Ion Channels as Targets in Drug Discovery

Gary Stephens • Edward Stevens
Editors

Ion Channels as Targets in Drug Discovery

Springer

Editors
Gary Stephens
School of Pharmacy
University of Reading
Reading, UK

Edward Stevens
Metrion Biosciences Ltd.
Cambridge, UK

ISBN 978-3-031-52196-6 ISBN 978-3-031-52197-3 (eBook)
https://doi.org/10.1007/978-3-031-52197-3

This Springer imprint is published by the registered company Springer Nature Switzerland AG
The registered company address is: Gewerbestrasse 11, 6330 Cham, Switzerland

The editors would like to dedicate this book to the memory of Professor William 'Bill' Catterall (1946–2024). It was with great sadness that we heard of the passing of Bill earlier this year, initially from colleagues attending the Calcium Channel Network event, at which Bill was a regularly attendee and leading light. It was unsurprising to realise that the first words of our introductory chapter are a definition of an "ion channel" from Bill Catterall; it was the best one we could find when writing the chapter. Bill was clearly an eminent and respected researcher who contributed, perhaps more than any other scientist in the modern era, to our general knowledge of ion channels. Importantly, both the voltage-gated calcium channel and voltage-gated sodium channel field are indebted to Bill's research. It is further appreciated that pioneering work on ion channel structure was always supported by relating these advances to function, a link sometimes only applied belated by other leaders of technical innovation. Bill was also generous and supportive in our careers. The first symposium GJS organised was at the International Brain Research Organization (IBRO) 2011 in Florence, Italy. It was at this meeting that GJS first met Bill, and his wife Tina, who often accompanied Bill to meetings. Bill also spoke at symposia at both the 18th World Congress of Basic and Clinical Pharmacology (WCP) in Kyoto, Japan, in 2018 and the more recent WCP2023 meeting in Glasgow. At the latter meeting, Bill was also honoured by a IUPHAR Lifetime Achievement Award to add to several similar awards. It was a pleasure to invite Bill and Tina to a symposium dinner and Bill's untimely passing made this time all the more appreciated.

Contents

About the Editors

Gary Stephens is Professor of Pharmacology at the School of Pharmacy, University of Reading, United Kingdom. He is an in vitro electrophysiologist with an interest in voltage-gated calcium channels and G protein-coupled receptors. He is a Senior Editor with the *British Journal of Pharmacology* and is on the Editorial Board of *Pharmaceuticals*.

Edward Stevens is Chief Scientific Officer at Metrion Biosciences and has over 20 years' experience of ion channel drug discovery within the biotech/pharma field.

Ion Channels as Targets in Drug Discovery: Outlook and Perspectives

Edward B. Stevens and Gary J. Stephens

Abstract Ion channels are ancient proteins, fundamental to basic biology and important targets for disease treatment, in particular, in conditions where imbalance of ionic flux disrupts homeostasis. The remit of the present volume is to present and discuss selected ion channel drugs with potential for clinical translation. Around one-fifth of all drugs act on ion channels, with over 170 available ion channel drugs, 27 of which are listed as WHO Essential Medicines. Such drugs were typically introduced in the "golden age" of modern drug discovery, which covers the latter half of the twentieth century. More recently, however, ion channel drug discovery has somewhat stalled, and it is this issue onto which contributors to this volume seek to shed a potential ray of light. Thus, there are currently over 60 ion channel drugs undergoing clinical trials. Such agents include the small molecules that have dominated ion channel drugs but also newer "biologic" agents, such as peptides and larger antibodies. Alongside this, new technological advances, including in drug engineering strategies, structural biology, and related computational modelling techniques, are now starting to make significant contributions. Moreover, combination of techniques whereby existing drugs are screened against specific phenotypes has utility for all drugs, including those against ion channel diseases with unmet clinical need. Overall, this introductory chapter and the broader volume will discuss and provide evidence that ion channel drug discovery is poised for a renaissance.

Keywords Voltage-gated ion channels · Ligand-gated ion channels · Small molecule drugs · Biologic drugs · Patch clamp electrophysiology · Structural biology

---

E. B. Stevens
Metrion Biosciences Ltd., Cambridge, UK
e-mail: edward.stevens@metrionbiosciences.com

G. J. Stephens (✉)
School of Pharmacy, University of Reading, Reading, UK
e-mail: g.j.stephens@reading.ac.uk

G. Stephens, E. Stevens (eds.), *Ion Channels as Targets in Drug Discovery*,
https://doi.org/10.1007/978-3-031-52197-3_1

1 Ion Channel Diversity and Difficulties in Targeting Ion Channels for Therapeutic Indications

1.1 Introduction

> Ion channels are a major class of membrane signaling proteins that produce changes in transmembrane electrical potential in response to many different stimuli… Activation of the ion channels opens a selective transmembrane pore through which specific ions may diffuse down their electrochemical gradient into or out of the cell. (Catterall 1996)

The above definition captures the concept of ion channels as proteins that are subject to regulation by different signaling pathways and, by extension, are modifiable and regulatable by exogenous drug molecules. Ion channels are ancient proteins, being widely expressed in prokaryotic cell membranes long before the evolution of multicellular organisms, where ion channel functionality is the basis for excitable neural networks. Evolutionary precursors such as bacterial channel homologs have provided source material for earlier structural studies and allowed initial predictive modelling of human drug-channel interactions. The human genome has over 400 predicted ion channel family genes, and the Guide to Pharmacology (curated by the International Union of Basic and Clinical Pharmacology and the British Pharmacological Society) divides target ion channel genes into ~145 voltage-gated ion channels (VGICs; Fig. 1), ~82 ligand-gated ion channels (LGICs; Fig. 2), and >50 genes for "other channels" including chloride channels, mechanosensitive Piezo channels, aquaporins, and store-operated channels such as Orai. Many of these targets await selective therapeutic targeting.

The gene products summarized in Figs. 1 and 2 encode proteins that form (or contribute directly to) the principal, ion-conducting channel pore; for VGICs, the pore-forming element is often termed the α subunit. Among VGICs, potassium channels represent the basic domain structure, containing segments S1-S6, that can form homo-, or hetero-, tetramers; by contrast, sodium and calcium VGICs consist of four such S1-S6 domains encoded by a single gene. LGICs typically consist of trimeric, tetrameric or pentameric assemblies of component subunits. Many channels also require auxiliary subunits for full functionality. These structural considerations increase diversity of ion channel subtypes and, potentially, drug targeting opportunities. For example, there are LGIC drugs designed to target specific subunits (or combinations thereof) preferentially, as discussed for nicotinic acetylcholine receptors by Wonnacott (chapter "Nicotinic Receptors of the Neuronal and Non-neuronal Cholinergic Systems as Therapeutic Targets: Opportunities and Limitations", this volume).

There are numerous volumes on ion channels; our remit is to consider how fundamental research can best translate to industrial drug development and thus supply clinicians with the tools to benefit the desired patient population. In this regard, the distinction and/or divide between academic and pharmaceutical industry research and development has narrowed considerably over the last decade, and translatability is now, rightly, a key goal within the global ion channel community. A healthy

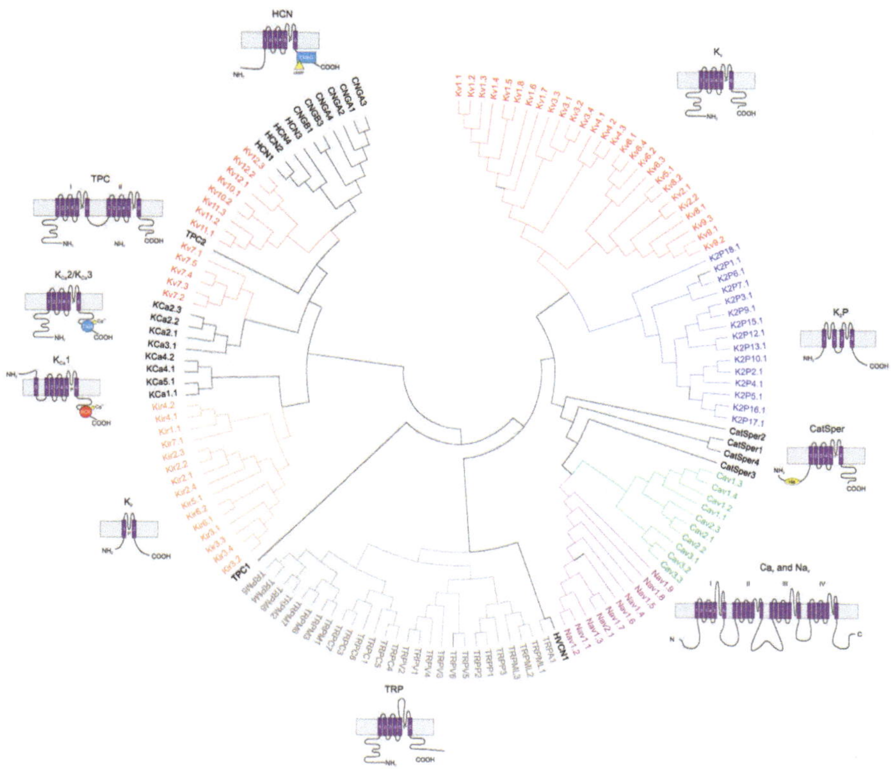

Fig. 1 Overview of voltage-gated ion channels. Schematic representation of phylogenetic relationship between voltage-gated ion channel superfamilies based on minimal pore regions. Insets indicated basic structure topology models. From: https://www.metrionbiosciences.com/knowledgelibrary/ion-channel-overviews

realization of the importance of drug discovery to global well-being has come from the scientific response to the COVID-19 pandemic. In particular, the exemplary academic/industry partnership between the Oxford Vaccine Group and AstraZeneca in the UK led the introduction of the first, fully tested severe acute respiratory syndrome coronavirus-2 (SARS-CoV-2) vaccine, ChAdOx1 nCoV-19. These and other vaccines clearly saved millions of lives worldwide. However, as emphasized by Sir Andrew Pollard (Oxford Vaccine Group) at the 19th World Congress of Basic and Clinical Pharmacology in 2023, the apparently rapid "translation" of the ChAdOx1 nCoV-19 vaccine to clinical use came on the back of around 20 years of existing academic research on the coronaviruses associated with previous severe acute respiratory syndrome (SARS) and Middle East respiratory syndrome (MERS) outbreaks (Stephens and Bell 2023). Together with efforts to repurpose existing antiviral drugs, such advances serve to illustrate not only the need for extensive basic research but also the advantages of a joint academic/industry approach to drug discovery and ultimate translation, as advocated herein.

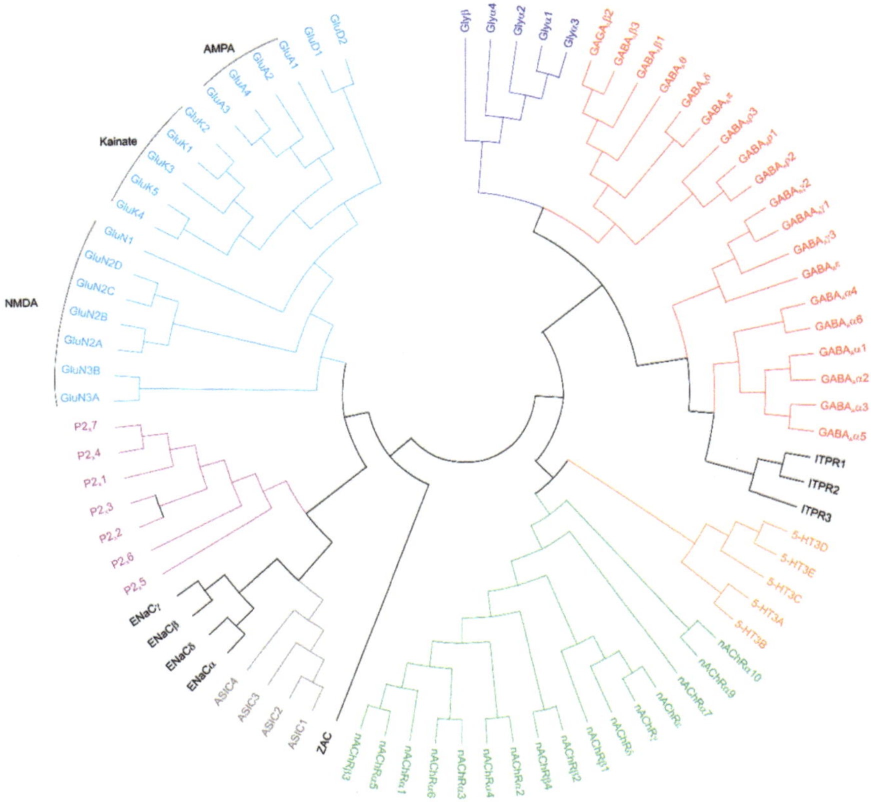

Fig. 2 Overview of ligand-gated ion channels. Schematic representation of phylogenetic relationship between ligand-gated ion channel superfamilies based on minimal pore regions. From: https://www.metrionbiosciences.com/knowledge-library/ion-channel-overviews

1.2 Milestones in Ion Channel Research

A landmark achievement in ion channel research was the award of the Nobel Prize in Physiology or Medicine in 1963 to Sir John Eccles, Alan Hodgkin, and Andrew Huxley for their initial descriptions of ion channel mechanisms. This work formed the basis for research on the function of single ion channels in excitable cells by Erwin Neher and Bert Sakmann, who were awarded the Nobel Prize in Physiology or Medicine in 1991. The history of ion channel research is intimately linked with technological developments enabling the measurement of picoamps (and occasionally nanoamps) of current flowing through ion channels and/or measurement of millivolt changes across the cell membrane. These advances were achieved largely by the development of patch clamp electrophysiology recordings which arose from Neher and Sakmann's work (see Hill and Stephens 2021). Many of the chapters herein describe ion channels at the plasma membrane; however, it should be appreciated that ion channels are also expressed on intracellular compartments, although targeting such conductances therapeutically awaits fuller exploitation (see Selezneva et al., chapter "Immune Cell Ion Channels as Therapeutic Targets", this volume). Equally, the rise of reliable automated patch clamp technology and other complementary techniques (as discussed further below) is now starting to realize the potential for higher throughput screening of candidate ion channel drugs.

Other more recent Nobel laureates in the ion channel field have made significant advances in revealing structural details, with the clear potential to benefit drug discovery. Thus, Peter Agre and Rod MacKinnon were awarded the Nobel Prize for Chemistry in 2003 for work on aquaporin and potassium ion channels, respectively. In 2021, David Julius and Ardem Patapoutian won the Nobel Prize in Physiology or Medicine for structural and mechanistic work on transient receptor potential (TRP) and mechanosensitive "Piezo" channels. These achievements have supported the more widespread use of gold-standard techniques such as X-ray crystallography and cryogenic electron microscopy (cryo-EM) for ion channel proteins and, as discussed further below and throughout this volume, are starting to inform new drug discovery.

1.3 The Problems with Ion Channels as Drug Targets

Ion channels are, in actuality, low abundance proteins that are difficult to target selectively. The latter concern likely relates most simply to the common ancestry of ion channel proteins, in particular for VGIC subtypes, which often have highly conserved structural motifs, including interaction/binding sites on extracellular residues close to the ion-conducting pore or voltage-sensing regions within the pore. This is a primary reason why small molecules must be tested against different ion channels to determine potential off-target effects, for example, the delayed rectifier potassium channel, $K_V11.1$, encoded by the human *ether-a-go-go* related (hERG) gene, where nonselective block of this conductance is potentially fatal. This regulatory step has inevitably led to attrition in approval of new ion channel drugs. Moreover, binding sites for different drug classes can overlap within an ion channel family; for example, voltage-gated sodium channels possess a pore binding site considered to be common to local anesthetic, antiarrhythmic, and anticonvulsant drugs (Tikhonov and Zhorov 2017). Also, ion channels typically have a paucity of extracellular epitopes available for targeting. This is particularly true for VGICs, where it is widely considered that the low percentage of exposed amino acids in extracellular domains and loops (e.g., ~15% for voltage-gated sodium channels) serves to limit targeting opportunities (Wulff et al. 2019). In general, good structural data (i.e., at low Angstrom (Å) resolution) for ion channels has lagged behind that of other major drug targets such as G protein-coupled receptors (GPCRs); however, ion channels are now starting to catch up in this area. A recent *British Journal of Pharmacology* Themed Issue "Structure Guided Pharmacology of Membrane

Proteins" discussed advances in structural knowledge of proteins including VGICs and LGICs (see Stephens and Shukla 2022). It is also illuminating that many of the chapters in this volume contain dedicated sections to structural biology and describe how recent advances are informing drug discovery.

A further general issue for drugs is the state dependence of ion channels, in particular for VGICs. Thus, ion channels shuttle between closed/resting, open, and inactivated states. It may be advantageous for drugs to target specific states; for example, there is evidence that different effective ion channel blockers preferentially target the inactivated state of sodium and calcium channels or the open state of potassium channels. Ion channels are also use-dependent, meaning that the more the channel is activated, the more channels move from rest to the open and then inactivated states. Use-dependency can allow some of the issues with drug selectivity to be overcome; for example, localized ion channels affected by hyperexcitability disorders (such as sodium channels in hippocampal neurons in epilepsy or sodium/calcium channels in nociceptive pathways in pain) can be blocked preferentially. However, an issue for drug discovery is replicating state-dependency and/or use-dependency when evaluating affinity/efficacy in preclinical studies. Here, techniques such as voltage-clamp, current-clamp and/or dynamic clamp electrophysiology can replicate ion channel dependency states to best measure drug effects.

The search for state-dependent blockers as a major drug discovery strategy is exemplified by the voltage-gated calcium channel $Ca_V2.2$ subunit. Here, the argument is that state-independent pore-blocking agents, designed to act on nociceptive pathways as analgesic agents, may be associated with unwanted effects on cardiovascular $Ca_V2.2$ channels. Moreover, for $Ca_V2.2$, specific splice variants have been shown to be preferentially expressed in dorsal root ganglion (DRG) neurons; thus, analgesic drug discovery for $Ca_V2.2$ has focused on the e37a isoform in screening campaigns (DuBreuil et al. 2021). The use of the $Ca_V2.2$ e37a splice variant raises another important concept, that of "functional selectivity." The e37a isoform functionally allows greater action potential-mediated Ca^{2+} influx, and thus the search for functionally selective, state-dependent inhibitors was initially prioritized. Such strategies may benefit drug discovery for other ion channel targets. It should be added that agents such as peptides and antibodies and inhibitors of protein-protein interactions and/or proteolysis-targeting chimeras (PROTACs) targeting $Ca_V2.2$ (as discussed further below) are unlikely to act in any state-dependent fashion and so different targeting strategies can be considered.

1.4 Current State of Play

This volume seeks to highlight the most promising ion channel drugs in preclinical development in terms of potential progress through clinical trials. In this spirit, it is illuminating to highlight existing ion channel drugs. It is estimated variously that around one-fifth of all small molecule drugs target ion channels, with over 170 ion channel drugs currently on the market. Although some of these drugs have common targets and/or similar indications to first-line drugs, on initial inspection this is an impressive list. Perhaps most importantly, 27 ion channel drugs are on the World Health Organization (WHO) List of Essential Medicines. Many of these drugs act on the cardiovascular system (such as the L-type voltage-gated calcium channel blockers amlodipine and verapamil and the voltage-gated potassium channel blocker amiodarone) or are anticonvulsant (including voltage-gated sodium channel blockers carbamazepine, phenytoin, and lamotrigine and the $GABA_A$ activator lorazepam). A further prominent example is gliclazide, which targets the $K_{ir}6.2$/sulfonylurea receptor 1 (SUR1) (K_{ATP} channel) complex in the treatment of type 2 diabetes. These ion channel drugs are all to be found in the "top 100" UK drug formulary (Audi et al. 2018).

As discussed further below, approved ion channel drugs are predominantly small molecules, typically orally active compounds that adhere well to Lipinski's rule of 5 or Veber variation of these rules for "druggability." The current, almost total, domination of small molecules is exemplified by the fact that only common "biologic" peptide drug is ziconotide, approved for chronic pain, although leconotide (ω-conotoxin CVID) and the muscle relaxant XEP-018 (μ-conotoxin CnIIIC) peptide toxins are also approved ion channel drugs. However, the search is very much on for biologic drugs such as peptide toxins and antibody drugs. Moreover, technological advances have potential to produce hybrid molecules that combine the advantages of different ion channel drug classes.

In general, small molecules tend to be hydrophilic compounds that bind close to the external vestibule and/or ion selectivity filter and effectively block ion conduction through the integral channel pore. Larger peptide toxins can similarly insert into the pore; for VGICs, so-called gating modifier toxins can also bind close to/at the voltage sensor within the pore and stabilize closed or open states to act as inhibitors or activators, respectively. Hydrophobic molecules can also affect intracellular sites via allosteric interactions. In general, ion channel drugs that act on voltage-gated sodium or calcium channels are blocking agents; by contrast, openers/activators include ion channel drugs that act to open different potassium channels or those that activate the cystic fibrosis transmembrane conductance regulator (CFTR); the latter drugs can also "correct" deficit CFTR function.

Against this, recent ion channel drug development has been disappointing and is often seen as a poor relation to advances seen for GPCRs, the other major molecular drug class. If we take the decade prior to 2019 as a not-so-arbitrary "pre-COVID pandemic" cutoff, analysis by Wulff et al. (2019) detailed 13 new approvals for ion channel drugs by the US Food and Drug Administration (FDA) during this period. To this, we can only really add the approval in 2019 of elexacaftor/ivacaftor/tezacaftor as the first triple combination therapy for cystic fibrosis, with further approvals for children in 2021 and 2023 (as discussed by Veselu et al., chapter "Cystic Fibrosis Transmembrane Conductance Regulator (CFTR)-Targeted Therapeutics for Cystic Fibrosis", this volume).

However, encouragement comes from the fact that there are currently over 60 ion channel drugs in different stages of clinical trials. One prominent area of unmet clinical need is that of pain (as an umbrella term), the subject of the *British Journal of Pharmacology* Themed Issue "Targeting Ion Channels in Chronic Pain" (see Stevens and Stephens 2018). There are several examples of ion channel blockers that have been marketed but subsequently discontinued for treatment of pain states; these include small molecules targeting $Na_V1.7$ (see Eagles et al. 2022), $Na_V1.8$, and $Ca_V2.2$. However, $Na_V1.7$ and, in particular, $Na_V1.8$ are far from being abandoned targets, as discussed further below. One new avenue in attempts to overcome the poor translatability of potential analgesics is illustrated by the recent pursuit of small molecules or venom-derived peptide toxins that dually affect Na_V1 and Ca_V3 channels, discussed by Finol-Urdaneta et al. (chapter "Conservation of Ligand Binding Between Voltage-Gated Sodium and T-Type Calcium Channels", this volume). Given the proof-of-concept for cancer pain treatment offered by the peptide toxin ziconotide, the potential for different natural venoms and toxins to affect ion channel targets is a tantalizing prospect, as discussed by Trim et al. (chapter "Ion Channel Tools and Therapeutics from Venoms and Toxins", this volume). Other notable molecular drug targets for pain therapy that are followed up in this volume include acid-sensitive ion channels (Senescall et al., chapter "Targeting Acid-Sensing Ion Channels in Disease", this volume), transient receptor potential channels (Rubaiy, chapter "Transient Receptor Potential Canonical (TRPC) as a Therapeutic Drug Target", this volume), and nicotinic acetylcholine receptor subtypes (Wonnacott, chapter "Nicotinic Receptors of the Neuronal and Non-neuronal Cholinergic Systems as Therapeutic Targets: Opportunities and Limitations", this volume).

1.5 Beyond the Pore: Targeting Auxiliary Subunits and Interacting Proteins

While ion channel drugs typically target pore-forming subunits to modify channel conduction or gating, there is also considerable interest in ion channel auxiliary subunits as drug targets. A useful definition of an auxiliary subunit can be adapted from the criteria described for ionotropic glutamate receptors by Yan and Tomita (2012). Thus, an auxiliary subunit should (i) be non-pore forming, (ii) interact in a direct and stable manner with the pore-forming subunit, (iii) modulate ion channel biophysical properties and/or trafficking, (iv) have functional necessity in vivo. In particular, criterion (iii) affords the potential for ligands at auxiliary subunits to have therapeutic utility. In this regard, gabapentinoids represent the most successful drug class that targets ion channel auxiliary subunits (as discussed by Ablinger et al., chapter "The Presynaptic $\alpha_2\delta$ Protein Family and Their Therapeutic Potential", this volume). Thus, the first-in-class gabapentin was found (somewhat unexpectedly for a GABA mimetic) to bind to voltage-gated calcium channel $\alpha_2\delta$ auxiliary subunits and to be an anticonvulsant agent, with important additional indications in neuropathic pain. The higher affinity $\alpha_2\delta$ ligand pregabalin is another commonly prescribed drug. More recently, the $\alpha_2\delta$ ligand mirogabalin has been approved in Japan, and new agents such as crisugabalin (HSK16149) are the subject of clinical interest/development (Gou et al. 2021).

A recent and novel advance is the targeting of the $Ca_V\beta$ subunit to reduce activity of high-voltage-activated (HVA) calcium channel $\alpha 1$ subunit activity. Here, the laboratory of Henry Colecraft (Columbia University Irving Medical Center, New York) has pioneered several approaches to develop potential new ways to modulate ion channel function. An elegant PROTAC approach was taken to create a nanobody linked to a ligase, termed "Ca_V-aβlator," which binds $Ca_V\beta$ and increases ubiquitination (Morgenstern et al. 2019). Further, a more specific functionalized nanobody, Chisel-1, which targets $Ca_V\beta 1$ selectively was shown to effectively abolish HVA Ca^{2+} current (Morgenstern et al. 2022). The future development of such technologies is another area of clear therapeutic interest.

A final possibility here are ligands acting at ion channel-interacting proteins that may exert downstream functional effects on ion channels. A notable example is the collapsin response mediator protein 2 (CRMP2), which is reportedly part of the functional ion channel proteome of $Ca_V2.2$ and $Na_V1.7$. Here, the academic laboratory of Rajesh Khanna (NYU Pain Research Center) together with an associated industry spinout has developed a synthetic peptide based on the CRMP2-$Ca_V2.2$ interaction site (the 15-mer peptide CBD3). Another approved drug, the anticonvulsant agent levetiracetam, binds to the synaptic vesicle protein SV2A; the exact mechanism by which this interaction causes hypoexcitability remains somewhat unclear, but there is evidence that levetiracetam acts to reduce ion channel conductance(s) (Contreras-García et al. 2022), including via a putative intracellular site on $Ca_V2.2$ (Vogl et al. 2012).

2 Small Molecule Ion Channel Drug Discovery Programs: Case Studies

The following section will provide brief case study examples of the current major small molecules targeting ion channels, with different pharmacological modalities (including inhibitors, activators, correctors, allosteric modulators) in a range of specified therapeutic areas. These different case studies highlight areas of significant industry activity and some of the most exciting ongoing research in targeting ion channels for treatment of disease.

2.1 $Na_V1.8$ Blockers in Pain

Existing analgesics for treatment of chronic pain act at the level of the CNS, e.g., antidepressants (serotonin-norepinephrine reuptake inhibitors or selective serotonin reuptake inhibitors), gabapentinoids (gabapentin or pregabalin), and opioids (e.g., oxycodone). These drugs are all associated with unwanted CNS side effects (such as sedation, respiratory depression, or addiction). The problems associated with addiction of centrally acting opioid analgesics has created the "opioid crisis," whereby since the 1990s widespread poor prescribing practices alongside overuse, misuse, and abuse (predominantly in North America and Europe) have led to a significant societal and medical impact (most notably a large number of deaths) (Skolnick 2018). Development of peripherally acting analgesics provides a route to pain relief for chronic conditions without the risks of centrally acting agents. Directly targeting ion channels on sensory neurons in the periphery (with resulting decrease in excitability) provides a clear opportunity for development of peripherally acting analgesics.

Small diameter sensory neurons highly express the $Na_V1.8$ voltage-gated sodium channel (Akopian et al. 1999; Coward et al. 2000). Genetic ablation of $Na_V1.8$ channels in rodents leads to deficits in nociception (Akopian et al. 1999). Gain-of-function mutations in the *SCN10A* gene that encode $Na_V1.8$ channels contribute to painful peripheral neuropathy (Faber et al. 2012). Thus, selective blockers of $Na_V1.8$ channels have the potential to treat a broad range of chronic pain states, including cancer pain, lower back pain, and neuropathic pain such as painful diabetic neuropathy. Many companies have pursued $Na_V1.8$ drug discovery programs over the last few decades (e.g., Vertex, Pfizer, Abbott) and have identified highly selective compound acting at the pore region (despite the high level of amino acid conservation in S6 domains across $Na_V1.1$-$Na_V1.8$). Vertex has recently made a breakthrough with development of the selective $Na_V1.8$ inhibitor, VX-548 (suzetrigine), which has shown efficacy in two phase II trials for acute pain in bunionectomy (surgery to remove bunions) (NCT04977336) and abdominoplasty (tummy tuck) (NCT05034952) (Jones et al. 2023); these areas are being extended to phase III with a further phase II study for neuropathic pain (diabetic neuropathy) planned. VX-548 received "breakthrough therapy" status from the FDA, reflecting substantial treatment advantages over existing options and allowing priority review for clinic progression. Orion Corporation (Finland) are also developing a selective $Na_V1.8$ blocker (ODM-111/JMKX000623), which is currently being positioned for treatment of neuropathic pain.

2.2 $K_V1.3$ Blockers in Autoimmune Disease

There has been an ongoing industry interest in targeting ion channel in immune cells, for example, the $K_V1.3$ voltage-gated potassium channel in T-cells and microglia (see Selezneva et al., chapter "Immune Cell Ion Channels as Therapeutic Targets", this volume), TRPA1 in astrocytes and microglia (Feske et al. 2019), and calcium release-activated calcium (CRAC) channels in T-cells (and other immune cell types including macrophages). For example, the CRAC inhibitor Auxora (Calcimedica) has completed phase II trials for treatment of COVID patients with pneumonia (Bruen et al. 2022) and is currently in phase II studies for treatment of acute respiratory distress syndrome in ventilated COVID patients (NCT04661540).

$K_V1.3$ has a key role in setting resting membrane potential (RMP) in T-cells (Lin et al. 1993) and upon T-cell activation causes a hyperpolarization of RMP, creating a large driving force for Ca^{2+} entry through CRAC channels (modelled by Hou et al. 2014). In autoimmune diseases (such as rheumatoid arthritis and type 1 diabetes), chronic self-antigen exposure leads to autoreactive T-cells associated with high $K_V1.3$ expression (Beeton et al. 2006). DE Shaw Research licensed a $K_V1.3$ compound DES-7114 to Lilly in 2022, which was identified using molecular dynamics and machine learning paradigms. DES-7114 is under development for ulcerative colitis, Crohn's disease, and atopic dermatitis. $K_V1.3$ has a key role in microglia-mediated neuronal cell death and inflammatory cytokine production by microglia (Wang et al. 2020). There is also upregulation of $K_V1.3$ in human microglia from Alzheimer's disease and Parkinson's disease (Fomina et al. 2021); based on these types of observations, Muna Therapeutics, Denmark, is currently targeting $K_V1.3$ for treatment of neuroinflammatory diseases. As described in further detail below, the synthetic peptide dalazatide is a potent $K_V1.3$ inhibitor under investigation to treat autoimmune conditions.

2.3 *$Ca_V2.2$ Blockers in Pain*

In addition to $Na_V1.7$ and $Na_V1.8$, the $Ca_V2.2$ voltage-gated calcium channel arose as a prominent rational drug target for pain (Patel et al. 2018). Thus, $Ca_V2.2$ is widely expressed in nociceptive pathways, including small diameter DRG neurons where they play a key role in nociception and there is good evidence of plastic $Ca_V2.2$ upregulation in pain states. Importantly, the $Ca_V2.2$ blocker ziconotide provides a clear proof-of-concept for targeting this specific subunit, as does use of gabapentinoids, which are $\alpha_2\delta$ ligands but with prominent effects on $Ca_V2.2$. However, these existing agents have issues: ziconotide must be delivered intrathecally; therapeutic index is low and compliance often compromised; gabapentinoids can also be poorly tolerated. Common to other targets, the high resolution of $Ca_V2.2$ cryo-EM structures has offered potential for refinements in drug discovery. The search for small molecule $Ca_V2.2$ blockers has prioritized functional selectivity through state-dependent inhibition, targeting the inactivated state driven by the use-dependent nature of $Ca_V2.2$ channel gating, and the promotion of such states in pathophysiological conditions. The search for $Ca_V2.2$ blockers has been intense but, despite several promising leads, ultimately has failed to lead to the introduction of new agents. Thus, a number of small molecule $Ca_V2.2$ blockers have been reported to have improved selectivity and appropriate preclinical profiles. An early example was the compound CNV2197944, initially developed by GlaxoSmithKline and then by a spinout, Convergence Pharmaceuticals. However, CNV2197944 failed to demonstrate any published phase II efficacy against postherpetic neuralgia or diabetic peripheral neuropathy, and, at present, there is no clear development of this compound. A more recent notable example is TROX-1, developed initially by Merck and then taken on by Grünenthal; however, as with all other $Ca_V2.2$ blockers, promising preclinical data (Patel et al. 2015) have failed to translate to human clinical trials. While there is potential for $Ca_V2.2$ to achieve a similar re-emergence as that of the $Na_V1.8$ channel, at present the jury remains out.

2.4 TRPA1 Modulators in Pain

TRPA1 is a nonselective cation channel with multiple modes of activation including by temperature, reactive oxygen species, inflammatory mediators, and exogenous irritants. TRPA1 is expressed in nociceptors and upregulated in rodent models of neuropathic and inflammatory pain. Human gain-of-function TRPA1 mutations are associated with familial episodic pain syndrome, an autosomal dominant neurological disorder characterized by onset in infancy of episodic debilitating upper body pain and abdominal guarding (a symptom of visceral pain) (Kremeyer et al. 2010). Exogenous TRPA1 activators, such as mustard oil, cause pain in rats and humans, while increased concentrations of endogenous activators in human diseases such as osteoarthritis and diabetic neuropathy are associated with pain (Andersson et al. 2008). Genetic knockout/knockdown of TRPA1 in mice reveal a key role in pain processing (Reese et al. 2020), while TRPA1 antagonists have been shown to reverse pain scores in multiple models of pain (Chen et al. 2011).

Glenmark's TRPA1 inhibitor GRC 17536 reported positive data in a phase IIa diabetic neuropathy study in 2014 (NCT01726413), but more recent progress has been restricted to conference communications. Lilly assessed LY-3526318 (from their Hydra collaboration) using the Chronic Pain Master Protocol designed to accelerate the development of chronic pain drug candidates, which included parallel clinical studies for osteoarthritis (NCT05080660), chronic low back pain (NCT05086289), and diabetic neuropathy (NCT05177094) in 2022. TRPA1 has been shown to be a key chemosensor involved in induction of airway inflammation and hyperreactivity (Balestrini et al. 2021). Based on this concept, Genentech has positioned TRPA1 antagonists for the treatment of respiratory diseases, where the clinical compound GDC-6599 is currently recruiting to phase II for chronic cough.

2.5 P_2X_3 Blockers in Cough

Ion channel drugs have likely utility in the treatment of persistent cough, a troubling condition associated with unmet clinical need. The discovery of the role of ionotropic P_2X_3 receptors in cough and the several receptor antagonists currently in advanced clinical trials is discussed by Stokes et al. (chapter "Purinergic P_2X Receptors as Therapeutic Targets", this volume). The first-in-class compound gefapixant (named famously after the purinergic research pioneer Geoff Burnstock) is approved currently in Japan, and it will be of interest if this approval can be replicated by the FDA, European Medicines Agency (EMA), and/or UK Medicines and Healthcare products Regulatory Agency (MHRA).

2.6 K_V7 Activators in Epilepsy and Beyond

$K_V7.2$, $K_V7.4$, or $K_V7.5$ voltage-gated potassium channels can co-assemble with $K_V7.3$ channels to form heterotetrameric voltage-gated channels, where $K_V7.2$/$K_V7.3$ channels are the principal channels responsible for the "M-current" that regulates neuronal excitability (Brown and Passmore 2009). Retigabine, a nonselective opener of the K_V7 subfamily (and derivative of the less potent analgesic flupirtine), was developed for the treatment of epilepsy as an adjunctive therapy for partial seizures (see also Gamper et al., chapter "Therapeutic Targeting of Potassium Channels", this volume). Retigabine was discontinued in 2016 due to side effects (urinary retention, retinal abnormalities, skin discoloration), limited use, and associated drop in demand. Retigabine and the fellow activator ML213, which show limited selectivity across K_V7 subfamily ($K_V7.1$ sparing), have been reported to bind at the pore region of K_V7 (through a conserved tryptophan residue in the pore-forming S5 helix, which is absent from $K_V7.1$) (Kim et al. 2015), while the selective $K_V7.2/7.3$ openers ICA-069673 and ICA-27243 bind in the voltage-sensor domain with no direct interaction with the pore region (Wang et al. 2018). Much of the industry focus of recent $K_V7.2/7.3$ opener programs has been around epilepsy, where Xenon's selective agent XEN1101 has demonstrated efficacy as an adjunctive treatment in adult patients with focal epilepsy in an ongoing phase IIb study (NCT03796962) (French et al. 2022), with phase III studies for focal-onset seizures currently also being recruited for. XEN1101 is also being positioned for primary generalized tonic-clonic seizures and depressive disorders.

Biohaven's selective $K_V7.2/7.3$ opener BHV-7000 is also being developed for epilepsy, including treatment of the rare disease KCNQ2 Developmental and Epileptic Encephalopathy (DEE). This disease gives rise to early-onset seizures and encephalopathy in infants and children due pathogenic mutations in $K_V7.2$. Biohaven are also investigating mood and pain disorders as potential indications for this mechanism. Biohaven released phase I data for BHV-7000 in Q1 2023 that demonstrated mild adverse events and no serious or severe adverse events or dose-limiting toxicities. Biohaven will progress BHV-7000 to phase II studies initially in focal epilepsy and bipolar disorder.

Retigabine has demonstrated efficacy in amyotrophic lateral sclerosis (ALS) in a phase II study in 2021 (NCT02450552, Wainger et al. 2021). In addition, potassium channel RNA and conductances (including $K_V7.2$) have been shown to be reduced in ALS (Kanai et al. 2006). Building on these data, Qualis are focused on treating motor neuron hyperexcitability-induced disease progression of ALS (present in approximately 50% of ALS patients). The lead compound QRL-101 demonstrated a better safety profile than retigabine in preclinical studies measuring fatigue/somnolence, dizziness, and bladder retention. Phase I studies for QRL-101 commenced at the beginning of 2023 (NCT05667779).

2.7 $K_V3.1/K_V3.2$ Activators in Schizophrenia and Fragile X Syndrome

Dysfunctional fast-spiking parvalbumin-positive GABAergic interneurons are thought to contribute to the symptoms of schizophrenia. $K_V3.1$ and $K_V3.2$ voltage-gated potassium channels have been shown to control firing properties of these interneurons, where small molecule $K_V3.1/K_V3.2$ activators restore the rapid firing frequency (Taskin et al. 2015; Boddum et al. 2017). Autifony's small molecule K_V3 activator, AUT00206, has been shown to modulate gamma oscillations (a marker of cortical excitation-inhibition balance) and reward-related striatal abnormalities in schizophrenia patients (Kaar et al. 2022). AUT00206 has completed phase I trials (NCT03164876) and has FDA Orphan Drug Designation for treatment of Fragile X syndrome. A collaboration between Lundbeck and LeadXPro has generated a high resolution cryo-EM structure suitable for structure-based discovery, which has identified a novel binding pocket for a novel K_V3 activator Lu AG00563 (Botte et al. 2022).

2.8 CFTR Agents in Cystic Fibrosis

Small molecules acting at CFTR are often termed "correctors" or "potentiators." These drugs overcome deficits in CFTR trafficking and gating, respectively, to restore channel function. Such drugs represent a high watermark for successful ion channel drug discovery for the life-changing treatment of cystic fibrosis. However, ongoing effort is needed to treat all people with cystic fibrosis due to the existence of disease variants. This work is reviewed by Veselu et al. (chapter "Cystic Fibrosis Transmembrane Conductance Regulator (CFTR)-Targeted Therapeutics for Cystic Fibrosis", this volume) and a case study chapter on some of the novel TMEM16A potentiators in development by Enterprise Therapeutics (Danahay and Gosling, chapter "Validation of TMEM16A Modulation as a Therapeutic Approach for the Treatment of Cystic Fibrosis: The Discovery of Novel TMEM16A Potentiators", this volume). As above, the most recent new ion channel drug is the elexacaftor/ivacaftor/tezacaftor a triple combination therapy, with work in this area ongoing.

2.9 Ligand-Gated Ion Channels: Partial Agonists and Allosteric Modulators

By contrast to VGICs where, as above, small molecules may be characterized as blockers, openers, correctors, or potentiators, action at LGICs may be instigated by ligands employing a wider range of pharmacological mechanisms. Full or partial agonists typically act at the externally facing residues comprising the orthosteric binding site. Partial agonists may have therapeutic benefits over full agonists due to their lower efficacy/submaximal response profile affording the potential to reduce ion channel desensitization, permit subunit selectivity, and improve safety profile. For example, clobazam is a $GABA_A$ partial agonist approved for treatment of anxiety and seizures. LGIC receptor antagonists typically block such orthosteric sites. Inverse agonists are class of drugs that promote actions in the opposite direction to that of full/partial agonist, typically by reducing constitutive LGIC activity via the orthosteric binding site. In general, there are no clear examples of clinically proven ion channel inverse agonists; however, compounds such as $GABA_A$ inverse agonist β-carbolines are useful research tools to investigate potential anxiolytic drugs.

The identification of distinct allosteric, noncanonical binding sites has prompted the search for new ligands that may activate or inhibit signal transduction pathways indirectly, so-called positive allosteric modulators (PAMs) or negative allosteric modulators (NAMs), respectively. The potential development of PAMs and NAMs against nicotinic acetylcholine receptors is considered by Wonnacott (chapter "Nicotinic Receptors of the Neuronal and Non-neuronal Cholinergic Systems as Therapeutic Targets: Opportunities and Limitations", this volume). Other prominent examples of allosteric modulators currently in clinical trials include those against $GABA_A$ receptors (predominantly PAMs) and NMDA receptors (PAMs and NAMs). With regard to $GABA_A$ receptors, the WHO Essential Medicine diazepam is a $GABA_A$ PAM, as are several approved drugs for epilepsies, such as clonazepam, clorazepate, primidone, and stiripentol, and for insomnia, such as triazolam. A number of $GABA_A$ PAMs are currently in clinical trials, including EVT 201 (Evotec Neurosciences) with positive phase I/II results for treatment of insomnia (NCT00380003); PRAX-114 (Praxis Precision Medicines) has completed phase II/III trials for major depressive disorder (MDD) (NCT04832425); SAGE-217 (Sage Therapeutics) has completed phase III trials for adults with MDD (NCT03864614) and severe postpartum depression (NCT04442503). SAGE-324 (Sage Therapeutics) has completed phase II trials in subjects with essential tremor (NCT04305275). $GABA_A$ NAMs have also been proposed to have therapeutic utility, and a trial for RO 7017773 (also known as RG 7816; Roche) against autism spectrum disorders is currently recruiting.

Another major target for allosteric modulation is the glutamatergic NMDA receptor. NMDA PAMs seek to combat NMDAR hypofunction, and this class includes plazinemdor (formerly CAD 9303; Cadent Therapeutics) with a completed phase I trial for schizophrenia (NCT04306146); zelquistinel (GATE-251, formerly AGN-241751; Gate Neurosciences) completed a phase II trial in adults with MDD (NCT03586427); SAGE-718 (Sage Therapeutics) represents a promising neuroactive steroid lead and has Fast Track Designation by the FDA and Orphan Designation by the EMA. SAGE-718 has completed phase II trials against Parkinson's disease (NCT04476017) and Alzheimer's disease (NCT04602624) and has completed a phase I trial for consideration in subjects with Huntington's disease (NCT03787758) and treatment of cognitive impairment associated with neurodegenerative disorders (NCT03771586).

NMDA NAMs currently in human trials include MIJ821 (Novartis) with a completed phase 2 trial in subjects with treatment-resistant depression (TRD) (NCT03756129) and current recruitment for treatment of MDD (NCT04722666). Radiprodil is an orally active NMDA NR2B subunit NAM; a trial in subjects with drug-resistant infantile spasms was terminated, but the drug is currently being evaluated for effects on seizure and behavior in children with GRIN (glutamate receptor ionotropic NMDA)-related disorder (NCT05818943). These small molecule allosteric modulators point to mechanisms beyond those at the "traditional" agonist orthosteric binding site. In further relation to NMDA receptors, AXS-05 (dextromethorphan-bupropion; Axsome Therapeutics) is an oral receptor antagonist (and also a sigma-1 receptor agonist) that has progressed through phase II trials for TRD (NCT04634669) and phase III trials for MDD (NCT04019704 and NCT02741791) and control of agitation in subjects with Alzheimer's disease (NCT04797715).

A further concept, at present largely associated with other receptor classes such as GPCRs, is that of "biased signaling," whereby an agonist favors one of (potentially) multiple signaling pathways. In general, this is an area that has yet to be exploited to achieve a selective therapeutic benefit; however, there is the potential for ion channel ligands to demonstrate bias in ion selectivity and/or activation of signaling pathways (Herrington and Arey 2014). A recent example of such ion channel bias signaling comes from elegant bioluminescence resonance energy transfer (BRET) measurements for different ligands at TRPV1 and ionotropic P_2X_5 and P_2X_7 receptors (Chappe et al. 2022).

3 Beyond Small Molecules as Ion Channel Drugs

The vast majority of approved ion channel drugs are small molecules, with ziconotide being the only peptide among common approved ion channel drugs. The pharmaceutical industry is actively looking to advance "next generation therapeutics" and associated technologies; such advances include peptide/protein therapeutics, antibody therapeutics, nucleotide-based therapeutics, and cell-based therapeutics. A major interest for ion channel drug discovery is the potential for advances in each of these areas or, as discussed below, combinations of such therapeutic areas.

3.1 Peptide/Protein Therapeutics

In addition to the aforementioned ziconotide, there are a number of other peptides in clinical trials as ion channel drugs. A prominent example is dalazatide (κ-stichotoxin-She3a, ShK-186; Kineta), a 37 amino acid synthetic peptide derived from sea anemone toxin, which is a potent $K_V1.3$ inhibitor under investigation to treat autoimmune conditions. The peptide was discovered at University of California, Irvine, then further progressed by Kineta and KPI Therapeutics. Dalazatide completed phase I trials for psoriasis in 2015 (NCT02435342, as discussed by Gamper et al. (chapter "Therapeutic Targeting of Potassium Channels", this volume), Trim et al. (chapter "Ion Channel Tools and Therapeutics from Venoms and Toxins", this volume) and Tarcha et al. (2017)). The 36-mer scorpion peptide chlorotoxin (TM-601; TransMolecular) is a chloride channel blocker that completed phase I trials for solid tumor cancer in 2008 (NCT00379132), as discussed by Trim et al. (chapter "Ion Channel Tools and Therapeutics from Venoms and Toxins", this volume). SOR-C13 (Soricimed Biopharma) is a synthetic peptide TRPV6 antagonist designed to treat solid tumor cancer that has completed phase I trials (NCT01578564). Granexin gel contains the synthetic peptide ACT1 (Xequel Bio) that targets the connexin Cx43 carboxyl terminus and has proposed anti-inflammatory properties. A phase III trial for Granexin gel treatment of diabetic foot ulcers was terminated; however, a phase II/III study for the indication of reducing surgery scar formation (NCT04331080) was completed in 2022. Such data provide hope that the next peptide ion channel drug is imminent.

3.2 *Antibody Therapeutics*

Antibodies offer clear potential advantages over small molecules for improved selectivity (including reduced off-target effects), potency, and long-lasting modulation of ion channels. Alongside other molecular targets, and in context of an era when biologic drugs such as monoclonal antibodies are increasingly dominating the list of blockbuster drugs, interest in ion channel antibodies has grown over the last 20 years, as summarized by a number of excellent reviews (Haustrate et al. 2019; Hutchings et al. 2019; Wulff et al. 2019). In common with other areas of preclinical development, several antibodies have been raised as potential drugs against a range of ion channels. Some such agents have shown both promising in vitro activity and in vivo disease model efficacy, including targeting $K_V1.3$ in autoimmune disease and $Na_V1.7/Na_V1.8$ in pain; however, as yet, no antibodies agoinst ion channels have become marketed drugs.

At present, there is only one example of an ion channel antibody in clinical trials: an antibody against the nonfunctional P_2X_7 receptor variant (nfP_2X_7) (which does not gate current, but is expressed on cancer cells). The nfP_2X_7 antibody was formulated as a 10% ointment (BIL010t; Biosceptre) and completed phase I trials for skin cancer in 2014 (NCT02587819), as discussed (along with other preclinical P_2X receptor antibodies) by Stokes et al. (chapter "Purinergic P_2X Receptors as Therapeutic Targets", this volume). Although no follow-up studies have been reported to date, the nfP_2X_7 receptor has also been targeted by Biosceptre's BIL06v peptide conjugate vaccine, as reported in a conference communication (Oliphant et al. 2019). A humanized monoclonal antibody, DS-2741 (Daiichi Sankyo), a specific blocker of the calcium release-activated calcium channel Orai1, completed phase I trials for atopic dermatitis in 2020 (NCT04211415); however, further development appears to have been terminated.

A further recent advance is the potential therapeutic advantages of nanobody drugs. Nanobodies are small, single domain antibodies that represent the ~15 Kd fragment of antigen binding (fab) region and may represent a more potent and efficacious biologic drug, for example, by overcoming the limitation of larger, poorly penetrant conventional antibodies (Jovčevska and Muyldermans 2020). An additional strategy may be to combine a nanobody with other therapeutic agents to produce improved engineered drug compounds. An example here is the development of a nanobody against the human acid-sensitive ion channel ASIC1a, which has been coupled with the peptide toxin psalmotoxin 1 (PcTx1) to produce a potent channel inhibitor, as discussed by Senescall et al. (chapter "Targeting Acid-Sensing Ion Channels in Disease", this volume). Thus, strategies whereby nanobodies facilitate the action of small molecule ion channel drugs are attractive for future hybrid drug development. Nanobodies can also target intracellular sites as "intrabodies"; as mentioned above, nanobody conjugates such as Ca_V-aβlator and Chisel-1 have potential as future therapeutic agents.

Another technical innovation is in engineered antibody-based drugs using the so-called "KnotBodies" technology developed by Iontas and, more latterly, Maxion Therapeutics. Many animal venoms contain a conserved six cysteine motif that can contribute to a chemically stable "knottin" or "inhibitor cystine knot (ICK) mini-protein" motif; such scaffolds have attracted the interest in drug discovery efforts. KnotBodies incorporate the ICK into complementarity-determining regions (CDR) of an antibody to produce a fusion molecule that combines the potency of a toxin with the therapeutic advantages of an antibody and, thus, represents a useful novel blueprint for drug discovery. Preclinical efforts have investigated the potential of producing KnotBodies from $K_V1.3$ and $Na_V1.7$ peptide toxins, as discussed more fully by Trim et al. (chapter "Ion Channel Tools and Therapeutics from Venoms and Toxins", this volume).

For all classes of antibody drugs, it will be important to confirm that high affinity binding to ion channel leads to functional efficacy in vivo.

3.3 *Nucleotide-Based Therapeutics*

A recent publication has provided an interim report for a phase I/II study (NCT03375047, RESTORE-CF) for the mRNA therapeutic MRT5005 (Translate Bio), whereby inhalation of a nebulized mRNA is proposed to deliver functional CFTR protein (Rowe et al. 2023). This is the first-in-human inhaled mRNA replacement therapy for people with cystic fibrosis; however, although the safety profile was largely positive, no clear changes to lung function were reported thus far.

Another area of considerable therapeutic interest is the potential of antisense oligonucleotides (ASOs), due to their ability to act specifically to silence mRNA translation into deleterious disease-associated proteins "at source." Several ASOs have been approved by the FDA over the last two decades, with over 20 more in advanced clinical trials (see Thakur et al. 2022). To date, there is only example of use of ASOs against ion channels in clinical trials, as described in further detail by Wengert et al. (chapter "Voltage-Gated Sodium Channels as Drug Targets in Epilepsy-Related Sodium Channelopathies", this volume). Briefly, a Targeted Augmentation of Nuclear Gene Output (TANGO) approach is being used, whereby an ASO (STK-001, Stoke Therapeutics) is used to boost expression of $Na_V1.1$ for the treatment of the severe childhood epilepsy, Dravet syndrome; this is currently at phase II as an open label extension study (NCT04740476).

3.4 Cell-Based Therapies

The use of cell-based therapies has massive therapeutic potential. However, as with many other areas, this is very much in its infancy for ion channels. One example that has entered clinical trials is the adaptation to the scorpion peptide chlorotoxin, described above and discussed further by Trim et al. (chapter "Ion Channel Tools and Therapeutics from Venoms and Toxins", this volume). Patients with glioma are currently being recruited for phase I trials with chimeric antigen receptor T (CAR T) cells with a chlorotoxin tumor-targeting domain (CHM-1101; Chimeric Therapeutics). Of further interest here is that the only ion channel antibody target nfP2X7 (described above) is currently also being investigated by Biosceptre using a novel bivalent antibody-CAR T-cell therapeutic approach.

3.5 Approaches to Solve Pharmacokinetic Issues with Biologics

While small molecule drugs have proven successful in terms of target engagement, there are a number of long-appreciated issues that need to be considered. A common problem is that small molecules often lack selectivity between ion channels and thus have safety/toxicity concerns. Other potential issues are the conversion of small molecules into toxic metabolites, as well as dangerous drug-drug interactions. One of the drivers toward biologic drugs such as peptides and antibodies is the avoidance of such concerns, largely due to their high target selectivity and their predictable breakdown as part of the body's natural catabolism processes. The latter, however, can lead to undesirably short half-lives, and there are also concerns regarding low oral bioavailability, low volume of distribution, and the triggering of immune responses to foreign biological material. There are several general strategies to increase plasma half-lives of biologic drugs, for example, conjugation of larger molecules such as polyethylene glycol (PEG), plasma proteins, or engineered fusion proteins (Wulff et al. 2019); extended half-live biologic drugs have been termed "bio-betters" (Strohl 2015). PEGylation of the K_V1.3-blocking scorpion peptide HsTx1[R14A] has been shown to increase half-life (Tanner et al. 2017). More recently, Jin et al. (2016) have described encapsulation of HsTx1[R14A] as a controlled release microsphere, as discussed by Trim et al. (chapter "Ion Channel Tools and Therapeutics from Venoms and Toxins", this volume). Trim et al. further discuss methodologies such as cyclizing peptides (e.g., conotoxin Vc1.1) and introducing targeted mutations to peptides to avoid protease degradation as strategies to improve stability of biologic drugs. Such manipulations are designed to improve bioavailability, but it is important that they do not adversely affect affinity/efficacy.

Biologic drugs also suffer from poor penetration due to lack of ability to partition into lipid membranes. This is of particular concern for targeting CNS diseases; even if the blood-brain barrier (BBB) becomes comprised and allows entry of systemically applied biologic drugs, the amount of such entry is unpredictable. Routes such as intrathecal injection can be considered. It may be argued that drugs such as ziconotide in the treatment of chronic pain represent an important proof-of-concept for intrathecal delivery, but this route is not straightforward. It is likely that technological advances in delivery systems and/or strategies to facilitate passage across the BBB are required. An example of the latter is the design of the CRMP2-Ca_V2.2 interaction blocking CBD3 peptide, which has been conjugated to a HIV TAT protein or a 9mer polyarginine sequence to facilitate cell permeability (see Feldman and Khanna 2013). This methodological advance is part of the basis for therapeutic development of CBD3 peptide-based drugs as analgesics.

Finally, it is of note that many exciting new technological developments are seeking to combine advantageous properties associated with small molecules and biologics to achieve high affinity, long-lasting target engagement, and efficacy. Future development of such hybrid agents against ion channels is an area of clear interest.

4 Screening Technologies

Screening technologies for ion channel drugs can be broadly subdivided into non-electrophysiological and electrophysiological methodologies. As described below, the basis of high-throughput screening (HTS) regimes is typically fluorescence-based or automated electrophysiological assays (see also Yu et al. 2016). The majority of screening campaigns make use of cell lines (e.g., CHO or HEK293 cells) stably expressing or transiently transfected with the single ion channel target or transduced with a vector containing the ion channel such as use of baculovirus-mediated gene transduction of mammalian cells (*BacMam*). Recent advances in ion channel structural biology, including drug-bound structures, support modelling techniques in drug development. By contrast to such target-specific methodologies, a return to a more "unbiased" investigation scenario has become increasingly in vogue, namely, investigations of disease phenotypes to identify potential therapeutic agents. These aspects are discussed further below.

4.1 Fluorescence-Based Platforms

Ion channel HTS campaigns have been supported historically using fluorescence-based assays using fluorescent dye kits suitable for HTS platforms (e.g., FDSS/μCELL; Hamamatsu Photonics, Japan; FLIPR Penta, Molecular Devices, USA). Channels that permeate Ca^{2+} can be assayed using calcium-sensitive dyes (e.g., Fluo-8, AAT Bioquest, USA; FLIPR Calcium 6/6QF kits, Molecular Devices, USA); channels permeating K^{+} can be monitored using thallium-sensitive dyes (e.g., FluxOR II Green Potassium Ion Channel Assay, Invitrogen, USA; FLIPR Potassium Assay Kit, Molecular Devices, USA); channels permeating Na^{+} can be measured using sodium-sensitive dyes (e.g., Natrium Green-2, Ion Biosciences). Channel activity can also be measured indirectly through effects on RMP using membrane potential dyes (e.g., FLIPR membrane potential dye kits, Molecular Devices, USA). Fluorescence assays are generally considered to be preferable to "direct" ion flux-based assays, such as use of radioactive isotopes ($^{22}Na^{+}$, $^{45}Ca^{2+}$, and $^{86}K^{+}$), as they are more accurate and have higher throughput.

An alternative to use of high external K^{+} or of sodium channel gating modifiers to evoke neuronal excitability is the adaptation of high-throughput fluorescence readers to use of electrical field stimulation (e.g., FDSS/μCELL; Hamamatsu Photonics, Japan) (Daily et al. 2017). Electrical stimulation of recombinant ion channels can provide an opportunity to investigate pharmacological modulation in different channel states or during action potential activity, thus avoiding the use of potentially contaminating pharmacological activation of channels (Molokanova et al. 2017). Vertex Pharmaceuticals initiated electrical stimulation integration with a fluorescent plate reader using an in-house screening platform (E-VIPR) with fluorescence resonance energy transfer (FRET)-based voltage dyes (Huang et al. 2006). More recently, other companies have developed similar systems that integrate fluorescence readouts with field stimulation (e.g., Cellectricon, Sweden; Hamamatsu Photonics, Japan; Vala Sciences, USA).

Use of genetically encoded Ca^{2+} and voltage indicators provides another method of measuring excitability (Zhang and Cohen 2017); for example, use of blue-light-activated channelrhodopsin (CheRiff) alongside red-light-excited Archaerhodopsin-derived voltage indicator (QuasAr2) provides high fidelity optical electrophysiology "optipatch" recordings of neuronal excitability (Lou et al. 2016; Zhang and Cohen 2017). New generation genetically encoded voltage indicators can be expressed in specific cell populations, for example, in excitatory vs. inhibitory nerve terminals, to permit greater fidelity of investigation (Knöpfel and Song 2019). Optogenetics screening platforms have been developed by a variety of companies (e.g., Axxam, Italy; Biosciences, USA; and Vala Sciences, USA); for example, Q-State Biosciences have reported use of their 384-well "Swarm" platform where 200,000 compounds were screened against $Na_V1.7$ (Borja et al. 2022).

4.2 *Automated Patch Clamp Electrophysiology*

Electrophysiology provides a more precise and robust measure of ion channel activity than fluorometric assays and provides more accurate mechanistic data, such as facilitating the investigation of state-dependence of compound interaction. While manual electrophysiological patch clamp recordings have been regarded as gold-standard (see Hill and Stephens 2021), the need for higher throughput ion channel screening platforms capable of supporting medicinal chemistry programs has led to the development of automated (giga-seal) electrophysiology platforms by a number of companies including Molecular Devices (USA), Cellectricon (Sweden), Sophion (Denmark), Nanion Technologies (Germany), and Fluxion (USA). All of these systems are centered around miniaturized chip-based technologies for patch clamp recordings, with integrated microfluidics for addition of cells in suspension or test compounds (Fig. 3; Suk et al. 2019). Further development of this technology has included the implementation of temperature control at the patch clamp recording site and the introduction of a current-clamp recording mode. Electrophysiological recordings have focused predominantly on cell lines stably expressing ion channel proteins at the cell surface, although there have been significant recent efforts to record native/induced pluripotent stem cell (iPSC)-derived neurons or cardiomyocytes that are now beginning to bear fruit (Ghovanloo et al. 2023). Higher throughput 384-well systems, such as Qube (Sophion) and Synchropatch (Nanion), have become fully integrated in pharma/biotech ion channel screening programs with the capability to support screening of large compound libraries.

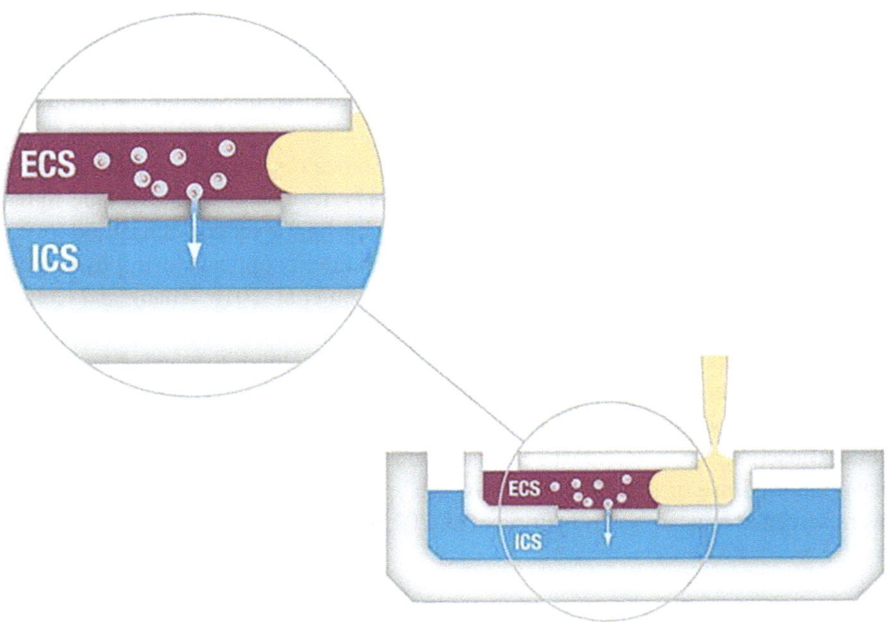

Fig. 3 Chip-based technologies for automated patch clamp recordings with integrated microfluidics. *ECS* extracellular solution, *ICS* intracellular solution. (From Prime et al. 2013)

4.3 Structure-Based Drug Design and Computational Modelling

As described above, a perceived deficit in ion channel drug discovery was the paucity of good structural data, in particular in comparison with GPCRs. However, it is fair to say that ion channel structural research is rapidly closing any existing gap. As eloquently covered by Trim et al. (chapter "Ion Channel Tools and Therapeutics from Venoms and Toxins", this volume), structural studies have progressed through broad epochs: with nuclear magnetic resonance representing the first wave (1994–2006); X-ray diffraction and crystallography representing the second wave (2007–2014); and the present, third wave, (2016–present) represented by cryo-EM.

The determination of ion channel structures by highest resolution X-ray crystallography techniques has been hindered by the generally low protein expression and complex multi-domain topology (leading to a highly dynamic nature and hence biochemical instability) associated with ion channel proteins (Singh et al. 2018). These limitations can be overcome using cryo-EM, where large transmembrane proteins can be observed in near-native conformations (also allowing identification of different channel states). Advancements in cryo-EM, in particular, advances in direct electron detector cameras and computational technologies (Callaway 2020; Nakane et al. 2020), alongside use of improved solubilization tools (such as nanodiscs, amphipols, and styrene-maleic acid copolymers (SMALPs) (Lees et al. 2021)), have led to significant improvements in structural resolution of ion channels within the accepted resolution of at least 2.5 Å required to model ligand-protein interactions (Renaud et al. 2018). For example, the cryo-EM structure of human β3 $GABA_A$ homopentamer (Nakane et al. 2020) was resolved at a resolution of 1.7 Å, allowing an understanding of the structure close to the atomic level. The technological advancements in cryo-EM (Van Drie and Tong 2020) have led to the publication of over 500 unique ligand-bound ion channel structures. The opportunities for exploitation of cryo-EM have been highlighted in the TRP ion channel family, where high resolution structures have been generated for several members (de Oliveira et al. 2021). For example, the binding site of TRPM8 antagonists AMTB and TC-I 2014 has been shown to be within the cavity S1-S4 and the TRP domain (a stretch of amino acids C-terminal to S6) of TRPM8 (Diver et al. 2019).

It is illuminating that many chapters in this volume include high resolution structural data as specific parts of the figures; thus, structures are reported for calcium channel $\alpha 1/\alpha_2\delta$ and CACHD1 subunits (Ablinger et al., chapter "The Presynaptic α2δ Protein Family and Their Therapeutic Potential", this volume), the HERG potassium channel (Gamper et al., chapter "Therapeutic Targeting of Potassium Channels", this volume), human CFTR with bound tezacaftor (Veselu et al., chapter "Cystic Fibrosis Transmembrane Conductance Regulator (CFTR)-Targeted Therapeutics for Cystic Fibrosis", this volume), ASIC channels (Senescall et al., chapter "Targeting Acid-Sensing Ion Channels in Disease", this volume), nicotinic acetylcholine receptors (Wonnacott, chapter "Nicotinic Receptors of the Neuronal and Non-neuronal Cholinergic Systems as Therapeutic Targets: Opportunities and

Limitations", this volume), and the P_2X_7 receptor with bound ATP (Stokes et al., chapter "Purinergic P2X Receptors as Therapeutic Targets", this volume), as well as providing improved resolution of more specialized regions such as the $K_V2.1$ tetramerization domain (Dallas et al., chapter "The Voltage-Gated Potassium Channel Kv2.1 as a Multicellular Drug Target", this volume). The specificity of biologic toxins facilitates useful structural data for toxin-bound receptors, such as bungarotoxin-bound acetylcholine receptors, as shown by Trim et al. (chapter "Ion Channel Tools and Therapeutics from Venoms and Toxins", this volume). In general, the wide availability of resources, such as the RCSB Protein Data Base, now allows easy access to structural information. Such access allows researchers to manipulate information further; for example, Senescall et al. (chapter "Targeting Acid-Sensing Ion Channels in Disease", this volume) model gating changes and drug binding to ASIC channels using PyMOL, an open-source molecular visualization system. This type of manipulation corresponds to the "fourth wave" of ion channel structural biology discussed by Trim et al. (chapter "Ion Channel Tools and Therapeutics from Venoms and Toxins", this volume). Moreover, in the past few years, this space is being increasingly filled with artificial intelligence (AI) technology platforms, exemplified by DeepMind's AlphaFold Protein Structure Database. For example, Ablinger et al. (chapter "The Presynaptic $\alpha_2\delta$ Protein Family and Their Therapeutic Potential", this volume) use AlphaFold technology to superimpose the human CACHD1 protein onto the cryo-EM structure of $Ca_V3.1$ available from the RCSB Protein Data Base. Such technology complements and potentially extends existing computational techniques such as homology modelling, molecular docking, and molecular dynamics simulations of drug-ion channel interactions (Menke et al. 2021). The contribution of machine learning paradigms to quantitative structure-activity relationship (QSAR) modelling is aimed to improve ion channel drug discovery (Zhu et al. 2022; Pliushcheuskaya and Künze 2023) .

4.4 Phenotypic Screening

As above, the majority of screening campaigns make use of cell lines stably expressing the single ion channel target of interest. An alternative approach, which can be considered more translational, is to measure an endpoint in a disease-relevant cell type, a process that is agnostic to target or mechanism. Patient-derived tissue or human iPSC-derived cells provide an attractive alternative to use of primary rodent cells due to their translational value, and that cell production can be scaled to support HTS (Stacey et al. 2018; Deng et al. 2023). For example, Pfizer Neusentis established a phenotypic screen with human iPSC-derived sensory neurons using a 384-well veratridine-evoked Ca^{2+} flux assay with a 2700-compound library (Stacey et al. 2018). As may be expected, use of patient-derived iPSCs in phenotypic screens is starting to identify ion channel targets (e.g., Huang et al. 2021). These studies use "standard" patch clamp electrophysiology and/or genetically encoded Ca^{2+} indicators as screening platforms, as described above, but screen based on disease phenotype.

5 Pharmacogenomics in Ion Channel Drug Discovery

Pharmacogenomics is the study of how genes affect a person's response to particular drugs. While this is an important aspect of pharmacokinetic handing of the drug, here we focus on how knowledge of inherited mutations in ion channel proteins that lead to excitability disorders has helped to reveal new ion channel targets and to consider therapeutic interventions. These diseases have their own specific term, the so-called ion channelopathies. Such genetic diseases can be associated with gain-of-function or loss-of-function mutations, typically leading to changes in biophysical properties of the ion channel protein or its modulation by incident signaling pathways and/or interacting proteome. Typically, gain-of-function mutations in voltage-gated sodium or calcium channels, or loss-of-function mutations in voltage-gated potassium or chloride, lead to diseases of hyperexcitability. There are several examples of prominent channelopathies, including *SCN9A* gene mutations underlying deficit $Na_V1.7$ function in pain disorders and *CFTR* gene mutations leading to cystic fibrosis (Veselu et al., chapter "Cystic Fibrosis Transmembrane Conductance Regulator (CFTR)-Targeted Therapeutics for Cystic Fibrosis", this volume). Other chapters herein discuss case studies of channelopathies in different sodium channels (Wengert et al., chapter "Voltage-Gated Sodium Channels as Drug Targets in Epilepsy-Related Sodium Channelopathies", this volume) and two pore domain potassium channels (Mathie et al., chapter "The Contribution of Genetic Sequencing Information to the Identification and Functional Characterization of Two-Pore Domain (K2P) Channels as Viable Therapeutic Targets", this volume) in more detail. As also discussed by Mathie et al., monogenic diseases are due to mutations in the ion channel protein itself, while polygenic diseases are associated with mutations that increase the likelihood of developing the disease. In terms of drug therapy, while some channelopathies can be treated by direct "targeted pharmacotherapy," for example, by blocking the ion channel encoded by the disease-causing mutation, the majority of therapeutic options offer only symptomatic therapy (Imbrici et al. 2016). The identification of people with a genetic channelopathy gives rise to the possibility of developing precision medicine (as discussed for sodium channel mutations by Wengert et al., chapter "Voltage-Gated Sodium Channels as Drug Targets in Epilepsy-Related Sodium Channelopathies", this volume) or personalized medicine (occasionally considered an interchangeable term, but one that infers treatment unique to an individual). The study of channelopathies has been advanced by the advent of new genetic study regimes. Most notably here are different genome-wide association studies (GWAS) that detail the effects of single nucleotide polymorphisms (SNPs) as genetic risk variants, and next generation sequencing methodologies, such as whole exome sequencing. GWAS technologies for ligand-gated ion channels are discussed in this volume for nicotinic acetylcholine receptors (Wonnacott, chapter "Nicotinic Receptors of the Neuronal and Non-neuronal Cholinergic Systems as Therapeutic Targets: Opportunities and Limitations", this volume) and for purinergic P_2X receptors (Stokes et al., chapter "Purinergic P2X Receptors as Therapeutic Targets", this volume). These technologies are supplemented by the increased availability of continually updating databases that facilitate in silico searching. Genetic studies are also supported by advances in structural biology, which can reveal detailed effects of mutations on ion channel function, and genetic "knock-in" techniques, whereby a targeted insertion of a human ion channel gene can lead to expression of mutant protein to create an animal model of disease (Carbone and Mori 2020). A couple of promising recent examples of drug development against specific channelopathies are highlighted below.

5.1 $Na_V1.2/Na_V1.6$ Inhibitors

Genetic association of voltage-gated sodium channel gain-of-function/loss-of-function mutations (in $Na_V1.1$, $Na_V1.2$, or $Na_V1.6$) with epilepsy represents the most common genetic epilepsy. Gain-of-function $Na_V1.2$ or $Na_V1.6$ variants are also associated with DEE, which are severe epilepsies characterized both by seizures and encephalopathy and a significant developmental delay or loss of developmental skills (see also Wengert et al., chapter "Voltage-Gated Sodium Channels as Drug Targets in Epilepsy-Related Sodium Channelopathies", this volume). Prax-562 is being developed by Praxis Precision Medicines for the treatment of pediatric patients carrying gain-of-function $Na_V1.2$ and $Na_V1.6$ variants underlying DEE (Anderson et al. 2023). Mechanistically, Prax-562 is a nonselective, use-dependent sodium channel blocker that shows a preference for persistent vs. peak currents, which contrasts with antiepileptic sodium channel inhibitors lamotrigine and carbamazepine. Preclinical studies demonstrate an improved therapeutic window compared to the poorly tolerated antiepileptic sodium channel drugs (Kahlig et al. 2022; Mahalingam et al. 2023). Prax-562 is currently in phase II studies (NCT05818553) for treatment of early-onset SCN2A-DEE and SCN8A-DEE patients.

An alternative strategy to treat SCN8A-DEE patients (and adult focal-onset seizures) was adopted by Xenon, who developed a highly selective $Na_V1.6$ inhibitor NBI-921352 (XEN901) (related to the arylsulfonide $Na_V1.7$ inhibitors) that interacts at voltage-sensor domain IV and stabilizes inactivation. This strategy of targeting subtype selectivity rather than gating mode (i.e., persistent vs. peak currents) also supports an improved therapeutic window, as shown in preclinical studies (employing a mouse model of SCN8-DEE) (Johnson et al. 2022). A phase II study for NBI-921352 as a treatment of SCN8A-DEE patents is currently being recruited.

5.2 $Na_V1.1$ Activators

$Na_V1.1$ voltage-gated sodium channels are thought to play a critical role in regulating excitability of fast-spiking parvalbumin-positive interneurons (through clustering at the axon initial segment) (Richards et al. 2018). Dravet syndrome is a pediatric epilepsy associated with cognitive impairment, psychomotor regression, autistic traits, and ataxia. The majority of Dravet syndrome patients carry de novo heterozygous loss-of-function mutations in the *SCN1A* gene that encodes $Na_V1.1$, thought to cause GABAergic interneuron dysfunction (see also Wengert et al., chapter "Voltage-Gated Sodium Channels as Drug Targets in Epilepsy-Related Sodium Channelopathies", this volume). Xenon has developed highly selective CNS-penetrant $Na_V1.1$ openers (a single series has been reported that includes XPC-8770 and XPC-7523), which cause a depolarizing shift in steady-state inactivation. XPC-8770 increases excitability of interneurons, has efficacy in the 6 Hz seizure model, and improves motor performance in the rotarod assay in *Scn1a*$^{+/-}$ mice (Goodchild et al. 2022).

6 Final Perspectives: How to Avoid Ion Channel Drugs Becoming Lost in Translation

This introductory, scene-setting chapter illustrates that ion channels are the basis for excitable neural networks, are fundamental to basic biology, and represent important targets for treatment of disease, in particular, in conditions where imbalance of excitability disrupts normal homeostasis. As with many initial therapeutic advantages seen since the advent of modern drug discovery from ~1950, ion channel drug discovery has slowed noticeably since the millennium. We propose that the new advances and refinements presented here suggest that we are potentially approaching a renaissance, as evidenced primarily by the number of new ion channel drugs in different stages of clinical trials. Such agents include representatives from the small molecule class that has dominated the ion channel drug field but also include newer biologic agents. Alongside this clinical impetus, higher throughput screening regimes including fluorescence-based assays and automated electrophysiology ideally suited to ion channel function, as well as newer phenotypic screening protocols, including those utilizing human iPSCs, are also making an impact. We also posit that the availability of improved structural biology techniques, including the solving of almost all ion channel structures at a requisite high resolution and the deposition of such information on accessible databases, offers real potential for meaningful structure-based drug design. Advances can also be achieved via virtual screening platforms with the help of computational and molecular modelling and docking studies. Moreover, new pathway analysis techniques can contribute further to ion channel drug discovery. No doubt these computational techniques will benefit, in common with many other areas, from the ubiquitous presence and likely quantum leaps in AI technology. Such points serve to illustrate the continued need for extensive basic research and full use of new enabling technologies. We also advocate the advantages of a joint academic/industry approach to drug discovery and ultimate translation. However, it is important to emphasize that such advances must be driven by biology and, therefore, must be empirical and proven by experimentation.

As discussed above, ion channels are low abundance proteins that are difficult to target selectively. It also appears clear that use of small molecule drugs will need to be complemented by biologic drugs. Although small molecules typically have properties conforming to the well-developed rules for "druglikeness" that appear well-suited for ion channel blockers or activators, they often suffer in terms of poor selectivity and, thus, unwanted off-target effects; such effects can also limit efficacy at concentrations predicted to be active using in vitro target development. Oral small molecules may also be subject to limited bioavailability due to first-pass metabolism and/or plasma protein binding; they may be metabolized to toxic compounds and/or undergo unwanted drug-drug interactions, among other potential limitations. In this regard, biologic drugs such as peptides and larger antibodies have advantages in high target selectivity and more predictable metabolism but have their own limitations, including short half-life, low oral bioavailability, low volume of distribution, triggering of immune responses to foreign biological material, and poor penetrability. Although the aim that biologic drugs can overcome such limitations is far from being achieved, there are enough positive indications to give us genuine hope that peptides and/or antibodies can start to make an increased contribution to the armory of ion channel drugs, as they have for other molecular drug targets. It is also hoped that the range of advances in next generation therapeutics, not only biologics, but also nucleotide-based therapeutics and cell-based therapeutics, will also extend the range of ion channel drugs. As stated above, drug engineering strategies are likely to be key in producing compounds that accentuate the positive and eliminate the negative properties associated with each drug class; an exciting development is the search for hybrid compounds that can meet this aim.

Due to the specialized nature of ion channels as proteins with conduction pathways that pass charged ions across a lipid bilayer, mechanistic studies typically involve fluorescence ion-sensitive dye assays and/or electrophysiology. Technical developments described above have allowed higher throughput screening and automation of patch clamp techniques that, certainly at early-stage discovery in industry, have largely superseded manual patch clamp. As automated patch clamp technology becomes more compact and/or more affordable, it is likely that academic labs also take advantage of such systems. The use of electrophysiology methodologies has advantages in allowing investigation of those agents designed to achieve functional selectivity via state-dependent or use-dependent actions. Although there have been successes with channels openers (e.g., with potassium channel openers (Gamper et al., chapter "Therapeutic Targeting of Potassium Channels", this volume) and CFTR (Veselu et al., chapter "Cystic Fibrosis Transmembrane Conductance Regulator (CFTR)-Targeted Therapeutics for Cystic Fibrosis", this volume)), it is fair to say strategies that have prioritized small molecule blockers that target the inactivated state of $Na_V1.7$ or $Na_V1.8$ or $Ca_V2.2$ have, to now, largely failed and that newer peptides/antibodies likely lack such dependency characteristics.

In addition to the issues inherent with targeting ion channels, there are also concerns regarding drug discovery for those diseases where ion channels are the predicted target of choice, but where an unmet clinical need still persists. The treatment of chronic pain represents one such condition (or group of conditions), and this is an area in which we have a research interest (Stevens and Stephens 2018). A principal concern is that the long and widely used animal models of pain typically measure drug effects on acute rather than chronic pain; moreover, reflexive responses to insults such as mechanical filaments, heat, or cold are the tests of choice. Such reflex responses typically fail to engage with higher brain function and poorly model the human condition, in particular the emotional aspects of pain. Even measuring these reflex responses has inherent concerns regarding operator bias, and it is essential that appropriate randomization and blinding procedures are in place. These specific concerns go alongside more general points regarding the typical use of inbred stains and use of younger, often male, animals that likely do not reflect the more heterologous human population. As alluded to, there is an ongoing issue of sex as a biological variable in drug research, this of particular relevance in pain research (see Shansky and Murphy 2021). There is also a move to develop more human-relevant models, such as those that reflect pain due to use of chemotherapeutic agents, viral infection, or visceral pain to improve clinical translatability.

However, advances in the area of pain research, whereby a number of recent developments are combined, may provide significant benefits. For example, Clifford Woolf's group (Boston's Children Hospital and Harvard Medical School, USA) has developed a range of innovative methodologies that have clear potential to improve development of analgesic drugs (including those targeting ion channels). A phenotypic screening regime uses a strategy involving the use of human iPSCs derived from patient pain (vs. other) pathways that are tested using ion flux assays, automated patch clamp, and multielectrode array platforms, supported by automated measurements of spontaneous pain in experimental animals (Zhang et al. 2022). The use of the latter technology is argued to better mirror the type of pain experienced in the human condition and is also designed to combat concerns about translatability, for example, in the use of filaments to quantifying tactile sensitivity as a measure of mechanical allodynia. It also remains to be seen how the recent adoption of the FDA Modernization Act 2.0 by the US Congress and Senate, which allows drugs to proceed to human testing using either animal or nonanimal tests, will have an impact. Thus, it will be of interest to follow the development of alternative techniques to animal models, such as those involving computer modelling and tissue or organ chip technology (Wadman 2023).

Finally here, there are a number of general points that can improve translation from preclinical drug discovery into human trials and onward. A general concern in drug discovery is the lack of efficacy at phase III. This may be of particular concern for small molecule ion channel drugs. Thus, the lack of selectivity/off-target effects of small molecule drugs likely further compromises efficacy in humans. A specific example here is that analgesia was typically not attained until plasma concentration of $Na_V1.7$ inhibitors exceeded in vitro IC_{50} values and that degree of block required may not be attainable in humans (as discussed by Eagles et al. 2022). Such points likely hold true for other ion channel blockers.

Improved stratification of clinical trial participants is also required. This could be achieved using appropriate biomarkers for a given indication, although for many conditions these are elusive. More specifically, knowledge of genetic channelopathies could be used to stratify participants to allow precision medicine to develop.

Overall, as the world emerges into a postpandemic landscape, we suggest that ion channel drug discovery is standing more at a threshold than a crossroad. The present volume aims to substantiate this proposal with important contributions from experts in the field of voltage-gated and ligand-gated ion channels, each outlining the current state of play for their specific ion channel of interest and offering insights into recent therapeutic development, as well as highlighting the challenges facing the translation of preclinical advances into clinical reality.

References

Akopian AN, Souslova V, England S, Okuse K, Ogata N, Ure J, et al. The tetrodotoxin-resistant sodium channel SNS has a specialized function in pain pathways. Nat Neurosci. 1999;2:541–8.

Anderson L, Jancovski N, Eckert W, Baker E, Wagley P, Maljevic S, et al. PRAX-562 is a well-tolerated, novel persistent sodium channel blocker with broad anticonvulsant activity in multiple DEE mouse models (S34.007). Neurology. 2023;100(17 Suppl 2):2118.

Andersson DA, Gentry C, Moss S, Bevan S. Transient receptor potential A1 is a sensory receptor for multiple products of oxidative stress. J Neurosci. 2008;28(10):2485–94.

Audi S, Burrage DR, Lonsdale DO, Pontefract S, Coleman JJ, Hitchings AW, et al. The 'top 100' drugs and classes in England: an updated 'starter formulary' for trainee prescribers. Br J Clin Pharmacol. 2018;84(11):2562–71.

Balestrini A, Joseph V, Dourado M, Reese RM, Shields SD, Rougé L, et al. A TRPA1 inhibitor suppresses neurogenic inflammation and airway contraction for asthma treatment. J Exp Med. 2021;218(4):e20201637.

Beeton C, Wulff H, Standifer NE, Azam P, Mullen KM, Pennington MW, et al. Kv1.3 channels are a therapeutic target for T cell-mediated autoimmune diseases. Proc Natl Acad Sci U S A. 2006;103(46):17414–9.

Boddum K, Hougaard C, Xiao-Ying Lin J, von Schoubye NL, Jensen HS, Grunnet M, et al. $K_v3.1$/$K_v3.2$ channel positive modulators enable faster activating kinetics and increase firing frequency in fast-spiking GABAergic interneurons. Neuropharmacology. 2017;118:102–12.

Borja GB, Zhang H, Harwood BN, Jacques J, Grooms J, Chantre RO, et al. Highly parallelized, multicolor optogenetic recordings of cellular activity for therapeutic discovery applications in ion channels and disease-associated excitable cells. Front Mol Neurosci. 2022;15:896320.

Botte M, Huber S, Bucher D, Klint JK, Rodríguez D, Tagmose L, et al. Apo and ligand-bound high resolution Cryo-EM structures of the human Kv3.1 channel reveal a novel binding site for positive modulators. PNAS Nexus. 2022;1(3):pgac083.

Brown DA, Passmore GM. Neural KCNQ (Kv7) channels. Br J Pharmacol. 2009;156(8):1185–95.

Bruen C, Al-Saadi M, Michelson EA, Tanios M, Mendoza-Ayala R, Miller J, et al. Auxora vs. placebo for the treatment of patients with severe COVID-19 pneumonia: a randomized-controlled clinical trial. Crit Care. 2022;26(1):101.

Callaway E. 'It opens up a whole new universe': revolutionary microscopy technique sees individual atoms for first time. Nature. 2020;582(7811):156–7.

Carbone E, Mori Y. Ion channelopathies to bridge molecular lesions, channel function, and clinical therapies. Pflugers Arch. 2020;472(7):733–8.

Catterall WA. Ion channels in plasma membrane signal transduction. J Bioenerg Biomembr. 1996;28(3):217–8.

Chappe YL, Pierredon S, Joushomme A, Molle P, Garenne A, Canovi A. Genetically-encoded BRET probes shed light on ligand bias-induced variable ion selectivity in TRPV1 and P2X5/7. Proc Natl Acad Sci U S A. 2022;119(46):e2205207119.

Chen J, Joshi SK, DiDomenico S, Perner RJ, Mikusa JP, Gauvin DM, et al. Selective blockade of TRPA1 channel attenuates pathological pain without altering noxious cold sensation or body temperature regulation. Pain. 2011;152(5):1165–72.

Contreras-García IJ, Cárdenas-Rodríguez N, Romo-Mancillas A, Bandala C, Zamudio SR, Gómez-Manzo S, et al. Levetiracetam mechanisms of action: from molecules to systems. Pharmaceuticals. 2022;15(4):475.

Coward K, Plumpton C, Facer P, Birch R, Carlstedt T, Tate S, et al. Immunolocalization of SNS/PN3 and NaN/SNS2 sodium channels in human pain states. Pain. 2000;85(1–2):41–50.

Daily NJ, Du ZW, Wakatsuki T. High-throughput phenotyping of human induced pluripotent stem cell-derived cardiomyocytes and neurons using electric field stimulation and high-speed fluorescence imaging. Assay Drug Dev Technol. 2017;15(4):178–88.

de Oliveira TM, van Beek L, Shilliday F, Debreczeni JÉ, Phillips C. Cryo-EM: the resolution revolution and drug discovery. SLAS Discov. 2021;26(1):17–31.

Deng T, Jovanovic VM, Tristan CA, Weber C, Chu PH, Inman J, et al. Scalable generation of sensory neurons from human pluripotent stem cells. Stem Cell Rep. 2023;18(4):1030–47.

Diver MM, Cheng Y, Julius D. Structural insights into TRPM8 inhibition and desensitization. Science. 2019;365(6460):1434–40.

DuBreuil DM, Chiang BM, Zhu K, Lai X, Flynn P, Sapir Y, et al. A high-content platform for physiological profiling and unbiased classification of individual neurons. Cell Rep Methods. 2021;1(1):100004.

Eagles DA, Chow CY, King GF. Fifteen years of $Na_V1.7$ channels as an analgesic target: why has excellent in vitro pharmacology not translated into in vivo analgesic efficacy? Br J Pharmacol. 2022;179(14):3592–611.

Faber CG, Lauria G, Merkies IS, Cheng X, Han C, Ahn HS, et al. Gain-of-function Nav1.8 mutations in painful neuropathy. Proc Natl Acad Sci U S A. 2012;109(47):19444–9.

Feldman P, Khanna R. Challenging the catechism of therapeutics for chronic neuropathic pain: targeting $Ca_V2.2$ interactions with CRMP2 peptides. Neurosci Lett. 2013;557 Pt A(0 0):27–36.

Feske S, Concepcion AR, Coetzee WA. Eye on ion channels in immune cells. Sci Signal. 2019;12(572):eaaw8014.

Fomina AF, Nguyen HM, Wulff H. Kv1.3 inhibition attenuates neuroinflammation through disruption of microglial calcium signaling. Channels. 2021;15(1):67–78.

French J, Porter R, Perucca E, Brodie M, Rogawski M, Pimstone S, et al. Phase 2b efficacy and safety of XEN1101, a novel potassium channel opener, in adults with focal epilepsy (X-TOLE) (P12-8.006). Neurology. 2022;98(18 Suppl):415.

Ghovanloo MR, Tyagi S, Zhao P, Kiziltug E, Estacion M, Dib-Hajj SD, et al. High-throughput combined voltage-clamp/current-clamp analysis of freshly isolated neurons. Cell Rep Methods. 2023;3(1):100385.

Goodchild SJ, Burford K, Dube C, Thouta S, Parrish R, Williams AD, et al. Molecularly selective $Na_V1.1$ potentiators increase PV+ fast-spiking interneuron excitability and restore motor performance in a mouse model of Dravet syndrome. Poster No. 3.049, American Epilepsy Society Annual Meeting, Nashville, TN, December 2–6, 2022; 2022.

Gou X, Yu X, Bai D, Tan B, Cao P, Qian M, et al. Pharmacology and mechanism of action of HSK16149, a selective ligand of $\alpha 2\delta$ subunit of voltage-gated calcium channel with analgesic activity in animal models of chronic pain. J Pharmacol Exp Ther. 2021;376(3):330–7.

Haustrate A, Hantute-Ghesquier A, Prevarskaya N, Lehen'kyi V. Monoclonal antibodies targeting ion channels and their therapeutic potential. Front Pharmacol. 2019;10:606.

Herrington J, Arey BJ. Conformational mechanisms of signaling bias of ion channels. In: Arey BJ, editor. Biased signaling in physiology, pharmacology and therapeutics. 1st ed. Academic; 2014. p. 173–207.

Hill CL, Stephens GJ. An introduction to patch clamp recording. Methods Mol Biol. 2021;2188:1–19.

Hou P, Zhang R, Liu Y, Feng J, Wang W, Wu Y, et al. Physiological role of Kv1.3 channel in T lymphocyte cell investigated quantitatively by kinetic modeling. PLoS One. 2014;9(3):e89975.

Huang CJ, Harootunian A, Maher MP, Quan C, Raj CD, McCormack K, et al. Characterization of voltage-gated sodium-channel blockers by electrical stimulation and fluorescence detection of membrane potential. Nat Biotechnol. 2006;24(4):439–46.

Huang X, Roet KCD, Zhang L, Brault A, Berg AP, Jefferson AB, et al. Human amyotrophic lateral sclerosis excitability phenotype screen: target discovery and validation. Cell Rep. 2021;35(10):109224.

Hutchings CJ, Colussi P, Clark TG. Ion channels as therapeutic antibody targets. MAbs. 2019;11(2):265–96.

Imbrici P, Liantonio A, Camerino GM, De Bellis M, Camerino C, Mele A, et al. Therapeutic approaches to genetic ion channelopathies and perspectives in drug discovery. Front Pharmacol. 2016;7:121.

Jin L, Boyd BJ, Larson IC, Pennington MW, Norton RS, Nicolazzo JA. Enabling noninvasive systemic delivery of the Kv1.3-blocking peptide HsTX1[R14A] via the buccal mucosa. J Pharm Sci. 2016;105(7):2173–9.

Johnson JP, Focken T, Khakh K, Tari PK, Dube C, Goodchild SJ, et al. NBI-921352, a first-in-class, $Na_V1.6$ selective, sodium channel inhibitor that prevents seizures in *Scn8a* gain-of-function mice, and wild-type mice and rats. elife. 2022;11:e72468.

Jones J, Correll DJ, Lechner SM, Jazic I, Miao X, Shaw D, et al. VX21-548-101 and VX21-548-102 trial groups. Selective inhibition of $Na_V1.8$ with VX-548 for acute pain. N Engl J Med. 2023;389(5):393–405.

Jovčevska I, Muyldermans S. The therapeutic potential of nanobodies. BioDrugs. 2020;34(1):11–26.

Kaar SJ, Angelescu I, Nour MM, Marques TR, Sharman A, Sajjala A, et al. The effects of AUT00206, a novel Kv3.1/3.2 potassium channel modulator, on task-based reward system activation: a test of mechanism in schizophrenia. Psychopharmacology. 2022;239(10):3313–23.

Kahlig KM, Scott L, Hatch RJ, Griffin A, Martinez Botella G, Hughes ZA, et al. The novel persistent sodium current inhibitor PRAX-562 has potent anticonvulsant activity with improved protective index relative to standard of care sodium channel blockers. Epilepsia. 2022;63(3):697–708.

Kanai K, Kuwabara S, Misawa S, Tamura N, Ogawara K, Nakata M, et al. Altered axonal excitability properties in amyotrophic lateral sclerosis: impaired potassium channel function related to disease stage. Brain. 2006;129(4):953–62.

Kim RY, Yau MC, Galpin JD, Seebohm G, Ahern CA, Pless SAYD, et al. Atomic basis for therapeutic activation of neuronal potassium channels. Nat Commun. 2015;6:8116.

Knöpfel T, Song C. Optical voltage imaging in neurons: moving from technology development to practical tool. Nat Rev Neurosci. 2019;20(12):719–27.

Kremeyer B, Lopera F, Cox JJ, Momin A, Rugiero F, Marsh SYD, et al. A gain-of-function mutation in TRPA1 causes familial episodic pain syndrome. Neuron. 2010;66(5):671–80.

Lees JA, Dias JM, Han S. Applications of Cryo-EM in small molecule and biologics drug design. Biochem Soc Trans. 2021;49(6):2627–38.

Lin CS, Boltz RC, Blake JT, Nguyen M, Talento A, Fischer PA, et al. Voltage-gated potassium channels regulate calcium-dependent pathways involved in human T lymphocyte activation. J Exp Med. 1993;177(3):637–45.

Lou S, Adam Y, Weinstein EN, Williams E, Williams K, Parot VYD, et al. Genetically targeted all-optical electrophysiology with a transgenic cre-dependent optopatch mouse. J Neurosci. 2016;36(43):11059–73.

Mahalingam R, Oldham M, Puryear C, Bansal P, Sriram B, Patel D, et al. PRAX-562-102: a phase 1 trial evaluating the safety, tolerability, pharmacokinetics and pharmacodynamics of PRAX-562 in healthy volunteers (P4-9.011). Neurology. 2023;100(17 Supplement 2):3192.

Menke J, Maskri S, Koch O. Computational ion channel research: from the application of artificial intelligence to molecular dynamics simulations. Cell Physiol Biochem. 2021;55(S3):14–45.

Molokanova E, Mercola M, Savchenko A. Bringing new dimensions to drug discovery screening: impact of cellular stimulation technologies. Drug Discov Today. 2017;22(7):1045–55.

Morgenstern TJ, Park J, Fan QR, Colecraft HM. A potent voltage-gated calcium channel inhibitor engineered from a nanobody targeted to auxiliary $Ca_V\beta$ subunits. elife. 2019;8:e49253.

Morgenstern TJ, Nirwan N, Hernández-Ochoa EO, Bibollet H, Choudhury P, Laloudakis YD, et al. Selective posttranslational inhibition of $Ca_V\beta 1$-associated voltage-dependent calcium channels with a functionalized nanobody. Nat Commun. 2022;13(1):7556.

Nakane T, Kotecha A, Sente A, McMullan G, Masiulis S, Brown PMGE, et al. Single-particle cryo-EM at atomic resolution. Nature. 2020;587(7832):152–6.

Oliphant CJ, Freeman E, Gilbert S, Lara R, Huang E, Barden J, et al. BIL06v immunizations generate antibodies specific for non-functional P2X7 (nfP2X7) that target solid tumors in vivo. Cancer Res. 2019;79(13_Suppl):LB-201.

Patel R, Rutten K, Valdor M, Schiene K, Wigge S, Schunk S, et al. Electrophysiological characterization of activation state-dependent Ca(v)2 channel antagonist TROX-1 in spinal nerve injured rats. Neuroscience. 2015;297:47–57.

Patel R, Montagut-Bordas C, Dickenson AH. Calcium channel modulation as a target in chronic pain control. Br J Pharmacol. 2018;175(12):2173–84.

Pliushcheuskaya P, Künze G. Recent advances in computer-aided structure-based drug design on ion channels. Int J Mol Sci. 2023;24(11):9226.

Prime R, Brown R, Kammonen J, Kirkup AJ, Stevens E. Automated patch clamp: advantages and limitations. In: Roberts GCK, editor. Encyclopedia of biophysics. Berlin/Heidelberg: Springer; 2013.

Reese RM, Dourado M, Anderson K, Warming S, Stark KL, Balestrini A, et al. Behavioral characterization of a CRISPR-generated TRPA1 knockout rat in models of pain, itch, and asthma. Sci Rep. 2020;10(1):979.

Renaud JP, Chari A, Ciferri C, Liu WT, Rémigy HW, Stark H, et al. Cryo-EM in drug discovery: achievements, limitations and prospects. Nat Rev Drug Discov. 2018;17(7):471–92.

Richards KL, Milligan CJ, Richardson RJ, Jancovski N, Grunnet M, Jacobson LH, et al. Selective $Na_V1.1$ activation rescues Dravet syndrome mice from seizures and premature death. Proc Natl Acad Sci U S A. 2018;115(34):E8077–85.

Rowe SM, Zuckerman JB, Dorgan D, Lascano J, McCoy K, Jain M, et al. Inhaled mRNA therapy for treatment of cystic fibrosis: interim results of a randomized, double-blind, placebo-controlled phase 1/2 clinical study. J Cyst Fibros. 2023;22(4):656–64.

Shansky RM, Murphy AZ. Considering sex as a biological variable will require a global shift in science culture. Nat Neurosci. 2021;24(4):457–64.

Singh AK, McGoldrick LL, Saotome K, Sobolevsky AI. X-ray crystallography of TRP channels. Channels. 2018;12(1):137–52.

Skolnick P. The opioid epidemic: crisis and solutions. Annu Rev Pharmacol Toxicol. 2018;58:143–59.

Stacey P, Wassermann AM, Kammonen L, Impey E, Wilbrey A, Cawkill D. Plate-based phenotypic screening for pain using human iPSC-derived sensory neurons. SLAS Discov. 2018;23(6):585–96.

Stephens GJ, Bell DC. 19th world congress of basic and clinical pharmacology, July 2nd–7th, 2023, Glasgow. Bioelectricity. 2023;5(3):227–30.

Stephens GJ, Shukla AK. Targeting the cell's gatekeepers for novel drug discovery. Br J Pharmacol. 2022;179(14):3485–6.

Stevens EB, Stephens GJ. Recent advances in targeting ion channels to treat chronic pain. Br J Pharmacol. 2018;175(12):2133–7.

Strohl WR. Fusion proteins for half-life extension of biologics as a strategy to make biobetters. BioDrugs. 2015;29(4):215–39.

Suk HJ, Boyden ES, van Welie I. Advances in the automation of whole-cell patch clamp technology. J Neurosci Methods. 2019;326:108357.

Tanner MR, Tajhya RB, Huq R, Gehrmann EJ, Rodarte KE, Atik MA, et al. Prolonged immunomodulation in inflammatory arthritis using the selective Kv1.3 channel blocker HsTX1[R14A] and its PEGylated analog. Clin Immunol. 2017;180:45–57.

Tarcha EJ, Olsen CM, Probst P, Peckham D, Muñoz-Elías EJ, Kruger JG, et al. Safety and pharmacodynamics of dalazatide, a Kv1.3 channel inhibitor, in the treatment of plaque psoriasis: a randomized phase 1b trial. PLoS One. 2017;12(7):e0180762.

Taskin B, von Schoubye NL, Sheykhzade M, Bastlund JF, Grunnet M, Jespersen T. Biophysical characterization of KV3.1 potassium channel activating compounds. Eur J Pharmacol. 2015;758:164–70.

Thakur S, Sinhari A, Jain P, Jadhav HR. A perspective on oligonucleotide therapy: approaches to patient customization. Front Pharmacol. 2022;13:1006304.

Tikhonov DB, Zhorov BS. Mechanism of sodium channel block by local anesthetics, antiarrhythmics, and anticonvulsants. J Gen Physiol. 2017;149(4):465–81.

Van Drie JH, Tong L. Cryo-EM as a powerful tool for drug discovery. Bioorg Med Chem Lett. 2020;30(22):127524.

Vogl C, Mochida S, Wolff C, Whalley BJ, Stephens GJ. The synaptic vesicle glycoprotein 2A ligand levetiracetam inhibits presynaptic Ca^{2+} channels through an intracellular pathway. Mol Pharmacol. 2012;82(2):199–208.

Wadman M. FDA no longer has to require animal testing for new drugs. Science. 2023;379(6628):127–8.

Wainger BJ, Macklin EA, Vucic S, McIlduff CE, Paganoni S, Maragakis NJ, et al. Effect of ezogabine on cortical and spinal motor neuron excitability in Amyotrophic Lateral Sclerosis: a randomized clinical trial. JAMA Neurol. 2021;78(2):186–96.

Wang CK, Lamothe SM, Wang AW, Yang RY, Kurata HT. Pore- and voltage sensor-targeted KCNQ openers have distinct state-dependent actions. J Gen Physiol. 2018;150(12):1722–34.

Wang X, Li G, Guo J, Zhang Z, Zhang S, Zhu Y, et al. Kv1.3 channel as a key therapeutic target for neuroinflammatory diseases: state of the art and beyond. Front Neurosci. 2020;13:1393.

Wulff H, Christophersen P, Colussi P, Chandy KG, Yarov-Yarovoy V. Antibodies and venom peptides: new modalities for ion channels. Nat Rev Drug Discov. 2019;18(5):339–57.

Yan D, Tomita S. Defined criteria for auxiliary subunits of glutamate receptors. J Physiol. 2012;590(1):21–31.

Yu HB, Li M, Wang WP, Wang XL. High throughput screening technologies for ion channels. Acta Pharmacol Sin. 2016;37(1):34–43.

Zhang H, Cohen AE. Optogenetic approaches to drug discovery in neuroscience and beyond. Trends Biotechnol. 2017;35(7):625–39.

Zhang Z, Roberson DP, Kotoda M, Boivin B, Bohnslav JP, González-Cano R, et al. Automated preclinical detection of mechanical pain hypersensitivity and analgesia. Pain. 2022;163(12):2326–36.

Zhu Z, Deng Z, Wang Q, Wang Y, Zhang D, Xu R, et al. Simulation and machine learning methods for ion-channel structure determination, mechanistic studies and drug design. Front Pharmacol. 2022;13:939555.

Conservation of Ligand Binding Between Voltage-Gated Sodium and T-Type Calcium Channels

Rocio K. Finol-Urdaneta, Jeffrey R. McArthur, and David J. Adams

Abstract In vertebrates, two types of ion channels are responsible for amplifying electrical signals and facilitating neurotransmission: voltage-gated sodium (Na_v1) and calcium (Ca_v3) channels. When the cell membrane is moderately depolarized, these channels allow the rapid influx of their respective ions – Na^+ and Ca^{2+}. Na_v1 and Ca_v3 channels are structurally related, and their expression overlaps in various excitable tissues, such as muscles and neurons. Dysfunction of these depolarization-activated cation channels can lead to various diseases including neuropathic pain, epilepsy, atrial fibrillation, and cancer. Hence, compounds that can modulate the function of both Na_v1 and Ca_v3 channels have the potential to be used as therapeutic agents. This chapter will primarily focus on venom-derived peptide toxins and small molecules that dually affect Na_v1 and Ca_v3 channels.

Keywords Gating modifier peptides · Pharmacophore · Venom-derived peptides · Voltage-gated ion channels · Voltage sensor

Voltage-gated sodium (Na_v) and calcium (Ca_v) ion channels are the primary mediators of electrical signal amplification and neurotransmission in the nervous system. Malfunction of these channels can lead to various disease states, such as neuropathic pain, epilepsy, atrial fibrillation, and cancer. Therefore, compounds capable of modulating Na_v1 and Ca_v3 channel function are considered prospective therapeutic agents (Cain and Snutch 2011; Simms and Zamponi 2014).

Na_v and Ca_v channels are large transmembrane proteins present in central and peripheral neurons, as well as skeletal, cardiac, and smooth muscle. As their name suggests, alterations in the plasma membrane electrical potential trigger conformational changes that lead to the opening of a permeation pathway, allowing the flow of ions that underlie electrical signaling in all excitable tissues.

R. K. Finol-Urdaneta · J. R. McArthur · D. J. Adams (✉)
Faculty of Science Medicine and Health, Illawarra Health and Medical Research Institute (IHMRI), University of Wollongong, Wollongong, NSW, Australia
e-mail: rfinolu@uow.edu.au; djadams@uow.edu.au

G. Stephens, E. Stevens (eds.), *Ion Channels as Targets in Drug Discovery*,
https://doi.org/10.1007/978-3-031-52197-3_2

Na_v channels play a crucial role in the generation and propagation of action potentials in neurons and muscle. When these are stimulated, Na_v channels on their membrane open, allowing a rapid influx of sodium ions into the cell. This influx of positively charged ions depolarizes the membrane and triggers the opening of adjacent Na_v channels, leading to a self-propagating wave of depolarization that constitutes the action potential. Na_v channels thus serve as the initial trigger for the electrical signals that underlie communication between excitable cells and the functioning of the nervous system as a whole (Hille 2001).

Ca_v channels support the normal functioning of neurons and other excitable cells. Calcium channels are important in pain management due to their regulatory role in neurotransmitter release at nociceptive synapses, which are the connections between nerve cells that transmit pain signals. When these synapses are activated, they release neurotransmitters such as glutamate, which stimulate pain receptors in the nervous system and contribute to the sensation of pain. Ca_vs plays a key role in modulating this process by controlling the influx of calcium ions, which in turn regulates neurotransmitter release and the intensity of pain signals (Catterall et al. 2005; Simms and Zamponi 2014).

Voltage-activated cation channels consist of a ~260 kD pore-forming α-subunit. Na_v and Ca_v subtypes are diverse, with different biophysical and pharmacological properties and varying tissue distribution. The canonical voltage-gated Na_v channels belong to the Na_v1 family, of which there are nine identified mammalian subtypes ($Na_v1.1$–$Na_v1.9$). In turn, Ca_v channels have been classified based on their biophysical properties into low-voltage-activated (LVA) channels consisting of T-type channels ($Ca_v3.1$ to $Ca_v3.3$) that require moderate membrane depolarization to open and high-voltage-activated (HVA) channels that require stronger depolarization for activation (Catterall et al. 2005). With the help of pharmacological tools, different types of HVA channels can be further distinguished into L-type ($Ca_v1.1$ to $Ca_v1.4$), P/Q-type ($Ca_v2.1$), N-type ($Ca_v2.2$), and R-type ($Ca_v2.3$) channels.

The α-subunits of all channel members in these four channel families share a common overall architecture. They consist of four distinct domains (DI–DIV), with intracellular N- and C-termini. Each domain is composed of six transmembrane segments (S1–S6). The voltage-sensing module (VSM) is made up of S1–S4, while the pore module (PM) is formed by S5, the pore loop, and S6. The PM is where ionic permeation occurs, and ion selectivity is determined at this site. The conformational changes that underlie ion channel gating and function have been thoroughly defined for Na_v1 channels and are assumed to be generally conserved for all Ca_v channels. In the resting state, channels are closed, and the intracellular gate in DIV prevents ion flow. Upon depolarization, positively charged residues in the S4 segments from each domain cause the sequential outward movement of the VSMs, in a motion that is coupled to the opening and closing of the permeation pathway. Channel opening is caused by VSM activation of DI–III, whereas in Na_v channels, activation of DIV's VSM uncovers the site where the fast inactivation particle binds.

The α-subunits of Na_v1 and Ca_v3 channels are sufficient to confer full voltage-activated cation channel functionality. Nevertheless, Na_v1 α-subunits may be associated with one or two auxiliary β-subunits, of which there are four known subtypes: β1 (36 kD), β2 (33 kD), β3 (25 kD), and β4 (38 kD). The β-subunits comprise a single transmembrane segment, short intracellular C-terminus, and large extracellular N-terminal domain that harbors two β-sheets in an immunoglobulin-like fold. The association of β-subunits to channel-forming α-subunits changes channel kinetics and voltage-dependent gating to that normally seen in native cells. Modulation of Na_v α-subunits by β-subunits is thought to mainly occur via interaction between the extracellular domains of both subunits (Isom 2001). However, two studies suggest that the intracellular domains of voltage-gated sodium channel (VGSC) α- and β-subunits also interact and that their interplay may be important for modulating certain VGSC subtypes (Meadows et al. 2001; Zimmer and Benndorf 2002). Functional activity can still be observed in the absence of modulatory β-subunits. In stark contrast with Na_v1s and Ca_v3s, Ca_v1 and Ca_v2 channels require obligatory association with auxiliary subunits for their function. Therefore, HVA calcium channels consist of a principal functional pore-forming α1-subunit that associates with an α2δ dimer (disulfide-linked), an intracellular β-subunit (phosphorylated), and often a transmembrane γ-subunit (Curtis and Catterall 1984; Takahashi and Catterall 1987). It was recently demonstrated that CACHD1 is structurally an α2δ-like protein that functionally modulates Ca_v3 activity (Cottrell et al. 2018).

Importantly, Na_v1 and Ca_v3 channels share functional properties, such as similar voltage activation thresholds and pharmacology. Therefore, this chapter will focus on peptide toxins and small molecules that act on Na_v1 and Ca_v3 channels expressed in mammalian sensory neurons. Multiple studies have identified that the binding sites of venom-derived peptide toxins and small molecules are restricted to well-defined and structurally conserved areas or domains (Stevens et al. 2011; Van Theemsche et al. 2020).

1 Toxin/Drug Binding Sites

Venom-derived peptides and small molecule inhibitors of voltage-gated ion channels, including Na_v1 subtypes, which are the best-studied group, have been shown to interact with eight distinct binding sites. These compounds can be grouped into "Site 1" drugs that interact with the extracellular pore domain; "Site 2 and 5 and LA binding site" drugs that interact with the inner vestibule; and "Site 3, 4, and 6 and ICA/PF" drugs that interact with the voltage-sensing domains of Na_v1 subunits (de Lera Ruiz and Kraus 2015). Venom-derived peptides typically interact with the voltage-sensing domains and extracellular pore domain, as they are unable to cross the membrane. On the other hand, most of the small molecules interact with the extracellular pore domain or inner vestibule due to their hydrophobic nature.

2 Venom-Derived Peptides Dually Targeting Na_v1 and Ca_v3 Channels

Venomous animals and plants devote substantial resources to producing and optimizing molecules that aid their survival, whether for defence or as a means to capture prey (Finol-Urdaneta et al. 2022). Among these, venom-derived peptides (VDPs) constitute a treasure trove of tried and tested powerful tools against intended and unintended targets.

Cysteine-rich VDPs are abundant in nature, and thus far, only a handful of spider venom peptides that target Na_v1 and Ca_v3 channels have been identified (Fig. 1). The fact that this dual activity has been observed only in spiders' venoms may be a reflection of biases in activity screening campaigns.

The overarching trends among Na_v/Ca_v3 active VDPs' current inhibitors include a ubiquitous inhibitory cystine knot (ICK) motif, peptide-induced modification of the target's voltage-dependent functions (gating modification), and typically higher inhibitory potency against Na_v than Ca_v3 channels. Importantly, the gating modifier peptides (GMPs) identified so far have provided unprecedented and much-needed interfamily selectivity to distinguish between Ca_v3 family members.

The presence of hydrophobic residues in VDPs is thought to direct the docking of key charged amino acids toward the VSM of voltage-gated ion channels, specifically targeting the "paddle motif". The paddle motif is formed by the

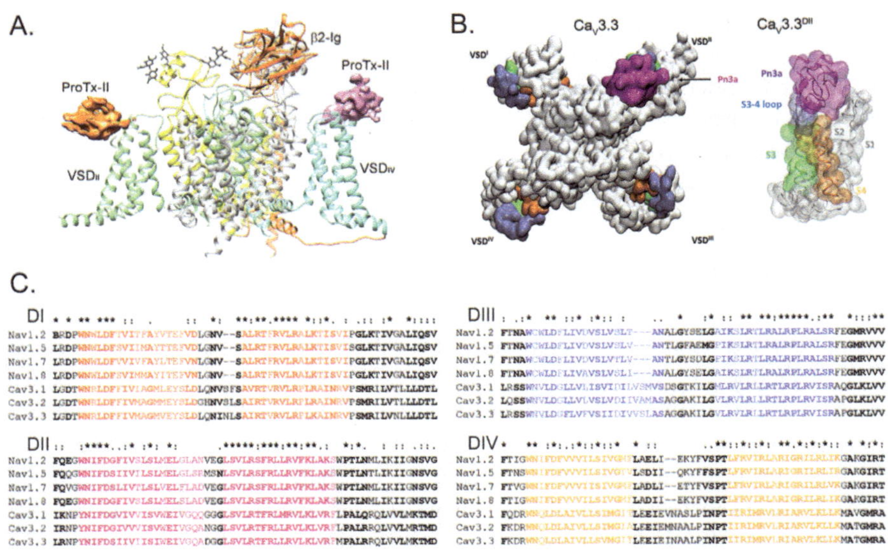

Fig. 1 Gating modifier peptides interacting with domain II of Na_v1 and Ca_v3 channels. (**a**) Protoxin II binding to the $Na_v1.7$ (Shen et al. 2019). (**b**) Pn3a binding to $Ca_v3.3$ (McArthur et al. 2022). (**c**) Partial sequence alignment of the voltage sensor modules from DI to DIV of selected Na_v1 and Ca_v3 channels. Colored text corresponds to S3 and S4 and conserved residues are marked (*, :, .) and bolded

extracellular-facing region of transmembrane segment 3 (S3b), transmembrane segment 4 (S4), and the linker adjoining them (S3–S4 linker) (Fig. 1c) (Takahashi et al. 2000; Lee and MacKinnon 2004; Mihailescu et al. 2014; Lawrence et al. 2019; Shen et al. 2019). These electrostatic interactions with the VDP constrain the movement of the VSM (VSM trapping, (Sokolov et al. 2008)), hindering transitions toward the channel's conducting states, as evidenced in electrophysiological studies as current inhibition. This mechanism of action contrasts with that of pore blockers that decrease macroscopic conductance by the hindrance of the permeation pathway (pore occlusions) and other functional modifications such as shielding of the channels' total gating charges in the presence of these GMPs.

The Swartz lab led the way in engineering voltage-sensitive proteins, such as potassium channels and membrane phosphatases, with the paddle motifs of other voltage-gated ion channels (Alabi et al. 2007). $K_v2.1$ was shown to be remarkably amenable to paddle "xenografting," and given the size and complexity of Na_v and Ca_v proteins, this K_v channel has been the classical template used to investigate the binding of GMPs to Na_v/Ca_v VSM (Bosmans et al. 2008). Nevertheless, several GMPs regulate the wild-type $K_v2.1$ channel (i.e., ProTx-I, Pn3a, etc.), and thus, alternative K_v templates like $K_v1.7$ have proven to be robust in suitability for similar chimeric approaches (McArthur et al. 2022).

Although several VDPs are active against Na_v and Ca_v3 channels (Table 1), for the purposes of this chapter, the discussion will focus on those peptides for which a detailed investigation of the molecular basis of their dual activity is available.

2.1 *Protoxins*

Early in the study of the peptide components in the venom of the Peruvian green velvet tarantula, *Thrixopelma pruriens*, the dual modulatory activity of the "protoxins" over the voltage-dependent gating of sodium and T-type calcium channels was observed (Middleton et al. 2002).

2.1.1 β/ω-Theraphotoxin-Tp1a (Protoxin I, ProTx-I, TXPR1_THRPR, P83480)

ProTx-1 promiscuously modulates several Na_v1 channels (Priest et al. 2007), $Ca_v3.1$ (Middleton et al. 2002), $K_v2.1$, and the transient receptor potential cation channel, subfamily A, member 1 (TRPA1) (Gui et al. 2014). Research using $K_v2.1/Na_v1.7$ paddle chimeras revealed that ProTx-I primarily interacts with $Na_v1.7$'s DII and DIV voltage sensor domains (Rupasinghe et al. 2020), whereas in $Ca_v3.1$, DIII appears to be the preferred site, with further evidence showing the contribution of the $Ca_v3.1$DIV in this interaction. Using $K_v2.1$ chimeric approaches, Bosmans et al. (2008) identified the pharmacophore of ProTx-I to be within the paddle motifs of domains II and IV of rat $Na_v1.2$ (Bosmans et al. 2008). This was consistent with

Table 1 Summary of toxins and small molecules that target both voltage-gated sodium ($Na_v1.x$) and T-type calcium ($Ca_v3.x$) channels

Compound	Target $hNa_V1.x$ (IC_{50} M)	Target $hCa_V3.x$ (IC_{50} M)
Protoxin I β/ω-Theraphotoxin-Tp1a	hNav1.7 (5.1×10^{-8}) (Middleton et al. 2002)	$hCa_v3.1$ (2×10^{-7}) (Ohkubo et al. 2010)
Protoxin II β/ω-Theraphotoxin-Tp2a	$hNa_v1.7$ (3.1×10^{-10}) (Schmalhofer et al. 2008)	$hCa_v3.2$ ($>1 \times 10^{-6}$) (Bladen et al. 2014)
μ-Theraphotoxin-Pn3a	$hNa_V1.7$ (9×10^{-10}) (Deuis et al. 2017)	$hCa_V3.3$ (9.6×10^{-7}) (McArthur et al. 2022)
μ-Theraphotoxin-Df1a	$hNa_v1.2/1.7$ (2×10^{-9}) (Cardoso et al. 2017)	$hCa_v3.1$ (4.5×10^{-8}) (Cardoso et al. 2017)
GpTx-I	$hNa_v1.7$ (4×10^{-9}) (Murray et al. 2015)	$rCa_v3.1$ (2.3×10^{-8}) (Ono et al. 2011)
μ/ω-Theraphotoxin-Tap1a	$hNa_v1.7$ (8×10^{-8}) (Cardoso et al. 2021)	$hCa_v3.1$ (9.9×10^{-8}) (Cardoso et al. 2021)
μ/ω-Theraphotoxin-Tap2a	$hNa_v1.7$ (1.7×10^{-7}) (Cardoso et al. 2021)	$hCa_v3.1$ (5.6×10^{-7}) (Cardoso et al. 2021)
μ/ω-Theraphotoxin-Pmu1a	$hNa_v1.7$ (7×10^{-9}) (Giribaldi et al. 2023)	$hCa_v3.2$ (9.5×10^{-7}) (Giribaldi et al. 2023)
[a]A-803467	$hNa_v1.8$ (8×10^{-9}) (Jarvis et al. 2007)	$hCa_v3.1$ (4.8×10^{-6}) (Bladen and Zamponi 2012)
Tetrodotoxin (TTX)	$hNa_v1.1$ (4.1×10^{-9}) (Tsukamoto et al. 2017)	$hCa_v3.3$ ($\sim 1.6 \times 10^{-6}$) (Sun et al. 2008)
Saxitoxin (STX)	$rNa_v1.2$ (1.7×10^{-9}) (Bricelj et al. 2005)	$hCa_v3.2$ ($\sim 3.0 \times 10^{-6}$) (Sun et al. 2008)
Cannabidiol	$hNa_v1.4$ (1.9×10^{-6}) (Ghovanloo et al. 2018)	$hCa_v3.2$ (7.8×10^{-7}) (Ross et al. 2008)
Mibefradil	$hNa_v1.5$ (9.8×10^{-7}) (Strege et al. 2005)	$hCa_v3.2$ (1.4×10^{-7}) (Martin et al. 2000)
Argentatin C	$hNa_v1.7$ ($\sim 6 \times 10^{-7}$) (Duran et al. 2023)	$hCa_v3.2$ ($\sim 1.4 \times 10^{-6}$) (Duran et al. 2023)

The half-maximal inhibitory concentration (IC_{50}) of each compound is for the highest affinity Na_v1 and Ca_v3 channel isoform

h = human; r = rat

[a]A-803467 = (5-(4-chlorophenyl)-*N*-(3,5-dimethoxyphenyl)-2-furancarboxamide)

ProTx-I binding to chimeric $K_v2.1/Na_v1.7$ channels DII and DIV but with unanticipated additional interactions with $Na_v1.7$ DI (Rupasinghe et al. 2020).

A remarkable aspect of ProTx-I is its potency and selectivity in inhibiting the T-type calcium channel $Ca_v3.1$ (half-maximal inhibitory concentration, $IC_{50} = 53$ nM) (Middleton et al. 2002), without evident effects on other T-type family members (Ohkubo et al. 2010; Bladen et al. 2014).

Differences in experimental approaches make it difficult to reach a definite conclusion about the gating modifying properties of ProTx-I over $Ca_v3.1$ channels. One study showed a small but significant 5 mV depolarizing shift in $Ca_v3.1$'s voltage dependence of activation in the presence of 1 μM ProTx-I (Ohkubo et al. 2010), whereas another study verified current inhibition but failed to observe significant changes to the half-activation potential of $Ca_v3.1$ in the presence of 1 μM ProTx-I (Bladen et al. 2014). Nevertheless, $Ca_v3.2$-mediated currents displayed shifts in both the half-activation (−6 mV) and half-inactivation (−11 mV) potentials in the presence of ProTx-I. In the case of $Ca_v3.1$, chimeras of ProTx-I-sensitive ($Ca_v3.1$) and ProTx-I-insensitive ($Ca_v3.2$) channels were used to explore ProTx-I's binding partners. With this approach, $Ca_v3.1$ DIV appeared to be relevant to the selectivity of ProTx-I for $Ca_v3.1$ over $Ca_v3.2$ (Ohkubo et al. 2010).

In a different study, it was described that ProTx-I also preferentially inhibited $Ca_v3.1$ channels with little inhibitory effect on $Ca_v3.2$ and low potency inhibition of $Ca_v3.3$-mediated currents (Bladen et al. 2014). In this report, exposure to ProTx-I of chimeric constructs between $Ca_v3.1$ (sensitive) and $Ca_v3.3$ (lower sensitivity) channels corroborated the role of $Ca_v3.1$ DIV in the observed selectivity within Ca_v3 channels, but the contribution of $Ca_v3.1$DII was also reported. Nevertheless, a clear binding site for ProTx-I could not be delineated through mutation of individual $Ca_v3.1$ paddle residues, as those did not result in measurable changes in toxin affinity nor substantial changes in the voltage dependence of activation/inactivation (Bladen et al. 2014).

2.1.2 β/ω-Theraphotoxin-Tp2a (Protoxin II, ProTx-II, TXPR2_THRPR, P83476)

ProTx-II is known to decrease the macroscopic current, modify the kinetics of activation (slowing) and deactivation (speeding), and shift the voltage dependence of activation of Na_v and Ca_v3 channels to more positive potentials, without affecting steady-state availability or recovery from inactivation (Edgerton et al. 2008; Edgerton et al. 2010).

ProTx-II targets $Na_v1.7$ channels with high potency (IC_{50} = 1–1.5 nM) (Park et al. 2014; Nicolas et al. 2019), at which shifts to the voltage dependence of activation and inactivation appear (Xiao et al. 2010). For $Na_v1.7$, it has been shown that ProTx-II interacts with pharmacological Site 4 by docking on top of the DII's S3 helix (Fig. 1a) (Shen et al. 2019; Xu et al. 2019) where it modulates $Na_v1.7$ activation (IC_{50} = 0.26–3 μM) and, by interacting with $Na_v1.7$ DIV, it shifts inactivation (IC_{50} = 250 nM) more potently than other Na_vs (approximately 4.6 μM for all channels reported) (Xiao et al. 2010).

Bladen et al. (2014) reported that 1 μM ProTx-II inhibited $Ca_v3.1$ (8.9%), $Ca_v3.2$ (29%), and $Ca_v3.3$ (8.9%). In the same study, an activation $V_{0.5}$ shift of −14 mV was observed for $Ca_v3.1$, and for $Ca_v3.2$, an inactivation $V_{0.5}$ shift of −11 mV was shown (Bladen et al. 2014).

Salari et al. (2016) reported a Kd of ~1.3 μM for the inhibition of $Ca_v3.1$-mediated currents by ProTx-II (in oocytes at −40 mV, with 1.8 mM Ca^{2+} as charge carrier; (Salari et al. 2016)). ProTx-II was shown to exert a large inhibitory effect on K^+ currents mediated by $K_v2.1/Ca_v3.1$ DIII chimeras, indicating more potent toxin binding to the domain III voltage sensor paddle of $Ca_v3.1$. Within this domain, alanine neutralization of the acidic residue D1372 of the $Ca_v3.1$ channel results in a modest decrease in ProTx-II potency, but, perhaps more interestingly, a 7 mV toxin-induced depolarizing shift in the voltage dependence of activation. Neutralization of equivalent amino acids in $Ca_v3.1$'s DI and DII (E198 and E855, respectively) chimeras also leads to decreases in toxin potency and toxin-concomitant shifts in activation $V_{0.5}$, suggesting that these paddle acidic residues may also participate in the interaction of ProTx-II with $Ca_v3.1$ channels. However, the same acidic residues are conserved in all T-type channels, obscuring the mechanisms of ProTx-II's selectivity over other T-types.

Both ProTx-I and ProTx-II preferentially modulate $Ca_v3.1$ channels, but the delineation of their pharmacophores appears to be concealed under experimental idiosyncrasies.

2.2 *μ-Theraphotoxin-Pn3a*

μ-Theraphotoxin-Pn3a (Pn3a, μ-TRTX-Pn3, PN3A_PAMNI, P0DM12) was identified in the venom of the giant blue bloom tarantula, *Pamphobeteus nigricolor*. Pn3a inhibits VGSCs with potencies ranging in the single-digit nanomolar range ($Na_v1.7$ $IC_{50} \cong 1$ nM) and with low-to-mid nanomolar potency against other tetrodotoxin-sensitive (TTX-S) Na_vs ($Na_v1.1$: 37 nM, $Na_v1.2$: 124 nM, $Na_v1.3$: 210 nM, $Na_v1.4$: 144 nM, and $Na_v1.6$: 126 nM). In contrast, Pn3a displays micromolar potencies inhibiting currents mediated by tetrodotoxin-resistant (TTX-R) Na_vs ($Na_v1.5$: 0.8 μM; $Na_v1.9$: 2.4 μM) (Deuis et al. 2017) as well as HVA Ca_v1 and Ca_v2 channels (McArthur et al. 2020).

While Pn3a seemingly spares $Na_v1.8$ (~50 μM), as well as $Ca_v3.1$- and $Ca_v3.2$-mediated currents ($IC_{50} > 100$ μM), the inhibitory activity of Pn3a against $Ca_v3.3$ channels ($IC_{50} = 1$ μM) was reported (McArthur et al. 2022). This highlights a remarkable >100-fold selectivity for $Ca_v3.3$ over $Ca_v3.1$ and $Ca_v3.2$ for this spider VDP.

As for $Na_v1.7$, Pn3a's inhibition of $Ca_v3.3$ is voltage-dependent, with greater inhibition at more negative potentials. Pn3a inhibits $Ca_v3.3$-mediated currents by inducing a depolarizing shift in the voltage dependence of activation, in a manner analogous to kurtoxin, ProTx-I, and ProTx-II actions on $Ca_v3.1$ channels (Chuang et al. 1998; Edgerton et al. 2010). This is consistent with Pn3a's modulation of $Na_v1.7$ (Deuis et al. 2017). Additionally, Pn3a modification induces $Ca_v3.3$ currents to recover ~1.5-fold faster from inactivation, suggestive of a toxin-induced destabilization of the inactivated state (McArthur et al. 2022). This observation appears at odds with the reported effects of Pn3a on $Na_v1.7$ channels, where it delays recovery from inactivation (Deuis et al. 2017). DI–DIII and DIV are important for Na_v activation and inactivation, respectively (Ahern et al. 2016). Similarly, the predominant effects of Pn3a on $Ca_v3.3$ activation are consistent with its interaction with DII of this channel. Pn3a's inhibitory actions on $Na_v1.7$ occur by targeting $Na_v1.7$ DII and $Na_v1.7$ DIV (Deuis et al. 2017). Chimeras of $Ca_v3.3$ and $K_v1.7$ indicated that Pn3a interacts with domains II and III. This was further verified by domain swap experiments between all T-type isoforms and alanine mutagenesis of $Ca_v3.3$'s aspartate 689, which identified this residue as the key interaction site in $Ca_v3.3$ DII (Fig. 1b) (McArthur et al. 2022).

2.3 *Other Spider VDPs with Dual Na_v1/Ca_v3 Activity*

2.3.1 μ-Theraphotoxin-Df1a

μ-Theraphotoxin-Df1a (Df1a, TXDF1_DAFVA, P0DPC2) was identified in the venom of the *Davus fasciatus* (Costa Rican tiger rump). The synthetic peptide Df1a has both -NH_2 and free -OH C-termini, and it inhibits Na_v1 and Ca_v3 channels. Both forms of Df1a inhibit sodium channels with higher potency (IC_{50} = 2–125 nM) than Ca_v3 channels (IC_{50} = 45–450 nM) (Cardoso et al. 2017). Df1a appears to interact with the DII, DIII, and DIV domains of $Na_v1.7$, but information on its molecular determinants and its effects on other equally sensitive Na_v1 isoforms and Ca_v3 channels is not available.

2.3.2 GpTx-I

Isolated from *Grammostola rosea* venom, GpTx-1 (toxin GTx1–15, TX15_GRARO, P0DJA9) inhibits $Na_v1.7$ (IC_{50} = 0.5–10 nM), $Na_v1.3$, $Na_v1.4$, and $Ca_v3.1$ and displays lower potency against TTX-R channels $Na_v1.5$ and $Na_v1.8$ (Ono et al. 2011; Murray et al. 2015). It is worth noting that the primary structure of mature GpTx-1 is identical to that of GpTx-1 from *Grammostola porteri* (AC P0DL72) and Gtx1–15 from *Paraphysa scrofa* (AC P0DL73). Thus, it may be inferred that this type of promiscuous peptide may have general physiological relevance to several large spiders.

2.3.3 Tap1a and Tap2a

μ/ω-Theraphotoxin-Tap1a (Tap1a, TAP1A_THEAO, P0DQO3) and μ/ω--theraphotoxin-Tap2a (Tap2a, TAP2A_THEAO, P0DQO4) are gating modifier toxins derived from *Theraphosa apophysis* (goliath pink foot tarantula). These toxins inhibit Na_v and Ca_v3 channels by inducing a hyperpolarizing shift in the voltage dependence of activation and steady-state inactivation.

Tap1a and Tap2a display higher potency in inhibiting $Na_v1.2$- and $Na_v1.7$- than $Ca_v3.1$- or $Ca_v3.2$-mediated currents (Cardoso 2020). Tap1a inhibits $Na_v1.1$, $Na_v1.2$, $Na_v1.3$, $Na_v1.6$, $Na_v1.7$, and $Ca_v3.1$ sodium and calcium channels at nanomolar concentrations (IC_{50} = 81–301 nM). Tap1a moderately inhibits $Ca_v3.2$ calcium channels (IC_{50} = 1233 nM). Tap2a inhibits $Na_v1.1$, $Na_v1.2$, $Na_v1.6$, $Na_v1.7$, and $Ca_v3.1$

channels with nanomolar potency (IC_{50} = 169–621 nM). Interestingly, Tap1a and Tap2a display weak modulation of muscle Na_vs ($Na_v1.4$ and $Na_v1.5$) while sparing $Ca_v3.3$ currents (Cardoso 2020).

2.3.4 μ/ω-Theraphotoxin-Pmu1a

μ/ω-TRTX-Pmu1a (Pmu1a, PmuTx1, PDB8FEY) was identified in a crude fraction of the venom of the orange baboon tarantula *Pterinochilus murinus*. Pmu1a exhibits inhibitory effects against Na_v1 and Ca_v3 channels (Giribaldi et al. 2023). Pmu1a was shown to have potent inhibitory effects against $Na_v1.7$ (IC_{50} ~ 7 nM)-, $Na_v1.6$ (IC_{50} ~ 10 nM)-, and $Na_v1.4$ (IC_{50} ~ 65 nM)-mediated currents. In addition, Pmu1a (at 10 μM) displayed modest potency at inhibiting $Ca_v3.1$ (95%), $Ca_v3.2$ (64%), and $Ca_v3.3$ (91%) without overt selectivity against T-type calcium channel isoform. Interestingly, Pmu1a (at 10 μM) preferentially inhibits Ca_v3 currents at negative potentials eliciting substantial depolarizing shifts in the voltage dependence of activation of $Ca_v3.1$ (~21 mV)-, $Ca_v3.2$ (~12 mV)-, and $Ca_v3.3$ (~26 mV)-mediated currents in the presence of this VDP (Giribaldi et al. 2023). To date, neither the effects of Pmu1a on the voltage-dependent parameters of Na_v1-mediated currents nor its pharmacophore on Na_v1 or Ca_v3 channels has been reported.

3 Small Molecules Dually Targeting Na_v1 and Ca_v3 Channels

Unlike peptide modulators, which predominantly bind extracellular pharmacophores (outer pore region and voltage sensor paddles), depending on their chemistry and charge, small molecules can also target nonexposed channel sites that can be accessed through the permeation pathway or via fenestrations, thus reaching the intracellular vestibule. Furthermore, members of both the Na_v1 and Ca_v3 families are susceptible to indirect modulation by G protein-coupled receptor (GPCR) activation (Tao et al. 2008; Perez-Reyes 2010; Rangel et al. 2010; Mattheisen et al. 2018). Additionally, changes in intracellular calcium concentrations can also lead to dual inhibition of Na_v1 and Ca_v3 by small molecule activation of GPCRs or calcium permeable membrane proteins (McArthur et al. 2019). The focus of this section is on the direct modulation of Na_v1 and Ca_v3 channels by small molecules.

3.1 A-803467

A-803467 was first identified as a highly selective inhibitor of the TTX-R sodium channel $Na_v1.8$ (Jarvis et al. 2007), with an IC_{50} affinity at $hNa_V1.8$ of 8 nM, which was >100-fold more selective than for other Na_V channels. The cryo-electron microscopy (cryo-EM) structure of $Na_V1.8$ showed A-803467 binding to the channel's intracellular vestibule directly below the selectivity filter, establishing interactions with the fenestration between DI and DIV (Huang et al. 2022). A-803467 has demonstrated efficacy in a variety of pain models; however, the observed effects of A-803467 are clouded by the low oral bioavailability, and concentrations of A-803467 in plasma needed to evidence clinical effects being 1.5–3-fold above the IC_{50} for rat TTX-R sodium channel inhibition. This raised questions about the therapeutic target of A-803467.

A-803467 was later shown to inhibit Ca_v3 channels (IC_{50} = 4.8–7 μM) close to its therapeutic dose range (plasma concentration of 3–5 μM) (Bladen and Zamponi 2012). The inner vestibule of Na_v channels possesses multiple drug interaction sites, including the binding of local anesthetics (LAs) and anticonvulsive drugs (Lipkind and Fozzard 2005). While there is no high-resolution structural data for the binding of A-803467 on Ca_v3 channels, mutational data suggests it binds to the inner vestibule of $Ca_v3.2$ at a site similar to that of LAs, as well as A-803467 in Na_v channels (Bladen and Zamponi 2012). Four residues were identified (I403, V1551, Q1848, and V1855, highlighted as purple residues in Fig. 2), which altered A-803467 binding to $Ca_v3.2$. Analogous residues in $Na_v1.8$ are all located within the binding pocket of the bound cryo-EM structure of $Na_v1.8$/A-804367 (F383, V1414, F1710, and Y1717; (Huang et al. 2022)) (see purple residues; Fig. 2). Despite the potential similarity in the A-803467 binding site on $Na_v1.8$ and $Ca_v3.2$, the effect of use dependence was observed to be the opposite for $Ca_v3.2$. Use dependence is critical for drugs that combat hyperexcitable neurons as expected in pain syndromes, as it will preferentially target these neurons. A-803467 showed increased affinity during tonic block of $Ca_v3.2$ (Bladen and Zamponi 2012); however, the opposite is true for $Na_v1.8$ where A-803467 showed use-dependent relief of block (Jo et al. 2023), which could lead to reduced affinity/efficacy during pathophysiological firing.

3.2 *Tetrodotoxin and Saxitoxin*

TTX and saxitoxin (STX) are heterocyclic guanidinium marine toxins initially described as inhibitors of Na_v channels via binding to the extracellular pore near the selectivity filter to occlude the permeation pathway (Henderson et al. 1974; Kao and Walker 1982; Kao 1986; Hille 2001). For Na_v channels, the selectivity filter sequence of DI-IV is DEKA (Catterall 2000), whereas, for Ca_v3 channels, it is EEDD (selectivity filter residues in red; Fig. 2 (Perez-Reyes 2003)). Mutation of DIII-Lys or DIV-Ala to glutamate conferred Ca^{2+} and Ba^{2+} selectivity to Na_v channels (Heinemann et al. 1992). TTX blocks six Na_v channels ($Na_v1.1$–1.4 and 1.6–1.7) in the nanomolar range (TTX-S), but it is unable to modify $Na_v1.5$, 1.8, and 1.9 (TTX-R) at this concentration. Previous studies have identified a tyrosine, one residue external to the DEKA selectivity filter in domain I, as the critical determinant of TTX inhibition of Na_v channels (Satin et al. 1992) (green residue; Fig. 2). In all TTX-S channels, this residue is a Tyr. However, in $Na_v1.5$, this residue is a cysteine

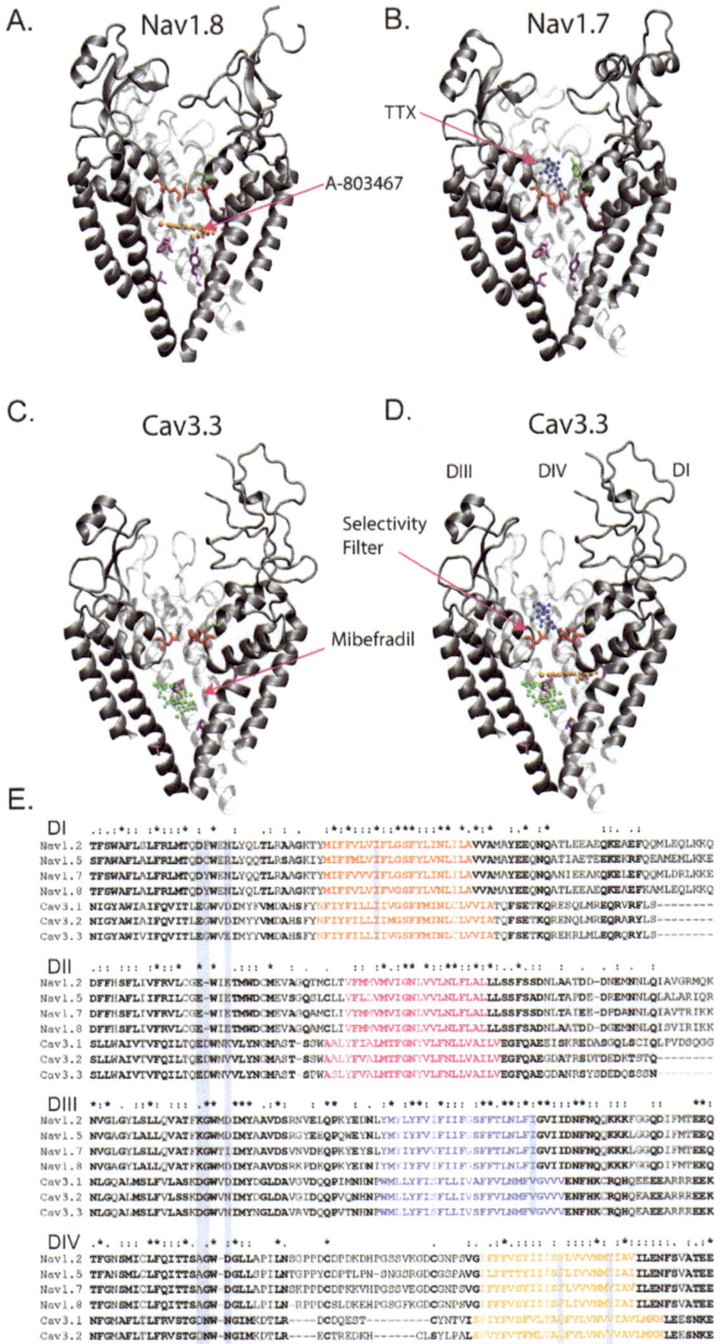

Fig. 2 Small molecules interacting with Na_v1 and Ca_v3 channels with A–D showing the S5-P loop-S6 of DI (right), DIII (left), and DIV (back, transparent). (**a**) A-803467 binding to $Na_v1.8$ (PDB: 7WE4; (Huang et al. 2022)). (**b**) TTX binding to $Na_v1.7$ (PDB: 6J8I; (Shen et al. 2019)). (**c**) Mibefradil binding to $Ca_v3.3$ (PDB: 7WLJ; (He et al. 2022)). (**d**) Overlay of TTX, mibefradil, and A-803467 on the $Ca_v3.3$ structure. (**e**) Partial sequence alignment of the pore domain and S6 from DI to DIV of selected Na_v1 and Ca_v3 channels. The colored text corresponds to conserved regions in each domain with conserved residues marked (*, :, .) and bolded. Shaded residues mark the selectivity filter and outer ring residues, as well as residues identified as critical for A-803467 inhibition of Ca_v3 channels

residue, whereas, in $Na_v1.8$ and $Na_v1.9$, it is a serine. When exchanged with tyrosine, the TTX-R channels' affinity for TTX increased (Sivilotti et al. 1997).

A LVA, TTX-S calcium current has been identified under Na^+-free conditions in human atrial (Lemaire et al. 1995), rat (Aggarwal et al. 1997), and guinea pig (Cole et al. 1997) ventricular myocytes. In guinea pig ventricular myocytes, TTX showed an IC_{50} value of 2.4 μM (Heubach et al. 2000), whereas other studies suggested that there were, in fact, two distinct LVA currents (ICa_T and ICa_{TTX}) distinguished by their sensitivity to the blockers Ni^{2+} and TTX. Both currents were inhibited by mibefradil, a classical Ca_v3 channel inhibitor. To examine these observations in more detail, the effect of TTX and STX was evaluated on heterologously expressed Ca_v3s. TTX (30 μM) displayed no inhibitory effects on $hCa_v3.1$ or $hCa_v3.2$ in HEK-293 cells, whereas the same concentration directly inhibited $hCa_v3.3$ by >60%, while 1 μM STX directly inhibited >80% of all Ca_v3s (Sun et al. 2008). Despite no direct inhibition of $Ca_v3.1$ and $Ca_v3.2$, TTX attenuated the block of Ni^{2+}, reducing its apparent affinity in both heterologously expressed channels and canine atrial myocytes. Ni^{2+} inhibition of $Ca_v3.2$ is voltage-dependent and has been suggested to interact with a histidine residue, His191, in the S3–S4 linker of DI (Kang et al. 2006). Despite the cross-talk between Ni^{2+} and TTX inhibition of $Ca_v3.2$, it is not clear if the observed interactions are due to competitive binding in the voltage sensor or if TTX binds to the pore of $Ca_v3.2$ without inhibiting Ca^{2+} permeation but is able to modulate the Ni^{2+} binding site. The fact that TTX and STX bind to an overlapping site in the extracellular pore region of TTX-S Na_v channels, and STX directly inhibits all Ca_v3s while TTX directly inhibits $Ca_v3.3$, suggests that both of these toxins are able to bind to the extracellular pore region of all Ca_v3 channels. This indicates that the pore domains of Na_v1 and Ca_v3 channels are structurally similar.

3.3 Cannabidiol

Cannabidiol is a hydrophobic compound produced from the cannabis plant and classified as a cannabinoid. It has been shown to have potential in treating epilepsy (Rosenberg et al. 2015; Rosenberg et al. 2017). Interestingly, Na_v1 channels, especially $Na_v1.1$, 1.2, and 1.6, (Mulley et al. 2005; Striano et al. 2006; Herlenius et al. 2007; Shi et al. 2009; O'Brien and Meisler 2013; Sawyer et al. 2014) as well as the Ca_v3 channels (Marescaux and Vergnes 1995; Chen et al. 2003; Becker et al. 2008; Ernst et al. 2009), have been linked to epilepsy.

Cannabidiol has been found to be a nonselective, direct inhibitor of Ca_v3 channels, with an IC_{50} of 1–4 μM, producing a hyperpolarizing shift in Ca_v3 inactivation and recovery from inactivation (Ross et al. 2008). More recently, cannabidiol was shown to nonselectively inhibit peak $Na_v1.1$–$Na_v1.7$ currents with an IC_{50} of 1–4 μM (Ghovanloo et al. 2018), as well as resurgent $Na_v1.1$ and $Na_v1.6$ currents (Patel et al. 2016). A crystal structure of the prokaryotic Na_vMs VGSC was solved in the presence of cannabidiol to elucidate its structural binding site on sodium channels (Sait et al. 2020). The study identified the intracellular vestibule, just below the selectivity filter, and the fenestrations as its binding site, similar to the LA binding site. This finding suggests a potential overlap with the A-803467 binding site discovered on $Na_v1.8$ and $Ca_v3.2$.

3.4 *Mibefradil*

Mibefradil (Ro 40–5967) was identified as a calcium channel antagonist that targeted Ca_v3 channels over Ca_v1 calcium channels. Mibefradil was marketed briefly as Posicor (Roche Pharmaceuticals, Nutley, NJ) to treat hypertension (Mehrke et al. 1994; Mishra and Hermsmeyer 1994; Ertel and Clozel 1997). Mibefradil inhibits heterologously expressed Ca_v3 channels at ~1 μM when 10 mM Ba^{2+} is used as the charge carrier, but its affinity is greatly increased when the charge carrier is 2 mM Ca^{2+} (IC_{50} $Ca_v3.1$ = 270 nM; $Ca_v3.2$ = 140 nM) (Martin et al. 2000), suggesting a potential binding site within the ion-conducting pathway.

Mibefradil has since been shown to target not only other calcium channels with an IC_{50} of approximately tenfold higher than Ca_v3 channels, including L-, N-, Q-, and R-type channels (Bezprozvanny and Tsien 1995), but also various potassium channels (Chouabe et al. 1998; Perchenet and Clement-Chomienne 2000), chloride channels (Nilius et al. 1997), and sodium channels, including $Na_v1.5$ (Eller et al. 2000; McNulty and Hanck 2004). The affinity for the $Na_v1.5$ channel is dependent on the holding potential and binds in the nanomolar range (~500 nM) when the holding potential is −130 mV, versus ~3 μM at a holding potential of −100 mV (McNulty and Hanck 2004). Mibefradil interacts with the LA site within the intracellular vestibule of sodium channels, as mutation of Asn406 ($Na_v1.5$) to cysteine or phenylalanine reduced mibefradil's binding affinity (McNulty et al. 2006). Interestingly, Asn406 of $Na_v1.5$ is conserved in Ca_v3 subunits. The recent cryo-EM structure of $Ca_v3.3$ was solved in the presence of mibefradil and displayed a density corresponding to the mibefradil molecule in the inner vestibule, in an analogous position to the LA site in $Na_v1.5$ (He et al. 2022). The overlap of the mibefradil binding site with the LA sites in both Na_v and Ca_v channels suggests that this binding site may be structurally conserved between these channels.

3.5 *Local Anesthetics and QX-314*

Today, the majority of clinically used LAs consist of a benzene ring linked via an amide or ester to an amine group. The target of LAs was first identified as Na_v channels (Taylor 1959); however, at clinically relevant concentrations, LAs can also modulate potassium (Olschewski et al. 1998; Komai and McDowell 2001; Slesinger 2001) and calcium channels (Sugiyama and Muteki 1994; Taga et al. 1997; Xiong and Strichartz 1998; Liu et al. 2001), which might explain some of their side effects.

LAs typically bind to all nine sodium channels in a state-dependent manner, with a low binding affinity (mM) to the resting/closed state and a high affinity toward the open/inactivated channel (μM) (Sheets et al. 2010). QX-314 is a quaternary derivative of the LA lidocaine that has a permanent positive charge, which does not allow it to cross the plasma membrane. This property allowed the identification of LA mechanism of blocking sodium channels via the intracellular vestibule, as QX-314 can only inhibit Na_v channels when applied to their intracellular side (Narahashi et al. 1970; Zamponi et al. 1993).

Only recently have LAs begun to be examined on Ca_v3s. While lidocaine was shown to be ineffective at 500 μM on both LVA and HVA Ca^{2+} currents from dog atrial cells (Sun et al. 2008), other LAs, including butamben at 200 μM and tetracaine at 300 μM, were able to inhibit LVA Ca^{2+} currents in small diameter mouse and rat DRG neurons, respectively (Sugiyama and Muteki 1994). QX-314 inhibited LVA calcium currents at 1 and 10 mM in CA1 pyramidal neurons (Talbot and Sayer 1996), a concentration similar to that shown for inhibition of $Na_v1.7$ (2 mM) by extracellular QX-314 (Klasfauseweh et al. 2022). Despite evidence that LAs are able to inhibit HVA Ca^{2+} currents in different cell types, there are currently no studies investigating the effect of LAs on heterologously expressed human Ca_v3 channels.

3.6 Argentatin C

Argentatin C (arg-C) is a natural plant product from *Pycnanthemum incanum*, which was identified from fractionation with activity at both sodium and LVA calcium influx in rat DRG neurons (Duran et al. 2023). Arg-C inhibited rat DRG neuron I_{Na} with an IC_{50} of 18.1 μM, and in heterologous expression systems, arg-C was able to acutely inhibit $Na_v1.7$ by ~75% at 20 μM (Duran et al. 2023). 20 μM arg-C also was able to inhibit LVA calcium channel currents from rat DRG neurons by 45% at 20 μM while blocking heterologous Ca_v3 channels between 42 and 65% (Duran et al. 2023). Predicted binding sites for arg-C based on computational modeling suggests it interacts with the fenestration and intracellular vestibule of both Na_v1 and Ca_v3 channels, similar to mibefradil and A-803467.

4 Closing Remarks

The sequence alignment shown in Fig. 1c highlights the remarkable conservation between Na_v1 and Ca_v3 channels. Although arthropod venom-derived Na_v1 modulators abound, only a handful have been shown to functionally modulate T-type calcium channels. A common theme between the spider peptides Pn3a and ProTx-II (Edgerton et al. 2010) appears to be the slowing of T-type channel deactivation, suggesting stabilization of the channel's closed state.

It is noteworthy that IC_{50}s for the inhibition of voltage-gated ion channels is highly dependent on experimental conditions such as the expression system (i.e., mammalian vs. nonmammalian cells), test pulse (duration and potential), stimulus frequency, holding potential, charge carrier ion, etc. Thus, a systematic comparison of a peptide's activity requires careful and consistent experimental design, which is paramount to the understanding of the structural and functional correlates of the pharmacology between these two related families of VGICs.

Given the evolutionary parallel between Na_v1 and Ca_v3 channels, it is not surprising that compounds targeting one may also inhibit the other. The inhibitory effects of TTX and STX suggest similarities within the outer vestibule between Na_v1 and Ca_v3, whereas A-803467, mibefradil, arg-C, and the LAs indicate a conserved site for small molecules within the inner vestibule. These observations are supported by cryo-EM structures. With more Na_v1 and Ca_v3 structures being solved, the question arises as to whether these Na_v1/Ca_v3 cross-talk small molecules bind to a similar pharmacophore site within these key channel families. Furthermore, it remains unclear to what extent Na_v1- or Ca_v3-selective compounds exhibit cross-reactivity when tested across both channel types. Investigating the potential benefits of developing drugs that target both Na_v1 and Ca_v3 may yield valuable insights.

Neuroexcitability disorders such as chronic pain and epilepsy have long been associated with the dysfunction of both Na_v1 and Ca_v3 channels, yet the vast majority of research to date has focused on the exploration of only one channel at a time. With the urgent need for new analgesic compounds, it is critical to explore the potential of both Na_v1- and Ca_v3-specific drugs, as well as dual-targeting molecules, to treat these debilitating conditions. By delving into the complex interplay between these channels in health and disease, new possibilities for developing novel, effective therapies that target multiple facets of neuroexcitability disorders may be unlocked.

References

Aggarwal R, et al. Tetrodotoxin-blockable calcium currents in rat ventricular myocytes; a third type of cardiac cell sodium current. J Physiol. 1997;505(Pt 2):353–69.

Ahern CA, et al. The hitchhiker's guide to the voltage-gated sodium channel galaxy. J Gen Physiol. 2016;147(1):1–24.

Alabi AA, et al. Portability of paddle motif function and pharmacology in voltage sensors. Nature. 2007;450(7168):370–5.

Becker AJ, et al. Transcriptional upregulation of $Ca_V3.2$ mediates epileptogenesis in the pilocarpine model of epilepsy. J Neurosci. 2008;28(49):13341–53.

Bezprozvanny I, Tsien RW. Voltage-dependent blockade of diverse types of voltage-gated Ca^{2+} channels expressed in Xenopus oocytes by the Ca^{2+} channel antagonist mibefradil (Ro 40-5967). Mol Pharmacol. 1995;48(3):540–9.

Bladen C, Zamponi GW. Common mechanisms of drug interactions with sodium and T-type calcium channels. Mol Pharmacol. 2012;82(3):481–7.

Bladen C, et al. Block of T-type calcium channels by protoxins I and II. Mol Brain. 2014;7:36.

Bosmans F, Martin-Eauclaire MF, Swartz KJ. Deconstructing voltage sensor function and pharmacology in sodium channels. Nature. 2008;456(7219):202–8.

Bricelj VM, et al. Sodium channel mutation leading to saxitoxin resistance in clams increases risk of PSP. Nature. 2005;434(7034):763–7.

Cain SM, Snutch TP. Voltage-gated calcium channels and disease. Biofactors. 2011;37(3):197–205.

Cardoso FC. Multi-targeting sodium and calcium channels using venom peptides for the treatment of complex ion channels-related diseases. Biochem Pharmacol. 2020;181:114107.

Cardoso FC, et al. Modulatory features of the novel spider toxin μ-TRTX-Df1a isolated from the venom of the spider *Davus fasciatus*. Br J Pharmacol. 2017;174(15):2528–44.

Cardoso FC, et al. A spider-venom peptide with multitarget activity on sodium and calcium channels alleviates chronic visceral pain in a model of irritable bowel syndrome. Pain. 2021;162(2):569–81.

Catterall WA. From ionic currents to molecular mechanisms: the structure and function of voltage-gated sodium channels. Neuron. 2000;26(1):13–25.

Catterall WA, et al. International Union of Pharmacology. XLVIII. Nomenclature and structure-function relationships of voltage-gated calcium channels. Pharmacol Rev. 2005;57(4):411–25.

Chen Y, et al. Association between genetic variation of CACNA1H and childhood absence epilepsy. Ann Neurol. 2003;54(2):239–43.

Chouabe C, et al. HERG and K_vLQT1/IsK, the cardiac K^+ channels involved in long QT syndromes, are targets for calcium channel blockers. Mol Pharmacol. 1998;54(4):695–703.

Chuang RS, et al. Inhibition of T-type voltage-gated calcium channels by a new scorpion toxin. Nat Neurosci. 1998;1(8):668–74.

Cole WC, et al. Ca^{2+} permeation through Na^+ channels in guinea pig ventricular myocytes. Am J Phys. 1997;273(Pt 2):H128–37.

Cottrell GS, et al. CACHD1 is an α2δ-like protein that modulates Ca_V3 voltage-gated calcium channel activity. J Neurosci. 2018;38(43):9186–201.

Curtis BM, Catterall WA. Purification of the calcium antagonist receptor of the voltage-sensitive calcium channel from skeletal muscle transverse tubules. Biochemistry. 1984;23(10):2113–8.

de Lera Ruiz M, Kraus RL. Voltage-gated sodium channels: structure, function, pharmacology, and clinical indications. J Med Chem. 2015;58(18):7093–118.

Deuis JR, et al. Pharmacological characterisation of the highly $Na_V1.7$ selective spider venom peptide Pn3a. Sci Rep. 2017;7:40883.

Duran P, et al. The natural product argentatin C attenuates postoperative pain via inhibition of voltage-gated sodium and T-type voltage-gated calcium channels. Br J Pharmacol. 2023;180:1267–85.

Edgerton GB, Blumenthal KM, Hanck DA. Evidence for multiple effects of ProTxII on activation gating in $Na_V1.5$. Toxicon. 2008;52(3):489–500.

Edgerton GB, Blumenthal KM, Hanck DA. Inhibition of the activation pathway of the T-type calcium channel $Ca_V3.1$ by ProTxII. Toxicon. 2010;56(4):624–36.

Eller P, et al. High affinity interaction of mibefradil with voltage-gated calcium and sodium channels. Br J Pharmacol. 2000;130(3):669–77.

Ernst WL, et al. Genetic enhancement of thalamocortical network activity by elevating alpha 1g-mediated low-voltage-activated calcium current induces pure absence epilepsy. J Neurosci. 2009;29(6):1615–25.

Ertel SI, Clozel JP. Mibefradil (Ro 40-5967): the first selective T-type Ca^{2+} channel blocker. Expert Opin Investig Drugs. 1997;6(5):569–82.

Finol-Urdaneta RK, et al. Multitarget nociceptor sensitization by a promiscuous peptide from the venom of the King Baboon spider. Proc Natl Acad Sci U S A. 2022;119(5):e2110932119.

Ghovanloo MR, et al. Inhibitory effects of cannabidiol on voltage-dependent sodium currents. J Biol Chem. 2018;293(43):16546–58.

Giribaldi J, et al. Pmu1a, a novel spider toxin with dual inhibitory activity at pain targets $hNa_V1.7$ and hCa_V3 voltage-gated channels. FEBS J. 2023;290(14):3688–702.

Gui J, et al. A tarantula-venom peptide antagonizes the TRPA1 nociceptor ion channel by binding to the S1-S4 gating domain. Curr Biol. 2014;24(5):473–83.

He L, et al. Structure, gating, and pharmacology of human $Ca_V3.3$ channel. Nat Commun. 2022;13(1):2084.

Heinemann SH, et al. Calcium channel characteristics conferred on the sodium channel by single mutations. Nature. 1992;356(6368):441–3.
Henderson R, Ritchie JM, Strichartz GR. Evidence that tetrodotoxin and saxitoxin act at a metal cation binding site in the sodium channels of nerve membrane. Proc Natl Acad Sci U S A. 1974;71(10):3936–40.
Herlenius E, et al. SCN2A mutations and benign familial neonatal-infantile seizures: the phenotypic spectrum. Epilepsia. 2007;48(6):1138–42.
Heubach JF, et al. T-type and tetrodotoxin-sensitive Ca^{2+} currents coexist in Guinea pig ventricular myocytes and are both blocked by mibefradil. Circ Res. 2000;86(6):628–35.
Hille B. Ion Channels of Excitable Membranes. 3rd ed. Sunderland: Sinauer Associates; 2001.
Huang X, et al. Structural basis for high-voltage activation and subtype-specific inhibition of human Nav1.8. Proc Natl Acad Sci U S A. 2022;119(30):e2208211119.
Isom LL. Sodium channel beta subunits: anything but auxiliary. Neuroscientist. 2001;7(1):42–54.
Jarvis MF, et al. A-803467, a potent and selective Nav1.8 sodium channel blocker, attenuates neuropathic and inflammatory pain in the rat. Proc Natl Acad Sci U S A. 2007;104(20):8520–5.
Jo S, Zhang HB, Bean BP. Use-dependent relief of inhibition of Nav1.8 channels by A-887826. Mol Pharmacol. 2023;103(4):221–29.
Kang HW, et al. A molecular determinant of nickel inhibition in Cav3.2 T-type calcium channels. J Biol Chem. 2006;281(8):4823–30.
Kao CY. Structure-activity relations of tetrodotoxin, saxitoxin, and analogues. Ann N Y Acad Sci. 1986;479:52–67.
Kao CY, Walker SE. Active groups of saxitoxin and tetrodotoxin as deduced from actions of saxitoxin analogues on frog muscle and squid axon. J Physiol. 1982;323:619–37.
Klasfauseweh T, et al. Low potency inhibition of $Na_V1.7$ by externally applied QX-314 via a depolarizing shift in the voltage-dependence of activation. Eur J Pharmacol. 2022;925:175013.
Komai H, McDowell TS. Local anesthetic inhibition of voltage-activated potassium currents in rat dorsal root ganglion neurons. Anesthesiology. 2001;94(6):1089–95.
Lawrence N, et al. Peptide-membrane interactions affect the inhibitory potency and selectivity of spider toxins ProTx-II and GpTx-1. ACS Chem Biol. 2019;14(1):118–30.
Lee SY, MacKinnon R. A membrane-access mechanism of ion channel inhibition by voltage sensor toxins from spider venom. Nature. 2004;430(6996):232–5.
Lemaire S, et al. Tetrodotoxin-sensitive Ca^{2+} and Ba^{2+} currents in human atrial cells. Recept Channels. 1995;3(2):71–81.
Lipkind GM, Fozzard HA. Molecular modeling of local anesthetic drug binding by voltage-gated sodium channels. Mol Pharmacol. 2005;68(6):1611–22.
Liu BG, et al. Effects of bupivacaine and ropivacaine on high-voltage-activated calcium currents of the dorsal horn neurons in newborn rats. Anesthesiology. 2001;95(1):139–43.
Marescaux C, Vergnes M. Genetic absence epilepsy in rats from Strasbourg (GAERS). Ital J Neurol Sci. 1995;16(1–2):113–8.
Martin RL, et al. Mibefradil block of cloned T-type calcium channels. J Pharmacol Exp Ther. 2000;295(1):302–8.
Mattheisen GB, Tsintsadze T, Smith SM. Strong G-protein-mediated inhibition of sodium channels. Cell Rep. 2018;23(9):2770–81.
McArthur JR, Finol-Urdaneta RK, Adams DJ. Analgesic transient receptor potential vanilloid-1-active compounds inhibit native and recombinant T-type calcium channels. Br J Pharmacol. 2019;176(13):2264–78.
McArthur JR, et al. Spider venom peptide Pn3a inhibition of primary afferent high voltage-activated calcium channels. Front Pharmacol. 2020;11:633679.
McArthur JR, et al. μ-Theraphotoxin Pn3a inhibition of $Ca_V3.3$ channels reveals a novel isoform-selective drug binding site. Elife. 2022;11:e74040.
McNulty MM, Hanck DA. State-dependent mibefradil block of Na^+ channels. Mol Pharmacol. 2004;66(6):1652–61.

McNulty MM, et al. An inner pore residue (Asn406) in the Nav1.5 channel controls slow inactivation and enhances mibefradil block to T-type Ca^{2+} channel levels. Mol Pharmacol. 2006;70(5):1514–23.

Meadows L, et al. The intracellular segment of the sodium channel beta 1 subunit is required for its efficient association with the channel alpha subunit. J Neurochem. 2001;76(6):1871–8.

Mehrke G, et al. The Ca(++)-channel blocker Ro 40-5967 blocks differently T-type and L-type Ca++ channels. J Pharmacol Exp Ther. 1994;271(3):1483–8.

Middleton RE, et al. Two tarantula peptides inhibit activation of multiple sodium channels. Biochemistry. 2002;41(50):14734–47.

Mihailescu M, et al. Structural interactions of a voltage sensor toxin with lipid membranes. Proc Natl Acad Sci U S A. 2014;111(50):E5463–70.

Mishra SK, Hermsmeyer K. Selective inhibition of T-type Ca^{2+} channels by Ro 40-5967. Circ Res. 1994;75(1):144–8.

Mulley JC, et al. SCN1A mutations and epilepsy. Hum Mutat. 2005;25(6):535–42.

Murray JK, et al. Engineering potent and selective analogues of GpTx-1, a tarantula venom peptide antagonist of the $Na_V1.7$ sodium channel. J Med Chem. 2015;58(5):2299–314.

Narahashi T, Frazier T, Yamada M. The site of action and active form of local anesthetics. I. Theory and pH experiments with tertiary compounds. J Pharmacol Exp Ther. 1970;171(1):32–44.

Nicolas S, et al. Chemical synthesis, proper folding, Na(v) channel selectivity profile and analgesic properties of the spider peptide Phlotoxin 1. Toxins (Basel). 2019;11(6)

Nilius B, et al. Inhibition by mibefradil, a novel calcium channel antagonist, of Ca^{2+}- and volume-activated Cl^- channels in macrovascular endothelial cells. Br J Pharmacol. 1997;121(3):547–55.

O'Brien JE, Meisler MH. Sodium channel SCN8A (Nav1.6): properties and de novo mutations in epileptic encephalopathy and intellectual disability. Front Genet. 2013;4:213.

Ohkubo T, Yamazaki J, Kitamura K. Tarantula toxin ProTx-I differentiates between human T-type voltage-gated Ca^{2+} channels Cav3.1 and Cav3.2. J Pharmacol Sci. 2010;112(4):452–8.

Olschewski A, et al. Blockade of Na^+ and K^+ currents by local anesthetics in the dorsal horn neurons of the spinal cord. Anesthesiology. 1998;88(1):172–9.

Ono S, Kimura T, Kubo T. Characterization of voltage-dependent calcium channel blocking peptides from the venom of the tarantula Grammostola rosea. Toxicon. 2011;58(3):265–76.

Park JH, et al. Studies examining the relationship between the chemical structure of protoxin II and its activity on voltage-gated sodium channels. J Med Chem. 2014;57(15):6623–31.

Patel RR, et al. Aberrant epilepsy-associated mutant $Na_V1.6$ sodium channel activity can be targeted with cannabidiol. Brain. 2016;139(Pt 8):2164–81.

Perchenet L, Clement-Chomienne O. Characterization of mibefradil block of the human heart delayed rectifier $hK_v1.5$. J Pharmacol Exp Ther. 2000;295(2):771–8.

Perez-Reyes E. Molecular physiology of low-voltage-activated T-type calcium channels. Physiol Rev. 2003;83(1):117–61.

Perez-Reyes E. G protein-mediated inhibition of Cav3.2 T-type channels revisited. Mol Pharmacol. 2010;77(2):136–8.

Priest BT, et al. ProTx-I and ProTx-II: gating modifiers of voltage-gated sodium channels. Toxicon. 2007;49(2):194–201.

Rangel A, Sanchez-Armass S, Meza U. Protein kinase C-mediated inhibition of recombinant T-type Cav3.2 channels by neurokinin 1 receptors. Mol Pharmacol. 2010;77(2):202–10.

Rosenberg EC, et al. Cannabinoids and epilepsy. Neurotherapeutics. 2015;12(4):747–68.

Rosenberg EC, Patra PH, Whalley BJ. Therapeutic effects of cannabinoids in animal models of seizures, epilepsy, epileptogenesis, and epilepsy-related neuroprotection. Epilepsy Behav. 2017;70(Pt B):319–27.

Ross HR, Napier I, Connor M. Inhibition of recombinant human T-type calcium channels by Δ^9-tetrahydrocannabinol and cannabidiol. J Biol Chem. 2008;283(23):16124–34.

Rupasinghe DB, et al. Mutational analysis of ProTx-I and the novel venom peptide Pe1b provide insight into residues responsible for selective inhibition of the analgesic drug target $Na_V1.7$. Biochem Pharmacol. 2020;181:114080.

Sait LG, et al. Cannabidiol interactions with voltage-gated sodium channels. Elife. 2020;9:e58593.
Salari A, et al. Molecular interactions between tarantula toxins and low-voltage-activated calcium channels. Sci Rep. 2016;6:23894.
Satin J, et al. A mutant of TTX-resistant cardiac sodium channels with TTX-sensitive properties. Science. 1992;256(5060):1202–5.
Sawyer NT, et al. *Scn8a* voltage-gated sodium channel mutation alters seizure and anxiety responses to acute stress. Psychoneuroendocrinology. 2014;39:225–36.
Schmalhofer WA, et al. ProTx-II, a selective inhibitor of $Na_V1.7$ sodium channels, blocks action potential propagation in nociceptors. Mol Pharmacol. 2008;74(5):1476–84.
Sheets MF, et al. Sodium channel molecular conformations and antiarrhythmic drug affinity. Trends Cardiovasc Med. 2010;20(1):16–21.
Shen H, et al. Structures of human Nav1.7 channel in complex with auxiliary subunits and animal toxins. Science. 2019;363(6433):1303–8.
Shi X, et al. Missense mutation of the sodium channel gene SCN2A causes Dravet syndrome. Brain and Development. 2009;31(10):758–62.
Simms BA, Zamponi GW. Neuronal voltage-gated calcium channels: structure, function, and dysfunction. Neuron. 2014;82(1):24–45.
Sivilotti L, et al. A single serine residue confers tetrodotoxin insensitivity on the rat sensory-neuron-specific sodium channel SNS. FEBS Lett. 1997;409(1):49–52.
Slesinger PA. Ion selectivity filter regulates local anesthetic inhibition of G-protein-gated inwardly rectifying K^+ channels. Biophys J. 2001;80(2):707–18.
Sokolov S, et al. Inhibition of sodium channel gating by trapping the domain II voltage sensor with protoxin II. Mol Pharmacol. 2008;73(3):1020–8.
Stevens M, Peigneur S, Tytgat J. Neurotoxins and their binding areas on voltage-gated sodium channels. Front Pharmacol. 2011;2:71.
Strege PR, et al. Effect of mibefradil on sodium and calcium currents. Am J Physiol Gastrointest Liver Physiol. 2005;289(2):G249–53.
Striano P, et al. A novel SCN2A mutation in family with benign familial infantile seizures. Epilepsia. 2006;47(1):218–20.
Sugiyama K, Muteki T. Local anesthetics depress the calcium current of rat sensory neurons in culture. Anesthesiology. 1994;80(6):1369–78.
Sun H, et al. Differential interactions of Na^+ channel toxins with T-type Ca^{2+} channels. J Gen Physiol. 2008;132(1):101–13.
Taga K, et al. Transient neuronal depolarization induces tolerance to subsequent forebrain ischemia in rats. Anesthesiology. 1997;87(4):918–25.
Takahashi M, Catterall WA. Identification of an alpha subunit of dihydropyridine-sensitive brain calcium channels. Science. 1987;236(4797):88–91.
Takahashi H, et al. Solution structure of hanatoxin1, a gating modifier of voltage-dependent K^+ channels: common surface features of gating modifier toxins. J Mol Biol. 2000;297(3):771–80.
Talbot MJ, Sayer RJ. Intracellular QX-314 inhibits calcium currents in hippocampal CA1 pyramidal neurons. J Neurophysiol. 1996;76(3):2120–4.
Tao J, et al. Activation of corticotropin-releasing factor receptor 1 selectively inhibits $Ca_V3.2$ T-type calcium channels. Mol Pharmacol. 2008;73(6):1596–609.
Taylor RE. Effect of procaine on electrical properties of squid axon membrane. Am J Phys. 1959;196(5):1071–8.
Tsukamoto T, et al. Differential binding of tetrodotoxin and its derivatives to voltage-sensitive sodium channel subtypes (Na(v) 1.1 to Na(v) 1.7). Br J Pharmacol. 2017;174(21):3881–92.
Van Theemsche KM, et al. Hydrophobic drug/toxin binding sites in voltage-dependent K^+ and Na^+ channels. Front Pharmacol. 2020;11:735.
Xiao Y, et al. The tarantula toxins ProTx-II and huwentoxin-IV differentially interact with human Nav1.7 voltage sensors to inhibit channel activation and inactivation. Mol Pharmacol. 2010;78(6):1124–34.

Xiong Z, Strichartz GR. Inhibition by local anesthetics of Ca^{2+} channels in rat anterior pituitary cells. Eur J Pharmacol. 1998;363(1):81–90.

Xu H, et al. Structural basis of Nav1.7 inhibition by a gating-modifier spider toxin. Cell. 2019;176(4):702–715 e14.

Zamponi GW, Doyle DD, French RJ. Fast lidocaine block of cardiac and skeletal muscle sodium channels: one site with two routes of access. Biophys J. 1993;65(1):80–90.

Zimmer T, Benndorf K. The human heart and rat brain IIA Na^{+} channels interact with different molecular regions of the beta1 subunit. J Gen Physiol. 2002;120(6):887–95.

The Presynaptic $\alpha_2\delta$ Protein Family and Their Therapeutic Potential

Cornelia Ablinger, Clarissa Eibl, Maria Roznovcova, Graeme S. Cottrell, Gary J. Stephens, and Gerald J. Obermair

Abstract $\alpha_2\delta$ proteins are well-established modulators of membrane expression and biophysical properties of voltage-gated calcium channels. Moreover, they are critical regulators of synapse formation and function and key players in transsynaptic signalling. The $\alpha_2\delta$ isoforms are highly glycosylated membrane-anchored proteins with distinct structural features, some of which are also observed in a human $\alpha_2\delta$-like protein termed CACHD1. Accumulating evidence has underpinned the involvement of $\alpha_2\delta$ proteins in neurological and neurodevelopmental disorders, making them attractive novel therapeutic targets. Also, CACHD1, through its modulation of T-type currents, is an emerging potential drug target, particularly for epilepsy and pain. Furthermore, $\alpha_2\delta$ proteins are targets of the widely prescribed gabapentinoids. Among these, gabapentin and pregabalin, which have a high binding affinity for $\alpha_2\delta$-1 and $\alpha_2\delta$-2, have been administered particularly in the treatment of neuropathic pain conditions, epilepsy, and restless leg syndrome. Extensive efforts have been and are being made to understand the structure and functions of $\alpha_2\delta$ proteins and how they interact with synaptic proteins. This is ultimately helping to understand the contribution of the $\alpha_2\delta$ protein family to neurological and neurodevelopmental disorders and provides insight into potential novel treatment options.

Keywords $\alpha_2\delta$ subunits · Voltage-gated calcium channels · CACHD1 · Synapse formation and transmission · Neurological disorders · Neurodevelopmental disorders · Gabapentin · Therapeutic targets · Structural homology modelling

---

Authors Cornelia Ablinger and Clarissa Eibl have equally contributed to this chapter.

C. Ablinger · C. Eibl · G. J. Obermair (✉)
Division of Physiology, Department of Pharmacology, Physiology, and Microbiology, Karl Landsteiner University of Health Sciences, Krems, Austria
e-mail: Cornelia.ablinger@kl.ac.at; Clarissa.Eibl@kl.ac.at; Gerald.Obermair@kl.ac.at

M. Roznovcova · G. S. Cottrell · G. J. Stephens
Reading School of Pharmacy, University of Reading, Reading, UK
e-mail: m.roznovcova@pgr.reading.ac.uk; g.s.cottrell@reading.ac.uk; g.j.stephens@reading.ac.uk

G. Stephens, E. Stevens (eds.), *Ion Channels as Targets in Drug Discovery*,
https://doi.org/10.1007/978-3-031-52197-3_3

1 $\alpha_2\delta$ Proteins

1.1 $\alpha_2\delta$ Isoforms and the $\alpha_2\delta$-Like Protein CACHD1

$\alpha_2\delta$ proteins were initially characterised as auxiliary components of the multi-subunit protein complex of voltage-gated calcium channels (VGCCs). More recent studies revealed that they are also critical regulators of synaptic functions, including synapse formation and synaptic transmission. $\alpha_2\delta$ proteins are membrane-anchored extracellular glycoproteins (reviewed in Davies et al. 2007), which are abundantly expressed in the brain and the peripheral nervous system (Obermair et al. 2005; Schlick et al. 2010). In 1988, the first $\alpha_2\delta$ isoform, $\alpha_2\delta$-1, was identified as a protein associated with the dihydropyridine receptor (DHPR, $Ca_V1.1$, or α_{1S}) complex in skeletal muscle, and the cDNA was cloned (Ellis et al. 1988). Since then, three additional isoforms have been discovered: $\alpha_2\delta$-2 (Klugbauer et al. 1999; Gao et al. 2000), $\alpha_2\delta$-3 (Klugbauer et al. 1999), and $\alpha_2\delta$-4 (Qin et al. 2002). All four isoforms are detected in multiple tissues, displaying both partially differential and overlapping expression patterns. Interestingly, while $\alpha_2\delta$-1, $\alpha_2\delta$-2, and $\alpha_2\delta$-3 are abundantly expressed in the brain (Schlick et al. 2010; Geisler et al. 2019; Cole et al. 2005), $\alpha_2\delta$-4 is mainly expressed in photoreceptor cells of the retina (Wang et al. 2017). The four $\alpha_2\delta$ isoforms ($\alpha_2\delta$-1, $\alpha_2\delta$-2, $\alpha_2\delta$-3, $\alpha_2\delta$-4) are encoded by the mammalian genes *Cacna2d1-4*. While mouse and human $\alpha_2\delta$ genes are highly conserved (Barclay and Rees 2000; Hanke et al. 2001), the amino acid sequence alignments of the individual isoforms show a relatively low similarity of approximately 60% ($\alpha_2\delta$-3 vs. $\alpha_2\delta$-4), 55% ($\alpha_2\delta$-1 vs. $\alpha_2\delta$-2), and 30% ($\alpha_2\delta$-1/$\alpha_2\delta$-2 vs. $\alpha_2\delta$-3/$\alpha_2\delta$-4) (Klugbauer et al. 1999; Qin et al. 2002). Nevertheless, they all share a similar protein domain organisation, including important common structural properties (see Fig. 1). CACHD1 [cache (Ca^{2+} channel and chemotaxis receptor) domain containing 1] protein was first identified to be related to $\alpha_2\delta$ auxiliary subunits by Cottrell et al. (2018) using bioinformatics approaches. Due to the similar structural properties, CACHD1 has been described as an $\alpha_2\delta$-like protein (Cottrell et al. 2018) and is now broadly considered a member of the $\alpha_2\delta$ protein family.

1.2 Structure of $\alpha_2\delta$ Proteins

The initial publication of the three-dimensional cryo-EM structure of the skeletal muscle calcium channel complex at 3.6 Å resolution delivered insights into the structural properties of $\alpha_2\delta$ proteins (Wu et al. 2016). All *CACNA2D* genes encode a single precursor protein that is proposed to be posttranslationally cleaved by the metalloprotease ADAM17 into an α_2 and a δ polypeptide (Fig. 2) (De Jongh et al. 1990; Jay et al. 1991; Kadurin et al. 2022). During posttranslational processing, α_2 and δ become linked by disulfide bonds (Wu et al. 2016) (reviewed in Andrade et al. 2007; Calderón-Rivera et al. 2012; Dolphin 2018). Posttranslational processing of

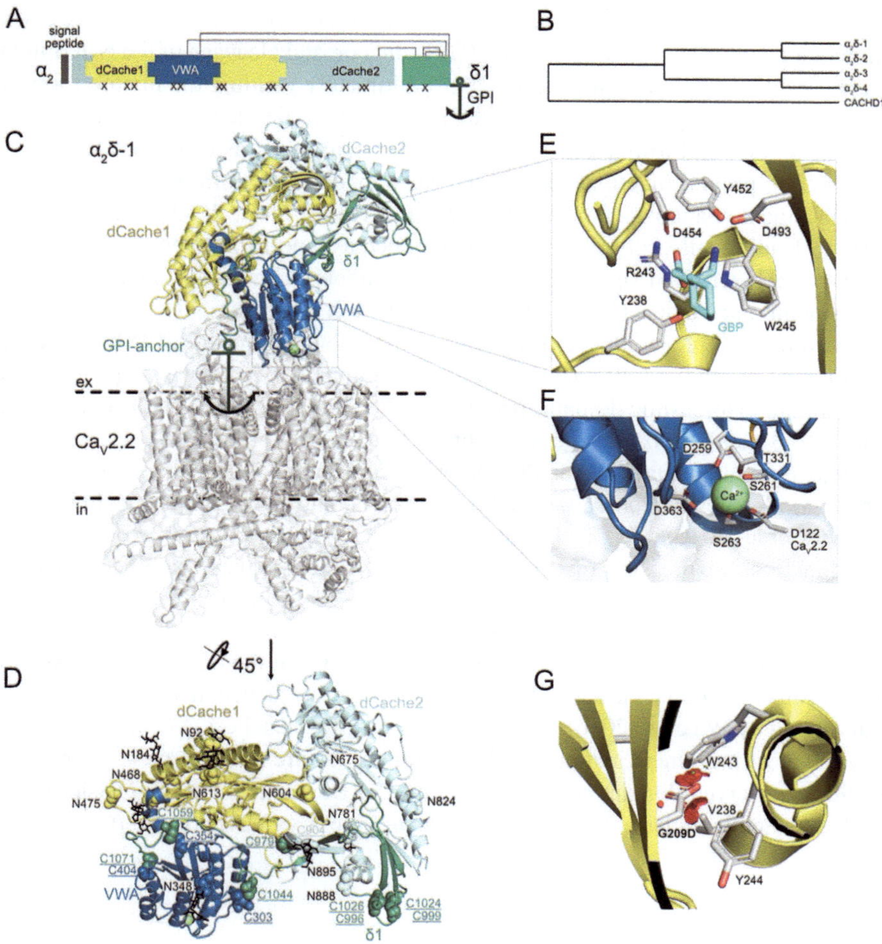

Fig. 1 Structure of $\alpha_2\delta$ proteins. (**a**) Protein domain structure of $\alpha_2\delta$-1. The two intertwined double Cache domains, dCache1 and dCache2 (yellow and turquoise), are interrupted by the von Willebrand factor type A (VWA) domain (blue). The proteolytically cleaved δ peptide (green) completes the dCache2 domain and forms five disulfide bonds (black brackets). The δ peptide harbours a GPI-anchor indicated by the C-terminal anchor motif. Sixteen predicted glycosylation sites are indicated by black Xs. (**b**) Phylogenetic tree of the individual $\alpha_2\delta$ proteins and CACHD1 created by the Simple Phylogeny, EBI Web Services. (**c**) Structure of the human $Ca_V2.2$ (grey, PDB code 7miy) bound to human $\alpha_2\delta$-1 (same colour code as in **a**) as inserted in the membrane (intracellular (in) and extracellular (ex) fractions are separated by dashed lines). (**d**) Top view onto the $\alpha_2\delta$-1 subunit showing the glycosylated asparagine residues (shown as spheres and labelled black). Resolved N-linked glycosylation is shown as black sticks. The δ peptide (green) covalently links several domains, creating rigidity in the overall structure. Disulfide-forming cysteines are labelled, underlined, and shown as spheres. (**e**) The inset shows the gabapentin (GBP, cyan) binding site within the first dCache1 domain (PDB code 8fd7). Under native conditions, leucine is bound at this site (Chen et al. 2023). Residues engaging in the interaction are shown as grey sticks. (**f**) The inset shows the Ca^{2+}-ion coordination by the metal ion-dependent adhesion site (MIDAS) motif within the VWA domain (residues are labelled and shown as grey sticks). The Asp122 of the $Ca_V2.2$ α_1 subunit completes the Ca^{2+} ion coordination. (**g**) Model of the $\alpha_2\delta$-1 disease mutant Gly209Asp. The mutated aspartate creates steric interference (indicated by red discs) with the residues of the gabapentin binding site

$\alpha_2\delta$ proteins also includes glycosylation at several N-glycosylation sites, the addition of a glycosylphosphatidylinositol (GPI)-anchor, and cleavage of the N-terminal signal peptide (reviewed in Dolphin 2018), ultimately yielding the mature form of $\alpha_2\delta$ proteins (Fig. 2). All these modifications (proteolytic cleavage, glycosylation, and disulfide bond formation) are essential for mature forms of $\alpha_2\delta$ proteins and, hence, for proper functions, including calcium channel trafficking and current enhancement (Sandoval et al. 2004; Calderón-Rivera et al. 2012; Kadurin et al. 2017). Traditionally, $\alpha_2\delta$ subunits were classified as type-I transmembrane proteins due to the N-terminal signal peptide, the hydrophobic C-terminus, and a potential transmembrane region (Jay et al. 1991; Gurnett et al. 1996; Wiser et al. 1996). However, more recent studies suggest that $\alpha_2\delta$ subunits are linked to the plasma membrane through a GPI-anchor, as all $\alpha_2\delta$ subunits have either a short or absent C-terminal cytoplasmic domain (reviewed in Davies et al. 2007) and $\alpha_2\delta$-1, $\alpha_2\delta$-2, and $\alpha_2\delta$-3 (and probably $\alpha_2\delta$-4 by prediction) display characteristics of GPI-anchored proteins (Davies et al. 2010). One such characteristic is the inclusion of ω (omega) residues, which act as anchoring motifs, and their delivery to lipid rafts (Davies et al. 2010) (Fig. 2). Furthermore, the GPI-anchored prion protein blocks the functional expression of calcium channels, which suggests competition with $\alpha_2\delta$-1 for GPI-anchoring or trafficking pathways (Senatore et al. 2012; Alvarez-Laviada et al. 2014). Moreover, the structure reported by Wu et al. (2016) shows that the $\alpha_2\delta$ protein is entirely extracellular and indicates that the δ peptide is linked to the plasma membrane via a glycophosphatidyl moiety attached to a cysteine residue (discussed by Wu et al. 2016), strongly supporting GPI-anchoring (Fig. 1c).

The most prominent and outstanding common feature of $\alpha_2\delta$ proteins, however, is their conserved domain structure (reviewed in Davies et al. 2007; Geisler et al. 2015), which includes a von Willebrand factor type A (VWA) domain (Cantí et al.

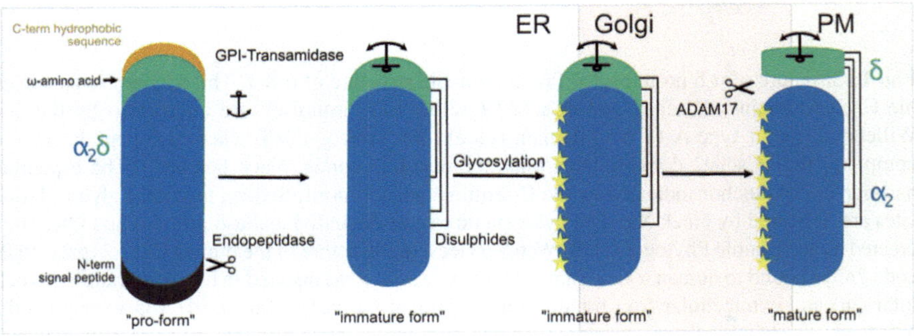

Fig. 2 Posttranslational modifications of $\alpha_2\delta$ proteins. The pro-form of $\alpha_2\delta$ is posttranslationally modified within the endoplasmic reticulum (ER), including disulfide formations during translation, replacing the C-terminal hydrophobic sequence (orange) with a GPI-anchor by the GPI-transamidase, cleavage of the N-terminal signal peptide by an endopeptidase, and addition of several glycosylations (yellow stars). The glycosylations are modified in the Golgi apparatus. The metalloprotease ADAM17 is proposed to cleave $\alpha_2\delta$ into the mature disulfide-linked α_2 and δ subunit found at the plasma membrane (PM)

2005) and two double Cache domains, dCache1 and dCache2 (Fig. 1a) (Wu et al. 2016). The VWA domain is generally found in integrins and is implicated in divalent cation-dependent interactions with the extracellular matrix and is crucial for the $\alpha_2\delta$-1/Ca$_V$1.1 association (Cantí et al. 2005; Wu et al. 2016), while the cache domains (homologous to bacterial chemotaxis receptors) bind small-molecule ligands such as amino acids (Anantharaman and Aravind 2000). Proximal to the VWA domain of $\alpha_2\delta$-1 and $\alpha_2\delta$-2 within the dCache1 domain is a triple arginine (RRR) motif (Davies 2006; Field et al. 2006), which facilitates the binding of the antiepileptic and anti-allodynic drugs gabapentin and pregabalin (Fig. 1e) (Gee et al. 1996; Gong et al. 2001; Bian et al. 2006; Field et al. 2006; Li et al. 2011). Another critical feature of $\alpha_2\delta$ proteins is the metal ion-dependent adhesion site (MIDAS) in the VWA domain. This site facilitates protein interactions by binding divalent cations (e.g., Ca^{2+} or Mg^{2+}) through a ring of up to five polar or charged residues (Whittaker and Hynes 2002) and is essential for calcium channel trafficking (Fig. 1f) (Cantí et al. 2005; Schöpf et al. 2021). Exclusively, $\alpha_2\delta$-1 and $\alpha_2\delta$-2 comprise a "perfect" MIDAS motif (DxSxS) (Cantí et al. 2005), while $\alpha_2\delta$-3 and $\alpha_2\delta$-4 lack a polar residue (Whittaker and Hynes 2002).

1.3 The $\alpha_2\delta$-Like Protein CACHD1

Due to its predicted structural similarities to members of the $\alpha_2\delta$ family, the CACHD1 protein was reported as an $\alpha_2\delta$-like protein (Whittaker and Hynes 2002) and shown to functionally modulate calcium channel currents (T-type by Cottrell et al. (2018) and N-type by Dahimene et al. (2018)) (Fig. 3) (Cottrell et al. 2018; Dahimene et al. 2018; Stephens and Cottrell 2019). Despite low sequence similarity (only a 13–16% gene homology and a < 21% protein identity) to $\alpha_2\delta$ proteins (Fig. 1b), there are multiple structural similarities between CACHD1 and $\alpha_2\delta$ proteins (Cottrell et al. 2018). For example, CACHD1, similar to $\alpha_2\delta$-1, contains an N-terminal signal peptide, a VWA domain, two double bacterial chemosensory-like cache domains with similar overall alignment (Gumerov et al. 2022), and a short hydrophobic transmembrane domain (Cottrell et al. 2018) (Fig. 3). Despite this striking similarity, CACHD1 and $\alpha_2\delta$ proteins differ in several aspects. First, $\alpha_2\delta$ proteins are single gene products that contain a critical six amino acid motif for proteolytic cleavage (Andrade et al. 2007), which is absent in CACHD1 (Cottrell et al. 2018). Second, the dCache2 domain of $\alpha_2\delta$ proteins contains a distinctive long insertion between the α9 helix and β9 sheet as opposed to the reportedly shorter insertion in CACHD1 (Gumerov et al. 2022). Third, CACHD1 has up to seven predicted posttranslational N-glycosylation sites (Stephens and Cottrell 2019), while $\alpha_2\delta$-1 can be glycosylated at multiple potential sites (Douglas et al. 2006). Fourth, CACHD1 was initially reported to contain a variant gabapentinoid binding motif (RSR amino acid sequence) as well as a variant MIDAS motif (DxGxS) (Cottrell et al. 2018). Recent work has described an additional amino acid binding motif that is also targeted by gabapentinoids located within the first double dCache1 domain

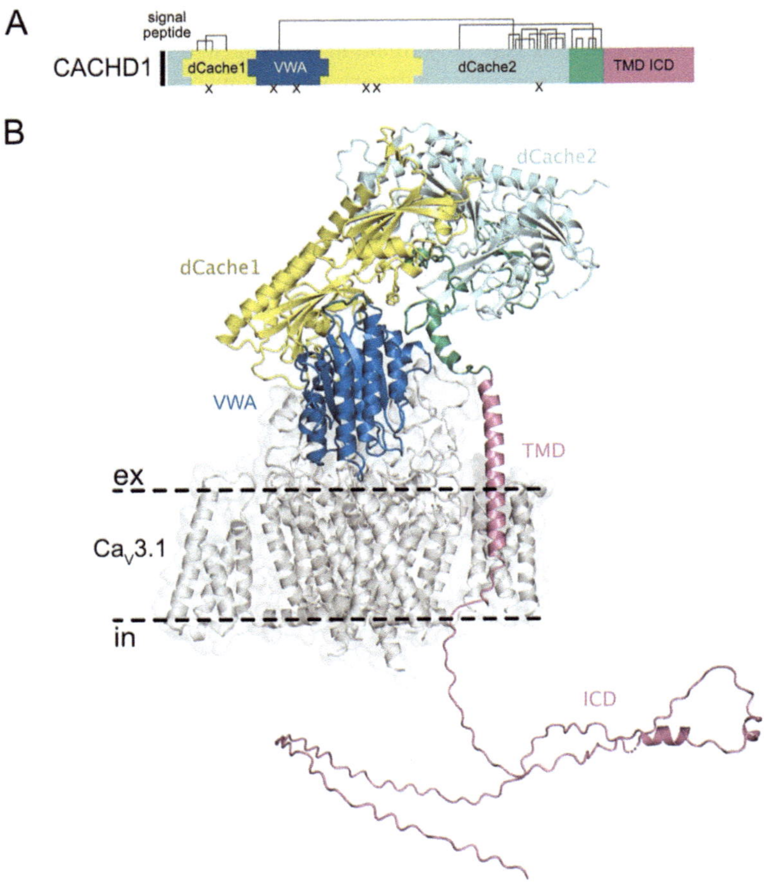

Fig. 3 Domain architecture and structural model of CACHD1 in complex with $Ca_V3.1$. (**a**) Protein domain structure of CACHD1. CACHD1 shows similarities to $\alpha_2\delta$-1 in its protein architecture, such as the two intertwined dCache domains (dCache1, yellow, and dCache2, turquoise) that are interrupted by the von Willebrand factor type A (VWA) domain (blue). The overall structure is highly stabilised by 14 predicted disulfide bonds (black brackets). In contrast to $\alpha_2\delta$-1, CACHD1 possesses a transmembrane domain (TMD), a primarily unstructured intracellular domain (ICD), and seven predicted glycosylation sites (black X). (**b**) Structural model of the Alphafold (Jumper et al. 2021) predicted structure of human CACHD1 (same colour code as above) modelled onto the cryo-EM structure of the T-type channel $Ca_V3.1$ (grey, PDB code: 6kzo)

of $\alpha_2\delta$ proteins (Gumerov et al. 2022); in this regard, a suggested consensus motif (YRWYD) is present entirely within the dCache2 domain in CACHD1, in contrast to $\alpha_2\delta$-1, where the N-terminal (YRW) and C-terminal (YD) parts of the amino acid binding motif straddle the VWA domain. Finally, contrary to $\alpha_2\delta$ subunits, CACHD1 has no predicted GPI-anchoring site but possesses a predicted transmembrane domain and a larger intracellular C-terminal domain (Cottrell et al. 2018).

2 Physiological Roles of $\alpha_2\delta$ Proteins

As canonical regulators of the trafficking and functions of Ca_V1 and Ca_V2 channels, $\alpha_2\delta$ subunits increase the channel density at the plasma membrane (Cassidy et al. 2014). Particularly in neurons, $\alpha_2\delta$ isoforms further direct trafficking of channels to subcellular sites such as dendrites (Obermair et al. 2005; Cassidy et al. 2014; Kadurin et al. 2017) and influence biophysical channel properties (Davies et al. 2007; Dolphin 2012, 2013). Moreover, $\alpha_2\delta$ proteins are targets of the widely prescribed antiepileptic and anti-allodynic gabapentinoid drugs (gabapentin, pregabalin), which find applications as treatment of neuropathic pain conditions (see Sects. 3.3 and 3.4) and have been strongly linked to synapse formation (elaborated on in Sect. 2.2.1).

2.1 *Calcium Channel-Dependent Functions*

A wealth of heterologous co-expression studies in nonneuronal cells revealed that $\alpha_2\delta$ proteins operate in a rather promiscuous manner by increasing the surface expression of Ca_V channels [Ca_V1 and Ca_V2 together with distinct β isoforms (for review, see Davies et al. 2007; Klugbauer et al. 2003; Geisler et al. 2015; Dolphin 2016)]. Expression of $\alpha_2\delta$-2 with $Ca_V2.1$ and β_4 in COS-7 cells increases calcium currents threefold (Ludwig et al. 1997; Schlick et al. 2010; Barclay et al. 2001), while expression of other $\alpha_2\delta$s further influenced the biophysical properties of L-type channels (kinetics and voltage dependence of activation) (Felix et al. 1997; Klugbauer et al. 1999; Lee et al. 2015). The proposed mechanism underlying the $\alpha_2\delta$-induced increase in maximum current occurs via an enhancement of plasma membrane expression of channel complexes concomitant with a decrease in their turnover (Dolphin 2012) but may also include increased forward trafficking (Cassidy et al. 2014) or transport from recycling endosomes (Fig. 5) (Tran-Van-Minh and Dolphin 2010). Moreover, studies on channel complexes in native cellular systems suggested isoform specificity. Ablation of $\alpha_2\delta$-1 in mice abolished $Ca_V2.2$ cell surface expression in dorsal root ganglion (DRG) neurons and strongly reduced expression in the dorsal horn (Nieto-Rostro et al. 2018). Overexpression of distinct isoforms differentially modulated presynaptic calcium transients in primary hippocampal neurons (Ablinger et al. 2022). For example, overexpression of $\alpha_2\delta$-1 increased presynaptic calcium transients and endogenous $Ca_V2.1$ expression (Ablinger et al. 2022). $\alpha_2\delta$-1 in sensory neurons was reported to increase currents and alter the kinetics and voltage dependence of activation (Li et al. 2006). Moreover, recombinant expression in rat cervical ganglion neurons of $\alpha_2\delta$-1 and $\alpha_2\delta$-2 affected $Ca_V2.2$ expression differently compared to expression of $\alpha_2\delta$-3 (Scott and Kammermeier 2017), and loss of $\alpha_2\delta$-3 reduced P/Q- and R-type currents, while N- and L-type currents remain unchanged (Stephani et al. 2019). Furthermore, loss of $\alpha_2\delta$-3 blocks homeostatic modulation of neurotransmitter release due to a failure in the potentiation of presynaptic calcium influx (Wang et al. 2016a). These different effects may also imply that $\alpha_2\delta$ proteins can exert specific and distinct properties on different Ca_V isoforms (see discussion in Obermair et al. 2008; Ablinger et al. 2022). A recent study provided evidence that $\alpha_2\delta$-1, like $\alpha_2\delta$-2, is recycled to the plasma membrane via a Rab11a-dependent endosomal recycling pathway and increases the plasma membrane expression of $Ca_V2.2$ (Meyer and Dolphin 2021).

The affinity and strength of the $\alpha_1/\alpha_2\delta$ interaction are under debate, despite numerous identified binding sites and the importance of the MIDAS motif for α_1 subunit trafficking (Cantí et al. 2005) due to a quantitative study of the neuronal Ca_V2 channel proteome (Müller et al. 2010). Furthermore, it was shown recently that the interaction sites between specific $\alpha_1/\alpha_2\delta$ proteins differ in distinct residues, which might be crucial for fine-tuning protein-protein interactions in native cells and tissues where more than one $\alpha_2\delta$ protein is expressed (Ablinger et al. 2022). In addition, multiple other studies suggest a relatively weak association of the $\alpha_1/\alpha_2\delta$-complex. For instance, subunits are believed to be GPI-anchored (Davies et al. 2010; Wu et al. 2016), which contradicts a possible tight transmembrane interaction between α_1 and $\alpha_2\delta$ (Robinson et al. 2011). Moreover, $Ca_V2.1$ solubilised from the cerebellum was found to be mainly uncomplexed in the membrane fraction, and only a tiny fraction was associated with $\alpha_2\delta$-2 in lipid rafts (Davies 2006), suggesting a rather weak interaction. Indeed, the comparison of the interaction interface of $\alpha_2\delta$-1 with $Ca_V1.1$ or $Ca_V2.2$ reveals more contact points (Fig. 4; pink and blue coloured residues) with $Ca_V1.1$, indicating a more robust interaction than with

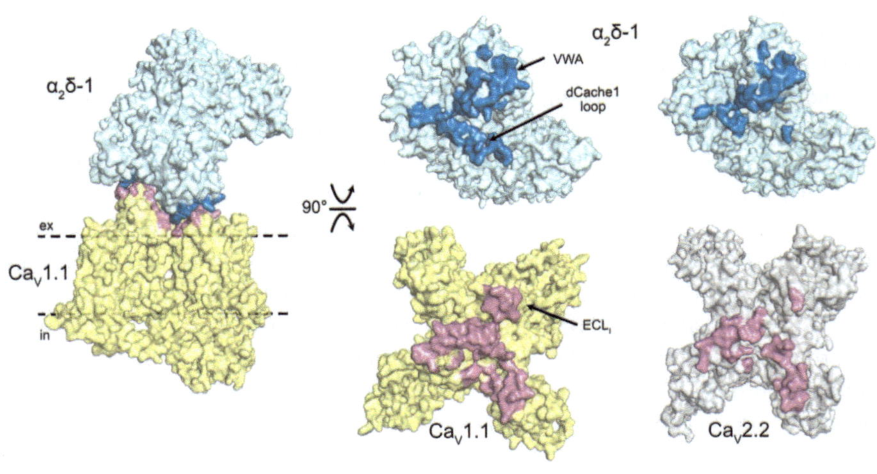

Fig. 4 Structures of distinct $\alpha_1/\alpha_2\delta$-1 complexes reveal differences in the interaction interfaces. Published structures of $Ca_V1.1$ (yellow, PDB code: 5gjv) and $Ca_V2.2$ (grey, PDB code: 7miy) in complex with $\alpha_2\delta$-1 (light blue) are opened to allow the view on the interaction interface (e.g., revealing the position of the VWA domain and the loop of dCache1). Interface areas engaging in the complex interaction are coloured pink for the channel subunit and blue in $\alpha_2\delta$-1. Negative charges in the extracellular loop I (ECL_I), present in $Ca_V1.1$ but not $Ca_V2.2$, allow tighter interactions with the $\alpha_2\delta$-1 loop in the dCache1 domain

$Ca_V2.2$ (Fig. 4), which is supported by the initial identification of the stoichiometric expression of calcium channel subunit proteins in skeletal muscle (Leung et al. 1987).

Both kinetics and voltage dependence of activation/inactivation can be differentially modulated by $\alpha_2\delta$ proteins, depending on the type and cellular environment of the channel (Wakamori et al. 1999; Obermair et al. 2005, 2008; Tuluc et al. 2007). Generally, $\alpha_2\delta$ subunits hyperpolarize the voltage dependence of the steady-state inactivation (Felix et al. 1997). However, a shift to more positive potentials was also observed in individual cases (Qin et al. 1998), and, in particular, $\alpha_2\delta$-1 facilitates channel activation by increasing the voltage sensitivity of the voltage-sensing domains (Savalli et al. 2016). Although it is still unknown by which exact molecular mechanism $\alpha_2\delta$ subunits exert their function, the MIDAS motif seems to be essential, as a MIDAS mutant of $\alpha_2\delta$-2 resulted in intracellular retention of α_1 subunits and failed to enhance $Ca_V1.2$, $Ca_V2.1$, and $Ca_V2.2$ currents (Cantí et al. 2005). Moreover, the integrity of the perfect MIDAS motif of $\alpha_2\delta$-1 and $\alpha_2\delta$-2 is further necessary for the trafficking and abundance of calcium channels in nonneuronal (Cantí et al. 2005; Hoppa et al. 2012) and neuronal cells (Hoppa et al. 2012; Schöpf et al. 2021). The extracellular localisation of $\alpha_2\delta$ proteins protruding into the synaptic cleft may allow a direct link between the calcium channel complex and extracellular signalling entities (Fig. 5).

In neurons, calcium channel function may be altered by aberrant interaction between $\alpha_2\delta$ proteins and other interaction partners (Ablinger et al. 2020). Thrombospondin (TSP)4, for example, increases calcium currents after depolarization at low voltages and contributes to changed excitatory synaptogenesis linked with neuropathic pain development, which can be blocked by gabapentin (Pan et al. 2016; Yu et al. 2018). α-Neurexins, together with $\alpha_2\delta$-1, modulate presynaptic calcium influx through $Ca_V2.1$ channels (Brockhaus et al. 2018), while prion protein exerts an inhibitory effect (Senatore et al. 2012), most probably by competing with $\alpha_2\delta$ for GPI-anchoring pathways (Alvarez-Laviada et al. 2014). The low-density lipoprotein receptor-related protein 1 (LRP1) reduces $\alpha_2\delta$-1 trafficking and, thus, $Ca_V2.2$ trafficking to the cell surface (Kadurin et al. 2017). The α subunit of the large conductance calcium-activated potassium (BK) channel (Slo1) is suggested to influence VGCC function by interaction with $\alpha_2\delta$-1 (Zhang et al. 2018).

Human CACHD1 and rat Cachd1 modulate Ca_V3 and $Ca_V2.2$ channels, respectively (Cottrell et al. 2018; Dahimene et al. 2018). Contrary to the classical auxiliary $\alpha_2\delta$- and β-subunits of VGCCs, which are believed to interact with and modulate exclusively high voltage-activated VGCCs of the Ca_V1 (L-type) and Ca_V2 (N-, P/Q-, and R-type) family, human CACHD1 showed modulation of Ca_V3 (T-type) in heterologous expression systems and native neurons by increasing the (i) peak current density with a corresponding increase in maximal conductance of human $Ca_V3.1$, $Ca_V3.2$, and $Ca_V3.3$, (ii) channel open probability of $Ca_V3.1$, and (iii) neuronal and rebound firing frequency (property mediated predominantly by T-type currents), which is altered by the selective T-type currents blocker TTA-P2 (Cottrell et al. 2018; Stephens and Cottrell 2019). Furthermore, proximity ligation assays showed that CACHD1 and $Ca_V3.1$ are <40 nm apart on the cell surface, strongly suggesting the formation of CACHD1-$Ca_V3.1$ protein complexes at the

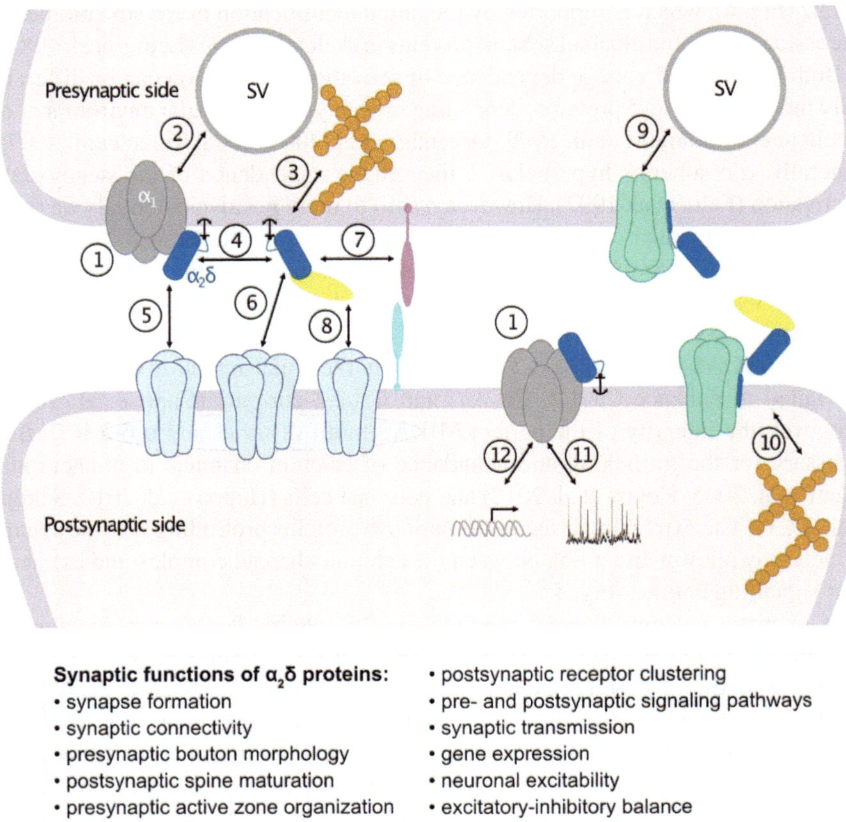

Fig. 5 Summary of proposed synaptic functions of $\alpha_2\delta$ proteins. The function of $\alpha_2\delta$ proteins (blue; numbers refer to the respective mechanisms in the figure) include (1) plasma membrane trafficking and current modulation of the Ca_V channel (α_1 subunit, grey), (2) accumulation of synaptic vesicles (SV), (3) regulating active zone architecture (AZ) and bouton morphogenesis either channel independent or (4) via the VGCC complex, (5) transsynaptic organisation of the presynaptic AZ with postsynaptic AMPAR and $GABA_AR$ (receptors in light blue) either partly or (6) entirely channel independent presumably via interactions with (7) cell adhesion molecules such as presynaptic neurexins (pink) and postsynaptic neuroligins (cyan), or (8) via interactions with extracellular matrix or secreted proteins (e.g., BDNF, TSP; yellow ellipses). Transmembrane $\alpha_2\delta$-1 bound to TSP is suggested to stabilise NMDAR (receptors in green) on the presynaptic (9) and postsynaptic (10) surface and might further regulate neuronal excitability (11) and gene expression (12). (Figure adapted from Ablinger et al. 2020)

plasma membrane, which further supports the interaction of CACHD1 with Ca_V3 channels (Cottrell et al. 2018).

As a modulator of $Ca_V2.2$, rat Cachd1 caused a significant increase in $Ca_V2.2$ currents while promoting $Ca_V2.2$ cell surface expression by reducing endocytosis (Dahimene et al. 2018). However, the increase in $Ca_V2.2$ currents caused by Cachd1 was not as profound as for $\alpha_2\delta$-1-induced current increases. Mutagenesis of the

Asp122 interaction site utilised by $\alpha_2\delta$-1 showed no effects on the Cachd1-mediated $Ca_V2.2$ current increase, suggesting that Cachd1 may engage different domains to interact with $Ca_V2.2$, perhaps via cache domains (Dahimene et al. 2018). Another study showed that the complex interaction between $\alpha_2\delta$ proteins or Cachd1 and the α_1 subunit is rather similar, although with differences in the affinity of the interaction (Ablinger et al. 2022). Moreover, rat Cachd1 did not modulate $Ca_V2.1$ (Dahimene et al. 2018), suggesting that $\alpha_2\delta$ proteins and CACHD1 may modulate VGCCs via differential modes of action.

In the future, it will be of interest to determine if different CACHD1 domains perform any equivalent functions to corresponding domains in $\alpha_2\delta$ proteins, for example, in terms of channel trafficking. Moreover, it will be of interest to determine the functionality of unique CACHD1 structural domains, such as the large intracellular C-terminus.

2.2 Calcium Channel-Independent Functions

Apart from their classical role as calcium channel subunits, $\alpha_2\delta$ proteins regulate synaptic functions. This has been suggested by overexpression or genetic ablation of specific $\alpha_2\delta$ isoforms in neuronal cell systems (e.g., Eroglu et al. 2009; Geisler et al. 2019; Schöpf et al. 2021) or knockout/mutant animals (e.g., Barclay and Rees 2000; Kurshan et al. 2009; Pirone et al. 2014; Fell et al. 2016; Wang et al. 2017). These studies show that $\alpha_2\delta$ isoforms regulate synaptic morphology and function by influencing synapse formation, synaptic connectivity and postsynaptic receptor abundance, presynaptic architecture, and synaptic transmission. However, the exact molecular mechanisms by which $\alpha_2\delta$ proteins convey synaptogenic properties remain elusive.

2.2.1 Functions Associated with Specific $\alpha_2\delta$ Isoforms

The following part will provide an overview of the current knowledge on the involvement of each $\alpha_2\delta$ subunit in synapse formation (also confer Fig. 5).

$\alpha_2\delta$-1 and Excitatory Synaptogenesis

The presynaptic importance of $\alpha_2\delta$ subunits and their synaptogenic potential were highlighted in an extensive study using $\alpha_2\delta$ double knockout mice and $\alpha_2\delta$ triple loss-of-function neurons (Schöpf et al. 2021). Double knockout mice not only showed a reduced survival rate, but cultured hippocampal neurons from these mice also showed reduced calcium currents (Schöpf et al. 2021). Presynaptic triple loss of $\alpha_2\delta$ isoforms resulted in a severe failure in presynaptic differentiation. This was evidenced by defective presynaptic calcium channel clustering and calcium influx, smaller presynaptic active zones, and a strongly reduced accumulation of presynaptic vesicle-associated proteins (synapsin and vGLUT) and was associated with the downscaling of postsynaptic AMPA receptors and the postsynaptic density (Schöpf et al. 2021). However, postsynaptic triple knockout neurons could still form dendritic spines and receive input from double knockout neurons, suggesting that postsynaptic differentiation is secondary to presynaptic differentiation (Schöpf et al. 2021). Interestingly, reintroduction of any $\alpha_2\delta$ isoform ($\alpha_2\delta$-1 to -3 (Schöpf et al. 2021), $\alpha_2\delta$-4 and Cachd1 (Ablinger et al. 2022)) was capable of rescuing presynaptic calcium channel trafficking and expression of synaptic proteins, suggesting that $\alpha_2\delta$ isoforms are highly redundant in their role as synaptic organisers. Importantly, mutated MIDAS variants of $\alpha_2\delta$-2 and $\alpha_2\delta$-3 were effective in restoring presynaptic synapsin expression but only partially restored calcium channel trafficking, highlighting that the regulatory role of $\alpha_2\delta$ proteins is independent of their role as calcium channel subunits (Schöpf et al. 2021).

The postsynaptic effect of $\alpha_2\delta$-1 on synapse formation has been described in retinal ganglion cells, in which the interaction of postsynaptic $\alpha_2\delta$-1, which acts as a receptor for its astrocyte-secreted ligand TSP, mediates excitatory synapse formation (Eroglu et al. 2009). The study suggests that the $\alpha_2\delta$-1/TSP-induced synaptogenesis does not rely on the presence or function of an α_1 subunit, which further supports the idea that synaptogenesis is a calcium channel-independent function of $\alpha_2\delta$ subunits. A subsequent study in a conditional knockout mouse strain (Park et al. 2016) demonstrated that loss of $\alpha_2\delta$-1 impairs excitatory synaptogenesis and dendritic spine morphology in the cortex (Risher et al. 2018) and that this defect could not be rescued by the expression of an $\alpha_2\delta$-1 mutant variant (p.Arg351Thr) associated with autism (Iossifov et al. 2014). Moreover, the p.Arg351Thr mutation in the VWA domain further prevented membrane expression (Risher et al. 2018), suggesting that the disease mutation might impair excitatory synapse formation and synaptic transmission. The presynaptic role of $\alpha_2\delta$-1 has also been investigated in the light of neuropathic pain (Chen et al. 2018) and hyperalgesia (Deng et al. 2019a) and in cultured hippocampal neurons (Bikbaev et al. 2020). Indeed, increased expression levels of $\alpha_2\delta$-1 through lentiviral injection (Chen et al. 2018) or opioid application (Deng et al. 2019a) potentiated presynaptic NMDAR trafficking in spinal cord synapses. Furthermore, gabapentin or an $\alpha_2\delta$-1 interfering peptide could normalise the enhanced excitatory synaptic transmission, which caused pain hypersensitivity. Interestingly, the interaction of $\alpha_2\delta$-1 with distinct NMDAR subtypes (GluN1, GluN2A, and GluN2B) appears to be isoform-specific as neither $\alpha_2\delta$-2 nor $\alpha_2\delta$-3 co-immunoprecipitated with NMDARs in spinal cord and HEK cells (Chen et al. 2018). Further analysis showed that the NMDAR interaction site is located after the C-terminal GPI-anchor cleavage site, which – considering that $\alpha_2\delta$ subunits have no cytoplasmic domain (Davies et al. 2010) – might argue for an NMDAR binding of a transmembrane form of $\alpha_2\delta$-1 (Fig. 5). Pre- and postsynaptic $\alpha_2\delta$-1/NMDAR complexes are expressed in the hypothalamus in disease states (hypertension), where they boost NMDAR-mediated glutamate release (Ma et al. 2018a, b), as well as in the striatum under normal physiological conditions, such as during cortico-striatal long-term potentiation in learning and memory formation (Zhou et al. 2018).

The levels of neuronal $\alpha_2\delta$-1 may be regulated by the brain-derived neurotrophic factor (BDNF) (Cordeira et al. 2014; Bowling et al. 2016; Qin et al. 2019) as BDNF deficiency in mice resulted in reduced $\alpha_2\delta$-1 cell surface expression in the ventromedial hypothalamus (Cordeira et al. 2014). Furthermore, BDNF mutant mice showed increased levels of $\alpha_2\delta$-1 and TSP2 and an enhanced abundance of glutamatergic synapses within the cortico-striatal pathway, after recovering from a stroke (Qin et al. 2019). Moreover, the same study further provided data that imply a regulation of the excitation-inhibition balance by synaptic $\alpha_2\delta$-1 expression levels, as they identified a reduction of GABAergic innervation in distinct cortical layers (Qin et al. 2019).

$\alpha_2\delta$-2 and Synaptic Connectivity and Postsynaptic Receptor Abundance

Synaptic functions of $\alpha_2\delta$-2, particularly the transsynaptic action of a splice variant lacking exon 23 (Geisler et al. 2019), are most likely independent of their role as a calcium channel subunit. A recent study investigated the consequences of presynaptic $\alpha_2\delta$ subunit expression in glutamatergic and GABAergic synapses, showing that presynaptic overexpression of $\alpha_2\delta$-2 in GABAergic synapses strongly increases the clustering of postsynaptic $GABA_A$Rs (Fig. 5) (Geisler et al. 2019). Surprisingly, presynaptic $\alpha_2\delta$-2 exerted the same effect in glutamatergic synapses, leading to a mismatched localisation of $GABA_A$Rs due to aberrant wiring of glutamatergic presynaptic boutons to GABAergic postsynaptic positions (Geisler et al. 2019). Furthermore, the same study showed that alternatively spliced exon 23 of $\alpha_2\delta$-2 is essential for the transsynaptic mechanism (Geisler et al. 2019). The amino acid sequence of exon 23 is located on the surface of the $\alpha_2\delta$ protein facing the synaptic cleft, opposite the channel interacting interface, suggesting that the transsynaptic effect is channel independent (Geisler et al. 2019).

Further supporting evidence for the channel-independent action of splice variants is provided by a heterologous co-expression study in which expression of distinct $\alpha_2\delta$-2 splice variants with various α_1 subunits resulted in similar effects on calcium current densities and activation/inactivation kinetics (Hobom et al. 2000). *Cacna2d2* displays a rather broad expression in the CNS (Ivanov et al. 2004; Cole et al. 2005). However, cerebellar Purkinje cells of different $\alpha_2\delta$-2 null mouse models show abnormal dendrite morphology and reduced calcium currents (Meier 1968; Barclay et al. 2001; Donato et al. 2006; Brodkin 2007; Wang et al. 2013). It has been shown that $\alpha_2\delta$-2 is also required for functional coupling of postsynaptic voltage-gated calcium channels, controlling depolarization-induced suppression of excitation and spike after hypolarization in Purkinje cells (Beeson et al. 2022). Moreover, $\alpha_2\delta$-2 was shown to regulate the structure and function of cerebellar climbing fibre synapses (Beeson et al. 2020). Consistently, in inner hair cells (IHC), $\alpha_2\delta$-2 is required for the proper spatial alignment of presynaptic calcium channels and postsynaptic glutamate receptors (AMPAR) and postsynaptic densities (PSD-95) (Fell et al. 2016). Importantly, mutations in the calcium channel partners of $\alpha_2\delta$-2 ($Ca_V2.1$ in the cerebellum and $Ca_V1.3$ in IHC) cause similar disease characteristics (Fletcher et al. 1996; Brandt et al. 2003; Mark et al. 2015), which suggests that either $\alpha_2\delta$-1 or $\alpha_2\delta$-3 might not be sufficient to compensate for the loss $\alpha_2\delta$-2 or that there is a preferential complex formation. $\alpha_2\delta$-2 potently regulates synaptic connectivity and postsynaptic receptor abundance as exclusively an $\alpha_2\delta$ splice variant that lacks a single exon (exon 23) induces wiring of presynaptic axons to GABAergic postsynaptic sites and increases postsynaptic $GABA_A$ receptor abundance (Geisler et al. 2019). Specifically, $\alpha_2\delta$-2 was implicated in establishing neuronal circuits (Tedeschi et al. 2016; Tedeschi and Popovich 2019) as increased $\alpha_2\delta$-2 and $Ca_V2.1$ expression was correlated with the developmental switch from growth-competent to transmitting axonal projections of maturing DRG neurons. In line with this, axon growth and regeneration after spinal cord injury were also observed in adult DRG neurons after genetic ablation or siRNA silencing of *Cacna2d2* (Tedeschi et al. 2016; Tedeschi and Popovich 2019). Distinct postnatal expression patterns of $\alpha_2\delta$-2 in corticospinal layer V neurons (Sun et al. 2019) suggest increased $\alpha_2\delta$-2 expression levels during developmental time points, which are concomitant with increased spontaneous firing almost at termination of corticospinal axon growth and the beginning of synaptogenesis. Multiple data suggest that distinct spatiotemporal expression patterns of $\alpha_2\delta$-2 and $\alpha_2\delta$-2 splice variants regulate synapse connectivity and specificity (Geisler et al. 2019) and that aberrant expression levels of $\alpha_2\delta$-2 in pathological conditions (e.g., spinal cord injury) might be involved in dysfunctional or maladaptive synaptogenesis or plasticity resulting in abnormal neuronal networks (Tedeschi et al. 2016; Sun et al. 2019).

$\alpha_2\delta$-3 and Size, Morphology, and Architecture of Presynaptic Boutons

The synaptic functions of $\alpha_2\delta$-3 were well characterised by its homologs in invertebrate synapses, where it was shown that $\alpha_2\delta$-3 is involved in synaptic transmission and synapse formation. For example, in *Caenorhabditis elegans*, the homologs of Ca_V2 (UNC-2), $\alpha_2\delta$ (UNC-36), and CaMKII (UNC-43) regulate the size and morphology of GABAergic motoneuron terminals in neuromuscular junctions (NMJs) (Caylor et al. 2013). Consistently, the *Drosophila melanogaster* homolog of $\alpha_2\delta$-3, *straightjacket* (stj), is necessary for the development of presynaptic boutons in motoneuron terminals (Kurshan et al. 2009). Kurshan et al. (2009) showed that the cytoskeleton of terminals at the NMJ in stj mutants was disrupted, and the Ca_V2 homolog *cacophony* was omitted from active zones. Consequently, synaptic release fails to be evoked but, together with bouton formation, can be rescued by the expression of an $\alpha_2\delta$-3 mutant lacking the δ peptide (Kurshan et al. 2009). Another study provided a mechanistic explanation of how the extracellular α_2 peptide modulates the development of *Drosophila* NMJ, suggesting that presynaptic $\alpha_2\delta$-3 induces the function of an activity-dependent autocrine bone morphogenetic protein (BMP) signalling pathway by regulating membrane retention of the glass bottom boat (Gbb) protein (Hoover et al. 2019). The study of Pirone et al. (2014) provides evidence that similar mechanisms might exist in vertebrate synapse formation, as alterations in the expression of mammalian $\alpha_2\delta$-3 affected the size and morphology of presynaptic boutons of auditory nerve fibres.

Furthermore, in central neuron synapses, overexpression of $\alpha_2\delta$-3 affected GABAergic synapses and postsynaptic receptor abundance (Geisler et al. 2019) and facilitated spontaneous GABA release with a concomitant increase in the density of inhibitory synapses in hippocampal neurons (Bikbaev et al. 2020). Importantly, contrary to overexpression of $\alpha_2\delta$-2, which modulated postsynaptic $GABA_AR$ abundance without affecting presynaptic bouton size, overexpression of $\alpha_2\delta$-3 may regulate synaptic differentiation and postsynaptic receptor abundance by a different mechanism as it induced smaller synapses, associated with reduced content of both presynaptic and postsynaptic proteins (Geisler et al. 2019). Moreover, loss of $\alpha_2\delta$-3 in *Drosophila* further blocked the rapid induction and sustained expression of presynaptic homeostatic potentiation (PHP) due to a failure to potentiate presynaptic calcium influx and the RIM-dependent readily releasable vesicle pool (Wang et al. 2016b). As PHP is a fundamental type of neuromodulation that is conserved from *Drosophila* to humans and required for stabilising information transfer at synaptic connections throughout the nervous system, the defect in PHP could contribute to the function of $\alpha_2\delta$-3 in neurological disorders (Wang et al. 2016b). These data imply that $\alpha_2\delta$-3 is involved in the function, formation, and maintenance of presynaptic structures and homeostatic potentiation, which in *Drosophila melanogaster* is independent of the calcium channel complex.

$\alpha_2\delta$-4 and Synaptic Transmission and Wiring of Photoreceptor Synapses

In retinal photoreceptor cells, the main $\alpha_2\delta$ isoform is $\alpha_2\delta$-4 (Knoflach et al. 2013). Hence, it is less surprising that mutations in the human, mouse, and zebrafish genes (Wycisk et al. 2006a, b; Wang et al. 2017; Kerov et al. 2018; Schlegel et al. 2019) of $\alpha_2\delta$-4 and the resulting link to pathological conditions are mainly resting upon proper rod and cone synaptogenesis and physiology. As $\alpha_2\delta$-4 regulates membrane expression of $Ca_V1.4$ channels and synaptic transmission of rods (Wang et al. 2017) and cones (Kerov et al. 2018; Schlegel et al. 2019), several studies suggest that abnormal $Ca_V1.4$ expression may be the leading cause of synaptic abnormalities of retinal photoreceptor cells. For example, in $\alpha_2\delta$-4 knockout mice, the critical synaptogenic molecule for rod synaptogenesis (Elfn1) is not recruited to rod synaptic terminals, which prevents rods from establishing proper contacts with their postsynaptic targets, the ON rod bipolar cells, thereby disrupting postsynaptic clustering of metabotropic glutamate receptor 6 (mGLUR6) (Wang et al. 2017). Moreover, Wang et al. showed that synaptic transmission was also severely impaired in cones of $\alpha_2\delta$-4 knockout mice (Wang et al. 2017). A follow-up study in a different $\alpha_2\delta$-4 knockout mouse (Kerov et al. 2018) expanded these findings by unravelling that impaired cone transmission through ON and OFF bipolar pathways underlies aberrant wiring of cone synapses. Nevertheless, rods show more severe abnormalities of ribbon synapses concomitant with a faster loss of $Ca_V1.4$ channels in terminals than cones. Since Elfn1 is exclusively expressed in rod synapses (Cao et al. 2015), the reports from both studies (Wang et al. 2017; Kerov et al. 2018) imply that loss of $Ca_V1.4$ might be the primary cause for synaptic abnormalities in rods and cones. An alternative possibility could be that $\alpha_2\delta$-4 regulates rod and cone synapse structure by two distinct mechanisms: an Elfn1-dependent and an Elfn1-independent mechanism. This idea has been supported by a recent study in zebrafish, showing a functional divergence and different developmental expression patterns of two $\alpha_2\delta$-4 variants in cone photoreceptors (Schlegel et al. 2019). As exclusively the loss of one variant, *Cacna2d4b*, leads to the specific occurrence of mislocalised synapses during larval development (Schlegel et al. 2019), it is tempting to speculate that spatio-temporal expression patterns of distinct $\alpha_2\delta$-4 variants might regulate synapse connectivity and specificity of rod photoreceptors (discussed in Bikbaev et al. 2020). Recently, the role of $\alpha_2\delta$-4 in central synapses has become more relevant. Previously, the protein and mRNA levels of human and murine $\alpha_2\delta$-4 were considered negligible in all previously examined CNS regions (Qin et al. 2002; Schlick et al. 2010; Geisler et al. 2019). However, a study reported increased levels of $\alpha_2\delta$-4 mRNA in human hippocampal biopsies obtained from epileptic patients, an exciting finding despite the absence of significant mRNA levels (van Loo et al. 2019). Moreover, another study supported this idea and illustrated the importance of $\alpha_2\delta$-4 in motor behaviour and cognition (Klomp et al. 2022). Interestingly, a recent study further suggested a functional relevance of $\alpha_2\delta$-4 in the brain, as $\alpha_2\delta$-4 not only differentially modulated presynaptic calcium transients compared to $\alpha_2\delta$-1 but was also capable of restoring glutamatergic synapse formation and function in $\alpha_2\delta$ triple knockout neurons (Ablinger et al. 2022). Furthermore, several studies link *CACNA2D4* with psychiatric disorders (Van Den Bossche et al. 2012; Klomp et al. 2022), suggesting that $\alpha_2\delta$-4 might play a more prominent role in the central neurons of the brain.

Cachd1 and Synaptic Function

A recent study detailed the effects of mouse Cachd1 on synaptic function for the first time (Ablinger et al. 2022). First, this study shows that expression of mouse Cachd1 in cultured hippocampal neurons resulted in clear somatodendritic protein expression. The study further demonstrated that Cachd1 was able to rescue severe synaptic deficits under conditions of triple knockout/knockdown of $\alpha_2\delta$ proteins in hippocampal neurons. Thus, mouse Cachd1 was able to rescue the formation of glutamatergic synapses by rescuing the expression of the synaptic vesicle-associated protein synapsin, and the clustering of $Ca_V2.1$ calcium channels, and restoring presynaptic calcium signalling (as measured by the calcium sensor SynGCaMP6f) (Ablinger et al. 2022). These effects replicate those seen for $\alpha_2\delta$-1 to $\alpha_2\delta$-3 isoforms (Schöpf et al. 2021). Of further interest is the fact that Cachd1 was able to rescue presynaptic calcium signalling in $\alpha_2\delta$ triple knockout/knockdown with improved efficiency compared to $\alpha_2\delta$-4 and at similar levels to those seen with $\alpha_2\delta$-1 (Ablinger et al. 2022). These effects suggest that Cachd1 modulates synaptic calcium channel signalling, further augmenting the description of CACHD1 as a functional member of the $\alpha_2\delta$ family.

3 $\alpha_2\delta$ Proteins and Disorders

3.1 $\alpha_2\delta$ Proteins Are Associated with a Variety of Neurological and Neurodevelopmental Disorders

Dysfunctions of voltage-gated calcium channels have been linked to various diseases, particularly neurological disorders (for review, see, e.g., Striessnig 2016; Zamponi 2016; Andrade et al. 2019; Hessenberger et al. 2023). Hence, it is not surprising that, likewise, $\alpha_2\delta$ proteins have been associated with many of these, summarised in Table 1. *CACNA2D1* is mainly associated with epilepsy and autism, while *CACNA2D2* is strongly linked to epilepsy, *CACNA2D3* to autism, and

Table 1 Human disorders associated with the $\alpha_2\delta$ protein family

Gene	Disorder	References
CACNA2D1	Brugada syndrome	Burashnikov et al. (2010)
	Short QT syndrome	Templin et al. (2011)
	Epilepsy	Mefford et al. (2011)
	Autism	Iossifov et al. (2014)
	West syndrome	Hino-Fukuyo et al. (2015)
CACNA2D2	Epilepsy	Edvardson et al. (2013), Pippucci et al. (2013), Butler et al. (2018), Punetha et al. (2019) and Valence et al. (2019)
	Schizophrenia	Rodríguez-López et al. (2018)
CACNA2D3	Central pain processing	Neely et al. (2010)
	Autism	Iossifov et al. (2012), The DDD Study, Homozygosity Mapping Collaborative for Autism, UK10K Consortium, The Autism Sequencing Consortium, et al. (2014), Wang et al. (2016c, 2020), Redin et al. (2017), Guo et al. (2018) and Satterstrom et al. (2020)
CACNA2D4	Night blindness	Wycisk et al. (2006b) and Ba-Abbad et al. (2016)
	Retinitis pigmentosa	Cheng et al. (2021)
	Inherited retinal dystrophy	Huang et al. (2015)
	Bipolar disorders	Van Den Bossche et al. (2012)
CACHD1	Type 1 diabetes	Rassi et al. (2008)

The table highlights disorders that are associated with $\alpha_2\delta$ (*CACNA2D*) and *CACHD1* genes

CACNA2D4 to retinal disorders and bipolar disorders. Currently, the number of genetic disease associations continuously expands by the wealth of genome-wide association studies that reveal novel mutations in *CACNA2D* genes in patients suffering from brain disorders (for more information on mutations, see Ablinger et al. 2020). Thus, $\alpha_2\delta$ proteins are attractive targets for treating a multitude of conditions. However, understanding their potential disease mechanism is crucial for developing treatment strategies. So far, the common understanding is that $\alpha_2\delta$ proteins act either on the calcium channel itself or via a channel-independent mechanism, which will be discussed in the following section.

By contrast to the knowledge of $\alpha_2\delta$-1 to $\alpha_2\delta$-4 proteins described, much less is known at present about the involvement of CACHD1 in disease phenotypes. A spontaneously arising mouse mutant with abnormal "tilt-o-whirl" (*tow*) circling movement was identified as a recessive missense mutation in the *Cachd1* gene; a study characterising such $Cachd1^{tow}$ mice and CRISPR/cas9 generated $Cachd1^{-/-}$ mice showed that the animals exhibit hearing and balance deficits associated with a disruption of calcium homeostasis in the inner ear (Tian et al. 2021). However, abnormal vestibular phenotype and any balance deficits showed high variability in *Cachd1* mutant mice, and it was suggested that the *tow* mutation is not fully penetrant and so behavioural deficits associated with these mice remain to be characterised fully (Tian et al. 2021).

Recent studies link CACHD1 to a number of human diseases, including (i) Alzheimer's disease, where CACHD1 was identified as an in vivo substrate for beta-site amyloid precursor protein (APP) cleaving enzyme 1 (BACE1) protease (Rudan Njavro et al. 2020) and for γ-secretase/presenilin complex (Güner and Lichtenthaler 2020) involved in the generation of pathogenic amyloid-β peptide; (ii) Parkinson's disease (Aguiar PM de C and Severino 2010), (iii) malignant peripheral nerve sheath tumours (MPNSTs) in BL1391 cell line established from a primary tumour, where a *YAP1-CACHD1* in-frame fusion resulted in a putative protein (1325 amino acids, 1-188 YAP1 + 88-1224 CACHD1) (Tolomeo et al. 2021); and (iv) nonalcoholic steatohepatitis (NASH)-associated liver preneoplastic and neoplastic lesions (Kakehashi et al. 2021) where CACHD1 was identified as a novel biomarker. Modulation of *CACHD1* gene expression was also found in patients with type 1 diabetes (Rassi et al. 2008). These findings suggest that CACHD1 has multiple functions in the human nervous system and, therefore, may provide a potential drug target for these disorders.

Cachd1 also shows strong expression in mouse dorsal root ganglia (Ray et al. 2018) – critical transducers of nociceptive information – further implicating CACHD1 as a potential therapeutic target for pain. Evidence from human brain atlas databases further indicates that CACHD1 is highly co-expressed with $Ca_V3.2$ [a key mediator of neuropathic and inflammatory pain, which can be regulated by de-ubiquitination (García-Caballero et al. 2014)] in adolescent peripheral sensory neurofilament 1 low threshold mechanoreceptors, which are important in mechanical allodynia (Adelman et al. 2019). Moreover, Ca_V3 channels are the targets for antiepileptic drugs (Powell et al. 2014). Thus, data from Cottrell et al. (2018)

strongly supports that human CACHD1, as a modulator of Ca_V3 channels, may offer a possible therapeutic target in hyperexcitability disorders associated with Ca_V3 dysfunction, such as pain and epilepsy.

3.2 *The Pathophysiological Mechanisms of $\alpha_2\delta$ Proteins*

3.2.1 Calcium Channel-Dependent Disease Mechanisms

As $\alpha_2\delta$ proteins have been linked to various neurological disorders (reviewed in Ablinger et al. 2020), it is tempting to speculate that a similar mechanism might be necessary for counteracting neuronal hyperexcitability in pathological conditions, including epilepsy. Therefore, mutations in $\alpha_2\delta$ proteins, which may affect expression levels, glycosylation, or attachment of the GPI-anchor, can affect calcium channel surface expression and calcium current modulation, potentially contributing to pathology. In line with this idea, a mutation in $\alpha_2\delta$-2 (p.Leu1040Pro), linked to epilepsy (Edvardson et al. 2013), was shown to reduce current densities and slow the inactivation of $Ca_V2.2$ and $Ca_V1.2$ upon co-expression in *Xenopus laevis* oocytes. Accordingly, a mutation in the VWA domain of $\alpha_2\delta$-1 (p.Arg351Thr), identified in a patient with autism (Iossifov et al. 2014), was reported to prevent stable membrane expression of $\alpha_2\delta$-1 in a heterologous expression system (Risher et al. 2018). Furthermore, a recent study characterised a biallelic loss-of-function variant in *CACNA2D1* (p.Leu9Alafs*5 and p.Gly209Asp) found in an epileptic patient, which showed reduced trafficking of the $\alpha_2\delta$-1^{G209D} mutant to the cell surface and a complete inability to increase the trafficking and function of Ca_V2 channels (Dahimene et al. 2022). Although the structure-function consequences of most human $\alpha_2\delta$ mutations are not yet determined, the limited data suggest that mutations can result in altered expression of pre- and postsynaptic calcium channels or defective calcium currents. Thus, these mutations may result in an aberrant excitatory-inhibitory balance and potentially underlie the pathophysiology of the associated diseases (discussed in Ablinger et al. 2020).

3.2.2 Calcium Channel-Independent Disease Mechanisms

Synaptic dysfunctions (synaptopathies) are a common feature of neurological disorders (reviewed in Torres et al. 2017). Hence, the synaptic functions of $\alpha_2\delta$ proteins, such as synaptogenesis (discussed in the following part), are particularly relevant for brain disorders. As such, the correct assembly of glutamatergic and GABAergic synapses and potential changes in the expression of individual synaptic proteins might cause perturbations in synapse physiology and morphology. Ultimately, changes in synaptic composition or synapse formation and differentiation could lead to synaptic dysfunction, including excitatory-inhibitory imbalance and aberrant wiring of neuronal circuits, and, finally, result in the development of neurodevelopmental disorders (discussed in Ablinger et al. 2020).

3.3 *Examples of Potential Disease Mechanisms and Treatment Options*

Considering that extensive research and revision processes must be conducted before clinical trials are approved, it is particularly promising to repurpose already approved drugs. Being regulators of calcium channel properties, $\alpha_2\delta$ proteins are attractive drug targets. On the other hand, distinct disorders might require particular modes of action, such as fine-tuning of synaptic transmission. As such, $\alpha_2\delta$ and CACHD1 proteins and their splice variants offer a vast toolbox to fine-tune synaptic activity.

From in vitro studies, we know, for example, that loss-of-function of all expressed $\alpha_2\delta$ proteins results in severe defects in synapse formation (Schöpf et al. 2021; Ablinger et al. 2022). Significantly, these severe defects can be rescued by reintroducing any protein of the $\alpha_2\delta$ family. On the other hand, overexpression of distinct isoforms in wild-type neurons leads to differential, albeit more subtle, consequences on calcium signalling (Ablinger et al. 2022). These results show that $\alpha_2\delta$ proteins have both redundant and isoform-specific functions. Hence, mutations in $\alpha_2\delta$ proteins, which affect isoform-specific functions, might affect the tuning of signalling complexes. As such, mutations that reduce expression levels or weaken the interaction with the principle α_1 subunit might lead to decreased sensitivity of signalling properties. In contrast, mutations that strengthen the interaction or increase expression levels might result in increased or more specific synaptic signaling.

As detailed structure-function studies about the consequences of most human $\alpha_2\delta$ mutations are not yet available, a selection of mutations with known and unknown functional consequences to exemplify potential disease mechanisms was made.

The rigid structural arrangements of the $\alpha_2\delta$ protein domains, such as their two intertwined double Cache domains that are further tied together with the VWA domain via the δ subunit, make alterations in the primary sequence more susceptible to functional consequences. The mutation Gly209Asp in $\alpha_2\delta$-1 was recently described in a patient with developmental and epileptic encephalopathy (Dahimene et al. 2022). This Gly209 is positioned at the end of the beta-sheet adjacent to the gabapentin binding site within the Cache1 domain. The substitution of the evolutionary highly conserved Gly209 by the bulkier and charged aspartate disrupts the structure by introducing steric interferences with Trp243 of the gabapentin binding site within the dCache1 domain (Fig. 1g). This mutant thus likely fails quality control in the ER, which would explain the observed decreased membrane expression of $\alpha_2\delta$-1 Gly209Asp mutant and the resulting loss of $Ca_V2.2$ current modulation (Dahimene et al. 2022). Of note, this mutation may mimic the mechanism of the drug gabapentin, which results in similar effects on membrane expression and modulatory functions of $\alpha_2\delta$-1 and $\alpha_2\delta$-2 (Hendrich et al. 2008; Tran-Van-Minh and Dolphin 2010).

Similarly, $\alpha_2\delta$-1 with the mutation Arg351Thr within the VWA domain, found in an autistic patient, does not reach the plasma membrane, while Leu1040Pro associated with epilepsy also showed reduced current densities (Edvardson et al. 2013; Iossifov et al. 2014; Risher et al. 2018). These are examples of identified $\alpha_2\delta$ disease mutations showing decreased membrane trafficking that finally may lead to the disease phenotypes. However, other disease-associated mutations might have a different mechanism, such as weaker interactions with the pore-forming channel subunit or extra-synaptic proteins. Thus, research on each identified $\alpha_2\delta$ mutant is essential to provide mechanism-specific treatments.

Other potential disease mechanisms of $\alpha_2\delta$ proteins might include autoantibodies. Autoantibodies against neuronal structures have a high potential to impair neuronal functions by killing neurons, thereby inducing behavioural, cognitive, and psychiatric abnormalities. Such autoantibodies directed against $\alpha_2\delta$ proteins were isolated in patients suffering from neuroendocrine carcinoma or meningitis (Lee et al. 2021). These patients experienced memory loss, seizures, psychosis, and altered mentality. Interestingly, autoantibodies against $\alpha_2\delta$ proteins do not alter the channel density in cultivated cells. Still, they inhibit neurotransmitter release in both excitatory and inhibitory synapses by interfering with the tight coupling of Ca^{2+} channels and Ca^{2+} sensors (Lee et al. 2021). Further studies must be conducted to determine the significance of these autoantibodies in immune encephalitis and whether they influence channel-independent $\alpha_2\delta$ functions. Moreover, the potential of subtype-specific autoantibodies against, e.g., $\alpha_2\delta$-1 or $\alpha_2\delta$-2, and their effects on the systemic organs require further investigations.

An early diagnosis is crucial for successful treatment options for neurodevelopmental and neuropsychiatric disorders and acute brain disorders. As such, biomarkers have turned out to be very helpful in early diagnostics. Plasmatic $\alpha_2\delta$-1 has a high potential as a biomarker for intracerebral haemorrhage and acute ischemic stroke. In complex with TSP1, an extracellular calcium-binding multidomain protein, $\alpha_2\delta$-1 promotes synaptogenesis (Christopherson et al. 2005; Eroglu et al. 2009; Risher et al. 2018). After an ischemic injury, the plasmatic levels of TSP1/2 and $\alpha_2\delta$-1 are increased, which is critical to induce synaptic regeneration in terms of numbers and structure of synapses and to recover neuronal functions. If one of the players, TSP1/2 or $\alpha_2\delta$-1, is missing or inhibited by gabapentin, the recovery is impaired (Liauw et al. 2008; Risher et al. 2018).

In contrast, another study connects the enriched levels of $\alpha_2\delta$-1 and TSP after cortical malformation in mice with increased neuronal cell death, which gabapentin can partially antagonize (Hendrich et al. 2008). It is proposed that the increase in synapses, due to $\alpha_2\delta$-1 and TSP, results in neuronal excitotoxicity by enhanced Ca^{2+} currents that finally lead to the activation of caspase3 and the subsequent neuronal cell death (Lau et al. 2017). The discrepancy in findings might arise from the linkage of increased postsynaptic $\alpha_2\delta$-1 to neuronal recovery, while increased presynaptic $\alpha_2\delta$-1 is linked to neuronal excitotoxicity. A recent study also points toward a beneficial excitatory over inhibitory signalling for stroke recovery (Qin et al. 2019). However, a complementary analysis of pre- and postsynaptic $\alpha_2\delta$-1 is needed to determine whether increased $\alpha_2\delta$-1 levels after ischemic stroke have a positive or negative effect and whether gabapentin treatment might be beneficial in that regard.

3.4 Implemented Treatment Paradigms

Both $\alpha_2\delta$-1 and $\alpha_2\delta$-2 are the therapeutic targets of the widely prescribed gabapentinoids. This class of drugs comprises the prototypical gabapentinoid, gabapentin, and its advanced forms, pregabalin and, more recently, mirogabalin. Gabapentinoids are anticonvulsants and antiepileptic or anti-seizure drugs that inhibit the excessive neuronal firing during seizures and prevent their spreading within the brain (Calandre et al. 2016). Furthermore, gabapentinoids treat neuropathic pain, postherpetic neuralgia (Deng et al. 2019b), diabetic neuropathy, fibromyalgia (Deng et al. 2019b), generalized anxiety disorder, insomnia, and restless legs syndrome. Off-label use of gabapentin includes the treatment of non-neuropathic pain, migraine, bipolar disorders, social phobia, panic disorder, and alcohol withdrawal (Goodman and Brett 2019). Gabapentinoids have considerable side effects, such as fatigue, ataxia, peripheral oedema, nystagmus, dizziness, and somnolence as well as abuse potential. Thus, other gabapentin analogues with higher specificity are needed.

While gabapentin has a similar low-nanomolar affinity for $\alpha_2\delta$-1 (K_D = 7.6 nM) and $\alpha_2\delta$-2 (19.4 nM), its descendent pregabalin has slightly improved affinity for $\alpha_2\delta$-1 (0.21 nM) over $\alpha_2\delta$-2 (4.9 nM) (Gee et al. 1996; Gong et al. 2001), offering a potentially more specific targeting of individual subunits. From pregabalin, mirogabalin was derived, which showed higher potency than pregabalin (Deeks 2019). High hopes were put into mirogabalin as a treatment for fibromyalgia; however, this was shown ultimately not to be effective. However, in 2019 mirogabalin was approved in Japan, later in South Korea and Taiwan, for the treatment of postherpetic neuralgia and diabetic peripheral neuropathic pain. In Europe and the USA, mirogabalin is currently in clinical trials (Baba et al. 2019; Kato et al. 2019). Gabapentinoids are effective and high-selling drugs; thus, many more potent and specific gabapentin analogues are being developed. Also, alternative application forms of gabapentinoids are developed, such as a skin patches and subcutaneous formulations of gabapentin.

In in vitro experiments gabapentinoids have no acute effect and must be applied chronically (Schumacher et al. 1998; Hendrich et al. 2008). This speaks for a mechanism that reduces the protein levels at their points of action. Indeed, gabapentinoids were shown to inhibit the trafficking of $\alpha_2\delta$-1 and $\alpha_2\delta$-2 to the plasma membrane as well as anterograde transport (Hendrich et al. 2008; Bauer et al. 2009; Domon et al. 2023). Simultaneously, the co-trafficking of the VGCC is inhibited under gabapentin treatment, leading to reduced calcium currents (Hendrich et al. 2008) and glutamate release at cortical synapses (Cunningham et al. 2004). Thus, gabapentinoids reduce excitatory neurotransmission, most likely explaining their therapeutic mechanism and associated side effects. The recent cryo-EM structure of gabapentin bound $\alpha_2\delta$-1 in complex with Ca_V1 provides the exact binding mode of gabapentin (Fig. 1e) and adds a possible explanation for the missing acute effect of gabapentin (Chen et al. 2023). However and surprisingly, the overall complex structure and the interaction between $\alpha_2\delta$-1 and α_1 do not change between complexes bound to the native L-Lys or gabapentin (RMSD of 0.155 Å) (Chen et al.

2023, EMC holdase). This speaks against the hypothesis that gabapentin binding induces a conformational change in $\alpha_2\delta$-1, leading to the dissociation of the complex. It, however, on the one hand, explains why functional Ca_V complexes on the membrane are not acutely affected by gabapentin. On the other hand, it suggests that gabapentin might interfere with the trafficking of newly synthesised $\alpha_2\delta$-1 and/or the recycling of complexes that lead/s to decreased protein levels on the plasma membrane, effects that are seen in functional experiments after chronic gabapentin application (Calandre et al. 2016; Deng et al. 2019b; Kakehashi et al. 2021).

It has yet to be confirmed if CACHD1 function can be regulated by gabapentinoids. While $\alpha_2\delta$-1 and $\alpha_2\delta$-2 comprise a perfect gabapentin binding site, $\alpha_2\delta$-3 and $\alpha_2\delta$-4 harbour a tyrosine that occludes the binding pocket, explaining the specificity for $\alpha_2\delta$-1 and $\alpha_2\delta$-2. Also, CACHD1 harbours the occluding tyrosine in the dCache1 domain (Chen et al. 2023). However, the dCachd2 domain might bind gabapentin as predicted from structural modelling (Gumerov et al. 2022). Future functional and structural studies should clarify if gabapentin can bind to CACHD1 and what the functional consequences might be.

4 Future Perspective

Recent advances in whole genome and whole exome sequencing have detected an increased number of disease-associated mutations and variations in $\alpha_2\delta$ proteins. Together with continuously improving structural information and homology modelling, this offers a novel approach to understanding the pathophysiological mechanisms underlying the various associated disorders and paves the way for developing new treatment options. Although functional data for mutations in $\alpha_2\delta$ proteins are largely missing, several features of $\alpha_2\delta$ proteins speak for them being highly promising therapeutic targets in the near future. First, $\alpha_2\delta$ proteins are located extracellularly on the cell surface, making them easily accessible for modulation, e.g., by targeted nanobodies. Besides their high specificity, a considerable advantage of using nanobodies in disorder treatment is their small size, which enables them to enter the narrow space in the synaptic cleft, which is a significant issue for regular antibodies. A potentially important proof of concept here is the description of the so-called Ca_V-aβlator, a llama nanobody directed against auxiliary $Ca_V\beta$-subunits fused to the catalytic HECT domain of Nedd4L, an E3 ubiquitin ligase (Morgenstern et al. 2019). Ca_V-aβlator inhibited a range of high voltage-activated calcium channels by degrading the β-subunit important for trafficking. Second, the fact that all four $\alpha_2\delta$ isoforms have been implicated in specific disorders and isoform-specific functions are continuously discovered suggests that each isoform might offer treatment potential for distinct functions. Hence, targeting nanobodies against individual isoforms might be a potential approach to manipulate functions of specific $\alpha_2\delta$ isoforms. It is tempting to speculate that by modulating overall expression levels of individual $\alpha_2\delta$ isoforms, one might be able to offer a treatment strategy for disorders such as schizophrenia, epilepsy, or autism. Third, the existence of exon-specific functions, such as the transsynaptic recruitment of $GABA_A$ receptors due to the lack of exon 23, suggests that transsynaptic interactions might be targeted with small molecules that interfere with these specific exon structures. This idea is particularly tempting in the light of disorders related to an excitatory-inhibitory imbalance. Fourth, as rescue experiments showed that reintroduction of any isoform was capable of rescuing the severe $\alpha_2\delta$ triple knockout/knockdown phenotype, it might be possible to apply $\alpha_2\delta$ proteins ectopically or to use viral gene therapy. Moreover, as autoantibodies against $\alpha_2\delta$ have been detected in human patients, it might be possible to use simple antibody-mediated treatments in certain conditions. Fifth, more structural information on $\alpha_2\delta$ proteins will allow the design of selective alternatives to gabapentin, which might discriminate between individual isoforms (such as, e.g., $\alpha_2\delta$-3 or $\alpha_2\delta$-4). And lastly, $\alpha_2\delta$ isoforms and CACHD1 modulate calcium channel properties; therefore, modulating their expression levels and functions might provide a potent alternative to traditional calcium channel blockers in the therapy of channelopathies.

As described above, identifying CACHD1 as a new member of the $\alpha_2\delta$ family, with a mix of structural and functional similarities and differences, provides a potential new target in future disease therapy. An attractive proposition is to target CACHD1 in T-type channel-mediated deficits, most notably in epilepsy and pain.

Taken together, the $\alpha_2\delta$ protein family is far more than simply channel subunits. They are synaptic organisers, modulators of synaptic transmission, and very potent novel therapeutic targets. Although research on $\alpha_2\delta$ proteins has greatly advanced in the last years, future studies will have to answer how exactly $\alpha_2\delta$ proteins convey their synaptogenic potential, what are isoform-specific functions, and which functional effects mutations in $\alpha_2\delta$ proteins have. Answering these will significantly advance our understanding of how the $\alpha_2\delta$ protein family causes and contributes to neurological and neurodevelopmental disorders and help progress innovative treatment options.

References

Ablinger C, Geisler SM, Stanika RI, Klein CT, Obermair GJ. Neuronal α2δ proteins and brain disorders. Pflugers Arch. 2020;472(7):845–63.

Ablinger C, Eibl C, Geisler SM, Campiglio M, Stephens GJ, Missler M, et al. α2δ-4 and Cachd1 proteins are regulators of presynaptic functions. Int J Mol Sci. 2022;23(17):9885.

Adelman PC, Baumbauer KM, Friedman R, Shah M, Wright M, Young E, et al. Single-cell q-PCR derived expression profiles of identified sensory neurons. Mol Pain. 2019;15:174480691988449.

Aguiar PM de C, Severino P. Biomarkers in Parkinson Disease: global gene expression analysis in peripheral blood from patients with and without mutations in PARK2 and PARK8. Einstein São Paulo. 2010;8(3):291–7.

Alvarez-Laviada A, Kadurin I, Senatore A, Chiesa R, Dolphin AC. The inhibition of functional expression of calcium channels by prion protein demonstrates competition with α2δ for GPI-anchoring pathways. Biochem J. 2014;458(2):365–74.

Anantharaman V, Aravind L. Cache – a signaling domain common to animal Ca^{2+}-channel subunits and a class of prokaryotic chemotaxis receptors. Trends Biochem Sci. 2000;25(11):535–7.

Andrade A, Sandoval A, Oviedo N, Waard MD, Elias D, Felix R. Proteolytic cleavage of the voltage-gated Ca^{2+} channel α2δ subunit: structural and functional features: proteolytic cleavage of the voltage-gated Ca^{2+} channel α2δ subunit. Eur J Neurosci. 2007;25(6):1705–10.

Andrade A, Brennecke A, Mallat S, Brown J, Gomez-Rivadeneira J, Czepiel N, et al. Genetic associations between voltage-gated calcium channels and psychiatric disorders. Int J Mol Sci. 2019;20(14):3537.

Ba-Abbad R, Arno G, Carss K, Stirrups K, Penkett CJ, Moore AT, et al. Mutations in CACNA2D4 cause distinctive retinal dysfunction in humans. Ophthalmology. 2016;123(3):668–71.

Baba M, Matsui N, Kuroha M, Wasaki Y, Ohwada S. Mirogabalin for the treatment of diabetic peripheral neuropathic pain: a randomized, double-blind, placebo-controlled phase III study in Asian patients. J Diabetes Investig. 2019;10(5):1299–306.

Barclay J, Rees M. Genomic organization of the mouse and human alpha2delta2 voltage-dependent calcium channel subunit genes. Mamm Genome Off J Int Mamm Genome Soc. 2000;11(12):1142–4.

Barclay J, Balaguero N, Mione M, Ackerman SL, Letts VA, Brodbeck J, et al. Ducky mouse phenotype of epilepsy and ataxia is associated with mutations in the Cacna2d2 gene and decreased calcium channel current in cerebellar Purkinje cells. J Neurosci. 2001;21(16):6095–104.

Bauer CS, Nieto-Rostro M, Rahman W, Tran-Van-Minh A, Ferron L, Douglas L, et al. The increased trafficking of the calcium channel subunit 2–1 to presynaptic terminals in neuropathic pain is inhibited by the 2 ligand pregabalin. J Neurosci. 2009;29(13):4076–88.

Beeson KA, Beeson R, Westbrook GL, Schnell E. α2δ-2 protein controls structure and function at the cerebellar climbing fiber synapse. J Neurosci. 2020;40(12):2403–15.

Beeson KA, Westbrook GL, Schnell E. α2δ-2 is required for depolarization-induced suppression of excitation in Purkinje cells. J Physiol. 2022;600(1):111–22.

Bian F, Li Z, Offord J, Davis MD, McCormick J, Taylor CP, et al. Calcium channel alpha2-delta type 1 subunit is the major binding protein for pregabalin in neocortex, hippocampus, amygdala, and spinal cord: an ex vivo autoradiographic study in alpha2-delta type 1 genetically modified mice. Brain Res. 2006;1075(1):68–80.

Bikbaev A, Ciuraszkiewicz-Wojciech A, Heck J, Klatt O, Freund R, Mitlöhner J, et al. Auxiliary α2δ1 and α2δ3 subunits of calcium channels drive excitatory and inhibitory neuronal network development. J Neurosci. 2020;40(25):4824–41.

Bowling H, Bhattacharya A, Zhang G, Lebowitz JZ, Alam D, Smith PT, et al. BONLAC: a combinatorial proteomic technique to measure stimulus-induced translational profiles in brain slices. Neuropharmacology. 2016;100:76–89.

Brandt A, Striessnig J, Moser T. $Ca_V1.3$ channels are essential for development and presynaptic activity of cochlear inner hair cells. J Neurosci. 2003;23(34):10832–40.

Brockhaus J, Schreitmüller M, Repetto D, Klatt O, Reissner C, Elmslie K, et al. α-Neurexins together with α2δ-1 auxiliary subunits regulate Ca^{2+} influx through $Ca_v2.1$ channels. J Neurosci. 2018;38(38):8277–94.

Brodkin ES. BALB/c mice: low sociability and other phenotypes that may be relevant to autism. Behav Brain Res. 2007;176(1):53–65.

Burashnikov E, Pfeiffer R, Barajas-Martinez H, Delpón E, Hu D, Desai M, et al. Mutations in the cardiac L-type calcium channel associated with inherited J-wave syndromes and sudden cardiac death. Heart Rhythm. 2010;7(12):1872–82.

Butler KM, Holt PJ, Milla SS, da Silva C, Alexander JJ, Escayg A. Epileptic encephalopathy and cerebellar atrophy resulting from compound heterozygous *CACNA2D2* variants. Case Rep Genet. 2018;2018:1–4.

Calandre EP, Rico-Villademoros F, Slim M. $Alpha_2$ delta ligands, gabapentin, pregabalin and mirogabalin: a review of their clinical pharmacology and therapeutic use. Expert Rev Neurother. 2016;16(11):1263–77.

Calderón-Rivera A, Andrade A, Hernández-Hernández O, González-Ramírez R, Sandoval A, Rivera M, et al. Identification of a disulfide bridge essential for structure and function of the voltage-gated Ca(2+) channel α(2)δ-1 auxiliary subunit. Cell Calcium. 2012;51(1):22–30.

Cantí C, Nieto-Rostro M, Foucault I, Heblich F, Wratten J, Richards MW, et al. The metal-ion-dependent adhesion site in the Von Willebrand factor-A domain of alpha2delta subunits is key to trafficking voltage-gated Ca^{2+} channels. Proc Natl Acad Sci U S A. 2005;102(32):11230–5.

Cao Y, Sarria I, Fehlhaber KE, Kamasawa N, Orlandi C, James KN, et al. Mechanism for selective synaptic wiring of rod photoreceptors into the retinal circuitry and its role in vision. Neuron. 2015;87(6):1248–60.

Cassidy JS, Ferron L, Kadurin I, Pratt WS, Dolphin AC. Functional exofacially tagged N-type calcium channels elucidate the interaction with auxiliary α2δ-1 subunits. Proc Natl Acad Sci U S A. 2014;111(24):8979–84.

Caylor RC, Jin Y, Ackley BD. The Caenorhabditis elegans voltage-gated calcium channel subunits UNC-2 and UNC-36 and the calcium-dependent kinase UNC-43/CaMKII regulate neuromuscular junction morphology. Neural Dev. 2013;8(1):10.

Chen J, Li L, Chen SR, Chen H, Xie JD, Sirrieh RE, et al. The α2δ-1-NMDA receptor complex is critically involved in neuropathic pain development and gabapentin therapeutic actions. Cell Rep. 2018;22(9):2307–21.

Chen Z, Mondal A, Minor DL. Structural basis for $Ca_V\alpha_2\delta$:gabapentin binding. Nat Struct Mol Biol. 2023;30:735.

Cheng J, Zhou Q, Fu J, Wei C, Zhang L, Khan MSS, et al. Novel compound heterozygous missense variants (cG955A and cA1822C) of CACNA2D4 likely causing autosomal recessive retinitis pigmentosa in a Chinese patient. 3 Biotech. 2021;11(5):208.

Christopherson KS, Ullian EM, Stokes CCA, Mullowney CE, Hell JW, Agah A, et al. Thrombospondins are astrocyte-secreted proteins that promote CNS synaptogenesis. Cell. 2005;120(3):421–33.

Cole RL, Lechner SM, Williams ME, Prodanovich P, Bleicher L, Varney MA, et al. Differential distribution of voltage-gated calcium channel alpha-2 delta (alpha2delta) subunit mRNA-containing cells in the rat central nervous system and the dorsal root ganglia. J Comp Neurol. 2005;491(3):246–69.

Cordeira JW, Felsted JA, Teillon S, Daftary S, Panessiti M, Wirth J, et al. Hypothalamic dysfunction of the thrombospondin receptor 2–1 underlies the overeating and obesity triggered by brain-derived neurotrophic factor deficiency. J Neurosci. 2014;34(2):554–65.

Cottrell GS, Soubrane CH, Hounshell JA, Lin H, Owenson V, Rigby M, et al. CACHD1 is an α2δ-like protein that modulates CaV3 voltage-gated calcium channel activity. J Neurosci. 2018;38(43):9186–201.

Cunningham MO, Woodhall GL, Thompson SE, Dooley DJ, Jones RSG. Dual effects of gabapentin and pregabalin on glutamate release at rat entorhinal synapses in vitro. Eur J Neurosci. 2004;20(6):1566–76.

Dahimene S, Page KM, Kadurin I, Ferron L, Ho DY, Powell GT, et al. The α2δ-like protein Cachd1 increases N-type calcium currents and cell surface expression and competes with α2δ-1. Cell Rep. 2018;25(6):1610–21.

Dahimene S, von Elsner L, Holling T, Mattas LS, Pickard J, Lessel D, et al. Biallelic *CACNA2D1* loss-of-function variants cause early-onset developmental epileptic encephalopathy. Brain. 2022;145(8):2721–9.

Davies A. The calcium channel 2-2 subunit partitions with CaV2.1 into lipid rafts in cerebellum: implications for localization and function. J Neurosci. 2006;26(34):8748–57.

Davies A, Hendrich J, Van Minh AT, Wratten J, Douglas L, Dolphin AC. Functional biology of the alpha(2)delta subunits of voltage-gated calcium channels. Trends Pharmacol Sci. 2007;28:220–8.

Davies A, Kadurin I, Alvarez-Laviada A, Douglas L, Nieto-Rostro M, Bauer CS, et al. The alpha2delta subunits of voltage-gated calcium channels form GPI-anchored proteins, a posttranslational modification essential for function. Proc Natl Acad Sci U S A. 2010;107(4):1654–9.

De Jongh KS, Warner C, Catterall WA. Subunits of purified calcium channels. Alpha 2 and delta are encoded by the same gene. J Biol Chem. 1990;265(25):14738–41.

Deeks ED. Mirogabalin: first global approval. Drugs. 2019;79(4):463–8.

Deng M, Chen SR, Chen H, Pan HL. α2δ-1–bound *N* -methyl- D -aspartate receptors mediate morphine-induced hyperalgesia and analgesic tolerance by potentiating glutamatergic input in rodents. Anesthesiology. 2019a;130(5):804–19.

Deng M, Chen SR, Pan HL. Presynaptic NMDA receptors control nociceptive transmission at the spinal cord level in neuropathic pain. Cell Mol Life Sci. 2019b;76(10):1889–99.

Dolphin AC. Calcium channel auxiliary α2δ and β subunits: trafficking and one step beyond. Nat Rev Neurosci. 2012;13(8):542–55.

Dolphin AC. The α2δ subunits of voltage-gated calcium channels. Biochim Biophys Acta Biomembr. 2013;1828(7):1541–9.

Dolphin AC. Voltage-gated calcium channels and their auxiliary subunits: physiology and pathophysiology and pharmacology: voltage-gated calcium channels. J Physiol. 2016; 594(19):5369–90.

Dolphin AC. Voltage-gated calcium channel α 2δ subunits: an assessment of proposed novel roles. F1000Research. 2018;7:F1000 Faculty Rev-1830.

Domon Y, Kobayashi N, Kubota K, Kitano Y, Ueki H, Shimojo Y, et al. The novel gabapentinoid mirogabalin prevents upregulation of α2δ-1 subunit of voltage-gated calcium channels in spinal dorsal horn in a rat model of spinal nerve ligation. Drug Res. 2023;73(01):54–60.

Donato R, Page KM, Koch D, Nieto-Rostro M, Foucault I, Davies A, et al. The ducky(2J) mutation in Cacna2d2 results in reduced spontaneous Purkinje cell activity and altered gene expression. J Neurosci. 2006;26(48):12576–86.

Douglas L, Davies A, Wratten J, Dolphin AC. Do voltage-gated calcium channel α2δ subunits require proteolytic processing into α2 and δ to be functional? Biochem Soc Trans. 2006;34(5):894–8.

Edvardson S, Oz S, Abulhijaa FA, Taher FB, Shaag A, Zenvirt S, et al. Early infantile epileptic encephalopathy associated with a high voltage gated calcium channelopathy. J Med Genet. 2013;50(2):118–23.

Ellis SB, Williams ME, Ways NR, Brenner R, Sharp AH, Leung AT, et al. Sequence and expression of mRNAs encoding the alpha 1 and alpha 2 subunits of a DHP-sensitive calcium channel. Science. 1988;241(4873):1661–4.

García-Caballero A, Gadotti VM, Stemkowski P, Weiss N, Souza IA, Hodgkinson V, et al. The deubiquitinating enzyme USP5modulates neuropathic and inflammatory pain by enhancing Cav3.2 channel activity. Neuron. 2014; 83(5):1144–58.

Eroglu C, Allen NJ, Susman MW, O'Rourke NA, Park CY, Ozkan E, et al. Gabapentin receptor alpha2delta-1 is a neuronal thrombospondin receptor responsible for excitatory CNS synaptogenesis. Cell. 2009;139(2):380–92.

Felix R, Gurnett CA, De Waard M, Campbell KP. Dissection of functional domains of the voltage-dependent Ca^{2+} channel alpha2delta subunit. J Neurosci. 1997;17(18):6884–91.

Fell B, Eckrich S, Blum K, Eckrich T, Hecker D, Obermair GJ, et al. α2δ2 controls the function and trans-synaptic coupling of Cav1.3 channels in mouse inner hair cells and is essential for normal hearing. J Neurosci. 2016;36(43):11024–36.

Field MJ, Cox PJ, Stott E, Melrose H, Offord J, Su TZ, et al. Identification of the α_2-δ-1 subunit of voltage-dependent calcium channels as a molecular target for pain mediating the analgesic actions of pregabalin. Proc Natl Acad Sci. 2006;103(46):17537–42.

Fletcher CF, Lutz CM, O'Sullivan TN, Shaughnessy JD, Hawkes R, Frankel WN, et al. Absence epilepsy in tottering mutant mice is associated with calcium channel defects. Cell. 1996;87(4):607–17.

Gao B, Sekido Y, Maximov A, Saad M, Forgacs E, Latif F, et al. Functional properties of a new voltage-dependent calcium channel alpha(2)delta auxiliary subunit gene (CACNA2D2). J Biol Chem. 2000;275(16):12237–42.

Gee NS, Brown JP, Dissanayake VU, Offord J, Thurlow R, Woodruff GN. The novel anticonvulsant drug, gabapentin (Neurontin), binds to the alpha2delta subunit of a calcium channel. J Biol Chem. 1996;271(10):5768–76.

Geisler S, Schöpf CL, Obermair GJ. Emerging evidence for specific neuronal functions of auxiliary calcium channel α2δ subunits. Gen Physiol Biophys. 2015;34(02):105–18.

Geisler S, Schöpf CL, Stanika R, Kalb M, Campiglio M, Repetto D, et al. Presynaptic $\alpha_2\delta$-2 calcium channel subunits regulate postsynaptic $GABA_A$ receptor abundance and axonal wiring. J Neurosci. 2019;39(14):2581–605.

Gong HC, Hang J, Kohler W, Li L, Su TZ. Tissue-specific expression and gabapentin-binding properties of calcium channel α2δ subunit subtypes. J Membr Biol. 2001;184(1):35–43.

Goodman CW, Brett AS. A clinical overview of off-label use of gabapentinoid drugs. JAMA Intern Med. 2019;179(5):695.

Gumerov VM, Andrianova EP, Matilla MA, Page KM, Monteagudo-Cascales E, Dolphin AC, et al. Amino acid sensor conserved from bacteria to humans. Proc Natl Acad Sci. 2022;119(10):e2110415119.

Güner G, Lichtenthaler SF. The substrate repertoire of γ-secretase/presenilin. Semin Cell Dev Biol. 2020;105:27–42.

Guo H, Wang T, Wu H, Long M, Coe BP, Li H, et al. Inherited and multiple de novo mutations in autism/developmental delay risk genes suggest a multifactorial model. Mol Autism. 2018;9(1):64.

Gurnett CA, De Waard M, Campbell KP. Dual function of the voltage-dependent Ca^{2+} channel alpha 2 delta subunit in current stimulation and subunit interaction. Neuron. 1996;16(2):431–40.

Hanke S, Bugert P, Chudek J, Kovacs G. Cloning a calcium channel alpha2delta-3 subunit gene from a putative tumor suppressor gene region at chromosome 3p21.1 in conventional renal cell carcinoma. Gene. 2001;264(1):69–75.

Hendrich J, Van Minh AT, Heblich F, Nieto-Rostro M, Watschinger K, Striessnig J, et al. Pharmacological disruption of calcium channel trafficking by the $\alpha_2\delta$ ligand gabapentin. Proc Natl Acad Sci. 2008;105(9):3628–33.

Hessenberger M, Haddad S, Obermair GJ. Pathophysiological roles of auxiliary calcium channel α2δ subunits. In: Handbook of experimental pharmacology. Berlin/Heidelberg: Springer Berlin Heidelberg; 2023.

Hino-Fukuyo N, Kikuchi A, Arai-Ichinoi N, Niihori T, Sato R, Suzuki T, et al. Genomic analysis identifies candidate pathogenic variants in 9 of 18 patients with unexplained West syndrome. Hum Genet. 2015;134(6):649–58.

Hobom M, Dai S, Marais E, Lacinova L, Hofmann F, Klugbauer N. Neuronal distribution and functional characterization of the calcium channel $\alpha_2\delta$-2 subunit: calcium channel $\alpha_2\delta$-2 subunit. Eur J Neurosci. 2000;12(4):1217–26.

Hoover KM, Gratz SJ, Qi N, Herrmann KA, Liu Y, Perry-Richardson JJ, et al. The calcium channel subunit α2δ-3 organizes synapses via an activity-dependent and autocrine BMP signaling pathway. Nat Commun. 2019;10(1):5575.

Hoppa MB, Lana B, Margas W, Dolphin AC, Ryan TA. α2δ expression sets presynaptic calcium channel abundance and release probability. Nature. 2012;486(7401):122–5.

Huang XF, Huang F, Wu KC, Wu J, Chen J, Pang CP, et al. Genotype–phenotype correlation and mutation spectrum in a large cohort of patients with inherited retinal dystrophy revealed by next-generation sequencing. Genet Med. 2015;17(4):271–8.

Iossifov I, Ronemus M, Levy D, Wang Z, Hakker I, Rosenbaum J, et al. De novo gene disruptions in children on the autistic spectrum. Neuron. 2012;74(2):285–99.

Iossifov I, O'Roak BJ, Sanders SJ, Ronemus M, Krumm N, Levy D, et al. The contribution of de novo coding mutations to autism spectrum disorder. Nature. 2014;515(7526):216–21.

Ivanov SV, Ward JM, Tessarollo L, McAreavey D, Sachdev V, Fananapazir L, et al. Cerebellar ataxia, seizures, premature death, and cardiac abnormalities in mice with targeted disruption of the Cacna2d2 gene. Am J Pathol. 2004;165(3):1007–18.

Jay SD, Sharp AH, Kahl SD, Vedvick TS, Harpold MM, Campbell KP. Structural characterization of the dihydropyridine-sensitive calcium channel alpha 2-subunit and the associated delta peptides. J Biol Chem. 1991;266(5):3287–93.

Jumper J, Evans R, Pritzel A, Green T, Figurnov M, Ronneberger O, et al. Highly accurate protein structure prediction with AlphaFold. Nature. 2021;596(7873):583–9.

Kadurin I, Rothwell SW, Lana B, Nieto-Rostro M, Dolphin AC. LRP1 influences trafficking of N-type calcium channels via interaction with the auxiliary α2δ-1 subunit. Sci Rep. 2017;7(1):43802.

Kadurin I, Dahimene S, Page KM, Ellaway JIJ, Chaggar K, Troeberg L, et al. ADAM17 mediates proteolytic maturation of voltage-gated calcium channel auxiliary α2δ subunits, and enables calcium current enhancement. Function. 2022;3(3):zqac013.

Kakehashi A, Chariyakornkul A, Suzuki S, Khuanphram N, Tatsumi K, Yamano S, et al. Cache domain containing 1 is a novel marker of non-alcoholic steatohepatitis-associated hepatocarcinogenesis. Cancer. 2021;13(6):1216.

Kato J, Matsui N, Kakehi Y, Murayama E, Ohwada S, Sugihara M. Mirogabalin for the management of postherpetic neuralgia: a randomized, double-blind, placebo-controlled phase 3 study in Asian patients. Pain. 2019;160(5):1175–85.

Kerov V, Laird JG, M-ling J, Knecht S, Soh D, Hagen J, et al. $\alpha_2\delta$-4 is required for the molecular and structural organization of rod and cone photoreceptor synapses. J Neurosci. 2018;38(27):6145–60.

Klomp A, Omichi R, Iwasa Y, Smith RJ, Usachev YM, Russo AF, et al. The voltage-gated Ca^{2+} channel subunit α2δ-4 regulates locomotor behavior and sensorimotor gating in mice. PLoS One. 2022;17(3):e0263197.

Klugbauer N, Lacinová L, Marais E, Hobom M, Hofmann F. Molecular diversity of the calcium channel alpha2delta subunit. J Neurosci. 1999;19(2):684–91.

Klugbauer N, Marais E, Hofmann F. Calcium channel alpha2delta subunits: differential expression, function, and drug binding. J Bioenerg Biomembr. 2003;35(6):639–47.

Knoflach D, Kerov V, Sartori SB, Obermair GJ, Schmuckermair C, Liu X, et al. Cav1.4 IT mouse as model for vision impairment in human congenital stationary night blindness type 2. Channels. 2013;7(6):503–13.

Kurshan PT, Oztan A, Schwarz TL. Presynaptic alpha2delta-3 is required for synaptic morphogenesis independent of its Ca^{2+}–channel functions. Nat Neurosci. 2009;12(11):1415–23.

Lau LA, Noubary F, Wang D, Dulla CG. α2δ-1 signaling drives cell death, synaptogenesis, circuit reorganization, and gabapentin-mediated neuroprotection in a model of insult-induced cortical malformation. eNeuro. 2017;4(5):ENEURO.0316-17.2017.

Lee A, Wang S, Williams B, Hagen J, Scheetz TE, Haeseleer F. Characterization of Cav1.4 complexes (α11.4, β2, and α2δ4) in HEK293T cells and in the retina. J Biol Chem. 2015; 290(3):1505–21.

Lee S, Lee BJ, Bae J, Kim YS, Han D, Shin H, et al. Ca_V α2δ autoimmune encephalitis: a novel antibody and its characteristics. Ann Neurol. 2021;89(4):740–52.

Leung AT, Imagawa T, Campbell KP. Structural characterization of the 1,4-dihydropyridine receptor of the voltage-dependent Ca^{2+} channel from rabbit skeletal muscle. Evidence for two distinct high molecular weight subunits. J Biol Chem. 1987;262(17):7943–6.

Li CY, Zhang XL, Matthews EA, Li KW, Kurwa A, Boroujerdi A, et al. Calcium channel alpha2delta1 subunit mediates spinal hyperexcitability in pain modulation. Pain. 2006;125(1–2):20–34.

Li Z, Taylor CP, Weber M, Piechan J, Prior F, Bian F, et al. Pregabalin is a potent and selective ligand for α(2)δ-1 and α(2)δ-2 calcium channel subunits. Eur J Pharmacol. 2011;667(1–3):80–90.

Liauw J, Hoang S, Choi M, Eroglu C, Choi M, Sun G-h, et al. Thrombospondins 1 and 2 are necessary for synaptic plasticity and functional recovery after stroke. J Cereb Blood Flow Metab. 2008;28(10):1722–32.

Ludwig A, Flockerzi V, Hofmann F. Regional expression and cellular localization of the alpha1 and beta subunit of high voltage-activated calcium channels in rat brain. J Neurosci. 1997;17(4):1339–49.

Ma H, Chen SR, Chen H, Li L, Li DP, Zhou JJ, et al. α2δ-1 is essential for sympathetic output and NMDA receptor activity potentiated by angiotensin II in the hypothalamus. J Neurosci. 2018a;38(28):6388–98.

Ma H, Chen SR, Chen H, Zhou JJ, Li DP, Pan HL. α2δ-1 couples to NMDA receptors in the hypothalamus to sustain sympathetic vasomotor activity in hypertension: α2δ-1-bound NMDA receptors in hypertension. J Physiol. 2018b;596(17):4269–83.

Mark MD, Krause M, Boele HJ, Kruse W, Pollok S, Kuner T, et al. Spinocerebellar ataxia type 6 protein aggregates cause deficits in motor learning and cerebellar plasticity. J Neurosci. 2015;35(23):8882–95.

Mefford HC, Yendle SC, Hsu C, Cook J, Geraghty E, McMahon JM, et al. Rare copy number variants are an important cause of epileptic encephalopathies. Ann Neurol. 2011;70(6):974–85.

Meier H. The neuropathology of ducky, a neurological mutation of the mouse: a pathological and preliminary histochemical study. Acta Neuropathol (Berl). 1968;11(1):15–28.

Meyer JO, Dolphin AC. Rab11-dependent recycling of calcium channels is mediated by auxiliary subunit α2δ-1 but not α2δ-3. Sci Rep. 2021;11(1):10256.

Morgenstern TJ, Park J, Fan QR, Colecraft HM. A potent voltage-gated calcium channel inhibitor engineered from a nanobody targeted to auxiliary CaVβ subunits. elife. 2019;8:e49253.

Müller CS, Haupt A, Bildl W, Schindler J, Knaus HG, Meissner M, et al. Quantitative proteomics of the Cav2 channel nano-environments in the mammalian brain. Proc Natl Acad Sci. 2010;107(34):14950–7.

Neely GG, Hess A, Costigan M, Keene AC, Goulas S, Langeslag M, et al. A genome-wide drosophila screen for heat nociception identifies α2δ3 as an evolutionarily conserved pain gene. Cell. 2010;143(4):628–38.

Nieto-Rostro M, Ramgoolam K, Pratt WS, Kulik A, Dolphin AC. Ablation of $\alpha_2\delta$-1 inhibits cell-surface trafficking of endogenous N-type calcium channels in the pain pathway in vivo. Proc Natl Acad Sci. 2018;115(51).

Obermair GJ, Kugler G, Baumgartner S, Tuluc P, Grabner M, Flucher BE. The Ca^{2+} channel alpha2delta-1 subunit determines Ca^{2+} current kinetics in skeletal muscle but not targeting of alpha1S or excitation-contraction coupling. J Biol Chem. 2005;280(3):2229–37.

Obermair G, Tuluc P, Flucher B. Auxiliary Ca^{2+} channel subunits: lessons learned from muscle. Curr Opin Pharmacol. 2008;8(3):311–8.

Pan B, Guo Y, Wu HE, Park J, Trinh VN, Luo ZD, et al. Thrombospondin-4 divergently regulates voltage-gated Ca^{2+} channel subtypes in sensory neurons after nerve injury. Pain. 2016;157(9):2068–80.

Park J, Yu YP, Zhou CY, Li KW, Wang D, Chang E, et al. Central mechanisms mediating thrombospondin-4-induced pain states. J Biol Chem. 2016;291(25):13335–48.

Pippucci T, Parmeggiani A, Palombo F, Maresca A, Angius A, Crisponi L, et al. A novel null homozygous mutation confirms CACNA2D2 as a gene mutated in epileptic encephalopathy. Brusgaard K, editor. PLoS One. 2013;8(12):e82154.

Pirone A, Kurt S, Zuccotti A, Rüttiger L, Pilz P, Brown DH, et al. α2δ3 is essential for normal structure and function of auditory nerve synapses and is a novel candidate for auditory processing disorders. J Neurosci. 2014;34(2):434–45.

Powell KL, Cain SM, Snutch TP, O'Brien TJ. Low threshold T-type calcium channels as targets for novel epilepsy treatments: Tchannelsand epilepsy treatments. Br J Clin Pharmacol. 2014;77(5):729–39.

Punetha J, Karaca E, Gezdirici A, Lamont RE, Pehlivan D, Marafi D, et al. Biallelic *CACNA2D2* variants in epileptic encephalopathy and cerebellar atrophy. Ann Clin Transl Neurol. 2019;6(8):1395–406.

Qin N, Olcese R, Stefani E, Birnbaumer L. Modulation of human neuronal α_{1E} -type calcium channel by $\alpha_2\delta$-subunit. Am J Physiol-Cell Physiol. 1998;274(5):C1324–31.

Qin N, Yagel S, Momplaisir ML, Codd EE, D'Andrea MR. Molecular cloning and characterization of the human voltage-gated calcium channel $\alpha_2\delta$-4 subunit. Mol Pharmacol. 2002;62(3):485–96.

Qin L, Actor-Engel HS, Woo MS, Shakil F, Chen YW, Cho S, et al. An increase of excitatory-to-inhibitory synaptic balance in the contralateral cortico-striatal pathway underlies improved stroke recovery in BDNF Val66Met SNP mice. Neurorehabil Neural Repair. 2019;33(12):989–1002.

Rassi DM, Junta CM, Fachin AL, Sandrin-Garcia P, Mello S, Silva GL, et al. Gene expression profiles stratified according to type 1 diabetes mellitus susceptibility regions. Ann N Y Acad Sci. 2008;1150(1):282–9.

Ray P, Torck A, Quigley L, Wangzhou A, Neiman M, Rao C, et al. Comparative transcriptome profiling of the human andmouse dorsal root ganglia: an RNA-seq–based resource for pain and sensory neuroscience research. Pain. 2018;159(7):1325–45.

Redin C, Brand H, Collins RL, Kammin T, Mitchell E, Hodge JC, et al. The genomic landscape of balanced cytogenetic abnormalities associated with human congenital anomalies. Nat Genet. 2017;49(1):36–45.

Risher WC, Kim N, Koh S, Choi JE, Mitev P, Spence EF, et al. Thrombospondin receptor α2δ-1 promotes synaptogenesis and spinogenesis via postsynaptic Rac1. J Cell Biol. 2018;217(10):3747–65.

Robinson P, Etheridge S, Song L, Shah R, Fitzgerald EM, Jones OT. Targeting of voltage-gated calcium channel α2δ-1 subunit to lipid rafts is independent from a gpi-anchoring motif. Degtyar VE, editor. PLoS One. 2011;6(6):e19802.

Rodríguez-López J, Sobrino B, Amigo J, Carrera N, Brenlla J, Agra S, et al. Identification of putative second genetic hits in schizophrenia carriers of high-risk copy number variants and resequencing in additional samples. Eur Arch Psychiatry Clin Neurosci. 2018;268(6):585–92.

Rudan Njavro J, Klotz J, Dislich B, Wanngren J, Shmueli MD, Herber J, et al. Mouse brain proteomics establishes MDGA1 and CACHD1 as in vivo substrates of the Alzheimer protease BACE1. FASEB J. 2020;34(2):2465–82.

Sandoval A, Oviedo N, Andrade A, Felix R. Glycosylation of asparagines 136 and 184 is necessary for the alpha2delta subunit-mediated regulation of voltage-gated Ca^{2+} channels. FEBS Lett. 2004;576(1–2):21–6.

Satterstrom FK, Kosmicki JA, Wang J, Breen MS, De Rubeis S, An JY, et al. Large-scale exome sequencing study implicates both developmental and functional changes in the neurobiology of autism. Cell. 2020;180(3):568–84.

Savalli N, Pantazis A, Sigg D, Weiss JN, Neely A, Olcese R. The α2δ-1 subunit remodels CaV1.2 voltage sensors and allows Ca^{2+} influx at physiological membrane potentials. J Gen Physiol. 2016;148(2):147–59.

Schlegel DK, Glasauer SMK, Mateos JM, Barmettler G, Ziegler U, Neuhauss SCF. A new zebrafish model for CACNA2D4-dysfunction. Investig Opthalmol Vis Sci. 2019;60(15):5124.

Schlick B, Flucher BE, Obermair GJ. Voltage-activated calcium channel expression profiles in mouse brain and cultured hippocampal neurons. Neuroscience. 2010;167(3):786–98.

Schöpf CL, Ablinger C, Geisler SM, Stanika RI, Campiglio M, Kaufmann WA, et al. Presynaptic $\alpha_2\delta$ subunits are key organizers of glutamatergic synapses. Proc Natl Acad Sci. 2021;118(14):e1920827118.

Schumacher TB, Beck H, Steinhauser C, Schramm J, Elger CE. Effects of phenytoin, carbamazepine, and gabapentin on calcium channels in hippocampal granule cells from patients with temporal lobe epilepsy. Epilepsia. 1998;39(4):355–63.

Scott MB, Kammermeier PJ. Ca_V2 channel subtype expression in rat sympathetic neurons is selectively regulated by $\alpha_2\delta$ subunits. Channels. 2017;11(6):555–73.

Senatore A, Colleoni S, Verderio C, Restelli E, Morini R, Condliffe SB, et al. Mutant PrP suppresses glutamatergic neurotransmission in cerebellar granule neurons by impairing membrane delivery of VGCC α2δ-1 subunit. Neuron. 2012;74(2):300–13.

Stephani F, Scheuer V, Eckrich T, Blum K, Wang W, Obermair GJ, et al. Deletion of the Ca2+ channel subunit α2δ3 differentially affects Cav2.1 and Cav2.2 currents in cultured spiral ganglion neurons before and after the onset of hearing. Front Cell Neurosci. 2019;13:278.

Stephens GJ, Cottrell GS. CACHD1: a new activity-modifying protein for voltage-gated calcium channels. Channels. 2019;13(1):120–3.

Striessnig J. Voltage-gated calcium channels – from basic mechanisms to disease: editorial. J Physiol. 2016;594(20):5817–21.

Sun W, Larson MJE, Kiyoshi CM, Annett AJ, Stalker WA, Peng J, et al. Gabapentinoid treatment promotes corticospinal plasticity and regeneration following murine spinal cord injury. J Clin Invest. 2019;130(1):345–58.

Tedeschi A, Popovich PG. The application of omics technologies to study axon regeneration and CNS repair. F1000Research. 2019;8:311.

Tedeschi A, Dupraz S, Laskowski CJ, Xue J, Ulas T, Beyer M, et al. The calcium channel subunit alpha2delta2 suppresses axon regeneration in the adult CNS. Neuron. 2016;92(2):419–34.

Templin C, Ghadri JR, Rougier JS, Baumer A, Kaplan V, Albesa M, et al. Identification of a novel loss-of-function calcium channel gene mutation in short QT syndrome (SQTS6). Eur Heart J. 2011;32(9):1077–88.

The DDD Study, Homozygosity Mapping Collaborative for Autism, UK10K Consortium, The Autism Sequencing Consortium, De Rubeis S, He X, et al. Synaptic, transcriptional and chromatin genes disrupted in autism. Nature. 2014;515(7526):209–15.

Tian C, Johnson KR, Lett JM, Voss R, Salt AN, Hartsock JJ, et al. CACHD1-deficient mice exhibit hearing and balance deficits associated with a disruption of calcium homeostasis in the inner ear. Hear Res. 2021;409:108327.

Tolomeo D, Agostini A, Macchia G, L'Abbate A, Severgnini M, Cifola I, et al. BL1391: an established cell line from a human malignant peripheral nerve sheath tumor with unique genomic features. Hum Cell. 2021;34(1):238–45.

Torres VI, Vallejo D, Inestrosa NC. Emerging synaptic molecules as candidates in the etiology of neurological disorders. Neural Plast. 2017;2017:1–25.

Tran-Van-Minh A, Dolphin AC. The alpha2delta ligand gabapentin inhibits the Rab11-dependent recycling of the calcium channel subunit alpha2delta-2. J Neurosci. 2010;30(38):12856–67.

Tuluc P, Kern G, Obermair GJ, Flucher BE. Computer modeling of siRNA knockdown effects indicates an essential role of the Ca^{2+} channel alpha2delta-1 subunit in cardiac excitation-contraction coupling. Proc Natl Acad Sci U S A. 2007;104(26):11091–6.

Valence S, Cochet E, Rougeot C, Garel C, Chantot-Bastaraud S, Lainey E, et al. Exome sequencing in congenital ataxia identifies two new candidate genes and highlights a pathophysiological link between some congenital ataxias and early infantile epileptic encephalopathies. Genet Med. 2019;21(3):553–63.

Van Den Bossche MJ, Strazisar M, De Bruyne S, Bervoets C, Lenaerts AS, De Zutter S, et al. Identification of a CACNA2D4 deletion in late onset bipolar disorder patients and implications for the involvement of voltage-dependent calcium channels in psychiatric disorders. Am J Med Genet B Neuropsychiatr Genet. 2012;159B(4):465–75.

van Loo KMJ, Rummel CK, Pitsch J, Müller JA, Bikbaev AF, Martinez-Chavez E, et al. Calcium channel subunit α2δ4 is regulated by early growth response 1 and facilitates epileptogenesis. J Neurosci. 2019;39(17):3175–87.

Wakamori M, Mikala G, Mori Y. Auxiliary subunits operate as a molecular switch in determining gating behaviour of the unitary N-type Ca^{2+} channel current in *Xenopus* oocytes. J Physiol. 1999;517(3):659–72.

Wang X, Whalley BJ, Stephens GJ. The *du*^{2J} mouse model of ataxia and absence epilepsy has deficient cannabinoid CB_1 receptor-mediated signalling: deficient CB_1 signalling in the *du*^{2J} model. J Physiol. 2013;591(16):3919–33.

Wang X, Bey AL, Katz BM, Badea A, Kim N, David LK, et al. Altered mGluR5-Homer scaffolds and corticostriatal connectivity in a Shank3 complete knockout model of autism. Nat Commun. 2016a;7:11459.

Wang T, Jones RT, Whippen JM, Davis GW. α2δ-3 is required for rapid transsynaptic homeostatic signaling. Cell Rep. 2016b;16(11):2875–88.

Wang T, Guo H, Xiong B, Stessman HAF, Wu H, Coe BP, et al. De novo genic mutations among a Chinese autism spectrum disorder cohort. Nat Commun. 2016c;7(1):13316.

Wang Y, Fehlhaber KE, Sarria I, Cao Y, Ingram NT, Guerrero-Given D, et al. The auxiliary calcium channel subunit α2δ4 is required for axonal elaboration, synaptic transmission, and wiring of rod photoreceptors. Neuron. 2017;93(6):1359–74.

Wang T, Hoekzema K, Vecchio D, Wu H, Sulovari A, Coe BP, et al. Large-scale targeted sequencing identifies risk genes for neurodevelopmental disorders. Nat Commun. 2020;11(1):4932.

Whittaker CA, Hynes RO. Distribution and evolution of von Willebrand/integrin A domains: widely dispersed domains with roles in cell adhesion and elsewhere. Mol Biol Cell. 2002;13(10):3369–87.

Wiser O, Trus M, Tobi D, Halevi S, Giladi E, Atlas D. The alpha 2/delta subunit of voltage sensitive Ca^{2+} channels is a single transmembrane extracellular protein which is involved in regulated secretion. FEBS Lett. 1996;379(1):15–20.

Wu J, Yan Z, Li Z, Qian X, Lu S, Dong M, et al. Structure of the voltage-gated calcium channel Cav1.1 at 3.6 Å resolution. Nature. 2016;537(7619):191–6.

Wycisk KA, Budde B, Feil S, Skosyrski S, Buzzi F, Neidhardt J, et al. Structural and functional abnormalities of retinal ribbon synapses due to Cacna2d4 mutation. Invest Ophthalmol Vis Sci. 2006a;47(8):3523–30.

Wycisk KA, Zeitz C, Feil S, Wittmer M, Forster U, Neidhardt J, et al. Mutation in the auxiliary calcium-channel subunit CACNA2D4 causes autosomal recessive cone dystrophy. Am J Hum Genet. 2006b;79(5):973–7.

Yu YP, Gong N, Kweon TD, Vo B, Luo ZD. Gabapentin prevents synaptogenesis between sensory and spinal cord neurons induced by thrombospondin-4 acting on pre-synaptic Ca_v α_2 δ_1 subunits and involving T-type Ca^{2+} channels: early gabapentin blocks abnormal synaptogenesis. Br J Pharmacol. 2018;175(12):2348–61.

Zamponi GW. Targeting voltage-gated calcium channels in neurological and psychiatric diseases. Nat Rev Drug Discov. 2016;15(1):19–34.

Zhang FX, Gadotti VM, Souza IA, Chen L, Zamponi GW. BK potassium channels suppress Cav$\alpha2\delta$ subunit function to reduce inflammatory and neuropathic pain. Cell Rep. 2018;22(8):1956–64.

Zhou JJ, Li DP, Chen SR, Luo Y, Pan HL. The $\alpha2\delta$-1-NMDA receptor coupling is essential for corticostriatal long-term potentiation and is involved in learning and memory. J Biol Chem. 2018;293(50):19354–64.

Voltage-Gated Sodium Channels as Drug Targets in Epilepsy-Related Sodium Channelopathies

Eric R. Wengert, Raquel M. Miralles, and Manoj K. Patel

Abstract Voltage-gated sodium channels (VGSCs) drive cellular excitability in various cells including the neurons of the central nervous system. Genetic variants in the genes encoding the brain-expressed VGSCs have been identified in patients with severe epilepsy syndromes making clear that proper VGSC activity is required for normal neurological development and function. Given this central role in determining neuronal physiology, VGSCs are one of the most common molecular targets for currently available antiseizure medications. Yet the significant limitations of existing medications indicate a need for novel therapeutic strategies, including more precise modulation of VGSC function. In this chapter, we discuss ongoing precision medicine approaches targeting VGSCs in the treatment of epilepsy-related sodium (Na) channelopathies. Notwithstanding significant technical hurdles, modulators of VGSCs are likely to remain a mainstay in the treatment of epilepsy-related channelopathies and epilepsy more generally.

Keywords Voltage-gated sodium channel · Antiseizure medication · Developmental and epileptic encephalopathy · Antisense oligonucleotide

1 Introduction

Voltage-gated sodium channels (VGSCs) are central determinants of cellular excitability in a variety of contexts including skeletal muscle cells, cardiomyocytes, as well as neurons of the central and peripheral nervous systems (Hille 2001). They

E. R. Wengert
Division of Neurology, Department of Pediatrics, The Children's Hospital of Philadelphia, Philadelphia, PA, USA

R. M. Miralles · M. K. Patel (✉)
Department of Anesthesiology, University of Virginia Health System, Charlottesville, VA, USA
e-mail: mkp5u@uvahealth.org

G. Stephens, E. Stevens (eds.), *Ion Channels as Targets in Drug Discovery*,
https://doi.org/10.1007/978-3-031-52197-3_4

contribute to the initiation and propagation of action potentials (APs), which are the basic unit of intracellular transmission of information required for nervous system function (Bean 2007). VGSCs show numerous features that are critical to proper functioning. They are typically nonconducting at relatively hyperpolarized membrane potentials. However, in response to membrane depolarization, they rapidly activate, opening the pore to the highly selective flow of sodium (Na) ions into the cell. Within milliseconds, the channel population inactivates and ceases to conduct Na ions regardless of continued depolarization, and upon returning to relatively hyperpolarized membrane potentials, the population recovers its voltage sensitivity to complete this sequential progression again. This stereotypical cycling of VGSCs to produce the action potential waveform, the subject of intense scientific investigation for decades, is essential for normal neuronal excitability. Therefore, any alteration in VGSC function that disturbs the conductance, kinetic rates, or voltage dependence of Na channel behavior will likely have profound effects on intrinsic neuronal physiology and overall brain function, leading to their contribution to a wide array of neurological diseases.

Structurally, VGSCs are comprised of alpha-subunits (~260 kDa) associated with beta-subunits (β1–β4) (Hartshorne and Catterall 1981; Hartshorne et al. 1982, 1985; Talvenheimo et al. 1984; Isom et al. 1994; Morgan et al. 2000; Yu et al. 2003); with more recent work suggesting that alpha-subunits form functional dimers (Clatot et al. 2017). The beta-subunits have important roles in the voltage dependence and kinetics of Na channel gating and channel trafficking (Isom et al. 1992, 1994); however, the alpha-subunit is functional independently of beta-subunits, able to produce Na currents on its own (Goldin et al. 1986). The alpha-subunit contains four homologous domains (DI–DIV) each with six α-helical transmembrane segments (S1–S6) that together form a complete channel with voltage sensor and pore domains. Previous physiological investigation and more recent structural experiments have provided evidence that the S1–S4 regions comprise the voltage sensor, with four arginine residues that, in response to depolarization, experience the electrostatic force of repulsion which changes channel conformation into the open state (Payandeh et al. 2011; Catterall et al. 2020). The pore of the VGSC is formed by the S5–S6 region, which is significantly larger than its counterparts in other voltage-gated channels, allowing coordinated passage of a hydrated Na ion when the channel is opened. This allows for high selectivity for Na ions, as the hydrated form of Na is unable to pass through other voltage-gated channels and other relevant ions (potassium, calcium, magnesium) pass through VGSCs at much lower rates (Ulmschneider et al. 2013; Naylor et al. 2016). At the base of the S6 region is the local anesthetic site, which is the structural location to which numerous small molecules bind and influence VGSC function. The basic structure of a VGSC alpha-subunit is shown in Fig. 1.

There are nine different isoforms for the alpha-subunit for VGSCs ($Na_V1.1$–$Na_V1.9$; Fig. 2) that have slight, yet important, differences in structure, functional channel properties, subcellular and cellular expression patterns, developmental time course of expression, and overall contribution to the vast repertoire of physiological behavior of individual neurons and neural circuits (Goldin 1999; Lorincz and Nusser

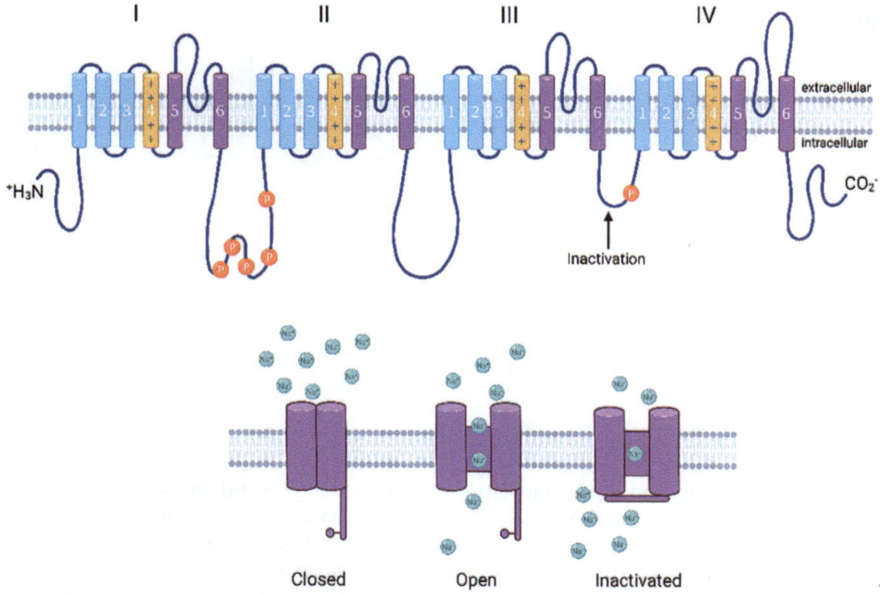

Fig. 1 Voltage-gated Na channel (VGSC) structure. Annotated structure of a VGSC showing DI–DIV and S1–S6 in each domain. Pore-lining segments in purple; S4 part of the voltage-sensing segments in yellow with gating charges. At rest, Na channels are typically nonconducting and rapidly open in response to membrane depolarization, allowing the selective flow of Na ions into the cell. Within milliseconds, the channel population inactivates and ceases to conduct Na ions regardless of continued depolarization. Upon repolarization, Na channels recover from inactivation and enter the closed state to complete this sequential progression again

2008). The predominant VGSC isoforms expressed in the central nervous system are $Na_V1.1$, $Na_V1.2$, $Na_V1.3$, and $Na_V1.6$ (Fig. 2a). These VGSC isoforms have differential expression in various types of neurons and other central nervous system cells (Fig. 2b). Additionally, their expression varies throughout nervous system development. Variants in each of the genes encoding these channels, *SCN1A*, *SCN2A*, *SCN3A*, and *SCN8A*, respectively, have been associated with developmental and epileptic encephalopathy (DEE), a class of severe and treatment-resistant early-onset epilepsy syndromes with a range of clinical manifestations. These DEEs have also provided patient-derived variants from which transgenic mouse models have been generated, which have greatly facilitated efforts to better understand VGSC function and potentially target them for treating epilepsy. Patient-derived variants have also been identified in $Na_V1.4$, $Na_V1.5$, $Na_V1.7$, $Na_V1.8$, and $Na_V1.9$ and are associated with periodic paralysis, cardiac arrhythmias, and changes to pain sensation (Dib-Hajj et al. 2015; Han et al. 2016, 2018; Baker and Nassar 2020; Mantegazza et al. 2021).

In the following sections, we briefly review the scientific basis of existing VGSC modulators as antiseizure medications and further describe ongoing efforts to generate novel precision medicine approaches corresponding to VGSC isoform-specific

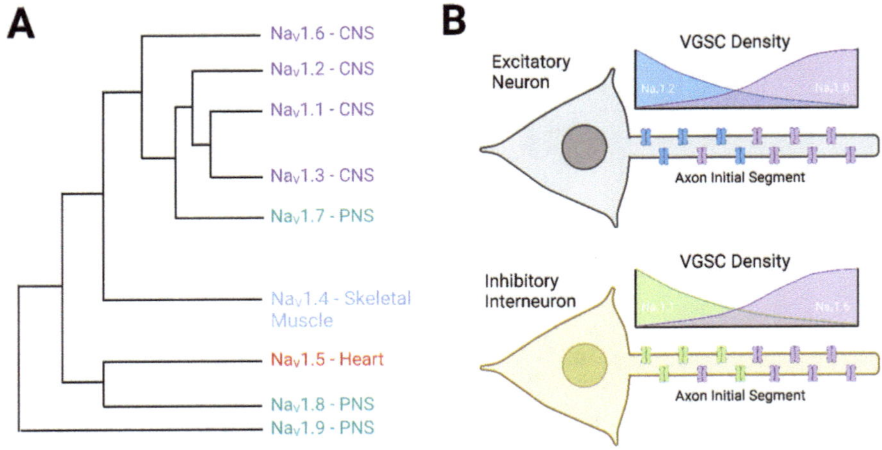

Fig. 2 VGSC diversity and expression pattern in neuronal subtypes. (**a**) Phylogenetic tree illustrating the evolution of different VGSC isoforms. VGSCs have evolved to have many isoforms predominantly expressed in various tissues, namely, the central nervous system (purple), peripheral nervous system (green), skeletal muscle (blue), and cardiac muscle (red) are indicated. (**b**) Shows the relative expression pattern of $Na_V1.1$, $Na_V1.2$, and $Na_V1.6$ in adult excitatory and inhibitory neurons

disease(s). Considering the significant advances being made in understanding VGSCs in structural detail alongside the strides in tools for controlled manipulation of channel function, there is reason for cautious optimism regarding the potential of targeting VGSCs selectively to help patients suffering with epilepsy.

2 Voltage-Gated Sodium Channels as Drug Targets in Epilepsy

2.1 *Epilepsy and Antiseizure Medications*

Epilepsy is a neurological disorder characterized by the occurrence of unprovoked seizures caused by aberrantly hyperexcitable and/or hypersynchronous brain excitability. Although around one-third of patients are refractory to existing pharmacological interventions, for many patients, chronic use of antiseizure medications provides significant relief of the risk for spontaneous seizures. Most existing antiseizure medications fit into categories of molecular mechanisms that are intuitively satisfying considering the general paradigm of hyperexcitable neural circuits as the underlying mechanism. Antiseizure medications (1) inhibit voltage-gated Na or calcium channels in order to reduce overall depolarization of neurons, (2) enhance voltage-gated potassium channel function as a means to accentuate the repolarization of neuronal membrane potential limiting intrinsic excitability, (3) augment

GABAergic inhibitory synaptic neurotransmission to increase inhibitory drive within the neuronal network, or (4) attenuate excitatory neurotransmission to reduce overall network excitation. Among the most common antiseizure medications are those that modulate VGSC function as their primary molecular mechanism (Mantegazza et al. 2010; Brodie and Sills 2011).

Antiseizure medications have been available as early as the 1850s, as shown in Fig. 3. Early antiseizure medications, phenytoin and carbamazepine, were developed as inhibitors of VGSCs through their preferential binding of channels in the inactivated state. These early VGSC blockers produced numerous cellular-level physiological effects which likely contributed to their aid in diminishing risk for the runaway excitation that underlies seizures, including facilitation of steady-state inactivation, use-dependent block of Na current, attenuation of the subthreshold non-inactivating persistent Na current, and reduction of the resurgent Na current known to contribute to high-frequency action potential firing (Segal and Douglas 1997; Kuo 1998; Sun et al. 2007; Colombo et al. 2013). In the past few decades, there are many additional VGSC-targeting antiseizure medications that share these

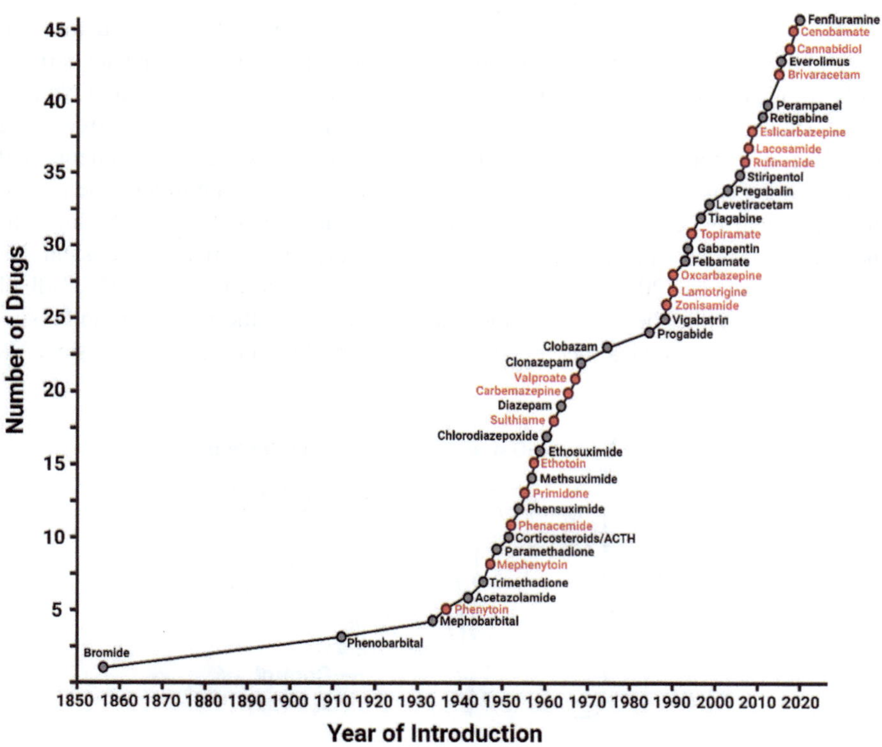

Fig. 3 Introduction of antiseizure medications to the market. Graph shows the developmental timeline for antiseizure medications approved for use in Europe, the USA, or Japan from 1853 to 2020. Not all antiseizure medications are shown. Antiseizure medications that modulate voltage-gated Na channels are shown in red

primary features but differ slightly in functional impact on Na channel biophysics, pharmacokinetics, and pharmacodynamics, including oxcarbazepine, lamotrigine, lacosamide, eslicarbazepine, and cenobamate. Site-directed mutagenesis studies have revealed that these conventional antiseizure medications, like other local anesthetics, bind to a small region in the S6 segment of domain IV termed the local anesthetic site (Catterall 1986; Ragsdale et al. 1994). All VGSC isoforms possess this local anesthetic site, which is helpful as it leads to blockade in a use-dependent manner. However, these antiseizure medications largely do not differentiate between the different Na channel isoforms, leading to relatively indiscriminate block of various Na channel isoforms at similar plasma and brain concentrations. This lack of specificity likely accounts for the high incidence of unwanted significant side effects experienced by patients, including sedation, motor impairment, and cardiac arrhythmias (French et al. 2021), and limits the ability of these traditional VGSC blockers to suppress seizures in refractory patients. Recent efforts have focused on developing subtype-selective blockers (Johnson et al. 2022). Unlike the conventional antiseizure medications, subtype-selective blockers bind to the domain IV voltage sensor domain (VSD4) when the channel is in the activated conformation, stabilizing the inactivated state (McCormack et al. 2013).

More recently, efforts have been made to generate novel VGSC modulators that seek to preferentially target the components of aberrant Na currents that are particularly implicated in epilepsy including the persistent (Fig. 4) and resurgent Na currents (Stafstrom 2007; Wengert and Patel 2021). Both of these currents are considered important in facilitating increased neuronal excitability. Riluzole and GS967 (also called Prax-330) were developed as specific inhibitors of the persistent/resurgent Na currents, although other antiseizure medications, including cannabidiol, may also preferentially block these currents relative to transient Na currents (Patel et al. 2016; Wengert et al. 2019a; Mason and Cummins 2020; Tidball et al. 2020). While the molecular mechanism underlying the persistent and resurgent Na currents remains incompletely understood, the block of the persistent Na

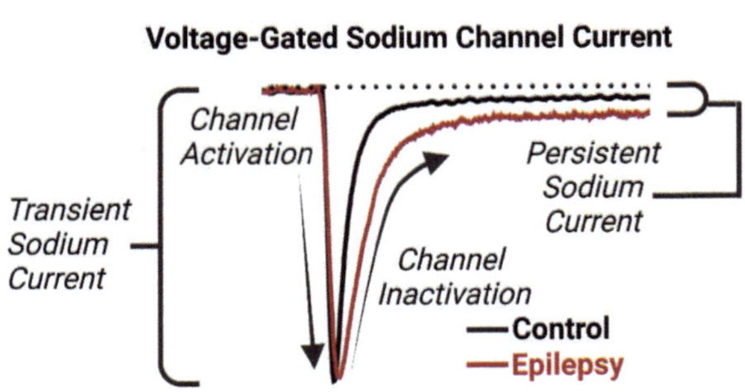

Fig. 4 Persistent Na current. VGSCs cycle between closed, open, and inactivated states. In response to depolarization, channels are activated and conduct a large transient Na current, which ceases following channel inactivation. Often, in epilepsy (red), a non-inactivating persistent Na current (I_{NaP}) follows the transient Na current. This typically leads to an increase in cellular excitability and susceptibility to seizures

current by GS967 seems to be driven, in part, via drug-binding to a distinct channel site other than the local anesthetic site – an engineered mutation in the local anesthetic site was able to reduce the impact of the drug on use-dependent block, but the inhibition of the persistent Na current was unaffected in $Na_V1.5$-expressing cells (Potet et al. 2016). Similarly, cannabidiol was recently shown to bind to a novel site in VGSCs, indicating a possible candidate for the modification of the persistent/resurgent currents (Sait et al. 2020). Future studies will likely shed additional light on the molecular mechanism(s) underlying the inhibition of the persistent and resurgent Na current observed by these antiseizure medications, and although mouse studies are encouraging regarding their potential value, more clinical studies are needed to validate their use in human epilepsy patients.

As mentioned earlier, the predominant VGSC isoforms expressed in the central nervous system are $Na_V1.1$, $Na_V1.2$, $Na_V1.3$ and $Na_V1.6$, and variants in each of the genes encoding these channels have been associated with developmental and epileptic encephalopathy (DEE), a class of severe and treatment-resistant early-onset epilepsy syndromes with a range of clinical manifestations. Generally, VGSC variants associated with DEE have been classed as either (1) gain-of-function (GoF), where increases in Na channel activity as a result of a genetic mutation in typically $Na_V1.2$ or $Na_V1.6$ lead to an increase in excitatory neuron activity and subsequent network excitability, or (2) loss-of-function (LoF), where decreases in VGSC activity in mainly $Na_V1.1$ can result in a decrease in inhibitory neuron activity and reduced inhibitory drive onto the neuronal network (Fig. 5). It is important to note that there are exceptions to this: GOF mutations in $Na_V1.1$ and LOF mutations in both $Na_V1.2$ and $Na_V1.6$ have been associated with seizure activity (de Kovel et al. 2014; Berecki et al. 2019; Spratt et al. 2021).

In the following sections, we focus on each of these channelopathies and precision medicine advances in treating these genetic disorders.

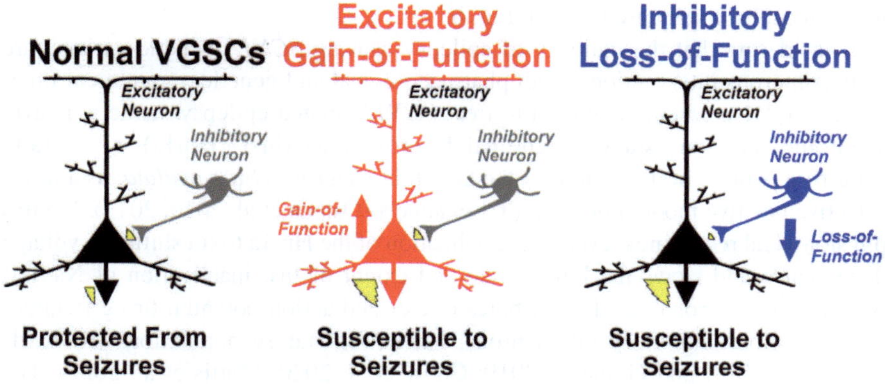

Fig. 5 Network mechanisms of epilepsy caused by GoF or LoF mutations in Na channels. Na channels in the CNS are expressed in excitatory and inhibitory neurons, illustrated in Fig. 2b. Here, we show a simplified view of the effects of gain- or loss-of-function mutations expressed in either excitatory or inhibitory neurons. This indicates simplified potential mechanisms of overall network excitability leading to susceptibility to seizures

2.2 *SCN1A*

The *SCN1A* gene encodes the $Na_V1.1$ VGSC isoform which is expressed at relatively high levels in forebrain GABAergic neuron populations including parvalbumin-positive (PV), somatostatin-positive (SST), and vasoactive intestinal peptide-positive (VIP) interneurons, among others (Ogiwara et al. 2007; Lorincz and Nusser 2008; Tai et al. 2014; Goff and Goldberg 2019). The majority of *SCN1A* variants identified in epilepsy patients have been functionally characterized as physiologically LoF, either due to hypofunction of the resulting protein or due to haploinsufficiency in response to a nonfunctional protein produced by the variant allele (Lossin et al. 2002; Rhodes et al. 2005). Reports for the incidence of variant-dependent Dravet syndrome vary between 1:15,700 births and 1:40,900 births (Brunklaus et al. 2012; Bayat et al. 2015; Wu et al. 2015). Due to the expression of $Na_V1.1$ primarily in inhibitory interneurons, LoF variants lead to a deficit in functional inhibition and, thus, an epilepsy phenotype through an overall loss of inhibition in the network (Fig. 5). Patients with LoF variants are typically diagnosed with genetic epilepsy with febrile seizures plus (GEFS+) or the more severe Dravet syndrome which both encompass a range of symptoms, including febrile and treatment-resistant afebrile seizures as well as cognitive and motor impairments (Claes et al. 2001; Ohmori et al. 2006; Catterall 2018). Incidence of sudden unexplained death in epilepsy (SUDEP) is high in patients with Dravet syndrome (Skluzacek et al. 2011). More recently, *SCN1A* GoF variants have been identified and cause a clinically distinct and mechanistically different epileptic encephalopathy that is more severe than Dravet syndrome (Sadleir et al. 2017; Berecki et al. 2019). This indicates that $Na_V1.1$ levels and functional properties must be tightly regulated given that strong deviations from proper $Na_V1.1$ function (i.e., hypofunction and hyperfunction) can both result in severe epilepsy syndromes (Wengert et al. 2022). Several relevant *SCN1A* mutations leading to Dravet syndrome or other *SCN1A*-related epileptic encephalopathy are noted in Fig. 6.

Considering that the majority of epilepsy-causing *SCN1A* genetic variants are LoF, numerous interventions, both pharmacological and genetic, to augment functional $Na_V1.1$ are being explored to treat *SCN1A*-related epilepsy, namely, Dravet syndrome. Previous studies identified heteroscodratoxin-1 (Hm1a), a naturally occurring peptide in the venom of the tarantula *Heteroscodra maculata*, as a novel selective positive modulator of $Na_V1.1$ channels (Osteen et al. 2016, 2017). In vitro physiological recordings reveal that application of the Hm1a toxin shifts the voltage dependence and slows the kinetic rate and extent of fast inactivation of $Na_V1.1$, which at the neuronal level contributes to elevated action potential firing frequencies in various inhibitory interneurons, but not excitatory populations (Richards et al. 2018; Goff and Goldberg 2019; Chow et al. 2020; Mattis et al. 2022). The therapeutic value of Hm1a is supported by in vivo studies in which acute and long-term intracerebroventricular application of Hm1a reduced epileptic discharges, reduced seizure frequency, and prolonged survival in a mouse model of Dravet syndrome (Richards et al. 2018). Significant considerations remain regarding the

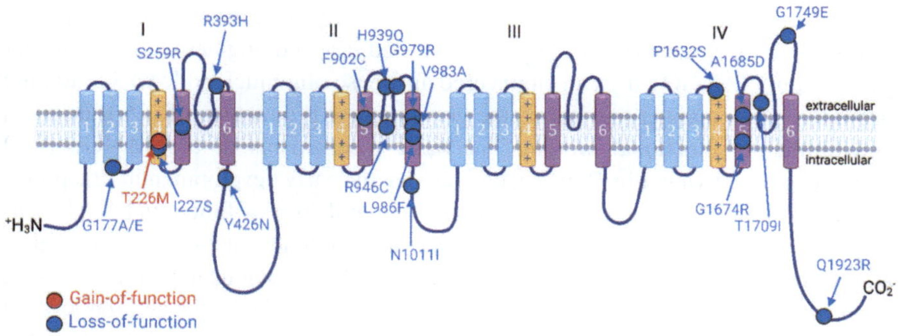

Fig. 6 Structure and mutations of $Na_V1.1$. Annotated structure of $Na_V1.1$ including several physiologically characterized *SCN1A* mutations that lead to a Dravet syndrome or severe epileptic encephalopathy phenotype. Structure shows domains DI–DIV and S1–S6 in each domain, with pore-lining segments in purple and voltage-sensing segments in yellow. A GoF mutation is noted with a red circle, whereas LoF mutations are noted with a blue circle

therapeutic value of Hm1a including delivery method as well as the significant potential off-target effects on the body.

Cannabidiol (CBD) is approved as an add-on therapy for the treatment of seizures associated with Dravet syndrome. CBD treatment in an open label trial resulted in a reduction in total seizure frequency for up to 48 weeks with acceptable side effects (Devinsky et al. 2019; Scheffer et al. 2021). Although the full mechanism of action of CBD is unknown, it has been shown to block VGSCs (Patel et al. 2016; Ghovanloo and Ruben 2022).

Using a screening approach for small compound $Na_V1.1$ positive modulators, Frederiksen and colleagues identified AA43279, which elevated the $Na_V1.1$-generated Na current largely through slowing fast inactivation (Frederiksen et al. 2017). In recordings of PV-positive fast-spiking inhibitory interneurons, they found that application of AA43279 hyperpolarized AP threshold and elevated spiking frequencies without having any effect on the physiological behavior of CA1 pyramidal neurons. Additionally, AA43279 displayed in vivo efficacy against seizures by significantly raising the current threshold for electrically induced convulsions (Frederiksen et al. 2017). AA43279 was also shown to be effective in a zebrafish model of Dravet syndrome in which treatment led to fewer bouts of abnormal burst movements and reduced the frequency of epileptiform events (Weuring et al. 2020). These results encourage further direct investigation into AA43279 and other small molecule activators of $Na_V1.1$ as a treatment for Dravet syndrome and epilepsy more generally.

Considering the mechanism of Dravet syndrome (haploinsufficiency of $Na_V1.1$), VGSC modulators have generally been avoided in favor of drugs targeting alternative mechanisms. However, despite numerous available medications working through a range of molecular mechanisms, seizures in Dravet syndrome remain refractory for many patients, emphasizing the potential benefit of modulating VGSCs directly to rescue the core mechanism of disease. In view of this, methods to augment the levels of the *SCN1A* transcript directly are currently being explored to treat Dravet syndrome. One approach, termed targeted augmentation of nuclear gene output (TANGO), uses an engineered antisense oligonucleotide (ASO) to bind pre-mRNA in a manner that boosts the generation of productive mRNA transcripts and ultimately functional protein. The basic mechanisms of the TANGO-ASO approach are shown in Fig. 7. When delivered at the early developmental timepoints of postnatal day 2 or 14, TANGO-mediated upregulation of functional *Scn1a* was able to increase levels of $Na_V1.1$, reduce seizure frequency, and elevate survival of a transgenic mouse model of Dravet syndrome without obvious negative effects on the health of the mice (Han et al. 2020; Wengert et al. 2022). At the cellular level, it was shown that TANGO-ASO is able to rescue the hypoexcitability of cortical PV interneurons (Wengert et al. 2022). TANGO-ASO is currently undergoing phase II clinical trials.

Another novel approach to elevate functional $Na_V1.1$ at the level of the gene expression is through the use of the CRISPR-Cas9 system in which specific guide-RNA traffics a nuclease-dead Cas9 protein (dCas9) fused to multiple transcriptional activator domains to proximal regions relative to the *Scn1a* gene. Using this technique, two separate groups were able to selectively potentiate *Scn1a* gene transcription and $Na_V1.1$ levels in mouse models of Dravet syndrome, which attenuated the severity and elevated the temperature threshold for hyperthermia-induced seizures in vivo (Colasante et al. 2020; Yamagata et al. 2020). Additional evidence for rescue of aberrant PV interneuron excitability and other Dravet syndrome-related behaviors was provided in these studies, supporting further investigation of this precision medicine approach to treat Dravet syndrome (Colasante et al. 2020; Yamagata et al.

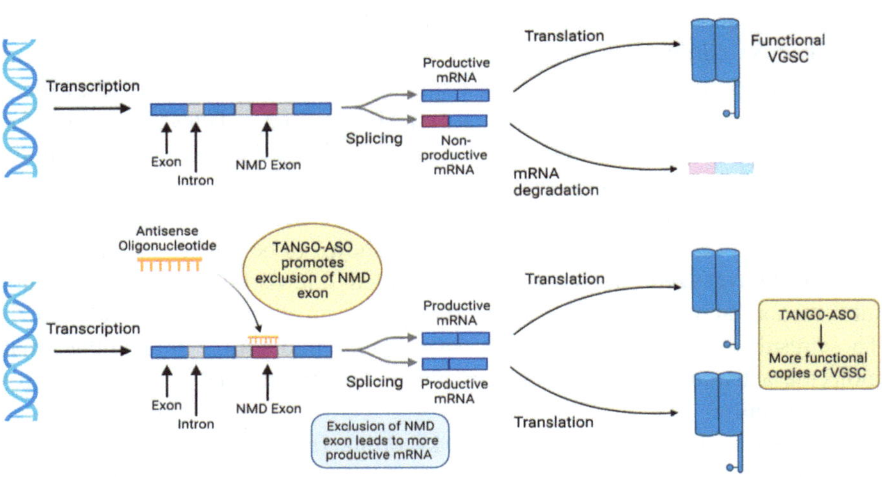

Fig. 7 TANGO-ASO. Mechanism of TANGO-ASO for use in epilepsy-related Na channelopathies. An ASO promotes the exclusion of the nonsense-mediated decay exon, which typically makes one copy of mRNA a target for degradation. With this exclusion, splicing leads to two productive copies of mRNA, allowing the production of more productive protein (i.e., $Na_V1.1$ channels)

2020). Still, other approaches to upregulate endogenous *Scn1a* expression are being explored. In one study, viral delivery (AAV9) of an engineered transcription factor designed to specifically upregulate *Scn1a* levels in GABAergic inhibitory interneurons led to amelioration of Dravet syndrome phenotype in mice and further, was well-tolerated in nonhuman primates (Tanenhaus et al. 2022).

While these novel gene therapy approaches are exciting and hold significant promise, a number of key issues remain in order for them to be translated into efficacious treatments for human patients with Dravet syndrome. One challenge for these approaches is to properly address the mechanism of disease in Dravet syndrome – a mechanism still incompletely understood. For example, although impairments in intrinsic excitability are observed at early timepoints in mouse models of Dravet syndrome, the chronic phase of the disease likely has an alternative mechanism considering that axonal dysfunction and synaptic deficits are observed alongside normalization of the intrinsic excitability of PV interneurons (Favero et al. 2018; Kaneko et al. 2022). For this reason, boosting $Na_V1.1$ function after the onset of symptoms might be predicted to be ineffective or even deleterious, although recent evidence suggests restoration of $Na_V1.1$ function in adulthood is therapeutically valuable in a mouse model of Dravet syndrome (Valassina et al. 2022). For each of the studies discussed, treatment of the mice was initiated presymptomatically, often within the first few postnatal days. Presymptomatic treatment of human patients with *SCN1A* genetic variants poses numerous clinical (not to mention ethical) challenges. Other challenges facing gene therapies targeting *SCN1A* include method of delivery of the antiseizure medications, reaching adequate levels of expression, ensuring cell-type specificity and limiting off-target effects, and determining the best clinical practices for long-term use of these medications. Yet, with ongoing scientific and clinical investigation, many of these issues will likely be addressed.

2.3 *SCN2A*

SCN2A encodes the VGSC isoform $Na_V1.2$, which is expressed in primarily excitatory pyramidal cells at the axon initial segment, a critical region of the axon that contributes to the initiation, generation, and backpropagation of action potentials in various neuronal subpopulations (Hu et al. 2009). Developmentally, $Na_V1.2$ is expressed throughout the entire axon initial segment in early postnatal development, but is later replaced by another VGSC ($Na_V1.6$) at the distal portion of this segment (Kaplan et al. 2001; Boiko et al. 2003); this is potentially important to disorders involving the dysfunction of either VGSC. Due to its early postnatal and widespread expression, it is unsurprising that $Na_V1.2$ dysfunction has been linked with a range of neurodevelopmental disorders including autism, intellectual disability, schizophrenia, and epilepsy (Wolff et al. 2017, 2019). Both GoF and LoF genetic variants have been associated with developmental and epileptic encephalopathy (Fig. 8); patients exhibiting the earliest onset of seizures (<3 months) have GoF variants, whereas later onset is typically observed for patients with mutations

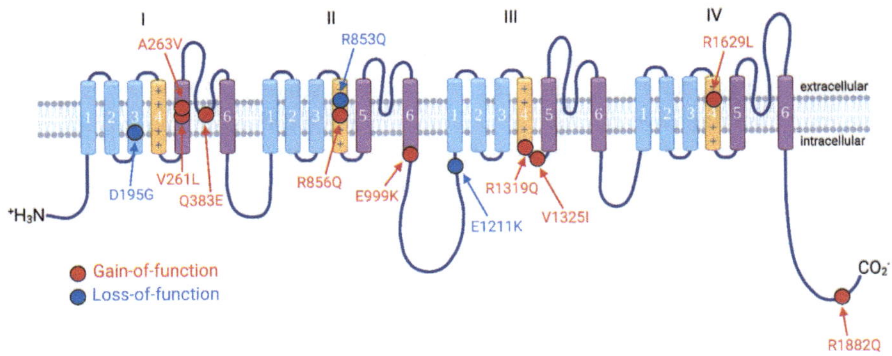

Fig. 8 Structure and mutations of $Na_V1.2$. Annotated structure of $Na_V1.2$ including several physiologically characterized *SCN2A* mutations leading to a severe epilepsy phenotype. Structure shows domains DI–DIV and S1–S6 in each domain, with pore-lining segments in purple and voltage-sensing segments in yellow. GoF mutations are in red, with LoF mutations noted in blue

leading to LoF at the channel level (Wolff et al. 2017). While GoF *SCN2A* variants have a somewhat direct relationship to an epilepsy phenotype via excitatory cell hyperexcitability, the relationship between LoF $Na_V1.2$ variants and network hyperexcitability/seizures appears to be more complicated. With LoF *SCN2A* mutations causing epilepsy, excitatory cells exhibit abnormal dendritic excitability and seem to be hypoexcitable early in development, with a paradoxical hyperexcitability occurring in adulthood (Spratt et al. 2019, 2021). Importantly, in cases with either GoF or LoF mutations, patients with *SCN2A*-related developmental and epileptic encephalopathy have a poor clinical prognosis and are often refractory to available antiseizure medications.

Subtype-specific blockade of $Na_V1.2$ would be predicted to be therapeutic in patients harboring GoF genetic variants in *SCN2A*. To date, a few $Na_V1.2$ subtype-specific blockers have been identified. One is the naturally occurring μ-conotoxin KIIIA, of which the detailed binding structure was recently determined in human $Na_V1.2$ (Wilson et al. 2011; Khoo et al. 2012; Pan et al. 2019). Due to its potency (dissociation constant; K_d is ~5 nM) and other factors limiting its translation for preclinical and clinical investigation, KIIIA is unlikely to be directly useful for the treatment of epilepsy, yet the structural insight gained will likely facilitate the development of small molecule compounds targeting $Na_V1.2$.

Small molecules that preferentially inhibit $Na_V1.2$ have also been identified through chemical synthesis of analogs of the VGSC inhibitor mexiletine (van der Peet et al. 2019). Through the rational design of compounds with relative enhanced lipophilicity over mexiletine to achieve better CNS penetration, the possibility of using these $Na_V1.2$-preferential inhibitors to treat *SCN2A*-related epilepsy is more feasible. Indeed, the authors identified that delivery of one of the N-(4′-hydroxy-3′,5′-ditert-butylbenzyl) mexiletine analogs was effective in suppressing seizures in the 6-Hz electrical stimulation model (van der Peet et al. 2019). Given this efficacy, future studies should examine this compound in mouse models of *SCN2A*-related epilepsy.

Mentioned previously, ASOs have been developed to upregulate $Na_V1.1$ to normalize effects of mutations causing LoF or haploinsufficiency in Dravet syndrome; whereas in the context of GoF *SCN2A* epileptic encephalopathy, ASOs have been developed to specifically downregulate functional $Na_V1.2$ and potentially normalize cellular and circuit excitability. In one study, Li and colleagues generated a novel mouse model of *SCN2A* developmental and epileptic encephalopathy that expressed the R1882Q GoF variant and well recapitulated the disease severity through frequent spontaneous seizures and premature lethality (Li et al. 2021). Intracerebroventricular delivery of the novel ASO targeting *Scn2a* at postnatal day 1 was able to strongly extend survival relative to treatment with a control ASO. Importantly, survival and relative life span were still increased when delivery of the ASO occurred after the symptomatic onset of spontaneous seizures, between mouse postnatal day 14 and 16. This is more clinically relevant, considering that patients would likely need to display DEE symptoms to identify a *SCN2A* variant before commencing treatment with an ASO. ASOs for the treatment of seizures associated with GoF *SCN2A* mutations are currently in development/phase II clinical trials (EMBRAVE: PRAX-222). Taken together, the preclinical evidence for the efficacy of ASO-mediated knockdown of *Scn2a* in a mouse model of *SCN2A* epileptic encephalopathy is encouraging for future studies.

A few important considerations remain for targeted manipulation of *SCN2A*/$Na_V1.2$. Given that hypofunction and hyperfunction in $Na_V1.2$ have both been associated with autism and epilepsy, issues related to dosing must be taken into account. As mentioned previously, LoF in $Na_V1.2$ has been paradoxically linked with hyperexcitability, and therefore excessive knockdown of a GoF channel may also be deleterious (Spratt et al. 2021). Despite these considerations, the ability to genetically up- and downregulate expression of other VGSCs in an effort to treat their related epileptic encephalopathies gives hope to the future treatment of *SCN2A* epilepsy.

2.4 *SCN3A*

The *SCN3A* gene encodes the $Na_V1.3$ subunit, which is highly expressed early in development with relatively reduced expression levels in adulthood compared to other VGSCs (Beckh et al. 1989; Whitaker et al. 2000; Cheah et al. 2013). Genetic variants in *SCN3A* have been found in patients with various types of epilepsy, ranging from the more mild partial epilepsy to the most severe epileptic encephalopathy (Holland et al. 2008; Zaman et al. 2018). Although functional characterization of genetic variants has revealed that both LoF and GoF variants are among those associated with epilepsy (Fig. 9), the majority of patients show some form of GoF channel behavior via increased persistent Na current, hyperpolarizing shift in voltage dependence of activation, or impairment of the kinetic rate of VGSC inactivation (Vanoye et al. 2014; Lamar et al. 2017; Zaman et al. 2018, 2020). Potentially due to

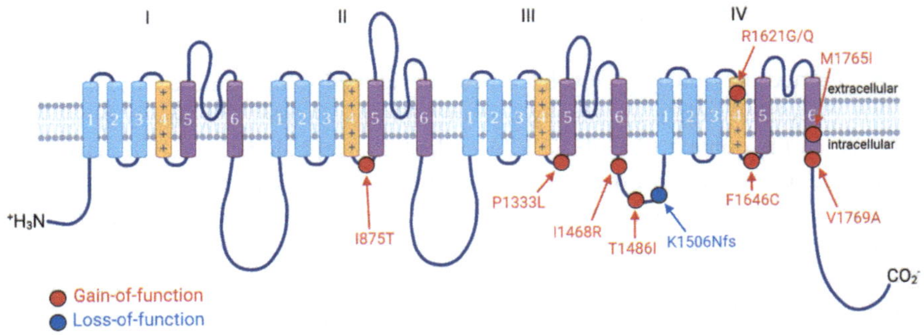

Fig. 9 Structure and mutations of $Na_V1.3$. Annotated structure of $Na_V1.3$ including some physiologically characterized *SCN3A* mutations that lead to an epilepsy phenotype. Structure shows Na channel domains DI–DIV and S1–S6 in each domain, with pore-lining segments in purple and voltage-sensing segments in yellow. GoF mutations are in red, with a single LoF mutation noted in blue. Some mutations (GoF only) also lead to brain malformations

the high expression of $Na_V1.3$ in early cortical development, patients with epilepsy-related mutations in *SCN3A* often also possess brain malformations, which are not typically present in patients with epilepsy-related variants in other Na channel genes. Interestingly, these malformations were only observed in patients with functionally GoF variants (Zaman et al. 2020). Additionally, while transgenic mice haploinsufficient for *Scn3a* exhibit a relatively mild neurological phenotype and do not exhibit spontaneous seizures, *Scn3a* mouse models expressing the epileptic encephalopathy patient-derived variants are needed to clarify the physiological impact of abnormal $Na_V1.3$ function on single neurons and neural circuits (Lamar et al. 2017). However, due to potential differences in key progenitor cells between mice and humans (along with nonhuman primates), this may prove quite difficult: *SCN3A* is expressed in outer radial glia, a type of progenitor cell only seen in humans and nonhuman primates that possess Na currents and are important in brain gyrification (Hansen et al. 2010; Smith et al. 2018; Rash et al. 2019).

Considering the predominantly GoF phenotype of the majority of epilepsy-causing *SCN3A* variants, a selective inhibitor for $Na_V1.3$ would be advantageous for precision treatment. To date, the only preferential $Na_V1.3$-selective small molecule antagonist is ICA-121431, which inhibits $Na_V1.3$ at nanomolar concentrations (~20 nM) with significantly less effect on $Na_V1.2$ and $Na_V1.6$ and has been explored as an antiseizure medication (McCormack et al. 2013; Masala et al. 2021). However, due to structural similarity, ICA-121431 inhibits $Na_V1.1$ to a similar degree as $Na_V1.3$ (Osteen et al. 2017; Li et al. 2022). Since hypofunction of $Na_V1.1$ is itself associated with GABAergic interneuron impairment, leading to imbalanced network excitability, ICA-121431's potential efficacy as an antiseizure medication may be limited to certain early developmental times when there is greater relative presence of $Na_V1.3$ compared to $Na_V1.1$. Nonetheless, alternative approaches to selectively target and modify *SCN3A* expression and/or $Na_V1.3$ function should be explored.

2.5 *SCN8A*

$Na_V1.6$, the VGSC encoded by the *SCN8A* gene, is highly expressed in various neuronal subpopulations including both forebrain excitatory and inhibitory interneuron populations (Lorincz and Nusser 2008; Li et al. 2014). Through expression at the axon initial segment and nodes of Ranvier, $Na_V1.6$ is a critical contributor to action potential generation and propagation, particularly in excitatory pyramidal cells through which it is crucial to the overall excitability of the network (Caldwell et al. 2000; Hu et al. 2009). Comparison of $Na_V1.6$ to other VGSCs indicates that it has a substantially large relative contribution to the total persistent and resurgent Na current implicated in repetitive firing and other aspects of cellular physiology (Rush et al. 2005; Chen et al. 2008; Katz et al. 2018). Considering that the $Na_V1.6$-derived persistent and resurgent Na currents have been strongly implicated in seizures (Stafstrom 2007; Alexandrou et al. 2016; Ottolini et al. 2017; Wengert et al. 2019a; Pan and Cummins 2020; Wengert and Patel 2021), it is unsurprising that genetic variants in *SCN8A* are a cause of severe DEE (Veeramah et al. 2012; Ohba et al. 2014; Wengert et al. 2019b; Talwar and Hammer 2021). Although there are numerous reports of epilepsy-causing *SCN8A* variants that are physiologically LoF (Blanchard et al. 2015; Wengert et al. 2019b), those epilepsy cases seem to be relatively more mild in their symptomatic severity (Wagnon et al. 2017, 2018; Talwar and Hammer 2021). Many of these patients suffer from intellectual disability (ID), autism spectrum disorder (ASD), and absence seizures (Wengert et al. 2019b; Johannesen et al. 2022). Meanwhile, patient-derived mutations classified as GoF are proexcitatory due to shifts in the voltage dependence of activation, steady-state inactivation, and increase in the magnitude of persistent and/or resurgent currents, among other changes, and often lead to a severe epileptic encephalopathy phenotype (Veeramah et al. 2012; Patel et al. 2016; Liu et al. 2019; Zaman et al. 2019). Epilepsy-causing *SCN8A* variants are shown in Fig. 10.

Given the predominant role that $Na_V1.6$ plays in controlling the excitability of forebrain excitatory neuron populations, evidenced by the fact that aberrant $Na_V1.6$ function specifically in forebrain excitatory neurons is sufficient for spontaneous seizures and premature lethality in mice (Bunton-Stasyshyn et al. 2019), development of an $Na_V1.6$-selective blocker is thought to better restore balanced network excitability in the context of epilepsy. Original efforts to identify $Na_V1.6$-selective inhibitors revealed that the tetrodotoxin metabolite 4,9-anhydro-TTX is preferential for $Na_V1.6$ relative to $Na_V1.7$ and $Na_V1.5$ (Rosker et al. 2007). Indeed, 4,9-anhydro-TTX is able to reduce hyperexcitability in neuronal populations (Barker et al. 2017). However, the selectivity of 4,9-anhydro-TTX has been questioned (Tsukamoto et al. 2017; Denomme et al. 2020), limiting the outlook of its therapeutic value.

More recently, the small molecule $Na_V1.6$-selective inhibitor, NBI-921352 (previously XEN901), has been developed to treat *SCN8A* epileptic encephalopathy and potentially other types of epilepsy (Johnson et al. 2022). That study demonstrated that NBI-921352 is over 100 times more selective for $Na_V1.6$ relative to other VGSC isoforms and functions through preferential binding to inactivated

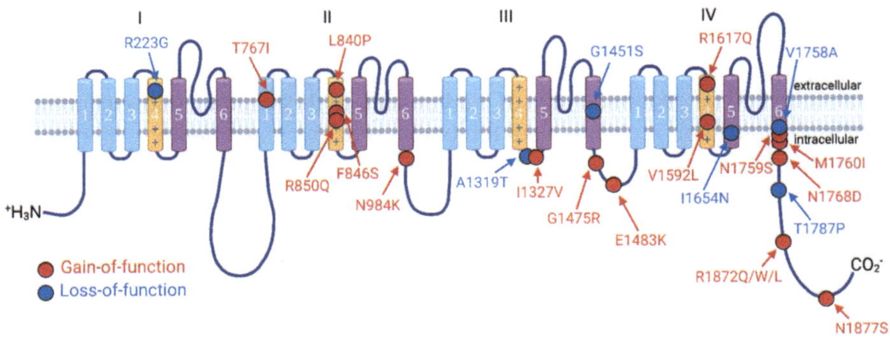

Fig. 10 Structure and mutations of $Na_V1.6$. Annotated structure of $Na_V1.6$ including several physiologically characterized *SCN8A* mutations that lead to an *SCN8A*-related epilepsy phenotype. Structure shows Na channel domains DI–DIV and S1–S6 in each domain, with pore-lining segments in purple and voltage-sensing segments in yellow. GoF mutations are in red, with LoF mutations noted in blue

channel populations (Johnson et al. 2022). As anticipated from a $Na_V1.6$-selective inhibitor, NBI-921352 reduced the magnitude of the persistent and resurgent Na currents and reduced excitability of excitatory neurons without overtly impacting fast-spiking interneuron function (Johnson et al. 2022). In vivo, NBI-921352 inhibited electrically induced seizures in both rats and mice at significantly lower doses than other non-isoform selective VGSC blockers, suggesting that NBI-921352 will be therapeutically valuable without the significant off-target effects contributing to the difficult tolerability of other VGSC-targeting antiseizure medications. While additional studies investigating the more detailed physiological impact of selective inhibition of $Na_V1.6$ at the cellular, circuit, and behavioral level are needed, these early results are supportive of the overall promise of targeting $Na_V1.6$ as a treatment for epilepsy. Phase II studies (KAYAK) of NBI-921352 as an adjunctive therapy for children with *SCN8A* DEE are currently in progress.

As for other VGSCs, approaches are being developed to modulate $Na_V1.6$ function through the direct manipulation of gene expression as a means to treat epilepsy. Considering that most epilepsy-causing variants in *SCN8A* are GoF, reduction in $Na_V1.6$ gene expression is predicted to be therapeutically valuable. Indeed, Lenk and colleagues developed an ASO which selectively reduced relative *Scn8a* expression, protected against spontaneous seizures, and extended survival in a mouse model of *SCN8A* epileptic encephalopathy (Lenk et al. 2020). Importantly, in these studies, the ASO was first delivered intracerebroventricularly at postnatal day 2, well before the onset of seizures in this mouse model. Future studies utilizing ASO-mediated manipulation of VGSC gene expression and function will likely clarify the potential efficacy of the ASO when given after the onset of spontaneous seizures and other comorbid symptoms.

3 Outlook

VGSCs have been targeted by drugs to treat epilepsy for decades, yet scientific and technological progress has continued to put forth additional approaches to potentially better help epileptic patients reach adequate seizure control. In the previous sections, we have outlined a number of prior and ongoing strategies to target and modulate VGSC function to help achieve balanced network excitation and prevent seizures. Although there is reason for optimism regarding novel pharmacological agents that preferentially target VGSC isoforms and/or functional features of Na currents (i.e., persistent Na current) as well as gene therapy approaches to selectively alter gene expression, there remain numerous important limitations and hurdles that must be considered if these treatments are going to be effectively translated into clinical use.

For one, the molecular-level details of the structure-function relationship for VGSCs and the drugs with which they interact are still incompletely understood. Substantial strides have been made in this area within the past decade, and a deeper structural understanding of general features of VGSC function including voltage-dependent activation, fast inactivation, the selectivity filtering for hydrated Na ions, etc. has come into view (Payandeh et al. 2011; Catterall and Swanson 2015; Shen et al. 2017; Pan et al. 2019; Sait et al. 2020; Noreng et al. 2021; Li et al. 2022). Solved structures now exist for various VGSC in conformations bound to classic pore blockers, toxins, and small compounds, which all serve to inform a structural perspective in pharmacology and subtype-selective modulation. Further, advances in the field of protein engineering has provided the ability to generate chimeric Na channels in which a bacterial Na channel is engineered to express key structural features of human Na channels (i.e., the voltage-sensing module of a human channel provided onto the bacterial Na channel structure) (Ahuja et al. 2015; Männikkö et al. 2018; Clairfeuille et al. 2019; Myshkin et al. 2019; Xu et al. 2019; Gao et al. 2020; Wisedchaisri et al. 2021). This strategy bypasses the significant challenges involved in producing full-length human Na channel proteins and opens new avenues for high-throughput screening of compound libraries for Na channel-selective modulating properties (Noreng et al. 2021). Given that advancements in structural biology and molecular pharmacology are mutually reinforcing, it is likely that both of these subfields will make ever-accelerating progress in the years to come to produce a highly detailed understanding of Na channel conformational states and the associated pharmacological properties. Nonetheless, important questions remain regarding how epilepsy-causing genetic variants alter the conformational structure of the channel and subsequently how these alterations might influence the impact of various toxins and pharmacological agents on the variant channel (Sait et al. 2020; Pan et al. 2019; Shen et al. 2017; Yan et al. 2017). This consideration is likely important given that genetic variant-specific responses to antiseizure medications have been reported (Barker et al. 2016; Tidball et al. 2020). Gaining a more detailed understanding of VGSC structure and function will inform the design and validation of clinical decision support algorithms and potentially support patient-specific precision medicine interventions as is already actively being explored (Brünger et al. 2022; Brunklaus et al. 2022).

One additional consideration in developing targeted VGSC-based treatments for patients with DEE is the wide heterogeneity of functional impacts of VGSC genetic variants. Although the general terms LoF and GoF are useful categorical descriptions for ion channels, the details of how any single variant influences channel, cellular, and neural circuit function are likely to be divergent, even for variants placed in the same general category. The therapeutic logic of the isoform-specific channel modulators and gene therapy-based approaches is such that up- or downregulation of channel expression or function is intended to address the core channel dysfunction described as gain- or loss-of-function – a logical view that may be complicated when the details of the variant phenotype are taken into account. For example, would ASO-mediated knockdown be similarly efficacious for treating patients with variants causing hyperpolarizing shifts in the voltage dependence of activation, depolarized steady-state inactivation, and/or elevated persistent Na currents, all of which are classified as GoF? This same question is relevant for cases in which the genetic variant results in a mixed functional phenotype of GoF and LoF such as the N1768D variant in the *SCN8A* gene which causes a large elevation in the magnitude of the persistent Na current (GoF), but a ~50% reduction in transient current density (LoF) as well as aspects of both GoF and LoF at the level of cellular physiology (Veeramah et al. 2012; Bunton-Stasyshyn et al. 2019; Wengert et al. 2019a, 2021). Ongoing experimental investigation is needed to facilitate a more nuanced understanding of VGSC variant function and clarification on how these various therapeutic interventions work in the full range of contexts.

The complexity of the brain and its development over time poses yet another challenge for treating severe genetic epilepsy syndromes in pediatric populations. Growing evidence indicates that the mechanism of seizures caused by VGSC dysfunction evolves over time as expression levels change, compensation occurs, and neurons mature and age. As an illustrative example, consider PV interneurons in a mouse model of Dravet syndrome, in which the intrinsic excitability of PV interneurons is impaired only at early developmental time points before apparently normalizing in adulthood (Favero et al. 2018; Kaneko et al. 2022). Would TANGO-mediated upregulation of $Na_V1.1$ in adult *Scn1a*$^{+/-}$ mice be efficacious? Limited evidence provided by Han and colleagues suggested that the efficacy of TANGO was reduced when delivered to *Scn1a*$^{+/-}$ mice at postnatal day 14 compared to postnatal day 2 (Han et al. 2020). Overall, it is clear that the developmental course of VGSC expression and overall neurological function must be considered as an element of precision medicine. Treatment should not merely be selected for particular patients and their underlying genetic cause of disease, but the precision must also take into account developmental time point.

Looking forward, it is likely that VGSCs will continue to be one of the most prominent drug targets for treating epileptic seizures. Although the future of modulating VGSCs to treat epilepsy is one of great promise, the path also contains numerous challenges which must be overcome. Facilitated by technological and scientific advancement, it is feasible that patients will eventually receive treatment highly tailored to their genetic and developmental context and adapted to meet the ever-evolving mechanistic underpinning of the disease. With the advent of various subtype-selective blockers and activators as well as genetic tools to bidirectionally influence gene expression, the ability to experimentally control VGSC expression and function will likely contribute more detailed insight into the underlying mechanisms of epilepsy and, hopefully, better available treatments for patients.

References

Ahuja S, et al. Structural basis of Nav1.7 inhibition by an isoform-selective small-molecule antagonist. Science. 2015;350:aac5464.

Alexandrou AJ, et al. Subtype-selective small molecule inhibitors reveal a fundamental role for Nav1.7 in nociceptor electrogenesis, axonal conduction and presynaptic release. PLoS One. 2016;11:e0152405.

Baker MD, Nassar MA. Painful and painless mutations of SCN9A and SCN11A voltage-gated sodium channels. Pflugers Arch. 2020;472:865–80.

Barker BS, et al. The SCN8A encephalopathy mutation p.Ile1327Val displays elevated sensitivity to the anticonvulsant phenytoin. Epilepsia. 2016;57:1458–66.

Barker BS, et al. Pro-excitatory alterations in sodium channel activity facilitate subiculum neuron hyperexcitability in temporal lobe epilepsy. Neurobiol Dis. 2017;108:183–94.

Bayat A, Hjalgrim H, Møller RS. The incidence of SCN1A-related Dravet syndrome in Denmark is 1:22,000: a population-based study from 2004 to 2009. Epilepsia. 2015;56:e36–9.

Bean BP. The action potential in mammalian central neurons. Nat Rev Neurosci. 2007;8:451–65.

Beckh S, Noda M, Lubbert H, Numa S. Differential regulation of three sodium channel messenger RNAs in the rat central nervous system during development. EMBO J. 1989;8:3611–6.

Berecki G, et al. SCN1A gain of function in early infantile encephalopathy. Ann Neurol. 2019;85:514–25.

Blanchard MG, et al. De novo gain-of-function and loss-of-function mutations of SCN8A in patients with intellectual disabilities and epilepsy. J Med Genet. 2015;52:330–7.

Boiko T, et al. Functional specialization of the axon initial segment by isoform-specific sodium channel targeting. J Neurosci. 2003;23:2306–13.

Brodie MJ, Sills GJ. Combining antiepileptic drugs – rational polytherapy? Seizure. 2011;20:369–75.

Brünger T, et al. Conserved patterns across ion channels correlate with variant pathogenicity and clinical phenotypes. Brain. 2022;146:923–34.

Brunklaus A, Ellis R, Reavey E, Forbes GH, Zuberi SM. Prognostic, clinical and demographic features in SCN1A mutation-positive Dravet syndrome. Brain. 2012;135:2329–36.

Brunklaus A, et al. Gene variant effects across sodium channelopathies predict function and guide precision therapy. Brain. 2022;145:4275–86.

Bunton-Stasyshyn RKA, et al. Prominent role of forebrain excitatory neurons in SCN8A encephalopathy. Brain. 2019;142:362–75.

Caldwell JH, Schaller KL, Lasher RS, Peles E, Levinson SR. Sodium channel Nav1.6 is localized at nodes of Ranvier, dendrites, and synapses. Proc Natl Acad Sci U S A. 2000;97:5616–20.

Catterall WA. Molecular properties of voltage-sensitive sodium channels. Annu Rev Biochem. 1986;55:953–85.

Catterall WA. Dravet syndrome: a sodium channel interneuronopathy. Curr Opin Physio. 2018;2:42–50.

Catterall WA, Swanson TM. Structural basis for pharmacology of voltage-gated sodium and calcium channels. Mol Pharmacol. 2015;88:141–50.

Catterall WA, Wisedchaisri G, Zheng N. The conformational cycle of a prototypical voltage-gated sodium channel. Nat Chem Biol. 2020;16:1314–20.
Cheah CS, et al. Correlations in timing of sodium channel expression, epilepsy, and sudden death in Dravet syndrome. Channels. 2013;7:468–72.
Chen Y, et al. Functional properties and differential neuromodulation of Nav1.6 channels. Mol Cell Neurosci. 2008;38:607–15.
Chow CY, et al. A selective NaV1.1 activator with potential for treatment of Dravet syndrome epilepsy. Biochem Pharmacol. 2020;181:113991.
Claes L, et al. De novo mutations in the sodium-channel gene SCN1A cause severe myoclonic epilepsy of infancy. Am J Hum Genet. 2001;68:1327–32.
Clairfeuille T, et al. Structural basis of α-scorpion toxin action on Nav channels. Science. 2019;363:eaav8573.
Clatot J, et al. Voltage-gated sodium channels assemble and gate as dimers. Nat Commun. 2017;8:2077.
Colasante G, et al. dCas9-based Scn1a gene activation restores inhibitory interneuron excitability and attenuates seizures in Dravet syndrome mice. Mol Ther. 2020;28:235–53.
Colombo E, Franceschetti S, Avanzini G, Mantegazza M. Phenytoin inhibits the persistent sodium current in neocortical neurons by modifying its inactivation properties. PLoS One. 2013;8:55329.
de Kovel CGF, et al. Characterization of a de novo SCN8A mutation in a patient with epileptic encephalopathy. Epilepsy Res. 2014;108:1511–8.
Denomme N, et al. The voltage-gated sodium channel inhibitor, 4,9-anhydrotetrodotoxin, blocks human Nav1.1 in addition to Nav1.6. Neurosci Lett. 2020;724:134853.
Devinsky O, et al. Long-term cannabidiol treatment in patients with Dravet syndrome: an open-label extension trial. Epilepsia. 2019;60:294–302.
Dib-Hajj SD, Black JA, Waxman SG. NaV1.9: a sodium channel linked to human pain. Nat Rev Neurosci. 2015;16:511–9.
Favero M, Sotuyo NP, Lopez E, Kearney JA, Goldberg EM. A transient developmental window of fast-spiking interneuron dysfunction in a mouse model of Dravet syndrome. J Neurosci. 2018;38:7912.
Frederiksen K, et al. A small molecule activator of Nav1.1 channels increases fast-spiking interneuron excitability and GABAergic transmission in vitro and has anti-convulsive effects in vivo. Eur J Neurosci. 2017;46:1887–96.
French JA, et al. FDA safety warning on the cardiac effects of lamotrigine: an advisory from the Ad Hoc ILAE/AES Task Force. Epilepsia Open. 2021;6:45–8.
Gao S, et al. Employing NaChBac for cryo-EM analysis of toxin action on voltage-gated Na+ channels in nanodisc. Proc Natl Acad Sci U S A. 2020;117:14187–93.
Ghovanloo M-R, Ruben PC. Cannabidiol and sodium channel pharmacology: general overview, mechanism, and clinical implications. Neuroscientist. 2022;28:318–34.
Goff KM, Goldberg EM. Vasoactive intestinal peptide-expressing interneurons are impaired in a mouse model of Dravet syndrome. eLife. 2019;8:e46846.
Goldin AL. Diversity of mammalian voltage-gated sodium channels. Ann N Y Acad Sci. 1999;868:38–50.
Goldin AL, et al. Messenger RNA coding for only the α subunit of the rat brain Na channel is sufficient for expression of functional channels in Xenopus oocytes. Proc Natl Acad Sci U S A. 1986;83:7503–7.
Han C, Huang J, Waxman SG. Sodium channel Nav1.8: emerging links to human disease. Neurology. 2016;86:473–83.
Han D, Tan H, Sun C, Li G. Dysfunctional Nav1.5 channels due to SCN5A mutations. Exp Biol Med (Maywood). 2018;243:852–63.
Han Z, et al. Antisense oligonucleotides increase Scn1a expression and reduce seizures and SUDEP incidence in a mouse model of Dravet syndrome. Sci Transl Med. 2020;12:eaaz6100.
Hansen DV, Lui JH, Parker PRL, Kriegstein AR. Neurogenic radial glia in the outer subventricular zone of human neocortex. Nature. 2010;464:554–61.

Hartshorne RP, Catterall WA. Purification of the saxitoxin receptor of the sodium channel from rat brain. Proc Natl Acad Sci U S A. 1981;78:4620–4.

Hartshorne RP, Messner DJ, Coppersmith JC, Catterall WA. The saxitoxin receptor of the sodium channel from rat brain. Evidence for two nonidentical β subunits. J Biol Chem. 1982;257:13888–95.

Hartshorne RP, Keller BU, Talvenheimo JA, Catterall WA, Montal M. Functional reconstitution of the purified brain sodium channel in planar lipid bilayers. Proc Natl Acad Sci U S A. 1985;82:240–4.

Hille B. Ion channels of excitable membranes. 3rd ed. Sunderland: Sinauer; 2001.

Holland KD, et al. Mutation of sodium channel SCN3A in a patient with cryptogenic pediatric partial epilepsy. Neurosci Lett. 2008;433:65–70.

Hu W, et al. Distinct contributions of Nav1.6 and Nav1.2 in action potential initiation and backpropagation. Nat Neurosci. 2009;12:996–1002.

Isom LL, et al. Primary structure and functional expression of the β1 subunit of the rat brain sodium channel. Science. 1992;256:839–42.

Isom LL, De Jongh KS, Catterall WA. Auxiliary subunits of voltage-gated ion channels. Neuron. 1994;12:1183–94.

Johannesen KM, et al. Genotype-phenotype correlations in SCN8A-related disorders reveal prognostic and therapeutic implications. Brain. 2022;145:2991–3009.

Johnson JP, et al. NBI-921352, a first-in-class, NaV1.6 selective, sodium channel inhibitor that prevents seizures in Scn8a gain-of-function mice, and wild-type mice and rats. eLife. 2022;11:e72468.

Kaneko K, et al. Developmentally regulated impairment of parvalbumin interneuron synaptic transmission in an experimental model of Dravet syndrome. Cell Rep. 2022;38:110580.

Kaplan MR, et al. Differential control of clustering of the sodium channels Nav1.2 and Nav1.6 at developing CNS nodes of Ranvier. Neuron. 2001;30:105–19.

Katz E, et al. Role of sodium channel subtype in action potential generation by neocortical pyramidal neurons. Proc Natl Acad Sci U S A. 2018;115:E7184–92.

Khoo KK, et al. Distinct disulfide isomers of μ-conotoxins KIIIA and KIIIB block voltage-gated sodium channels. Biochemistry. 2012;51:9826–35.

Kuo CC. A common anticonvulsant binding site for phenytoin, carbamazepine, and lamotrigine in neuronal Na+ channels. Mol Pharmacol. 1998;54:712–21.

Lamar T, et al. SCN3A deficiency associated with increased seizure susceptibility. Neurobiol Dis. 2017;102:38–48.

Lenk GM, et al. Scn8a antisense oligonucleotide is protective in mouse models of SCN8A encephalopathy and Dravet syndrome. Ann Neurol. 2020;87:339–46.

Li T, et al. Action potential initiation in neocortical inhibitory interneurons. PLoS Biol. 2014;12:e1001944.

Li M, et al. Antisense oligonucleotide therapy reduces seizures and extends life span in an SCN2A gain-of-function epilepsy model. J Clin Invest. 2021;131:e152079.

Li X, et al. Structural basis for modulation of human NaV1.3 by clinical drug and selective antagonist. Nat Commun. 2022;13:1286.

Liu Y, et al. Neuronal mechanisms of mutations in SCN8A causing epilepsy or intellectual disability. Brain. 2019;142:376–90.

Lorincz A, Nusser Z. Cell-type-dependent molecular composition of the axon initial segment. J Neurosci. 2008;28:14329–40.

Lossin C, Wang DW, Rhodes TH, Vanoye CG, George AL. Molecular basis of an inherited epilepsy. Neuron. 2002;34:877–84.

Männikkö R, et al. Spider toxin inhibits gating pore currents underlying periodic paralysis. Proc Natl Acad Sci U S A. 2018;115:4495–500.

Mantegazza M, Curia G, Biagini G, Ragsdale DS, Avoli M. Voltage-gated sodium channels as therapeutic targets in epilepsy and other neurological disorders. Lancet Neurol. 2010;9:413–24.

Mantegazza M, Cestèle S, Catterall WA. Sodium channelopathies of skeletal muscle and brain. Physiol Rev. 2021;101:1633–89.
Masala N, et al. Targeting aberrant dendritic integration to treat cognitive comorbidities of epilepsy. bioRxiv 2020.11.23.393694. 2021.
Mason ER, Cummins TR. Differential inhibition of human Nav1.2 resurgent and persistent sodium currents by cannabidiol and GS967. Int J Mol Sci. 2020;21:2454.
Mattis J, et al. Corticohippocampal circuit dysfunction in a mouse model of Dravet syndrome. eLife. 2022;11:e69293.
McCormack K, et al. Voltage sensor interaction site for selective small molecule inhibitors of voltage-gated sodium channels. Proc Natl Acad Sci U S A. 2013;110:E2724–32.
Morgan K, et al. Beta3: an additional auxiliary subunit of the voltage-sensitive sodium channel that modulates channel gating with distinct kinetics. Proc Natl Acad Sci U S A. 2000;97:2308–13.
Myshkin MY, et al. Cell-free expression of sodium channel domains for pharmacology studies. Noncanonical spider toxin binding site in the second voltage-sensing domain of human Nav1.4 channel. Front Pharmacol. 2019;10:953.
Naylor CE, et al. Molecular basis of ion permeability in a voltage-gated sodium channel. EMBO J. 2016;35:820–30.
Noreng S, Li T, Payandeh J. Structural pharmacology of voltage-gated sodium channels. J Mol Biol. 2021;433:166967.
Ogiwara I, et al. Nav1.1 localizes to axons of parvalbumin-positive inhibitory interneurons: a circuit basis for epileptic seizures in mice carrying an Scn1a gene mutation. J Neurosci. 2007;27:5903–14.
Ohba C, et al. Early onset epileptic encephalopathy caused by de novo SCN8A mutations. Epilepsia. 2014;55:994–1000.
Ohmori I, Kahlig KM, Rhodes TH, Wang DW, George AL. Nonfunctional SCN1A is common in severe myoclonic epilepsy of infancy. Epilepsia. 2006;47:1636–42.
Osteen JD, et al. Selective spider toxins reveal a role for the Nav1.1 channel in mechanical pain. Nature. 2016;534:494–9.
Osteen JD, Sampson K, Iyer V, Julius D, Bosmans F. Pharmacology of the Nav1.1 domain IV voltage sensor reveals coupling between inactivation gating processes. Proc Natl Acad Sci U S A. 2017;114:6836–41.
Ottolini M, Barker BS, Gaykema RP, Meisler MH, Patel MK. Aberrant sodium channel currents and hyperexcitability of medial entorhinal cortex neurons in a mouse model of SCN8A encephalopathy. J Neurosci. 2017;37:7643–55.
Pan Y, Cummins TR. Distinct functional alterations in SCN8A epilepsy mutant channels. J Physiol. 2020;598:381–401.
Pan X, et al. Molecular basis for pore blockade of human Na^+ channel $Na_v1.2$ by the μ-conotoxin KIIIA. Science. 2019;363:1309–13.
Patel RR, Barbosa C, Brustovetsky T, Brustovetsky N, Cummins TR. Aberrant epilepsy-associated mutant Nav1.6 sodium channel activity can be targeted with cannabidiol. Brain. 2016;139:2164–81.
Payandeh J, Scheuer T, Zheng N, Catterall WA. The crystal structure of a voltage-gated sodium channel. Nature. 2011;475:353–9.
Potet F, Vanoye CG, George AL. Use-dependent block of human cardiac sodium channels by GS967. Mol Pharmacol. 2016;90:52–60.
Ragsdale DS, McPhee JC, Scheuer T, Catterall WA. Molecular determinants of state-dependent block of Na+ channels by local anesthetics. Science. 1994;265:1724–8.
Rash BG, et al. Gliogenesis in the outer subventricular zone promotes enlargement and gyrification of the primate cerebrum. Proc Natl Acad Sci. 2019;116:7089–94.
Rhodes TH, et al. Sodium channel dysfunction in intractable childhood epilepsy with generalized tonic-clonic seizures. J Physiol. 2005;569:433–45.
Richards KL, et al. Selective NaV1.1 activation rescues Dravet syndrome mice from seizures and premature death. Proc Natl Acad Sci U S A. 2018;115:E8077–85.

Rosker C, et al. The TTX metabolite 4,9-anhydro-TTX is a highly specific blocker of the Nav1.6 voltage-dependent sodium channel. Am J Physiol Cell Physiol. 2007;293:C783–9.

Rush AM, Dib-Hajj SD, Waxman SG. Electrophysiological properties of two axonal sodium channels, Nav1.2 and Nav1.6, expressed in mouse spinal sensory neurones. J Physiol. 2005;564:803–15.

Sadleir LG, et al. Not all SCN1A epileptic encephalopathies are Dravet syndrome. Neurology. 2017;89:1035–42.

Sait LG, et al. Cannabidiol interactions with voltage-gated sodium channels. eLife. 2020;9:e58593.

Scheffer IE, et al. Add-on cannabidiol in patients with Dravet syndrome: results of a long-term open-label extension trial. Epilepsia. 2021;62:2505–17.

Segal MM, Douglas AF. Late sodium channel openings underlying epileptiform activity are preferentially diminished by the anticonvulsant phenytoin. J Neurophysiol. 1997;77:3021–34.

Shen H, et al. Structure of a eukaryotic voltage-gated sodium channel at near-atomic resolution. Science. 2017;355:eaal4326.

Skluzacek JV, Watts KP, Parsy O, Wical B, Camfield P. Dravet syndrome and parent associations: the IDEA League experience with comorbid conditions, mortality, management, adaptation, and grief. Epilepsia. 2011;52(Suppl 2):95–101.

Smith RS, et al. Sodium channel SCN3A (NaV1.3) regulation of human cerebral cortical folding and oral motor development. Neuron. 2018;99:905–913.e7.

Spratt PWE, et al. The autism-associated gene Scn2a contributes to dendritic excitability and synaptic function in the prefrontal cortex. Neuron. 2019;103:673–685.e5.

Spratt PWE, et al. Paradoxical hyperexcitability from NaV1.2 sodium channel loss in neocortical pyramidal cells. Cell Rep. 2021;36:109483.

Stafstrom CE. Persistent sodium current and its role in epilepsy. Epilepsy Curr. 2007;7:15–22.

Sun GC, Werkman TR, Battefeld A, Clare JJ, Wadman WJ. Carbamazepine and topiramate modulation of transient and persistent sodium currents studied in HEK293 cells expressing the Nav1.3 α-subunit. Epilepsia. 2007;48:774–82.

Tai C, Abe Y, Westenbroek RE, Scheuer T, Catterall WA. Impaired excitability of somatostatin- and parvalbumin-expressing cortical interneurons in a mouse model of Dravet syndrome. Proc Natl Acad Sci U S A. 2014;111:E3139–48.

Talvenheimo JA, et al. Structure and functional reconstitution of the sodium channel from rat brain. Biophys J. 1984;45:37–40.

Talwar D, Hammer MF. SCN8A epilepsy, developmental encephalopathy, and related disorders. Pediatr Neurol. 2021;122:76–83.

Tanenhaus A, et al. Cell-selective adeno-associated virus-mediated SCN1A gene regulation therapy rescues mortality and seizure phenotypes in a Dravet syndrome mouse model and is well tolerated in nonhuman primates. Hum Gene Ther. 2022;33:579–97.

Tidball AM, et al. Variant-specific changes in persistent or resurgent sodium current in SCN8A-related epilepsy patient-derived neurons. Brain. 2020;143:3025–40.

Tsukamoto T, et al. Differential binding of tetrodotoxin and its derivatives to voltage-sensitive sodium channel subtypes (Nav1.1 to Nav1.7). Br J Pharmacol. 2017;174:3881–92.

Ulmschneider MB, et al. Molecular dynamics of ion transport through the open conformation of a bacterial voltage-gated sodium channel. Proc Natl Acad Sci U S A. 2013;110:6364–9.

Valassina N, et al. Scn1a gene reactivation after symptom onset rescues pathological phenotypes in a mouse model of Dravet syndrome. Nat Commun. 2022;13:161.

van der Peet PL, Sandanayake S, Jarrott B, Williams SJ. Discovery of N-aryloxypropylbenzylamines as voltage-gated sodium channel NaV1.2-subtype-selective inhibitors. ChemMedChem. 2019;14:570–82.

Vanoye CG, Gurnett CA, Holland KD, George AL, Kearney JA. Novel SCN3A variants associated with focal epilepsy in children. Neurobiol Dis. 2014;62:313–22.

Veeramah KR, et al. De novo pathogenic SCN8A mutation identified by whole-genome sequencing of a family quartet affected by infantile epileptic encephalopathy and SUDEP. Am J Hum Genet. 2012;90:502–10.

Wagnon JL, et al. Loss-of-function variants of SCN8A in intellectual disability without seizures. Neurol Genet. 2017;3:e170.
Wagnon JL, et al. Partial loss-of-function of sodium channel SCN8A in familial isolated myoclonus. Hum Mutat. 2018;39:965–9.
Wengert ER, Patel MK. The role of the persistent sodium current in epilepsy. Epilepsy Curr. 2021;21:40–7.
Wengert ER, Saga AU, Panchal PS, Barker BS, Patel MK. Prax330 reduces persistent and resurgent sodium channel currents and neuronal hyperexcitability of subiculum neurons in a mouse model of SCN8A epileptic encephalopathy. Neuropharmacology. 2019a;158:107699.
Wengert ER, et al. Biallelic inherited SCN8A variants, a rare cause of SCN8A-related developmental and epileptic encephalopathy. Epilepsia. 2019b;60:2277–85.
Wengert ER, et al. Somatostatin-positive interneurons contribute to seizures in SCN8A epileptic encephalopathy. J Neurosci. 2021;41:9257–73.
Wengert ER, et al. Targeted Augmentation of Nuclear Gene Output (TANGO) of Scn1a rescues parvalbumin interneuron excitability and reduces seizures in a mouse model of Dravet syndrome. Brain Res. 2022;1775:147743.
Weuring WJ, et al. NaV1.1 and NaV1.6 selective compounds reduce the behavior phenotype and epileptiform activity in a novel zebrafish model for Dravet syndrome. PLoS One. 2020;15:e0219106.
Whitaker WRJ, et al. Distribution of voltage-gated sodium channel α-subunit and β-subunit mRNAs in human hippocampal formation, cortex, and cerebellum. J Comp Neurol. 2000;422:123–39.
Wilson MJ, et al. μ-Conotoxins that differentially block sodium channels NaV1.1 through 1.8 identify those responsible for action potentials in sciatic nerve. Proc Natl Acad Sci. 2011;108:10302–7.
Wisedchaisri G, et al. Structural basis for high-affinity trapping of the NaV1.7 channel in its resting state by tarantula toxin. Mol Cell. 2021;81:38–48.e4.
Wolff M, et al. Genetic and phenotypic heterogeneity suggest therapeutic implications in SCN2A-related disorders. Brain. 2017;140:1316–36.
Wolff M, Brunklaus A, Zuberi SM. Phenotypic spectrum and genetics of SCN2A-related disorders, treatment options, and outcomes in epilepsy and beyond. Epilepsia. 2019;60:S59–67.
Wu YW, et al. Incidence of Dravet syndrome in a US population. Pediatrics. 2015;136:e1310-5.
Xu H, et al. Structural basis of Nav1.7 inhibition by a gating-modifier spider toxin. Cell. 2019;176:702–715.e14.
Yamagata T, et al. CRISPR/dCas9-based Scn1a gene activation in inhibitory neurons ameliorates epileptic and behavioral phenotypes of Dravet syndrome model mice. Neurobiol Dis. 2020;141:104954.
Yan Z, et al. Structure of the Nav1.4-β1 complex from electric eel. Cell. 2017;170:470–482.e11.
Yu FH, et al. Sodium channel beta4, a new disulfide-linked auxiliary subunit with similarity to beta2. J Neurosci. 2003;23:7577–85.
Zaman T, et al. Mutations in SCN3A cause early infantile epileptic encephalopathy. Ann Neurol. 2018;83:703–17.
Zaman T, Abou Tayoun A, Goldberg EM. A single-center SCN8A-related epilepsy cohort: clinical, genetic, and physiologic characterization. Ann Clin Transl Neurol. 2019;6:1445–55.
Zaman T, et al. SCN3A-related neurodevelopmental disorder: a spectrum of epilepsy and brain malformation. Ann Neurol. 2020;88:348–62.

Therapeutic Targeting of Potassium Channels

Nikita Gamper, Dongyang Huang, Jinlong Qi, Lilong Dong, and Hailin Zhang

Abstract Potassium (K^+) channels are the largest group of ion channels in mammals. The immediate function of these channels is electrogenic transport of K^+ ions across the cellular membranes, but the physiological processes that such K^+ conductance regulate in cells are very diverse, ranging from the establishment and maintenance of the resting membrane potential and control of excitability in neurons, muscle, and neuroendocrine cells to metabolic control, mechanotransduction, epithelial transport, and cell volume regulation. Mutations within K^+ channel genes or abnormal activity of K^+ channel proteins cause multiple severe disorders, including epilepsies, ataxias, deafness, cardiac arrhythmias, diabetes, and developmental and cognitive disorders. Consequently, K^+ channels are validated drug targets for many

N. Gamper (✉)
Faculty of Biological Sciences, University of Leeds, Leeds, UK

Department of Pharmacology, Hebei Medical University, Shijiazhuang, Hebei, China
e-mail: n.gamper@leeds.ac.uk

D. Huang
Institute of Chinese Integrative Medicine, Hebei Medical University, Shijiazhuang, Hebei, China

The Key Laboratory of Neural and Vascular Biology, Ministry of Education of China, Hebei Medical University, Shijiazhuang, Hebei, China

J. Qi · H. Zhang (✉)
Department of Pharmacology, Hebei Medical University, Shijiazhuang, Hebei, China

The Key Laboratory of Neural and Vascular Biology, Ministry of Education of China, Hebei Medical University, Shijiazhuang, Hebei, China

The Key Laboratory of New Drug Pharmacology and Toxicology, Hebei Medical University, Shijiazhuang, Hebei, China
e-mail: zhanghl@hebmu.edu.cn

L. Dong
Department of Pharmaceutical Analysis, School of Pharmaceutical Science, Hebei Medical University, Shijiazhuang, Hebei, China

G. Stephens, E. Stevens (eds.), *Ion Channels as Targets in Drug Discovery*,
https://doi.org/10.1007/978-3-031-52197-3_5

of these disorders. The aim of the present chapter is to provide an overview of the therapeutic targeting of K^+ channels, paying attention to new developments in the field.

Keywords Potassium channel · Voltage-gated potassium channel · Inwardly rectifying potassium channel · Two-pore domain potassium channel · Calcium-activated potassium channel · hERG · Therapeutic target

1 Introduction

Potassium (K^+) channels are a class of ion channels with characteristically high permeability to K^+ ions over all other physiologically relevant ions found in biological solutions (Hille 2001; Taura et al. 2021). K^+ channels are present in virtually all living organisms, and all types of mammalian cells express a complement of different K^+ channels; there are about 80 K^+-channel-encoding genes in humans (Gonzalez et al. 2012). Due to the highly conserved selectivity filter in their pore, K^+ channels allow electrogenic diffusion of K^+ across biological membranes, down an electrochemical gradient. A variety of gating mechanisms, developed throughout evolution, allow tight control over the precise circumstances at which such diffusion can occur. The functions of K^+ channels include the establishment and maintenance of resting membrane potential, control of excitability parameters in neurons and muscle cells, metabolic control, mechanotransduction, secretion, cell volume regulation, etc. It comes as no surprise that mutations within K^+ channel genes often cause severe disorders, including epilepsies, ataxias, deafness, cardiac arrhythmias, diabetes, and many, many others. Consequently, K^+ channels are validated drug targets for many of the above disorders, and some are also screening benchmarks for safety pharmacology (Gamper and Wang 2021). K^+ channels are commonly subdivided into four main families: inwardly rectifying (K_{ir}), voltage-gated (K_v), Ca^{2+}-activated (K_{Ca}), and two-pore domain (K_{2P}) K^+ channels (Fig. 1). K_{ir} channels were first reported by Bernard Katz in skeletal muscle fibers (Katz 1949); he observed an inward K^+ current with "anomalous" (as compared to outwardly rectifying K^+ currents present in the squid axon) inward rectification. This inward rectification arises from the asymmetric blockade of the K_{ir} channel pore by intracellular Mg^{2+} and polyamines (Matsuda 1988; Lopatin et al. 1994), a property that allows K_{ir} channels to conduct K^+ currents at highly negative membrane potentials and play a key role in the maintenance of the negative resting membrane potential in excitable cells (Hibino et al. 2010; Cui et al. 2021). There are 15 genes encoding K_{ir} channel subunits in mammals; these are divided into 7 subfamilies ($K_{ir}1$–$K_{ir}7$) (Alexander et al. 2021) (Fig. 1). K_{ir} channels can also be classified into four functional groups: (1) K^+ transport channels ($K_{ir}1.1$, $K_{ir}4.1$–$K_{ir}4.2$, $K_{ir}5.1$, and $K_{ir}7.1$); (2) "classical" K_{ir} channels ($K_{ir}2.1$–$K_{ir}2.4$); (3) G-protein-gated K_{ir} channels (GIRK; $K_{ir}3.1$–$K_{ir}3.4$ or GIRK1–4); and (4) ATP-sensitive K+ channels (K_{ATP}; $K_{ir}6.1$–$K_{ir}6.2$) (Hibino et al. 2010; Alexander et al. 2021; Cui et al. 2021). K_{ir} channel subunits have two

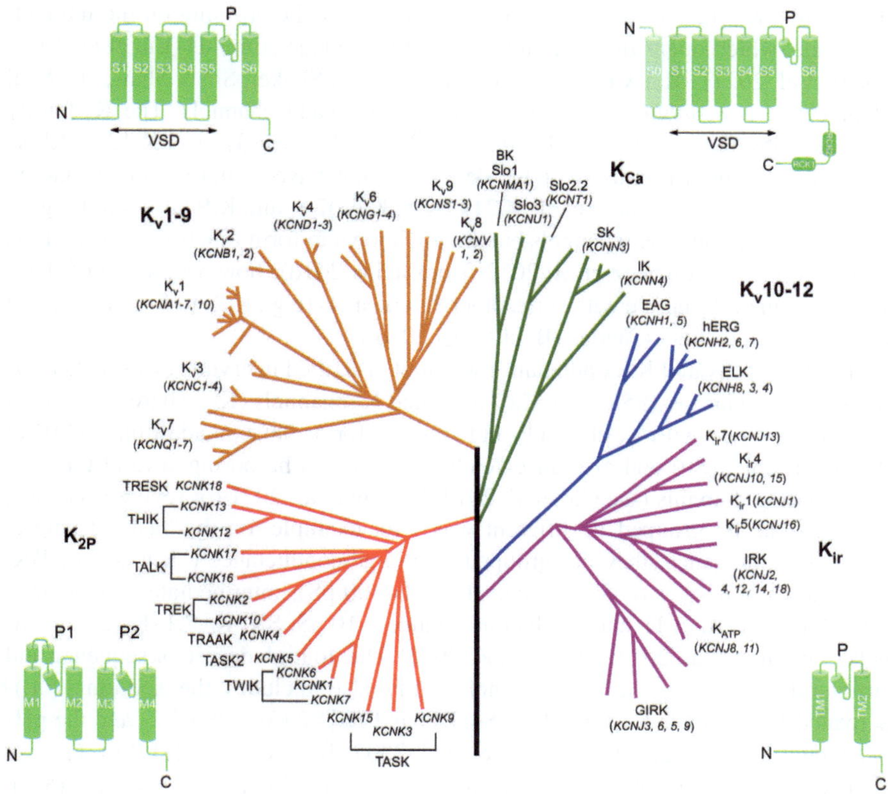

Fig. 1 Dendrogram of the potassium channel families. (Based on Taura et al. (2021))

transmembrane domains (TMDs), separated by an intracellular pore-forming P-loop that includes the K^+ selectivity filter; the cytoplasmic N- and C-terminal domains form an extended cytoplasmic pore structure (Nishida et al. 2007). The diversity of K_{ir} channels is further increased by heterodimerization of members of different subfamilies and by alternative splicing (Hibino et al. 2010; Gonzalez et al. 2012; Cui et al. 2021).

The largest mammalian K^+ channel family is the family of the voltage-gated K^+ channels (K_v). These channels are extensively studied and modeled as they underlie the repolarizing phase of an action potential. Their existence was first theorized in the mid-twentieth century by Hodgkin and Huxley in their classical work on squid axon biophysics (Hodgkin and Huxley 1952), with the first K_v channel cloned in the 1980s from the *Drosophila* genome by Lily Jan's group (Papazian et al. 1987). Currently, K_v channels are divided into 12 families encompassing 40 genes encoding pore-forming (α) subunits (Gonzalez et al. 2012; Taura et al. 2021) (Fig. 1). Voltage-gated K^+ channels share a six-TMD topology (S1–S6) with a characteristic voltage-sensor domain (VSD; S1–S4) and a pore domain formed by S5 and S6; a functional K_v channel assembles as a tetramer formed by the same or different α-subunits. K_v channels also contain a gate at the bundle crossing on the intracellular side, which opens the pore in response to a voltage stimulus (Liu et al. 2015; Taura et al. 2021) . K_v1.x to K_v4.x (corresponding to Shaker, Shab, Shaw, and Shal channels in *Drosophila*) form "classical" voltage-gated channels. The K_v family also contains K_v7.x (KCNQ), K_v10.x (EAG), K_v11.x (ERG), and K_v12.x (ELK) subfamilies, which are also voltage-dependent but have a more complex gating (Gonzalez et al. 2012; Taura et al. 2021) . K_v5, K_v6, K_v8, and K_v9 have similar general structure to other K_v channels but are not known to form functional channels as homotetramers (Gonzalez et al. 2012; Bocksteins 2016); however, some of these subunits can multimerize with some other K_v channels (e.g., K_v2) to form functional channels with distinct kinetics (Bocksteins 2016).

The Ca^{2+}-activated K^+ conductance was first described in erythrocytes by Gardos in the 1950s (Gardos 1958). The Ca^{2+}-activated K^+ channels (K_{Ca}) share the general 6-TMD topology with K_v channels, but some isoforms have an additional TMD at the N-terminus (S0) and also an extended C-terminus harboring several types of Ca^{2+}-binding domains (see Cui et al. 2014, Dudem et al. 2021 for review). The K_{Ca} channels can be grouped in different ways; for example, on the basis of single-channel conductance, they are grouped into large-conductance ("Big K^+" or BK, also known as K_{Ca}1.1 or Slo1) channels (250–300 pS); intermediate conductance (30–80 pS; IK; K_{Ca}3.1); and small conductance (~10 pS; SK; K_{Ca}2.1–K_{Ca}2.3) channels (Adelman et al. 2012; Dudem et al. 2021). Due to high degree of sequence and structural homology, the K_{Ca} channel family also includes the sodium (Na^+)-activated K^+ channels K_{Ca}4.1 (Slack/Slo2.1) and K_{Ca}4.2 (Slick/Slo2.2) and the pH-dependent K_{Ca}5.1 channels (Slo3) (Wei et al. 2005; Taura et al. 2021) (Fig. 1).

The fourth family of K^+ channels are the two-pore domain K^+ channels (K_{2P}), also sometimes referred to as "leak" K^+ channels. The first K_{2P} channel, TOK1, was cloned from yeast in 1995 by the Goldstein laboratory (Ketchum et al. 1995) and soon thereafter 15 related K_{2P} channels were identified in human genome (Kamuene et al. 2021). K_{2P} channels contribute to a background or "leak" K^+ conductance in many types of neurons, including sensory and motoneurons (Talley et al. 2000, 2001; Kang and Kim 2006; Marsh et al. 2012; Acosta et al. 2014), cerebellar granule neurons (Plant et al. 2002), and many other neurons, as well as in the non-neuronal tissues, such as the kidney (Reyes et al. 1998). The neuronal background K^+ conductance was first observed in the 1940s and is included in the Hodgkin and Huxley formalism describing neuronal excitability (Hodgkin and Huxley 1952). Due to weak or absent voltage dependence of K_{2P} channels, they are fractionally open at membrane potentials, positive to E_K, hence contributing to maintenance of negative resting membrane potentials in neurons, as well as to setting action potential firing threshold and other excitability parameters. In humans, the K_{2P} family is composed of 15 channel subunits (K_{2P}1–15) and encoded by *KCNK1–18* genes (with *KCNK8*, *KCNK11*, and *KCNK14* not expressing functionally) (Taura et al. 2021). These K^+ channels are historically divided into six subfamilies: the TREK subfamily (TREK1/K_{2P}2/KCNK2; TREK2/K_{2P}10/KCNK10; TRAAK/K_{2P}4/

KCNK4), the TALK subfamily (TALK1/K_{2P}16/KCNK16; TALK2/K_{2P}17/KCNK17); the TWIK subfamily (TWIK1/K_{2P}1/KCNK1; TWIK2/K_{2P}6/KCNK6; K_{2P}7/KCNK7), the TASK subfamily (TASK1/K_{2P}3/KCNK3; TASK3/K_{2P}9/KCNK9; TASK5/K_{2P}15/KCNK15), the THIK subfamily (THIK1/K_{2P}12/KCNK12; THIK2/K_{2P}13/KCNK13), and the TRESK channel (K_{2P}18/KCNK18) (Fig. 1).

All four K^+ channel families of are highly selective for K^+ over other physiologically relevant ions, particularly Na^+, which is conferred by the selectivity filter within the K^+ channel pore (Hille 1986). The selectivity filter is comprised of a highly conserved sequence, containing the T-I/V-G-Y/F-G motif (Heginbotham et al. 1994; Doyle et al. 1998; Zhou et al. 2001a; Varma et al. 2011). At the selectivity filter, the pore diameter, geometry, and the polarity profile are such that the pore environment mimics the water dipoles surrounding K^+ ions in an aqueous solution. Hence, when entering the selectivity filter, K^+ ions lose the surrounding water molecules and pass though the filter along the electrochemical gradient, while passing of smaller Na^+ ion is unfavorable since it remains bound to water dipoles (Aqvist and Luzhkov 2000; Berneche and Roux 2001; Shrivastava et al. 2002; Noskov et al. 2004).

Given that nearly 20% of validated human protein drug targets are ion channels (Santos et al. 2017), with K^+ channels being the largest ion channel family in humans, the scope for therapeutic targeting of K^+ channels for treatment of human diseases is vast. The pharmacology and therapeutic targeting of the individual K^+ channel families has been extensively reviewed, including in the recent volume the *Handbook of Experimental Pharmacology* (Gamper and Wang 2021); hence, for the in-depth coverage of these individual K^+ channel families and subfamilies, the reader is advised to refer to these recent articles (e.g., Cui et al. 2021; Dudem et al. 2021; Jones et al. 2021; Kamuene et al. 2021; Liu et al. 2021; Li et al. 2021a; Su et al. 2021). The aim of the present chapter is to provide a concise overview of the therapeutic targeting of K^+ channels, paying attention to the most important topics and recent developments in the field. In the interest of this focus, we do not specifically consider therapeutic potential of targeting auxiliary subunits and other K^+ channel interacting partners; interested reader can refer to an excellent recent review on this topic (Abbott 2021).

2 Inwardly Rectifying K^+ Channels as Drug Targets

K_{ir} channels are important K^+ transport proteins in both excitable and non-excitable cells, such as cardiac myocytes, neurons, blood cells, osteoclasts, endothelial cells, glial cells, epithelial cells, and oocytes (Cui et al. 2021); accordingly, K_{ir} channels likely represent valuable new drug targets for cardiovascular, metabolic, immune, and neurological diseases. Functional classification of K_{ir} channels is shown in Fig. 2a.

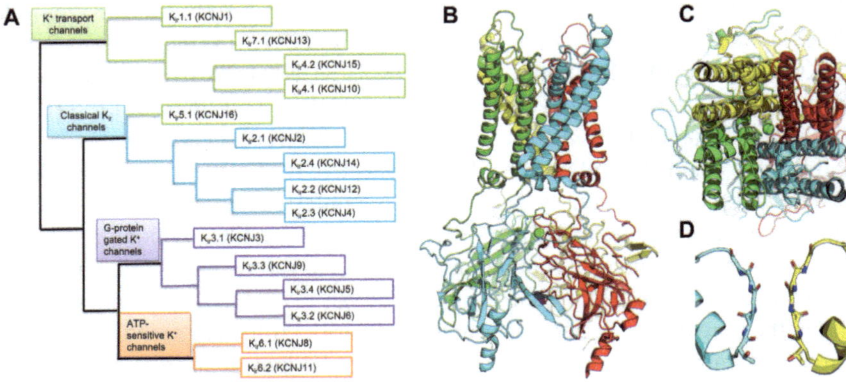

Fig. 2 Inwardly rectifying K^+ channels. (**a**) Phylogenetic and functional classification of human K_{ir} channels. (**b**) Side view of a tetrameric inward inward-rectifier potassium channel 2 (K_{ir}2.1, PDB ID: 7ZDZ), colored by subunit in four colors. (**c**) The view of the K_{ir}2.1 from the extracellular side (top view). (**d**) A side view of selectivity filter of K_{ir}2.1. In (**b**), (**c**), potassium and strontium ions are shown as green and purple balls, respectively

2.1 K^+ Transport Channels

K_{ir}1.1: K_{ir}1.1, also known as KCNJ1 and, formerly, ROMK, is primarily expressed in non-excitable epithelial cells of the thick ascending limb (TAL) of the loop of Henle, the distal convoluted tubule (DCT), as well as the cortical collecting duct of the kidney (Lee and Hebert 1995; Fakler et al. 1996; Hebert et al. 2005). K_{ir}1.1 dysfunction causes Bartter's syndrome, which is characterized by hypokalemic metabolic alkalosis, renal salt wasting, hyperreninemia, and hyperaldosteronism. This is consistent with the channel's critical role in salt absorption in TAL (Weaver and Denton 2021). K_{ir}1.1 has been investigated as a novel mechanism for diuretic drugs to manage hypertension and edema (Weaver and Denton 2021). The channel is inhibited by the nonspecific K_{ir} channel blockers such as Ba^{2+} and Cs^+, as well as by some more specific drugs, such as β-dendrotoxin, tertiapin, and some others (Table 1).

K_{ir}4.1 is expressed in the CNS glia, cochlea, gastric parietal cells, and kidney (Takumi et al. 1995). In the brain, K_{ir}4.1 is involved in K^+ buffering by glial cells, which is pivotal for normal neuronal functions; deficiencies in such K^+ buffering were linked to major depression (Cui et al. 2018). Additionally, mutations in the K_{ir}4.1 gene (*KCNJ10*) are associated with autism and epilepsy spectrum disorder (AED) (Cucchiara et al. 2020). In gastric parietal cells, K_{ir}4.1 colocalizes with and modulates the function of H^+-K^+-ATPase at the apical membrane (Fujita et al. 2002; Kaufhold et al. 2008). K_{ir}4.1 is the only K_{ir} subunit expressed in the stria vascularis of cochlear duct, where it regulates K^+ transport, similar to its role in renal epithelial cells (Hibino et al. 1997). Deletion of K_{ir}4.1 results in a complete loss of membrane polarization of endocochlear cells and a 50% reduction of [K^+] in the endolymph, eventually leading to hearing loss in mice (Hibino et al. 1997). Some

Table 1 Modulators of K_{ir} channels of K^+ transport channel group

Channel	Compound	IC_{50}/EC_{50}	Class	Cell type/ species	Ref.
$K_{ir}1.1$, $K_{ir}7.1$	VU590	$K_{ir}1.1$ $IC_{50} = 0.2\mu M$, $K_{ir}7.1$ $IC_{50} = 8\mu M$	Inhibitor	C1 HEK293	Lewis et al. (2009)
$K_{ir}7.1$	VU714	$IC_{50} = 1.5\mu M$	Inhibitor	HEK293	Swale et al. (2016)
$K_{ir}7.1$	ML418	$IC_{50} = 0.31\mu M$	Inhibitor	HEK293	Swale et al. (2016)
$K_{ir}1.1$	VU591	IC_{50} = 240–300 nM	Inhibitor	HEK293, rat	Bhave et al. (2011)
$K_{ir}1.1$	BNBI	$IC_{50} = 8\mu M$	Inhibitor	C1 HEK293	Bhave et al. (2011)
$K_{ir}1.1$	Tertiapin-Q	$K_{ir}1.1$ Ki = 1.3 nM	Inhibitor	Xenopus oocytes	Jin and Lu (1999)
$K_{ir}3.1$, $K_{ir}3.2$, $K_{ir}3.4$	*n*-Alcohols	$EC_{50} > 100\mu M$	Activator	Xenopus oocytes	Kobayashi et al. (1999); Lewohl et al. (1999)
$K_{ir}3$, $K_{ir}4.1$	Fluoxetine	$K_{ir}3$ $IC_{50} = 7–13\mu M$, $K_{ir}4.1$ $IC_{50} = 15.2\mu M$	Inhibitor	Xenopus oocytes	Kobayashi et al. (2003); Ohno et al. (2007)
$K_{ir}1.1$, $K_{ir}3.1/3.2$, $K_{ir}7.1$	VU573	$K_{ir}1.1$ $IC_{50} = 19\mu M$, $K_{ir}3.1/3.2$ $IC_{50} = 1.3\mu M$, $K_{ir}7.1$ $IC_{50} = 0.9\mu M$	Inhibitor	Xenopus oocytes	Raphemot et al. (2011)
$K_{ir}1.1$, $K_{ir}3.1/3.4$	Tertiapin	$K_{ir}3.1/3.4$ IC_{50} = 8 nM, $K_{ir}1.1$ IC_{50} = 2 nM, $K_{ir}3.1/3.4$ IC_{50} = 8 nM (rabbit cardiac myocytes)	Inhibitor	Xenopus oocytes	Jin and Lu (1998); Kitamura et al. (2000)
$K_{ir}3$, $K_{ir}4.1$	Tricyclic antidepressants	$K_{ir}3$ $IC_{50} = 18–71\mu M$, $K_{ir}4.1$ $IC_{50} = 16–38\mu M$	Inhibitor	Xenopus oocytes, HEK293	Kobayashi et al. (2004); Su et al. (2007)
$K_{ir}4.1$, $K_{ir}5.1$	VU0134992	$K_{ir}4.1$ IC50 = $0.9\mu M$, $K_{ir}4.1/K_{ir}5.1$ $IC_{50} = 9\mu M$	Inhibitor	HEK293T, mouse	Kharade et al. (2018)

antidepressants, such as tricyclics and fluoxetine, can inhibit $K_{ir}4.1$ with low-micromolar potency (Table 1).

$K_{ir}4.2$ (formerly $K_{ir}1.3$; *KCNJ15*) was first cloned from a human kidney cDNA library (Shuck et al. 1997). $K_{ir}4.2$ expresses in the kidney, liver, microvascular endothelial cells, and embryonic fibrocytes (Shuck et al. 1997; Pearson et al. 1999; deHart et al. 2008; Lachheb et al. 2008). $K_{ir}4.2$ can form functional homomeric channels or functional heteromeric channels with $K_{ir}5.1$, but with different single-channel conductance and whole-cell current (Pearson et al. 1999; Pessia et al. 2001).

$K_{ir}5.1$ channels are expressed in the kidney, liver, spleen adrenal gland, and several brain regions, but it appears to be nonfunctional as a homomeric channel (Cui et al. 2021). However, when $K_{ir}5.1$ heterodimerizes with $K_{ir}4.1$, a functional channel is formed with biophysical properties that differ from the $K_{ir}4.1$ homomer (Bond et al. 1994; Pessia et al. 1996; Tanemoto et al. 2000; Tucker et al. 2000). Heterotetrameric $K_{ir}4.1/5.1$ channels have been detected in various tissues, including the kidney and the brain (Hibino et al. 2004; Wang and Lin 2022). In the renal distal nephron's basolateral membrane, $K_{ir}4.1/5.1$ and/or $K_{ir}4.2/5.1$ channels are key players in the reabsorption of Na^+ in conjunction with the function of epithelial Na^+ channels and basolateral Na^+-K^+-ATPase (Wang and Lin 2022).

$K_{ir}7.1$ expresses in various regions of the retina, including the retinal pigmental epithelia and the internal limiting membrane, as well as in some other epithelial cells (Kusaka et al. 2001; Hejtmancik et al. 2008; Cui et al. 2021). Malfunction of $K_{ir}7.1$ may cause the degeneration of cells in these tissues, which is related to the etiology of snowflake vitreoretinal degeneration (SVD). Compound ML418 is reported to be a selective blocker for $K_{ir}7.1$ (Table 1).

2.2 *Classical K_{ir} Channels ($K_{ir}2.x$)*

The first member of this family was cloned from a mouse macrophage cell line and named IRK1 ($K_{ir}2.1$, *KCNJ2*) (Kubo et al. 1993). Soon after, three additional subunits were identified: $K_{ir}2.2$ (IRK2, *KCNJ12*) (Koyama et al. 1994; Takahashi et al. 1994), $K_{ir}2.3$ (IRK3, BIR11, *KCNJ4*) (Bond et al. 1994; Morishige et al. 1994; Bredt et al. 1995), and $K_{ir}2.4$ (IRK4, *KCNJ14*) (Töpert et al. 1998). $K_{ir}2$ families are constitutively active and exhibit strong inward rectification. These channels contribute to the establishment of a highly negative resting membrane potential (E_{rest}) and long-lasting action potential plateau in the excitable cells. An example of classical K_{ir} channel ($K_{ir}2.1$) structure is given in Fig. 2b–d.

$K_{ir}2.x$ ($K_{ir}2.1$, $K_{ir}2.2$, $K_{ir}2.3$) channels are abundant in cardiac myocytes, including Purkinje fibers and ventricular and atrial tissues, and they play a critical role in determining the shape of the cardiac action potential (Schimpf et al. 2005; Schreiber and Seebohm 2021). They set the E_{rest} and contribute to the rapid final stages of cardiac action potential repolarization (see hERG channel section and Fig. 4 therein). The reduction of $K_{ir}2.1$ activity prolongs the plateau phase of the action potential and depolarizes E_{rest}, causing the symptoms of Andersen's syndrome in the heart. On the other hand, evidence shows that a gain-of-function (GOF) mutations in the $K_{ir}2.1$ gene *KCNJ2* produces an abrupt increase in the rate of final repolarization of the action potential and shortens the duration of ventricular depolarization, thus shortening the Q-T interval (Priori et al. 2005); this process can result in short Q-T syndrome associated with sudden cardiac death, syncope, and/or atrial fibrillation (Priori et al. 2005; Schimpf et al. 2005; Schreiber and Seebohm 2021).

$K_{ir}2.x$ channels are also expressed in the vasculature, including endothelial and smooth muscle cells (Nilius et al. 1993; von Beckerath et al. 1996; Nilius and Droogmans 2001). In vascular endothelial cells, they are considered to be the most prominent K^+ channels (Nilius et al. 1993; von Beckerath et al. 1996; Nilius and Droogmans 2001). Functional expression of $K_{ir}2.1$ sets the E_{rest} of endothelial cells to negative potentials, providing the driving force for Ca^{2+} influx through Ca^{2+}-permeable channels (Yamamura et al. 2016).

Classical K_{ir} channels are functionally expressed in the nervous systems, including hippocampal neurons (Brown et al. 1990) and spinal motor neurons (Takahashi 1990). Currents generated by $K_{ir}2.x$ subunits are critically involved in the maintenance of E_{rest} and regulation of the excitability of these neurons. Classical K_{ir} currents were recorded from the Schwann cells myelinating peripheral nerve fibers (Wilson and Chiu 1990). Modulation of $K_{ir}2$ channels by acetylcholine muscarinic M_1 receptor signaling is suggested to be involved in neuronal phenotypes of Parkinson's disease (Shen et al. 2007).

Various modulators (activators and inhibitors) of classical K_{ir} channels have been reported (Table 2), but these are usually not very selective. In the expression systems, the homomeric $K_{ir}2.x$ channels differ substantially in their sensitivity to Ba^{2+} and Cs^+ (Table 2). Recent pharmacological and electrophysiological studies have identified several inhibitors of $K_{ir}2.x$ channels, but these are not very specific. Thus, first-generation histamine H_1 receptor blockers, mepyramine and diphenhydramine, can inhibit $K_{ir}2.3$ current by approximately 25% and 17%, respectively (Liu et al. 2007). $K_{ir}2.3$ current is also partially reduced by genistein, an inhibitor of protein tyrosine kinases, by approximately 50% (Zhao et al. 2008). These two types of reagents have little effect on $K_{ir}2.1$ currents (Liu et al. 2007; Zhao et al. 2008) (Table 2). The antimalarial drug, chloroquine, inhibits $K_{ir}2.1$ channels, which may contribute to the side effects of the drug, namely, lethal ventricular arrhythmias (Benavides-Haro and Sánchez-Chapula 2000; Sánchez-Chapula et al. 2001). Quinidine, a class Ia antiarrhythmic, is also known to block cardiac $K_{ir}2.x$ currents (Hiraoka et al. 1986; Nenov et al. 1998), which may also contribute to the clinical effects of the drug.

2.3 G-Protein-Gated K_{ir} Channels

$K_{ir}3.x$ channels ($K_{ir}3.1$–$K_{ir}3.4$), also known as G-protein-gated, inwardly rectifying K^+ channels (GIRKs), are primarily regulated through interactions with $G_{i/o}$-coupled G-protein-coupled receptors (GPCRs). Activation of $G_{i/o}$-coupled GPCRs results in an increase in channel activity, primarily via Gβγ dimer binding (reviewed in Cui et al. (2021)). GIRK/$K_{ir}3$ channels are common effectors for $G_{i/o}$-dependent signaling pathways in the heart and nervous system (Cui et al. 2021; Luo et al. 2022; Campos-Rios et al. 2022). $K_{ir}3$ channels expressed in the CNS are critical effectors

Table 2 Modulators of classical K_{ir} channels

Channel	Compound	IC_{50}/EC_{50}	Class	Cell type/ species	Ref.
$K_{ir}2.1$	Flecainide	EC_{50} = 1μM	Activator	Human atrial myocytes, guinea pig cardiomyocytes	Caballero et al. (2010)
$K_{ir}2.3$	Mepyramine	IC_{50} = 306.4μM	Inhibitor	Xenopus oocytes	Liu et al. (2007)
$K_{ir}2.3$	Diphenhydramine	IC_{50} = 689.2μM	Inhibitor	Xenopus oocytes	Liu et al. (2007)
$K_{ir}2.3$	Genistein	IC_{50} = 14 ~ 22μM	Inhibitor	Xenopus oocytes, HEK293	Zhao et al. (2008)
$K_{ir}2.3$	Tenidap	EC_{50} = 1.3μM	Activator	CHO	Liu et al. (2002a)
$K_{ir}2.3$	Verapamil	IC_{50} = 58.1μM	Inhibitor	Xenopus oocytes	Xynogalos et al. (2023)
$K_{ir}2.1$, $K_{ir}2.3$	Terfenadine	$K_{ir}2.3$ IC50 = 1.06 ± 0.11μM $K_{ir}2.1$ IC50 = 27.8 ± 4.8μM	Inhibitor	HEK293	Delgado-Ramírez et al. (2021)
$K_{ir}2.1$, $K_{ir}2.2$, $K_{ir}2.4$	Ba^{2+}	$K_{ir}2.1$ IC_{50} = 8μM IC_{50} = 6μM $K_{ir}2.2$ IC_{50} = 10.3μM $K_{ir}2.4$ IC_{50} = 390μM	Inhibitor	Xenopus oocytes, various mammalian cells	Töpert et al. (1998); Preisig-Müller et al. (2002)
$K_{ir}2.1$, $K_{ir}2.4$	Cs^{+}	$K_{ir}2.1$ IC_{50} = 420μM $K_{ir}2.4$ IC_{50} = 8 mM	Inhibitor	Xenopus oocytes, mammalian cells	Töpert et al. (1998); Preisig-Müller et al. (2002)
$K_{ir}2.1$	Chloroquine	≈9μM	Inhibitor	HEK293, feline ventricular myocytes	Sánchez-Chapula et al. (2001); Rodríguez-Menchaca et al. (2008)
$K_{ir}2.x$	Chlorpromazine	~9μM	Inhibitor	Cat ventricular myocytes	Sánchez-Chapula et al. (2001)
$K_{ir}2.x$	Quinidine	IC_{50}, ~57μM	Inhibitor	Various mammalian cells	Fremstad et al. (1979); Yang et al. (2009); Koepple et al. (2017);

(continued)

Table 2 (continued)

Channel	Compound	IC_{50}/EC_{50}	Class	Cell type/ species	Ref.
$K_{ir}2.x$	PA-6	IC_{50}, ~50–200 nM	Inhibitor	Human, mouse, canine ventricular cardiomyocytes	Takanari et al. (2013)
$K_{ir}2.x$	ML133	1.8μM	Inhibitor	HEK293	Wang et al. (2011)

of numerous signal transduction pathways underlying analgesia, addiction, and respiratory-depressive effects of opioids. Accordingly, dysregulation of GIRK signaling has been linked to various neurological and cardiovascular human disorders (Cui et al. 2021; Campos-Rios et al. 2022; Luo et al. 2022).

$K_{ir}3.1$ (GIRK1, *KCNJ3*) cDNA was initially isolated from a rat heart cDNA library, and it was the first K_{ir} subunit demonstrated to contribute to the formation of functional GIRK channels in native cells and tissues (Dascal et al. 1993). $K_{ir}3.1$ generally heteromerizes with other $K_{ir}3.x$ subunits to form GIRK channels in native cells and tissues (Cui et al. 2021).

There are at least four different variants ($K_{ir}3.2a$–d) of $K_{ir}3.2$ (GIRK2, *KCNJ6*) subunit. $K_{ir}3.2$ variants exhibit differential expression patterns in various tissues, such as the brain, pituitary, pancreas, and testis (Kobayashi et al. 1995; Karschin et al. 1996; Liao et al. 1996; Inanobe et al. 1999). $K_{ir}3.3$ was cloned from mouse brain cDNA library (Lesage et al. 1994). $K_{ir}3.4$ (GIRK4 or CIR, *KCNJ5*) was identified in bovine atrial membranes by immunoprecipitation with $K_{ir}3.1$ (Krapivinsky et al. 1995).

$K_{ir}3.1$ and $K_{ir}3.4$ subunits are the main contributors of the cardiac $I_{K\text{-}Ach}$ current, the K^+ current activated by acetylcholine released by the vagus nerve. Activation of $I_{K\text{-}Ach}$ slows down the heart rate, which is an important physiological mechanism for heart rate adaptability (Mesirca et al. 2013; Cui et al. 2021). Dysfunction of $K_{ir}3.4$ does not significantly affect the resting heart rate but disrupts the balance between sympathetic and parasympathetic activity, leading to a decrease in heart rate in response to parasympathetic stimulation (Mesirca et al. 2013; Cui et al. 2021). Vagal deprivation (or $K_{ir}3.4$ knockdown) prevents induction of atrial fibrillation (AF) (Kovoor et al. 2001; Cui et al. 2021) , one of the most common types of cardiac arrhythmia. Hence, $I_{K\text{-}ACh}$ may represent a therapeutic target for preventing and treating AF.

$K_{ir}3.1/3.2$ (Liao et al. 1996; Inanobe et al. 1999), $K_{ir}3.2/3.3$ (Inanobe et al. 1999; Jelacic et al. 2000; Cruz et al. 2004), and $K_{ir}3.2/3.4$ (Lesage et al. 1995) heteromers have been shown to be expressed in various brain regions. Numerous GPCRs can activate $K_{ir}3$ channels at both postsynaptic and presynaptic sites (Misgeld et al. 1989; Thompson et al. 1992; Thompson and Gähwiler 1992; Yi et al. 2001; Kurachi and Ishii 2004). Postsynaptic $K_{ir}3$ channels generate slow inhibitory postsynaptic potentials (sIPSP) and suppress neuronal excitability. Several studies suggest that $K_{ir}3.2$ is a major subunit involved in the formation of IPSPs in hippocampal and cerebellar neurons (Luscher et al. 1997; Signorini et al. 1997). Loss of $K_{ir}3.2$ can lead to the development of spontaneous seizures (Signorini et al. 1997). In the locus coeruleus, $K_{ir}3.1$, $K_{ir}3.2$, and $K_{ir}3.3$ are involved in the hyperpolarization of neurons induced by opioid agonists such as met-enkephalin (Torrecilla et al. 2002; Matsui and Williams 2010). $K_{ir}3$ currents activated in hippocampal CA1 neurons by $GABA_B$ receptor agonists mediate postsynaptic inhibitory effects of GABA (Koyrakh et al. 2005). $K_{ir}3.4$-deficient mice showed impaired performance in spatial learning and memory tasks (Wickman et al. 2000). Studies have demonstrated that neuronal $K_{ir}3$ channels are involved in drug abuse and addiction due to the release of dopamine in response to activation of the mesocorticolimbic system (Robbins and Everitt 1999). A heteromeric complex of $K_{ir}3.1/K_{ir}3.4$ has been identified in the secretory thyrotropin-containing vesicles in the rat pituitary gland, where it is involved in an effective negative-feedback loop that controls blood plasma hormone concentrations (Morishige et al. 1999).

Some most common modulators of GIRK channels are aggregated in Table 3. Tertiapin is a potent inhibitor of $K_{ir}3.1/3.4$ and $K_{ir}1.1$ channels with nanomolar potency (Jin and Lu 1998; Kitamura et al. 2000). n-Alcohols such as methanol, ethanol, and 1-propanol can activate GIRK channels made up by a different combinations of $K_{ir}3.1$, $K_{ir}3.2$, and $K_{ir}3.4$ subunits (Kobayashi et al. 1999; Lewohl et al. 1999). Channels containing $K_{ir}3.2$ show the greatest enhancement by ethanol. $I_{K\text{-}ACh}$/$K_{ir}3.1/K_{ir}3.4$ channels are blocked by quinidine and quinine with IC_{50} in the range of ~10μM (Kurachi et al. 1987; Nakajima et al. 1989; Katayama et al. 1997), while verapamil inhibits these channels with IC_{50} of ~1μM (Ito et al. 1989). On the other hand, GIRKs are insensitive to many other general K^+ channel blockers (Lacey et al. 1987, 1988; Inoue et al. 1988; Nakajima et al. 1989; Katayama et al. 1997). $K_{ir}3.x$ subunits can be modulated by many other compounds, including (i) antipsychotic drugs such as haloperidol, thioridazine, pimodine, and clozapine (Kobayashi et al. 2000); (ii) antidepressants (Kobayashi et al. 2003, 2004); (iii) volatile anesthetics (Weigl and Schreibmayer 2001; Yamakura et al. 2001; Milovic et al. 2004); and (iv) local anesthetics, such as bupivacaine (Zhou et al. 2001b). Additionally, the selective dopamine D_1 receptor antagonist R-(+)-7-chloro-8-hydroxy-3-methyl-1-phenyl-2,3,4,5-tetrahydro-1H-3-benzazepine hydrochloride (Kuzhikandathil and Oxford 2002), the NMDA receptor antagonists MK-801 (Kofuji et al. 1996) and ifenprodil (Kobayashi et al. 2006), the inhaled drug of abuse toluene (Del Re et al. 2006), the Ca^{2+} channel blocker verapamil, and the classical cation channel blocker QX-314 (Kofuji et al. 1996) also inhibit $K_{ir}3$ channels. Further details are provided in Table 3.

Table 3 Modulators of G-protein-gated K^+ channels

Channel	Compound	IC_{50}/EC_{50}	Class	Cell type/ species	Ref.
$K_{ir}3.x$	Quinidine and quinine	IC_{50} ~ 10μM	Inhibitor	Guinea pig atrial cells	Kurachi et al. (1987)
$K_{ir}3.x$	Naringin	EC_{50} > 100μM	Activator	Xenopus oocytes	Yow et al. (2011)
$K_{ir}3.x$	SCH23390	IC_{50} = 0.3–8μM	Inhibitor	AtT-20 cells	Kuzhikandathil and Oxford (2002)
$K_{ir}3.1/3.2$	ML297 (VU0456810)	$K_{ir}3.1/3.2$ EC_{50} = 160 nM, $K_{ir}3.1/3.4$ EC_{50} = 1.8μM	Activator	HEK293 cells, mice	Kaufmann et al. (2013)
$K_{ir}3.1/3.2$, $K_{ir}3.1/3.4$	VU0032230	$K_{ir}3.1/3.2$ EC_{50} = 1.4μM, $K_{ir}3.1/3.4$ EC_{50} = 0.69μM	Activator	HEK293	Ramos-Hunter et al. (2013)
$K_{ir}3.1/3.2$, $K_{ir}3.1/3.4$	VU0259369	$K_{ir}3.1/3.2$ EC_{50} = 1.1μM, $K_{ir}3.1/3.4$ EC_{50} = 1.1μM	Activator	HEK293	Ramos-Hunter et al. (2013)
$K_{ir}3.2$, $K_{ir}3.1/3.2$, $K_{ir}3.1/3.4$	VU0529331	EC_{50} = 5μM	Activator	HEK293	Kozek et al. (2019)
$K_{ir}3.1/3.2$	VU0466551	$K_{ir}3.1/3.2$ EC_{50} = 49 nM	Activator	HEK293	Kozek et al. (2019)
$K_{ir}3.x$	Thioridazine	IC_{50} = 50μM	Inhibitor	Xenopus oocyte	Kobayashi et al. (2000)
$K_{ir}3.x$	Bupivacaine	IC_{50} = 22–170μM Gβγ and voltage dependent	Inhibitor	Xenopus oocyte	Zhou et al. (2001b)
$K_{ir}3.x$	Ethosuximide	IC_{50} = 0.4–3.5 mM (inhibition rate 61–76%)	Inhibitor	Xenopus oocytes	Kobayashi et al. (2009)
$K_{ir}3.x$, $K_{ir}4.1$	Fluoxetine	$K_{ir}3$ IC_{50} = 7–13μM, $K_{ir}4.1$ IC50 = 15.2μM	Inhibitor	Xenopus oocytes	Kobayashi et al. (2003); Ohno et al. (2007)
$K_{ir}3.x$	Halothane	IC_{50} = 60μM	Inhibitor	Xenopus oocytes	Weigl and Schreibmayer (2001)
$K_{ir}2.3$, $K_{ir}3.4$	Mepyramine	$K_{ir}2.3$ IC_{50} = 306μM, $K_{ir}3.4$ IC_{50} < 300μM	Inhibitor	Xenopus oocytes	Liu et al. (2007)
$K_{ir}3.1/3.4$	NIP-142	IC_{50} = 0.64μM	Inhibitor	HEK293	Matsuda et al. (2006)
$K_{ir}3.1/3.4$	NIP-151	IC_{50} = 1.6 nM	Inhibitor	HEK293	Hashimoto et al. (2008)

(continued)

Table 3 (continued)

Channel	Compound	IC_{50}/EC_{50}	Class	Cell type/ species	Ref.
$K_{ir}3.x$, $K_{ir}4.1$	Tricyclic antidepressants (nortriptyline)	$K_{ir}3$ $IC_{50} = 18–71\mu M$, $K_{ir}4.1$ $IC_{50} = 16–38\mu M$	Inhibitor	Xenopus oocytes, HEK293	Kobayashi et al. (2004); Su et al. (2007)
$K_{ir}3.1$/ $K_{ir}3.2$	U50488H	$IC_{50} = 70\mu M$	Inhibitor	Xenopus oocytes	Ulens et al. (1999)
$K_{ir}3.2$/ $K_{ir}3.1$	MK801	$IC_{50} = 3.1\mu M$	Inhibitor	Xenopus oocytes, mouse	Kofuji et al. (1996)
$K_{ir}3.2$/ $K_{ir}3.1$	QX314	$IC_{50} = 3.5\mu M$	Inhibitor	Xenopus oocytes, mouse	Kofuji et al. (1996)
$K_{ir}3.x$	Ifenprodil	IC_{50} at submicromolar concentrations	Inhibitor	Xenopus oocytes	Kobayashi et al. (2006)

2.4 ATP-Sensitive K^+ Channels

K_{ATP} channels were initially identified in cardiac myocytes (Noma 1983; Trube and Hescheler 1984). These channels are also found in pancreatic β-cells (Ashcroft et al. 1984; Cook and Hales 1984), skeletal muscle (Spruce et al. 1985, 1987), vascular smooth muscle (Standen et al. 1989; Kajioka et al. 1991; Beech et al. 1993; Kamouchi and Kitamura 1994), and some neurons (Ashford et al. 1988, 1990). K_{ATP} channels are formed by the α-subunits from the $K_{ir}6$ subfamily, forming hetero-octameric complexes with sulfonylurea receptors (SUR1, SUR2A, SUR2B). In these complex channels, K_{ir} subunits are responsible for ATP inhibition, and the SUR molecules are responsible for MgADP activation (Cui et al. 2021). K_{ATP} channels function to couple cellular metabolic state (i.e., ATP/ADP ratio) to cellular activity (Cui et al. 2021).

In pancreatic insulin-secreting β-cells, K_{ATP} channels, composed of $K_{ir}6.2$ and SUR1 (Inagaki et al. 1995, 1996), function to control the E_{rest} (~70 mV) and couple blood glucose concentration to insulin secretion (Ashcroft and Kakei 1989; Ashcroft and Rorsman 1989). Impairment of K_{ATP} channel activity leads to loss of control of insulin secretion by any physiological process, including hypoglycemia. GOF mutations in either $K_{ir}6.2$ or SUR1 genes or pathological enhancement of channel function cause diabetes mellitus with reduced insulin secretion and resultant hyperglycemia (Ashcroft 2005).

Studies suggest that cardiac K_{ATP} channels play a cardioprotective role in ischemic conditions (Noma 1983; Nichols and Lederer 1991; Terzic et al. 1995; Isomoto et al. 1997; Kane et al. 2005). Unlike pancreatic K_{ATP} channels, cardiac K_{ATP} channels are closed under physiological conditions due to high cellular ATP concentrations (Nichols and Lederer 1991; Terzic et al. 1995; Kane et al. 2005). However, the channels can be opened by metabolic insult, such as increased cardiac workload, hypoxia, or ischemia. $K_{ir}6.1$ knockout mice typically die between 5 and 6 weeks after birth because of a high rate of sudden cardiac death syndrome (Miki et al. 2002). Dysfunction of $K_{ir}6.2$ channels increases vulnerability to sympathetic stress and vigorous sympathetic challenge, leading to arrhythmia and sudden death.

$K_{ir}6.1$/SUR2B channels resemble native K_{ATP} channels observed in vascular smooth muscle throughout the body (Teramoto 2006). Opening of vascular K_{ATP} channels hyperpolarizes the membrane potential toward the E_K, resulting in the closure of the voltage-gated Ca^{2+} channels and, hence, relaxation of the smooth muscle of the blood vessels, especially veins (Cui et al. 2021).

Various pharmacological agents can either stimulate or inhibit K_{ATP} channels, many of these act by binding to SUR subunits. Inhibitory agents include sulfonylureas, such as chlorpropamide, tolbutamide, and glibenclamide (Edwards and Weston 1993; Wickenden 2002; Doyle and Egan 2003), which are used therapeutically to treat type 2 diabetes (Dean and Matthews 1968; Schmid-Antomarchi et al. 1987a, b; Wickenden 2002; Doyle and Egan 2003). Stimulatory agents (activators or openers) include pinacidil, nicorandil, and diazoxide (Ashcroft 1988; Terzic et al. 1995), which are clinically used as hypotensive agents. Mitiglinide, nateglinide, repaglinide, and meglitinide are non-sulfonylurea acyl-amino-alkyl benzoic acid derivatives, nonselective K_{ATP} channel blockers that represent a new chemical class of drugs for treating type 2 diabetes (Landgraf 2000; Malaisse 2003; Blicklé 2006). There are also many more recent compounds that can regulate K_{ATP} channels; the details of these are given in Table 4. For further detail on the pharmacology of K_{ir} channels, the reader is referred to the recent comprehensive review from the Diomedes Logothetis laboratory (Cui et al. 2021).

3 Voltage-Gated K^+ Channels (Excluding ERG and K_v7) as Drug Targets

Voltage-gated potassium (K_v) channels are the largest superfamily of $K^{\pm}$ channels with the common properties of strong dependence of their open probability (P_o) on the membrane voltage. Such channels are typically closed at resting voltages of −70 to −90 mV, and open when the membrane depolarizes (Taura et al. 2021). K_v channels have a wide range of regional expression, pharmacological, and biophysical features and are found in the nervous and cardiovascular systems, as well as in many other cell types throughout the body (Taura et al. 2021). As a general principle, the function of $K^{\pm}$ channels in excitable cells (e.g., neurons, myocytes) is to keep the excitation within limits, to shape action potentials, and to control excitability parameters. Thus, K_vs generally oppose the excitatory action of voltage-gated Na^+ and $Ca^{2\pm}$ channels, whose basic roles are to initiate and conduct action potentials and release Ca^{2+} in presynaptic nerve terminals (Anderson and Greenberg 2001; Kohling and Wolfart 2016).

Table 4 Modulatory compounds of K_{ATP} channels

Channel	Compound	IC_{50}/EC_{50}	Class	Cell type/species	Ref.
$K_{ir}6.2$/SUR1	*Diazoxide*	$EC_{50} = 30 \sim 60 \mu M$	Activator	Xenopus oocyte	Schwanstecher et al. (1998); Dabrowski et al. (2003)
K_{ATP}	*Nicorandil*	$EC_{50} = 500$ mM	Activator	Mouse, rat, rabbit	Kajioka et al. (1990); Giblin et al. (2002); Mannhold (2004)
SUR2	P1075	$EC_{50} \leq 50$ nM	Activator	HEK293, isolated vascular preparations from rat, rabbit, and dog	Higdon et al. (1997); Schwanstecher et al. (1998)
SUR2	Levcromakalim	$EC_{50} = 0.4–3 \mu M$	Activator	Human, mouse, rat	Randall and Griffith (1993); Schwanstecher et al. (1998)
SUR1	NN414	$EC_{50} = 450$ nM	Activator	HEK cells	Dabrowski et al. (2003)
K_{ATP}	*Minoxidil*	$EC_{50} = 182.6 \mu M$ (sarcK_{ATP}) $EC_{50} = 7.3 \mu M$ (mitoK_{ATP})	Activator	Guinea pig ventricular myocytes	Sato et al. (2004)
$K_{ir}6.2$/SUR2	Pinacidil	$EC_{50} \sim 10$ nM	Activator	HEK293	Gribble et al. (1998); Okuyama et al. (1998)
$K_{ir}6.2$/SUR1	Iptakalim	$EC_{50} = 10 \sim 100 \mu M$	Activator	Rat, mouse, human	Misaki et al. (2007); Costa (2009)
K_{ATP}	Cromakalim	$EC_{50} = 100 \sim 500$ nM	Activator	Coronary endothelial cells, coronary smooth muscle cells	Daut et al. (1990)
$K_{ir}6.2$/SUR1	VU0071063	$EC_{50} = 7 \mu M$	Activator	HEK293	Raphemot et al. (2014)
SUR2	Levosimendan	$EC_{50} = 0.4 \sim 3 \mu M$	Activator	COS-7, in hypoxic smooth muscle cells	Randall and Griffith (1993); Schwanstecher et al. (1998)

Channel	Compound	IC_{50}/EC_{50}	Class	Cell type/species	Ref.
$K_{ir}6.2$	Phentolamine	$IC_{50} = 106 \mu M$	Inhibitor	Xenopus oocytes	Proks and Ashcroft (1997)
$K_{ir}6.2$/SUR2	Glibenclamide	$IC_{50} \sim 150$ nM	Inhibitor	Rat, human, rabbit	Gribble et al. (1998); Okuyama et al. (1998)
$K_{ir}6.2$/SUR2	Tolbutamide	IC_{50}, $2 \sim 7 \mu M$	Inhibitor	HEK293	Gribble et al. (1998); Okuyama et al. (1998)
SURl	Gliclazide	$IC_{50} = \sim 50$ nM	Inhibitor	Xenopus oocytes	Gribble and Ashcroft (1999)
SUR1	Glipizide	$IC_{50} = 4$ nM	Inhibitor	COS-7	Dörschner et al. (1999)
SURl ~ SUR2	Mitiglinide	SUR1 $IC_{50} = \sim 0.3 \mu M$ SUR2A $IC_{50} = \sim 0.5 \mu M$	Inhibitor	COS-1, MIN6 cells	Sunaga et al. (2001)
SUR1 ~ SUR2B ≫ SUR2A	Nateglinide	SUR1, SUR2B $IC_{50} = 100$ nM SUR2A $IC_{50} > 1 \mu M$	Inhibitor	COS-1, MIN6 cells	Sunaga et al. (2001)
SUR1 > SUR2A	Glyburide	SUR1 $IC_{50} \sim 4$ nM SUR2A $IC_{50} \sim 27$ nM	Inhibitor	Xenopus oocytes	Gribble et al. (1998)
$K_{ir}6.1$/SUR2B $K_{ir}6.2$/SURx	Rosiglitazone	$K_{ir}6.1$/SUR2B $IC_{50} \sim 12 \mu M$ $K_{ir}6.2$/SURx $IC_{50} \sim 45 \mu M$	Inhibitor	HEK293	Yu et al. (2012) Wang et al. (2013)
$K_{ir}6.1$/SUR2B	AS-252424	$IC_{50} = 4 \mu M$	Inhibitor	HEK293	Wang et al. (2013)
$K_{ir}6.1$/SUR2B	Englitazone	$IC_{50} = 7 \mu M$	Inhibitor	HEK293	Wang et al. (2013)
$K_{ir}6.1$/SUR2B	A6355	$IC_{50} = 8 \mu M$	Inhibitor	HEK293	Wang et al. (2013)
$K_{ir}6.1$/SUR2B	cay10415	$IC_{50} = 15 \mu M$	Inhibitor	HEK293	Wang et al. (2013)
$K_{ir}6.1$	PNU-37883A	IC50 ~ 50μM	Inhibitor	RINm5F cells	Humphrey (2006)

K_v channels are encoded by 40 genes in humans and are classified into the K_v1–K_v12 subfamilies based on sequence homology (Fig. 1) (Gutman et al. 2005; Taura et al. 2021). Each K_v subfamily contains multiple members, and members of each K_v subfamily feature distinct gating modes and drug sensitivity. In this section we will consider K_v channels with the exception of ERG (K_v11) and K_v7 (KCNQ) channels; due to their particular clinical relevance and wealth of literature, these be considered in separate sections.

A functional K_v channel is assembled from four subunits (in a homo- or hetero-tetrameric fashion), which encircle a central ion conduction pathway. Each subunit comprises six helices (TMDs) commonly labeled S1–S6, with both an amino and a carboxyl termini located on the intracellular side of the plasma membrane (Fig. 3). The voltage-sensor domain (VSD) is composed of the first four transmembrane helices (S1–S4), while the pore domain is made up of the final two transmembrane helices (S5–S6) with a reentrant P-loop in between (Fig. 3) (Sands et al. 2005). The VSD mediates voltage sensing, with S4 carrying the majority of the gating charges. Changes in the transmembrane electric field cause conformational changes in the VSD, which are coupled to the pore/gate module via the cytosolic S4–S5 linker. The

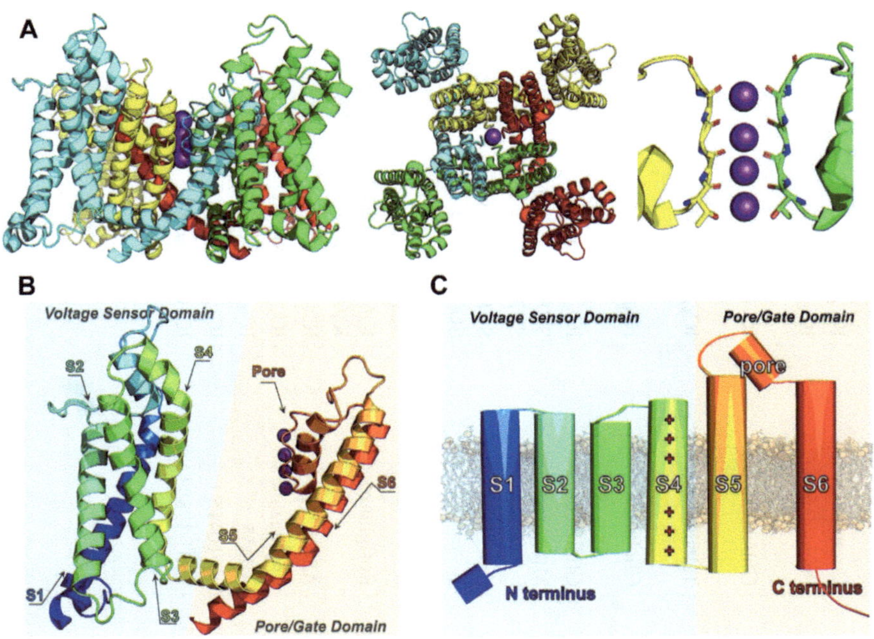

Fig. 3 General structure of voltage-gated K⁺ (K_v) channels. (**a**) Structure of the mammalian K_v1.2/K_v2.1 chimeric K_v channel (PDB ID: 2R9R). Left: the side view of the tetrameric channel assembly. Middle: a top view of the channel from the extracellular side. Right: a close-up view of the selectivity filter in the pore region. Potassium ions are shown as purple balls. (**b**) Structural elements covering S1–S6 of a single subunit of the Kv1.2/2.1 chimera. (**c**) Topological model of a 6TM α-subunit of voltage-gated K+ channels illustrating the voltage-sensor domain (S1–S4), the pore/gate domain (S5–S6), and an N-terminal domain that functions in some channel inactivation "ball" domain

latter then causes the S6 segments to move, allowing the pore to open on the intracellular side. The fundamental gating characteristics of voltage-gated K^+ channels are defined by the combination of VSD and pore domain. Extended N- and C-terminal protein domains, on the other hand, contribute many additional features, such as inactivation and modulation by second messengers such as Ca^{2+} or cyclic nucleotides. The conserved selectivity filter region comprised of an extensive polypeptide chain with a highly conserved (among all K^+ channels) sequence pattern TVGYG and a small pore helix in the reentrant loop (Sands et al. 2005; Jensen et al. 2012; Kim and Nimigean 2016). Depending on the G-Y-G or G-F-G sequence in the selectivity filter region, neighboring carbonyl oxygens in the pore loops are guided into the center of the channel to bestow selectivity for K^+. A water-filled cavity that extends toward the cytosolic side completes the channel pore. Different types of drugs attach to K_v channels at this site to either block the pore or interfere with gating (Long et al. 2007; Kim and Nimigean 2016).

Mutations in multitude of K_v channel genes can lead to a range of severe channelopathies; hence, K_v channels are one the most well-studied ion channel families. These mutations not only underscore the physiological importance of K_v channels but also highlight their potential as therapeutic targets. Indeed, numerous studies have demonstrated that K_v channels represent confirmed or potential therapeutic targets for a wide range of diseases, including neurological disorders, cardiovascular diseases, and even cancer (see section on K^+ channels in cancer below). The therapeutic potential of K_v channels is summarized in Table 5.

3.1 K_v1 Channels

Eight members of the Shaker-related Kv1 family (Kv1.1–Kv1.8) are encoded by the corresponding *KCNA1-KCNA8* genes. K_v1 channels are expressed as homo- or heterotetrameric complexes in a variety of brain regions and other tissues (Coetzee et al. 1999; Fountain et al. 2004; Hoegg and Meyer 2007). These channels are sensitive to "classical" K^+ channel blocker, 4-aminopiridine (4-AP); some K_v1 channels are also inhibited by the mamba snake toxin, dendrotoxin (DTX) (Coetzee et al. 1999). Mutations in various K_v1 subtypes can cause several channelopathies, ranging from ataxias and epilepsies to autoimmune diseases (see Table 5). K_v1.1, K_v1.3, and K_v1.5 channels of the K_v1 subfamily have received particular attention, and these show promising potential as therapeutic targets (Table 5).

K_v1.1, which is mainly expressed in the nervous system, is essential for regulating neuronal excitability (Wang et al. 1994). When K_v1.1 α-subunits are expressed alone, they display slow, delayed-rectifier kinetics with slow and only partial inactivation, but once co-expressed with β1.1-auxiliary subunit, they become rapidly inactivating, A-type channels (Rettig et al. 1994; Heinemann et al. 1996; Jow et al. 2004). The K_v1.1 channel complex is a potential drug target for treatment of pain and epilepsy. Several substances that activate K_v1.1 channel (or attenuate inactivation) were found to have anti-excitatory effects in vitro and anticonvulsant effects in

Table 5 Channelopathies and potential therapeutic targets associated with K_v channels (excluding K_v7 and K_v11 channels)

Channel	Channelopathy	Ref	Therapeutic potential	Ref
$K_v1.1$	Myokymia; episodic ataxia type 1; acquired neuromyotonia	Adelman et al. (1995); Zuberi et al. (1999); Kleopa et al. (2006); Chen et al. (2007); Irani et al. (2010); Tristan-Clavijo et al. (2016);	Epilepsy (anti-seizure); multiple sclerosis and spinal damage; cancer	Hayes (2004); Lu et al. (2008a); Jang et al. (2011)
$K_v1.2$	Epileptic encephalopathy; ataxia and myoclonic epilepsy; episodic ataxia and pharmacoresponsive epilepsy	Syrbe et al. (2015); Corbett et al. (2016); Hundallah et al. (2016);	Neuropathic pain	Zhang et al. (2021)
$K_v1.3$	Polymorphism associated with autoimmune pancreatic diabetes; polymorphism associated with hyperglycemia and reduced olfaction	Tschritter et al. (2006); Guthoff et al. (2009); Ota et al. (2011)	Multiple sclerosis, rheumatoid arthritis, other autoimmune disorders, neuroinflammatory diseases, diabetes, cancer	Beraud et al. (2006); Bielanska et al. (2009); Lam and Wulff (2011); Comes et al. (2013); Prentis et al. (2018)
$K_v1.5$	Atrial fibrillation	Olson et al. (2006)	Atrial fibrillation	Wettwer and Terlau (2014); Borrego et al. (2021)
$K_v1.6$	Acquired neuromyotonia	Antozzi et al. (2005)	NA	
$K_v2.1$	Non-syndromic intellectual disability; epileptic encephalopathy	Torkamani et al. (2014); Thiffault et al. (2015); Latypova et al. (2017)	Epilepsy; hyperactivity and defects in spatial learning	Jedrychowska and Korzh (2019); Yeh et al. (2020)
$K_v2.2$	Nonfamilial Brugada syndrome	Juang et al. (2014)	NA	
$K_v3.1$	Intellectual disability; myoclonus epilepsy and ataxia	Muona et al. (2015); Nascimento and Andrade (2016); Poirier et al. (2017)	Epilepsy; cancer metastasis	Muona et al. (2015); Song et al. (2018); Li et al. (2021b)
$K_v3.3$	Spinocerebellar ataxia 13	Waters et al. (2006)	NA	

(continued)

Table 5 (continued)

Channel	Channelopathy	Ref	Therapeutic potential	Ref
$K_v3.4$	NA		Alzheimer's disease; spinal cord injury	Angulo et al. (2004); Ritter et al. (2015); Boscia et al. (2017); Yeap et al. (2022)
$K_v4.2$	Temporal lobe epilepsy	Singh et al. (2006)	Ischemic stroke impairments; epilepsy	Tiwari et al. (2020); Xiao et al. (2021)
$K_v4.3$	Lone atrial fibrillation; intellectual disability; early-onset cerebellar ataxia; oral apraxia; epilepsy	Olesen et al. (2013); Smets et al. (2015)	Pain	Conner et al. (2016); Kanda et al. (2021)
$K_v8.2$	Epilepsy susceptibility	Jorge et al. (2011)	NA	
$K_v10.1$	Temple-Baraitser and Zimmermann-Laband syndromes; epilepsy	Simons et al. (2015); Fukai et al. (2016); Mastrangelo et al. (2016)	Chemotherapy, cancer	Cazares-Ordonez and Pardo (2017); Bauer and Schwarz (2018); Galan-Vidal et al. (2022); Luis et al. (2022)
$K_v10.2$	Epilepsy and autism	Yang et al. (2013); Galan-Vidal et al. (2022)	Epilepsy	Liu et al. (2020); Hu et al. (2022)

animal models (Lu et al. 2008a; D'Adamo et al. 2020), although the small-molecule $K_v1.1$ openers remain few in number. Among the activators are the resin acids produced by many plants such as pimaric acid (PiMA) and dehydroabietic acid (DHAA) (D'Adamo et al. 2020). $K_v1.1$ channel inhibitors may potentially also be helpful, particularly in multiple sclerosis and spinal nerve injury (Hayes 2004). Additionally, $K_v1.1$ was suggested to be associated with several cancer types (Jang et al. 2011; Zhao and Chen 2016; D'Adamo et al. 2020; Finol-Urdaneta et al. 2020).

$K_v1.3$ is another widely studied member of the K_v1 subfamily; it was first functionally identified in lymphocytes (DeCoursey et al. 1984) and then independently cloned from brain tissue (Swanson et al. 1990) and T-lymphocytes (Grissmer et al. 1990); overall, the body-wide distribution of its expression is remarkable: this channel is abundantly expressed in lymphocytes, adipose tissue, skeletal muscle, and both smooth and skeletal muscles, as well as in the olfactory bulb and the hippocampus (Swanson et al. 1990; Zhao and Chen 2016). $K_v1.3$ function is crucial for lymphocyte physiology and for immune responses. It is necessary for maintenance of Ca^{2+} driving force and Ca^{2+} signaling integrity; blockade of $K_v1.3$ channels in T cells suppresses Ca^{2+} signaling, cytokine production, and proliferation (Wulff et al.

2003; Chandy et al. 2004). Thus, it is not surprising that $K_v1.3$ is a validated target for immunosuppression (Lam and Wulff 2011), as well as for the treatment of multiple sclerosis, rheumatoid arthritis, and other autoimmune disorders (Beeton et al. 2005, 2006). Due to the presence in microglia, $K_v1.3$ is a potential therapeutic target for treating neuroinflammatory diseases (Saxena et al. 2021). $K_v1.3$ is implicated in cancer development (see section on K^+ channels and cancer below); thus, some tumors, such as melanoma, pancreatic ductal adenocarcinoma (PDAC), and B-type chronic lymphocytic leukemia (B-CLL), may respond to therapy better when the $K_v1.3$ channels are inhibited (Bielanska et al. 2009; Comes et al. 2013; Teisseyre et al. 2019). $K_v1.3$ channel is also a prospective pharmaceutical target for diabetes and obesity because it plays a role in regulating insulin action in insulin-sensitive tissues (Xu et al. 2003, 2004; Tucker et al. 2008). The main types of $K_v1.3$ channel blockers currently in use are small molecular inhibitors and toxins from animal venoms. Examples of $K_v1.3$ inhibitors are given in Table 6; some of these compounds progressed to clinical trials.

Table 6 Examples of Kv1.3 channel-specific inhibitors

Compound	Type	Source	IC_{50} (nM)	Therapeutic potential	Ref
HsTX1 (scorpion toxin)	Peptide	*Heterometrus spinifer*	0.012	Arthritis; neuroinflammation	Tanner et al. (2017); Nicolazzo et al. (2022)
Hg1	Peptide	*Hadrurus gertschi*	6.2	NA	Chen et al. (2012)
ShK-186 (anemone toxin)	Peptide	*Stichodactyla helianthus*	0.07	Insulin resistance and obesity; influenza and chlamydia; psoriasis; autoimmune diseases	Nilius et al. (1993); von Beckerath et al. (1996); Nilius and Droogmans (2001); Matheu et al. (2008); Tarcha et al. (2012); Rashid et al. (2013); Upadhyay et al. (2013); Pennington et al. (2015); Yamamura et al. (2016)
OsK1	Peptide	*Orthochirus scrobiculosus*	0.014	Autoimmune diseases	Mouhat et al. (2005); Chen and Chung (2012)
ADWX-1	Peptide	*Buthus martensii*	0.002	NA	Yin et al. (2008)
PAP-1	Small molecule	Organic synthesis	2.0	Psoriasis; chronic lymphocytic leukemia; colitis; reperfusion injury in stroke	Schmitz et al. (2005); Leanza et al. (2013); Mei et al. (2019); Chen et al. (2021)
Psora-4	Small molecule	Organic synthesis	2.9	Chronic lymphocytic leukemia; glomerular basement membrane glomerulonephritis	Hyodo et al. (2010); Leanza et al. (2013)

$K_v1.4$ is one of the isoforms forming characteristic transient A-type current in neurons (Johnston et al. 2010; Johnston 2021). $K_v1.4$ channels are relatively insensitive to tetraethylammonium (TEA) and have low sensitivity ($IC_{50} > 10$ mM) to 4-AP (Coetzee et al. 1999; Johnston 2021). Small molecules CP-339,818 (Nguyen et al. 1996) and UK-78,282 (Hanson et al. 1999) were reported as relatively selective inhibitors of $K_v1.4$, although these also block $K_v1.3$.

$K_v1.5$ channels are also ubiquitously expressed in the mammalian body, including the immune system, kidney, skeletal and smooth muscle, and the brain (Tamkun et al. 1991; Overturf et al. 1994). In the heart, $K_v1.5$ contributes to the ultrarapid delayed-rectifier K^+ current (I_{Kur}) of atrial myocytes (Wang et al. 1993), which is one of the major repolarizing currents in human atrium (Wettwer et al. 2004). Mutations of this channel have been linked to hereditary forms of atrial fibrillation (AF), and the channel is an appealing target for AF treatment (Wettwer et al. 2004; Olson et al. 2006). Some effective small-molecule $K_v1.5$ inhibitors are considered for the treatment of AF (Wettwer and Terlau 2014; Borrego et al. 2021).

3.2 *K_v2 Channels*

$K_v2.1$ (*KCNB1*) was cloned from rat brain in 1989 (Frech et al. 1989). Originally known as drk1, this channel forms a classic delayed-rectifier K^+ channel with an activation threshold of around −15 mV (Frech et al. 1989). Three years later, $K_v2.2$ (*KCNB2*), originally named cdrk1, was cloned from rat taste buds (Hwang et al. 1992). In terms of tissue distribution, K_v2 channels are very widely expressed: cortical, hippocampal, and α-motoneurons contain them, as do the retinal bipolar cells, vascular and cardiac myocytes, and gastrointestinal smooth muscle and pancreatic beta cells (Johnson et al. 2019). Mutation of $K_v2.1$ has been linked to epilepsy in humans, and $K_v2.1$ knockout mice are epileptic, hyperactive and show defects in spatial learning (Jedrychowska and Korzh 2019; Yeh et al. 2020). K_v2 channels are sensitive to 4-AP and TEA (Coetzee et al. 1999). A subpopulation of non-conducting $K_v2.1$ channels localize to and stabilize junctions between the endoplasmic reticulum (ER) and plasma membrane (ER-PM junctions) in neurons (Fox et al. 2015; Yeh et al. 2020). $K_v2.1$-dirrected targeted disruption of these microdomains was shown to exert antiapoptotic effect (Justice et al. 2017) and provide neuroprotection against ischemic stroke (Schulien et al. 2020; Yeh et al. 2020).

3.3 *K_v3 Channels*

The K_v3 family includes four channels: $K_v3.1$–$K_v3.4$ encoded by the *KCNC1-KCNC4* genes (Kaczmarek and Zhang 2017). One feature that distinguishes these channels from those in other mammalian K_v subfamilies is that they typically open only at positive potentials and that activation and deactivation in response to voltage changes are extremely fast (Wei et al. 1990). The discovery of a recurrent de novo mutation of K_v3.1(R320H), which results in progressive myoclonus epilepsy (Muona et al. 2015; Nascimento and Andrade 2016), as well as ataxia and action myoclonus (Poirier et al. 2017), has provided direct evidence in humans that K_v3.1 constitutes a valid target for preventing excitability disorders, such as epileptic seizures. Further evidence demonstrates that low levels of neuronal K_v3 (primarily K_v3.1 and K_v3.2) channels may predispose an animal to seizures (Poirier et al. 2017). K_v3.3 and K_v3.4, together with K_v1.4 and K_v4 subunits, form neuronal A-type channels (Johnston 2021). Mutations within *KCNC3* gene, which codes for the K_v3.3 subunit, cause spinocerebellar ataxia type 13 (SCA13) (Waters et al. 2006). Furthermore, dysregulations in the K_v3.4 channel expression/activity were linked to Alzheimer's disease and spinal cord injury (Angulo et al. 2004; Ritter et al. 2015; Boscia et al. 2017; Yeap et al. 2022).

3.4 K_v4 Channels

K_v4.1 (*KCND1*), K_v4.2 (*KCND2*), and two splice variants of K_v4.3 (*KCND3*) comprise the K_v4-subfamily or Shal-related subfamily of K_v channels, with an overall sequence identity of about 60% (Isbrandt et al. 2000; Birnbaum et al. 2004). K_v4 channels are abundant in smooth muscles, the heart, and the brain. In the heart, K_v4.2 and K_v4.3 contribute to the transient outward current, I_{to} (Schreiber and Seebohm 2021), and in neurons these channels contribute to A-type current (e.g., somatodendritic A-type current, I_{SA}, in the hippocampal neurons) (Serodio et al. 1994, 1996). K_v4 dysfunctions in the heart are linked to Brugada syndrome, atrial fibrillation, hypertrophy, and heart failure. K_v4.x channelopathies in the hippocampus have been linked to epilepsy, schizophrenia, and Alzheimer's disease (Huo et al. 2014; Cercos et al. 2021). K_v4.1 and K_v4.3 are expressed in peripheral somatosensory neurons, and K_v4.3 downregulation was linked to neuropathic pain development (Uchida et al. 2010; Zhang et al. 2019). Pharmacology of K_v4 channels has been reviewed in Cercos et al. (2021), and recent review of the A-type K^+ channel pharmacology is also available (Johnston 2021).

3.5 K_v5, K_v6, K_v8, and K_v9 Channels (Electrically Silent K_v Channels)

Members of the K_v5, K_v6, K_v8, and K_v9 subfamilies do not form functional ion channels as homomers; they are referred to collectively as the electrically silent K_v (K_vS) channel subunits (Bocksteins 2016). When expressed alone in heterologous expression systems, K_vS subunits, unlike other K_v subunits, do not form functional homotetrameric channels. However, some K_vS subunits heterotetramerize with K_v2 subunits to form functional channels with distinct biophysical properties; K_vS are not known to co-assemble with K_v1, K_v3, or K_v4 channels (Bocksteins and Snyders 2012; Bocksteins 2016). K_vS subunits are found in a variety of tissues and organs, including the brain, spinal cord, internal organs, and male and female reproductive systems (Bocksteins and Snyders 2012; Bocksteins 2016). The $K_v8.2$ coding gene (*KCNV2*) was identified as a potential epilepsy modifier gene, and it influences the epilepsy phenotype in the *Scn2a*^{Q54} transgenic mouse model (Bergren et al. 2009). Likewise, $K_v6.1$ may also be linked to epilepsy (Friedman et al. 2013). Downregulation of the $K_v9.1$ was reported to contribute to neuropathic pain following nerve injury (Tsantoulas et al. 2012). K_vS subunits may also be involved in other central nervous system diseases such as schizophrenia (Georgiev et al. 2014) and migraine (Lafreniere and Rouleau 2012). However, the level of understanding of the K_vS channel function, perhaps related to its electrically silent phenotype, is currently insufficient to pursue this class fully for therapeutic targeting.

3.6 K_v10, K_v11, and K_v12 Channels (Ether-à-Go-Go K^+ Channels)

The ether-à-go-go K^+ channel (EAG) subfamily is divided into three subfamilies based on sequence similarity: ether-à-go-go (EAG, including $K_v10.1$, $K_v10.2$), ether-à-go-go-related-gene (ERG, including $K_v11.1$, $K_v11.2$, $K_v11.3$), and ether-à-go-go-like (ELK, including $K_v12.1$, $K_v12.2$, $K_v12.3$) (Wimmers et al. 2001; Schonherr et al. 2002; Lin et al. 2014). As mentioned, ERG channels will be considered in a separate section.

The $K_v10.1$ channel is encoded by the *KCNH1* gene located on chromosome 1, band q32.2 in humans, and comprises 457,343 bases and 12 exons (Occhiodoro et al. 1998). $K_v10.1$ is virtually undetectable in peripheral tissues in healthy or physiological conditions, with enriched expression only in specific brain regions, including the olfactory bulb, hippocampus, cerebral cortex, striatum, hypothalamus, and parallel fiber-Purkinje cell synapses (Occhiodoro et al. 1998; Martin et al. 2008). *KCNH1* mutations have been associated with Temple-Baraitser and Zimmermann-Laband syndromes (Fukai et al. 2016; Mastrangelo et al. 2016), as well as other forms of developmental deficits characterized by mental retardation and epilepsy (Simons et al. 2015), indicating that $K_v10.1$ is important for cognitive development in humans. Low expression of $K_v10.1$ in peripheral tissues is contrasted with frequent ectopic expression of this channel in tumor biopsies and cancer cell lines (Pardo et al. 2012; Pardo and Stuhmer 2014). The enriched expression of $K_v10.1$ in cancer cells, its limited expression in other tissues, and its membrane localization make $K_v10.1$ a promising protein in cancer research. K_v10 channel activity can impact cell cycle, proliferation, migration, and microenvironment of tumor cells (Ganser et al. 2021; Pardo and Stuhmer 2014). Furthermore, ectopic $K_v10.1$ expression has been linked to chemoresistance cell phenotypes. Thus, inhibiting the

$K_v10.1$ channel may improve cell response to therapeutic drugs commonly used in chemotherapy (Cazares-Ordonez and Pardo 2017; Galan-Vidal et al. 2022; Hu et al. 2022; Luis et al. 2022) (see section on K^+ channels in cancer).

$K_v10.2$ (*KCNH5*) is found in many organs and nervous systems, including the skeletal muscle, placenta, heart, lung, and retina. Loss-of-function (LOF) mutations of $K_v10.2$ in humans are also linked to epilepsy and autism (Veeramah et al. 2013); Yang et al. 2013; Galan-Vidal et al. 2022). Consistently, viral overexpression of $K_v10.2$ in the CA3 region of the hippocampus of epileptic rats ameliorated some cognitive deficits caused by kainic acid-induced temporal lobe epilepsy (Liu et al. 2020).

4 Therapeutic Implications of hERG Channels

Contraction of the heart is governed by the action potentials generated by cardiac myocytes. The specific shape, duration, and frequency of these action potentials is, in turn, controlled by the specific set of ion channels expressed by the myocytes and is different across different heart regions (i.e., sinoatrial, atrial, or the ventricular myocardium) (Fig. 4) (Schreiber and Seebohm 2021). Abnormalities in the expression and function of these cardiac ion channels often lead to aberrations in heart function, such as those leading to arrhythmias and sudden cardiac death (Keating and Sanguinetti 1996; Su et al. 2021). As an example, we can consider the ventricular action potential, which is subdivided into five stages (phases 0–4): phase 0 is a rapid rise of the membrane potential (E_m) due to opening of voltage-gated $Na_v1.5$ channels (I_{Na}). During phase 1 E_m reaches its maximum and starts to reverse due to the opening of voltage-gated $K_v4.2$, $K_v4.3$/KChIP (I_{Tof}), and $K_v1.4$ (I_{Tos}) channels. Meanwhile, depolarized membrane E_m leads to the opening of voltage-gated $Ca_v1.2$ (L-Type) Ca^{2+} channels, as well slow-kinetic K^+ channels $K_v7.1$/KCNE1 (I_{Ks}) and $K_v11.1$/KCNE2 (I_{Kr}) channels, together generating a plateau phase (phase 2). The return toward the resting potential (phase 3) is driven by the inactivation of T-type Ca^{2+} channels, while I_{Ks} and I_{Kr} channels are still active. Finally, opening of inwardly rectifying K^+ channels $K_{ir}2.1$, $K_{ir}2.2$, $K_{ir}2.3$ (I_{K1}) returns E_m to the resting potential (phase 4) (Schreiber and Seebohm 2021).

The pore-forming K^+ channel subunit ($K_v11.1$) of I_{Kr} is encoded by the human *ether-á-go-go related* gene (*hERG*, *KCNH2*). I_{Kr} is critical for the repolarization (phase 3) of the human ventricle and effectively controls the QT interval of the electrocardiogram (Sanguinetti et al. 1995). Inhibition of I_{Kr} results in the prolongation of repolarization (Fig. 4a), which can cause arrhythmias in certain circumstances (Su et al. 2021). A wide range of unrelated compounds displays a potential to inhibit hERG channels, posing a threat of the drug-induced form of acquired long QT syndrome (LQTS) and risk of an unusual life-threatening form of arrhythmia, torsades de pointes (TdP) (Sanguinetti and Tristani-Firouzi 2006; Vandenberg et al. 2012). As a result, profiling prospective drug candidates for potential hERG inhibition has become a major requirement in drug development (Sanguinetti and Mitcheson 2005; Hancox et al. 2008).

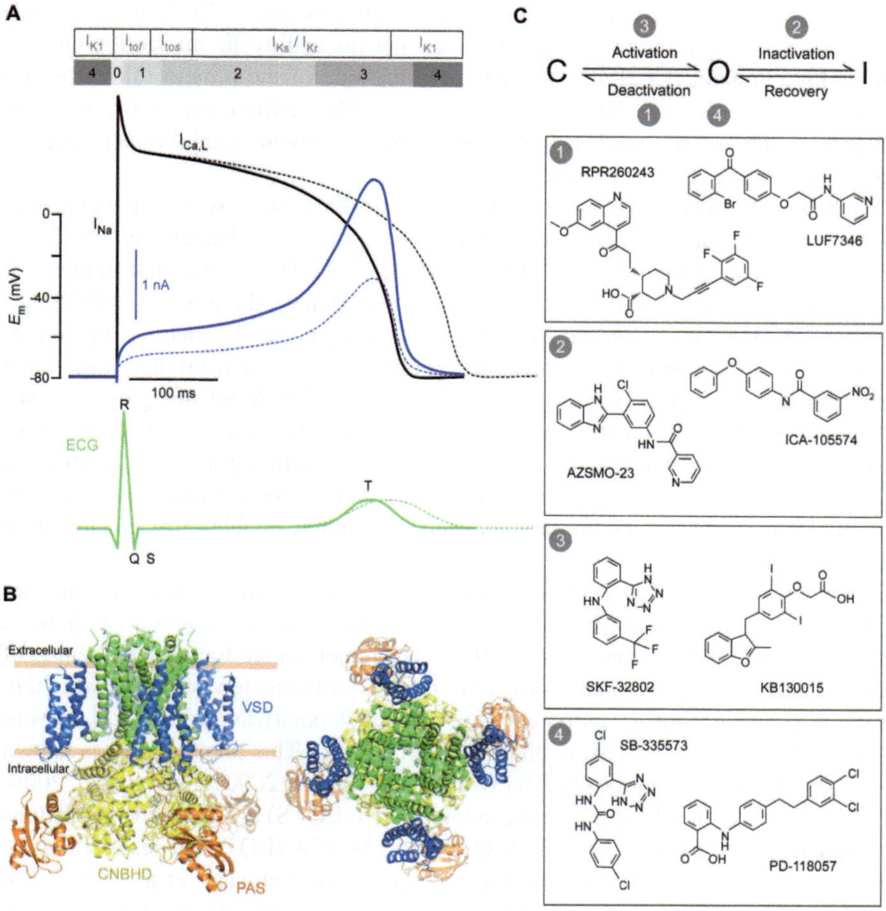

Fig. 4 hERG channels and the cardiac function. (**a**) Schematic overlay of the ventricular action potential (AP) waveform (black) and hERG current kinetics during the AP (blue). Corresponding action potential phases and K^+ currents active during the AP development are indicated by the horizontal bars on the top. I_{K1}, inwardly rectifying current ($K_{ir}2.1/K_{ir}2.2/K_{ir}2.3$); I_{tof}, fast transient current ($K_v4.2/K_v4.3$/KChIP2/KCNE3); I_{tos}, slow transient current ($K_v1.4$); I_{Ks}, slow outwardly rectifying current ($K_v7.1$/KCNE1); I_{Kr}, rapid outwardly rectifying current (hERG). Schematic depiction of the electrocardiogram (ECG) is shown in green. Dotted lines depict changes to the AP and ECG induced by reduction in hERG channel activity. (**b**) Cryo-EM structure of hERG channel (PDB: 5VA2); side view and top views of the ribbon representation of the channel. Voltage-sensor domain (VSD) is in blue, PAS domain is in orange, and CNBHD domain is in yellow (Modified from Asai et al. (2021)). (**c**) Examples of the four types of hERG activators; simple hERG gating model with the main sites of action of each activator type is shown on the top; C, closed state; O, open state; I, inactivated state

A cryo-EM structure of hERG has provided new level of structural understanding of hERG channel gating and pharmacology (Wang and MacKinnon 2017; Butler et al. 2019; Asai et al. 2021). Like other K_v channels, hERG assembles as a tetramer; S1–S4 helices of each α-subunit form a voltage sensor, while S5–S6 TMDs and the intervening pore loop form the pore domain (Piper et al. 2003; Subbiah et al. 2004;

Jiang et al. 2005). In addition, hERG channels possess a Per-Arnt-Sim (PAS) domain at the N-terminal side and C-linker helices and cyclic nucleotide-binding homology domain (CNBHD) at the C-terminal side of the protein (all cytosolic) (Wang and MacKinnon 2017; Asai et al. 2021). The hERG channel also has four deep hydrophobic side pockets that likely contribute to binding inhibitors just below the K^+ selectivity filter (Fig. 4b).

The hERG channel has unique kinetics: relatively slow activation and deactivation, but very fast, voltage-dependent inactivation (Vandenberg et al. 2004; Sanguinetti and Tristani-Firouzi 2006; Su et al. 2021). During the plateau phase of the cardiac action potential (phase 2; usually near 0 mV), there is little hERG current, as channels inactivate rapidly. However, during phase 3, when E_m repolarizes toward the resting membrane potential, the channels recover from inactivation and contribute more current, increasing rate of repolarization (Vandenberg et al. 2004). This unusual kinetics creates an apparent inward rectification, which is crucial for the maintenance of the plateau phase of the cardiac action potential (Smith et al. 1996; Sanguinetti and Tristani-Firouzi 2006), which, in turn, is necessary for sufficient calcium release from the sarcoplasmic reticulum to enable cardiac contraction (Su et al. 2021).

Long QT syndrome (LQTS) is a relatively rare form of arrhythmia characterized by a prolongation of the QT interval of the surface electrocardiogram and T-wave abnormalities (Witchel and Hancox 2000). The major source for the lengthening of the QT interval is delayed repolarization (phase 2 of the action potential) which, in turn, may result in the development of oscillating depolarizations of the E_m of ventricular cardiomyocytes during phase 2 or 3 (Fig. 4a). This phenomenon is known as early afterdepolarization (EAD) (Witchel and Hancox 2000; Hancox et al. 2008) . Three quarters of all cases of congenital LQTS (cLQTS) are caused by mutations within *KCNQ1* (LQT1), *KCNH2* (LQT2), and *SCN5A* (LQT3) genes, coding for $K_v7.1$, hERG, and $Na_v1.5$ ion channels, respectively (Schwartz et al. 2012). The acquired form of the syndrome aLQTS can be produced by the electrolyte imbalance, thyroid disease, or, most commonly, the inhibition of hERG by the prescribed medications (drug-induced long QT syndrome, diLQTS) (Hancox et al. 2008; Kallergis et al. 2012; Vandenberg et al. 2012). Ionic mechanisms of LQTS are described in details elsewhere (Witchel and Hancox 2000; Hancox et al. 2008; Schreiber and Seebohm 2021).

4.1 *hERG Inhibitors*

hERG channels appear to be uniquely susceptible to inhibition by a multitude of compounds of varying chemical structure and properties. Accordingly, hERG inhibition turned out to be a ubiquitous molecular mechanism for many drugs that produce TdP associated with QT prolongation (Rampe and Brown 2013). Consequently, testing for hERG inhibition has become an obligatory step in drug discovery process (Rampe and Brown 2013; Su et al. 2021). Those interested in general mechanisms of arrhythmias are directed to excellent recent reviews on the topic (Schreiber and Seebohm 2021; Krijger Juarez et al. 2023). Here, we will briefly review pro- and anti-arrhythmic hERG channel pharmacology.

Clinical observations of cardiac side effects of unrelated drugs have a long history. Thus, syncope and sudden death were reported in patients treated with anti-arrhythmic agent, quinidine (Levy 1922); these effects were later attributed to hERG inhibition (Po et al. 1999; Witchel and Hancox 2000). Since then, many other drugs and candidate compounds of different structures and properties were identified as hERG inhibitors, including anti-arrhythmic (e.g., disopyramide, dronedarone), antidepressant (e.g., citalopram), antipsychotic (e.g., mesoridazine), antibiotic (e.g., ciprofloxacin), antihistamine (e.g., astemizole), and many other drugs. hERG inhibitor activity flags these as posing proarrhythmic, TdP risk (Rampe and Brown 2013; Su et al. 2021). Indeed, many of such drugs were withdrawn from clinical use, or issued with strict limitations. Examples of these include astemizole, cisapride, droperidol, sertindole, levomethadyl, lidoflazine, and terfenadine, among others (see Su et al. (2021) for more examples). Due to high susceptibility to inhibition by unrelated small-molecule drugs, assessing the hERG liability is an obligatory and early stage of most drug discovery programs with appropriate requirements put in place by national and international regulators. Thus, the International Conference on Harmonisation of Technical Requirements for Registration of Pharmaceuticals for Human Use (ICH) adopted a guideline S7B for assessing hERG blocking risk for new drugs in 2005. The cardiac safety screening approaches continue to evolve and now include non-hERG targets as well (e.g., new cardiac safety paradigm, the Comprehensive in Vitro Proarrhythmia Assay (CiPA) (Sager et al. 2014; Li et al. 2020a)). Nevertheless, low hERG inhibition risk remains a benchmark requirement for novel pharmaceuticals, and commercial high-throughput hERG safety screening services are widely available worldwide. Because of this increased scrutiny, hundreds of potential hERG inhibitors were reported recently; moreover, machine learning and AI-assisted approaches are being deployed to assess hERG risk in yet untested compounds (Ogura et al. 2019; Creanza et al. 2021; Zhang et al. 2022). An integrated, searchable database for hERG blocking small molecules is available here: http://drugdesign.riken.jp/hERGdb/ (Sato et al. 2018).

4.2 hERG Activators

Since *KCNH2* LOF mutations or pharmacological hERG inhibitions are capable of producing LQT, it was logical to suggest hERG activation/potentiation as a plausible therapeutic strategy for treatment of some arrhythmias. Indeed, over 20 distinct small-molecule hERG activators were identified in recent years (Su et al. 2021; El Harchi and Brincourt 2022), and some of these were reported to produce anti-arrhythmic effects in various models of cLQTS and aLQTS (Asayama et al. 2013; Zhou et al. 2005; Kang et al. 2005a; Hansen et al. 2006a; Peitersen et al. 2008; Potet et al. 2012; Zhang et al. 2012a; Meng et al. 2013; Perissinotti et al. 2015; Yu et al.

2015; Sala et al. 2016; Sale et al. 2017; Qile et al. 2019; Perry et al. 2020; Qile et al. 2020; Shi et al. 2020; Kemp et al. 2021, reviewed in Su et al. 2021; El Harchi and Brincourt (2022)). These activators can be categorized into four main types by the primary mechanism of hERG potentiation: (1) slowing of the channel deactivation; (2) attenuation of channel inactivation or shift of the voltage dependence of inactivation toward more positive potentials; (3) negative shift of voltage dependence of activation; and (4) general increase of channel P_o (Sanguinetti 2014; Su et al. 2021; El Harchi and Brincourt 2022) (Fig. 4c). Yet, many hERG activators have multiple mechanisms of action (e.g., VU0405601, A-935142, HW-0168) and cannot be easily classified (El Harchi and Brincourt 2022). It is important to point out that a critical requirement for a hERG channel opener to display efficacy in any of the LQTS is that I_{Kr} current is not completely lost.

Examples of compounds that slow hERG deactivation kinetics (type-1) are compound RPR260243 (one of the first hERG activators discovered) (Kang et al. 2005a; Perry et al. 2007), ginsenoside Rg3 (Choi et al. 2011), and compound LUF7346 (Sala et al. 2016). Mutagenesis strategy revealed that RPR260243 may bind to the cytosolic areas of S5, such that binding of the compound may directly constrain movement of the S6 to slow the rate of channel closure (Perry et al. 2007). RPR260243 has shown efficacy against dofetilide-induced arrhythmia in both guinea pig (Kang et al. 2005a) and zebrafish (Shi et al. 2020).

Type-2 compounds enhance hERG activity mainly by attenuating C-type (selectivity filter) inactivation; examples include ICA-105574 (Gerlach et al. 2010), AZSMO-23 (Mannikko et al. 2015), NS3623 (Hansen et al. 2006b), MC-450 (Gualdani et al. 2017), PD118057 (Zhou et al. 2005), PD307243 (Gordon et al. 2008), and some others. Although classified as type-2, most of these compounds have multiple mechanisms of action. ICA-105574 is one of the first type-2 activators reported and is one of the most potent hERG activators discovered to date. ICA-105574 prevented moxifloxacin-induced ventricular tachycardia and fibrillation in guinea pig hearts, presumably acting to reverse IhERG inhibition by moxifloxacin (Meng et al. 2013). Importantly, when applied alone ICA-105574 caused shortening of the ventricular action potential to the point of triggering ventricular fibrillation (Meng et al. 2013; Qiu et al. 2019). Experimental evidence suggests that many type-2 activators bind near the selectivity filter of hERG (Perry et al. 2009; Gerlach et al. 2010; Garg et al. 2011). Thus, PD118057 interacts with residues within the pore helix of one α-subunit and the N-terminal half of the S6 helix of an adjacent subunit (Perry et al. 2009). Similarly, ICA-105574 binds to residues within the pore helix at the base of the selectivity filter and S6 segments (Garg et al. 2011); these interactions may directly stabilize the pore in its active state (Schewe et al. 2019; Zangerl-Plessl et al. 2020). Whether this is indeed a generalized mechanism for other type-2 hERG activators remains to be established.

Examples of predominantly type-3 compounds are mallotoxin (MTX) (Zeng et al. 2006), KB130015 (Gessner et al. 2010), and SKF-32802 (Donovan et al. 2018). These were shown to induce a hyperpolarizing shift in the voltage dependence of hERG activation, thus amplifying the channel activity at positive voltages.

Additionally, MTX was shown also to speed up the activation and slow the deactivation kinetics (Zeng et al. 2006).

An example of a type-4 opener is SB-335573, which is structurally related to SKF-32802. It potentiates hERG current by increasing P_o across the voltage range, without affecting the voltage dependence of activation (Donovan et al. 2018). Another drug that increases P_o with only a small effect on voltage dependence of activation is PD-118057 (Perry et al. 2009). As mentioned above, this compound also attenuates channel inactivation.

4.3 *Limitations for Clinical Use of hERG Activators*

There are several factors limiting potential utility of hERG openers as LQTS drugs. Firstly, shortening of cardiac action potential may lead to short QT syndromes (SQTS). Congenital SQTS are associated with severe ventricular arrhythmias and early cardiac death (Schreiber and Seebohm 2021), which raises concerns that QT-shortening drugs could also lead to arrhythmias (Crotti et al. 2010). Indeed, experimental and in silico evidence revealed a potential proarrhythmic risk of some hERG activators including MTX, NS1643, ICA-105574, and PD-118057 (Peitersen et al. 2008; Lu et al. 2008b; Bentzen et al. 2011; Schewe et al. 2019). Additionally, some hERG agonists may have narrow therapeutic windows, e.g., due to blocking hERG at higher concentrations (Zhou et al. 2005; Kang et al. 2005a; Hansen et al. 2006b; Casis et al. 2006; Su et al. 2009; Gerlach et al. 2010; Perissinotti et al. 2015) and/or due to lack of selectivity and effects on other ion channels. For example, the type 1–2 agonist, NS1643, was suggested to carry a proarrhythmic risk due to Nav1.5 channel inhibition (Peitersen et al. 2008). Likewise, KB130015 affects multiple ion channels, including K_{ir}, L-type Ca^{2+} channels, and voltage-gated Na^+ channels (Mubagwa et al. 2003).

5 K_v7 Channels and Excitability Disorders

The K_v7 subfamily of voltage-gated K^+ channels, also known as the KCNQ- or M-type channels, plays a critical role in regulating the membrane potential and firing properties of numerous excitable cell types; they also play an important role in epithelial transport and cell volume regulation (Delmas and Brown 2005; Gamper and Shapiro 2015; Du et al. 2018; Jones et al. 2021). Due to their well-understood role in human health and disease and due to their rich pharmacological profile, like in the case with hERG, we will consider these channels in a separate section. K_v7 channel physiology, pharmacology, and associated channelopathies have been extensively reviewed elsewhere; hence, here we will only consider some major topics and examples; for in-depth coverage of this fascinating subfamily, the reader is referred to the topical reviews (Brown and Passmore 2009; Maljevic et al. 2010; Brueggemann et al. 2012; Gamper and Shapiro 2015; Barrese et al. 2018; Du et al. 2018; Jones et al. 2021; Liu et al. 2021).

The K_v7 channels ($K_v7.1$–$K_v7.5$) conduct K^+ currents with slow activation and deactivation kinetics, little inactivation, and a notably low activation threshold (approximately −60 mV), which makes them a unique tool for controlling excitability at the threshold (Delmas and Brown 2005; Gamper and Shapiro 2015; Du et al. 2018; Jones et al. 2021). The α-subunits of K_v7 subfamily are encoded by the corresponding *KCNQ* genes (KCNQ1–5) that are positioned on chromosomes 11p15, 20q13, 8q24, 1p34, and 6q13, respectively. Each *KCNQ* gene expresses a K_v7 protein with 650–940 amino acids, which assembles into a tetrameric K_v channel. Homo- or heterotetramers of K_v7 proteins form channels with distinct tissue localizations and functions (Maljevic et al. 2003; Gutman et al. 2005; Schwake et al. 2003, 2006). The discovery of numerous mutations in the *KCNQ* genes that are associated with human excitability disorders, including epilepsy, hearing loss, cardiac arrhythmia, and some others, underscores the physiological significance of these channels (reviewed in Gamper and Shapiro 2015; Jones et al. 2021). Recently, the cryo-EM structures of the K_v7 family, including the human $K_v7.1$, $K_v7.2$, and $K_v7.4$ channels, have been resolved, providing further insight into the gating mechanisms and drug action of these channels (Sun and MacKinnon 2017, 2020; Li et al. 2020b, c).

5.1 *Tissue Distribution, Function and Channelopathies (see Table 7 for Summary)*

The K_v7 subfamily is unevenly split into the predominantly neuronal M channels ($K_v7.2$–$K_v7.5$) and a cardiac channel, $K_v7.1$ (this division is not absolute; see below). $K_v7.1$, also known as K_vLQT1, differs from the other four members of the K_v7 family in that it is not believed to form heterotetramers with any other K_v7 channel subtypes and is not typically expressed in neurons. The $K_v7.1$ channel is present throughout the body, including in the heart, airways, uterus, pancreas, gastrointestinal tract, kidneys, and inner ear (Barhanin et al. 1996; Sanguinetti et al. 1996). In the heart, $K_v7.1$ tetramers form functional channels in complex with KCNE1 (formerly known as MinK) β-subunits. These channels are the basis for the I_{Ks}, a slow delayed-rectifier K^+ current involved in repolarization of the cardiac action potential and an important determinant of the QT interval in the electrocardiogram (Fig. 3) (Barhanin et al. 1996; Sanguinetti et al. 1996; Nicolas et al. 2008). In the inner ear, $K_v7.1$/KCNE1 regulate auditory function, and in the pancreatic acinar cells, activation of the $K_v7.1$/KCNE1 drives Cl^- secretion (Kottgen et al. 1999; Robbins 2001; Wang and Li 2016; van der Horst et al. 2020). Mutations in this channel have been linked to a form of inherited LQTS as well as to a form of deafness, indicating that KCNQ1 plays a role in K^+ recycling in the inner ear (Kapplinger et al. 2015;

Table 7 Overview on location, properties, channelopathies, and potential therapeutic potential of K_v7 channels

Subunit	Tissue/organ expression	Current properties	Functions	Channelopathies	Therapeutic potential
$K_v7.1$	Heart, inner ear, epithelia, smooth muscle	Delayed-rectifier I_{Ks}; epithelial K^+ current	Repolarizes action potential; potassium transport	LQTS; JLNS; AF; type 2 diabetes; deafness	Arrhythmias
$K_v7.2$	Brain, spinal cord, autonomic and sensory ganglia, skeletal muscle	M-current $I_{K(M)}$	Controls excitability	BFNE; KCNQ2-DEE; RE; KS; myokymia; neuromyotonia	Epilepsy; pain; anxiety; schizophrenia; learning, memory; neurodevelopmental disorders
$K_v7.3$	Brain, autonomic and sensory ganglia, gut	M-current $I_{K(M)}$	Controls excitability	BFNE; BFIE; IGE; ASD	Depression; schizophrenia; learning, memory, neurodevelopmental disorders; epilepsy
$K_v7.4$	Cochlea, auditory nuclei, mechanosensitive neurons, smooth muscle	Auditory hair cell K^+ current I_{Ks}; M-like current	Potassium transport; regulation of smooth muscle contractility	DFNA2	Depression; hypertension; airway hyperconstriction; ED
$K_v7.5$	Brain, autonomic and sensory ganglia, smooth muscle, retinal pigment epithelium	M-like current	Controls excitability	EE; ID	Epilepsy

Aggregated data from Leppert and Singh (1999); Mackie and Byron (2008); Peroz et al. (2008); Brown and Passmore (2009); Maljevic et al. (2010); Brueggemann et al. (2012); Maljevic and Lerche (2014); Gamper and Shapiro (2015); Barrese et al. (2018); Allen et al. (2020); Nappi et al. (2020); Jones et al. (2021); Kruger et al. (2022)

AF atrial fibrillation, *ASD* autism spectrum disorder, *BFNE* benign familial neonatal epilepsy, *BFIE* benign familial infantile epilepsy, *DFNA2* autosomal dominant hearing loss 2, *EE* epileptic encephalopathy, *KCNQ2-DEE* KCNQ2-encephalopathy, *KS* Kleefstra syndrome, *IGE* idiopathic generalized epilepsy, *ID* intellectual disability, *JLNS* Jervell and Lange-Nielsen syndrome (LQTS + deafness), *LQTS* long QT syndrome, *OHC* outer hair cells, *RE* rolandic epilepsy, *ED* erectile dysfunction

Mousavi Nik et al. 2015; Oertli et al. 2021). $K_v7.1$ channels in pancreatic ß-cells help to regulate insulin secretion; consequently, LOF mutations in the *KCNQ1* gene cause impaired ß-cell function (Ullrich et al. 2005) and are linked to type 2 diabetes (Unoki et al. 2008; Zhang et al. 2020). In many non-excitable cells $K_v7.1$ co-assemble with KCNE3, which results in a constitutively open channel (Sun and MacKinnon 2020; Abbott 2021). In the colon, small intestine, and airway epithelial tissues, $K_v7.1$/KCNE3 regulates water and salt transport (Schroeder et al. 2000a; Vallon et al. 2005).

$K_v7.2$ channel is found in both the peripheral and central nervous systems, with a high expression level in the neocortex, hippocampus, and peripheral sensory neurons (Cooper et al. 2001; Klinger et al. 2011; Du et al. 2018; Zheng et al. 2019). $K_v7.2$ is also reportedly expressed in the skeletal muscle (Iannotti et al. 2010). $K_v7.3$ is highly expressed in the hippocampus, thalamus (Kanaumi et al. 2008), and cerebellum (Geiger et al. 2006; Klinger et al. 2011). The heteromultimers of $K_v7.2$ and $K_v7.3$ are the molecular basis of the "classical" neuronal M-type current, a crucial regulator of neuronal excitability (Wang et al. 1998; Delmas and Brown 2005). LOF mutations within *KCNQ2* (predominantly) and *KCNQ3* genes in humans often result in an inherited benign form of epilepsy, benign familial neonatal epilepsy (BFNE), and this has led to the identification and cloning of these channels (Biervert et al. 1998; Singh et al. 1998). Variations in the *KCNQ2* gene are responsible for a broad range of phenotypes associated with hyperexcitability, ranging from mild and self-limiting epilepsy to severe epileptic encephalopathy with cognitive impairment, as well as pharmacoresistant seizures, such as neonatal epileptic encephalopathy (NEE) (Miceli et al. 1993; Neubauer et al. 2008; Goto et al. 2019; Gao et al. 2022; Liu et al. 2022a). Epileptogenic *KCNQ3* variants have also been described (Leppert and Singh 1999; Nappi et al. 2020).

$K_v7.4$ is highly expressed in the cochlear hair cells, which are essential for normal hearing. Mutations in *KCNQ4* are a significant cause of congenital deafness (Kubisch et al. 1999; Kharkovets et al. 2000, 2006; Nouvian et al. 2003). Dopaminergic neurons, as well as the vestibular and auditory systems, primarily express the homomeric form of the $K_v7.4$ channel, while smooth muscle cells express $K_v7.4/7.5$ heteromers (Chadha et al. 2014; Gamper and Shapiro 2015; Mani et al. 2016; Jones et al. 2021). This variant is present in various smooth muscle tissues, including blood vessels, bladder, gastrointestinal tract, arteries, uterus, airways, and the corpus cavernosum of penis. In these smooth muscles, $K_v7.4$ plays a significant role in regulating contractility and vascular tone (McCallum et al. 2009, 2011; Ipavec et al. 2011; Evseev et al. 2013; Svalo et al. 2015; Jepps et al. 2016). Additionally, skeletal muscle differentiation has been linked to upregulation of $K_v7.4$ expression (Zavaritskaya et al. 2013).

The *KCNQ5* gene, which was identified in 2000, is highly expressed in neural tissues, including the retinal pigment epithelium (Schroeder et al. 2000b). The $K_v7.5$ protein, which has a comparable distribution in the brain to that of $K_v7.2$ and $K_v7.3$, is expressed in different parts of the brain in both homomeric form and as $K_v7.3/7.5$ heteromer, and is also believed to be a component of the M-type currents (Lerche et al. 2000). Mutation of $K_v7.5$ results in retinopathy (Caminos et al. 2015) and epileptic encephalopathy (Kruger et al. 2022; Wei et al. 2022). $K_v7.5$ is also expressed in the vascular smooth muscle, and regulates vascular tone and blood pressure, often in heteromeric form with $K_v7.4$ (Mani et al. 2016; Zimmer et al. 2017).

5.2 Pharmacology of K_v7 Channels

Pharmacology of K_v7 channels has been considered in detail elsewhere (Du et al. 2018; Borgini et al. 2021; Jones et al. 2021; Liu et al. 2021). Briefly, $K_v7.2$ but not $K_v7.3$ is highly sensitive to TEA (Wang et al. 1998; Shapiro et al. 2000). Two closely related compounds, linopirdine and XE991, are potent inhibitors of all K_v7 channels (IC_{50}s in the range of 1–5μM for $K_v7.1$–$K_v7.4$ (Wang et al. 1998, 2000; Sogaard et al. 2001) and ~60μM for $K_v7.5$ (Schroeder et al. 2000b; Jensen et al. 2005)). Chromanol 293B inhibits $K_v7.1$ and I_{Ks} (Suessbrich et al. 1996; Busch et al. 1997). A great variety of endogenous and exogenous substances have the ability to activate K_v7 channels. Among the endogenous ones is the obligatory cofactor, phosphatidylinositol 4,5-bisphosphate (PIP_2), which is needed for channel activity (Suh and Hille 2002; Zhang et al. 2003; Li et al. 2005). In turn, GPCR-induced PIP_2 depletion mediates physiological M channel inhibition (and neuronal excitation) by acetylcholine, bradykinin, angiotensin II, and many other such neurotransmitters and neuromodulators (reviewed in Gamper and Shapiro, 2015; Gamper and Shapiro 2015; Jones et al. 2021). K_v7 channels are also activated by GABA and related compounds (Manville et al. 2018; Abbott 2020), polyunsaturated fatty acids (PUFAs) (Liin et al. 2016; Frampton et al. 2022), endogenous and exogenous reactive oxygen species such as hydrogen peroxide (Gamper et al. 2006; Linley et al. 2012; Nunez et al. 2023) and intracellular zinc (Gao et al. 2017; Yang et al. 2023), among other substances (Abbott 2020). A great variety of small-molecule K_v7 openers were also developed; some examples are considered below.

5.3 K_v7 Channels and Nervous System Diseases

Due to the prominent and well-understood inhibitory action of K_v7/M currents in both central and peripheral neurons, K_v7 channel activators are attracting close attention as potential anticonvulsants and analgesics (reviewed in Du et al. 2018; Liu et al. 2021; Jones et al. 2021). The "classical" M channel activators/openers, retigabine (RTG, ezogabine) and flupirtine, have been used clinically as an anticonvulsant (retigabine) and an analgesic (flupirtine), but both are currently restricted due to a range of side effects (Clark et al. 2015; Bock and Link 2019). Both are potent but relatively nonselective K_v7 openers that activate all K_v7 subunits except $K_v7.1$, with $K_v7.3$ being the most sensitive subunit (Tatulian and Brown 2003; Schenzer et al. 2005; Wuttke et al. 2005). Multiple K_v7 activators are being investigated for treatments of epilepsy and diseases related to the neuronal hyperexcitability including depression, anxiety, mania, attention-deficit/hyperactivity disorder (ADHD), addiction to psychostimulants, migraines, and pain (Wickenden and McNaughton-Smith 2009; Du et al. 2018; Greene and Hoshi 2016; Borgini et al. 2021; Naffaa and Al-Ewaidat 2021). Additionally, recent studies demonstrate that K_v7 channel openers may have potential neuroprotective benefits in animal models of traumatic brain injury (TBI), induced acute cerebral infarction, and catastrophic stroke (Gamper et al. 2006; Bierbower et al. 2015; Vigil et al. 2020a, b); the general idea here is that hyperpolarization produced by K_v7 activation would limit excitotoxicity by dampening the release of glutamate (and other excitatory substances) by the damaged or stressed neurons (Gamper et al. 2006). On the other hand, blockade of K_v7 may mitigate the cognitive loss seen in neurodegenerative illnesses (such as Parkinson's disease, PD), by raising neuronal excitability and enhancing the release of various neurotransmitters, including serotonin, dopamine, and glutamate (Drion et al. 2010; Shi et al. 2013). In the somatosensory system, K_v7 is one of the dominant regulators of excitability of peripheral nociceptive neurons; M-like channel activity is confirmed at the peripheral and central terminals of nociceptive neurons, as well as in the cell bodies of these neurons in the dorsal root and trigeminal ganglia, where K_v7s is one of the major regulator of E_{rest} and excitability (Passmore et al. 2003, 2012; Linley et al. 2008; Liu et al. 2010; Du et al. 2014, as reviewed in Du et al. 2018). Overall, strong biological and pharmacological evidence for the causative link between $K_v7.2$ and 7.5 channel activity and both neurological and mental illnesses is very strong. Based on this, we expect to see greater clinical recognition of the potential benefits of K_v7 modulation in treating nervous system diseases, further expanding the benefits of the two already approved (even though with restricted use) K_v7 activators, retigabine and flupirtine. Fortunately, K_v7 channels lend themselves very well to ligand-mediated opening, with hundreds of compounds identified to date (see Du and Gamper 2013; Du et al. 2018; Borgini et al. 2021; Liu et al. 2021; for review; examples are given in Table 8). Many of these modulators demonstrated antiepileptic and analgesic effects in animal models, and some have undergone additional preclinical studies and clinical trials.

5.4 K_v7 Channels and Cardiopathies

I_{Ks}, which is the slow component of the delayed-rectifier current observed during the late repolarizing phase of the cardiac action potential (Fig. 4), is generated by $K_v7.1$ in complex with KCNE1 proteins. The slow deactivation rate of I_{KS} causes current to accumulate with increased cardiac pacing, indicating that inhibiting I_{Ks} may have a scope in managing some arrhythmias (Nicolas et al. 2008; Seebohm 2009; Wu and Larsson 2020; Schreiber and Seebohm 2021). However, mutations in *KCNQ1* are linked to LQTS (Mousavi Nik et al. 2015; Oertli et al. 2021), and, accordingly, inhibiting I_{KS} in the ventricular muscle reduces the repolarization reserve, unfortunately increasing the likelihood of a reentrant arrhythmias, such as LQTS. As I_{KS} inhibition is likely to precipitate ventricular fibrillation, experts argue that not only hERG but also $K_v7.1$ safety testing should be mandatory for new drug development (Towart et al. 2009; Huang et al. 2017). Although targeting the plasmalemmal $K_v7.1$/KCNE1 channels may not offer many clinical advantages, doing so for heart mitochondrial $K_v7.4$ could have therapeutic benefits (Testai et al. 2016). Additionally, $K_v7.1$-specific activators may have the potential to treat familial atrial

Table 8 Examples of K_v7 openers and their physiological action

Compound	Selectivity	EC_{50} (μM) or efficacy (fold)	In vivo evidence for potential therapeutic use
Flupirtine	Pan $K_v7.2$–$K_v7.5$ activator	4.6 ($K_v7.2/7.3$) 1.6 ($K_v7.3/7.5$)	Analgesic (marketed); depression; ischemic brain injury; memory impairment; cocaine addiction
Retigabine (ezogabine)	Pan $K_v7.2$–$K_v7.5$ activator	2.5 ($K_v7.2$) 0.34 ($K_v7.2/7.3$) 5.2 ($K_v7.4$)	Anticonvulsant (marketed); pain (including neuropathic, musculoskeletal, postoperative pain, migraine, and inflammatory pain); depression; anxiety
BMS-204352 (MaxiPost)	Pan $K_v7.2$–$K_v7.5$ activator, inhibits $K_v7.1$; also activates BK channels	At 10μM: 2.3-fold (Kv7.2) 1.3-fold (Kv7.2/7.3) 2.0-fold (Kv7.4) >12-fold (Kv7.5)	Depression; anxiety; acute ischemic stroke
ICA-27243	$K_v7.2/7.3$ > $K_v7.4$ and $K_v7.5$ activator	0.4 ($K_v7.2/7.3$) 4.8 ($K_v7.3$) 9.7 ($K_v7.4$)	Inflammatory pain; L-DOPA-induced dyskinesia; mania; epilepsy
ML213	Selective $K_v7.2$, $K_v7.3$ and $K_v7.5$ activator	0.23 ($K_v7.2$) 1.6 ($K_v7.4$) 3.8 ($K_v7.5$)	Epilepsy
NS15370	Pan $K_v7.2$–$K_v7.5$ activator	0.1 ($K_v7.2$) 0.13 ($K_v7.2/7.3$) 0.04 ($K_v7.4$) 0.15 ($K_v7.5$)	Epilepsy; mania and psychosis
SF0034	Selective $K_v7.2$, $K_v7.3$, and $K_v7.2/7.3$ activator	1.3 ($K_v7.2/7.3$)	Epilepsy
QO-58	Pan $K_v7.2$–$K_v7.5$ activator	0.06 ($K_v7.2/7.3$) 1.3 ($K_v7.2$) 0.6 ($K_v7.4$)	Pain

Aggregated data from Block et al. (1997); Schroder et al. (2001); Dupuis et al. (2002); Jensen (2002); Korsgaard et al. (2005); Roeloffs et al. (2008); Wickenden et al. (2008); Redrobe and Nielsen (2009); Klinger et al. (2012); Dalby-Brown et al. (2013); Sander et al. (2013); Zhang et al. (2013); Brueggemann et al. (2014); Hayashi et al. (2014); Kalappa et al. (2015); Huang et al. (2015a); Friedman et al. (2016); Teng et al. (2016); Mooney and Rawls (2017); Kanyo et al. (2020); Borgini et al. (2021)

fibrillation brought on by mutations that impair function of *KCNQ1* or KCNE1 (Xu et al. 2015; Hou et al. 2019). In addition, $K_v7.4$ and $K_v7.5$ channels are present in the smooth muscle of the coronary arteries (Khanamiri et al. 2013; Morales-Cano et al. 2015); thus, increasing $K_v7.4$ may be advantageous in lowering coronary resistance and minimizing infarct size after an ischemic episode (Hedegaard et al. 2014).

5.5 K_v7 Channels and Auditory Diseases

Mutations within *KCNQ1* (Wei et al. 2000; Wang et al. 2002) and *KCNQ4* (Kubisch et al. 1999; Nouvian et al. 2003) genes are associated with the congenital forms of deafness. Thus, K_v7 activators may provide hope for treatment of some hearing conditions, e.g., tinnitus. Although the cause of tinnitus is not fully understood, it appears to be associated with neuronal hyperexcitability within the auditory pathways. Tinnitus can arise from various causes, such as ear infections, noise-induced hearing loss, aging, brain tumors, and neurodegeneration, as well as other potential factors. In a model of noise-induced hearing loss, tinnitus is often accompanied by feelings of sadness and anxiety (Henry et al. 2014). Several K_v7 activators were found to prevent the onset of tinnitus without impairing temporal processing or hearing thresholds (Kalappa et al. 2015). There is a growing interest in identification of additional *KCNQ4* variants linked to genetic hearing loss (e.g., autosomal dominant hearing loss 2, DFNA2) and the possibility to restore their function with the use of K_v7 openers (Oliver et al. 2003; Leitner et al. 2012; Xia et al. 2020), with some success demonstrated in vitro and in vivo (Kalappa et al. 2015). This lays the groundwork for a potential strategy to treat DFNA2; however, $K_v7.4$-selective openers without efficacy on other K_v7 subunits are desirable for such a strategy. Fasudil was recently identified as one such $K_v7.4$-selective opener (Zhang et al. 2016), but it also has other pharmacological actions, such as RhoA/Rho kinase (ROCK) inhibition (Nagumo et al. 2000).

5.6 K_v7 Channels and Smooth Muscle-Related Disorders

As mentioned, $K_v7.1$, $K_v7.4$, and $K_v7.5$ are functionally expressed in many types of smooth muscle. As in neurons, K_v7 channels hyperpolarize the E_m in smooth muscle cells, suppressing action potentials and voltage-gated Ca^{2+} channel opening, which provide Ca^{2+} necessary for contractile activity of the muscle (Jin and Lu 1999; Bhave et al. 2011). It is believed that the amount of membrane hyperpolarization produced by pharmacological activation of vascular K_v7 channels could be sufficient to decrease the chance of Ca^{2+} channel opening and, as a result, attenuate vasoconstrictor reactions (Greenwood and Ohya 2009; Rode et al. 2010; Haick and Byron 2016; Takagi and Hashitani 2016). In several rodent and human arteries, including the mesenteric, gracilis, cerebral, coronary, penile and renal, as well as in the visceral adipose smooth muscle, gastrointestinal tract, corpus cavernosum, uterus, and airways, transcripts for *KCNQ* genes have been found (*KCNQ1*, *KCNQ4*, *KCNQ5* appear to be dominant) (Greenwood and Ohya 2009; Haick and Byron 2016). K_v7 channel activity maintains the relaxed state of smooth muscles; in addition, these channels are among the functional endpoints for several receptor vasorelaxants (Jepps et al. 2009; Evseev et al. 2013; Mani et al. 2016). K_v7 channel activators can effectively relax precontracted arteries and reduce the effects of vasoconstrictors, although it is important to point out that due to broad expression in the nervous system, nonselective K_v7 openers are likely to produce CNS side effects (e.g., sedation). It is tempting to speculate that developing molecules that could selectively target $K_v7.4/K_v7.5$ heteromers (which hitherto were only described in the vasculature (Chadha et al. 2014; Mani et al. 2016) could provide a more specific avenue to modulate vascular tone via K_v7 channel activity.

The discovery of K_v7 functional expression in the penile arteries and corpus cavernosum sparked interest in the role of K_v7 activators in erectile dysfunction (ED) (Diniz et al. 2020; Lee et al. 2020). Additionally, identification of expression of K_v7 (primarily $K_v7.3$) channels in mouse nodose neurons innervating the lung may bring additional benefit for using K_v7 openers for some airway diseases such as hypersecretion disorders, dyspnea, bronchoconstriction, and chronic cough (Mondejar-Parreno et al. 2020).

In contrast to K_v7 channel activators, K_v7 channel inhibition increases the spontaneous contractility of phasic tissues in the gastrointestinal tract, bladder, and uterus (Jackson 2018; Malysz and Petkov 2020; Wray and Arrowsmith 2021). Since K_v7 channels are physiologically inhibited by many GPCR (e.g., via PIP_2 depletion, PKC activation, or Ca^{2+} transients; reviewed in Delmas and Brown 2005; Gamper and Shapiro 2015; Jones et al. 2021), GPCR-mediated K_v7 inhibition may contribute to vasoconstriction by some vasoactive agents, such as angiotensin II (Clement-Chomienne et al. 1996; Jackson 2018). Taken together, there is significant evidence that K_v7 channels may play a role of "master regulators" of smooth muscle activity, which could potentially lead to the treatment of hypertension, overactive bladder syndrome, airway hyperconstriction, and other conditions (Brueggemann et al. 2012; Haick and Byron 2016; Jepps 2021; Jones et al. 2021). However, the lack of mechanisms for selective/specific targeting of vascular K_v7 channels hitherto hinders translational efforts and will have to be overcome in order to develop successful future therapies.

6 Targeting Ca^{2+}-Activated K^+ Channels

The K_{Ca} channels encompass large-, intermediate-, and small conductance K_{Ca} channels (BK, IK, and SK), respectively, as well as K_{Na} (Slo2.1 and Slo2.2) and the pH-dependent Slo3 channel (Fig. 1). Mutations in most of these channels are linked to various excitability disorders, and, hence, these channels offer plausible therapeutic opportunities.

6.1 BK Channels

BK channels are widely expressed in most mammalian tissue types. The channels are cooperatively activated by both depolarization and intracellular Ca^{2+} elevation, enabling them to play an important regulatory role in cells where depolarization is linked to Ca^{2+} influx (e.g., neurons and muscle cells). Hence, BK channels are important for control of neuronal excitability (Adams et al. 1982) neurotransmitter release (Robitaille et al. 1993), muscle contraction (Brayden and Nelson 1992), endothelial function (Nilius and Droogmans 2001), etc. In non-excitable (epithelial, glial, etc.) cells, BK channels can regulate Ca^{2+} entry (e.g., via P2X, CRAC, or TRP channels) by clamping membrane voltage and, hence, Ca^{2+} driving force (reviewed in Nilius and Droogmans 2001; Dudem et al. 2021). BK channels have a general architecture similar to K_v channels, but with a number of important distinctions; thus, BK has an additional N-terminal TMD, the S0, and two C-terminal RCK ("regulator of conductance for K^+") domains crucial for Ca^{2+} binding. The channels are regulated by a number of auxiliary subunits (reviewed in Dudem et al. (2021)).

Both gain- and loss-of-function mutations within the BK α-subunit-encoding gene, *KCNMA1*, were linked to neurological disorders, including epilepsy, paroxysmal dyskinesia, and other movement abnormalities (Sausbier et al. 2004; Du et al. 2005; Yang et al. 2010). BK channels are expressed in smooth muscles and represent a valid therapeutic target for urinary incontinence and overactive bladder (Meredith et al. 2004). Mice lacking the BK auxiliary subunit-β1 displayed increased arterial tone and blood pressure (Brenner et al. 2000); hence, BK channel modulators may have a role in treatment of hypertension. BK channels were also implicated in autism and mental retardation (Laumonnier et al. 2006) and cerebellar ataxia (Sausbier et al. 2004) (reviewed in Dudem et al. 2021).

BK channel activators are being developed for treatments of a number of diseases, such as epilepsy, stroke, asthma, hypertension, ischemic heart disease, erectile dysfunction, bladder overactivity, etc. (Nardi and Olesen 2008; Dudem et al. 2021). Despite a large number of compounds identified, none have yet passed phase III clinical trials, likely due to their poor selectivity and off-target effects, limited bioavailability, or lack of efficacy. Yet, the search is continuing, and we are hopeful that clinically successful BK channel openers are not long coming (see the last section of this chapter).

For a comprehensive listing of BK channel activators, the reader is directed to the excellent recent review (Dudem et al. 2021); here we will highlight some examples. BK activators can be subdivided into endogenous, synthetic, and naturally accruing compounds.

Among endogenous activators are metabolites of cytochrome 450, like epoxygenase, lipoxygenase, and arachidonic acid (Hou et al. 2009), hemoxigenase (Williams et al. 2004), the omega-3 fatty acid DHA (Hoshi et al. 2013), and steroidal hormones such as 17β-estradiol (E2) (Valverde et al. 1999). Action of many of the endogenous modulators often depends on the presence of specific β-subunits (Dudem et al. 2021).

Natural compounds with BK opener activity include dehydrosoyasaponin-1 (DHS-1), a compound extracted from the medicinal herb *Desmodium adscendens*, which is historically used in Ghana for the treatment of asthma (Addy and Awumey 1984; McManus et al. 1993). Pimaric acid, purified from the resin of *Pinus* plants, is also a potent BK activator (Imaizumi et al. 2002). Trans-resveratrol, a polyphenol from *Polygonum cuspidatum*, was shown to activate the BK cannels in vascular endothelial cells (Li et al. 2000). A number of other plant-derived BK openers were also discovered (Dudem et al. 2021).

There are many synthetic BK activators reported to date (Bentzen et al. 2014). Among the first discovered were benzimidazoles NS004 and NS1619 (Olesen et al. 1994). These compounds induced strong negative shift of BK channel voltage dependence, making them to open more easily. These compounds lacked selectivity (e.g., also blocked voltage-gated Ca^{2+} channels (Edwards et al. 1994; Holland et al. 1996)), but did provide a useful template for the subsequent structure-activity relationship studies. Thus, NS11021, developed later by the same group, was more potent and efficacious than the original compounds and had little effect on T- or L-type Ca^{2+} channels, Na^+ channels, K_{ir}, or K_v channels, although it did activate $K_v7.4$ (Bentzen et al. 2007). NS11021 showed efficacy in improving erectile responses in rats (Kun et al. 2009; Kiraly et al. 2013) and reduced contractility in the guinea pig bladder (Layne et al. 2010).

Some dyes, such as the voltage-sensitive fluorescent dye, bis-(1,3-dibutylbarbituric acid) trimethine oxonol [DiBAC$_4$(3)] (Morimoto et al. 2007; Scornik et al. 2013) and aniline compound, Cibacron blue (Cotton et al. 1996), were shown to activate BK channels. The latter compound was used to synthesize a series of GoSlo compounds (Roy et al. 2012) with the aim to develop therapy for the overactive bladder (OAB). An in-depth review of the GoSlo family compounds can be found in recent literature (Hoshi and Heinemann 2016; Dudem et al. 2021).

Compound BMS204352 with BK opener activity reduced regional cerebral edema and motor impairment after brain surgery in mice (Cheney et al. 2001), although it also a K_v7 channel opener (Jensen 2002) (Table 8). Another BK opener, andolast, showed promise in the treatment of asthma (Malerba et al. 2015). None of the above passed full clinical trials yet (Dudem et al. 2021).

6.2 SK Channels

SK channels ($K_{Ca}2.1$–$K_{Ca}2.3$) are activated by relatively low intracellular Ca^{2+} concentrations (~500 nM), which binds to the calmodulin (CaM), which is constitutively associated with the channel via its C-terminal CaM-binding domain (CaMBD) (Schumacher et al. 2001; Wei et al. 2005; Li et al. 2009; Zhang et al. 2012b). SK subunits are broadly expressed in the nervous and cardiovascular systems and regulate various parameters of neuronal excitability; in particular, SK channels are involved in afterhyperpolarization of the neuronal action potential (Faber and Sah 2007). Mutations in $K_{Ca}2.3$-encoding *KCNN3* gene were associated with schizophrenia (Miller et al. 2001; Soden et al. 2013) and hypertension (Koot et al. 2016). Consequently, these channels are considered as therapeutic targets for these diseases, as well as for Parkinson's and Alzheimer's diseases (reviewed in Cui et al. 2014; Kshatri et al. 2018; Dudem et al. 2021). Modulators of SK channels include blockers, as well as positive and negative gating modulators (Lam et al. 2013; Christophersen and Wulff 2015; Dudem et al. 2021). The most well-known blocker of SK channels is the bee venom toxin, apamin (Kawai and Watanabe 1986; Maylie et al. 2004), which binds to the outer vestibule of the SK channel pore and inhibits all three SK channels ($K_{Ca}2.2 > K_{Ca}2.3 > K_{Ca}2.1$) with picomolar to low-nanomolar potency (Nolting et al. 2007; Lamy et al. 2010; Kuzmenkov et al. 2022), having little effect on most other ion channels (Kuzmenkov et al. 2022). Another toxin (scorpion venom), tamapin, is more selective for $K_{Ca}2.2$ over other two subunits (Pedarzani et al. 2002).

Small-molecule pan-SK inhibitors can be exemplified by methyl-laudanosine (Scuvee-Moreau et al. 2002) and methyl-noscapine (Scuvee-Moreau et al. 2004), a series of bis-quinolinium cyclophanes, e.g., UCL1684 and UCL1848 (Conejo-Garcia and Campos 2008), and an aminobenzimidazole derivative NS8593 (Strobaek et al. 2006). NS11757, on the other hand, exhibits selectivity towards $K_{Ca}2.3$ channels and inhibits the channel with nanomolar potency (Sorensen et al. 2008). The antidepressant fluoxetine also blocks all three SK subunits with low-micromolar potencies (Terstappen et al. 2003). A common feature of the SK channel inhibitors is that they attenuate afterhyperpolarization in many types of CNS neurons.

Positive gating modulators (activators) of SK channels include 1-EBIO (1-ethyl-2-benzimidazolinone) (Devor et al. 1996) and a related, more potent compound, NS309. Another compound developed by the same group, NS13001, selectively activates $K_{Ca}2.2$ and $K_{Ca}2.3$ with low-micromolar potency. The latter compound was shown to have a neuroprotective effect on Purkinje neurons in a mouse model of spinocerebellar ataxia type 2 (Kasumu et al. 2012). Purkinje neurons from the ataxic mice fired irregularly and NS13001 restored regular pacemaker activity of these neurons (Kasumu et al. 2012). Another SK activator, SKA-31, induced smooth muscle hyperpolarization and decrease contractility in human and guinea pig detrusor, identifying SK channel activators as a potential treatment for bladder dysfunctions (Parajuli et al. 2012; Soder et al. 2013). GW542573X, a $K_{Ca}2.1$-selective activator, was shown to activate the channel even in the absence of Ca^{2+} (Hougaard et al. 2009).

6.3 IK Channels

IK channels ($K_{Ca}3.1$, *KCNN4*) are activated by Ca^{2+}/CaM in a manner similar to the SK channels. This channel was cloned in 1997 by three different groups (Ishii et al. 1997; Joiner et al. 1997; Logsdon et al. 1997), but was functionally first described by Gardos in human red blood cells in the 1950s (Gardos 1958; Hoffman et al.

2003). IK channels play a major role in red blood cell shrinkage via KCl and water loss that occurs in sickle cell anemia; hence, it has long been recognized as a potential target for treatment of this debilitating disease (Maher and Kuchel 2003). Besides erythrocytes, the channel is found in the bone marrow, lymph nodes, spleen, thymus, prostate, colon, lung, placenta, etc. (Ishii et al. 1997; Joiner et al. 1997; Logsdon et al. 1997), where it plays a role in proliferation, secretion, and volume regulation (Wulff et al. 2007). In the peripheral nervous system, IK channels are found in enteric, sensory, and sympathetic neurons (Furness et al. 2004; Bahia et al. 2005; Mongan et al. 2005), while in the CNS, these channels are mainly found in microglia and endothelial cells (Pedarzani and Stocker 2008).

IK channels are insensitive to apamin, but a number of other selective inhibitors, activators, and gating modulators of IK channels are reported in the literature (Maher and Kuchel 2003; Cui et al. 2014; Christophersen and Wulff 2015). Thus, IK channels are inhibited by the scorpion peptide toxins maurotoxin (Castle et al. 2003) and charybdotoxin (Miller et al. 1985), both with low-nanomolar potency. The K_v1-blocking toxins, margatoxin (MgTX) and stichodactyla toxin (ShK), also inhibit IK channels, but with somewhat lower potency (IC_{50}s ~ hundreds nM) (Jensen et al. 1998). Small-molecule inhibitors of IK channels include clotrimazole, nitrendipine, cetiedil, miconazole, econazole, and promethazine (Roxburgh et al. 1996; Jensen et al. 1998; Maher and Kuchel 2003; Wittekindt et al. 2006). Another potent (low nanomolar) and selective IK channel inhibitor is TRAM-34 (1-[(2-chlorophenyl)diphenylmethyl]-1H-pyrazole) (Wulff et al. 2000). It was shown to reduce astrogliosis and microglial activation and attenuated memory loss in a mouse model of Alzheimer's disease (Yi et al. 2016). TRAM-34 also decreased microglia activation, reduced infarction size, and improved neurological deficits in a mouse model of ischemic stroke (Chen et al. 2016). The compound also improved symptoms of experimental autoimmune encephalomyelitis in a mouse model of multiple sclerosis (Reich et al. 2005).

Among IK activators are the muscle relaxants chlorzoxazone and zoxazolamine (Syme et al. 2000). It must be noted that SK and IK channels are structurally close and many SK channel modulators also affect IK channels. Thus, the SK activator 1-EBIO and other related compounds (DCEBIO, NS309) also activate IK channels (Singh et al. 2001; Strobaek et al. 2004). Compounds SKA-111 and SKA-121 are more selective positive gating modulators of the IK channels with about 100-fold selectivity over the SKs (Coleman et al. 2014). SKA-121 significantly lowered arterial blood pressure in normotensive and hypertensive wild-type but not in $K_{Ca}3.1$ knockout mice (Coleman et al. 2014). The role of IK and SK channels in human diseases is covered in detail in several recent reviews (Cui et al. 2014; Kshatri et al. 2018; Dudem et al. 2021).

6.4 K_{Na} Channels

K_{Na} channels (Slack, *KCNT1* and Slick, *KCNT2*) were functionally first reported in the 1980s in mammalian cardiomyocytes (Kameyama et al. 1984) and avian sensory neurons (Bader et al. 1985). K_{Na} channels are structurally related to BK channels and also have two C-terminal RCK domains, but lack the N-terminal S0 domain. However, in contrast to BK channels, K_{Na} are activated by the intracellular Na^+, rather than Ca^{2+} (reviewed in Cole et al. 2021; Liu et al. 2022b). In excitable cells, K_{Na} channels are activated following Na^+ influx via voltage-gated Na^+ channels during action potential firing (especially during high-frequency firing). Accordingly, the physiological role of K_{Na} channels is in generating the slow afterhyperpolarization phase of an action potential (Franceschetti et al. 2003; Liu and Stan Leung 2004). They were also reported to contribute to resting membrane potential, control excitability and regulate precision of burst firing in CNS neurons (Yang et al. 2007; Lee et al. 2019; Reijntjes et al. 2019), and control intrinsic excitability of arterial smooth muscle cells (Li et al. 2019). The channels are highly, but often differentially, expressed in several brain areas, including all areas of the hippocampus and cortical layers II, III, and V (Bhattacharjee et al. 2002, 2005); K_{Na} channels are also expressed in somatosensory neurons of dorsal root ganglion (Gao et al. 2008; Tamsett et al. 2009; Nuwer et al. 2010). These channels are also found in cardiac and smooth muscles (reviewed in Kaczmarek 2013).

Typically, LOF mutations in neuronal K^+ channels result in hyperexcitability disorders, such as epilepsy and pain. Somewhat paradoxically, it is largely the GOF mutations within the *KCNT1* and *KCNT2* that were linked to such disorders. Thus far, more than 50 distinct pathogenic GOF variants of *KCNT1* and several more mutations in *KCNT2* were identified and linked to a number of developmental brain excitability disorders, including autosomal dominant nocturnal frontal lobe epilepsy (ADNFLE), Ohtahara syndrome (OS), West syndrome, early infantile epileptic encephalopathies (EOEE), epilepsy of infancy with migrating focal seizures (EIMFS), leukoencephalopathy, and other severe epilepsies (reviewed in Cole et al. 2021; Liu et al. 2022b). It is still being debated how exactly an increase in K_{Na} channel activity can result in epilepsy. The main ideas are as follows: (i) mostly inhibitory interneurons are affected and increased K_{Na} current in those neurons would result in disinhibition (Shore et al. 2020; Kuchenbuch et al. 2021); (ii) enhanced K_{Na} current may increase afterhyperpolarization, which, in turn, could help in synchronizing excitatory networks (Fernandez de Sevilla et al. 2006; Liu et al. 2022b); and (iii) potential pro-epileptic alterations in synaptic connectivity during development (Barcia et al. 2012; Liu et al. 2022b).

Regardless of the mechanistic explanation for the pro-epileptic effect of *KCNT* GOF, the K_{Na} channels represent an attractive target for treatment of the *KCNT*-associated encephalopathies, especially given the fact that these are often resistant to currently available treatments. Importantly, anti-arrhythmic drugs quinidine, bepridil and clofilium, are all efficacious K_{Na} channel inhibitors (reviewed in Kaczmarek 2013; Cole et al. 2021; Liu et al. 2022b), adding more to the promise. While these older anticonvulsants mentioned above are rather nonselective, the search for new, more potent, and selective K_{Na} inhibitors is ongoing. A recent study used cryo-EM structure of Slack (Hite et al. 2015) and mutational analysis to identify quinidine binding site within the channel pore region (Cole et al. 2020). The authors then performed virtual high-throughput screen of a library of 100,000 commercially available drug-like compounds in order to identify more potent inhibitors. The authors identified six putative inhibitors ("BC series") with low- to submicromolar potencies and confirmed their efficacy electrophysiologically. Another study used a high-throughput thallium flux assay to screen another large compound library and identified VU0606170, a selective Slack channel inhibitor with low-micromolar potency, which decreased the firing rate in overexcited, spontaneously firing cortical neuron cultures (Spitznagel et al. 2020). Yet another study used high-throughput methods to develop a series of oxadiazole derivatives as potent and selective Slack inhibitors (Griffin et al. 2021). They identified "compound 31" as an orally available Slack inhibitor with nanomolar potency, which reduced hippocampal neuron firing frequency and exerted anticonvulsant activity in vivo. These exciting latest developments still await translation into clinic.

7 Targeting Two-Pore K^+ Channels

K_{2P} channels have a unique topology among K^+ channels: each α-subunit of a K_{2P} channel contains four TMDs (M1–M4) (except of the TOC channel, which has eight TMDs) and two pore-loops (Fig. 5); the functional channel is formed by two α-subunits (Goldstein et al. 2001; Kollewe et al. 2009; Brohawn et al. 2012; Miller and Long 2012; Lolicato et al. 2017). Also unique among K^+ channel superfamily, K_{2P} channels form extracellular "cap" structure, which is formed by the M1–P1 extracellular linkers of each subunit in the dimer (Brohawn et al. 2012; Miller and Long 2012). The cap forms two tunnellike side portals which serve as extracellular ion pathway and shield the pore opening from many classical K^+ channel inhibitors (TEA, 4-AP, peptide toxins) (Piechotta et al. 2011; Miller and Long 2012; Zuniga and Zuniga 2016; Lolicato et al. 2017).

In humans, the K_{2P} family contains 15 subunits, based on sequence similarity and physiological and pharmacological properties; these are further subdivided into 6 subfamilies (see Introduction). K_{2P} channels do not possess a "classical" voltage-sensor domain (as in voltage-gated channels) and, hence, are not primarily gated by voltage (although different secondary voltage-dependence mechanisms do exist in many K_{2P} channels (Mathie et al. 2010; Schewe et al. 2016)). Hence, K_{2P} channels are open across the range of physiologically relevant membrane voltages, mediating background (or "leak") K^+ currents, which are important to maintain negative membrane potential and control excitability parameters in excitable cells. Consequently, deficiencies in K_{2P} channel expression and/or activity contribute to a number of excitability disorders. For comprehensive coverage of K_{2P} channel physiology and pharmacology, readers are directed to excellent recent reviews (Decher et al. 2021; Kamuene et al. 2021); here, we will focus on some key examples and latest developments.

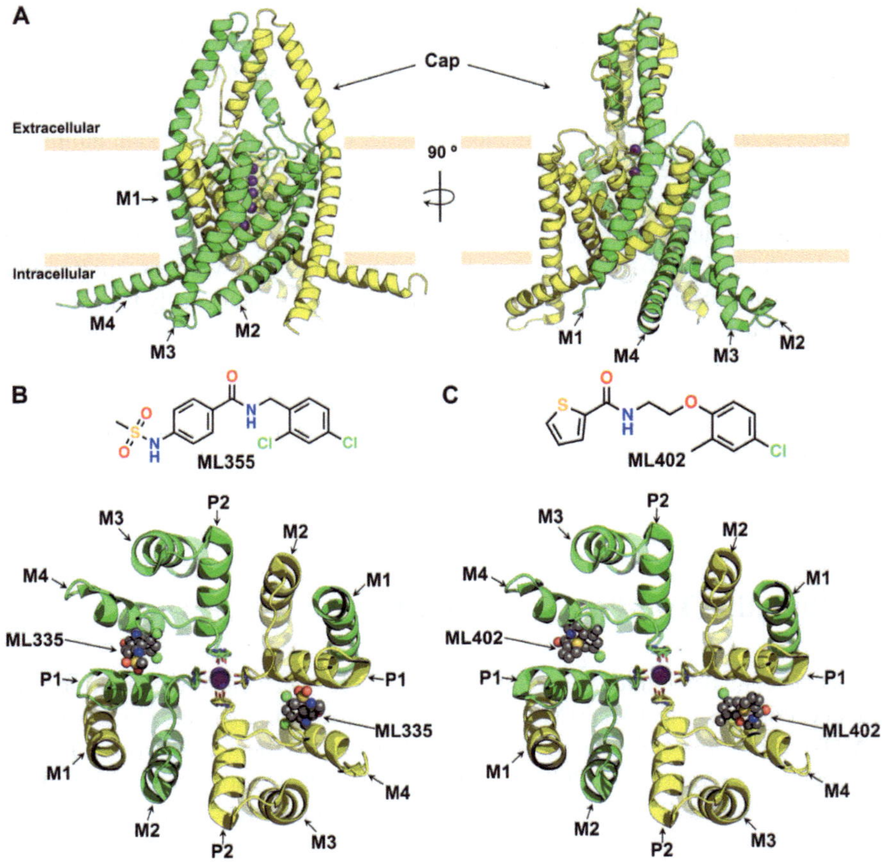

Fig. 5 K2P channel structure exemplified by the crystal structure of TREK1. (**a**) Two side views of TREK1 structure (PDB: 6CQ6). (**b**, **c**) Extracellular views (excluding the cap domain) of channel with two activators bound, ML335 (**b**, PDB: 6CQ9) and ML402 (**c**, PDB: 6CQ8); the channel represented as the ribbon, potassium and magnesium ions as purple and lime balls, respectively; ML335 or ML402 are shown as ball and stick, colored by element: C in green, N in blue, O in red, P in orange, respectively. (Based on Lolicato et al. (2017))

7.1 Pain and Migraine

Many K_{2P} subunits are expressed in the dorsal root ganglion (DRG) and trigeminal ganglion sensory neurons, including the damage-sensing (nociceptive) neurons; these subunits include TASK1–3, TREK1–2, TRAAK, TWIK1–2, and TRESK (Talley et al. 2001; Marsh et al. 2012; Acosta et al. 2014; Zheng et al. 2019). Pharmacological inhibition or genetic knockdown of K_{2P} channels significantly depolarizes nociceptors (Acosta et al. 2014; Du et al. 2014), while K_{2P} activators, such as riluzole (Duprat et al. 2000), hyperpolarize the E_m (Du et al. 2014). Hence, decreased activity/expression of K_{2P} channel subunits in nociceptors may result in pain (Acosta et al. 2014), while K_{2P} activators may produce analgesia. Electrophysiological recordings supported a prominent role of TREK-2 and TRESK in nociceptors (Kang and Kim 2006; Acosta et al. 2014).

Specifically, TRESK ($K_{2P}18$) channels are expressed in sensory neurons of the trigeminal ganglia (Lafreniere et al. 2010; Royal et al. 2019) and truncations or other LOF mutations in TRESK-encoding *KCNK18* gene have been associated with familial migraine with aura (Lafreniere et al. 2010). Moreover, one frameshift *KCNK18* mutation (F139WfsX24) was later shown to induce an alternative translation initiation site, leading not only to TRESK LOF but also to the production of second protein fragment, TRESK-MT2, which co-assembles with and inhibits two other K_{2P} subunits expressed in the trigeminal neurons: TREK1 and TREK2 (Royal et al. 2019). Such comprehensive disabling of K_{2P} current increased trigeminal sensory neuron excitability and produced a migraine-like phenotype in rodents. These observations support the idea that not only TRESK but TREK1 and TREK2 could emerge as potential therapeutic targets for migraine.

There is some evidence of downregulation of certain K_{2P} subunits in chronic pain models. Thus, TWIK1 and, to a lesser extent, TASK3 were found to be downregulated in a neuropathic pain model (spared nerve injury, SNI) (Pollema-Mays et al. 2013). TREK2 was found to be downregulated in another neuropathic pain model, spinal nerve axotomy (SNA), and after the hind-paw injection of complete Freund's adjuvant (inflammatory pain model) (Acosta et al. 2014). Consistently, in vivo overexpression of a K_{2P} subunit ($K_{2P}18$ or TRESK) by intrathecal delivery of a DRG-targeting adenoviral construct significantly reduced SNI-associated hyperalgesia (Zhou et al. 2013).

Many K_{2P} channels are modulated by temperature, mechanical, and chemical stimuli. Hence, these channels may contribute to or modulate primary sensory responses of nociceptors (reviewed in Plant 2012; Kamuene et al. 2021; Avalos Prado et al. 2022). For example, TREK1, TREK2, and TRAAK were reported to respond to thermal and mechanical stimuli (Patel et al. 1998; Maingret et al. 1999, 2001; Kang et al. 2005b; Honore 2007); TASK1 and TASK3 are sensitive to acidic pH (Kim et al. 2000; Lopes et al. 2000, 2001). TASK1, TASK3, and TRESK are inhibited by pungent compound of Szechuan peppers, hydroxyl-α-sanshool and, hence, may contribute to sensory response to this compound (Bocksteins 2016).

An important difference with other sensory ion channels (e.g., TRPs, Piezos) is that in order to excite a neuron, K_{2P} channels must be inhibited, rather than activated, by the noxious stimulation. Inversely, an analgesic strategy targeting K_{2P} channels would require channel activation, not inhibition. After discovery of the K_{2P}-opener activity of riluzole (Duprat et al. 2000), identification of new small-molecule K_{2P} activators is on the rise. For example, compound GI-530139 was identified, which hyperpolarized DRG neurons by activating TRAAK and TREK1 (Loucif et al. 2018); several caffeate esters were reported to activate TREK1 and exhibit analgesic activity in vivo (Rodrigues et al. 2014). A high-throughput, yeast-based screen for novel modulators performed by the Daniel Minor's group identified a series of K_{2P} activators; the most potent of these, ML67-33, activated TREK1 channels in the low-micromolar range by activating the extracellular selectivity filter-based gating apparatus (Bagriantsev et al. 2013). ML67-33 was later shown to inhibit trigeminal ganglion nociceptors and reverses migraine phenotype in rodents in an experimental migraine model (Avalos Prado et al. 2021). The same group later reported two further derivatives, ML335 and ML402, which stabilized TRAAK and TREK1 channel opening (Lolicato et al. 2017) (Fig. 5b, c). Aristolochic acid (AristA), a natural alkaloid from plants of Aristolochiaceae family, is used in traditional medicines to treat pain. A recent study has found that AristA activates TREK1 and TREK2 while inhibiting TRESK (Veale and Mathie 2016). While it is tempting to speculate that analgesic activity of AristA may be mediated by TREK1/2 potentiation, inhibitory action on TRESK complicates the matter and further investigation is required. Another natural compound that was shown to activate K_{2P} channels is mycolactone, a toxin produced by *Mycobacterium ulcerans*, a bacterium causing a topical skin disease called Buruli ulcer (Marion et al. 2014; Song et al. 2017). Skin ulceration produced by this disease (which is often extensive) is remarkably not painful; the endogenous analgesia of Buruli ulcer was suggested to result from the activation of TRAAK channels downstream of mycolactone action, although it is apparently not a direct effect but rather one mediated by the angiotensin II receptor signaling cascade (Marion et al. 2014; Song et al. 2017). Finally, it is worth noting here that some K_{2P} channels are activated by volatile general anesthetics, such as halothane, chloroform, and isoflurane. Thus, TASK channels are activated by clinically relevant concentrations of such anesthetics, which hyperpolarize neurons and reduce excitability, potentially contributing to anesthesia (Patel et al. 1999). Such action on brainstem motoneurons and thalamocortical neurons may contribute to loss of consciousness and motor ability under general anesthesia (Sirois et al. 2002; Meuth et al. 2003). Accordingly, sedation, hypnosis, and immobilization, as well as halothane-/isoflurane-induced motoneuron hyperpolarization, were reduced in TASK-1/TASK-3 KO mice (Lazarenko et al. 2010).

7.2 Other Neurological Disorders

A genome-wide linkage analysis identified a LOF G236R mutation in *KCNK9* (TASK3) to be associated with the maternally inherited Birk-Barel syndrome, a rare genetic disorder characterized with mental retardation, hypotonia, hyperactivity, and syndromic facies (Barel et al. 2008). Another mutation in *KCNK9* has been linked to absence epilepsy in a rat model, but the exact causative relationship with channel activity is yet to be elucidated since the mutation produces no discernible TASK3 loss-of-function (Holter et al. 2005). Regardless of this, pharmacological activation of K_{2P} channels does look like a plausible strategy to quench seizures. Growing evidence links various K_{2P} channels with other neurological disorders, such as depression, schizophrenia/psychosis, etc., but the causative links between the K_{2P} channel expression/activity and the disease pathophysiology are, as yet, less firmly established (reviewed in Aggarwal et al. 2021; Luo et al. 2021).

7.3 Cardiovascular Disorders

A heterozygous point mutation in *KCNK2* (TREK-1; I267T) has been identified in a patient with idiopathic right ventricular outflow tract tachycardia. The substitution is located just above the selectivity filter of the second pore loop of TREK-1 and results in enhanced stretch sensitivity and increased Na^+ permeability. This reduction in selectivity can cause the depolarization of ventricular cardiomyocytes and result in arrhythmias, especially in parts of the myocardium that are exposed to mechanical stress (Decher et al. 2017a, b). TREK1 surface expression is regulated by the Popeye domain-containing proteins, POPDC1 and POPDC2, which are expressed in the heart and skeletal muscle (Froese et al. 2012; Schindler et al. 2016). A homozygous missense variant in POPDC1 (S201F) was identified in a family with cardiac arrhythmia and limb-girdle muscular dystrophy (Schindler et al. 2016). The mutation, located in the cAMP binding site of POPDC1, reduced POPDC1-mediated TREK1 surface expression, but increased channel activity; it was hypothesized that these deregulations may result in arrhythmia (Schindler et al. 2016). Mutations in POPDC2 have been linked to atrioventricular block without a skeletal muscle phenotype (Decher et al. 2017a). Additionally, cardiomyocyte-specific TREK1 knockout in mice displayed stress-induced sick sinus syndrome (Unudurthi et al. 2016). LOF mutations within *KCNK3* (TASK1) are linked to pulmonary artery hypertension (PAH), a devastating pulmonary disease (Lambert et al. 2019). Loss of TASK1 was reported to contribute to the pulmonary vascular smooth muscle cell depolarization, leading to vasoconstriction of pulmonary arteries and PAH (Olschewski et al. 2006; Cunningham et al. 2019; Lambert et al. 2019; Fan et al. 2022). Several LOF mutations in *KCNK3* were identified in PAH patients, and these are considered the major causative factor of PAH in these patients (Ma et al. 2013; Best et al. 2017; Higasa et al. 2017; Navas Tejedor et al. 2017; Rosenzweig et al. 2019; Southgate et al. 2020).

8 Potential for Targeting Potassium Channels in Cancer Therapy

There is a growing interest to the role of ion channels, and, particularly K^+ channels, in cancer development and therapy. Since the early 1990s, it becomes apparent that the repertoire of ion channels expressed by a cell changes significantly when it becomes cancerous (Nilius and Wohlrab 1992). A specific set of K^+ channels came to light later as these often found overexpressed in the cancer cells (Huang and Jan 2014; Pardo and Stuhmer 2014; Bates 2015; Ganser et al. 2021). Genetic or pharmacological manipulations with some of these K^+ channels have been successfully used to counter cancer onset/progression in several preclinical models (Steudel et al. 2017; Mohr et al. 2022), rising hopes for potential future therapies.

More recent literature identified three groups of K^+ channels most consistently found overexpressed in cancer cells:

(i) hEAG ($K_v10.1$, *KCNH1*) and hERG ($K_v11.1$, *KCNH2*) (reviewed in Lastraioli et al. 2015; Cazares-Ordonez and Pardo 2017)
(ii) $K_v1.3$ (*KCNA3*), which is found overexpressed in many cancer types, including breast and colon cancers; smooth muscle (leiomyosarcoma), skeletal muscle (alveolar rhabdomyosarcoma), and lymph node cancer; and B cells in chronic lymphocytic leukemia (reviewed in Teisseyre et al. 2019), although downregulation of $K_v1.3$ was also reported in some cases (Teisseyre et al. 2019)
(iii) BK ($K_{Ca}1.1$) and IK ($K_{Ca}3.1$) Ca^{2+}-activated K^+ channels, which were found to be overexpressed in multiple cancer cells, including carcinomas and glioblastomas (BK) as well as sarcomas, leukemias, and some others (IK), reviewed in Liu et al. 2002b; Mohr et al. 2019, 2022; Ganser et al. 2021

There is no single theory for the precise role of K^+ channels in tumorigenesis; it is even unclear if a unifying mechanism exists. One concept which is being put forward is that cell proliferation (including abnormal) depends on the oscillations of membrane potential and cell volume, which, in turn, are regulated by K^+ channel activity (Wonderlin and Strobl 1996; Ganser et al. 2021). Maintaining the E_m at highly negative voltages was suggested to lock cells in the G_0 phase of the cell cycle, while depolarization may promote proliferation in nonterminally differentiated cells (Pardo and Stuhmer 2014; Ganser et al. 2021). The E_m in fast-cycling cancer cells exhibits oscillations, with the repolarization phase depending on K^+ channel activity (similar to their role in an action potential generation). Thus, control of such oscillations by K^+ channel activity may become a feature of tumorigenic transformation (Ganser et al. 2021). hERG was one of the first cancer-related K^+ channels identified (Arcangeli et al. 1995). Specifically, it was shown that the E_m of neuroblastoma cells is related to their proliferative capabilities and that it is controlled by the expression of an inwardly rectifying K^+ channel, later identified as hERG (Arcangeli et al. 1997).

Beyond cell cycle regulation, aberrant expression of K^+ channels in cancers cells was linked to changes in cell volume regulation, adhesion, migration, energy metabolism, and many other cellular function (reviewed in Pardo and Stuhmer 2014; Ganser et al. 2021).

In clinical terms, K^+ channel expression often correlates with poorer outcome. For example, EAG expression correlates with poor prognosis in colonic carcinoma (Ousingsawat et al. 2007), fibrosarcoma (Mello de Queiroz et al. 2006), acute myeloid leukemia (Agarwal et al. 2010), ovarian carcinoma (Asher et al. 2010), and head and neck cancer (Menendez et al. 2012a). hERG channel activity has been linked to the chemotherapy resistance of tumor cells (reviewed in Pardo and Stuhmer (2014)); hERG activity/expression has prognostic significance in colon, head and neck, ovarian, and gastric cancers (Cicek et al. 2013; Lastraioli et al. 2004, 2019; Menendez et al. 2012b). There are a few examples where agents with K^+ channel-modulating activity were shown to exert antitumorigenic activity. For example, the vitamin D metabolite calcitriol (a compound that decreases EAG1 expression) and the antineoplastic drug astemizole (which inhibits EAG1 activity) both suppressed growth of human breast cancer tumors xenografted into mice (Garcia-Quiroz et al.

2014). Likewise, blockage of EAG2 by thioridazine reduced growth and metastasis of medulloblastoma cell xenografts (Huang et al. 2015b). Yet, to the best of our knowledge, no K^+-channel-based therapy has yet progressed to the clinical trial stage. The reason for this is at least twofold. Firstly, compared to the relatively well-understood mechanistic function of K^+ channels in excitable cells, the exact role of K^+ channels in tumorigenesis is only beginning to emerge. Secondly, there is a great heterogeneity among tumors and cell types involved, and K^+ channel expression and activity are not always uniformly affected in different tumor types. Nevertheless, strong prognostic links identified for some K^+ channels (such as EAG and hERG) do suggest that future therapies could arise.

9 Concluding Remarks: Current Challenges and Future Directions

This chapter has considered recent advances in K^+ channel pharmacology with particular attention to potential treatment of human diseases. The scope for therapeutic targeting of K^+ channels is vast: K^+ channels constitute the largest mammalian ion channel family and represent the third largest class of protein targets (after GPCRs and kinases) (Rogers 2019).

Yet, K^+ channels are not an easy target. One of the issues is the relatively high degree of structural similarity between the K^+ channels, especially in some key domains, such as the selectivity filter and voltage sensor (Gonzalez et al. 2012; Taura et al. 2021). Additionally, most K^+ channel subfamilies contain several closely related ion channel isoforms (Fig. 1), often with very different cell and tissue expression profiles. This makes selective targeting of specific K^+ channel in a specific cell type by a small-molecule drug an uphill battle. One example of this issue is targeting K_v7s for treatment of epilepsy and pain. Five K_v7 subunits are expressed in cardiovascular and nervous systems and epithelia. Two closely related and relatively nonselective (within the K_v7 subfamily) compounds, flupirtine and retigabine, had reasonable success in the clinic, but were recently restricted due to side effects, some of which probably arise from on-target non-neuronal effects of the drugs (reviewed in Du et al. 2018; Liu et al. 2021). Thus, a new generation of more specific drugs that would only target relevant K_v7 subunits in specific cells is being highly sought after.

There is also the notable issue of hERG and cardiac safety. As discussed above, in order to progress anywhere close to clinical use, any K^+ channel modulator drug must spare the hERG channel. One direction toward the improving selectivity and specific cell targeting is development of ion channel-targeting biologics, such as peptide toxins, antibodies, nanobodies, affimers, etc. (see Bednenko et al. 2021; Matsumura et al. 2021 for review).

Even though K^+ channels are indeed a difficult target at present, several technological and conceptual advances of last few decades are positively transforming the field and leaving ample scope for optimism. Among these advances the following appear to be the most critical:

(i) "Structural revolution," the advancement of X-ray crystallography and cryo-EM has resulted in a rapid (and ongoing) accumulation of high-resolution structures of all major ion channel types (including the key K^+ channels).
(ii) Rapid development and AI-assisted folding and ligand-binding algorithms (in combination with wealth of structural information mentioned above) are revealing structure-functional relationships and major drug binding sites for the therapeutically relevant K^+ channels; computerized docking simulations now allow screening of large compound libraries in silico (see, e.g., recent in silico screen to identify K_{Na} channel modulators for treatment of epilepsy (Cole et al. 2020, 2021)).
(iii) Advent of automated patch-clamp and other advanced screening technologies has made possible a (relatively) rapid in vitro screening of large compound libraries (Milligan et al. 2009; Bell and Dallas 2021). This, in combination with the above two advances, makes it possible to perform structure-informed, focused screens (Cole et al. 2020).
(iv) Continuing progress in genetics, genomics, and next-generation sequencing brings about a rapid advancement in ion channel target validation. There are now hundreds of disease-linked genetic mutations in K^+ channel genes identified, categorized, and catalogued.
(v) Another rapid development that affects ion channel drug discovery is the development of the new in vitro human disease models using human cells or organoids derived from the induced pluripotent stem cell (iPSC) from patients carrying disease-associated mutations. Although not without its own limitations, this promising new development may breach the gap between the animal-based preclinical drug discovery research and clinical trials. This is exactly what is needed as this "translatability gap" is currently resulting in many high-profile failures and costing the industry a fortune (Sun et al. 2022).

Therefore, we hope that we are finally close to accumulating the "critical mass" of the basic, preclinical K^+ channel research and methodology for the rapid explosion into the new and improved therapies.

Acknowledgements Work in our laboratories is supported by the Wellcome Trust Investigator Award 212302/Z/18/Z (NG), Medical Research Council project grant MR/V012738/1 (NG), Biotechnology and Biological Sciences Research Council project grant BB/V010344/1, and International Partnering Award BB/R02104X/1 (NG); NG received funding from the European Union's Horizon 2020 research and innovation program under the Marie Sklodowska-Curie grant agreement No. 956477. DH is supported by the Scientific Research Project of Colleges and Universities of Hebei Province (ZD2021075). JQ is supported by the National Natural Science Foundation of Hebei Province, China (H2022206201), and Projects Guided by Central Government of Hebei Province, China (226Z2602G). HZ is supported by National Natural Science Foundation of China (81871075, 82071533), Science Fund for Creative Research Groups of Natural Science Foundation of Hebei Province (H2020206474), and Basic Research Fund for Provincial Universities (JCYJ2021010).

References

Abbott GW. KCNQs: ligand- and voltage-gated potassium channels. Front Physiol. 2020;11:583.

Abbott GW. Control of biophysical and pharmacological properties of potassium channels by ancillary subunits. Handb Exp Pharmacol. 2021;267:445–80.

Acosta C, et al. TREK2 expressed selectively in IB4-binding C-fiber nociceptors hyperpolarizes their membrane potentials and limits spontaneous pain. J Neurosci. 2014;34:1494–509.

Adams PR, Constanti A, Brown DA, Clark RB. Intracellular Ca2+ activates a fast voltage-sensitive K+ current in vertebrate sympathetic neurones. Nature. 1982;296:746–9.

Addy ME, Awumey EM. Effects of the extracts of Desmodium adscendens on anaphylaxis. J Ethnopharmacol. 1984;11:283–92.

Adelman JP, Bond CT, Pessia M, Maylie J. Episodic ataxia results from voltage-dependent potassium channels with altered functions. Neuron. 1995;15:1449–54.

Adelman JP, Maylie J, Sah P. Small-conductance Ca^{2+}-activated K^{+} channels: form and function. Annu Rev Physiol. 2012;74:245–69.

Agarwal JR, Griesinger F, Stuhmer W, Pardo LA. The potassium channel Ether a go-go is a novel prognostic factor with functional relevance in acute myeloid leukemia. Mol Cancer. 2010;9:18.

Aggarwal P, Singh S, Ravichandiran V. Two-pore domain potassium channel in neurological disorders. J Membr Biol. 2021;254:367–80.

Alexander SP, et al. THE CONCISE GUIDE TO PHARMACOLOGY 2021/22: ion channels. Br J Pharmacol. 2021;178(Suppl 1):S157–245.

Allen NM, Weckhuysen S, Gorman K, King MD, Lerche H. Genetic potassium channel-associated epilepsies: clinical review of the K(v) family. Eur J Paediatr Neurol. 2020;24:105–16.

Anderson PA, Greenberg RM. Phylogeny of ion channels: clues to structure and function. Comp Biochem Physiol B Biochem Mol Biol. 2001;129:17–28.

Angulo E, et al. Up-regulation of the Kv3.4 potassium channel subunit in early stages of Alzheimer's disease. J Neurochem. 2004;91:547–57.

Antozzi C, et al. Sequential antibodies to potassium channels and glutamic acid decarboxylase in neuromyotonia. Neurology. 2005;64:1290–3.

Aqvist J, Luzhkov V. Ion permeation mechanism of the potassium channel. Nature. 2000;404:881–4.

Arcangeli A, et al. A novel inward-rectifying K^{+} current with a cell-cycle dependence governs the resting potential of mammalian neuroblastoma cells. J Physiol. 1995;489(Pt 2):455–71.

Arcangeli A, et al. HERG- and IRK-like inward rectifier currents are sequentially expressed during neuronal development of neural crest cells and their derivatives. Eur J Neurosci. 1997;9:2596–604.

Asai T, et al. Cryo-EM structure of K^{+}-bound hERG channel complexed with the blocker astemizole. Structure. 2021;29:203–212.e4.

Asayama M, et al. Effects of an hERG activator, ICA-105574, on electrophysiological properties of canine hearts. J Pharmacol Sci. 2013;121:1–8.

Ashcroft FM. Adenosine 5′-triphosphate-sensitive potassium channels. Annu Rev Neurosci. 1988;11:97–118.

Ashcroft FM. ATP-sensitive potassium channelopathies: focus on insulin secretion. J Clin Invest. 2005;115:2047–58.

Ashcroft FM, Kakei M. ATP-sensitive K^{+} channels in rat pancreatic beta-cells: modulation by ATP and Mg2+ ions. J Physiol. 1989;416:349–67.

Ashcroft FM, Rorsman P. Electrophysiology of the pancreatic beta-cell. Prog Biophys Mol Biol. 1989;54:87–143.

Ashcroft FM, Harrison DE, Ashcroft SJ. Glucose induces closure of single potassium channels in isolated rat pancreatic beta-cells. Nature. 1984;312:446–8.

Asher V, et al. The Eag potassium channel as a new prognostic marker in ovarian cancer. Diagn Pathol. 2010;5:78.

Ashford ML, Sturgess NC, Trout NJ, Gardner NJ, Hales CN. Adenosine-5′-triphosphate-sensitive ion channels in neonatal rat cultured central neurones. Pflugers Arch. 1988;412:297–304.

Ashford ML, Boden PR, Treherne JM. Glucose-induced excitation of hypothalamic neurones is mediated by ATP-sensitive K^+ channels. Pflugers Arch. 1990;415:479–83.

Avalos Prado P, et al. TREK channel activation suppresses migraine pain phenotype. iScience. 2021;24:102961.

Avalos Prado P, Chassot AA, Landra-Willm A, Sandoz G. Regulation of two-pore-domain potassium TREK channels and their involvement in pain perception and migraine. Neurosci Lett. 2022;773:136494.

Bader CR, Bernheim L, Bertrand D. Sodium-activated potassium current in cultured avian neurones. Nature. 1985;317:540–2.

Bagriantsev SN, et al. A high-throughput functional screen identifies small molecule regulators of temperature- and mechano-sensitive K2P channels. ACS Chem Biol. 2013;8:1841–51.

Bahia PK, et al. A functional role for small-conductance calcium-activated potassium channels in sensory pathways including nociceptive processes. J Neurosci. 2005;25:3489–98.

Barcia G, et al. De novo gain-of-function KCNT1 channel mutations cause malignant migrating partial seizures of infancy. Nat Genet. 2012;44:1255–9.

Barel O, et al. Maternally inherited Birk Barel mental retardation dysmorphism syndrome caused by a mutation in the genomically imprinted potassium channel KCNK9. Am J Hum Genet. 2008;83:193–9.

Barhanin J, et al. K(V)LQT1 and lsK (minK) proteins associate to form the I(Ks) cardiac potassium current. Nature. 1996;384:78–80.

Barrese V, Stott JB, Greenwood IA. KCNQ-encoded potassium channels as therapeutic targets. Annu Rev Pharmacol Toxicol. 2018;58:625–48.

Bates E. Ion channels in development and cancer. Annu Rev Cell Dev Biol. 2015;31:231–47.

Bauer CK, Schwarz JR. Ether-a-go-go K^+ channels: effective modulators of neuronal excitability. J Physiol. 2018;596:769–83.

Bednenko J, Colussi P, Hussain S, Zhang Y, Clark T. Therapeutic antibodies targeting potassium ion channels. Handb Exp Pharmacol. 2021;267:507–45.

Beech DJ, Zhang H, Nakao K, Bolton TB. K channel activation by nucleotide diphosphates and its inhibition by glibenclamide in vascular smooth muscle cells. Br J Pharmacol. 1993;110:573–82.

Beeton C, et al. Targeting effector memory T cells with a selective peptide inhibitor of Kv1.3 channels for therapy of autoimmune diseases. Mol Pharmacol. 2005;67:1369–81.

Beeton C, et al. Kv1.3 channels are a therapeutic target for T cell-mediated autoimmune diseases. Proc Natl Acad Sci U S A. 2006;103:17414–9.

Bell DC, Dallas ML. Advancing ion channel research with automated patch clamp (APC) electrophysiology platforms. Adv Exp Med Biol. 2021;1349:21–32.

Benavides-Haro DE, Sánchez-Chapula JA. Chloroquine blocks the background potassium current in guinea pig atrial myocytes. Naunyn Schmiedeberg's Arch Pharmacol. 2000;361:311–8.

Bentzen BH, et al. The small molecule NS11021 is a potent and specific activator of Ca^{2+}-activated big-conductance K^+ channels. Mol Pharmacol. 2007;72:1033–44.

Bentzen BH, et al. Pharmacological activation of Kv11.1 in transgenic long QT-1 rabbits. J Cardiovasc Pharmacol. 2011;57:223–30.

Bentzen BH, Olesen SP, Ronn LC, Grunnet M. BK channel activators and their therapeutic perspectives. Front Physiol. 2014;5:389.

Beraud E, et al. Block of neural Kv1.1 potassium channels for neuroinflammatory disease therapy. Ann Neurol. 2006;60:586–96.

Bergren SK, Rutter ED, Kearney JA. Fine mapping of an epilepsy modifier gene on mouse Chromosome 19. Mamm Genome. 2009;20:359–66.

Berneche S, Roux B. Energetics of ion conduction through the K^+ channel. Nature. 2001;414:73–7.

Best DH, et al. EIF2AK4 mutations in patients diagnosed with pulmonary arterial hypertension. Chest. 2017;151:821–8.

Bhattacharjee A, Gan L, Kaczmarek LK. Localization of the Slack potassium channel in the rat central nervous system. J Comp Neurol. 2002;454:241–54.

Bhattacharjee A, von Hehn CA, Mei X, Kaczmarek LK. Localization of the Na^+-activated K^+ channel Slick in the rat central nervous system. J Comp Neurol. 2005;484:80–92.

Bhave G, et al. Development of a selective small-molecule inhibitor of Kir1.1, the renal outer medullary potassium channel. Mol Pharmacol. 2011;79:42–50.

Bielanska J, et al. Voltage-dependent potassium channels Kv1.3 and Kv1.5 in human cancer. Curr Cancer Drug Targets. 2009;9:904–14.

Bierbower SM, Choveau FS, Lechleiter JD, Shapiro MS. Augmentation of M-type (KCNQ) potassium channels as a novel strategy to reduce stroke-induced brain injury. J Neurosci. 2015;35:2101–11.

Biervert C, et al. A potassium channel mutation in neonatal human epilepsy. Science. 1998;279:403–6.

Birnbaum SG, et al. Structure and function of Kv4-family transient potassium channels. Physiol Rev. 2004;84:803–33.

Blicklé JF. Meglitinide analogues: a review of clinical data focused on recent trials. Diabetes Metab. 2006;32:113–20.

Block F, Pergande G, Schwarz M. Flupirtine reduces functional deficits and neuronal damage after global ischemia in rats. Brain Res. 1997;754:279–84.

Bock C, Link A. How to replace the lost keys? Strategies toward safer K_V7 channel openers. Future Med Chem. 2019;11:337–55.

Bocksteins E. Kv5, Kv6, Kv8, and Kv9 subunits: no simple silent bystanders. J Gen Physiol. 2016;147:105–25.

Bocksteins E, Snyders DJ. Electrically silent Kv subunits: their molecular and functional characteristics. Physiology (Bethesda). 2012;27:73–84.

Bond CT, et al. Cloning and expression of a family of inward rectifier potassium channels. Receptors Channels. 1994;2:183–91.

Borgini M, Mondal P, Liu R, Wipf P. Chemical modulation of K_v7 potassium channels. RSC Med Chem. 2021;12:483–537.

Borrego J, et al. Peptide inhibitors of Kv1.5: an option for the treatment of atrial fibrillation. Pharmaceuticals (Basel). 2021;14:1303.

Boscia F, et al. The expression and activity of $K_V3.4$ channel subunits are precociously upregulated in astrocytes exposed to Abeta oligomers and in astrocytes of Alzheimer's disease Tg2576 mice. Neurobiol Aging. 2017;54:187–98.

Brayden JE, Nelson MT. Regulation of arterial tone by activation of calcium-dependent potassium channels. Science. 1992;256:532–5.

Bredt DS, Wang TL, Cohen NA, Guggino WB, Snyder SH. Cloning and expression of two brain-specific inwardly rectifying potassium channels. Proc Natl Acad Sci U S A. 1995;92:6753–7.

Brenner R, et al. Vasoregulation by the beta1 subunit of the calcium-activated potassium channel. Nature. 2000;407:870–6.

Brohawn SG, del Marmol J, MacKinnon R. Crystal structure of the human K2P TRAAK, a lipid- and mechano-sensitive K^+ ion channel. Science. 2012;335:436–41.

Brown DA, Passmore GM. Neural KCNQ (Kv7) channels. Br J Pharmacol. 2009;156:1185–95.

Brown DA, Gähwiler BH, Griffith WH, Halliwell JV. Membrane currents in hippocampal neurons. Prog Brain Res. 1990;83:141–60.

Brueggemann LI, et al. Kv7 potassium channels in airway smooth muscle cells: signal transduction intermediates and pharmacological targets for bronchodilator therapy. Am J Physiol Lung Cell Mol Physiol. 2012;302:L120–32.

Brueggemann LI, Haick JM, Cribbs LL, Byron KL. Differential activation of vascular smooth muscle Kv7.4, Kv7.5, and Kv7.4/7.5 channels by ML213 and ICA-069673. Mol Pharmacol. 2014;86:330–41.

Busch AE, et al. The role of the IsK protein in the specific pharmacological properties of the IKs channel complex. Br J Pharmacol. 1997;122:187–9.

Butler A, Helliwell MV, Zhang Y, Hancox JC, Dempsey CE. An update on the structure of hERG. Front Pharmacol. 2019;10:1572.

Caballero R, et al. Flecainide increases Kir2.1 currents by interacting with cysteine 311, decreasing the polyamine-induced rectification. Proc Natl Acad Sci U S A. 2010;107:15631–6.

Caminos E, Vaquero CF, Martinez-Galan JR. Relationship between rat retinal degeneration and potassium channel KCNQ5 expression. Exp Eye Res. 2015;131:1–11.

Campos-Rios A, Rueda-Ruzafa L, Lamas JA. The relevance of GIRK channels in heart function. Membranes (Basel). 2022;12:1119.

Casis O, Olesen SP, Sanguinetti MC. Mechanism of action of a novel human ether-a-go-go-related gene channel activator. Mol Pharmacol. 2006;69:658–65.

Castle NA, et al. Maurotoxin: a potent inhibitor of intermediate conductance Ca^{2+}-activated potassium channels. Mol Pharmacol. 2003;63:409–18.

Cazares-Ordonez V, Pardo LA. Kv10.1 potassium channel: from the brain to the tumors. Biochem Cell Biol. 2017;95:531–6.

Cercos P, et al. Pharmacological approaches for the modulation of the potassium channel $K_V4.x$ and KChIPs. Int J Mol Sci. 2021;22:1419.

Chadha PS, et al. Contribution of $K_V7.4/K_V7.5$ heteromers to intrinsic and calcitonin gene-related peptide-induced cerebral reactivity. Arterioscler Thromb Vasc Biol. 2014;34:887–93.

Chandy KG, et al. K^+ channels as targets for specific immunomodulation. Trends Pharmacol Sci. 2004;25:280–9.

Chen R, Chung SH. Engineering a potent and specific blocker of voltage-gated potassium channel Kv1.3, a target for autoimmune diseases. Biochemistry. 2012;51:1976–82.

Chen H, et al. Functional analysis of a novel potassium channel (KCNA1) mutation in hereditary myokymia. Neurogenetics. 2007;8:131–5.

Chen ZY, et al. Hg1, novel peptide inhibitor specific for Kv1.3 channels from first scorpion Kunitz-type potassium channel toxin family. J Biol Chem. 2012;287:13813–21.

Chen YJ, et al. The potassium channel KCa3.1 constitutes a pharmacological target for neuroinflammation associated with ischemia/reperfusion stroke. J Cereb Blood Flow Metab. 2016;36:2146–61.

Chen YJ, Cui Y, Singh L, Wulff H. The potassium channel Kv1.3 as a therapeutic target for immunocytoprotection after reperfusion. Ann Clin Transl Neurol. 2021;8:2070–82.

Cheney JA, et al. The maxi-K channel opener BMS-204352 attenuates regional cerebral edema and neurologic motor impairment after experimental brain injury. J Cereb Blood Flow Metab. 2001;21:396–403.

Choi SH, et al. Ginsenoside Rg(3) decelerates hERG K^+ channel deactivation through Ser631 residue interaction. Eur J Pharmacol. 2011;663:59–67.

Christophersen P, Wulff H. Pharmacological gating modulation of small- and intermediate-conductance Ca^{2+}-activated K^+ channels (KCa2.x and KCa3.1). Channels (Austin). 2015;9:336–43.

Cicek MS, et al. Epigenome-wide ovarian cancer analysis identifies a methylation profile differentiating clear-cell histology with epigenetic silencing of the HERG K^+ channel. Hum Mol Genet. 2013;22:3038–47.

Clark S, Antell A, Kaufman K. New antiepileptic medication linked to blue discoloration of the skin and eyes. Ther Adv Drug Saf. 2015;6:15–9.

Clement-Chomienne O, Walsh MP, Cole WC. Angiotensin II activation of protein kinase C decreases delayed rectifier K^+ current in rabbit vascular myocytes. J Physiol. 1996;495(Pt 3):689–700.

Coetzee WA, et al. Molecular diversity of K^+ channels. Ann N Y Acad Sci. 1999;868:233–85.

Cole BA, et al. Structure-based identification and characterization of inhibitors of the epilepsy-associated K(Na)1.1 (KCNT1) potassium channel. iScience. 2020;23:101100.

Cole BA, Clapcote SJ, Muench SP, Lippiat JD. Targeting K(Na)1.1 channels in KCNT1-associated epilepsy. Trends Pharmacol Sci. 2021;42:700–13.

Coleman N, et al. New positive Ca^{2+}-activated K^+ channel gating modulators with selectivity for KCa3.1. Mol Pharmacol. 2014;86:342–57.

Comes N, et al. The voltage-dependent K^+ channels Kv1.3 and Kv1.5 in human cancer. Front Physiol. 2013;4:283.

Conejo-Garcia A, Campos JM. Bis-quinolinium cyclophanes: highly potent and selective non-peptidic blockers of the apamin-sensitive Ca^{2+}-activated K^+ channel. Curr Med Chem. 2008;15:1305–15.

Conner LB, Alvarez P, Bogen O, Levine JD. Role of Kv4.3 in vibration-induced muscle pain in the rat. J Pain. 2016;17:444–50.

Cook DL, Hales CN. Intracellular ATP directly blocks K^+ channels in pancreatic B-cells. Nature. 1984;311:271–3.

Cooper EC, Harrington E, Jan YN, Jan LY. M channel KCNQ2 subunits are localized to key sites for control of neuronal network oscillations and synchronization in mouse brain. J Neurosci. 2001;21:9529–40.

Corbett MA, et al. Dominant KCNA2 mutation causes episodic ataxia and pharmacoresponsive epilepsy. Neurology. 2016;87:1975–84.

Costa AD. Iptakalim: a new or just another KCO? Cardiovasc Res. 2009;83:417–8.

Cotton KD, Hollywood MA, Thornbury KD, McHale NG. Effect of purinergic blockers on outward current in isolated smooth muscle cells of the sheep bladder. Am J Phys. 1996;270:C969–73.

Creanza TM, et al. Structure-based prediction of hERG-related cardiotoxicity: a benchmark study. J Chem Inf Model. 2021;61:4758–70.

Crotti L, Taravelli E, Girardengo G, Schwartz PJ. Congenital short QT syndrome. Indian Pacing Electrophysiol J. 2010;10:86–95.

Cruz HG, et al. Bi-directional effects of GABA(B) receptor agonists on the mesolimbic dopamine system. Nat Neurosci. 2004;7:153–9.

Cucchiara F, et al. Electrophysiological features of sleep in children with Kir4.1 channel mutations and Autism-Epilepsy phenotype: a preliminary study. Sleep. 2020;43:zsz255.

Cui M, Qin G, Yu K, Bowers MS, Zhang M. Targeting the small- and intermediate-conductance Ca2+-activated potassium channels: the drug-binding pocket at the channel/calmodulin interface. Neurosignals. 2014;22:65–78.

Cui Y, et al. Astroglial Kir4.1 in the lateral habenula drives neuronal bursts in depression. Nature. 2018;554:323–7.

Cui M, Cantwell L, Zorn A, Logothetis DE. Kir channel molecular physiology, pharmacology, and therapeutic implications. Handb Exp Pharmacol. 2021;267:277–356.

Cunningham KP, et al. Characterization and regulation of wild-type and mutant TASK-1 two pore domain potassium channels indicated in pulmonary arterial hypertension. J Physiol. 2019;597:1087–101.

D'Adamo MC, Liantonio A, Rolland JF, Pessia M, Imbrici P. Kv1.1 channelopathies: pathophysiological mechanisms and therapeutic approaches. Int J Mol Sci. 2020;21:2935.

Dabrowski M, Larsen T, Ashcroft FM, Bondo Hansen J, Wahl P. Potent and selective activation of the pancreatic beta-cell type K_{ATP} channel by two novel diazoxide analogues. Diabetologia. 2003;46:1375–82.

Dalby-Brown W, et al. Characterization of a novel high-potency positive modulator of K_v7 channels. Eur J Pharmacol. 2013;709:52–63.

Dascal N, et al. Atrial G protein-activated K^+ channel: expression cloning and molecular properties. Proc Natl Acad Sci U S A. 1993;90:10235–9.

Daut J, et al. Hypoxic dilation of coronary arteries is mediated by ATP-sensitive potassium channels. Science. 1990;247:1341–4.

Dean PM, Matthews EK. Electrical activity in pancreatic islet cells. Nature. 1968;219:389–90.

Decher N, Kiper AK, Rinne S. Stretch-activated potassium currents in the heart: focus on TREK-1 and arrhythmias. Prog Biophys Mol Biol. 2017a;130:223–32.

Decher N, et al. Sodium permeable and "hypersensitive" TREK-1 channels cause ventricular tachycardia. EMBO Mol Med. 2017b;9:403–14.

Decher N, Rinne S, Bedoya M, Gonzalez W, Kiper AK. Molecular pharmacology of K(2P) potassium channels. Cell Physiol Biochem. 2021;55:87–107.

DeCoursey TE, Chandy KG, Gupta S, Cahalan MD. Voltage-gated K^+ channels in human T lymphocytes: a role in mitogenesis? Nature. 1984;307:465–8.

deHart GW, Jin T, McCloskey DE, Pegg AE, Sheppard D. The alpha9beta1 integrin enhances cell migration by polyamine-mediated modulation of an inward-rectifier potassium channel. Proc Natl Acad Sci U S A. 2008;105:7188–93.

Del Re AM, Dopico AM, Woodward JJ. Effects of the abused inhalant toluene on ethanol-sensitive potassium channels expressed in oocytes. Brain Res. 2006;1087:75–82.

Delgado-Ramírez M, et al. Inhibitory effect of terfenadine on Kir2.1 and Kir2.3 channels. Acta Pharm. 2021;71:317–24.

Delmas P, Brown DA. Pathways modulating neural KCNQ/M (Kv7) potassium channels. Nat Rev Neurosci. 2005;6:850–62.

Devor DC, Singh AK, Frizzell RA, Bridges RJ. Modulation of Cl- secretion by benzimidazolones. I. Direct activation of a Ca^{2+}-dependent K^+ channel. Am J Phys. 1996;271:L775–84.

Diniz AFA, Ferreira RC, de Souza ILL, da Silva BA. Ionic channels as potential therapeutic targets for erectile dysfunction: a review. Front Pharmacol. 2020;11:1120.

Donovan BT, et al. Discovery and electrophysiological characterization of SKF-32802: a novel hERG agonist found through a large-scale structural similarity search. Eur J Pharmacol. 2018;818:306–27.

Dörschner H, Brekardin E, Uhde I, Schwanstecher C, Schwanstecher M. Stoichiometry of sulfonylurea-induced ATP-sensitive potassium channel closure. Mol Pharmacol. 1999;55:1060–6.

Doyle ME, Egan JM. Pharmacological agents that directly modulate insulin secretion. Pharmacol Rev. 2003;55:105–31.

Doyle DA, et al. The structure of the potassium channel: molecular basis of K^+ conduction and selectivity. Science. 1998;280:69–77.

Drion G, et al. M-type channels selectively control bursting in rat dopaminergic neurons. Eur J Neurosci. 2010;31:827–35.

Du X, Gamper N. Potassium channels in peripheral pain pathways: expression, function and therapeutic potential. Curr Neuropharmacol. 2013;11:621–40.

Du W, et al. Calcium-sensitive potassium channelopathy in human epilepsy and paroxysmal movement disorder. Nat Genet. 2005;37:733–8.

Du X, et al. Control of somatic membrane potential in nociceptive neurons and its implications for peripheral nociceptive transmission. Pain. 2014;155:2306–22.

Du X, Gao H, Jaffe D, Zhang H, Gamper N. M-type K^+ channels in peripheral nociceptive pathways. Br J Pharmacol. 2018;175:2158–72.

Dudem S, Sergeant GP, Thornbury KD, Hollywood MA. Calcium-activated K^+ channels (K^{Ca}) and therapeutic implications. Handb Exp Pharmacol. 2021;267:379–416.

Duprat F, et al. The neuroprotective agent riluzole activates the two P domain K^+ channels TREK-1 and TRAAK. Mol Pharmacol. 2000;57:906–12.

Dupuis DS, et al. Activation of KCNQ5 channels stably expressed in HEK293 cells by BMS-204352. Eur J Pharmacol. 2002;437:129–37.

Edwards G, Weston AH. The pharmacology of ATP-sensitive potassium channels. Annu Rev Pharmacol Toxicol. 1993;33:597–637.

Edwards G, Niederste-Hollenberg A, Schneider J, Noack T, Weston AH. Ion channel modulation by NS 1619, the putative BKCa channel opener, in vascular smooth muscle. Br J Pharmacol. 1994;113:1538–47.

El Harchi A, Brincourt O. Pharmacological activation of the hERG K^+ channel for the management of the long QT syndrome: a review. J Arrhythm. 2022;38:554–69.

Evseev AI, et al. Functional effects of KCNQ K^+ channels in airway smooth muscle. Front Physiol. 2013;4:277.

Faber ES, Sah P. Functions of SK channels in central neurons. Clin Exp Pharmacol Physiol. 2007;34:1077–83.

Fakler B, et al. Identification of a titratable lysine residue that determines sensitivity of kidney potassium channels (ROMK) to intracellular pH. EMBO J. 1996;15:4093–9.

Fan X, Lu Y, Du G, Liu J. Advances in the understanding of two-pore domain TASK potassium channels and their potential as therapeutic targets. Molecules. 2022;27:8296.

Fernandez de Sevilla D, Garduno J, Galvan E, Buno W. Calcium-activated afterhyperpolarizations regulate synchronization and timing of epileptiform bursts in hippocampal CA3 pyramidal neurons. J Neurophysiol. 2006;96:3028–41.

Finol-Urdaneta RK, et al. Marine toxins targeting Kv1 channels: pharmacological tools and therapeutic scaffolds. Mar Drugs. 2020;18:173.

Fountain SJ, et al. Functional up-regulation of KCNA gene family expression in murine mesenteric resistance artery smooth muscle. J Physiol. 2004;556:29–42.

Fox PD, et al. Induction of stable ER-plasma-membrane junctions by Kv2.1 potassium channels. J Cell Sci. 2015;128:2096–105.

Frampton DJA, Choudhury K, Nikesjo J, Delemotte L, Liin SI. Subtype-specific responses of hKv7.4 and hKv7.5 channels to polyunsaturated fatty acids reveal an unconventional modulatory site and mechanism. eLife. 2022;11:e77672.

Franceschetti S, et al. Na^+-activated K^+ current contributes to postexcitatory hyperpolarization in neocortical intrinsically bursting neurons. J Neurophysiol. 2003;89:2101–11.

Frech GC, VanDongen AM, Schuster G, Brown AM, Joho RH. A novel potassium channel with delayed rectifier properties isolated from rat brain by expression cloning. Nature. 1989;340:642–5.

Fremstad D, Nilsen OG, Storstein L, Amlie J, Jacobsen S. Pharmacokinetics of quinidine related to plasma protein binding in man. Eur J Clin Pharmacol. 1979;15:187–92.

Friedman LK, et al. Transcriptome profiling of hippocampal CA1 after early-life seizure-induced preconditioning may elucidate new genetic therapies for epilepsy. Eur J Neurosci. 2013;38:2139–52.

Friedman AK, et al. KCNQ channel openers reverse depressive symptoms via an active resilience mechanism. Nat Commun. 2016;7:11671.

Froese A, et al. Popeye domain containing proteins are essential for stress-mediated modulation of cardiac pacemaking in mice. J Clin Invest. 2012;122:1119–30.

Fujita A, et al. Specific localization of an inwardly rectifying K^+ channel, Kir4.1, at the apical membrane of rat gastric parietal cells; its possible involvement in K^+ recycling for the H^+-K^+-pump. J Physiol. 2002;540:85–92.

Fukai R, et al. De novo KCNH1 mutations in four patients with syndromic developmental delay, hypotonia and seizures. J Hum Genet. 2016;61:381–7.

Furness JB, et al. Intermediate conductance potassium (IK) channels occur in human enteric neurons. Auton Neurosci. 2004;112:93–7.

Galan-Vidal J, et al. A novel loss-of-function mutation of the voltage-gated potassium channel Kv10.2 involved in epilepsy and autism. Orphanet J Rare Dis. 2022;17:345.

Gamper N, Shapiro MS. Chapter 20: KCNQ channels. In: Zheng J, Trudeau MC, editors. Handbook of ion channels. Boca Raton: CRC Press; 2015. p. 275–306.

Gamper N, Wang KW, editors. Pharmacology of potassium channels. Cham: Springer; 2021. p. 553.

Gamper N, et al. Oxidative modification of M-type K^+ channels as a mechanism of cytoprotective neuronal silencing. EMBO J. 2006;25:4996–5004.

Ganser K, et al. Potassium channels in cancer. Handb Exp Pharmacol. 2021;267:253–75.

Gao SB, et al. Slack and Slick K_{Na} channels are required for the depolarizing afterpotential of acutely isolated, medium diameter rat dorsal root ganglion neurons. Acta Pharmacol Sin. 2008;29:899–905.

Gao HX, et al. Intracellular zinc activates KCNQ channels by reducing their dependence on phosphatidylinositol 4,5-bisphosphate. Proc Natl Acad Sci U S A. 2017;114:E6410–9.

Gao K, Lin Z, Wen S, Jiang Y. Potassium channels and epilepsy. Acta Neurol Scand. 2022;146:699–707.

Garcia-Quiroz J, et al. In vivo dual targeting of the oncogenic Ether-a-go-go-1 potassium channel by calcitriol and astemizole results in enhanced antineoplastic effects in breast tumors. BMC Cancer. 2014;14:745.

Gardos G. The function of calcium in the potassium permeability of human erythrocytes. Biochim Biophys Acta. 1958;30:653–4.

Garg V, Stary-Weinzinger A, Sachse F, Sanguinetti MC. Molecular determinants for activation of human ether-à-go-go-related gene 1 potassium channels by 3-nitro-n-(4-phenoxyphenyl) benzamide. Mol Pharmacol. 2011;80:630–7.

Geiger J, Weber YG, Landwehrmeyer B, Sommer C, Lerche H. Immunohistochemical analysis of KCNQ3 potassium channels in mouse brain. Neurosci Lett. 2006;400:101–4.

Georgiev D, et al. Lower gene expression for KCNS3 potassium channel subunit in parvalbumin-containing neurons in the prefrontal cortex in schizophrenia. Am J Psychiatry. 2014;171:62–71.

Gerlach AC, Stoehr SJ, Castle NA. Pharmacological removal of human ether-a-go-go-related gene potassium channel inactivation by 3-nitro-N-(4-phenoxyphenyl) benzamide (ICA-105574). Mol Pharmacol. 2010;77:58–68.

Gessner G, Macianskiene R, Starkus JG, Schonherr R, Heinemann SH. The amiodarone derivative KB130015 activates hERG1 potassium channels via a novel mechanism. Eur J Pharmacol. 2010;632:52–9.

Giblin JP, Cui Y, Clapp LH, Tinker A. Assembly limits the pharmacological complexity of ATP-sensitive potassium channels. J Biol Chem. 2002;277:13717–23.

Goldstein SA, Bockenhauer D, O'Kelly I, Zilberberg N. Potassium leak channels and the KCNK family of two-P-domain subunits. Nat Rev Neurosci. 2001;2:175–84.

Gonzalez C, et al. K^+ channels: function-structural overview. Compr Physiol. 2012;2:2087–149.

Gordon E, et al. 2-[2-(3,4-dichloro-phenyl)-2,3-dihydro-1H-isoindol-5-ylamino]-nicotinic acid (PD-307243) causes instantaneous current through human ether-a-go-go-related gene potassium channels. Mol Pharmacol. 2008;73:639–51.

Goto A, et al. Characteristics of KCNQ2 variants causing either benign neonatal epilepsy or developmental and epileptic encephalopathy. Epilepsia. 2019;60:1870–80.

Greene DL, Hoshi N. Modulation of Kv7 channels and excitability in the brain. Cell Mol Life Sci. 2016;74:495–508.

Greenwood IA, Ohya S. New tricks for old dogs: KCNQ expression and role in smooth muscle. Br J Pharmacol. 2009;156:1196–203.

Gribble FM, Ashcroft FM. Differential sensitivity of beta-cell and extrapancreatic K_{ATP} channels to gliclazide. Diabetologia. 1999;42:845–8.

Gribble FM, Tucker SJ, Seino S, Ashcroft FM. Tissue specificity of sulfonylureas: studies on cloned cardiac and beta-cell K_{ATP} channels. Diabetes. 1998;47:1412–8.

Griffin AM, et al. Discovery of the first orally available, selective $K_{Na}1.1$ inhibitor: in vitro and in vivo activity of an oxadiazole series. ACS Med Chem Lett. 2021;12:593–602.

Grissmer S, et al. Expression and chromosomal localization of a lymphocyte K^+ channel gene. Proc Natl Acad Sci U S A. 1990;87:9411–5.

Gualdani R, Cavalluzzi MM, Tadini-Buoninsegni F, Lentini G. Discovery of a new mexiletine-derived agonist of the hERG K^+ channel. Biophys Chem. 2017;229:62–7.

Guthoff M, et al. Effect of genetic variation in Kv1.3 on olfactory function. Diabetes Metab Res Rev. 2009;25:523–7.

Gutman GA, et al. International Union of Pharmacology. LIII. Nomenclature and molecular relationships of voltage-gated potassium channels. Pharmacol Rev. 2005;57:473–508.

Haick JM, Byron KL. Novel treatment strategies for smooth muscle disorders: targeting Kv7 potassium channels. Pharmacol Ther. 2016;165:14–25.

Hancox JC, McPate MJ, El Harchi A, Zhang YH. The hERG potassium channel and hERG screening for drug-induced torsades de pointes. Pharmacol Ther. 2008;119:118–32.

Hansen RS, et al. Activation of human ether-a-go-go-related gene potassium channels by the diphenylurea 1,3-bis-(2-hydroxy-5-trifluoromethyl-phenyl)-urea (NS1643). Mol Pharmacol. 2006a;69:266–77.

Hansen RS, et al. Biophysical characterization of the new human ether-a-go-go-related gene channel opener NS3623 [N-(4-bromo-2-(1H-tetrazol-5-yl)-phenyl)-N′-(3′-trifluoromethylphenyl) urea]. Mol Pharmacol. 2006b;70:1319–29.

Hanson DC, et al. UK-78,282, a novel piperidine compound that potently blocks the Kv1.3 voltage-gated potassium channel and inhibits human T cell activation. Br J Pharmacol. 1999;126:1707–16.

Hashimoto N, Yamashita T, Tsuruzoe N. Characterization of in vivo and in vitro electrophysiological and antiarrhythmic effects of a novel IKACh blocker, NIP-151: a comparison with an IKr-blocker dofetilide. J Cardiovasc Pharmacol. 2008;51:162–9.

Hayashi H, Iwata M, Tsuchimori N, Matsumoto T. Activation of peripheral KCNQ channels attenuates inflammatory pain. Mol Pain. 2014;10:15.

Hayes KC. The use of 4-aminopyridine (fampridine) in demyelinating disorders. CNS Drug Rev. 2004;10:295–316.

Hebert SC, Desir G, Giebisch G, Wang WH. Molecular diversity and regulation of renal potassium channels. Physiol Rev. 2005;85:319–71.

Hedegaard ER, et al. KV 7 channels are involved in hypoxia-induced vasodilatation of porcine coronary arteries. Br J Pharmacol. 2014;171:69–82.

Heginbotham L, Lu Z, Abramson T, MacKinnon R. Mutations in the K^+ channel signature sequence. Biophys J. 1994;66:1061–7.

Heinemann SH, Rettig J, Graack HR, Pongs O. Functional characterization of Kv channel beta-subunits from rat brain. J Physiol. 1996;493(Pt 3):625–33.

Hejtmancik JF, et al. Mutations in KCNJ13 cause autosomal-dominant snowflake vitreoretinal degeneration. Am J Hum Genet. 2008;82:174–80.

Henry JA, Roberts LE, Caspary DM, Theodoroff SM, Salvi RJ. Underlying mechanisms of tinnitus: review and clinical implications. J Am Acad Audiol. 2014;25:5–22; quiz 126.

Hibino H, et al. An ATP-dependent inwardly rectifying potassium channel, KAB-2 (Kir4. 1), in cochlear stria vascularis of inner ear: its specific subcellular localization and correlation with the formation of endocochlear potential. J Neurosci. 1997;17:4711–21.

Hibino H, Fujita A, Iwai K, Yamada M, Kurachi Y. Differential assembly of inwardly rectifying K^+ channel subunits, Kir4.1 and Kir5.1, in brain astrocytes. J Biol Chem. 2004;279:44065–73.

Hibino H, et al. Inwardly rectifying potassium channels: their structure, function, and physiological roles. Physiol Rev. 2010;90:291–366.

Higasa K, et al. A burden of rare variants in BMPR2 and KCNK3 contributes to a risk of familial pulmonary arterial hypertension. BMC Pulm Med. 2017;17:57.

Higdon NR, Khan SA, Buchanan LV, Meisheri KD. Tissue and species variation in the vascular receptor binding of 3H-P1075, a potent KATP opener vasodilator. J Pharmacol Exp Ther. 1997;280:255–60.

Hille B. Ionic channels: molecular pores of excitable membranes. Harvey Lect. 1986;82:47–69.

Hille B. Ion channels of excitable membranes. 3rd ed. Sunderland: Sinauer; 2001. xviii, 814 p.

Hiraoka M, Sawada K, Kawano S. Effects of quinidine on plateau currents of guinea-pig ventricular myocytes. J Mol Cell Cardiol. 1986;18:1097–106.

Hite RK, et al. Cryo-electron microscopy structure of the Slo2.2 Na^+-activated K^+ channel. Nature. 2015;527:198–203.

Hodgkin AL, Huxley AF. The components of membrane conductance in the giant axon of Loligo. J Physiol. 1952;116:473–96.

Hoegg S, Meyer A. Phylogenomic analyses of KCNA gene clusters in vertebrates: why do gene clusters stay intact? BMC Evol Biol. 2007;7:139.

Hoffman JF, et al. The hSK4 (KCNN4) isoform is the Ca^{2+}-activated K^+ channel (Gardos channel) in human red blood cells. Proc Natl Acad Sci U S A. 2003;100:7366–71.

Holland M, Langton PD, Standen NB, Boyle JP. Effects of the BKCa channel activator, NS1619, on rat cerebral artery smooth muscle. Br J Pharmacol. 1996;117:119–29.

Holter J, Carter D, Leresche N, Crunelli V, Vincent P. A TASK3 channel (KCNK9) mutation in a genetic model of absence epilepsy. J Mol Neurosci. 2005;25:37–51.

Honore E. The neuronal background K2P channels: focus on TREK1. Nat Rev Neurosci. 2007;8:251–61.

Hoshi T, Heinemann SH. Modulation of BK channels by small endogenous molecules and pharmaceutical channel openers. Int Rev Neurobiol. 2016;128:193–237.

Hoshi T, et al. Omega-3 fatty acids lower blood pressure by directly activating large-conductance Ca^{2+}-dependent K^+ channels. Proc Natl Acad Sci U S A. 2013;110:4816–21.

Hou S, Heinemann SH, Hoshi T. Modulation of BKCa channel gating by endogenous signaling molecules. Physiology (Bethesda). 2009;24:26–35.

Hou P, Shi J, White KM, Gao Y, Cui J. ML277 specifically enhances the fully activated open state of KCNQ1 by modulating VSD-pore coupling. eLife. 2019;8:e48576.

Hougaard C, et al. Selective activation of the SK1 subtype of human small-conductance Ca^{2+}-activated K^+ channels by 4-(2-methoxyphenylcarbamoyloxymethyl)-piperidine-1-carboxylic acid tert-butyl ester (GW542573X) is dependent on serine 293 in the S5 segment. Mol Pharmacol. 2009;76:569–78.

Hu X, et al. Clinical feature, treatment, and KCNH5 mutations in epilepsy. Front Pediatr. 2022;10:858008.

Huang X, Jan LY. Targeting potassium channels in cancer. J Cell Biol. 2014;206:151–62.

Huang P, et al. Flupirtine attenuates chronic restraint stress-induced cognitive deficits and hippocampal apoptosis in male mice. Behav Brain Res. 2015a;288:1–10.

Huang X, et al. EAG2 potassium channel with evolutionarily conserved function as a brain tumor target. Nat Neurosci. 2015b;18:1236–46.

Huang H, et al. Cardiac voltage-gated ion channels in safety pharmacology: review of the landscape leading to the CiPA initiative. J Pharmacol Toxicol Methods. 2017;87:11–23.

Humphrey SJ. Pharmacology of the K-ATP channel blocking morpholinoguanidine PNU-37883A. Cardiovasc Drug Rev. 2006;17:295–328.

Hundallah K, Alenizi A, AlHashem A, Tabarki B. Severe early-onset epileptic encephalopathy due to mutations in the KCNA2 gene: expansion of the genotypic and phenotypic spectrum. Eur J Paediatr Neurol. 2016;20:657–60.

Huo R, Sheng Y, Guo WT, Dong DL. The potential role of Kv4.3 K^+ channel in heart hypertrophy. Channels (Austin). 2014;8:203–9.

Hwang PM, Glatt CE, Bredt DS, Yellen G, Snyder SH. A novel K^+ channel with unique localizations in mammalian brain: molecular cloning and characterization. Neuron. 1992;8:473–81.

Hyodo T, et al. Voltage-gated potassium channel Kv1.3 blocker as a potential treatment for rat anti-glomerular basement membrane glomerulonephritis. Am J Physiol Renal Physiol. 2010;299:F1258–69.

Iannotti FA, et al. Expression, localization, and pharmacological role of Kv7 potassium channels in skeletal muscle proliferation, differentiation, and survival after myotoxic insults. J Pharmacol Exp Ther. 2010;332:811–20.

Imaizumi Y, et al. Molecular basis of pimarane compounds as novel activators of large-conductance Ca^{2+}-activated K^+ channel alpha-subunit. Mol Pharmacol. 2002;62:836–46.

Inagaki N, et al. Reconstitution of IKATP: an inward rectifier subunit plus the sulfonylurea receptor. Science. 1995;270:1166–70.

Inagaki N, et al. A family of sulfonylurea receptors determines the pharmacological properties of ATP-sensitive K^+ channels. Neuron. 1996;16:1011–7.

Inanobe A, et al. Characterization of G-protein-gated K^+ channels composed of Kir3.2 subunits in dopaminergic neurons of the substantia nigra. J Neurosci. 1999;19:1006–17.

Inoue M, Nakajima S, Nakajima Y. Somatostatin induces an inward rectification in rat locus coeruleus neurones through a pertussis toxin-sensitive mechanism. J Physiol. 1988;407:177–98.

Ipavec V, Martire M, Barrese V, Taglialatela M, Currò D. K_V7 channels regulate muscle tone and nonadrenergic noncholinergic relaxation of the rat gastric fundus. Pharmacol Res. 2011;64:397–409.

Irani SR, et al. Antibodies to Kv1 potassium channel-complex proteins leucine-rich, glioma inactivated 1 protein and contactin-associated protein-2 in limbic encephalitis, Morvan's syndrome and acquired neuromyotonia. Brain. 2010;133:2734–48.

Isbrandt D, et al. Gene structures and expression profiles of three human KCND (Kv4) potassium channels mediating A-type currents I(TO) and I(SA). Genomics. 2000;64:144–54.

Ishii TM, et al. A human intermediate conductance calcium-activated potassium channel. Proc Natl Acad Sci U S A. 1997;94:11651–6.

Isomoto S, Yamada M, Horio Y, Kurachi Y. Molecular aspects of ATP-sensitive K^+ channels in the cardiovascular system. Jpn J Physiol. 1997;47(Suppl 1):S5–6.

Ito H, Takikawa R, Kurachi Y, Sugimoto T. Anti-cholinergic effect of verapamil on the muscarinic acetylcholine receptor-gated K^+ channel in isolated guinea-pig atrial myocytes. Naunyn Schmiedeberg's Arch Pharmacol. 1989;339:244–6.

Jackson WF. K_V channels and the regulation of vascular smooth muscle tone. Microcirculation. 2018;25:12421.

Jang SH, Ryu PD, Lee SY. Dendrotoxin-kappa suppresses tumor growth induced by human lung adenocarcinoma A549 cells in nude mice. J Vet Sci. 2011;12:35–40.

Jedrychowska J, Korzh V. Kv2.1 voltage-gated potassium channels in developmental perspective. Dev Dyn. 2019;248:1180–94.

Jelacic TM, Kennedy ME, Wickman K, Clapham DE. Functional and biochemical evidence for G-protein-gated inwardly rectifying K^+ (GIRK) channels composed of GIRK2 and GIRK3. J Biol Chem. 2000;275:36211–6.

Jensen BS. BMS-204352: a potassium channel opener developed for the treatment of stroke. CNS Drug Rev. 2002;8:353–60.

Jensen BS, et al. Characterization of the cloned human intermediate-conductance Ca^{2+}-activated K^+ channel. Am J Phys. 1998;275:C848–56.

Jensen HS, Callo K, Jespersen T, Jensen BS, Olesen SP. The KCNQ5 potassium channel from mouse: a broadly expressed M-current like potassium channel modulated by zinc, pH, and volume changes. Brain Res Mol Brain Res. 2005;139:52–62.

Jensen MO, et al. Mechanism of voltage gating in potassium channels. Science. 2012;336:229–33.

Jepps TA. K_v7 channel trafficking by the microtubule network in vascular smooth muscle. Acta Physiol (Oxf). 2021;232:e13692.

Jepps TA, Greenwood IA, Moffatt JD, Sanders KM, Ohya S. Molecular and functional characterization of K_v7 K^+ channel in murine gastrointestinal smooth muscles. Am J Physiol Gastrointest Liver Physiol. 2009;297:G107–15.

Jepps TA, Olesen SP, Greenwood IA, Dalsgaard T. Molecular and functional characterization of Kv 7 channels in penile arteries and corpus cavernosum of healthy and metabolic syndrome rats. Br J Pharmacol. 2016;173:1478–90.

Jiang M, et al. Dynamic conformational changes of extracellular S5-P linkers in the hERG channel. J Physiol. 2005;569:75–89.

Jin W, Lu Z. A novel high-affinity inhibitor for inward-rectifier K^+ channels. Biochemistry. 1998;37:13291–9.

Jin W, Lu Z. Synthesis of a stable form of tertiapin: a high-affinity inhibitor for inward-rectifier K^+ channels. Biochemistry. 1999;38:14286–93.

Johnson B, Leek AN, Tamkun MM. Kv2 channels create endoplasmic reticulum/plasma membrane junctions: a brief history of Kv2 channel subcellular localization. Channels (Austin). 2019;13:88–101.

Johnston J. Pharmacology of A-type K^+ channels. Handb Exp Pharmacol. 2021;267:167–83.

Johnston J, Forsythe ID, Kopp-Scheinpflug C. Going native: voltage-gated potassium channels controlling neuronal excitability. J Physiol. 2010;588:3187–200.

Joiner WJ, Wang LY, Tang MD, Kaczmarek LK. hSK4, a member of a novel subfamily of calcium-activated potassium channels. Proc Natl Acad Sci U S A. 1997;94:11013–8.

Jones F, Gamper N, Gao H. Kv7 channels and excitability disorders. Handb Exp Pharmacol. 2021;267:185–230.

Jorge BS, et al. Voltage-gated potassium channel KCNV2 (Kv8.2) contributes to epilepsy susceptibility. Proc Natl Acad Sci U S A. 2011;108:5443–8.

Jow F, Zhang ZH, Kopsco DC, Carroll KC, Wang K. Functional coupling of intracellular calcium and inactivation of voltage-gated Kv1.1/Kvbeta1.1 A-type K^+ channels. Proc Natl Acad Sci U S A. 2004;101:15535–40.

Juang JM, et al. Disease-targeted sequencing of ion channel genes identifies de novo mutations in patients with non-familial Brugada syndrome. Sci Rep. 2014;4:6733.

Justice JA, et al. Disruption of $K_V2.1$ somato-dendritic clusters prevents the apoptogenic increase of potassium currents. Neuroscience. 2017;354:158–67.

Kaczmarek LK. Slack, Slick and sodium-activated potassium channels. ISRN Neurosci. 2013;2013:354262.

Kaczmarek LK, Zhang Y. Kv3 channels: enablers of rapid firing, neurotransmitter release, and neuronal endurance. Physiol Rev. 2017;97:1431–68.

Kajioka S, Oike M, Kitamura K. Nicorandil opens a calcium-dependent potassium channel in smooth muscle cells of the rat portal vein. J Pharmacol Exp Ther. 1990;254:905–13.

Kajioka S, Kitamura K, Kuriyama H. Guanosine diphosphate activates an adenosine 5′-triphosphate-sensitive K^+ channel in the rabbit portal vein. J Physiol. 1991;444:397–418.

Kalappa BI, et al. Potent KCNQ2/3-specific channel activator suppresses in vivo epileptic activity and prevents the development of tinnitus. J Neurosci. 2015;35:8829–42.

Kallergis EM, Goudis CA, Simantirakis EN, Kochiadakis GE, Vardas PE. Mechanisms, risk factors, and management of acquired long QT syndrome: a comprehensive review. ScientificWorldJournal. 2012;2012:212178.

Kameyama M, et al. Intracellular Na^+ activates a K^+ channel in mammalian cardiac cells. Nature. 1984;309:354–6.

Kamouchi M, Kitamura K. Regulation of ATP-sensitive K^+ channels by ATP and nucleotide diphosphate in rabbit portal vein. Am J Phys. 1994;266:H1687–98.

Kamuene JM, Xu Y, Plant LD. The pharmacology of two-pore domain potassium channels. Handb Exp Pharmacol. 2021;267:417–43.

Kanaumi T, et al. Developmental changes in KCNQ2 and KCNQ3 expression in human brain: possible contribution to the age-dependent etiology of benign familial neonatal convulsions. Brain and Development. 2008;30:362–9.

Kanda H, et al. Kv4.3 channel dysfunction contributes to trigeminal neuropathic pain manifested with orofacial cold hypersensitivity in rats. J Neurosci. 2021;41:2091–105.

Kane GC, Liu XK, Yamada S, Olson TM, Terzic A. Cardiac KATP channels in health and disease. J Mol Cell Cardiol. 2005;38:937–43.

Kang D, Kim D. TREK-2 (K2P10.1) and TRESK (K2P18.1) are major background K^+ channels in dorsal root ganglion neurons. Am J Physiol Cell Physiol. 2006;291:C138–46.

Kang J, et al. Discovery of a small molecule activator of the human ether-a-go-go-related gene (HERG) cardiac K^+ channel. Mol Pharmacol. 2005a;67:827–36.

Kang D, Choe C, Kim D. Thermosensitivity of the two-pore domain K^+ channels TREK-2 and TRAAK. J Physiol. 2005b;564:103–16.

Kanyo R, et al. Functional and behavioral signatures of Kv7 activator drug subtypes. Epilepsia. 2020;61:1678–90.

Kapplinger JD, et al. Enhancing the predictive power of mutations in the C-terminus of the KCNQ1-encoded Kv7.1 voltage-gated potassium channel. J Cardiovasc Transl Res. 2015;8:187–97.

Karschin C, Dissmann E, Stühmer W, Karschin A. IRK(1-3) and GIRK(1-4) inwardly rectifying K^+ channel mRNAs are differentially expressed in the adult rat brain. J Neurosci. 1996;16:3559–70.

Kasumu AW, et al. Selective positive modulator of calcium-activated potassium channels exerts beneficial effects in a mouse model of spinocerebellar ataxia type 2. Chem Biol. 2012;19:1340–53.

Katayama J, Yakushiji T, Akaike N. Characterization of the K^+ current mediated by 5-HT1A receptor in the acutely dissociated rat dorsal raphe neurons. Brain Res. 1997;745:283–92.

Katz B. Les constantes electriques de la membrane du muscle. Arch Sci Physiol. 1949;2:285–99.

Kaufhold MA, et al. Localization, trafficking, and significance for acid secretion of parietal cell Kir4.1 and KCNQ1 K^+ channels. Gastroenterology. 2008;134:1058–69.

Kaufmann K, et al. ML297 (VU0456810), the first potent and selective activator of the GIRK potassium channel, displays antiepileptic properties in mice. ACS Chem Neurosci. 2013;4:1278–86.

Kawai T, Watanabe M. Blockade of Ca-activated K conductance by apamin in rat sympathetic neurones. Br J Pharmacol. 1986;87:225–32.

Keating MT, Sanguinetti MC. Pathophysiology of ion channel mutations. Curr Opin Genet Dev. 1996;6:326–33.

Kemp JM, et al. Electrophysiological characterization of the hERG R56Q LQTS variant and targeted rescue by the activator RPR260243. J Gen Physiol. 2021;153:e202112923.

Ketchum KA, Joiner WJ, Sellers AJ, Kaczmarek LK, Goldstein SA. A new family of outwardly rectifying potassium channel proteins with two pore domains in tandem. Nature. 1995;376:690–5.

Khanamiri S, et al. Contribution of Kv7 channels to basal coronary flow and active response to ischemia. Hypertension. 2013;62:1090–7.

Kharade SV, et al. Discovery, characterization, and effects on renal fluid and electrolyte excretion of the Kir4.1 potassium channel pore blocker, VU0134992. Mol Pharmacol. 2018;94:926–37.

Kharkovets T, et al. KCNQ4, a K^+ channel mutated in a form of dominant deafness, is expressed in the inner ear and the central auditory pathway. Proc Natl Acad Sci U S A. 2000;97:4333–8.

Kharkovets T, et al. Mice with altered KCNQ4 K^+ channels implicate sensory outer hair cells in human progressive deafness. EMBO J. 2006;25:642–52.

Kim DM, Nimigean CM. Voltage-gated potassium channels: a structural examination of selectivity and gating. Cold Spring Harb Perspect Biol. 2016;8:a029231.

Kim Y, Bang H, Kim D. TASK-3, a new member of the tandem pore K^+ channel family. J Biol Chem. 2000;275:9340–7.

Kiraly I, et al. Involvement of large-conductance Ca^{2+} -activated K^+ channels in both nitric oxide and endothelium-derived hyperpolarization-type relaxation in human penile small arteries. Basic Clin Pharmacol Toxicol. 2013;113:19–24.

Kitamura H, et al. Tertiapin potently and selectively blocks muscarinic K^+ channels in rabbit cardiac myocytes. J Pharmacol Exp Ther. 2000;293:196–205.

Kleopa KA, Elman LB, Lang B, Vincent A, Scherer SS. Neuromyotonia and limbic encephalitis sera target mature Shaker-type K^+ channels: subunit specificity correlates with clinical manifestations. Brain. 2006;129:1570–84.

Klinger F, Gould G, Boehm S, Shapiro MS. Distribution of M-channel subunits KCNQ2 and KCNQ3 in rat hippocampus. NeuroImage. 2011;58:761–9.

Klinger F, et al. Concomitant facilitation of GABAA receptors and K_V7 channels by the non-opioid analgesic flupirtine. Br J Pharmacol. 2012;166:1631–42.

Kobayashi T, et al. Molecular cloning of a mouse G-protein-activated K^+ channel (mGIRK1) and distinct distributions of three GIRK (GIRK1, 2 and 3) mRNAs in mouse brain. Biochem Biophys Res Commun. 1995;208:1166–73.

Kobayashi T, et al. Ethanol opens G-protein-activated inwardly rectifying K^+ channels. Nat Neurosci. 1999;2:1091–7.

Kobayashi T, Ikeda K, Kumanishi T. Inhibition by various antipsychotic drugs of the G-protein-activated inwardly rectifying K^+ (GIRK) channels expressed in xenopus oocytes. Br J Pharmacol. 2000;129:1716–22.

Kobayashi T, Washiyama K, Ikeda K. Inhibition of G protein-activated inwardly rectifying K^+ channels by fluoxetine (Prozac). Br J Pharmacol. 2003;138:1119–28.

Kobayashi T, Washiyama K, Ikeda K. Inhibition of G protein-activated inwardly rectifying K^+ channels by various antidepressant drugs. Neuropsychopharmacology. 2004;29:1841–51.

Kobayashi T, Washiyama K, Ikeda K. Inhibition of G protein-activated inwardly rectifying K^+ channels by ifenprodil. Neuropsychopharmacology. 2006;31:516–24.

Kobayashi T, et al. Inhibitory effects of the antiepileptic drug ethosuximide on G protein-activated inwardly rectifying K^+ channels. Neuropharmacology. 2009;56:499–506.

Koepple C, et al. Dual mechanism for inhibition of inwardly rectifying Kir2.x channels by quinidine involving direct pore block and PIP_2-interference. J Pharmacol Exp Ther. 2017;361:209–18.

Kofuji P, et al. Functional analysis of the weaver mutant GIRK2 K^+ channel and rescue of weaver granule cells. Neuron. 1996;16:941–52.

Kohling R, Wolfart J. Potassium channels in epilepsy. Cold Spring Harb Perspect Med. 2016;6:a022871.

Kollewe A, Lau AY, Sullivan A, Roux B, Goldstein SA. A structural model for K2P potassium channels based on 23 pairs of interacting sites and continuum electrostatics. J Gen Physiol. 2009;134:53–68.

Koot BG, Alders M, Verheij J, Beuers U, Cobben JM. A de novo mutation in KCNN3 associated with autosomal dominant idiopathic non-cirrhotic portal hypertension. J Hepatol. 2016;64:974–7.

Korsgaard MP, et al. Anxiolytic effects of Maxipost (BMS-204352) and retigabine via activation of neuronal Kv7 channels. J Pharmacol Exp Ther. 2005;314:282–92.

Kottgen M, et al. Carbachol activates a K^+ channel of very small conductance in the basolateral membrane of rat pancreatic acinar cells. Pflugers Arch. 1999;438:597–603.

Kovoor P, et al. Evaluation of the role of I(KACh) in atrial fibrillation using a mouse knockout model. J Am Coll Cardiol. 2001;37:2136–43.

Koyama H, et al. Molecular cloning, functional expression and localization of a novel inward rectifier potassium channel in the rat brain. FEBS Lett. 1994;341:303–7.

Koyrakh L, et al. Molecular and cellular diversity of neuronal G-protein-gated potassium channels. J Neurosci. 2005;25:11468–78.

Kozek KA, et al. Discovery and characterization of VU0529331, a synthetic small-molecule activator of homomeric G protein-gated, inwardly rectifying, potassium (GIRK) channels. ACS Chem Neurosci. 2019;10:358–70.

Krapivinsky G, et al. The G-protein-gated atrial K^+ channel IKACh is a heteromultimer of two inwardly rectifying K^+-channel proteins. Nature. 1995;374:135–41.

Krijger Juarez C, Amin AS, Offerhaus JA, Bezzina CR, Boukens BJ. Cardiac repolarization in health and disease. JACC Clin Electrophysiol. 2023;9:124–38.

Kruger J, et al. Loss-of-function variants in the KCNQ5 gene are implicated in genetic generalized epilepsies. EBioMedicine. 2022;84:104244.

Kshatri AS, Gonzalez-Hernandez A, Giraldez T. Physiological roles and therapeutic potential of Ca^{2+} activated potassium channels in the nervous system. Front Mol Neurosci. 2018;11:258.

Kubisch C, et al. KCNQ4, a novel potassium channel expressed in sensory outer hair cells, is mutated in dominant deafness. Cell. 1999;96:437–46.

Kubo Y, Baldwin TJ, Jan YN, Jan LY. Primary structure and functional expression of a mouse inward rectifier potassium channel. Nature. 1993;362:127–33.

Kuchenbuch M, et al. In silico model reveals the key role of GABA in KCNT1-epilepsy in infancy with migrating focal seizures. Epilepsia. 2021;62:683–97.

Kun A, et al. NS11021, a novel opener of large-conductance Ca^{2+} -activated K^+ channels, enhances erectile responses in rats. Br J Pharmacol. 2009;158:1465–76.

Kurachi Y, Ishii M. Cell signal control of the G protein-gated potassium channel and its subcellular localization. J Physiol. 2004;554:285–94.

Kurachi Y, Nakajima T, Sugimoto T. Quinidine inhibition of the muscarine receptor-activated K+ channel current in atrial cells of guinea pig. Naunyn Schmiedeberg's Arch Pharmacol. 1987;335:216–8.

Kusaka S, et al. Functional Kir7.1 channels localized at the root of apical processes in rat retinal pigment epithelium. J Physiol. 2001;531:27–36.

Kuzhikandathil EV, Oxford GS. Classic D1 dopamine receptor antagonist R-(+)-7-chloro-8-hydroxy-3-methyl-1-phenyl-2,3,4,5-tetrahydro-1H-3-benzazepine hydrochloride (SCH23390) directly inhibits G protein-coupled inwardly rectifying potassium channels. Mol Pharmacol. 2002;62:119–26.

Kuzmenkov AI, et al. Apamin structure and pharmacology revisited. Front Pharmacol. 2022;13:977440.

Lacey MG, Mercuri NB, North RA. Dopamine acts on D2 receptors to increase potassium conductance in neurones of the rat substantia nigra zona compacta. J Physiol. 1987;392:397–416.

Lacey MG, Mercuri NB, North RA. On the potassium conductance increase activated by GABAB and dopamine D2 receptors in rat substantia nigra neurones. J Physiol. 1988;401:437–53.

Lachheb S, et al. Kir4.1/Kir5.1 channel forms the major K^+ channel in the basolateral membrane of mouse renal collecting duct principal cells. Am J Physiol Renal Physiol. 2008;294:F1398–407.

Lafreniere RG, Rouleau GA. Identification of novel genes involved in migraine. Headache. 2012;52(Suppl 2):107–10.

Lafreniere RG, et al. A dominant-negative mutation in the TRESK potassium channel is linked to familial migraine with aura. Nat Med. 2010;16:1157–60.

Lam J, Wulff H. The lymphocyte potassium channels Kv1.3 and KCa3.1 as targets for immunosuppression. Drug Dev Res. 2011;72:573–84.

Lam J, Coleman N, Garing AL, Wulff H. The therapeutic potential of small-conductance KCa2 channels in neurodegenerative and psychiatric diseases. Expert Opin Ther Targets. 2013;17:1203–20.

Lambert M, et al. Characterization of Kcnk3-mutated rat, a novel model of pulmonary hypertension. Circ Res. 2019;125:678–95.

Lamy C, et al. Allosteric block of KCa2 channels by apamin. J Biol Chem. 2010;285:27067–77.

Landgraf R. Meglitinide analogues in the treatment of type 2 diabetes mellitus. Drugs Aging. 2000;17:411–25.

Lastraioli E, et al. hergl gene and HERG1 protein are overexpressed in colorectal cancers and regulate cell invasion of tumor cells. Cancer Res. 2004;64:606–11.

Lastraioli E, Lottini T, Bencini L, Bernini M, Arcangeli A. hERG1 potassium channels: novel biomarkers in human solid cancers. Biomed Res Int. 2015;2015:896432.

Lastraioli E, et al. The hERG1 potassium channel behaves as prognostic factor in gastric dysplasia endoscopic samples. Onco Targets Ther. 2019;12:9377–84.

Latypova X, et al. Novel KCNB1 mutation associated with non-syndromic intellectual disability. J Hum Genet. 2017;62:569–73.

Laumonnier F, et al. Association of a functional deficit of the BKCa channel, a synaptic regulator of neuronal excitability, with autism and mental retardation. Am J Psychiatry. 2006;163:1622–9.

Layne JJ, Nausch B, Olesen SP, Nelson MT. BK channel activation by NS11021 decreases excitability and contractility of urinary bladder smooth muscle. Am J Physiol Regul Integr Comp Physiol. 2010;298:R378–84.

Lazarenko RM, et al. Motoneuronal TASK channels contribute to immobilizing effects of inhalational general anesthetics. J Neurosci. 2010;30:7691–704.

Leanza L, et al. Clofazimine, Psora-4 and PAP-1, inhibitors of the potassium channel Kv1.3, as a new and selective therapeutic strategy in chronic lymphocytic leukemia. Leukemia. 2013;27:1782–5.

Lee WS, Hebert SC. ROMK inwardly rectifying ATP-sensitive K^+ channel. I. Expression in rat distal nephron segments. Am J Phys. 1995;268:F1124–31.

Lee JH, et al. The local translation of K_{Na} in dendritic projections of auditory neurons and the roles of K(Na) in the transition from hidden to overt hearing loss. Aging (Albany NY). 2019;11:11541–64.

Lee JH, et al. Characterization and functional roles of KCNQ-encoded voltage-gated potassium (K_v7) channels in human corpus cavernosum smooth muscle. Pflugers Arch. 2020;472:89–102.

Leitner MG, et al. Restoration of ion channel function in deafness-causing KCNQ4 mutants by synthetic channel openers. Br J Pharmacol. 2012;165:2244–59.

Leppert M, Singh N. Benign familial neonatal epilepsy with mutations in two potassium channel genes. Curr Opin Neurol. 1999;12:143–7.

Lerche C, et al. Molecular cloning and functional expression of KCNQ5, a potassium channel subunit that may contribute to neuronal M-current diversity. J Biol Chem. 2000;275:22395–400.

Lesage F, et al. Cloning provides evidence for a family of inward rectifier and G-protein coupled K^+ channels in the brain. FEBS Lett. 1994;353:37–42.

Lesage F, et al. Molecular properties of neuronal G-protein-activated inwardly rectifying K^+ channels. J Biol Chem. 1995;270:28660–7.

Levy RL. Clinical studies of quinidin: IV. The clinical toxicology of quinidin. J Am Med Assoc. 1922;79:1108–13.

Lewis LM, et al. High-throughput screening reveals a small-molecule inhibitor of the renal outer medullary potassium channel and Kir7.1. Mol Pharmacol. 2009;76:1094–103.

Lewohl JM, et al. G-protein-coupled inwardly rectifying potassium channels are targets of alcohol action. Nat Neurosci. 1999;2:1084–90.

Li HF, Chen SA, Wu SN. Evidence for the stimulatory effect of resveratrol on Ca^{2+}-activated K^+ current in vascular endothelial cells. Cardiovasc Res. 2000;45:1035–45.
Li Y, Gamper N, Hilgemann DW, Shapiro MS. Regulation of Kv7 (KCNQ) K^+ channel open probability by phosphatidylinositol (4,5)-bisphosphate. J Neurosci. 2005;25:9825–35.
Li W, Halling DB, Hall AW, Aldrich RW. EF hands at the N-lobe of calmodulin are required for both SK channel gating and stable SK-calmodulin interaction. J Gen Physiol. 2009;134:281–93.
Li P, et al. Sodium-activated potassium channels moderate excitability in vascular smooth muscle. J Physiol. 2019;597:5093–108.
Li Z, et al. General principles for the validation of proarrhythmia risk prediction models: an extension of the CiPA in silico strategy. Clin Pharmacol Ther. 2020a;107:102–11.
Li X, et al. Molecular basis for ligand activation of the human KCNQ2 channel. Cell Res. 2020b;31:52–61.
Li Y, Aziz Q, Tinker A. The pharmacology of ATP-sensitive K^+ channels (K_{ATP}). Handb Exp Pharmacol. 2021a;267:357–78.
Li X, et al. Kv3.1 channelopathy: a novel loss-of-function variant and the mechanistic basis of its clinical phenotypes. Ann Transl Med. 2021b;9:1397.
Li T, et al. Structural basis for the modulation of human KCNQ4 by small-molecule drugs. Mol Cell. 2021c;81:25–37.e4.
Liao YJ, Jan YN, Jan LY. Heteromultimerization of G-protein-gated inwardly rectifying K^+ channel proteins GIRK1 and GIRK2 and their altered expression in weaver brain. J Neurosci. 1996;16:7137–50.
Liin SI, Karlsson U, Bentzen BH, Schmitt N, Elinder F. Polyunsaturated fatty acids are potent openers of human M-channels expressed in Xenopus laevis oocytes. Acta Physiol (Oxf). 2016;218:28–37.
Lin TF, et al. The subfamily-specific assembly of Eag and Erg K^+ channels is determined by both the amino and the carboxyl recognition domains. J Biol Chem. 2014;289:22815–34.
Linley JE, et al. Inhibition of M current in sensory neurons by exogenous proteases: a signaling pathway mediating inflammatory nociception. J Neurosci. 2008;28:11240–9.
Linley JE, et al. Reactive oxygen species are second messengers of neurokinin signaling in peripheral sensory neurons. Proc Natl Acad Sci U S A. 2012;109:E1578–86.
Liu X, Stan Leung L. Sodium-activated potassium conductance participates in the depolarizing afterpotential following a single action potential in rat hippocampal CA1 pyramidal cells. Brain Res. 2004;1023:185–92.
Liu Y, et al. Tenidap, a novel anti-inflammatory agent, is an opener of the inwardly rectifying K^+ channel hKir2.3. Eur J Pharmacol. 2002a;435:153–60.
Liu X, Chang Y, Reinhart PH, Sontheimer H, Chang Y. Cloning and characterization of glioma BK, a novel BK channel isoform highly expressed in human glioma cells. J Neurosci. 2002b;22:1840–9.
Liu B, et al. Selective inhibition of Kir currents by antihistamines. Eur J Pharmacol. 2007;558:21–6.
Liu B, et al. The acute nociceptive signals induced by bradykinin in rat sensory neurons are mediated by inhibition of M-type K^+ channels and activation of Ca^{2+}-activated Cl^- channels. J Clin Invest. 2010;120:1240–52.
Liu S, et al. Ion-binding properties of a K^+ channel selectivity filter in different conformations. Proc Natl Acad Sci U S A. 2015;112:15096–100.
Liu Y, Duan Y, Du D, Chen F. Rescuing Kv10.2 protein changes cognitive and emotional function in kainic acid-induced status epilepticus rats. Epilepsy Behav. 2020;106:106894.
Liu Y, Bian X, Wang K. Pharmacological activation of neuronal voltage-gated Kv7/KCNQ/M--channels for potential therapy of epilepsy and pain. Handb Exp Pharmacol. 2021;267:231–51.
Liu HF, et al. A novel de novo heterozygous variant of the KCNQ2 gene: contribution to early-onset epileptic encephalopathy in a female infant. Mol Med Rep. 2022a;26:282.
Liu R, et al. Double-edged role of K_{Na} channels in brain tuning: identifying epileptogenic network micro-macro disconnection. Curr Neuropharmacol. 2022b;20:916–28.
Logsdon NJ, Kang J, Togo JA, Christian EP, Aiyar J. A novel gene, hKCa4, encodes the calcium-activated potassium channel in human T lymphocytes. J Biol Chem. 1997;272:32723–6.

Lolicato M, et al. K(2P)2.1 (TREK-1)-activator complexes reveal a cryptic selectivity filter binding site. Nature. 2017;547:364–8.
Long SB, Tao X, Campbell EB, MacKinnon R. Atomic structure of a voltage-dependent K^+ channel in a lipid membrane-like environment. Nature. 2007;450:376–82.
Lopatin AN, Makhina EN, Nichols CG. Potassium channel block by cytoplasmic polyamines as the mechanism of intrinsic rectification. Nature. 1994;372:366–9.
Lopes CM, Gallagher PG, Buck ME, Butler MH, Goldstein SA. Proton block and voltage gating are potassium-dependent in the cardiac leak channel Kcnk3. J Biol Chem. 2000;275:16969–78.
Lopes CM, Zilberberg N, Goldstein SA. Block of Kcnk3 by protons. Evidence that 2-P-domain potassium channel subunits function as homodimers. J Biol Chem. 2001;276:24449–52.
Loucif AJC, et al. GI-530159, a novel, selective, mechanosensitive two-pore-domain potassium (K(2P)) channel opener, reduces rat dorsal root ganglion neuron excitability. Br J Pharmacol. 2018;175:2272–83.
Lu Q, et al. Disruption of Kv1.1 N-type inactivation by novel small molecule inhibitors (disinactivators). Bioorg Med Chem. 2008a;16:3067–75.
Lu HR, et al. Predicting drug-induced changes in QT interval and arrhythmias: QT-shortening drugs point to gaps in the ICHS7B guidelines. Br J Pharmacol. 2008b;154:1427–38.
Luis E, Anaya-Hernandez A, Leon-Sanchez P, Duran-Pasten ML. The Kv10.1 channel: a promising target in cancer. Int J Mol Sci. 2022;23:8458.
Luo Y, Huang L, Liao P, Jiang R. Contribution of neuronal and glial two-pore-domain potassium channels in health and neurological disorders. Neural Plast. 2021;2021:8643129.
Luo H, de Velasco EMF, Wickman K. Neuronal G protein-gated K^+ channels. Am J Physiol Cell Physiol. 2022;323:C439–60.
Luscher C, Jan LY, Stoffel M, Malenka RC, Nicoll RA. G protein-coupled inwardly rectifying K^+ channels (GIRKs) mediate postsynaptic but not presynaptic transmitter actions in hippocampal neurons. Neuron. 1997;19:687–95.
Ma L, et al. A novel channelopathy in pulmonary arterial hypertension. N Engl J Med. 2013;369:351–61.
Mackie AR, Byron KL. Cardiovascular KCNQ (K_v7) potassium channels: physiological regulators and new targets for therapeutic intervention. Mol Pharmacol. 2008;74:1171–9.
Maher AD, Kuchel PW. The Gardos channel: a review of the Ca^{2+}-activated K^+ channel in human erythrocytes. Int J Biochem Cell Biol. 2003;35:1182–97.
Maingret F, Patel AJ, Lesage F, Lazdunski M, Honore E. Mechano- or acid stimulation, two interactive modes of activation of the TREK-1 potassium channel. J Biol Chem. 1999;274:26691–6.
Maingret F, Patel AJ, Lazdunski M, Honore E. The endocannabinoid anandamide is a direct and selective blocker of the background K^+ channel TASK-1. EMBO J. 2001;20:47–54.
Malaisse WJ. Pharmacology of the meglitinide analogs: new treatment options for type 2 diabetes mellitus. Treat Endocrinol. 2003;2:401–14.
Malerba M, et al. Efficacy of andolast in mild to moderate asthma: a randomized, controlled, double-blind multicenter study (the Andast trial). Curr Pharm Des. 2015;21:3835–43.
Maljevic S, Lerche H. Potassium channel genes and benign familial neonatal epilepsy. Prog Brain Res. 2014;213:17–53.
Maljevic S, et al. C-terminal interaction of KCNQ2 and KCNQ3 K^+ channels. J Physiol. 2003;548:353–60.
Maljevic S, Wuttke TV, Seebohm G, Lerche H. KV7 channelopathies. Pflugers Arch. 2010;460:277–88.
Malysz J, Petkov GV. Detrusor smooth muscle K_V7 channels: emerging new regulators of urinary bladder function. Front Physiol. 2020;11:1004.
Mani BK, et al. $K_v7.5$ potassium channel subunits are the primary targets for PKA-dependent enhancement of vascular smooth muscle Kv7 currents. Mol Pharmacol. 2016;89:323–34.
Mannhold R. K_{ATP} channel openers: structure-activity relationships and therapeutic potential. Med Res Rev. 2004;24:213–66.
Mannikko R, et al. Pharmacological and electrophysiological characterization of AZSMO-23, an activator of the hERG K^+ channel. Br J Pharmacol. 2015;172:3112–25.

Manville RW, Papanikolaou M, Abbott GW. Direct neurotransmitter activation of voltage-gated potassium channels. Nat Commun. 2018;9:1847.

Marion E, et al. Mycobacterial toxin induces analgesia in buruli ulcer by targeting the angiotensin pathways. Cell. 2014;157:1565–76.

Marsh B, Acosta C, Djouhri L, Lawson SN. Leak K^+ channel mRNAs in dorsal root ganglia: relation to inflammation and spontaneous pain behaviour. Mol Cell Neurosci. 2012;49:375–86.

Martin S, et al. Eag1 potassium channel immunohistochemistry in the CNS of adult rat and selected regions of human brain. Neuroscience. 2008;155:833–44.

Mastrangelo M, et al. Epilepsy in KCNH1-related syndromes. Epileptic Disord. 2016;18:123–36.

Matheu MP, et al. Imaging of effector memory T cells during a delayed-type hypersensitivity reaction and suppression by Kv1.3 channel block. Immunity. 2008;29:602–14.

Mathie A, Al-Moubarak E, Veale EL. Gating of two pore domain potassium channels. J Physiol. 2010;588:3149–56.

Matsuda H. Open-state substructure of inwardly rectifying potassium channels revealed by magnesium block in guinea-pig heart cells. J Physiol. 1988;397:237–58.

Matsuda T, et al. Blockade by NIP-142, an antiarrhythmic agent, of carbachol-induced atrial action potential shortening and GIRK1/4 channel. J Pharmacol Sci. 2006;101:303–10.

Matsui A, Williams JT. Activation of micro-opioid receptors and block of Kir3 potassium channels and NMDA receptor conductance by L- and D-methadone in rat locus coeruleus. Br J Pharmacol. 2010;161:1403–13.

Matsumura K, Yokogawa M, Osawa M. Peptide toxins targeting KV channels. Handb Exp Pharmacol. 2021;267:481–505.

Maylie J, Bond CT, Herson PS, Lee WS, Adelman JP. Small conductance Ca^{2+}-activated K^+ channels and calmodulin. J Physiol. 2004;554:255–61.

McCallum LA, Greenwood IA, Tribe RM. Expression and function of K_v7 channels in murine myometrium throughout oestrous cycle. Pflugers Arch. 2009;457:1111–20.

McCallum LA, Pierce SL, England SK, Greenwood IA, Tribe RM. The contribution of Kv7 channels to pregnant mouse and human myometrial contractility. J Cell Mol Med. 2011;15:577–86.

McManus OB, et al. An activator of calcium-dependent potassium channels isolated from a medicinal herb. Biochemistry. 1993;32:6128–33.

Mei Y, et al. PAP-1 ameliorates DSS-induced colitis with involvement of NLRP3 inflammasome pathway. Int Immunopharmacol. 2019;75:105776.

Mello de Queiroz F, Suarez-Kurtz G, Stuhmer W, Pardo LA. Ether a go-go potassium channel expression in soft tissue sarcoma patients. Mol Cancer. 2006;5:42.

Menendez ST, et al. Frequent aberrant expression of the human ether a go-go (hEAG1) potassium channel in head and neck cancer: pathobiological mechanisms and clinical implications. J Mol Med (Berl). 2012a;90:1173–84.

Menendez ST, et al. Role of HERG1 potassium channel in both malignant transformation and disease progression in head and neck carcinomas. Mod Pathol. 2012b;25:1069–78.

Meng J, Shi C, Li L, Du Y, Xu Y. Compound ICA-105574 prevents arrhythmias induced by cardiac delayed repolarization. Eur J Pharmacol. 2013;718:87–97.

Meredith AL, Thorneloe KS, Werner ME, Nelson MT, Aldrich RW. Overactive bladder and incontinence in the absence of the BK large conductance Ca^{2+}-activated K^+ channel. J Biol Chem. 2004;279:36746–52.

Mesirca P, et al. The G-protein-gated K^+ channel, IKACh, is required for regulation of pacemaker activity and recovery of resting heart rate after sympathetic stimulation. J Gen Physiol. 2013;142:113–26.

Meuth SG, et al. Contribution of TWIK-related acid-sensitive K^+ channel 1 (TASK1) and TASK3 channels to the control of activity modes in thalamocortical neurons. J Neurosci. 2003;23:6460–9.

Miceli F, Soldovieri MV, Weckhuysen S, Cooper E, Taglialatela M. Review KCNQ2-related disorders. In: Adam MP, et al., editors. GeneReviews((R)). Seattle: University of Washington, Seattle; 1993.

Miki T, et al. Mouse model of Prinzmetal angina by disruption of the inward rectifier Kir6.1. Nat Med. 2002;8:466–72.

Miller AN, Long SB. Crystal structure of the human two-pore domain potassium channel K2P1. Science. 2012;335:432–6.

Miller C, Moczydlowski E, Latorre R, Phillips M. Charybdotoxin, a protein inhibitor of single Ca^{2+}-activated K^+ channels from mammalian skeletal muscle. Nature. 1985;313:316–8.

Miller MJ, et al. Nuclear localization and dominant-negative suppression by a mutant SKCa3 N-terminal channel fragment identified in a patient with schizophrenia. J Biol Chem. 2001;276:27753–6.

Milligan CJ, et al. Robotic multiwell planar patch-clamp for native and primary mammalian cells. Nat Protoc. 2009;4:244–55.

Milovic S, Steinecker-Frohnwieser B, Schreibmayer W, Weigl LG. The sensitivity of G protein-activated K^+ channels toward halothane is essentially determined by the C terminus. J Biol Chem. 2004;279:34240–9.

Misaki N, et al. Iptakalim, a vascular ATP-sensitive potassium (K_{ATP}) channel opener, closes rat pancreatic beta-cell KATP channels and increases insulin release. J Pharmacol Exp Ther. 2007;322:871–8.

Misgeld U, Müller W, Brunner H. Effects of (−)baclofen on inhibitory neurons in the guinea pig hippocampal slice. Pflugers Arch. 1989;414:139–44.

Mohr CJ, et al. Cancer-associated intermediate conductance Ca^{2+}-activated K^+ channel K(Ca)3.1. Cancers (Basel). 2019;11:109.

Mohr CJ, et al. Subunits of BK channels promote breast cancer development and modulate responses to endocrine treatment in preclinical models. Br J Pharmacol. 2022;179:2906–24.

Mondejar-Parreno G, Perez-Vizcaino F, Cogolludo A. Kv7 channels in lung diseases. Front Physiol. 2020;11:634.

Mongan LC, et al. The distribution of small and intermediate conductance calcium-activated potassium channels in the rat sensory nervous system. Neuroscience. 2005;131:161–75.

Mooney J, Rawls SM. KCNQ2/3 channel agonist flupirtine reduces cocaine place preference in rats. Behav Pharmacol. 2017;28:405–7.

Morales-Cano D, et al. Kv7 channels critically determine coronary artery reactivity: left-right differences and down-regulation by hyperglycaemia. Cardiovasc Res. 2015;106:98–108.

Morimoto T, et al. Voltage-sensitive oxonol dyes are novel large-conductance Ca^{2+}-activated K^+ channel activators selective for beta1 and beta4 but not for beta2 subunits. Mol Pharmacol. 2007;71:1075–88.

Morishige K, et al. Molecular cloning and functional expression of a novel brain-specific inward rectifier potassium channel. FEBS Lett. 1994;346:251–6.

Morishige K, et al. Secretagogue-induced exocytosis recruits G protein-gated K^+ channels to plasma membrane in endocrine cells. J Biol Chem. 1999;274:7969–74.

Mouhat S, et al. K^+ channel types targeted by synthetic OSK1, a toxin from Orthochirus scrobiculosus scorpion venom. Biochem J. 2005;385:95–104.

Mousavi Nik A, Gharaie S, Jeong Kim H. Cellular mechanisms of mutations in Kv7.1: auditory functions in Jervell and Lange-Nielsen syndrome vs. Romano-Ward syndrome. Front Cell Neurosci. 2015;9:32.

Mubagwa K, et al. KB130015, a new amiodarone derivative with multiple effects on cardiac ion channels. Cardiovasc Drug Rev. 2003;21:216–35.

Muona M, et al. A recurrent de novo mutation in KCNC1 causes progressive myoclonus epilepsy. Nat Genet. 2015;47:39–46.

Naffaa MM, Al-Ewaidat OA. Ligand modulation of KCNQ-encoded (K_V7) potassium channels in the heart and nervous system. Eur J Pharmacol. 2021;906:174278.

Nagumo H, et al. Rho kinase inhibitor HA-1077 prevents Rho-mediated myosin phosphatase inhibition in smooth muscle cells. Am J Physiol Cell Physiol. 2000;278:C57–65.

Nakajima T, Kurachi Y, Ito H, Takikawa R, Sugimoto T. Anti-cholinergic effects of quinidine, disopyramide, and procainamide in isolated atrial myocytes: mediation by different molecular mechanisms. Circ Res. 1989;64:297–303.

Nappi P, et al. Epileptic channelopathies caused by neuronal Kv7 (KCNQ) channel dysfunction. Pflugers Arch. 2020;472:881–98.

Nardi A, Olesen SP. BK channel modulators: a comprehensive overview. Curr Med Chem. 2008;15:1126–46.

Nascimento FA, Andrade DM. Myoclonus epilepsy and ataxia due to potassium channel mutation (MEAK) is caused by heterozygous KCNC1 mutations. Epileptic Disord. 2016;18:135–8.

Navas Tejedor P, et al. An homozygous mutation in KCNK3 is associated with an aggressive form of hereditary pulmonary arterial hypertension. Clin Genet. 2017;91:453–7.

Nenov NI, Crumb WJ Jr, Pigott JD, Harrison LH Jr, Clarkson CW. Quinidine interactions with human atrial potassium channels: developmental aspects. Circ Res. 1998;83:1224–31.

Neubauer BA, et al. KCNQ2 and KCNQ3 mutations contribute to different idiopathic epilepsy syndromes. Neurology. 2008;71:177–83.

Nguyen A, et al. Novel nonpeptide agents potently block the C-type inactivated conformation of Kv1.3 and suppress T cell activation. Mol Pharmacol. 1996;50:1672–9.

Nichols CG, Lederer WJ. Adenosine triphosphate-sensitive potassium channels in the cardiovascular system. Am J Phys. 1991;261:H1675–86.

Nicolas CS, et al. IKs response to protein kinase A-dependent KCNQ1 phosphorylation requires direct interaction with microtubules. Cardiovasc Res. 2008;79:427–35.

Nicolazzo JA, et al. Blockade of microglial Kv1.3 potassium channels by the peptide HsTX1[R14A] attenuates lipopolysaccharide-mediated neuroinflammation. J Pharm Sci. 2022;111:638–47.

Nilius B, Droogmans G. Ion channels and their functional role in vascular endothelium. Physiol Rev. 2001;81:1415–59.

Nilius B, Wohlrab W. Potassium channels and regulation of proliferation of human melanoma cells. J Physiol. 1992;445:537–48.

Nilius B, Schwarz G, Droogmans G. Modulation by histamine of an inwardly rectifying potassium channel in human endothelial cells. J Physiol. 1993;472:359–71.

Nishida M, Cadene M, Chait BT, MacKinnon R. Crystal structure of a Kir3.1-prokaryotic Kir channel chimera. EMBO J. 2007;26:4005–15.

Nolting A, Ferraro T, D'Hoedt D, Stocker M. An amino acid outside the pore region influences apamin sensitivity in small conductance Ca^{2+}-activated K^+ channels. J Biol Chem. 2007;282:3478–86.

Noma A. ATP-regulated K^+ channels in cardiac muscle. Nature. 1983;305:147–8.

Noskov SY, Berneche S, Roux B. Control of ion selectivity in potassium channels by electrostatic and dynamic properties of carbonyl ligands. Nature. 2004;431:830–4.

Nouvian R, et al. Degeneration of sensory outer hair cells following pharmacological blockade of cochlear KCNQ channels in the adult guinea pig. Eur J Neurosci. 2003;17:2553–62.

Nunez E, et al. Redox regulation of K_v7 channels through EF3 hand of calmodulin. eLife. 2023;12:e81961.

Nuwer MO, Picchione KE, Bhattacharjee A. PKA-induced internalization of slack K_{Na} channels produces dorsal root ganglion neuron hyperexcitability. J Neurosci. 2010;30:14165–72.

Occhiodoro T, et al. Cloning of a human ether-a-go-go potassium channel expressed in myoblasts at the onset of fusion. FEBS Lett. 1998;434:177–82.

Oertli A, et al. Molecular mechanism of autosomal recessive long QT-syndrome 1 without deafness. Int J Mol Sci. 2021;22:1112.

Ogura K, Sato T, Yuki H, Honma T. Support Vector Machine model for hERG inhibitory activities based on the integrated hERG database using descriptor selection by NSGA-II. Sci Rep. 2019;9:12220.

Ohno Y, Hibino H, Lossin C, Inanobe A, Kurachi Y. Inhibition of astroglial Kir4.1 channels by selective serotonin reuptake inhibitors. Brain Res. 2007;1178:44–51.

Okuyama Y, et al. The effects of nucleotides and potassium channel openers on the SUR2A/Kir6.2 complex K^+ channel expressed in a mammalian cell line, HEK293T cells. Pflugers Arch. 1998;435:595–603.

Olesen SP, Munch E, Moldt P, Drejer J. Selective activation of Ca^{2+}-dependent K^+ channels by novel benzimidazolone. Eur J Pharmacol. 1994;251:53–9.

Olesen MS, et al. A novel KCND3 gain-of-function mutation associated with early-onset of persistent lone atrial fibrillation. Cardiovasc Res. 2013;98:488–95.
Oliver D, Knipper M, Derst C, Fakler B. Resting potential and submembrane calcium concentration of inner hair cells in the isolated mouse cochlea are set by KCNQ-type potassium channels. J Neurosci. 2003;23:2141–9.
Olschewski A, et al. Impact of TASK-1 in human pulmonary artery smooth muscle cells. Circ Res. 2006;98:1072–80.
Olson TM, et al. Kv1.5 channelopathy due to KCNA5 loss-of-function mutation causes human atrial fibrillation. Hum Mol Genet. 2006;15:2185–91.
Ota M, et al. Polymorphism in the KCNA3 gene is associated with susceptibility to autoimmune pancreatitis in the Japanese population. Dis Markers. 2011;31:223–9.
Ousingsawat J, et al. Expression of voltage-gated potassium channels in human and mouse colonic carcinoma. Clin Cancer Res. 2007;13:824–31.
Overturf KE, et al. Cloning and characterization of a Kv1.5 delayed rectifier K^+ channel from vascular and visceral smooth muscles. Am J Phys. 1994;267:C1231–8.
Papazian DM, Schwarz TL, Tempel BL, Jan YN, Jan LY. Cloning of genomic and complementary DNA from Shaker, a putative potassium channel gene from Drosophila. Science. 1987;237:749–53.
Parajuli SP, Soder RP, Hristov KL, Petkov GV. Pharmacological activation of small conductance calcium-activated potassium channels with naphtho[1,2-d]thiazol-2-ylamine decreases guinea pig detrusor smooth muscle excitability and contractility. J Pharmacol Exp Ther. 2012;340:114–23.
Pardo LA, Stuhmer W. The roles of K^+ channels in cancer. Nat Rev Cancer. 2014;14:39–48.
Pardo LA, et al. Approaches targeting $K_V10.1$ open a novel window for cancer diagnosis and therapy. Curr Med Chem. 2012;19:675–82.
Passmore GM, et al. KCNQ/M currents in sensory neurons: significance for pain therapy. J Neurosci. 2003;23:7227–36.
Passmore GM, et al. Functional significance of M-type potassium channels in nociceptive cutaneous sensory endings. Front Mol Neurosci. 2012;5:63.
Patel AJ, et al. A mammalian two pore domain mechano-gated S-like K^+ channel. EMBO J. 1998;17:4283–90.
Patel AJ, et al. Inhalational anesthetics activate two-pore-domain background K^+ channels. Nat Neurosci. 1999;2:422–6.
Pearson WL, Dourado M, Schreiber M, Salkoff L, Nichols CG. Expression of a functional Kir4 family inward rectifier K^+ channel from a gene cloned from mouse liver. J Physiol. 1999;514(Pt 3):639–53.
Pedarzani P, Stocker M. Molecular and cellular basis of small--and intermediate-conductance, calcium-activated potassium channel function in the brain. Cell Mol Life Sci. 2008;65:3196–217.
Pedarzani P, et al. Tamapin, a venom peptide from the Indian red scorpion (Mesobuthus tamulus) that targets small conductance Ca^{2+}-activated K^+ channels and afterhyperpolarization currents in central neurons. J Biol Chem. 2002;277:46101–9.
Peitersen T, et al. Computational analysis of the effects of the hERG channel opener NS1643 in a human ventricular cell model. Heart Rhythm. 2008;5:734–41.
Pennington MW, et al. Development of highly selective Kv1.3-blocking peptides based on the sea anemone peptide ShK. Mar Drugs. 2015;13:529–42.
Perissinotti LL, et al. Kinetic model for NS1643 drug activation of WT and L529I variants of Kv11.1 (hERG1) potassium channel. Biophys J. 2015;108:1414–24.
Peroz D, et al. Kv7.1 (KCNQ1) properties and channelopathies. J Physiol. 2008;586:1785–9.
Perry M, Sachse FB, Sanguinetti MC. Structural basis of action for a human ether-a-go-go-related gene 1 potassium channel activator. Proc Natl Acad Sci U S A. 2007;104:13827–32.
Perry M, Sachse FB, Abbruzzese J, Sanguinetti MC. PD-118057 contacts the pore helix of hERG1 channels to attenuate inactivation and enhance K^+ conductance. Proc Natl Acad Sci U S A. 2009;106:20075–80.

Perry MD, et al. Pharmacological activation of IKr in models of long QT type 2 risks overcorrection of repolarization. Cardiovasc Res. 2020;116:1434–45.
Pessia M, Tucker SJ, Lee K, Bond CT, Adelman JP. Subunit positional effects revealed by novel heteromeric inwardly rectifying K^+ channels. EMBO J. 1996;15:2980–7.
Pessia M, Imbrici P, D'Adamo MC, Salvatore L, Tucker SJ. Differential pH sensitivity of Kir4.1 and Kir4.2 potassium channels and their modulation by heteropolymerisation with Kir5.1. J Physiol. 2001;532:359–67.
Piechotta PL, et al. The pore structure and gating mechanism of K2P channels. EMBO J. 2011;30:3607–19.
Piper DR, Varghese A, Sanguinetti MC, Tristani-Firouzi M. Gating currents associated with intramembrane charge displacement in HERG potassium channels. Proc Natl Acad Sci U S A. 2003;100:10534–9.
Plant LD. A role for K2P channels in the operation of somatosensory nociceptors. Front Mol Neurosci. 2012;5:21.
Plant LD, Kemp PJ, Peers C, Henderson Z, Pearson HA. Hypoxic depolarization of cerebellar granule neurons by specific inhibition of TASK-1. Stroke. 2002;33:2324–8.
Po SS, et al. Modulation of HERG potassium channels by extracellular magnesium and quinidine. J Cardiovasc Pharmacol. 1999;33:181–5.
Poirier K, et al. Loss of function of KCNC1 is associated with intellectual disability without seizures. Eur J Hum Genet. 2017;25:560–4.
Pollema-Mays SL, Centeno MV, Ashford CJ, Apkarian AV, Martina M. Expression of background potassium channels in rat DRG is cell-specific and down-regulated in a neuropathic pain model. Mol Cell Neurosci. 2013;57:1–9.
Potet F, et al. Identification and characterization of a compound that protects cardiac tissue from human ether-a-go-go-related gene (hERG)-related drug-induced arrhythmias. J Biol Chem. 2012;287:39613–25.
Preisig-Müller R, et al. Heteromerization of Kir2.x potassium channels contributes to the phenotype of Andersen's syndrome. Proc Natl Acad Sci U S A. 2002;99:7774–9.
Prentis PJ, Pavasovic A, Norton RS. Sea anemones: quiet achievers in the field of peptide toxins. Toxins (Basel). 2018;10:36.
Priori SG, et al. A novel form of short QT syndrome (SQT3) is caused by a mutation in the KCNJ2 gene. Circ Res. 2005;96:800–7.
Proks P, Ashcroft FM. Phentolamine block of K_{ATP} channels is mediated by Kir6.2. Proc Natl Acad Sci U S A. 1997;94:11716–20.
Qile M, et al. LUF7244, an allosteric modulator/activator of K_v11.1 channels, counteracts dofetilide-induced torsades de pointes arrhythmia in the chronic atrioventricular block dog model. Br J Pharmacol. 2019;176:3871–85.
Qile M, et al. LUF7244 plus dofetilide rescues aberrant K_v11.1 trafficking and produces functional I(Kv11.1). Mol Pharmacol. 2020;97:355–64.
Qiu B, Wang Y, Li C, Guo H, Xu Y. Utility of the JT peak interval and the JT area in determining the proarrhythmic potential of QT-shortening agents. J Cardiovasc Pharmacol Ther. 2019;24:160–71.
Ramos-Hunter SJ, et al. Discovery and SAR of a novel series of GIRK1/2 and GIRK1/4 activators. Bioorg Med Chem Lett. 2013;23:5195–8.
Rampe D, Brown AM. A history of the role of the hERG channel in cardiac risk assessment. J Pharmacol Toxicol Methods. 2013;68:13–22.
Randall MD, Griffith TM. Modulation of vasodilatation to levcromakalim by hypoxia and EDRF in the rabbit isolated ear: a comparison with pinacidil, sodium nitroprusside and verapamil. Br J Pharmacol. 1993;109:386–93.
Raphemot R, et al. Discovery, characterization, and structure-activity relationships of an inhibitor of inward rectifier potassium (Kir) channels with preference for Kir2.3, Kir3.x, and Kir7.1. Front Pharmacol. 2011;2:75.
Raphemot R, et al. Direct activation of β-cell K_{ATP} channels with a novel xanthine derivative. Mol Pharmacol. 2014;85:858–65.

Rashid MH, et al. A potent and selective peptide blocker of the Kv1.3 channel: prediction from free-energy simulations and experimental confirmation. PLoS One. 2013;8:e78712.
Redrobe JP, Nielsen AN. Effects of neuronal K_v7 potassium channel activators on hyperactivity in a rodent model of mania. Behav Brain Res. 2009;198:481–5.
Reich EP, et al. Blocking ion channel KCNN4 alleviates the symptoms of experimental autoimmune encephalomyelitis in mice. Eur J Immunol. 2005;35:1027–36.
Reijntjes DOJ, et al. Sodium-activated potassium channels shape peripheral auditory function and activity of the primary auditory neurons in mice. Sci Rep. 2019;9:2573.
Rettig J, et al. Inactivation properties of voltage-gated K^+ channels altered by presence of beta-subunit. Nature. 1994;369:289–94.
Reyes R, et al. Cloning and expression of a novel pH-sensitive two pore domain K^+ channel from human kidney. J Biol Chem. 1998;273:30863–9.
Ritter DM, et al. Dysregulation of Kv3.4 channels in dorsal root ganglia following spinal cord injury. J Neurosci. 2015;35:1260–73.
Robbins J. KCNQ potassium channels: physiology, pathophysiology, and pharmacology. Pharmacol Ther. 2001;90:1–19.
Robbins TW, Everitt BJ. Drug addiction: bad habits add up. Nature. 1999;398:567–70.
Robitaille R, Garcia ML, Kaczorowski GJ, Charlton MP. Functional colocalization of calcium and calcium-gated potassium channels in control of transmitter release. Neuron. 1993;11:645–55.
Rode F, Svalo J, Sheykhzade M, Ronn LC. Functional effects of the KCNQ modulators retigabine and XE991 in the rat urinary bladder. Eur J Pharmacol. 2010;638:121–7.
Rodrigues N, et al. Synthesis and structure-activity relationship study of substituted caffeate esters as antinociceptive agents modulating the TREK-1 channel. Eur J Med Chem. 2014;75:391–402.
Rodríguez-Menchaca AA, et al. The molecular basis of chloroquine block of the inward rectifier Kir2.1 channel. Proc Natl Acad Sci U S A. 2008;105:1364–8.
Roeloffs R, et al. In vivo profile of ICA-27243 [N-(6-chloro-pyridin-3-yl)-3,4-difluoro-benzamide], a potent and selective KCNQ2/Q3 (Kv7.2/Kv7.3) activator in rodent anticonvulsant models. J Pharmacol Exp Ther. 2008;326:818–28.
Rogers M. Ion channel drug discovery and modern medicine. MedNous. 2019;November/December:9–11.
Rosenzweig EB, et al. Paediatric pulmonary arterial hypertension: updates on definition, classification, diagnostics and management. Eur Respir J. 2019;53:1801916.
Roxburgh CJ, et al. The synthesis and some pharmacological actions of the enantiomers of the K^+-channel blocker cetiedil. J Pharm Pharmacol. 1996;48:851–7.
Roy S, et al. Structure-activity relationships of a novel group of large-conductance Ca^{2+}-activated K^+ (BK) channel modulators: the GoSlo-SR family. ChemMedChem. 2012;7:1763–9.
Royal P, et al. Migraine-associated TRESK mutations increase neuronal excitability through alternative translation initiation and inhibition of TREK. Neuron. 2019;101:232–245.e6.
Sager PT, Gintant G, Turner JR, Pettit S, Stockbridge N. Rechanneling the cardiac proarrhythmia safety paradigm: a meeting report from the Cardiac Safety Research Consortium. Am Heart J. 2014;167:292–300.
Sala L, et al. A new hERG allosteric modulator rescues genetic and drug-induced long-QT syndrome phenotypes in cardiomyocytes from isogenic pairs of patient induced pluripotent stem cells. EMBO Mol Med. 2016;8:1065–81.
Sale H, et al. Modulation of $K_v11.1$ (hERG) channels by 5-(((1H-indazol-5-yl)oxy)methyl)-N-(4-(trifluoromethoxy)phenyl)pyrimidin-2-amine (ITP-2), a novel small molecule activator. Br J Pharmacol. 2017;174:2484–500.
Sánchez-Chapula JA, Salinas-Stefanon E, Torres-Jácome J, Benavides-Haro DE, Navarro-Polanco RA. Blockade of currents by the antimalarial drug chloroquine in feline ventricular myocytes. J Pharmacol Exp Ther. 2001;297:437–45.
Sander SE, Lambrecht C, Richter A. The $K_V7.2/3$ preferring channel opener ICA 27243 attenuates L-DOPA-induced dyskinesia in hemiparkinsonian rats. Neurosci Lett. 2013;545:59–63.
Sands Z, Grottesi A, Sansom MS. Voltage-gated ion channels. Curr Biol. 2005;15:R44–7.

Sanguinetti MC. HERG1 channel agonists and cardiac arrhythmia. Curr Opin Pharmacol. 2014;15:22–7.

Sanguinetti MC, Mitcheson JS. Predicting drug-hERG channel interactions that cause acquired long QT syndrome. Trends Pharmacol Sci. 2005;26:119–24.

Sanguinetti MC, Tristani-Firouzi M. hERG potassium channels and cardiac arrhythmia. Nature. 2006;440:463–9.

Sanguinetti MC, Jiang C, Curran ME, Keating MT. A mechanistic link between an inherited and an acquired cardiac arrhythmia: HERG encodes the IKr potassium channel. Cell. 1995;81:299–307.

Sanguinetti MC, et al. Coassembly of K_VLQT1 and minK (IsK) proteins to form cardiac I_{Ks} potassium channel. Nature. 1996;384:80–3.

Santos R, et al. A comprehensive map of molecular drug targets. Nat Rev Drug Discov. 2017;16:19–34.

Sato T, Li Y, Saito T, Nakaya H. Minoxidil opens mitochondrial K_{ATP} channels and confers cardioprotection. Br J Pharmacol. 2004;141:360–6.

Sato T, Yuki H, Ogura K, Honma T. Construction of an integrated database for hERG blocking small molecules. PLoS One. 2018;13:e0199348.

Sausbier M, et al. Cerebellar ataxia and Purkinje cell dysfunction caused by Ca^{2+}-activated K^+ channel deficiency. Proc Natl Acad Sci U S A. 2004;101:9474–8.

Saxena S, Kruys V, Vamecq J, Maze M. The role of microglia in perioperative neuroinflammation and neurocognitive disorders. Front Aging Neurosci. 2021;13:671499.

Schenzer A, et al. Molecular determinants of KCNQ (K_v7) K^+ channel sensitivity to the anticonvulsant retigabine. J Neurosci. 2005;25:5051–60.

Schewe M, et al. A non-canonical voltage-sensing mechanism controls gating in K2P K^+ channels. Cell. 2016;164:937–49.

Schewe M, et al. A pharmacological master key mechanism that unlocks the selectivity filter gate in K(+) channels. Science. 2019;363:875–80.

Schimpf R, Wolpert C, Gaita F, Giustetto C, Borggrefe M. Short QT syndrome. Cardiovasc Res. 2005;67:357–66.

Schindler RF, et al. POPDC1(S201F) causes muscular dystrophy and arrhythmia by affecting protein trafficking. J Clin Invest. 2016;126:239–53.

Schmid-Antomarchi H, de Weille J, Fosset M, Lazdunski M. The antidiabetic sulfonylurea glibenclamide is a potent blocker of the ATP-modulated K^+ channel in insulin secreting cells. Biochem Biophys Res Commun. 1987a;146:21–5.

Schmid-Antomarchi H, De Weille J, Fosset M, Lazdunski M. The receptor for antidiabetic sulfonylureas controls the activity of the ATP-modulated K^+ channel in insulin-secreting cells. J Biol Chem. 1987b;262:15840–4.

Schmitz A, et al. Design of PAP-1, a selective small molecule Kv1.3 blocker, for the suppression of effector memory T cells in autoimmune diseases. Mol Pharmacol. 2005;68:1254–70.

Schonherr R, Gessner G, Lober K, Heinemann SH. Functional distinction of human EAG1 and EAG2 potassium channels. FEBS Lett. 2002;514:204–8.

Schreiber JA, Seebohm G. Cardiac K^+ channels and channelopathies. Handb Exp Pharmacol. 2021;267:113–38.

Schroder RL, et al. KCNQ4 channel activation by BMS-204352 and retigabine. Neuropharmacology. 2001;40:888–98.

Schroeder BC, et al. A constitutively open potassium channel formed by KCNQ1 and KCNE3. Nature. 2000a;403:196–9.

Schroeder BC, Hechenberger M, Weinreich F, Kubisch C, Jentsch TJ. KCNQ5, a novel potassium channel broadly expressed in brain, mediates M-type currents. J Biol Chem. 2000b;275:24089–95.

Schulien AJ, et al. Targeted disruption of Kv2.1-VAPA association provides neuroprotection against ischemic stroke in mice by declustering Kv2.1 channels. Sci Adv. 2020;6:eaaz8110.

Schumacher MA, Rivard AF, Bachinger HP, Adelman JP. Structure of the gating domain of a Ca^{2+}-activated K^+ channel complexed with Ca^{2+}/calmodulin. Nature. 2001;410:1120–4.

Schwake M, Jentsch TJ, Friedrich T. A carboxy-terminal domain determines the subunit specificity of KCNQ K^+ channel assembly. EMBO Rep. 2003;4:76–81.

Schwake M, et al. Structural determinants of M-type KCNQ (Kv7) K^+ channel assembly. J Neurosci. 2006;26:3757–66.

Schwanstecher M, et al. Potassium channel openers require ATP to bind to and act through sulfonylurea receptors. EMBO J. 1998;17:5529–35.

Schwartz PJ, Crotti L, Insolia R. Long-QT syndrome: from genetics to management. Circ Arrhythm Electrophysiol. 2012;5:868–77.

Scornik FS, et al. $DiBAC_4(3)$ hits a "sweet spot" for the activation of arterial large-conductance Ca^{2+}-activated potassium channels independently of the β_1-subunit. Am J Physiol Heart Circ Physiol. 2013;304:H1471–82.

Scuvee-Moreau J, Liegeois JF, Massotte L, Seutin V. Methyl-laudanosine: a new pharmacological tool to investigate the function of small-conductance Ca^{2+}-activated K^+ channels. J Pharmacol Exp Ther. 2002;302:1176–83.

Scuvee-Moreau J, et al. Electrophysiological characterization of the SK channel blockers methyl-laudanosine and methyl-noscapine in cell lines and rat brain slices. Br J Pharmacol. 2004;143:753–64.

Seebohm G. Kv7.1 in atrial fibrillation. Heart Rhythm. 2009;6:1154–5.

Serodio P, Kentros C, Rudy B. Identification of molecular components of A-type channels activating at subthreshold potentials. J Neurophysiol. 1994;72:1516–29.

Serodio P, Vega-Saenz de Miera E, Rudy B. Cloning of a novel component of A-type K^+ channels operating at subthreshold potentials with unique expression in heart and brain. J Neurophysiol. 1996;75:2174–9.

Shapiro MS, et al. Reconstitution of muscarinic modulation of the KCNQ2/KCNQ3 K^+ channels that underlie the neuronal M current. J Neurosci. 2000;20:1710–21.

Shen W, et al. Cholinergic modulation of Kir2 channels selectively elevates dendritic excitability in striatopallidal neurons. Nat Neurosci. 2007;10:1458–66.

Shi L, et al. Peptide hormone ghrelin enhances neuronal excitability by inhibition of Kv7/KCNQ channels. Nat Commun. 2013;4:1435.

Shi YP, et al. The hERG channel activator, RPR260243, enhances protective I(Kr) current early in the refractory period reducing arrhythmogenicity in zebrafish hearts. Am J Physiol Heart Circ Physiol. 2020;319:H251–61.

Shore AN, et al. Reduced GABAergic neuron excitability, altered synaptic connectivity, and seizures in a KCNT1 gain-of-function mouse model of childhood epilepsy. Cell Rep. 2020;33:108303.

Shrivastava IH, Tieleman DP, Biggin PC, Sansom MS. K^+ versus Na^+ ions in a K channel selectivity filter: a simulation study. Biophys J. 2002;83:633–45.

Shuck ME, et al. Cloning and characterization of two K^+ inward rectifier (Kir) 1.1 potassium channel homologs from human kidney (Kir1.2 and Kir1.3). J Biol Chem. 1997;272:586–93.

Signorini S, Liao YJ, Duncan SA, Jan LY, Stoffel M. Normal cerebellar development but susceptibility to seizures in mice lacking G protein-coupled, inwardly rectifying K^+ channel GIRK2. Proc Natl Acad Sci U S A. 1997;94:923–7.

Simons C, et al. Mutations in the voltage-gated potassium channel gene KCNH1 cause Temple-Baraitser syndrome and epilepsy. Nat Genet. 2015;47:73–7.

Singh NA, et al. A novel potassium channel gene, KCNQ2, is mutated in an inherited epilepsy of newborns. Nat Genet. 1998;18:25–9.

Singh S, Syme CA, Singh AK, Devor DC, Bridges RJ. Benzimidazolone activators of chloride secretion: potential therapeutics for cystic fibrosis and chronic obstructive pulmonary disease. J Pharmacol Exp Ther. 2001;296:600–11.

Singh B, et al. A Kv4.2 truncation mutation in a patient with temporal lobe epilepsy. Neurobiol Dis. 2006;24:245–53.

Sirois JE, Lynch C 3rd, Bayliss DA. Convergent and reciprocal modulation of a leak K^+ current and I(h) by an inhalational anaesthetic and neurotransmitters in rat brainstem motoneurones. J Physiol. 2002;541:717–29.

Smets K, et al. First de novo KCND3 mutation causes severe Kv4.3 channel dysfunction leading to early onset cerebellar ataxia, intellectual disability, oral apraxia and epilepsy. BMC Med Genet. 2015;16:51.

Smith PL, Baukrowitz T, Yellen G. The inward rectification mechanism of the HERG cardiac potassium channel. Nature. 1996;379:833–6.

Soden ME, et al. Disruption of dopamine neuron activity pattern regulation through selective expression of a human KCNN3 mutation. Neuron. 2013;80:997–1009.

Soder RP, Parajuli SP, Hristov KL, Rovner ES, Petkov GV. SK channel-selective opening by SKA-31 induces hyperpolarization and decreases contractility in human urinary bladder smooth muscle. Am J Physiol Regul Integr Comp Physiol. 2013;304:R155–63.

Sogaard R, Ljungstrom T, Pedersen KA, Olesen SP, Jensen BS. KCNQ4 channels expressed in mammalian cells: functional characteristics and pharmacology. Am J Physiol Cell Physiol. 2001;280:C859–66.

Song OR, et al. A bacterial toxin with analgesic properties: hyperpolarization of DRG neurons by mycolactone. Toxins (Basel). 2017;9:E227.

Song MS, et al. Kv3.1 and Kv3.4, are involved in cancer cell migration and invasion. Int J Mol Sci. 2018;19:1061.

Sorensen US, et al. Synthesis and structure-activity relationship studies of 2-(N-substituted)-aminobenzimidazoles as potent negative gating modulators ofsmall conductance Ca^{2+}-activated K^+ channels. J Med Chem. 2008;51:7625–34.

Southgate L, Machado RD, Graf S, Morrell NW. Molecular genetic framework underlying pulmonary arterial hypertension. Nat Rev Cardiol. 2020;17:85–95.

Spitznagel BD, et al. VU0606170, a selective slack channels inhibitor, decreases calcium oscillations in cultured cortical neurons. ACS Chem Neurosci. 2020;11:3658–71.

Spruce AE, Standen NB, Stanfield PR. Voltage-dependent ATP-sensitive potassium channels of skeletal muscle membrane. Nature. 1985;316:736–8.

Spruce AE, Standen NB, Stanfield PR. Studies of the unitary properties of adenosine-5′-triphosphate-regulated potassium channels of frog skeletal muscle. J Physiol. 1987;382:213–36.

Standen NB, et al. Hyperpolarizing vasodilators activate ATP-sensitive K^+ channels in arterial smooth muscle. Science. 1989;245:177–80.

Steudel FA, et al. SK4 channels modulate Ca^{2+} signalling and cell cycle progression in murine breast cancer. Mol Oncol. 2017;11:1172–88.

Strobaek D, et al. Activation of human IK and SK Ca^{2+}-activated K^+ channels by NS309 (6,7-dichloro-1H-indole-2,3-dione 3-oxime). Biochim Biophys Acta. 2004;1665:1–5.

Strobaek D, et al. Inhibitory gating modulation of small conductance Ca^{2+}-activated K^+ channels by the synthetic compound (R)-N-(benzimidazol-2-yl)-1,2,3,4-tetrahydro-1-naphtylamine (NS8593) reduces afterhyperpolarizing current in hippocampal CA1 neurons. Mol Pharmacol. 2006;70:1771–82.

Su S, et al. Inhibition of astroglial inwardly rectifying Kir4.1 channels by a tricyclic antidepressant, nortriptyline. J Pharmacol Exp Ther. 2007;320:573–80.

Su Z, et al. Electrophysiologic characterization of a novel hERG channel activator. Biochem Pharmacol. 2009;77:1383–90.

Su S, Sun J, Wang Y, Xu Y. Cardiac hERG K^+ channel as safety and pharmacological target. Handb Exp Pharmacol. 2021;267:139–66.

Subbiah RN, et al. Molecular basis of slow activation of the human ether-a-go-go related gene potassium channel. J Physiol. 2004;558:417–31.

Suessbrich H, et al. Specific blockade of slowly activating I(sK) channels by chromanols – impact on the role of I(sK) channels in epithelia. FEBS Lett. 1996;396:271–5.

Suh B, Hille B. Recovery from muscarinic modulation of M current channels requires phosphatidylinositol 4,5-bisphosphate synthesis. Neuron. 2002;35:507–20.

Sun J, MacKinnon R. Cryo-EM structure of a KCNQ1/CaM complex reveals insights into congenital long QT syndrome. Cell. 2017;169:1042–1050.e9.

Sun J, MacKinnon R. Structural basis of human KCNQ1 modulation and gating. Cell. 2020;180:340–347.e9.

Sun D, Gao W, Hu H, Zhou S. Why 90% of clinical drug development fails and how to improve it? Acta Pharm Sin B. 2022;12:3049–62.

Sunaga Y, et al. The effects of mitiglinide (KAD-1229), a new anti-diabetic drug, on ATP-sensitive K^+ channels and insulin secretion: comparison with the sulfonylureas and nateglinide. Eur J Pharmacol. 2001;431:119–25.

Svalo J, Sheykhzade M, Nordling J, Matras C, Bouchelouche P. Functional and molecular evidence for Kv7 channel subtypes in human detrusor from patients with and without bladder outflow obstruction. PLoS One. 2015;10:e0117350.

Swale DR, et al. ML418: the first selective, sub-micromolar pore blocker of Kir7.1 potassium channels. ACS Chem Neurosci. 2016;7:1013–23.

Swanson R, et al. Cloning and expression of cDNA and genomic clones encoding three delayed rectifier potassium channels in rat brain. Neuron. 1990;4:929–39.

Syme CA, Gerlach AC, Singh AK, Devor DC. Pharmacological activation of cloned intermediate- and small-conductance Ca^{2+}-activated K^+ channels. Am J Physiol Cell Physiol. 2000;278:C570–81.

Syrbe S, et al. De novo loss- or gain-of-function mutations in KCNA2 cause epileptic encephalopathy. Nat Genet. 2015;47:393–9.

Takagi H, Hashitani H. Effects of K^+ channel openers on spontaneous action potentials in detrusor smooth muscle of the guinea-pig urinary bladder. Eur J Pharmacol. 2016;789:179–86.

Takahashi T. Inward rectification in neonatal rat spinal motoneurones. J Physiol. 1990;423:47–62.

Takahashi N, et al. Molecular cloning and functional expression of cDNA encoding a second class of inward rectifier potassium channels in the mouse brain. J Biol Chem. 1994;269:23274–9.

Takanari H, et al. Efficient and specific cardiac IK_1 inhibition by a new pentamidine analogue. Cardiovasc Res. 2013;99:203–14.

Takumi T, et al. A novel ATP-dependent inward rectifier potassium channel expressed predominantly in glial cells. J Biol Chem. 1995;270:16339–46.

Talley EM, Lei Q, Sirois JE, Bayliss DA. TASK-1, a two-pore domain K^+ channel, is modulated by multiple neurotransmitters in motoneurons. Neuron. 2000;25:399–410.

Talley EM, Solorzano G, Lei Q, Kim D, Bayliss DA. CNS distribution of members of the two-pore-domain (KCNK) potassium channel family. J Neurosci. 2001;21:7491–505.

Tamkun MM, et al. Molecular cloning and characterization of two voltage-gated K^+ channel cDNAs from human ventricle. FASEB J. 1991;5:331–7.

Tamsett TJ, Picchione KE, Bhattacharjee A. NAD^+ activates KNa channels in dorsal root ganglion neurons. J Neurosci. 2009;29:5127–34.

Tanemoto M, Kittaka N, Inanobe A, Kurachi Y. In vivo formation of a proton-sensitive K^+ channel by heteromeric subunit assembly of Kir5.1 with Kir4.1. J Physiol. 2000;525(Pt 3):587–92.

Tanner MR, et al. Prolonged immunomodulation in inflammatory arthritis using the selective Kv1.3 channel blocker HsTX1[R14A] and its PEGylated analog. Clin Immunol. 2017;180:45–57.

Tarcha EJ, et al. Durable pharmacological responses from the peptide ShK-186, a specific Kv1.3 channel inhibitor that suppresses T cell mediators of autoimmune disease. J Pharmacol Exp Ther. 2012;342:642–53.

Tatulian L, Brown DA. Effect of the KCNQ potassium channel opener retigabine on single KCNQ2/3 channels expressed in CHO cells. J Physiol. 2003;549:57–63.

Taura J, Kircher DM, Gameiro-Ros I, Slesinger PA. Comparison of K^+ channel families. Handb Exp Pharmacol. 2021;267:1–49.

Teisseyre A, Palko-Labuz A, Sroda-Pomianek K, Michalak K. Voltage-gated potassium channel Kv1.3 as a target in therapy of cancer. Front Oncol. 2019;9:933.

Teng BC, et al. Activation of neuronal Kv7/KCNQ/M-channels by the opener QO58-lysine and its anti-nociceptive effects on inflammatory pain in rodents. Acta Pharmacol Sin. 2016;37:1054–62.

Teramoto N. Physiological roles of ATP-sensitive K^+ channels in smooth muscle. J Physiol. 2006;572:617–24.

Terstappen GC, et al. The antidepressant fluoxetine blocks the human small conductance calcium-activated potassium channels SK1, SK2 and SK3. Neurosci Lett. 2003;346:85–8.

Terzic A, Jahangir A, Kurachi Y. Cardiac ATP-sensitive K^+ channels: regulation by intracellular nucleotides and K+ channel-opening drugs. Am J Phys. 1995;269:C525–45.
Testai L, et al. Expression and function of Kv7.4 channels in rat cardiac mitochondria: possible targets for cardioprotection. Cardiovasc Res. 2016;110:40–50.
Thiffault I, et al. A novel epileptic encephalopathy mutation in KCNB1 disrupts Kv2.1 ion selectivity, expression, and localization. J Gen Physiol. 2015;146:399–410.
Thompson SM, Gähwiler BH. Comparison of the actions of baclofen at pre- and postsynaptic receptors in the rat hippocampus in vitro. J Physiol. 1992;451:329–45.
Thompson SM, Haas HL, Gähwiler BH. Comparison of the actions of adenosine at pre- and postsynaptic receptors in the rat hippocampus in vitro. J Physiol. 1992;451:347–63.
Tiwari D, et al. The potassium channel Kv4.2 regulates dendritic spine morphology, electroencephalographic characteristics and seizure susceptibility in mice. Exp Neurol. 2020;334:113437.
Töpert C, et al. Kir2.4: a novel K^+ inward rectifier channel associated with motoneurons of cranial nerve nuclei. J Neurosci. 1998;18:4096–105.
Torkamani A, et al. De novo KCNB1 mutations in epileptic encephalopathy. Ann Neurol. 2014;76:529–40.
Torrecilla M, et al. G-protein-gated potassium channels containing Kir3.2 and Kir3.3 subunits mediate the acute inhibitory effects of opioids on locus ceruleus neurons. J Neurosci. 2002;22:4328–34.
Towart R, et al. Blockade of the I(Ks) potassium channel: an overlooked cardiovascular liability in drug safety screening? J Pharmacol Toxicol Methods. 2009;60:1–10.
Tristan-Clavijo E, et al. Dominant-negative mutation p.Arg324Thr in KCNA1 impairs Kv1.1 channel function in episodic ataxia. Mov Disord. 2016;31:1743–8.
Trube G, Hescheler J. Inward-rectifying channels in isolated patches of the heart cell membrane: ATP-dependence and comparison with cell-attached patches. Pflugers Arch. 1984;401:178–84.
Tsantoulas C, et al. Sensory neuron downregulation of the Kv9.1 potassium channel subunit mediates neuropathic pain following nerve injury. J Neurosci. 2012;32:17502–13.
Tschritter O, et al. A new variant in the human Kv1.3 gene is associated with low insulin sensitivity and impaired glucose tolerance. J Clin Endocrinol Metab. 2006;91:654–8.
Tucker SJ, Imbrici P, Salvatore L, D'Adamo MC, M. Pessia, pH dependence of the inwardly rectifying potassium channel, Kir5.1, and localization in renal tubular epithelia. J Biol Chem. 2000;275:16404–7.
Tucker K, Overton JM, Fadool DA. Kv1.3 gene-targeted deletion alters longevity and reduces adiposity by increasing locomotion and metabolism in melanocortin-4 receptor-null mice. Int J Obes. 2008;32:1222–32.
Uchida H, Sasaki K, Ma L, Ueda H. Neuron-restrictive silencer factor causes epigenetic silencing of Kv4.3 gene after peripheral nerve injury. Neuroscience. 2010;166:1–4.
Ulens C, Daenens P, Tytgat J. The dual modulation of GIRK1/GIRK2 channels by opioid receptor ligands. Eur J Pharmacol. 1999;385:239–45.
Ullrich S, et al. Effects of I(Ks) channel inhibitors in insulin-secreting INS-1 cells. Pflugers Arch. 2005;451:428–36.
Unoki H, et al. SNPs in KCNQ1 are associated with susceptibility to type 2 diabetes in East Asian and European populations. Nat Genet. 2008;40:1098–102.
Unudurthi SD, et al. Two-pore K^+ channel TREK-1 regulates sinoatrial node membrane excitability. J Am Heart Assoc. 2016;5:e002865.
Upadhyay SK, et al. Selective Kv1.3 channel blocker as therapeutic for obesity and insulin resistance. Proc Natl Acad Sci U S A. 2013;110:E2239–48.
Vallon V, et al. KCNQ1-dependent transport in renal and gastrointestinal epithelia. Proc Natl Acad Sci U S A. 2005;102:17864–9.
Valverde MA, et al. Acute activation of Maxi-K channels (hSlo) by estradiol binding to the beta subunit. Science. 1999;285:1929–31.
van der Horst J, Greenwood IA, Jepps TA. Cyclic AMP-dependent regulation of Kv7 voltage-gated potassium channels. Front Physiol. 2020;11:727.

Vandenberg JI, Torres AM, Campbell TJ, Kuchel PW. The HERG K^+ channel: progress in understanding the molecular basis of its unusual gating kinetics. Eur Biophys J. 2004;33:89–97.
Vandenberg JI, et al. hERG K^+ channels: structure, function, and clinical significance. Physiol Rev. 2012;92:1393–478.
Varma S, Rogers DM, Pratt LR, Rempe SB. Perspectives on: ion selectivity: design principles for K+ selectivity in membrane transport. J Gen Physiol. 2011;137:479–88.
Veale EL, Mathie A. Aristolochic acid, a plant extract used in the treatment of pain and linked to Balkan endemic nephropathy, is a regulator of K2P channels. Br J Pharmacol. 2016;173:1639–52.
Veeramah KR, et al. Exome sequencing reveals new causal mutations in children with epileptic encephalopathies. Epilepsia. 2013;54:1270–81.
Vigil FA, Carver CM, Shapiro MS. Pharmacological manipulation of K_v7 channels as a new therapeutic tool for multiple brain disorders. Front Physiol. 2020a;11:688.
Vigil FA, et al. Prevention of brain damage after traumatic brain injury by pharmacological enhancement of KCNQ (Kv7, "M-type") K^+ currents in neurons. J Cereb Blood Flow Metab. 2020b;40:1256–73.
von Beckerath N, Dittrich M, Klieber HG, Daut J. Inwardly rectifying K^+ channels in freshly dissociated coronary endothelial cells from guinea-pig heart. J Physiol. 1996;491(Pt 2):357–65.
Wang JJ, Li Y. KCNQ potassium channels in sensory system and neural circuits. Acta Pharmacol Sin. 2016;37:25–33.
Wang WA-O, Lin DA-O. Inwardly rectifying K^+ channels 4.1 and 5.1 (Kir4.1/Kir5.1) in the renal distal nephron. Am J Physiol Cell Physiol. 2022;323:C277–88.
Wang W, MacKinnon R. Cryo-EM structure of the open human ether-a-go-go-related K^+ channel hERG. Cell. 2017;169:422–430.e10.
Wang Z, Fermini B, Nattel S. Sustained depolarization-induced outward current in human atrial myocytes. Evidence for a novel delayed rectifier K^+ current similar to Kv1.5 cloned channel currents. Circ Res. 1993;73:1061–76.
Wang H, Kunkel DD, Schwartzkroin PA, Tempel BL. Localization of Kv1.1 and Kv1.2, two K channel proteins, to synaptic terminals, somata, and dendrites in the mouse brain. J Neurosci. 1994;14:4588–99.
Wang HS, et al. KCNQ2 and KCNQ3 potassium channel subunits: molecular correlates of the M-channel. Science. 1998;282:1890–3.
Wang HS, Brown BS, McKinnon D, Cohen IS. Molecular basis for differential sensitivity of KCNQ and I(Ks) channels to the cognitive enhancer XE991. Mol Pharmacol. 2000;57:1218–23.
Wang Z, et al. Compound heterozygous mutations in KvLQT1 cause Jervell and Lange-Nielsen syndrome. Mol Genet Metab. 2002;75:308–16.
Wang HR, et al. Selective inhibition of the K(ir)2 family of inward rectifier potassium channels by a small molecule probe: the discovery, SAR, and pharmacological characterization of ML133. ACS Chem Biol. 2011;6:845–56.
Wang Y, et al. Differential sensitivities of the vascular K_{ATP} channel to various PPAR activators. Biochem Pharmacol. 2013;85:1495–503.
Waters MF, et al. Mutations in voltage-gated potassium channel KCNC3 cause degenerative and developmental central nervous system phenotypes. Nat Genet. 2006;38:447–51.
Weaver CD, Denton JS. Next-generation inward rectifier potassium channel modulators: discovery and molecular pharmacology. Am J Physiol Cell Physiol. 2021;320:C1125–40.
Wei A, et al. K^+ current diversity is produced by an extended gene family conserved in Drosophila and mouse. Science. 1990;248:599–603.
Wei J, Fish FA, Myerburg RJ, Roden DM, George AL Jr. Novel KCNQ1 mutations associated with recessive and dominant congenital long QT syndromes: evidence for variable hearing phenotype associated with R518X. Hum Mutat. 2000;15:387–8.
Wei AD, et al. International Union of Pharmacology. LII. Nomenclature and molecular relationships of calcium-activated potassium channels. Pharmacol Rev. 2005;57:463–72.
Wei AD, et al. Human KCNQ5 de novo mutations underlie epilepsy and intellectual disability. J Neurophysiol. 2022;128:40–61.

Weigl LG, Schreibmayer W. G protein-gated inwardly rectifying potassium channels are targets for volatile anesthetics. Mol Pharmacol. 2001;60:282–9.

Wettwer E, Terlau H. Pharmacology of voltage-gated potassium channel Kv1.5--impact on cardiac excitability. Curr Opin Pharmacol. 2014;15:115–21.

Wettwer E, et al. Role of IKur in controlling action potential shape and contractility in the human atrium: influence of chronic atrial fibrillation. Circulation. 2004;110:2299–306.

Wickenden A. K^+ channels as therapeutic drug targets. Pharmacol Ther. 2002;94:157–82.

Wickenden AD, McNaughton-Smith G. K_v7 channels as targets for the treatment of pain. Curr Pharm Des. 2009;15:1773–98.

Wickenden AD, et al. N-(6-chloro-pyridin-3-yl)-3,4-difluoro-benzamide (ICA-27243): a novel, selective KCNQ2/Q3 potassium channel activator. Mol Pharmacol. 2008;73:977–86.

Wickman K, Karschin C, Karschin A, Picciotto MR, Clapham DE. Brain localization and behavioral impact of the G-protein-gated K^+ channel subunit GIRK4. J Neurosci. 2000;20:5608–15.

Williams SE, et al. Hemoxygenase-2 is an oxygen sensor for a calcium-sensitive potassium channel. Science. 2004;306:2093–7.

Wilson GF, Chiu SY. Ion channels in axon and Schwann cell membranes at paranodes of mammalian myelinated fibers studied with patch clamp. J Neurosci. 1990;10:3263–74.

Wimmers S, Wulfsen I, Bauer CK, Schwarz JR. Erg1, erg2 and erg3 K channel subunits are able to form heteromultimers. Pflugers Arch. 2001;441:450–5.

Witchel HJ, Hancox JC. Familial and acquired long qt syndrome and the cardiac rapid delayed rectifier potassium current. Clin Exp Pharmacol Physiol. 2000;27:753–66.

Wittekindt OH, Schmitz A, Lehmann-Horn F, Hansel W, Grissmer S. The human Ca^{2+}-activated K^+ channel, IK, can be blocked by the tricyclic antihistamine promethazine. Neuropharmacology. 2006;50:458–67.

Wonderlin WF, Strobl JS. Potassium channels, proliferation and G1 progression. J Membr Biol. 1996;154:91–107.

Wray S, Arrowsmith S. Uterine excitability and ion channels and their changes with gestation and hormonal environment. Annu Rev Physiol. 2021;83:331–57.

Wu X, Larsson HP. Insights into cardiac IKs (KCNQ1/KCNE1) channels regulation. Int J Mol Sci. 2020;21:9440.

Wulff H, et al. Design of a potent and selective inhibitor of the intermediate-conductance Ca^{2+}-activated K^+ channel, IKCa1: a potential immunosuppressant. Proc Natl Acad Sci U S A. 2000;97:8151–6.

Wulff H, et al. The voltage-gated Kv1.3 K^+ channel in effector memory T cells as new target for MS. J Clin Invest. 2003;111:1703–13.

Wulff H, Kolski-Andreaco A, Sankaranarayanan A, Sabatier JM, Shakkottai V. Modulators of small- and intermediate-conductance calcium-activated potassium channels and their therapeutic indications. Curr Med Chem. 2007;14:1437–57.

Wuttke TV, Seebohm G, Bail S, Maljevic S, Lerche H. The new anticonvulsant retigabine favors voltage-dependent opening of the Kv7.2 (KCNQ2) channel by binding to its activation gate. Mol Pharmacol. 2005;67:1009–17.

Xia X, et al. Molecular basis and restoration of function deficiencies of Kv7.4 variants associated with inherited hearing loss. Hear Res. 2020;388:107884.

Xiao F, et al. Voltage-dependent potassium channel Kv4.2 alleviates the ischemic stroke impairments through activating neurogenesis. Neurochem Int. 2021;150:105155.

Xu J, et al. The voltage-gated potassium channel Kv1.3 regulates energy homeostasis and body weight. Hum Mol Genet. 2003;12:551–9.

Xu J, et al. The voltage-gated potassium channel Kv1.3 regulates peripheral insulin sensitivity. Proc Natl Acad Sci U S A. 2004;101:3112–7.

Xu Y, et al. Probing binding sites and mechanisms of action of an I(Ks) activator by computations and experiments. Biophys J. 2015;108:62–75.

Xynogalos P, et al. Verapamil inhibits Kir2.3 channels by binding to the pore and interfering with PIP_2 binding. Naunyn Schmiedebergs Arch Pharmacol. 2023;396(4):659–67.

Yamakura T, Lewohl JM, Harris RA. Differential effects of general anesthetics on G protein-coupled inwardly rectifying and other potassium channels. Anesthesiology. 2001;95:144–53.

Yamamura H, Suzuki Y, Asai K, Imaizumi Y. Hypoxic stress up-regulates Kir2.1 expression and facilitates cell proliferation in brain capillary endothelial cells. Biochem Biophys Res Commun. 2016;476(4):386–92.

Yang B, Desai R, Kaczmarek LK. Slack and Slick K_{Na} channels regulate the accuracy of timing of auditory neurons. J Neurosci. 2007;27:2617–27.

Yang F, Hanon S, Lam P, Schweitzer P. Quinidine revisited. Am J Med. 2009;122:317–21.

Yang J, et al. An epilepsy/dyskinesia-associated mutation enhances BK channel activation by potentiating Ca^{2+} sensing. Neuron. 2010;66:871–83.

Yang Y, et al. Multistate structural modeling and voltage-clamp analysis of epilepsy/autism mutation Kv10.2-R327H demonstrate the role of this residue in stabilizing the channel closed state. J Neurosci. 2013;33:16586–93.

Yang X, et al. Intracellular zinc protects $K_{v7\,K}^+$ channels from Ca^{2+}calmodulin-mediated inhibition. J Biol Chem. 2023;299:102819.

Yeap J, et al. Reducing voltage-dependent potassium channel Kv3.4 levels ameliorates synapse loss in a mouse model of Alzheimer's disease. Brain Neurosci Adv. 2022;6:23982128221086464.

Yeh CY, Schulien AJ, Molyneaux BJ, Aizenman E. Lessons from recent advances in ischemic stroke management and targeting Kv2.1 for neuroprotection. Int J Mol Sci. 2020;21:6107.

Yi BA, Minor DL Jr, Lin YF, Jan YN, Jan LY. Controlling potassium channel activities: interplay between the membrane and intracellular factors. Proc Natl Acad Sci U S A. 2001;98:11016–23.

Yi M, et al. KCa3.1 constitutes a pharmacological target for astrogliosis associated with Alzheimer's disease. Mol Cell Neurosci. 2016;76:21–32.

Yin SJ, et al. Different residues in channel turret determining the selectivity of ADWX-1 inhibitor peptide between Kv1.1 and Kv1.3 channels. J Proteome Res. 2008;7:4890–7.

Yow TT, et al. Naringin directly activates inwardly rectifying potassium channels at an overlapping binding site to tertiapin-Q. Br J Pharmacol. 2011;163:1017–33.

Yu L, et al. Rosiglitazone selectively inhibits K_{ATP} channels by acting on the $K_{IR}6$ subunit. Br J Pharmacol. 2012;167:26–36.

Yu Z, et al. Synthesis and biological evaluation of negative allosteric modulators of the Kv11.1(hERG) channel. Eur J Med Chem. 2015;106:50–9.

Zangerl-Plessl EM, et al. Toward a structural view of hERG activation by the small-molecule activator ICA-105574. J Chem Inf Model. 2020;60:360–71.

Zavaritskaya O, et al. Role of KCNQ channels in skeletal muscle arteries and periadventitial vascular dysfunction. Hypertension. 2013;61:151–9.

Zeng H, et al. Mallotoxin is a novel human ether-a-go-go-related gene (hERG) potassium channel activator. J Pharmacol Exp Ther. 2006;319:957–62.

Zhang H, et al. PIP_2 activates KCNQ channels, and its hydrolysis underlies receptor-mediated inhibition of M currents. Neuron. 2003;37:963–75.

Zhang H, et al. Modulation of hERG potassium channel gating normalizes action potential duration prolonged by dysfunctional KCNQ1 potassium channel. Proc Natl Acad Sci U S A. 2012a;109:11866–71.

Zhang M, et al. Structural basis for calmodulin as a dynamic calcium sensor. Structure. 2012b;20:911–23.

Zhang F, et al. Modulation of K_v7 potassium channels by a novel opener pyrazolo[1,5-a]pyrimidin-7(4H)-one compound QO-58. Br J Pharmacol. 2013;168(4):1030–42.

Zhang X, et al. Selective activation of vascular Kv 7.4/Kv 7.5 K^+ channels by fasudil contributes to its vasorelaxant effect. Br J Pharmacol. 2016;173:3480–91.

Zhang F, et al. Repressor element 1-silencing transcription factor drives the development of chronic pain states. Pain. 2019;160:2398.

Zhang J, et al. Gain-of-function mutation in the voltage-gated potassium channel gene KCNQ1 and glucose-stimulated hypoinsulinemia – case report. BMC Endocr Disord. 2020;20:38.

Zhang J, et al. Epigenetic restoration of voltage-gated potassium channel Kv1.2 alleviates nerve injury-induced neuropathic pain. J Neurochem. 2021;156:367–78.

Zhang X, Mao J, Wei M, Qi Y, Zhang JZH. HergSPred: accurate classification of hERG blockers/nonblockers with machine-learning models. J Chem Inf Model. 2022;62:1830–9.

Zhao W, Chen Y. Progress in research of KV1.1 and KV1.3 channels as therapeutic targets. Curr Top Med Chem. 2016;16:1877–85.

Zhao Z, et al. Molecular basis for genistein-induced inhibition of Kir2.3 currents. Pflugers Arch. 2008;456:413–23.

Zheng Y, et al. Deep sequencing of somatosensory neurons reveals molecular determinants of intrinsic physiological properties. Neuron. 2019;103:598–616.e7.

Zhou Y, Morais-Cabral JH, Kaufman A, MacKinnon R. Chemistry of ion coordination and hydration revealed by a K^+ channel-Fab complex at 2.0 A resolution. Nature. 2001a;414:43–8.

Zhou W, Arrabit C, Choe S, Slesinger PA. Mechanism underlying bupivacaine inhibition of G protein-gated inwardly rectifying K^+ channels. Proc Natl Acad Sci U S A. 2001b;98:6482–7.

Zhou J, et al. Novel potent human ether-a-go-go-related gene (hERG) potassium channel enhancers and their in vitro antiarrhythmic activity. Mol Pharmacol. 2005;68:876–84.

Zhou J, Yang CX, Zhong JY, Wang HB. Intrathecal TRESK gene recombinant adenovirus attenuates spared nerve injury-induced neuropathic pain in rats. Neuroreport. 2013;24:131–6.

Zimmer J, Takahashi T, Hofmann AD, Puri P. Downregulation of KCNQ5 expression in the rat pulmonary vasculature of nitrofen-induced congenital diaphragmatic hernia. J Pediatr Surg. 2017;52:702–5.

Zuberi SM, et al. A novel mutation in the human voltage-gated potassium channel gene (Kv1.1) associates with episodic ataxia type 1 and sometimes with partial epilepsy. Brain. 1999;122(Pt 5):817–25.

Zuniga L, Zuniga R. Understanding the cap structure in K2P channels. Front Physiol. 2016;7:228.

The Contribution of Genetic Sequencing Information to the Identification and Functional Characterization of Two-Pore Domain Potassium (K2P) Channels as Viable Therapeutic Targets

Alistair Mathie, Samuel R. Bourne, Rachel Forfar, Walter E. Perfect, and Emma L. Veale

Abstract Genetic information can provide important information linking two-pore domain potassium (K2P) channels with disease states, both for diseases where the mutation causes the disease (monogenic diseases) and where the mutation is more subtly associated with the disease and changes the likelihood of a patient developing the disease (polygenic diseases).

In this review, we describe examples of monogenic diseases where both gain and loss of channel function can lead to the same disease state (TASK3/*KCNK9* in Birk-Barel mental retardation syndrome/KCNK9 imprinting syndrome), where mutations in one channel can lead to altered expression of another to cause the disease (TRESK/*KCNK18* in migraine), and where mutations in the same channel can lead to different diseases (TASK1/*KCNK3* in sleep apnea, pulmonary hypertension, and atrial fibrillation; TRAAK/*KCNK4* in FHEIG syndrome and Rolandic epilepsy).

While this functional genomic information is extremely useful in its own right for patients that are identified as having the mutations, it also has wider implications, when particular K2P channels have a role in physiological processes underlying disease states. As such, modifying the activity of the channels associated with both monogenic and polygenic diseases may be of more general therapeutic benefit even for patients where these proteins are functioning normally.

A. Mathie (✉)
Medway School of Pharmacy, University of Kent and University of Greenwich, Chatham Maritime, UK

School of Allied Health Sciences, University of Suffolk, Ipswich, UK
e-mail: a.mathie@uos.ac.uk

S. R. Bourne · R. Forfar
LifeArc, Stevenage, UK

W. E. Perfect · E. L. Veale
Medway School of Pharmacy, University of Kent and University of Greenwich, Chatham Maritime, UK

G. Stephens, E. Stevens (eds.), *Ion Channels as Targets in Drug Discovery*,
https://doi.org/10.1007/978-3-031-52197-3_6

Keywords K2P channel · KCNK · Monogenic diseases · Potassium channel function · Channelopathy

1 Introduction

Two-pore domain potassium (K2P) channels are widely expressed throughout the body and underlie background currents that stabilize the resting membrane potential of both excitable and non-excitable cells (Enyedi and Czirják 2010; Mathie et al. 2021). There are 15 mammalian K2P channels, which, based on their structural and functional properties, can be grouped into 6 families (Alexander et al. 2023, see also Table 1 in Mathie et al. 2021).

K2P channels have been shown to make important contributions in a number of physiological responses and pathological conditions (Mathie et al. 2021), including action potential propagation in myelinated axons (Brohawn et al. 2019; Kanda et al. 2019), microglial surveillance (Madry et al. 2018), cancer (Pei et al. 2003; Mu et al. 2003; Sun et al. 2016), inflammation (Bittner et al. 2009), cardiac arrhythmias (Decher et al. 2017), and pain (Alloui et al. 2006; Devilliers et al. 2013; Vivier et al. 2017). Additionally, mutations in specific K2P channels have been shown to underlie a number of rare diseases including pulmonary arterial hypertension (Ma et al. 2013), Birk-Barel mental retardation syndrome (BBMRS)/KCNK9 imprinting syndrome (KIS) (Graham et al. 2016), developmental delay with sleep apnea (DDSA, Sörmann et al. 2022) and facial dysmorphism, hypertrichosis, epilepsy, intellectual disability/developmental delay, and gingival overgrowth (FHEIG) syndrome (Bauer et al. 2018).

We intend to use this chapter as an opportunity to share our current thinking of the importance of genetic information in identifying ion channel drug targets, using K2P channels as an exemplar. In some cases, mutations in individual K2P channel genes have been proposed as the underlying cause of a disease (e.g., FHEIG or BBMRS/KIS), and so these conditions may be considered monogenic diseases or "channelopathies." However, more subtly, in other cases, mutations in K2P channels may be associated with or contribute to particular diseases or illnesses (e.g., pain, depression, autism, schizophrenia), so these diseases are considered polygenic diseases with a measurable contribution from a particular K2P channel or channels.

In this chapter, we will focus primarily on monogenic diseases because this is where the most compelling advances have been made recently. We have focused on four K2P channel case studies. These are TASK3 ($K_{2P}9.1$, *KCNK9*) channels in BBMRS/KIS; TASK1 ($K_{2P}3.1$, *KCNK3*) channels in sleep apnea, pulmonary hypertension, and atrial fibrillation; TRESK ($K_{2P}18.1$, *KCNK18*) channels in migraine; and TRAAK ($K_{2P}4.1$, *KCNK4*) channels in FHEIG syndrome. In each case, emerging evidence suggests that the role of these channels in the particular disease states is more complex than initially envisaged.

2 Case Study 1: *KCNK9* Imprinting Syndrome

Birk-Barel mental retardation syndrome (BBMRS) or *KCNK9* imprinting syndrome (KIS), BBMRS/KIS, is a maternally transmitted genetically imprinted disease, resulting from missense mutations in the maternal copy of the *KCNK9* gene, which encodes the K2P channel, TASK3. *KCNK9* is predominantly maternally transmitted (Cooper et al. 2020), as it is imprinted with paternal silencing (Luedi et al. 2007). The predominant features of this disorder include mental retardation, muscle hypotonia, and dysmorphic features including an elongated face but also include motor and speech delay, early feeding difficulties, microretrognathia, high arched eyebrows, and bitemporal narrowing (Barel et al. 2008; Graham et al. 2016; Cousin et al. 2022).

The original description of the syndrome by Barel et al. (2008) mapped the disease-associated locus to ~7.27 Mb on chromosome 8q24, an area previously associated with intellectual disabilities such as autism (Court et al. 2014), and demonstrated that the disease is caused by a missense mutation G236R (a glycine to arginine mutation at position 236) in *KCNK9* within this locus. Barel et al. (2008) stated that the G236R mutation fully abolished the TASK3 channel's current, both when functioning as a homodimer and as a heterodimer with $K_{2P}3.1$ (TASK1). A number of other patients have subsequently been identified with the same G236R mutation in *KCNK9* leading to similar clinical features (Graham et al. 2016) albeit with some variation in severity from patient to patient (Cousin et al. 2022).

Subsequent work by our laboratory (Veale et al. 2014) showed that, in mammalian expression systems, TASK3-G236R-mutated channels *did* give rise to a functional current; however, this current was significantly smaller in an outward direction when compared with wild-type (WT) TASK3 channels. In contrast to WT TASK3 channels, the current was inwardly rectifying. Furthermore, the current through mutated channels was significantly less sensitive to a number of regulators, such as extracellular acidification and activation of Gαq-coupled muscarinic (M3) receptors, compared with WT TASK3 channels.

Interestingly, we found that the reduced outward current through mutated TASK3-G236R channels could be overcome, at least in part, by a gain-of-function (GoF) additional mutation of TASK3 channels (A237T) or by application of both the nonsteroidal anti-inflammatory drug flufenamic acid (FFA) and the antifungal agent terbinafine (Wright et al. 2017). Furthermore, both FFA and terbinafine produced a significantly greater enhancement of current through mutated TASK3-G236R channels than through WT TASK3 channels. As such, we proposed that pharmacological enhancement of mutated TASK3-G236R channel current during development may, therefore, provide a potentially useful therapeutic strategy in the treatment of BBMRS/KIS. Garcia et al. (2019) demonstrated that intrathecal pretreatment with terbinafine reduced the formalin-induced flinching and allodynia/hyperalgesia in rat, further supporting the putative future clinical relevance of TASK activators (Garcia et al. 2019; Decher et al. 2021).

An alternative, related, therapeutic approach has been proposed by Cooper et al. (2020) who showed a partial rescue of the BBMRS/KIS behavioral and neuronal phenotypes mediated via residual expression from the paternal *KCNK9* allele. In an animal model of the disease, they further demonstrated that the histone deacetylase inhibitor, CI-994, induced enhanced expression from the paternally silenced *KCNK9* allele and led to a full rescue of the behavioral phenotype.

While the mapping of *KCNK9* to the 8q24 chromosomal locus might suggest a more widespread association between mutations affecting this gene and intellectual disability (Court et al. 2014; Yamada et al. 2020), to date, there is little concrete evidence to support this association. Indeed, a study of 120 individuals with intellectual disability, some of whom displayed many of the clinical symptoms of BBMRS/KIS, suggested that either mutations of *KCNK9* or even epigenetic disturbances within the wider imprinted cluster did not significantly contribute to the cause of the developmental disabilities observed in these patients (Sánchez Delgado et al. 2014). Intriguingly, Guo et al. (2019) describe a novel mutation of *KCNK9* (R303C) in a study of multiple deleterious variants in autism patients with more severe phenotypes. We have characterized the functional properties of this mutation (Fig. 1) and have found that although the current through the mutated channels is not different from WT TASK3 channels (Fig. 1a, b), the regulation of the channel by extracellular acidification is blunted (Fig. 1c) and the regulation by Gαq-coupled muscarinic (M3) receptors is accentuated (Fig. 1d).

More recently, mutations of the *KCNK9* gene additional to the G236R mutation have been identified and shown to lead to BBMRS/KIS phenotypes. Šedivá et al. (2020) reported a mutation in the adjacent amino acid (A237D) in a patient presenting with a similar clinical phenotype to previously described BBMRS/KIS patients. Our own research in collaboration with colleagues at the Mayo Clinic and the Medical College of Wisconsin and clinicians in a number of countries described 15 novel *KCNK9* mutations from 47 affected individuals (Cousin et al. 2022). We used clinical genetics and computer-assisted facial phenotyping to describe the spectrum

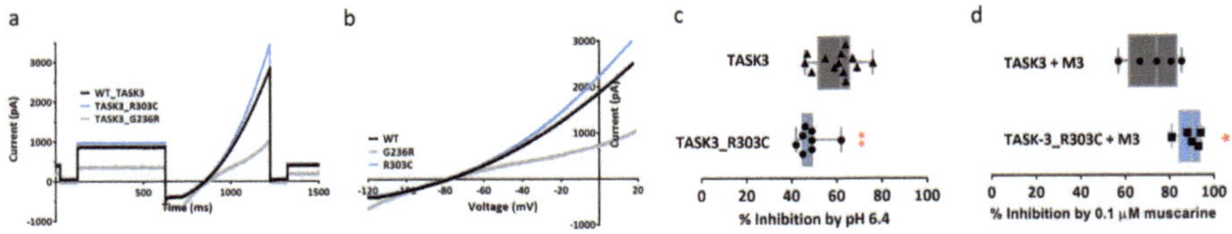

Fig. 1 Effect of autism-associated mutation R303C on TASK3 channel activity and its regulation by various modulators. (**a**) Average raw current traces from WT TASK3, TASK3_R303C autism mutation, and classical TASK3_G236R BBMRS/KIS mutation, representing current amplitude vs time. (**b**) Current-voltage relationships for WT TASK3, TASK3_R303C, and TASK3_G236R. (**c**) Box and Whisker plots of percentage inhibition of current in pH 6.4. (**d**) Box and Whisker plots of percentage inhibition of current by 0.1 micromolar (μM) muscarine for WT and R303C channels co-expressed with an M3, muscarinic receptor. Bars represent the min and max inhibition and the lines of the median inhibition, for each channel type. Triangle, circle, or square symbols represent the individual data points for each condition and channel. $^{**}P < 0.005$, $^{*}P < 0.05$ for between channel differences, determined using unpaired Student's *t*-tests. Abbreviations: *pA* picoamp, *ms* millisecond, *mV* millivolt, *WT* wildtype, *M3* muscarinic receptor 3

of clinical features observed. We also measured the functional consequences of the variants in the encoded TASK3 channel using sequence-based analysis, 3D molecular mechanic and dynamic protein modelling, and in vitro electrophysiological recordings. Our results strongly suggest that the large majority of these variants alter channel function, and we classified them as pathogenic or likely pathogenic for BBMRS/KIS (Cousin et al. 2022). We concluded that the constellation of clinical features observed in this study was not specific enough to allow for this disorder to be easily recognizable from other similar clinical syndromes (such as Prader-Willi syndrome) and therefore requires a confirmatory molecular or functional diagnosis to ascertain BBMRS/KIS diagnosis.

Our functional studies revealed a surprisingly large variability of the impact of these novel *KCNK9* variants on TASK3 channel function, identifying variants causing gain and others causing loss of conductance. Notably, however, the most consistent functional impact of the novel *KCNK9* genetic variants was altered channel regulation by extracellular pH and activation by Gαq, whereby the effect of these regulators on the channel was attenuated following these mutations.

Among the novel KCNK9 mutations identified in our cohort, eight families with nine affected individuals carried missense alterations at R131, including five families with R131H, and one each with R131C, R131P, and R131S in an affected sibling pair was identified. Thus, we have identified a novel mutational hotspot at R131 in addition to the original G236R mutation. Remarkably, each of the mutations at R131 caused a significant increase in channel current in direct contrast to that seen for the G236R mutation. However, all R131 mutated channels had a loss of regulation by external pH and Gαq activation in the same way as G236R-mutated channels (Veale et al. 2014; Cousin et al. 2022).

The observed variable effects on channel function caused by the variants in KCNK9 have ramifications on potential therapy for KIS. It was previously hypothesized that channel-stimulatory drugs, such as a subset of the nonsteroidal anti-inflammatory fenamic acid class of drugs, may be useful in treating BBMRS/KIS as FFA had been shown to partially rescue the reduced current in G236R-mutated channels (see above). Indeed, Graham et al. (2016) described the early results of treatment of two individuals with BBMRS/KIS with mefenamic acid suggesting positive effects. The data presented in this study suggest that at a minimum these medications may only be affective for individuals with a subset of variants causing reduced current, similar to that of G236R. These data could also suggest caution in treating an individual with normal or increased channel current as we may hypothesize exacerbated detrimental effects and that an opposite therapeutic approach is desirable. Paradoxical loss-of-function (LoF) and GoF of channel activity leading to similar clinical disorders have been seen for other potassium channels (such as $K_V1.2$, $K_V2.1$, $K_V7.2$, and BK channels) leading to neurodevelopmental disorders (Park et al. 2022; Veale et al. 2022) and hyperexcitability disorders such as epilepsy (Niday and Tzingounis 2018). As such, our data contribute to the growing recognition that individualized therapeutic strategies may be needed, depending on the particular functional consequences of the individual potassium channel mutations.

3 Case Study 2: *KCNK3* in Sleep Apnea, Pulmonary Hypertension, and Atrial Fibrillation

Another channel from the TASK family of K2P channels, TASK1 ($K_{2P}3.1$), encoded by *KCNK3* has been implicated in the condition, sleep apnea (Gurges et al. 2021). Sleep apnea is characterized by abnormal, interrupted breathing during sleep. It is a common disorder that represents a global public health burden. TASK1 channels are expressed in a wide variety of cells throughout the body including many chemosensitive regions and hypoglossal and spinal motoneurons (Olschewski et al. 2017). TASK1 channels are also found in the lung, heart, pulmonary arterial smooth muscle cells, and carotid bodies. Both TASK1 and TASK3 channels are expressed in the carotid body and brain stem regions and have been associated with respiratory control. Mice lacking these TASK channels have impaired carotid body function (Ortega-Saenz et al. 2010; Buckler 2015), and pharmacological blockers of these channels, such as the clinically used compound doxapram, are known respiratory stimulants (Cotton et al. 2006; O'Donohoe et al. 2018).

Given the proposed physiological role of TASK channels in breathing, it is not perhaps surprising that TASK1 channels have been linked with sleep apnea (Gurges et al. 2021). The mechanisms underlying the role of TASK1 channels in sleep apnea are, however, not well understood, and their importance in this condition has subsequently been questioned (Gurges et al. 2021).

Sörmann et al. (2022) described a new developmental disorder with associated sleep apnea (and termed it developmental delay with sleep apnea, or DDSA). They showed that DDSA is caused by six rare de novo missense mutations describing nine probands in *KCNK3*. Four of these mutations, L122V, L122P, G129D (two probands), and N133S (three probands), are in the second transmembrane helix (M2) of the TASK1 channel, while L239P and L241F are situated in the fourth transmembrane helix (M4). These mutations result in a GoF phenotype that appears to correlate with the clinical severity of the disorder, with G129D causing the most severe symptoms. Single channel data shows that the GoF is associated with an increase in channel open probability rather than any significant change in channel expression levels. These mutations are dominant positive in heteromers, including heteromers with TASK3 channel subunits. The mutations cluster around the lower "X-gate," a gating motif that controls channel opening (Rödström et al. 2020) and which leads to overactive channels. The six residues in M4 that form this constriction were previously identified as important for channel regulation by volatile anesthetics, Gαq-coupled pathways (Talley and Bayliss 2002; Chen et al. 2006), and the cannabinoid, methanandamide (Veale et al. 2007), although the mechanisms involved are unclear. Strikingly, these six mutated channels no longer respond to inhibition by Gαq activation, reminiscent of a similar loss of Gαq-mediated regulation of both LoF and GoF TASK3 channels underlying BBMRS/KIS syndrome (Veale et al. 2014, Cousin et al. 2022; see above).

However, despite their defective X-gating, these mutant channels can still be inhibited by a range of known TASK channel inhibitors including BAY-100493,

A293, A1899, PK-THPP, and doxapram with relatively similar IC_{50} values compared to those of WT TASK1 channels (Streit et al. 2011; Kiper et al. 2015). Interestingly, an X-ray crystal structure of the TASK1 channel has recently been reported in complex with a compound class of TASK1 inhibitors (related to BAY-100493) used in clinical trials for the treatment of sleep apnea (Rödström et al. 2020; see below). Thus, these results not only highlight an important new role for TASK1 K2P channels and their link with sleep apnea but also identify possible therapeutic strategies with which to manage the condition.

In one sense, it is somewhat surprising that these mutated channels can still be inhibited as normal by this range of inhibitory compounds, since a number of studies on both TASK1 and TASK3 channels have suggested that some of these residues notably L122 and L239 are critical for block by these compounds and mutation of these residues interferes with block (Streit et al. 2011; Chokshi et al. 2015; Kiper et al. 2015; Cunningham et al. 2020). One explanation for this may be the identity of the amino acid replacing the native residue. In the pharmacological experiments investigating putative binding sites, the mutations L122D and L239D were used. These interfered with compound block (e.g., Cunningham et al. 2020), whereas the mutations identified here (L122V, L122P, L239P) alter channel function but do not appear to affect pharmacological antagonism.

In addition to sleep apnea, the development of selective antagonists of TASK channels would also represent a useful therapeutic strategy for atrial fibrillation (AF, Burg and Attali 2021, Mondéjar-Parreño 2022) which may arise from overactive TASK1 channels (Schmidt et al. 2015) and where genetic ablation of TASK1 channels has been shown to prevent AF (Schmidt et al. 2019). Additionally, atrial TASK1 mRNA/protein levels are upregulated in a human chronic AF cohort compared with people in sinus rhythm (Schmidt et al. 2014). TASK1 channel upregulation causes a shortened AP duration in the right atrium which is restored using pharmacological TASK1 inhibitors (Schmidt et al. 2015; Mondéjar-Parreño 2022). It is of interest, however, that despite the reported overexpression of KCNK3 in AF (Schmidt et al. 2014), none of the DDSA probands described above exhibited any obvious cardiac phenotype (Sörmann et al. 2022).

The putative selective $K_V1.5$ channel blockers A1899 and A293, which have been proposed as antiarrhythmic drugs for AF treatment, are, in fact, about 70-fold more potent at blocking TASK1 channels, making them effective TASK channel selective antagonists at low doses (Decher et al. 2021). Furthermore, the TASK channel blocker A293 significantly reduced AF burden in a persistent animal model of AF (Wiedmann et al. 2020). Interestingly, the TASK1 and TASK3 channel blocker doxapram (Cunningham et al. 2020) is being used in ongoing clinical trials as a potential treatment for AF (Doctos trial, EudraCT 2018-002979-17, also Kraft et al. 2021).

On the other hand, other studies support the downregulation of the TASK1 channel or the absence of TASK1 current in perioperative AF and in AF associated with heart failure (Mondéjar-Parreño 2022). In 2015, Harleton et al. showed that TASK1 current was present in human and canine atrial myocytes with regular sinus rhythm but was absent in humans with AF undergoing cardiac surgery, while Wiedmann et al. (2018) found atrial TASK1 channel expression was markedly reduced in a mouse model of heart failure. Furthermore, mutations in TASK1 which reduce channel function or expression (Liang et al. 2014, Perfect, Mathie and Veale unpublished data) have been associated with AF. Therefore, any therapeutic strategies to alleviate AF would need to consider that the optimum level of TASK1 channel activity may vary from patient to patient.

A complication for the clinical use of TASK1 channel inhibitors for the treatment of DDSA or AF is the possible onset of pulmonary arterial hypertension (PAH) as there is strong evidence that this condition is either initiated or exacerbated by reduced TASK1 channel activity (see Mathie et al. 2021 and Fig. 2). Six heterozygous TASK1 mutations were first described in 2013 for patients suffering from PAH (Ma et al. 2013). These mutations resulted in LoF. Two additional TASK1 channel mutations, this time homozygous mutations, were identified in a cohort of PAH patients with an aggressive form of the disease (Navas et al. 2016), and these mutations also caused a LoF (Cunningham et al. 2019).

In rat models of PAH, TASK1 current decreased progressively during the development of the disease (Antigny et al. 2016). Furthermore, pulmonary expression of TASK1 is reduced in patients with PAH (Antigny et al. 2016), and a reduction in TASK1 channel function has been observed in right ventricular cardiomyocytes prior to the development of right ventricular hypertrophy related to pulmonary hypertension (Lambert et al. 2018). Thus, loss of TASK1 channel function is a key event in the pathogenesis of both heritable and idiopathic pulmonary hypertension.

A useful alternative to treat PAH is perhaps to seek to restore optimum/normal levels of activity of the channel target. For example, there is a strong genetic linkage

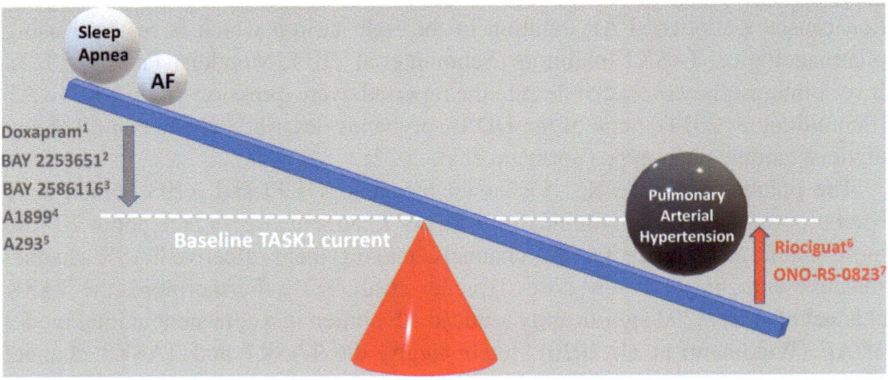

Fig. 2 Schematic representation of the membrane topology of TRESK and TRESK migraine-associated mutation generated protein fragments (MT1 and MT2) and their inhibition of wild-type TRESK (by MT1) and TREK (by MT2) channels (adapted from Royal et al. 2019). Schematic topology of a single α-subunit of TRESK consisting of four transmembrane domains, two-pore-forming domains, and intracellular N- and C-terminal domains. The location of the two-base pair frameshift mutation on the second transmembrane domain is represented by a blue oval. The insertion of a premature stop codon is represented by a red dot and frameshifted sequence generated, shown in blue

between LoF mutations of TASK1 channels and the development of PAH (Ma et al. 2013; Olschewski et al. 2017; Cunningham et al. 2019). More recent studies suggest that reduced TASK1 channel activity is a more general condition seen not just in genetic pulmonary hypertension but also in the idiopathic condition (Antigny et al. 2016). As such, restoration of "normal" channel function may be a more widely applicable therapeutic intervention in this and other situations (Fig. 2).

Clinical trials associated with the use of TASK1 antagonists in sleep apnea have provided equivocal evidence to date. Interestingly, in both trials, TASK1 channel blockers were applied topically to target TASK1 channels in airway muscle cells avoiding those TASK1 expressing neurons associated with breathing control or cardiac function.

A multicenter double-blind randomized controlled trial using the antagonist BAY 2253651 (the SANDMAN trial; Gaisl et al. 2021) has hypothesized that block of TASK1 channel activity in genioglossus muscle to amplify upper airway reflex activity may improve obstructive sleep apnea. Nasal application of the compound was well tolerated and safe, but, unfortunately, no significant therapeutic effect was observed in the small number of patients studied.

More recently, Osman et al. (2023) have shown that in a small number of patients suffering from obstructive sleep apnea, application of a related compound BAY 2586116 by nasal spray caused a significant improvement in pharyngeal collapsibility. A similar effect of the compound was seen in parallel experiments on pigs. These authors concluded that acute topical application of a localized TASK1 channel antagonist is a potential therapeutic approach to treat patients with obstructive sleep apnea.

4 Case Study 3: KCNK18 in Migraine

The human TRESK (K2P18.1) channel is encoded by the *KCNK18* gene (Sano et al. 2003). TRESK channels share most of the characteristics of the K2P channel family but are distinguished by having a large intracellular loop (120 amino acids) between the second and third transmembrane domain and a much shorter C terminal domain than other K2P channels (Enyedi and Czirják 2015). These features suggest that TRESK channels are regulated in different ways by intracellular messengers compared to other K2P channels, since the regulation for other K2P channels is predominantly mediated through the C terminus (Enyedi and Czirják 2015). Gαq-coupled receptor stimulation was found to activate heterologously expressed human TRESK channels, and this response could be mimicked by application of the calcium ionophore ionomycin (Czirják et al. 2004). This activation has proven to be a useful tool for investigating the functional properties of mutated TRESK channels. Overall sequence identity between different species of TRESK channel is much lower than for other K2P channels which hampers translational research from rodents to humans, emphasizing the need for studies in humans or at least with human version of the TRESK gene (Enyedi and Czirják 2015).

In terms of distribution, originally, human TRESK was cloned and exclusively found in the spinal cord (Sano et al. 2003). However, subsequently, more widespread expression was detected particularly in different areas of the nervous system including the dorsal root and trigeminal ganglion (TG) (Kang and Kim 2006; Bautista et al. 2008; Enyedi and Czirják 2015). The high level of expression in the dorsal root and TG suggests a role for TRESK channels in the regulation of pain. Indeed, the abundant expression of TRESK in TG neurons suggests an important role for these channels in regulating the resting membrane potential of these neurons and therefore that any reduction in TRESK current might lead to depolarization of these neurons suggesting the possibility that TRESK abnormalities might underlie at least some forms of migraine associated with dysfunction of TG neurons.

Migraine is a primary headache disorder affecting 10–15% of the population. Two major categories of the disease can be distinguished, based on the presence or absence of a preceding aura. When present, the aura phase can stimulate the release of inflammatory transmitters (such as CGRP and/or substance P), and this triggers the subsequent headache (Enyedi and Czirják 2015). Both forms of migraine often show multiple occurrences in families indicating the importance of a component of genetic predisposition.

A frameshift mutation (F139Wfsx24) in TRESK was identified in a large multigenerational pedigree, where it co-segregated perfectly with typical migraine with aura and had a significant genome-wide linkage score (Lafrenière et al. 2010). Furthermore, functional analysis revealed that this TRESK mutation, which comprised a truncated protein of 162 residues from the N terminus, incorporating the first transmembrane domain and the large extracellular loop between the first and second transmembrane domains (Fig. 3), caused a complete loss of TRESK function and that the truncated subunit was also capable of downregulating WT channel function as a dominant-negative action. A more detailed analysis over four generations of family members related to the original patient found that eight affected individuals had the frameshift mutation but that eight unaffected individuals from the same family group did not. This therefore highlighted *KCNK18* as potentially important candidate gene and suggested that TRESK dysfunction might play a role in the pathogenesis of familial migraine with visual aura. Thus, upregulation of TRESK current may be a useful more general therapeutic strategy in the treatment of migraine (see also Lafrenière and Rouleau 2012).

Liu et al. (2013) found that mutant TRESK subunits exhibited dominant-negative effects not only on the size of the whole-cell TRESK currents but also on the level of TRESK channels on the plasma membrane. Liu et al. (2013) went on to express mutant TRESK subunits in cultured TG neurons and observed a significant decrease in the lamotrigine-sensitive current (although lamotrigine also inhibits TREK channels with equal potency to inhibition of TRESK channels; Walsh et al. 2021), suggesting that the mutant TRESK subunits also have a dominant-negative effect on currents through endogenous TRESK channels. Current-clamp recordings showed that neurons expressing mutant TRESK subunits had a higher input resistance, a lower current threshold for action potential initiation, and a higher spike frequency

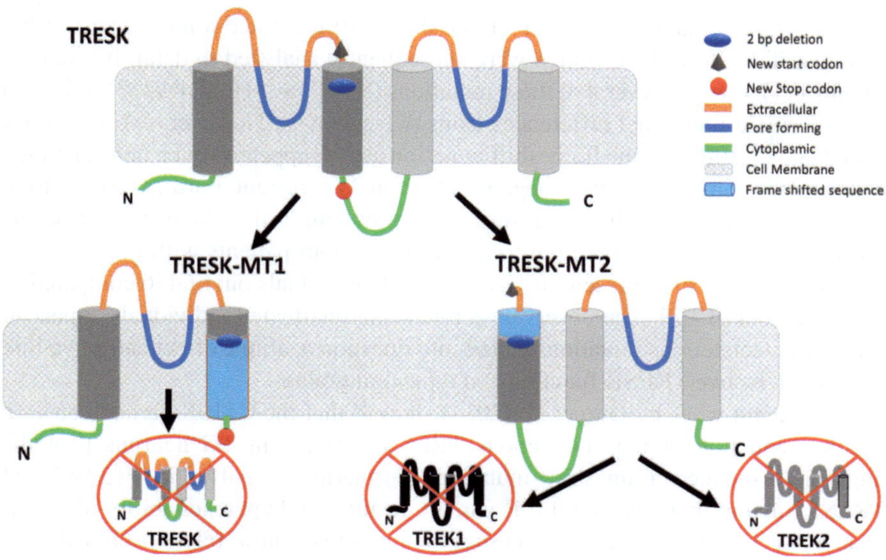

Fig. 3 Schematic representation of the disease-causing effects of TASK1 GoF and LoF mutations on channel activity. GoF mutations of TASK1 identified in some patients with sleep apnea or atrial fibrillation cause an increase of current passing through the channels, compared to wild-type channels. Drugs that act to decrease current through these channels such as the ventilatory stimulant doxapram could be useful in treating patients with these specific mutations, to restore current to baseline levels. Mutations of TASK1 channels that cause LoF of the channel, in conditions such as pulmonary arterial hypertension, could be therapeutically treated with channel activators to restore current to baseline. Abbreviations. AF, atrial fibrillation. [1]Cunningham et al. (2020), [2]Gaisl et al. (2021), [3]Osman et al. (2023), [4]Decher et al. (2021), [5]Wiedmann et al. (2020), [6]Cunningham et al. (2019), [7]Ma et al. (2013)

in response to suprathreshold stimuli, indicating that the mutation resulted in hyperexcitability of TG neurons.

More recently, an atypical clinical case has been reported (Imbrici et al. 2020) in which a male individual carrying a W101R missense mutation in the *KCNK18* gene was diagnosed with intellectual disability and migraine with brainstem aura. This mutation is located in a highly conserved residue close to the selectivity filter, and the study showed that although these mutant channels retain their potassium selectivity and calcineurin-dependent regulation, the variant caused a substantive reduction of TRESK channel function and a dominant-negative effect when co-expressed with WT channels.

Taken together, these results suggested that *KCNK18* is a clear candidate gene for migraine with aura and that TRESK dysfunction plays a role in the pathogenesis of this disease through trigeminal nociceptor sensitization in migraine patients. However, not all experimental evidence suggested that the picture is quite so clear cut.

The original analysis of Lafrenière et al. (2010) identified a number of further missense TRESK variants which were subsequently analyzed in detail by Andres-Enguix et al. (2012). Several of these mutations (R10G, S231P, and A233V) showed no discernible functional differences from WT channels. However, two mutations A34V and C110R, like the frameshift mutation above, appeared to be nonfunctional and exerted a similar dominant-negative effect on WT current. Further investigation of the C110R mutation, however, revealed a problem, in that the degree of occurrence of this mutation turned out to be similar in both patients with migraine and control subjects, being present in 12 unaffected individuals out of 496 compared to 8 affected out of 479. The observation that some unaffected individuals appear to carry quite deleterious mutations called into question a simple direct causative link between defective TRESK function and typical migraine.

Subsequent work by Guo et al. (2014) showed that the C110R variant causes a smaller reduction of the endogenous TRESK currents in TG neurons than the migraine-associated frameshift mutation. Furthermore, only the F139Wfsx24 TRESK subunit, but not the C110R variant, resulted in hyperexcitability of small-diameter TG neurons. These results suggest a possible causal relationship between the TRESK frameshift mutation and a higher susceptibility of migraine headache compared to other point mutations such as C110R. Additionally, in two further studies, a number of common polymorphisms of the *KCNK18* gene were analyzed in case control migraine populations, and no association was detected with the migraine phenotype (Maher et al. 2013; Nyholt et al. 2008).

If the influence of TRESK on TG neuronal excitability does indeed play a role in the pathogenesis of migraine, then its effects are undoubtedly complex. These results demonstrate that despite the significant genetic linkage of the F139Wfsx24 variant in an extended family, the presence of a single nonfunctional variant is probably not alone sufficient to determine whether an individual develops migraine. This further highlights the importance of understanding the broader genetic background of any affected individual and suggests perhaps that *KCNK18* mutations may be responsible for only a very small fraction of migraine cases (Enyedi and Czirják 2015). Nevertheless, the rare type of headache caused by *KCNK18* frameshift mutation leaves the possibility open that pharmacological stimulation of the channel will prevent or alleviate the symptoms also in the common forms of the disorder.

A number of genes have been implicated in rare familial syndromes which have migraine as part of their phenotype, but these genes have not previously been implicated in the common form of migraine. Markel and Curtis (2022), using data from the UK Biobank, identified 7194 migraine cases with the remaining 193,433 participants classified as controls. They investigated rare variants in ten genes including *KCNK18* previously reported to be implicated in conditions with migraine and carried out gene- and variant-based tests for association. They found that the frameshift variant in *KCNK18* (F139Wfsx24), which had been shown to segregate with migraine with aura in a multiple affected pedigree (see above), was found in 0.10% controls and in 0.14% cases with no significant difference between the two. Since there is no other reported evidence to implicate *KCNK18*, the authors concluded that this gene and its product, TRESK, should no longer be regarded as being involved in migraine etiology as it does not make a contribution to the commoner forms of migraine observed in this population.

This view is supported by the earlier work of Gormley et al. (2016) who carried out a large meta-analysis of 22 genome-wide association studies involving 375,000 individuals. They identified 38 distinct genomic loci associated with migraine, but only 2 of these harbored known ion channel genes and neither was *KCNK18*. They concluded that in common forms of migraine, ion channel dysfunction is not the major pathophysiological mechanism but that genes involved in ion homeostasis more generally may be a component of genetic susceptibility to the disease.

More recent pieces of work provide further evidence which may help understand these anomalies further (Cader 2020). The first study used induced pluripotent stem cells (iPSCs) to demonstrate how the TRESK frameshift mutation might contribute to migraine susceptibility (Pettingill et al. 2019). Stem cells were generated from members of the family with the TRESK F139Wfsx24 mutation, a non-migraine subject with the C110R mutation and several healthy control subjects. The study was able to show that the F139Wfsx24 mutation but not the C110R mutation caused a loss of TRESK leak current and significantly increased nociceptor excitability. Furthermore, Pettingill et al. were able to demonstrate that loss of TRESK alone, in a mouse model, increased susceptibility to nitroglycerin induced orofacial hypersensitivity compared to WT mice, an established rodent model of migraine (Pettingill et al. 2019).

In a much more surprising development, Royal et al. (2019) showed that TRESK-F139Wfsx24 promotes hyperexcitability by impeding the function of TREK potassium channels that regulate sensory neuron excitability. The authors found that the frameshift in TRESK not only produces a dominant-negative truncated protein (re-characterized as TRESK-MT1) but also a second protein product (TRESK-MT2) translated from a start site created by the frameshift. TRESK-F139Wfsx24 increases TG neuron excitability primarily because, remarkably, TRESK-MT2 inhibits TREK potassium currents, not because of loss of TRESK function (Hill and Bautista 2019; Fig. 3). In addition, Royal et al. (2019) found that TRESK physically interacts with TREK1 and TREK2, which are expressed within the same subsets of somatosensory neurons. As previously shown, overexpression of TRESK-MT or TRESK-C110R inhibited TRESK currents (Andres-Enguix et al. 2012). However, unlike previous studies, Royal et al. (2019) found that TRESK-MT, but not TRESK-C110R, inhibits TREK1/TREK2 currents in both heterologous cells and TG neurons. Interestingly, unlike WT TRESK channels (see also Lengyel et al. 2020; Prado et al. 2021), they showed that TRESK-C110R did not interact with TREK1 and TREK2. Electrophysiological studies of sensory neurons confirmed that TRESK-F139Wfsx24 increased neuronal excitability primarily via inhibition of TREK and not TRESK currents, whereas TRESK-C110R did not affect TREK currents or sensory neuron excitability.

Prado et al. (2021) hypothesized that enhancing TREK1 and TREK2 activity would reduce TG excitability, prevent the release of pro-inflammatory peptides, and therefore suppress migraine pain. They showed that ML67–33, a TREK1/TREK2 agonist (Bagriantsev et al. 2013), reduced TG sensory neuronal excitability, which provided important evidence that TREK channels regulate the excitability of these neurons. Furthermore, using behavioral tests in WT and double KO mice for *TREK1/TREK2,* they demonstrated that ML67–33 reversed a migraine-like phenotype in WT but not KO mice. These results demonstrate the importance of TREK1 and TREK2 channels in the regulation of TG excitability, and the authors suggest that targeting TREK channels to inhibit TG firing and prevent CGRP release underlying migraine should therefore be considered when developing molecules to treat migraine. It will be interesting to examine whether humans harboring TREK mutations also experience migraine susceptibility, consistent with the proposed model.

Migraine is not the only condition associated with KCNK18. Pavinato et al. (2021) reported two mutations in the long intracellular loop of TRESK (Y163D and S252L) in a family affected by mild to moderate intellectual disability, autism spectrum disorder, and other neurodevelopmental disorders. Neither variant, alone or in combination, affected the amplitude of current through the channels under resting conditions; however, both mutations severely attenuated the activation of TRESK channels by ionomycin in both homomeric and heteromeric mutant channels. These results add to the growing findings showing the importance of a number of different K channels in coordinating developmental signaling and therefore leading to neurodevelopment disorders, autism, and intellectual disability when their activity is altered.

5 Case Study 4: TRAAK (*KCNK4*) Gain-of-Function Variants in Neuronal Development Syndrome

The TRAAK (TWIK-related arachidonic acid-sensitive potassium) channel, also known as the $K_{2P}4.1$ (*KCNK4*), belongs to the TRAAK/TREK family of two-pore domain K-selective (K2P) channels that are commonly grouped by their collective sensitivity to lipids, temperature, and mechanical stimulation and insensitivity to classical K channel blockers such as TEA, quinidine, and barium (Fink et al. 1998; Lesage et al. 2000). TRAAK has been shown to be highly expressed in neurons of the central and peripheral nervous system and in the retina (Fink et al. 1998; Reyes et al. 2000; Talley et al. 2001) and more precisely located in the nodes of Ranvier of 80% of myelinated nerve fibers (Brohawn et al. 2019). Rapidly activated by tension transmitted through the lipid bilayer of the membrane gating the channel open, these unique properties relate closely to some of these channels' known physiological roles, such as allodynia, hyperalgesia, and inflammation identified through the use of TRAAK KO rodent models (Nóel et al. 2009). The identified location of TRAAK in the nodes of Ranvier uncovers another physiological role, underlying the K conductance that is responsible for maintaining a negative nodal resting potential, increasing nodal Na_V channel availability for action potential regeneration (Brohawn et al. 2019). Dysregulation of TRAAK channels through de novo heterozygous missense GoF mutations in the *KCNK4* gene, identified clinically in three unrelated children, is thought to underlie a rare, polymalformative, disorder, facial dysmorphism; hypertrichosis; epilepsy; intellectual disability/developmental delay; and gingival overgrowth (FHEIG) in humans (Bauer et al. 2018). FHEIG relates to the various developmental and neurological phenotypes underlying this complex disease. In two of the unrelated patients, a de novo heterozygous missense mutation c.515C > A (NM_001317090.1) results in a highly conserved alanine (A172) in transmembrane region 3 (M3) of the channel being substituted for a glutamic acid (A172E) in the *KCNK4* gene. In a third unrelated child, a de novo heterozygous c.730G > C (NM_001317090.1) missense mutation in the *KCNK4* gene results in a highly conserved alanine (A244) of transmembrane domain 4 (M4) being substituted for a proline (A244P). Similar to other neurodevelopmental pleiotropic disorders caused by dysregulated GoF ion channel mutations, such as BBMRS/KIS (Cousin et al. 2022), variable developmental and cognitive impairment varying from mild to severe was observed between the different individuals (Bauer et al. 2018). Functional expression studies showed that the individual mutations expressed homozygously caused a significant increase in basal activity of the TRAAK channels, compared to the low basal activity (<1%) of WT channels in the absence of any stimulation (Brohawn et al. 2014; Kang et al. 2005). Basal activity of the mutant channels could not be further enhanced by either voltage, pressure stimulation, or arachidonic acid (Bauer et al. 2018). Molecular modelling simulations predicted that the lateral fenestrations of the TRAAK channel were significantly less open for the mutant channels, than for WT, preventing membrane lipids from entering the conduction pathway and blocking the passage of K ions through the channel, gating the channel open (Bauer et al. 2018). This resulted in an increase in the resting basal activity of the mutant channels, in agreement with the lateral fenestration gating mechanism of Brohawn et al. (2014). A second, A244P missense point mutation of *KCNK4* was later reported for a 6-year-old Italian boy (Mariani et al. 2021), with described neurodevelopment phenotypes which overlapped with some (hypertrichosis, epilepsy, and gingival overgrowth) but not all of the physiological phenotypes described by Bauer et al. (2018).

An additional study (Yan et al. 2023) screening a cohort of patients with epilepsy using whole exon sequencing identified two further de novo likely pathogenic variants in *KCNK4*, although there is as yet no direct functional data to determine whether these mutations also cause a GoF. One of these mutations (G139E) led to the patient presenting with FHEIG syndrome as seen previously, but the other patient, with mutation P233L, presented with Rolandic epilepsy which was successfully treated with valproate and oxcarbazepine. The G139E, A172E, and A244P mutations are located, respectively, in membrane spanning regions TM2, TM3, and TM4; however, the P233L mutation, causing Rolandic epilepsy, is located on the P2/TM4 extracellular loop of the channel (Bauer et al. 2018; Yan et al. 2023).

Despite evidence for a novel neurodevelopmental disorder caused by these mutations in TRAAK, further cases would need to be identified in order to delineate the condition further and differentiate this as a TRAAK monogenic pleiotropic disease state. This is exemplified further by the fact that other GoF mutations, in other voltage-gated potassium channel families, *KCNH1* and *KCNQ1*, have been linked to similar overlapping phenotypic disorders, including Zimmermann-Laband, Temple-Baraitser and gingival overgrowth (Simons et al. 2015; Kortüm et al. 2020; Tommiska et al. 2017; Gripp et al. 2021), and Cantú syndrome, in two ATP-sensitive potassium channel subunits, *ABCC9* and *KCNJ8* (Harakalova et al. 2012; Cooper et al. 2014). Indeed, Gripp et al. 2021, who identified dominant variants of *KCNN3, KCNH1*, and *KCNK4* that share a number of clinical phenotypes as described for FHEIG syndrome, suggest combining them in a subgroup of potassium channelopathies, known as syndromic neurodevelopmental potassium channelopathies (Gripp et al. 2021).

6 Conclusion

The mutations identified in all four K2P channels considered in this study are illustrated in Fig. 4. The large majority (~70%) of the mutations are found in or close to the transmembrane domains of the channels, and most of these lead to either gain or

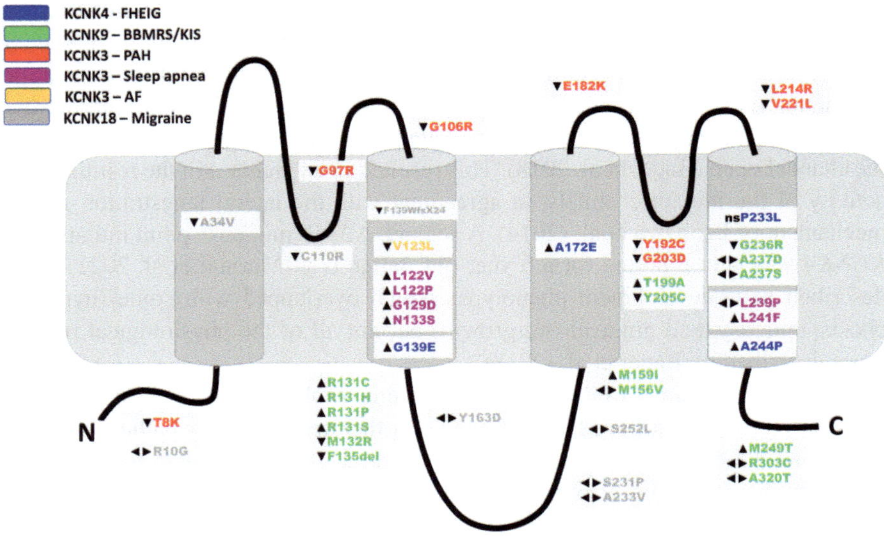

Fig. 4 A schematic representation of a K2P α-subunit showing the location of published mutations of KCNK3 (TASK1), KCNK9 (TASK3), KCNK4 (TRAAK), and KCNK18 (TRESK) channels linked to a number of clinical disease phenotypes. Color of the text, codes for a particular channel. Mutations indicated with ▲ correspond to a gain-of-function. Mutations indicated with ▼ correspond to a loss-of-function, and ◀▶ corresponds to no change in channel activity. ns – changes to channel activity are not stated. Abbreviations: *AF* atrial fibrillation, *BBMRS/KIS* Birk-Barel mental retardation syndrome/KCNK9 imprinting syndrome, *FHEIG* facial dysmorphism, hypertrichosis, epilepsy, intellectual disability/developmental delay, and gingival overgrowth, *PAH* pulmonary arterial hypertension. Amino acid codes: A, alanine; C, cysteine; D, aspartic acid; E, glutamic acid; F, phenylalanine; G, glycine; H, histidine; I, isoleucine; K, lysine; L, leucine; M, methionine; P, proline; R, arginine; S, serine; T, threonine; V, valine; Y, tyrosine

loss of channel function. Generally, mutations in the intracellular loops or N or C termini of the channels are less likely to alter channel current density. For *KCNK3*, identified mutations in the transmembrane domains lead to a GoF and are linked to sleep apnea, whereas mutations primarily in the extracellular loops or N terminus give a LoF and have been linked to pulmonary hypertension.

While this review has focused on rare diseases which result from mutations in K2P channels, interest is growing in common polygenic diseases (such as depression, autism, schizophrenia, pain, etc.) where one can look for associations between particular K2P channels and the disease state. In order to do this with any degree of confidence, one needs large numbers of patients with both appropriate genomic information and with suitable diagnoses or disease related phenotypes. The advent of large scale biobanks such as the UK Biobank (https://pan.ukbb.broadinstitute.org), the Finnish Biobank (FinnGen, https://www.finngen.fi/en), or deCODE (www.decode.com, based in Iceland) allows such associations to be made. Our initial work in this area has revealed, for the first time, associations between patients with a range of different pain related phenotypes and mutations in different K2P channels (Bourne, Forfar, Veale, and Mathie unpublished data).

Such studies can also reveal unexpected information. For example, the STAR*D study, a GWAS study of 1554 patients with nonpsychotic major depressive disorder being treated with the antidepressant, citalopram (Perlis et al. 2008), showed that patients with genetic variants in the *KCNK2* gene locus (which codes for TREK1, K2P channels) had a higher risk of resistance to antidepressant treatment than patients without these variants. These observations are consistent with earlier laboratory-based studies suggesting a link between depression, antidepressants, and TREK1 channels (Kennard et al. 2005; Heurteaux et al. 2006).

It is clear that genetic information can provide critical information linking K2P channels with disease states, both for diseases where the mutation causes the disease and for diseases where the mutation is more subtly associated with the disease and changes the likelihood of a patient developing the disease. Even for proposed monogenic diseases, however, it is also clear that the link between the channel mutation and the disease state is complicated. In this review of K2P channel monogenic diseases, we have shown examples where both gain and loss of channel function can lead to the same disease state (*KCNK9* and BBMRS/KIS), where mutations in one channel can lead to altered expression of another to cause the disease (*KCNK18* and migraine) and where mutations in the same channel can lead to different diseases (*KCNK3* and sleep apnea, pulmonary hypertension and atrial fibrillation, *KCNK4* and FHEIG syndrome and Rolandic epilepsy).

While this functional genomic information is extremely useful in its own right for patients that are identified as having the mutations, it also has wider implications for informing us when particular K2P channels have a role in the physiological processes underlying disease states. As such, modifying the activity of the channels associated with both monogenic and polygenic diseases may be of therapeutic benefit even for patients where these proteins are not mutated and functioning normally.

References

Alexander SPH, Mathie AA, Peters JA, et al. (2023) The Concise Guide to PHARMACOLOGY 2023/24: Ion channels. Br J Pharmacol. 180(S2).

Alloui A, Zimmermann K, Mamet J, et al. TREK-1, a K+ channel involved in polymodal pain perception. EMBO J. 2006;25(11):2368–76.

Andres-Enguix I, Shang L, Stansfeld PJ, et al. Functional analysis of missense variants in the TRESK (KCNK18) K channel. Sci Rep. 2012;2:237.

Antigny F, Hautefort MJ, et al. Potassium channel subfamily K member 3 (KCNK3) contributes to the development of pulmonary arterial hypertension. Circulation. 2016;133:1371–85.

Bagriantsev SN, Ang K-H, Gallardo-Godoy A, et al. A high-throughput functional screen identified small molecule regulators of temperature- and mechano-sensitive K2P channels. ACS Chem Biol. 2013;8(8):1841–51.

Barel O, Shalev SA, Ofir R, et al. Maternally inherited Birk Barel mental retardation dysmorphism syndrome cased by a mutation in the genomically imprinted potassium channel KCNK9. Am J Hum Genet A. 2008;83(3):193–9.

Bauer CK, Calligari P, Radio FC, et al. Mutations in KCNK4 that affect gating cause a recognizable neurodevelopmental syndrome. AJHG. 2018;103:621–30.

Bautista DM, Sigal YM, Milstein AD, et al. Pungent agents from Szechuan peppers excite sensory neurons by inhibiting two-pore potassium channels. Nat Neurosci. 2008;11(7):772–9.

Bittner S, Meuth SG, Göbel K, et al. TASK1 modulates inflammation and neurodegeneration in autoimmune inflammation of the central nervous system. Brain. 2009;132(Pt 9):2501–16.

Brohawn SG, Campbell EB, Mackinnon R. Physical mechanism for gating and mechanosensitivity of the human TRAAK K+ channel. Nature. 2014;516:126–30.

Brohawn SG, Wang W, Handler A, et al. The mechanosensitive ion channel TRAAK is localized to the mammalian node of Ranvier. eLife. 2019;8:e50403.

Buckler KJ. TASK channels in arterial chemoreceptors and their role in oxygen and acid sensing. Pflugers Arch. 2015;467(5):1013–25.

Burg S, Attali B. Targeting of potassium channels in cardiac arrhythmias. Trends Pharmacol Sci. 2021;42(6):491–506.

Cader MZ. The genetics of migraine and the path to precision medicine. Prog Brain Res. 2020;255:403–18.

Chen K, Talley EM, Patel N, et al. Inhibition of a background potassium channel by Gq protein α-subunits. Proc Natl Acad Sci U S A. 2006;103(9):3422–7.

Chokshi RH, Larsen AT, Bhayana B, et al. Breathing stimulant compounds inhibit TASK-3 potassium channel function likely by binding at a common site in the channel pore. Mol Pharmacol. 2015;88(5):926–34.

Cooper PE, Reutter H, Woelfle J, et al. Cantú syndrome resulting from activating mutation in the KCNJ8 gene. Hum Mutat. 2014;35(7):809–13.

Cooper A, ButtoT HN, et al. Inhibition of histone deacetylation rescues phenotype in a mouse model of Birk-Barel intellectual disability syndrome. Nat Commun. 2020;11:480.

Cotton JF, Keshavaprasad B, Laster MJ, et al. The ventilator stimulant doxapram inhibits TASK tandem pore (K2P) potassium channel function but does not affect minimum alveolar anesthetic concentration. Anesth Analg. 2006;102(3):779–85.

Court F, Camprubi C, Vicente Garcia C, et al. The PEG13-DMR and brain-specific enhancers dictate imprinted expression within the 8q24 intellectual disability risk locus. Epigenetics Chromatin. 2014;7:5.

Cousin MA, Veale EL, Dsouza NR, et al. Gain and loss of TASK3 channel function and its regulation by novel variation cause KCNK9 imprinting syndrome. Genome Med. 2022;14(1):62.

Cunningham KP, Holden RG, Escribano-Subias PM, et al. Characterization and regulation of wild-type and mutant TASK-1 two pore domain potassium channels indicated in pulmonary arterial hypertension. J Physiol. 2019;597(4):1087–101.

Cunningham KP, Macintyre DE, Mathie A, et al. Effects of the ventilatory stimulant, doxapram on human TASK-3 (KCNK9, K2P9.1) channels and TASK-1 (KCNK3, K2P3.1) channels. Acta Physiol. 2020;228(2):e13361.

Czirják G, Toth Z, Enyedi P. The two-pore domain K+ channel, TRESK, is activated by the cytoplasmic calcium signal through calcineurin. J Biol Chem. 2004;279(18):18550–8.

Decher N, Ortiz-Bonnin B, Friedrich C, et al. Sodium permeable and "hypersensitive" TREK-1 channels cause ventricular tachycardia. EMBO Mol Med. 2017;9(4):403–14.

Decher N, Rinne S, Bedoya M, et al. Molecular pharmacology of K2P potassium channels. Cell Physiol Biochem. 2021;55(53):87–107.

Devilliers M, Busserolles J, Lolignier S, et al. Activation of TREK-1 by morphine results in analgesia without adverse side effects. Nat Commun. 2013;4:2941.

Enyedi P, Czirják G. Molecular Background of Leak K + Currents: Two-Pore Domain Potassium Channels Physiological Reviews. 2010;90(2):559–605.

Enyedi P, Czirják G. Properties, regulation, pharmacology, and functions of the K2P channel TRESK. Pflugers Arch. 2015;467(5):945–58.

Fink M, Lesage F, Duprat F, et al. A neuronal two P domain K+ channel stimulated by arachidonic acid and polyunsaturated fatty acids. EMBO J. 1998;17:3297–308.

Gaisl T, Turnbull CD, Weimann G, et al. BAY 2253651 for the treatment of obstructive sleep apnoea: a multicentre, double-blind, randomised controlled trial (SANDMAN). Eur Respir J. 2021;58(5):2101937.

Garcia G, Noriega-Navarro R, Martínez-Rojas VA. Spinal TASK-1 and TASK-3 modulate inflammatory and neuropathic pain. Eur J Pharmacol. 2019;862:172631.

Gormley P, Anttila V, Winsvold BS, et al. Meta-analysis of 375,000 individuals identifies 38 susceptibility loci for migraine. Nat Genet. 2016;48(8):856–66.

Graham JM, Zadeh N, Kelley M, et al. KCNK9 imprinting syndrome-further delineation of a possible treatable disorder. Am J Med Genet A. 2016;170(10):2632–7.

Gripp KW, Smithson SF, Scurr IJ, et al. Syndromic disorders caused by gain-of-function variants in KCNH1, KCNK4, and KCNN3-a subgroup of K+ channelopathies. Eur J Hum Genet. 2021;29(9):1384–95.

Guo Z, Liu P, Ren F, et al. Nonmigraine-associated TRESK K+ channel variant C110R does not increase the excitability of trigeminal ganglion neurons. J Neurophysiol. 2014;112(3):568–79.

Guo H, Duyzend MH, Coe BP, et al. Genome sequencing identifies multiple deleterious variants in autism patients with more severe phenotypes. Genet Med. 2019;21(7):1611–20.

Gurges P, Liu H, Horner RL. Modulation of TASK-1/3 channels at the hypoglossal motoneuron pool and effects on tongue motor output and responses to excitatory inputs in vivo: implications for strategies for obstructive sleep apnea pharmacotherapy. SleepJ. 2021;44(1):zsaa144.

Harakalova M, van Harssel JJT, Terhal PA, et al. Dominant missense mutations in ABCC9 cause Cantó syndrome. Nat Genet. 2012;44(7):793–6.

Harleton E, Besana A, Chandra P, et al. TASK-1 current is inhibited by phosphorylation during human and canine chronic atrial fibrillation. Am J Physiol Heart Circ Physiol. 2015;308(2):H126–34.

Heurteaux C, Lucas G, Guy N, et al. Deletion of the background potassium channel TREK-1 results in a depression-resistant phenotype. Nat Neurosci. 2006;9(9):1134–41.

Hill RZ, Bautista DM. A TREK to Translate Genetics to Mechanisms of Migraine Neuron. 2019;101(2):193–195.

Imbrici P, Nematian-Ardestani E, Hasan S, et al. Altered functional properties of a missense variant in the TRESK K+ channel (KCNK18) associated with migraine and intellectual disability. Pflugers Arch. 2020;472(7):923–30.

Kanda H, Ling J, Tonomura S, et al. TREK-1 and TRAAK are principal K+ channels at the nodes of Ranvier for rapid action potential conduction on mammalian myelinated afferent nerves. Neuron. 2019;104(5):960–97.

Kang S, Kim D. TREK-2 (K2P10.1) and TRESK (K2P18.1) are major background K+ channels in dorsal root ganglion neurons. Am J Physiol Cell Physiol. 2006;291(1):C138–46.

Kang D, Choe C, Kim D. Thermosensitivity of the two-pore domain K^+ channels TREK-2 and TRAAK. J Physiol. 2005;564(1):103–16.

Kennard LE, Chumbley JR, Ranatunga KM, et al. Inhibition of the human two-pore domain potassium channel, TREK-1, by fluoxetine and its metabolite norfluoxetine. Br J Pharmcol. 2005;144(6):821–9.

Kiper AK, Rinne S, Rolfes C, et al. Kv1.5 blockers preferentially inhibit TASK-1 channels: TASK-1 as a target against atrial fibrillation and obstructive sleep apnea? Pflugers Arch. 2015;467(5):1081–90.

Kortüm F, Niceta M, Magliozzi M, et al. Cantú syndrome versus Zimmermann-Laband syndrome: report of nine individuals with ABCC9 variants. Eur J Med Genet. 2020;63(9):103996.

Kraft M, Buscher A, Wiedmann F, et al. Current drug treatment strategies for atrial fibrillation and TASK-1 inhibition as an emerging novel therapy option. Front Pharmacol. 2021;12:638445.

Lafrenière RG, Rouleau GA. Identification of novel genes involved in migraine. Headache. 2012;52(suppl 2):107–10.

Lafrenière RG, Cader MZ, Poulin J-F, et al. A dominant-negative mutation in the TRESK potassium channel is linked to familial migraine with aura. Nat Med. 2010;16(10):1157–60.

Lambert M, Boet A, Rucker-Martin C, et al. Loss of KCNK3 is a hallmark of RV hypertrophy/dysfunction associated with pulmonary hypertension. Cardiovasc Res. 2018;114:880–93.

Lengyel M, Czirják G, Jacobsen DA, Enyedi P. TRESK and TREK-2 two-pore-domain potassium channel subunits form functional heterodimers in primary somatosensory neurons. J Biol Chem. 2020;295(35):12408–25.

Lesage F, Maingret F, Lazdunski. Cloning and expression of human TRAAK, a polyunsaturated fatty acids-activated and mechano-sensitive K^+ channel. FEBS Lett. 2000;471:137–40.

Liang B, Soka M, Christensen AH, et al. Genetic variation in the two=pore domain potassium channel, TASK-1, may contribute to an atrial substrate in arrhythmogenesis. J Mol Cell Cardiol. 2014;67:69–76.

Liu P, Xiao Z, Ren F, et al. Functional analysis of a migraine-associated TRESK K+ channel mutation. J Neurosci. 2013;33(31):12810–24.

Luedi PP, Dietrich FS, Weidman JR, et al. Computation and experimental identification of novel human imprinted genes. Genome Res. 2007;17(12):1723–30.

Ma L, Roman-Campos D, Austin ED, et al. A novel channelopathy in pulmonary arterial hypertension. N Engl J Med. 2013;369:351–61.

Madry C, Kyrargyri V, Arancibia-Cárcamo, et al. Microglial ramification, surveillance, and interleukin-1B release are regulated by the two-pore domain K+ channel THIK-1. Neuron. 2018;97(2):299–312.e6.

Maher BH, Taylor M, Stuart S. Analysis of 3 common polymorphisms in the KCNK18 gene in an Australian Migraine case-control cohort. Gene. 2013;528(2):343–6.

Mariani P, Zhurakivska K, Santoro R. Hereditary gingival fibromatosis associated with the missense mutation of the KCNK4 gene. Oral Surg Oral Med Oral Pathol Oral Radiol. 2021;131(6):e175–82.

Markel KA, Curtis D. Study of variants in genes implicated in rare familial migraine syndromes and their association with migraine in 200,000 exome-sequenced UK Biobank participants. Ann Hum Genet. 2022;86(6):353–60.

Mathie A, Veale EL, Cunningham KP, et al. Two-pore domain potassium channels as drug targets: anesthesia and beyond. Annu Rev Pharmacol Toxicol. 2021;61:401–20.

Mondéjar-Parreño G. Unravelling the role of K2P channels in atrial fibrillation. Front Biosci. 2022;14:31.

Mu D, Chen L, Zhang X, et al. Genomic amplification and oncogenic properties of the KCNK9 potassium channel gene. Cancer Cell. 2003;3(3):297–302.

Navas P, Tenorio J, Quezada CA, et al. Molecular analysis of BMPR2, TBX4, and KCNK3 and genotype-phenotype correlations in Spanish patients and families with idiopathic and hereditary pulmonary arterial hypertension. Rev Esp Cardiol. 2016;69(11):1011–9.

Niday Z, Tzingounis AV. Potassium channel gain of function in epilepsy: an unresolved paradox. Neuroscientist. 2018;24(4):368–80.

Nóel J, Zimmermann K, Busserolles J, et al. The mechano-activated K+ channels TRAAK and TREK-1 control both warm and cold perception. EMBO J. 2009;28:1308–18.
Nyholt DR, LaForge KS, Kallela M. A high-density association screen of 155 ion transport genes for involvement with common migraine. Hum Mol Genet. 2008;17(21):3318–31.
O'Donohoe PB, Huskens N, Turner PJ. A1899, PK-THPP, ML365, and Doxapram inhibit endogenous TASK channels and excite calcium signalling in carotid body type-1 cells. Physiol Rep. 2018;6(19):e13876.
Olschewski A, Veale EL, Nagy BM, et al. TASK-1 (KCNK3) channels in the lung from cell biology to clinical implications. Eur Respir J. 2017;50:1700754.
Ortega-Saenz P, Levitsky KL, Marcos-Almarez MT. Carotid body chemosensory responses in mice deficient of TASK channels. J Gen Physiol. 2010;135(4):379–92.
Osman AM, Mukherje S, Altree TJ, et al. Topical potassium channel blockage improves pharyngeal collapsibility: a translational, placebo-controlled trial. Chest. 2023;163:P953–65.
Park SM, Roache CE, Iffland PH II, et al. BK channel properties correlate with neurobehavioral severity in three KCNMA1-linked channelopathy mouse models. eLife. 2022;11:e77953.
Pavinato L, Nematian-Ardestani E, Zonta A, et al. KCNK18 biallelic variants associated with intellectual disability and neurodevelopmental disorders alter TRESK channel actiivity. Int J Mol Sci. 2021;22(11):6064.
Pei L, Wiser O, Slavin A, et al. Oncogenic potential of TASK3 (Kcnk9) depends on K+ channel function. 2003;100(13):7803–7.
Perlis RH, Moorjani P, Fagerness J, et al. Pharmacogenic analysis of genes implicated in rodent models of antidepressant response: association of TREK1 and treatment resistance in the STAR*D study. Neuropsychopharmacology. 2008;33:2810–9.
Pettingill P, Weir GA, Wei T, et al. A causal role for TRESK loss of function in migraine mechanisms. Brain. 2019;142(12):3852–67.
Prado PA, Landra-Willm A, Verkest C, et al. TREK channel activation suppresses migraine pain phenotype. iScience. 2021;24(9):102961.
Reyes R, Lauritzen I, Lesage F, et al. Immunolocalization of the arachidonic acid and mechanosensitive baseline traak potassium channel in the nervous system. Neuroscience. 2000;95(3):893–901.
Rödström KEJ, Kiper AK, Zhang W, et al. A lower X-gate in TASK channels trap inhibitors within the vestibule. Nature. 2020;583(7812):443–7.
Royal P, Andres-Bilbe A, Prado PA, et al. Migraine-associated TRESK mutations increase neuronal excitability through alternative translation initiation and inhibition of TREK. Neuron. 2019;101(2):232–245.e6.
Sánchez Delgado M, Camprubi C, Tümer Z, et al. Screening individuals with intellectual disability, autism and Tourette's syndrome for KCNK9 mutations and aberrant DNA methylation within the 8q24 imprinted cluster. Am J Med Genet B Neuropsychiatr Genet. 2014;1658(6):472–8.
Sano Y, Inamura K, Miyake A, et al. A novel two-pore domain K+ channel, TRESK, is localized in the spinal cord. J Biol Chem. 2003;278(30):27406–12.
Schmidt C, Wiedmann F, Tristram F, et al. Cardiac expression and atrial fibrillation-associated remodeling of K2P2.1 (TREK-1) K+ channels in a porcine model. Life Sci. 2014;97(2):107–15.
Schmidt C, Wiedmann F, Voigt N, et al. Upregulation of K(2P)3.1 K+ current causes action potential shortening in patients with chronic atrial fibrillation. Circulation. 2015;132:82–92.
Schmidt C, Wiedmann F, Beyersdorf C, et al. Genetic ablation of TASK-1 (tandem of P domains in a weak inward rectifying K+ channel-related acid sensitive K+ channel-1) (K2P3.1) K+ channels suppresses atrial fibrillation and prevents electrical remodeling. Circ Arrhythm Electrophysiol. 2019;12(9):e007465.
Šedivá M, Laššuthová P, Zámečník J, et al. Novel variant in the KCNK9 gene in a girl with Birk Barel syndrome. Eur J Med Genet. 2020;63(1):103619.
Simons C, Rash LD, Crawford J, et al. Mutations in the voltage-gated potassium channel gene in KCNH1 cause Temple-Baraitser syndrome and epilepsy. Nat Genet. 2015;47:73–7.
Sörmann J, Schewe M, Proks P, et al. Gain-of-function mutations in KCNK3 cause a developmental disorder with sleep apnea. Nat Genet. 2022;54:1534–43.

Streit AK, Netter MF, Kempf F, et al. A specific two-pore domain potassium channel blocker defines the structure of the TASK-1 open pore. J Biol Chem. 2011;286(16):13977–84.

Sun H, Luo L, Lai B, et al. A monoclonal antibody against KCNK9 K+ channel extracellular domain inhibits tumour growth and metastasis. Nat Commun. 2016;7:10339.

Talley EM, Bayliss DA. Modulation of TASK-1 (Kcnk3) and TASK-3 (Kcnk9) potassium channels: volatile anesthetics and neurotransmitters share a molecular site of action. J Biol Chem. 2002;277(20):17733–42.

Talley EM, Solórzano G, Lei Q, et al. CNS distribution of members of the two-pore-domain (KCNK) potassium channel family. J Neurosci. 2001;21(19):7491–505.

Tommiska J, Känsäkoski J, Skibsbye L, et al. Two missense mutations in KCNQ1 cause pituitary hormone deficiency and maternally inherited gingival fibromatosis. Nat Commun. 2017;8:1289.

Veale EL, Buswell R, Clarke CE, Mathie A. Identification of a region in the TASK3 two pore domain potassium channel that is critical for its blockade by methanandamide. Brit J Pharmacol. 2007;152(5):778–86.

Veale EL, Hassan M, Walsh Y, et al. Recovery of current through mutated TASK3 potassium channels underlying Birk Barel syndrome. Mol Pharmacol. 2014;85(3):397–407.

Veale EL, Golluscio A, Grand K, et al. A KCNB1 gain of function variant causes developmental delay and speech apraxia but not seizures. Front Pharmacol. 2022;13.

Vivier D, Soussia IB, Rodrigues N, et al. Development of the first two-pore domain potassium channel TWIK-related K+ channel 1-selective agonist possessing in vivo Antinociceptive activity. J Med Chem. 2017;60(3):1076–88.

Walsh Y, Leach M, Veale EL, Mathie A. Block of TREK and TRESK K2P channels by lamotrigine and two derivatives sipatrigine and CEN-092. Biochem Biophys Rep. 2021;26:101021.

Wiedmann F, Schulte JS, Gomes B, et al. Atrial fibrillation and heart failure associated remodeling of two-pore -domain potassium (K2P) channels in murine disease models focus on TASK-1. Basic Res Cardiol. 2018;113(4):27.

Wiedmann F, Beyersdorf C, Zhou X-B, et al. The experimental TASK-1 potassium channel inhibitor A293- can be employed for rhythm control of persistent atrial fibrillation in a translational large animal model. Front Physiol. 2020;11:629421.

Wright PD, Veale EL, McCoull D, et al. Terbinafine is a novel and selective activator of the two-pore domain potassium channel TASK3. Biochem Biophys Res Commun. 2017;493(1):444–50.

Yamada M, Suzuki H, Watanabe A. Role of chimeric transcript formation in the pathogenesis of birth defects. Congenit Anom. 2020;61:76–81.

Yan H-J, He Y-Y, Jin L, et al. Expanding the phenotypic spectrum of KCNK4: from syndromic neurodevelopmental disorder to Rolandic epilepsy. Front Mol Neurosci. 2023;15:1081097.

The Voltage-Gated Potassium Channel $K_v2.1$ as a Multicellular Drug Target

Moza M. Al-Owais, Nagihan Ozsoy, and Mark L. Dallas

Abstract Voltage-gated potassium (K^+; K_v) channels are critical homeostatic regulators of ionic equilibrium. The diversity of channel subtypes and expression profiles makes them interesting drug targets. The $K_v2.1$ channel, encoded by the *KCNB1* gene, has a multitude of physiological roles from setting neuronal firing frequency to influencing viral replication. Here we look at the biophysical fingerprint of the $K_v2.1$ channel and its role in a variety of unique cellular systems. These cellular systems provide a wealth of evidence for modulating $K_v2.1$ to offer novel avenues to treat a range of disease from neurological disorders to diabetes. Existing pharmacological compounds that target $K_v2.1$ function are highlighted, along with presenting differential pharmacological approaches to modulate $K_v2.1$ function. This chapter reveals the $K_v2.1$ channel to be an exciting drug target that could afford treatment options in complex diseases affecting ionic homeostasis; future research should be dedicated to delivering the drug in the appropriate manner to provide novel compounds and repurpose existing medicines.

Keywords Potassium ion channel · $K_v2.1$ · Electrophysiology · Voltage-gated · Drug target

M. M. Al-Owais (✉)
School of Biomedical Sciences, University of Leeds, Leeds, UK
e-mail: m.al-owais@leeds.ac.uk

N. Ozsoy · M. L. Dallas (✉)
Reading School of Pharmacy, University of Reading, Reading, UK
e-mail: m.dallas@reading.ac.uk

G. Stephens, E. Stevens (eds.), *Ion Channels as Targets in Drug Discovery*,
https://doi.org/10.1007/978-3-031-52197-3_7

1 Introduction

1.1 Voltage-Gated K^+ Channels

Voltage-gated K^+ (K_v) channels, which belong to the six-transmembrane domain family of K^+ channels, activate upon depolarization of the cell membrane and deactivate upon repolarization, promoting the efflux of K^+ ions. Forty genes encoding K_v channels have been identified in humans, forming 12 families, each then divided further into subfamilies. Based on their sequence identity, the classification is usually given as K_vm.n where m denotes the family and n is the subfamily, which display multiple physiological roles (see Gamper et al., this volume).

The K_v2, or Shab-related, K^+ channel family consists of $K_v2.1$ and $K_v2.2$ (Gutman et al. 2003). The kinetics of these channels are distinct from other K_v channels, with slow activation and with inactivation that takes up to several seconds, characterizing the currents carried by the Shab-related channels as delayed rectifier (Frech et al. 1989; Hwang et al. 1992).

The Shab gene was initially identified in *Drosophila* (Butler et al. 1989), and its product has subsequently been shown to maintain the electrical excitability of neurons and muscle cells (Frech et al. 1989).

2 $K_v2.1$ Function and Structure

$K_v2.1$ is encoded by the *KCNB1* gene and consists of α-subunits that share the structural characteristic of six-transmembrane helices, designated as S1–S6, like other K_v channels, with two long intracellular domains known as the N-terminus and carboxyl terminus (Fig. 1). A highly conserved tetramerization T1 domain is located at the N-terminus, playing a major role in determining the compatibility of different K_v subunits. A recent study investigating the crystal structure of the human $K_v2.1$ T1 domain found that the domain forms a pentamer structure (Xu et al. 2022; Fig. 1).

The carboxy terminus tail of the $K_v2.1$ channel protein contains a motif representing the C-terminal activation (CTA) domain (Ju et al. 2003) and has been associated with the targeting and localization of the channel (Lim et al. 2000; Tamkun et al. 2007). Changes in the phosphorylation level have been attributed to changes in the channel's gating and surface expression (Grizel et al. 2014).

Highly conserved positively charged residues are located on the S4 domain, which acts as the main voltage sensor and contains the main components of the gating charge for the channel. The selectivity filter loop is characteristically located between the S5 and S6 transmembrane helices and includes the signature sequence of the K_v channel selectivity filter: TXXTXGYG (Heginbotham et al. 1994).

In its open state when the channel activated, the pore opening step is distinct. This step is weakly voltage-dependent leading to slow activation of $K_v2.1$ with

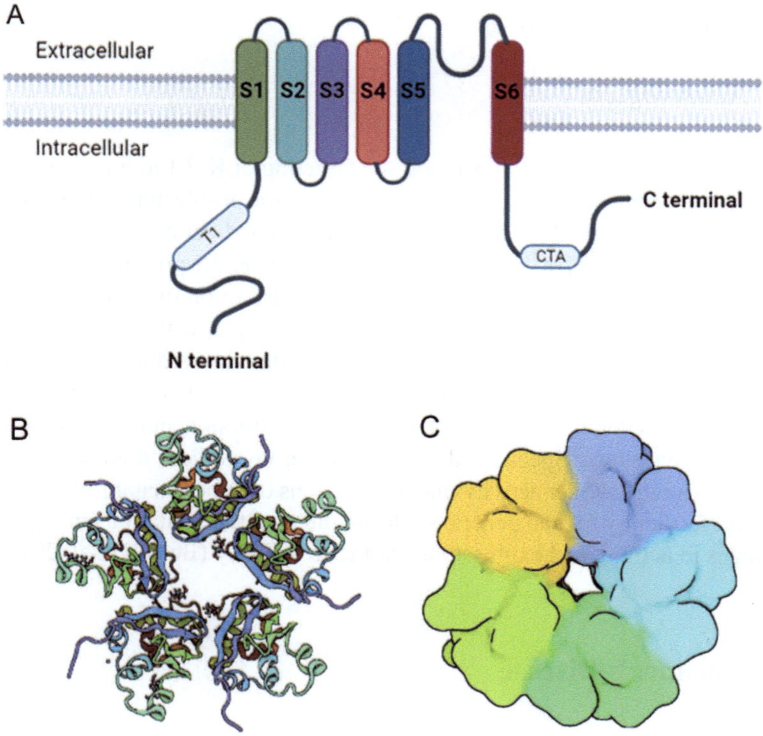

Fig. 1 Proposed architecture for K$_v$2.1. (**a**) K$_v$ α-subunits consist of six-transmembrane segments (S1–S6). S4 forms the voltage sensor domain of the channel and accounts for the channel gating where, in response to depolarization, movement of positively charged amino acids results in opening of the channel. S5–S6 form the electrical pore. The N- and C-termini that are located intracellularly contain a T1 tetramerization domain and CTA domain that is mainly responsible for activation of the channel with T1 domain. (**b**) Pentameric assembly of the K$_v$2.1 tetramerization domain viewed in cartoon form. (**c**) Pentameric assembly of the K$_v$2.1 tetramerization domain viewed in quick surface form. Figure made in Biorender.com and retrieved using PDB ID 7SPD

unique kinetics supporting K^+ ion flux and therefore forming its physiological role in membrane excitability. In the channel's inactivated state, the channel remains dormant via two separate inactivation pathways and does not response to voltage changes across the membrane. Firstly, a very rapid inactivation pathway that involves conformational change of the N-terminus in the main subunit, leading to pore blocking of the channels, is known as an N-type inactivation or "ball and chain" inactivation (Rettig et al. 1994). However, the channel also undergoes a slower C-type inactivation mechanism, regulated at the selectivity filter, which involves the narrowing of outer mouth of the channel. An additional U-type rapid inactivation mechanism, involving a voltage-dependent recovery, has also been proposed (Klemic et al. 2001).

3 $K_v2.1$ as a Drug Target

3.1 Neurological

There is an array of evidence to support the targeting of $K_v2.1$ to manage a range of neurological conditions, arising from its expression profile throughout the brain (Uhlén et al. 2015) and in discreet neuronal loci (Maletic-Savatic et al. 1995; Misonou and Trimmer 2004). This pattern allows the $K_v2.1$ protein to regulate neuronal excitability (Du et al. 1998, 2000) by its contribution to the delayed rectifier K^+ currents (Murakoshi and Trimmer 1999). For example, within the hippocampus, $K_v2.1$ channels are reported to carry 50% of recorded delayed rectifier K^+ currents (Murakoshi and Trimmer 1999). Genetic deletion of $K_v2.1$ leads to a hyperactive phenotype, characterized by accelerated seizure progression and increased sensitivity to anticonvulsants (Speca et al. 2014). Therefore, a clear therapeutic strategy would be to modify $K_v2.1$ activity under conditions of hyperexcitability. This role is supported further by human genetic studies indicating that mutations in $K_v2.1$ result in changes in activity and expression, leading to epilepsy (Bersell et al. 2014).

4 Alzheimer's Disease

The characteristic feature of Alzheimer's disease (AD) is the plaques which form in the brain by the accumulation of fragments of β-amyloid (Aβ) protein (Selkoe 1991). A result of this accumulation is the disruption of ionic homeostasis; while changes in Ca^{2+} have been extensively studied, a role for changes in K^+ homeostasis has also been proposed. In relation to $K_v2.1$, studies have identified altered levels of $K_v2.1$ channel mRNA expressed in cultured rat hippocampal neurons, which had been incubated in vitro with Aβ fragments (Jin et al. 2002); these studies are supported by comparable observations in vivo in drug-induced memory deficiency in the rat hippocampus (Zhong et al. 2005). While these are animal studies and link to accumulation of the amyloid beta peptide, they still demonstrate the sensitivity of K^+ channels to disease modification. More physiological research has demonstrated a link between the phosphorylation state of the $K_v2.1$ channel and neuronal hyperexcitability in hypoxia and ischemic stroke, where augmentation of the K^+ currents carried by the $K_v2.1$ channel results in suppression of hyperexcitability (Misonou et al. 2004, 2005). Interestingly, the phosphorylation of $K_v2.1$ is regulated by AMPK, which is a master energy regulator in cellular metabolism (Ikematsu et al. 2011).

5 Cell Death

Evidence suggests that the $K_v2.1$ channel has a distinct role in neuronal apoptosis, as demonstrated in vitro in cortical neurons, where the expression of $K_v2.1$ channels increased cellular susceptibility to apoptosis, when compared to dominant negative

$K_v2.1$-transfected neurons (Pal et al. 2003). Moreover, a surge in $K_v2.1$ channel activity was observed in both HEK293 cells and CHO cells undergoing apoptosis following induced oxidative stress (Pal et al. 2006; Al-Owais et al. 2012). This apoptotic surge is mediated through p38 phosphorylation at the serine 800 site of the $K_v2.1$ channel (Redman et al. 2007) following Zn^{2+} activation of signal-regulating kinase 1 (ASK-1) (Aras and Aizenman 2005). The requirement for Zn^{2+} mediated phosphorylation was also determined for the N-terminal residue tyrosine at position 124 of the channel (Redman et al. 2009). In fact, dual phosphorylation of S800 and Y124 aids trafficking of the $K_v2.1$ channel protein delivery to the plasma membrane (He et al. 2015), highlighting the channel's role in the apoptotic cascade.

Multiple signaling mechanisms are responsible for modulation of the $K_v2.1$ channel's activity: in hippocampal neurons, SUMOylation has been demonstrated to regulate the firing of cell excitability, by shifting the half-maximal activation voltage (Plant et al. 2011). Meanwhile, a reduction in the phosphorylation status of the channel alters the voltage dependence of activation and location of $K_v2.1$ (Murakoshi et al. 1997; Mohapatra et al. 2007). Several amino acid residues on the carboxy terminal domain of the $K_v2.1$ channel are highly phosphorylated in neurons, and rapid phosphorylation mediated by cyclin-dependent kinase 5 can restore neuronal excitability (Cerda and Trimmer 2011) or can indeed suppress excitability, influencing hyperpolarizing shifts in the voltage dependence of activation, when mediated by AMP-activated protein kinase (Ikematsu et al. 2011). Recently, S-sulfhydration, the posttranslational modification of proteins by hydrogen sulfide (H_2S), has been shown to play a role in neuronal excitability surges, by increasing the action potential firing rate (Dallas et al. 2021).

To assess the therapeutical potential of prompting a decrease in proapoptotic $K_v2.1$ surges, and K^+ loss, thus increasing neuronal survival, the in vitro inhibition of $K_v2.1$ channel activity following ischemic insult resulted in suppression of hyperexcitability in neurons (Misonou et al. 2005). Additionally, de novo experiments using syntaxin-1A-mimicking peptides, which target the control of membrane trafficking insertion of the $K_v2.1$ channel protein, preventing the channel's clustering and disrupting its location on the plasma membrane, reduced cell death and enhanced the neurological outcome in an ischemic stroke model (Yeh et al. 2017). Indeed, recent in vivo observations of therapeutic peptides designed to disrupt the association between $K_v2.1$ and the vesicle-associated membrane protein-associated protein-A (VAP-A), thus reducing the channel's trafficking to the membrane, have shown promising neuroprotection following administration in ischemic stroke mice. However, in vitro, these peptides do not bind to VAP-A directly, suggesting another trafficking delivery mechanism may be involved (Schulien et al. 2020).

Alternatively, K^+ channel blockers such as tetraethylammonium (TEA) and 4-aminopyridine (4-AP) could provide protection against apoptosis accelerated by a range of insults, including oxidative stress (Redman et al. 2007). Cholinesterase inhibitors, for example, donepezil, rivastigmine, and galantamine, are approved drugs for use in early and mild stages of AD. These drugs have been shown to inhibit $K_v2.1$ (Pan et al. 2003; Zhang et al. 2004; Yuan et al. 2011) with donepezil, exhibiting beneficial effects on neuronal damage and on cognitive deficits following ischemic insults (Yuan et al. 2011). However, those blockers can also act on other K^+ channels, such as $K_v1.5$ and hERG (Chae et al. 2015; Li et al. 2017), thus producing detrimental side effects. Clearly, the development of a therapeutic agent that selectively acts on the $K_v2.1$ channel, without affecting other ion channels, would be advantageous. Research on 48F10, a norbornyl catechol derivative, which selectively inhibits the $K_v2.1$ channel with an IC_{50} value of 60 μM, has shown to prevent oxidants-induced apoptosis in cortical neurons (Zaks-Makhina et al. 2004).

6 Metabolic

$K_v2.1$ is expressed in the pancreas (Yan et al. 2004), specifically in the β-cells, playing a role in regulating membrane excitability, with a demonstrated role in regulating membrane repolarization, consecutively controlling glucose-stimulated insulin secretion (Roe et al. 1996). Pancreatic β-cells respond to high concentrations of glucose through membrane depolarization, in response to K^+ induced Ca^{2+} influx (Dawson et al. 1984), followed by rapid repolarization when the K_v channels are active. The $K_v2.1$ channel has been found to be the main contributor to the voltage-dependent outward K^+ currents (MacDonald et al. 2001); thus, its impact on insulin secretion has been extensively investigated.

In isolated rat islets, dominant negative knockout of $K_v2.1$ resulted in extensive reduction of the delayed rectifier K^+ currents and caused an increase in insulin production, following stimulation by glucose (MacDonald et al. 2002). This role was further highlighted by use of the specific $K_v2.1$ channel blocker hanatoxin, a toxin peptide isolated from Chilean rose tarantula venom (Swartz and MacKinnon 1995). The inhibition of the $K_v2.1$ channel resulted in an increase in membrane potential, elevated intracellular Ca^{2+} oscillation in response to glucose, and improved glucose dose dependent insulin secretion (Tamarina et al. 2005). These results suggest that $K_v2.1$ could serve as a potential target in the treatment of type 2 diabetes, where the $K_v2.1$ channel is shown to be reduced in human β-cells isolated from type 2 diabetic patients (Fu et al. 2017).

Interestingly, the depolarization duration through outward K^+ currents can influence insulin exocytosis from cell plasma membrane (Fig. 2), through membrane clustering of the $K_v2.1$ channels (Greitzer-Antes et al. 2018; Fu et al. 2019).

$K_v2.1$ clustering aids in insulin exocytosis and contribute to rescuing dysfunctional human β-cell function (Fu et al. 2017). SNARE syntaxin1A and syntaxin-3 proteins are required for delivering these $K_v2.1$ clusters to the plasma membrane (Dai et al. 2012; Greitzer-Antes et al. 2018; Fu et al. 2019) and regulated through the SUMOylation-mediated $K_v2.1$ channels and thus regulation of pancreatic β-cell excitability (Plant et al. 2011). Glucose activation of β-cells results in action potential (AP) initiation. Following rapid depolarization, K_v channel activation facilitates the rapid repolarization phase of β-cell APs, restoring K^+ efflux via the opening of the channel, thus affecting the duration and the subsequent firing of APs (Jacobson and Philipson, 2007). Consequently, blocking the $K_v2.1$ channel in diabetic patients

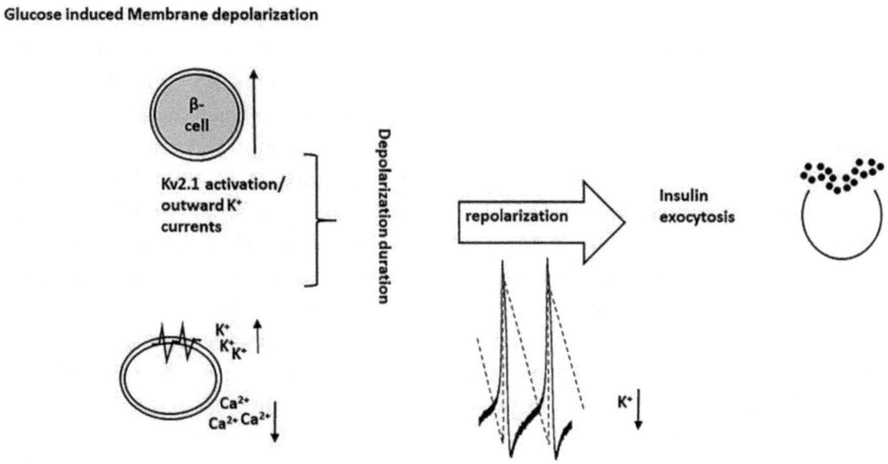

Fig. 2 Schematic representation of glucose-induced changes in β-cell membrane potential. Dynamic ionic flux resulting in an increase in membrane potential triggering rapid depolarization followed by rapid repolarization phase to restore the K^+ efflux via the opening of $K_v2.1$ channels influencing action potential duration and the subsequent firing of APs and insulin secretion

would contribute to inhibition of β-cell overexcitability and delay repolarization of the action potential, therefore minimizing K^+ loss and sustaining cytosolic Ca^{2+} (MacDonald and Wheeler 2003).

However, the precise role of $K_v2.1$ in repolarization of the AP in human β-cells is still unclear. It has been proposed that the large conductance Ca^{2+} activated K^+ channels (BK channels) are the main contributors to the repolarization phase in these cells, as stromatoxin inhibition of $K_v2.1$ had no effect on stimulated electrical activity in human β-cells obtained from nondiabetic donors (Braun et al. 2008). Nevertheless, the significance of the $K_v2.1$ channel's role in human β-cells was demonstrated using in vitro studies; tetramers of the $K_v2.1$ channel were found to be clustered on the plasma membrane of human β-cells, and $K_v2.1$ knockdown had an adverse effect on the depolarization-induced exocytosis in human β-cells (Fu et al. 2017).

In addition, examination of the structure of β-cells, obtained from type 2 diabetic human donors, revealed abnormalities in granule fusion, compared with nondiabetic donors. The density of these membrane-attached granules is regulated by clusters of K_v2, and the relative proportion of these granule fusions is determined by SUMOylation at two $K_v2.1$ residues: K145 and K470 (Fu et al. 2019). Moreover, dual blocking of BK and $K_v2.1$ channel activities, during glucose triggered AP, resulted in an increase in amplitude, duration, and latency of the AP, indicating that the repolarizing phase involves multiple K^+ channels and does indeed include the $K_v2.1$ channel (MacDonald et al. 2002).

In further studies, $K_v2.1$ gene disruption in animal models resulted in the modification of β-cell electrical activity, including reduced firing frequency and the prolongation of AP duration, as well as enhanced insulin secretion (Jacobson et al. 2007), confirming the important role of $K_v2.1$ in modulating glucose-induced APs.

Furthermore, application of the pan K_v2 channel inhibitor, RY796, which selectively inhibits $K_v2.1$ and $K_v2.2$ channels, leads to the enhancement of glucose-stimulated insulin secretion in vitro in both isolated mouse and human islets (Li et al. 2013). These effects were supported by the inhibition of the $K_v2.1$ channel using further pharmacological $K_v2.1$-specific blockers such as SP6616, and the small molecule inhibitor 4-ethoxy-N-{[6-(2-thienyl)-7H-[1,2,4]triazolo[3,4-b][1,3,4]thiadiazin-3-yl]methyl}aniline (ETA) ameliorates hyperglycemia and promotes β-cell survival in diabetic mice (Yao et al. 2013; Zhou et al. 2016, 2018). Even a modest inhibition of $K_v2.1$ channels can sustain cytosolic Ca^{2+} and stimulate insulin production, thus improving glucose tolerance in diabetic mice treated with glucagon-like peptide-1 based drugs; such observations have opened a new avenue to reduce the doses of such drugs and potentially the associated adverse effects while attaining the same glycemic control in type 2 diabetics (Sukma Rita et al. 2015).

In addition to improving glucose tolerance and enhancing insulin secretion, inhibition of $K_v2.1$ by the selective inhibitor, SP6616, in diabetic mice has been shown to improve the diabetic peripheral neuropathy-like pathology of the mice (Zhu et al. 2020), and a summary of selected $K_v2.1$ inhibitors and their proposed role is listed in Table 1. For a more comprehensive picture of $K_v2.1$ pharmacological tools, readers are signposted to the International Union of Basic and Clinical Pharmacology (IUPHAR) database (Attali et al. 2019).

7 Virology

Antiapoptotic strategies are also employed by viruses, and there is growing evidence that viruses manipulate K^+ homeostasis to their advantage for a number of processes from infection to replication (Gazzarrini et al. 2003; Gargan and Stevenson 2021). There is also a body of evidence that supports the idea that a viral-encoded K^+ protein might be the K^+ channel protein evolutionary progenitor (Thiel et al. 2013). Modulating K^+ efflux through manipulation of $K_v2.1$ has been proposed to be an effective strategy used by the hepatitis C virus to maintain infectivity (Mankouri et al. 2009). This original observation indicated that the $K_v2.1$ channel had viral antiapoptotic properties, and this has been developed further to indicate specific signaling mechanisms (Amako et al. 2013; Clemens et al. 2015). These observations have been used in generating a novel neuroprotective strategy involving the interaction of the hepatitis C protein (NS5A) and the $K_v2.1$ protein (Justice et al. 2018).

Table 1 Selected summary list of investigated $K_v2.1$ inhibitors and their proposed/reposted physiological role

Compound	Reported physiological role	Concentration	System tested	In vitro/ in vivo	Reference
Tetraethyl ammonium	Blocked excessive proapoptotic K^+ efflux	5 mM	Murine cortical cultures	In vitro	Yu et al. (1997) and Wei et al. (2003)
	Reduced cerebral infarct volume	5 μg/kg	Ischemic rats	In vivo	Wei et al. (2003)
Donepezil	Reduced oxygen-glucose deprivation-induced apoptosis	30 μM	HEK293 expressing $K_v2.1$	In vitro	Yuan et al. (2011)
Rivastigmine	Improved proteasome activity contributing to amyloid aggregation removal	2 mg/kg	Rat cortex and hippocampus	In vivo	Gupta et al. (2021)
MiDCA1 toxin	Induced hydrolysis of phosphatidylcholine	1 μM	Isolated murine dorsal root ganglion neurons	In vitro	Schütter et al. (2019)
			Xenopus oocyte expressing $K_v2.1$	In vitro	
Hanatoxin	Blocked K^+ channels	0.5 μM	Xenopus oocyte expressing $K_v2.1$	In vitro	Swartz and MacKinnon (1995)
	Reduced proliferation	0.2–1 μM	Uterine cancer cells	In vivo	Suzuki and Takimoto (2004)
48F10	Suppressed oxidative induced apoptosis	1–3 μM	Rat cortical neurons	In vitro	Zaks-Makhina et al. (2004)
		IC_{50} 1 μM	CHO cells expressing $K_v2.1$	In vitro	
		10 μM	IEC-6 enterocytes	In vitro	Grishin et al. (2005)
SP6616	Promoted glucose-stimulated insulin secretion (GSIS); promoted β-cell survival	1–10 μM	INS-1832/13 rat Insulinoma cell line	In vitro	Zhou et al. (2016)
	Ameliorated hyperglycemia and β-cell dysfunction	IC_{50} 2.58 μM	CHO cells expressing $K_v2.1$	In vitro	
	Ameliorated peripheral neuropathy	20 and 40 mg/kg	Diabetic mice	In vivo	Zhou et al. (2018)

(continued)

Table 1 (continued)

Compound	Reported physiological role	Concentration	System tested	In vitro/ in vivo	Reference
ETA	Ameliorated hyperglycemia and improved β-cell dysfunction	20 mg/kg/day	Type 2 diabetic mice	In vivo	Zhou et al. (2018)
		20 μM	INS-1832/13 rat insulinoma cell line	In vitro	
Vindoline	Stimulates insulin secretion and enhanced GSIS	20 mg/kg	Diabetic models (mice) type 2 diabetic (rats)	In vivo	Yao et al. (2013)
		IC_{50} 50 μM	MIN6 mouse insulinoma cell line and primary mice islets	In vitro	
Stromatoxin-1	Modulated insulin secretion	100 nM	Human β-cells	In vitro	Braun et al. (2008)
	Modulation of myogenic contraction in concert with BK channels	300 nM	Rabbit urethral smooth muscle cells	In vitro	Kyle et al. (2011)
	Blockade of channel pore	150 nM	HEK293 expressing $K_v2.1$	In vitro	
	Reduced proliferation	600 nM	Prostate cancer cells	In vitro	Park et al. (2021)
Perifosine	Inactivation gate modifier	0.3 μM	HEK293 expressing $K_v2.1$	In vitro	Delgado-Ramírez et al. (2016)

8 Pharmacological Approaches to Influence $K_v2.1$ Activity

Given the diverse array of physiological processes that $K_v2.1$ are involved in, there is a need for distinct pharmacological tools to modulate channel activity in an appropriate manner to offer therapeutic potential. Here we consider some options available that could be used to target diseases where $K_v2.1$ activity has been shown to regulate pathophysiology (see above).

9 Channel Blockers

K^+ channel blockers such as TEA and 4-AP have been used to inhibit $K_v2.1$ to achieve an array of therapeutic goals. For example, direct channel blockade offered protection against apoptosis accelerated by a range of insults, including oxidative stress (Redman et al. 2007). In addition, 4-AP has been used in the management of multiple sclerosis (Henney 3rd and Blight 2012), so there is a place for pan K_v blockers in the clinical, but how much of the 4-AP effect is mediated by $K_v2.1$ inhibition is unclear. So, while pan K_v channel blockers have their place in the clinical, they are likely to report a clinical benefit based on multiple drug-target interactions. Therefore, more selective tools are required to influence $K_v2.1$ channels to offer clinical potential. Here, more specific $K_v2.1$ tools have been used in highlighting discreet roles for the $K_v2.1$ protein (e.g., 48F10, Zaks-Makhina et al. 2004; Table 1). Of note here is that the 48F10 compound showed some selectivity against other K_v channels ($K_v1.5$). 48F10 demonstrated inhibitory actions on the Kir2.1 channel. However, most of these new chemical entities are still experimental compounds, and further research is needed to explore their full target profiles.

10 Channel Modulators

Channel blockade may afford therapeutic potential for broad spectrum conditions; however, another more targeted strategy would be the modulation of individual $K_v2.1$ channels. There are several approaches that have been tested that range from channel trafficking to channel gating modifications. $K_v2.1$ protein trafficking is an important determinant of channel assembly and insertion of operational channels into the plasma membrane. Removal of K^+ conductance pathways via $K_v2.1$ could be achieved by disrupting channel clustering that appears to act as supply platform for the channel's trafficking to the plasma membrane (Fox et al. 2015). Indeed, research has highlighted that this approach, targeted by calcineurin-mediated dephosphorylation, disrupted $K_v2.1$ channel clustering and afforded greater neuroprotection (Misonou et al. 2004; Shepherd et al. 2012).

Small molecule mediated gating modification has also been demonstrated to offer therapeutic potential. This has been reported for experimental compounds such as the tarantula guangxitoxin-1E (GxTx) which modulates $K_v2.1$ activity by increasing time in the closed state (Tilley et al. 2019). This has led to the development of high throughput screens looking to identify novel gating modifiers (Schmalhofer et al. 2009). While the search for novel gating modifiers is underway, existing medicines have been reported to act through $K_v2.1$ dependent mechanisms. Evidence suggests that the existing antitumor drug perifosine can alter the inactivation gate with the $K_v2.1$ channel complex leading to a hyperpolarizing shift in the voltage dependence of $K_v2.1$ (Delgado-Ramírez et al. 2016). This $K_v2.1$ modulation may underpin the antitumor effects in tandem with the more widely reported perifosine mediated Akt inhibition (Taniguchi et al. 2021).

In summary, the $K_v2.1$ channel is a promising drug target in a range of cell types to address distinct therapeutic areas. The work exploring modulation of $K_v2.1$ in diabetes has provided a pathway to research and development programs. Evidence indicates that blockade of $K_v2.1$ channels in β-cells could promote insulin release with a lower risk of hypoglycemic events compared with the more appreciated K_{ATP}

channel blockers (e.g., sulfonylureas). Suppression of channel activity to a level above the reported IC_{50} threshold would be more advantageous as opposed to full blockade and depending on the route of administration could mitigate other disease related pathology, such as diabetes associated neuropathy (Xu et al. 2020). This chapter also indicates interesting datasets about drug repurposing and some previously unreported effects on the $K_v2.1$ channel that could enhance clinical benefits of currently approved medicines. With respect to the other therapeutic areas highlighted in this chapter (e.g., neurological), more detailed structural and functional studies to provide the cell and channel specific reagents are needed to realize $K_v2.1$ targeted therapies.

References

Al-Owais MM, Scragg JL, Dallas ML, Boycott HE, Warburton P, Chakrabarty A, Boyle JP, Peers C. Carbon monoxide mediates the anti-apoptotic effects of heme oxygenase-1 in medulloblastoma DAOY cells via K+ channel inhibition. J Biol Chem. 2012;287:24754–64.

Amako Y, Igloi Z, Mankouri J, Kazlauskas A, Saksela K, Dallas M, Peers C, Harris M. Hepatitis C virus NS5A inhibits mixed lineage kinase 3 to block apoptosis. J Biol Chem. 2013;288(34):24753–63.

Aras MA, Aizenman E. Obligatory role of ASK1 in the apoptotic surge of K+ currents. Neurosci Lett. 2005;387(3):136–40.

Attali B, Chandy KG, Giese MH, Grissmer S, Gutman GA, Jan LY, Lazdunski M, Mckinnon D, Nerbonne J, Pardo LA, Robertson GA, Rudy B, Sanguinetti MC, Stühmer W, Trimmer JS, Wang X. Voltage-gated potassium channels (K_v) in the IUPHAR/BPS guide to pharmacology database. IUPHAR/BPS guide to pharmacology CITE. 2019;2019(3).

Bersell KR, Jorge BS, Kearney JA, George AL Jr. Epilepsy-associated point mutation in the pore domain of $K_v2.1$. Biophys J. 2014;106(2. Suppl):433a.

Braun M, Ramracheya R, Bengtsson M, Zhang Q, Karanauskaite J, Partridge C, Johnson PR, Rorsman P. Voltage-gated ion channels in human pancreatic beta-cells: electrophysiological characterization and role in insulin secretion. Diabetes. 2008;57:1618–28.

Butler A, Wei AG, Baker K, Salkoff LA. Family of putative potassium channel genes in drosophila. Science. 1989;243:943–7.

Cerda O, Trimmer JS. Activity-dependent phosphorylation of neuronal $K_v2.1$ potassium channels by CDK5. J Biol Chem. 2011;286:28738–48.

Chae YJ, Lee HJ, Jeon JH, Kim IB, Choi JS, Sung KW, Hahn SJ. Effects of donepezil on hERG potassium channels. Brain Res. 2015;1597:77–85.

Clemens K, Yeh CY, Aizenman E. Critical role of casein kinase 2 in hepatitis C NS5A-mediated inhibition of $K_v2.1$ K(+) channel function. Neurosci Lett. 2015;609:48–52.

Dai XQ, Manning Fox JE, Chikvashvili D, Casimir M, Plummer G, Hajmrle C, Spigelman AF, Kin T, Singer-Lahat D, Kang Y, Shapiro AM, Gaisano HY, Lotan I, Macdonald PE. The voltage-dependent potassium channel subunit $K_v2.1$ regulates insulin secretion from rodent and human islets independently of its electrical function. Diabetologia. 2012;55:1709–20.

Dallas ML, Al-Owais MM, Hettiarachchi NT, Vandiver MS, Jarosz-Griffiths HH, Scragg JL, Boyle JP, Steele D, Peers C. Hydrogen sulfide regulates hippocampal neuron excitability via S-sulfhydration of $K_v2.1$. Sci Rep. 2021;11:8194.

Dawson CM, Atwater I, Rojas E. The response of pancreatic beta-cell membrane potential to potassium-induced calcium influx in the presence of glucose. Q J Exp Physiol. 1984;69:819–30.

Delgado-Ramírez M, Morán-Zendejas R, Aréchiga-Figueroa IA, Toro-Castillo C, Ramírez-Martínez JF, Rodríguez-Menchaca AA. Modulation of the voltage-gated potassium channel $K_v2.1$ by the anti-tumor alkylphospholipid perifosine. Pharmacol Rep. 2016;68(2):457–61.

Du J, Tao-Cheng JH, Zerfas P, McBain CJ. The K+ channel, $K_v2.1$, is apposed to astrocytic processes and is associated with inhibitory postsynaptic membranes in hippocampal and cortical principal neurons and inhibitory interneurons. Neuroscience. 1998;84:37–48.

Du J, Haak LL, Phillips-Tansey E, Russell JT, McBain CJ. Frequency-dependent regulation of rat hippocampal somato-dendritic excitability by the K^+ channel subunit $K_v2.1$. J Physiol. 2000;522:19–31.

Fox PD, Haberkorn CJ, Akin EJ, Seel PJ, Krapf D, Tamkun MM. Induction of stable ER-plasma-membrane junctions by $K_v2.1$ potassium channels. J Cell Sci. 2015;128:2096–105.

Frech GC, VanDongen AM, Schuster G, Brown AM, Joho RH. A novel potassium channel with delayed rectifier properties isolated from rat brain by expression cloning. Nature. 1989;340:642–5.

Fu J, Dai X, Plummer G, Suzuki K, Bautista A, Githaka JM, Senior L, Jensen M, Greitzer-Antes D, Manning Fox JE, Gaisano HY, Newgard CB, Touret N, MacDonald PE. $K_v2.1$ clustering contributes to insulin exocytosis and rescues human β-cell dysfunction. Diabetes. 2017;66:1890–900.

Fu J, Githaka JM, Dai X, Plummer G, Suzuki K, Spigelman AF, Bautista A, Kim R, Greitzer-Antes D, Fox JEM, Gaisano HY, MacDonald PE. A glucose-dependent spatial patterning of exocytosis in human β-cells is disrupted in type 2 diabetes. JCI Insight. 2019;5:e127896.

Gargan S, Stevenson NJ. Unravelling the immunomodulatory effects of viral ion channels, towards the treatment of disease. Viruses 27. 2021;13(11):2165.

Gazzarrini S, Severino M, Lombardi M, Morandi M, DiFrancesco D, Van Etten JL, Thiel G, Moroni A. The viral potassium channel Kcv: structural and functional features. FEBS Lett 18. 2003;552(1):12–6.

Greitzer-Antes D, Xie L, Qin T, Xie H, Zhu D, Dolai S, Liang T, Kang F, Hardy AB, He Y, Kang Y, Gaisano HY. $K_v2.1$ clusters on β-cell plasma membrane act as reservoirs that replenish pools of newcomer insulin granule through their interaction with syntaxin-3. J Biol Chem. 2018;293:6893–904.

Grishin A, Ford H, Wang J, Li H, Salvador-Recatala V, Levitan ES, Zaks-Makhina E. Attenuation of apoptosis in enterocytes by blockade of potassium channels. Am J Physiol Gastrointest Liver Physiol. 2005;289(5):G815–21.

Grizel A, Popinako A, Kasimova MA, Stevens L, Karlova M, Moisenovich MM, Sokolova OS. Domain structure and conformational changes in rat $K_v2.1$ ion channel. J Neuroimmune Pharmacol. 2014;9:727–39.

Gupta P, Tiwari S, Singh A, Pal A, Mishra A, Singh S. Rivastigmine attenuates the Alzheimer's disease related protein degradation and apoptotic neuronal death signalling. Biochem J. 2021;478(7):1435–51.

Gutman GA, Chandy KG, Adelman JP, Aiyar J, Bayliss DA, Clapham DE, Covarriubias M, Desir GV, Furuichi K, Ganetzky B, Garcia ML, Grissmer S, Jan LY, Karschin A, Kim D, Kuperschmidt S, Kurachi Y, Lazdunski M, Lesage F, Lester HA, McKinnon D, Nichols CG, O'Kelly I, Robbins J, Robertson GA, Rudy B, Sanguinetti M, Seino S, Stuehmer W, Tamkun MM, Vandenberg CA, Wei A, Wulff H, Wymore RS. International Union of Pharmacology. XLI. Compendium of voltage-gated ion channels: potassium channels. Pharmacol Rev. 2003;55:583–6.

He K, McCord MC, Hartnett KA, Aizenman E. Regulation of pro-apoptotic phosphorylation of $K_v2.1$ K+ channels. PLoS One. 2015;10:e0129498.

Heginbotham L, Lu Z, Abramson T, MacKinnon R. Mutations in the K+ channel signature sequence. Biophys J. 1994;66:1061–7.

Henney HR 3rd, Blight AR. Walking impairment in patients with multiple sclerosis - a new therapeutic approach and clinical potential of dalfampridine extended release tablets. Degener Neurol Neuromuscul Dis. 2012;2:53–64.

Hwang PM, Glatt CE, Bredt DS, Yellen G, Snyder SH. A novel K+ channel with unique localizations in mammalian brain: molecular cloning and characterization. Neuron. 1992;8:473–81.

Ikematsu N, Dallas ML, Ross FA, Lewis RW, Rafferty JN, David JA, Suman R, Peers C, Hardie DG, Evans AM. Phosphorylation of the voltage-gated potassium channel $K_v2.1$ by AMP-activated protein kinase regulates membrane excitability. Proc Natl Acad Sci U S A. 2011;108:18132–7.

Jacobson DA, Philipson LH. Action potentials and insulin secretion: new insights into the role of K_v channels. Diabetes Obes Metab. 2007;9:89–98.

Jacobson DA, Kuznetsov A, Lopez JP, Kash S, Ammälä CE, Philipson LH. $K_v2.1$ ablation alters glucose-induced islet electrical activity, enhancing insulin secretion. Cell Metab. 2007;6:229–35.

Jin HW, Zhang W, Wang XL. Effects of chronic exposure to beta-amyloid-peptide25–35 on the mRNA expressions of voltage-gated outward potassium channel subunits in cultured rat hippocampal neurons. Acta Pharm Sin. 2002;37:598–602.

Ju M, Stevens L, Leadbitter E, Wray D. The roles of N- and C-terminal determinants in the activation of the $K_v2.1$ potassium channel. J Biol Chem. 2003;278:12769–78.

Justice JA, Manjooran DT, Yeh CY, Hartnett-Scott KA, Schulien AJ, Kosobucki GJ, Mammen S, Palladino MJ, Aizenman E. Molecular neuroprotection induced by zinc-dependent expression of hepatitis C-derived protein NS5A targeting $K_v2.1$ potassium channels. J Pharmacol Exp Ther. 2018;367(2):348–55.

Klemic KG, Kirsch GE, Jones SW. U-type inactivation of $K_v3.1$ and shaker potassium channels. Biophys J. 2001;81:814–26.

Kyle B, Bradley E, Ohya S, Sergeant GP, McHale NG, Thornbury KD, Hollywood MA. Contribution of $K_v2.1$ channels to the delayed rectifier current in freshly dispersed smooth muscle cells from rabbit urethra. Am J Physiol Cell Physiol. 2011;301(5):C1186–200.

Li X, Herrington J, Petrov A, Ge L, Eiermann G, Xiong Y, Jensen MV, Hohmeier HE, Newgard CB, Garcia ML, Wagner M, Zhang BB, Thornberry NA, Howard AD, Kaczorowski GJ, Zhou YP. The role of voltage-gated potassium channels $K_v2.1$ and $K_v2.2$ in the regulation of insulin and somatostatin release from pancreatic islets. J Pharmacol Exp Ther. 2013;344:407–16.

Li K, Cheng N, Li XT. Inhibitory effects of cholinesterase inhibitor donepezil on the $K_v1.5$ potassium channel. Sci Rep. 2017.

Lim ST, Antonucci DE, Scannevin RH, Trimmer JS. A novel targeting signal for proximal clustering of the $K_v2.1$ K+ channel in hippocampal neurons. Neuron. 2000;25:385–97.

MacDonald PE, Wheeler MB. Voltage-dependent K(+) channels in pancreatic beta cells: role, regulation and potential as therapeutic targets. Diabetologia. 2003;46:1046–62.

MacDonald PE, Ha XF, Wang J, Smukler SR, Sun AM, Gaisano HY, Salapatek AM, Backx PH, Wheeler MB. Members of the K_v1 and K_v2 voltage-dependent K(+) channel families regulate insulin secretion. Mol Endocrinol. 2001;15:1423–35.

MacDonald PE, Sewing S, Wang J, Joseph JW, Smukler SR, Sakellaropoulos G, Wang J, Saleh MC, Chan CB, Tsushima RG, Salapatek AM, Wheeler MB. Inhibition of $K_v2.1$ voltage-dependent K+ channels in pancreatic beta-cells enhances glucose-dependent insulin secretion. J Biol Chem. 2002;277:44938–45.

Maletic-Savatic M, Lenn NJ, Trimmer JS. Differential spatiotemporal expression of K+ channel polypeptides in rat hippocampal neurons developing in situ and in vitro. J Neurosci. 1995;15:3840–51.

Mankouri J, Dallas ML, Hughes ME, Griffin SD, Macdonald A, Peers C, Harris M. Suppression of a pro-apoptotic K+ channel as a mechanism for hepatitis C virus persistence. Proc Natl Acad Sci U S A. 2009;106(37):15903–8.

Misonou H, Trimmer JS. Determinants of voltage-gated potassium channel surface expression and localization in mammalian neurons. Crit Rev Biochem Mol Biol. 2004;39:125–45.

Misonou H, Mohapatra DP, Park EW, Leung V, Zhen D, Misonou K, Anderson AE, Trimmer JS. Regulation of ion channel localization and phosphorylation by neuronal activity. Nat Neurosci. 2004;7:711–8.

Misonou H, Mohapatra DP, Menegola M, Trimmer JS. Calcium- and metabolic state-dependent modulation of the voltage-dependent $K_v2.1$ channel regulates neuronal excitability in response to ischemia. J Neurosci. 2005;25:11184–93.

Mohapatra DP, Park KS, Trimmer JS. Dynamic regulation of the voltage-gated $K_v2.1$ potassium channel by multisite phosphorylation. Biochem Soc Trans. 2007;35:1064–8.

Murakoshi H, Trimmer JS. Identification of the $K_v2.1$ K+ channel as a major component of the delayed rectifier K+ current in rat hippocampal neurons. J Neurosci. 1999;19:1728–35.

Murakoshi H, Shi G, Scannevin RH, Trimmer JS. Phosphorylation of the $K_v2.1$ K+ channel alters voltage-dependent activation. Mol Pharmacol. 1997;52:821–8.

Pal S, Hartnett KA, Nerbonne JM, Levitan ES, Aizenman E. Mediation of neuronal apoptosis by $K_v2.1$-encoded potassium channels. J Neurosci. 2003;23:4798–802.

Pal SK, Takimoto K, Aizenman E, Levitan ES. Apoptotic surface delivery of K+ channels. Cell Death Differ. 2006;13:661–7.

Pan Y, Xu X, Wang X. Rivastigmine blocks voltage-activated K+ currents in dissociated rat hippocampal neurons. Br J Pharmacol. 2003;140:907–12.

Park HW, Song MS, Sim HJ, Ryu PD, Lee SY. The role of the voltage-gated potassium channel, $K_v2.1$ in prostate cancer cell migration. BMB Rep. 2021;54(2):130–5.

Plant LD, Dowdell EJ, Dementieva IS, Marks JD, Goldstein SA. SUMO modification of cell surface $K_v2.1$ potassium channels regulates the activity of rat hippocampal neurons. J Gen Physiol. 2011;137:441–54.

Redman PT, He K, Hartnett KA, Jefferson BS, Hu L, Rosenberg PA, Levitan ES, Aizenman E. Apoptotic surge of potassium currents is mediated by p38 phosphorylation of $K_v2.1$. Proc Natl Acad Sci U S A. 2007;104:3568–73.

Redman PT, Hartnett KA, Aras MA, Levitan ES, Aizenman E. Regulation of apoptotic potassium currents by coordinated zinc-dependent signaling. J Physiol. 2009;587:4393–404.

Rettig J, Heinemann SH, Wunder F, Lorra C, Parcej DN, Dolly JO, Pongs O. Inactivation properties of voltage-gated K+ channels altered by presence of beta-subunit. Nature. 1994;369(6478):289–94.

Roe MW, Worley JF III, Mittal AA, Kuznetsov A, DasGupta S, Mertz RJ, Witherspoon SM 3rd, Blair N, Lancaster ME, McIntyre MS, Shehee WR, Dukes ID, Philipson LH. Expression and function of pancreatic beta-cell delayed rectifier K+ channels. Role in stimulus-secretion coupling. J Biol Chem. 1996;271:32241–6.

Schmalhofer WA, Ratliff KS, Weinglass A, Kaczorowski GJ, Garcia ML, Herrington. A $K_v2.1$ gating modifier binding assay suitable for high throughput screening. J Channels (Austin). 2009;3(6):437–47.

Schulien AJ, Yeh CY, Orange BN, Pav OJ, Hopkins MP, Moutal A, Khanna R, Sun D, Justice JA, Aizenman E. Targeted disruption of $K_v2.1$-VAPA association provides neuroprotection against ischemic stroke in mice by declustering $K_v2.1$ channels. Sci Adv 1. 2020;6(27):eaaz8110.

Schütter N, Barreto YC, Vardanyan V, Hornig S, Hyslop S, Marangoni S, Rodrigues-Simioni L, Pongs O, Dal Belo CA. Inhibition of $K_v2.1$ potassium channels by MiDCA1, a pre-synaptically active PLA2-type toxin from Micrurus dumerilii carinicauda Coral Snake venom. Toxins (Basel). 2019;11(6):335.

Selkoe DJ. The molecular pathology of Alzheimer's disease. Neuron. 1991;6:487–98.

Shepherd AJ, Loo L, Gupte RP, Mickle AD, Mohapatra DP. Distinct modifications in $K_v2.1$ channel via chemokine receptor CXCR4 regulate neuronal survival-death dynamics. J Neurosci. 2012;32:17725–39.

Speca DJ, Ogata G, Mandikian D, Bishop HI, Wiler SW, Eum K, Wenzel HJ, Doisy ET, Matt L, Campi KL, Golub MS, Nerbonne JM, Hell JW, Trainor BC, Sack JT, Schwartzkroin PA, Trimmer JS. Deletion of the $K_v2.1$ delayed rectifier potassium channel leads to neuronal and behavioral hyperexcitability. Genes Brain Behav. 2014;13(4):394–408.

Sukma Rita R, Dezaki K, Kurashina T, Kakei M, Yada T. Partial blockade of $K_v2.1$ channel potentiates GLP-1's insulinotropic effects in islets and reduces its dose required for improving glucose tolerance in type 2 diabetic male mice. Endocrinology. 2015;156:114–23.

Suzuki T, Takimoto K. Selective expression of HERG and K_v2 channels influences proliferation of uterine cancer cells. Int J Oncol. 2004;25(1):153–9.

Swartz KJ, MacKinnon R. An inhibitor of the $K_v2.1$ potassium channel isolated from the venom of a Chilean tarantula. Neuron. 1995;15:941–9.

Tamarina NA, Kuznetsov A, Fridlyand LE, Philipson LH. Delayed-rectifier ($K_v2.1$) regulation of pancreatic beta-cell calcium responses to glucose: inhibitor specificity and modeling. Am J Physiol Endocrinol Metab. 2005;289:E578–85.

Tamkun MM, O'connell KM, Rolig AS. A cytoskeletal-based perimeter fence selectively corrals a sub-population of cell surface $K_v2.1$ channels. J Cell Sci. 2007;120:2413–23.

Taniguchi K, Suzuki T, Okamura T, Kurita A, Nohara G, Ishii S, Kado S, Takagi A, Tsugane M, Shishido Y. Perifosine, a bioavailable alkylphospholipid Akt inhibitor, exhibits antitumor activity in murine models of cancer brain metastasis through favorable tumor exposure. Front Oncol. 2021;11:754365.

Thiel G, Moroni A, Blanc G, Van Etten JL. Potassium ion channels: could they have evolved from viruses? Plant Physiol. 2013;162(3):1215–24.

Tilley DC, Angueyra JM, Eum KS, Kim H, Chao LH, Peng AW, Sack JT. The tarantula toxin GxTx detains K^+ channel gating charges in their resting conformation. J Gen Physiol. 2019;151(3):292–315.

Uhlén M, Fagerberg L, Hallström BM, Lindskog C, Oksvold P, Mardinoglu A, Sivertsson Å, Kampf C, Sjöstedt E, Asplund A, Olsson I, Edlund K, Lundberg E, Navani S, Szigyarto CA, Odeberg J, Djureinovic D, Takanen JO, Hober S, Alm T, Edqvist PH, Berling H, Tegel H, Mulder J, Rockberg J, Nilsson P, Schwenk JM, Hamsten M, von Feilitzen K, Forsberg M, Persson L, Johansson F, Zwahlen M, von Heijne G, Nielsen J, Pontén F. Proteomics. Tissue-based map of the human proteome. Science 23. 2015;347(6220):1260419.

Wei L, Yu SP, Gottron F, Snider BJ, Zipfel GJ, Choi DW. Potassium channel blockers attenuate hypoxia- and ischemia-induced neuronal death in vitro and in vivo. Stroke. 2003;34(5):1281–6.

Xu X, Xu X, Hao Y, Zhu X, Lu J, Ouyang X, Lu Y, Huang X, Li Y, Wang J, Shen X. Antispasmodic drug drofenine as an inhibitor of $K_v2.1$ channel ameliorates peripheral neuropathy in diabetic mice. iScience. 2020;23(10):101617.

Xu Z, Khan S, Schnicker NJ, Baker S. Pentameric assembly of the $K_v2.1$ tetramerization domain. Acta Crystallogr D Struct Biol. 2022;78:792–802.

Yan L, Figueroa DJ, Austin CP, Liu Y, Bugianesi RM, Slaughter RS, et al. Expression of voltage-gated potassium channels in human and rhesus pancreatic islets. Diabetes. 2004;53:597–607.

Yao XG, Chen F, Li P, Quan L, Chen J, Yu L, Ding H, Li C, Chen L, Gao Z, Wan P, Hu L, Jiang H, Shen X. Natural product vindoline stimulates insulin secretion and efficiently ameliorates glucose homeostasis in diabetic murine models. J Ethnopharmacol. 2013;150:285–97.

Yeh CY, Bulas AM, Moutal A, Saloman JL, Hartnett KA, Anderson CT, Tzounopoulos T, Sun D, Khanna R, Aizenman E. Targeting a potassium channel/syntaxin interaction ameliorates cell death in ischemic stroke. J Neurosci. 2017;37:5648–58.

Yu SP, Yeh CH, Sensi SL, Gwag BJ, Canzoniero LMT, Farhangrazi ZS, Ying HS, Tian M, Dugan LL, Choi DW. Mediation of neuronal apoptosis by enhancement of outward potassium current. Science. 1997;278:114–7.

Yuan H, Wang WP, Feng N, Wang L, Wang XL. Donepezil attenuated oxygen-glucose deprivation insult by blocking $K_v2.1$ potassium channels. Eur J Pharmacol. 2011;657:76–83.

Zaks-Makhina E, Kim Y, Aizenman E, Levitan ES. Novel neuroprotective K+ channel inhibitor identified by high-throughput screening in yeast. Mol Pharmacol. 2004;65:214–9.

Zhang HX, Zhang W, Jin HW, Wang XL. Galantamine blocks cloned $K_v2.1$, but not $K_v1.5$ potassium channels. Brain Res Mol Brain Res. 2004;131:136–40.

Zhong CB, Pan YP, Tong XY, Xu XH, Wang XL. Delayed rectifier potassium currents and $K_v2.1$ mRNA increase in hippocampal neurons of scopolamine-induced memory-deficient rats. Neurosci Lett. 2005;373:99–104.

Zhou TT, Quan LL, Chen LP, Du T, Sun KX, Zhang JC, Yu L, Li Y, Wan P, Chen LL, Jiang BH, Hu LH, Chen J, Shen X. SP6616 as a new $K_v2.1$ channel inhibitor efficiently promotes β-cell survival involving both PKC/Erk1/2 and CaM/PI3K/Akt signaling pathways. Cell Death Dis. 2016;7:e2216.

Zhou T, Du M, Zhao T, Quan L, Zhu Z, Chen J. ETA as a novel $K_v2.1$ inhibitor ameliorates β-cell dysfunction and hyperglycaemia. Clin Exp Pharmacol Physiol. 2018;45:1257–64.

Zhu X, Chen Y, Xu X, Xu X, Lu Y, Huang X, Zhou J, Hu L, Wang J, Shen X. SP6616 as a $K_v2.1$ inhibitor efficiently ameliorates peripheral neuropathy in diabetic mice. EBioMedicine. 2020;61:103061.

Cystic Fibrosis Transmembrane Conductance Regulator (CFTR)-Targeted Therapeutics for Cystic Fibrosis

Diana-Florentina Veselu, Han-I Yeh, Mayuree Rodrat, Jacob D. Bradbury, Yanira Manzanares Fornies, Hongyu Li, Deborah K. Shoemark, A. Sofia F. Oliveira, Tzyh-Chang Hwang, and David N. Sheppard

Abstract Treatment of the common life-limiting genetic disease cystic fibrosis (CF) has been transformed by small molecule therapeutics that target the root cause of the disease, pathogenic variants in the ATP-binding cassette transporter cystic fibrosis transmembrane conductance regulator (CFTR). Here, we selectively review understanding of CFTR function as an ATP-gated anion channel, its physiological role in epithelia and the molecular mechanisms of CFTR dysfunction, which paved the way for the development of CFTR modulators. These orally bioavailable small molecules include CFTR correctors, which improve CFTR folding and domain assembly to facilitate its transport to the plasma membrane and CFTR potentiators, which enhance channel gating to augment CFTR function. Elexacaftor-tezacaftor-ivacaftor, the combination of two CFTR correctors and one CFTR potentiator, is a highly effective therapeutic for most, but not all, people with CF. Ongoing research aims to provide highly effective therapeutics for all people with CF and ultimately cure the disease.

Keywords Cystic fibrosis · CFTR · CFTR correctors · CFTR potentiators · *CFTR* variants

D.-F. Veselu · D. K. Shoemark
School of Biochemistry, University of Bristol, Bristol, UK
e-mail: diana.veselu@bristol.ac.uk; Deb.Shoemark@bristol.ac.uk

H.-I. Yeh
Department of Pharmacology, National Yang Ming Chiao Tung University, Taipei, City, Taiwan

Dalton Cardiovascular Research Center, University of Missouri-Columbia, Columbia, MO, USA
e-mail: hyc3c@mail.missouri.edu

G. Stephens, E. Stevens (eds.), *Ion Channels as Targets in Drug Discovery*,
https://doi.org/10.1007/978-3-031-52197-3_8

1 Introduction

The last 10 years has witnessed a revolution in the treatment of the common life-limiting genetic disease cystic fibrosis (CF). Before 2012, evidence-based symptomatic therapy was the standard of care for people with CF (Ratjen et al. 2015; Elborn 2016; Ramsey and Welsh 2017). These treatments included time-consuming chest physiotherapy and an arsenal of drugs to clear mucus obstructed airways, fight lung infection, dampen airway inflammation and replace missing pancreatic enzymes. Together with specialist care from multidisciplinary CF teams, improvements in symptomatic therapy increased average life expectancy from ~10 years in the 1960s to ~40 years in 2010 in North America and Western Europe (Ramsey and Welsh 2017).

In 2012, ivacaftor (Kalydeco; VX-770; Vertex Pharmaceuticals, Boston, MA, USA) (Van Goor et al. 2009; Ramsey et al. 2011) (Fig. 1), an orally bioavailable small molecule targeting the root cause of CF, defective cystic fibrosis transmembrane conductance regulator (CFTR) (Riordan et al. 1989), was approved by the Food and Drug Administration (FDA) and the European Medicines Agency (EMA) for people with CF and the *CFTR* variant G551D (p.Gly551Asp; c.1652G>A). The clinical impact of ivacaftor was transformational with clinical trials reporting substantial health improvements compared with previous CF therapeutics (Ramsey et al. 2011). Further CFTR-targeted therapeutics quickly followed, culminating with the approval of elexacaftor-tezacaftor-ivacaftor (Trikafta/Kaftrio; VX-445–VX-661–VX-770; Vertex Pharmaceuticals) (Heijerman et al. 2019; Middleton et al. 2019) (Fig. 1) for people with CF and the predominant *CFTR* variant F508del (p.Phe508del; c.1521_1523delCTT) by the FDA in 2019 and the EMA in 2020.

M. Rodrat
School of Physiology, Pharmacology and Neuroscience, University of Bristol, Bristol, UK

Center of Research and Development for Biomedical Instrumentation, Institute of Molecular Biosciences, Mahidol University, Nakhon Pathom, Thailand
e-mail: tt18094@bristol.ac.uk

J. D. Bradbury · Y. Manzanares Fornies · H. Li · D. N. Sheppard (✉)
School of Physiology, Pharmacology and Neuroscience, University of Bristol, Bristol, UK
e-mail: jacob.bradbury@msdtc.ox.ac.uk; yanira.manzanaresfornies@bristol.ac.uk; H.Li@bristol.ac.uk; D.N.Sheppard@bristol.ac.uk

A. S. F. Oliveira
Centre for Computational Chemistry, School of Chemistry, University of Bristol, Bristol, UK
e-mail: sofia.oliveira@bristol.ac.uk

T.-C. Hwang
Department of Pharmacology, National Yang Ming Chiao Tung University, Taipei City, Taiwan

Dalton Cardiovascular Research Center and Department of Medical Pharmacology and Physiology, University of Missouri-Columbia, Columbia, MO, USA
e-mail: HwangT@health.missouri.edu

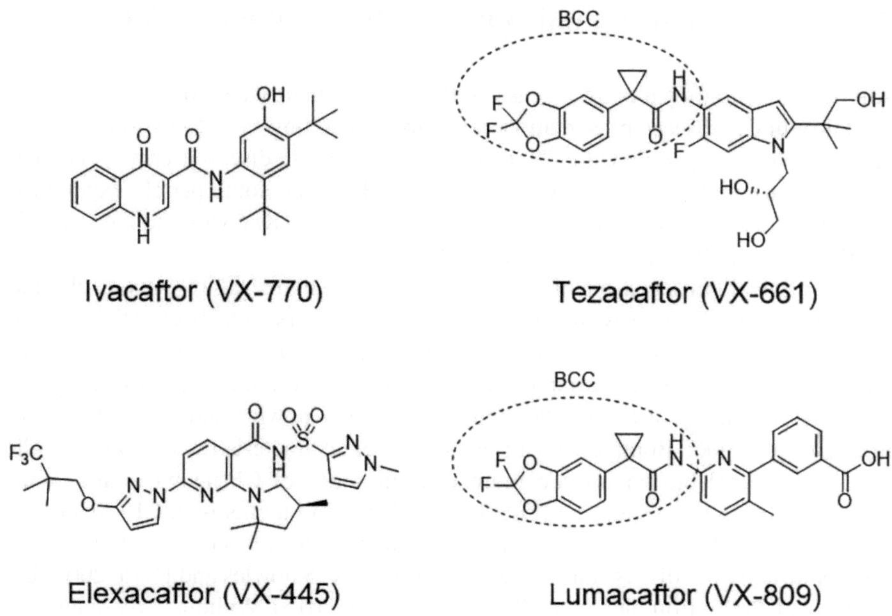

Fig. 1 Chemical structures of clinically approved CFTR modulators. Orkambi is the combination of lumacaftor and ivacaftor, Symdeko/Symkevi combines tezacaftor and ivacaftor, while Trikafta/Kaftrio combines elexacaftor, tezacaftor and ivacaftor. The 1,3-benzodioxol-5-yl-cyclopropane carboxamide (BCC) moiety of lumacaftor and tezacaftor is indicated

With the approval of elexacaftor-tezacaftor-ivacaftor, most people with CF now have an orally bioavailable, highly effective modulator therapy targeting the root cause of their disease.

For several reasons, the development of highly effective modulator therapy for CF is noteworthy (Sheppard et al. 2017). First, drug development benefited hugely from a wealth of basic science research, which identified the *CFTR* gene, demonstrated the function of the CFTR protein as an ATP-gated epithelial anion channel and elucidated how *CFTR* variants cause CFTR dysfunction. Second, drug development was assisted greatly by high-quality patient registries and clinical trial networks in North America and Europe. Third, the willingness of the CF community to provide Vertex Pharmaceuticals with advice and resources. Finally, the unique collaboration between the non-profit organisation Cystic Fibrosis Foundation (Bethesda, MD, USA) and Vertex Pharmaceuticals, which led to a new model of drug discovery and development termed venture philanthropy. Here, we review selectively the development of CFTR-targeted therapeutics for CF. For reviews of *CFTR* variant-independent strategies to develop therapeutics for all people with CF, we refer the reader to Alton et al. (2016), Li et al. (2017) and Harrison (2022).

2 The Spectrum of Cystic Fibrosis Disease Presentation

Although respiratory airway disease is the predominant cause of morbidity and mortality in CF, disease presentation is variable, including symptoms, severity and progression. Most people with CF are afflicted by a severe disease (termed classical CF) affecting multiple organs within the body. However, some people experience less severe disease (termed non-classical or atypical CF) (Knowles and Durie 2002; De Boeck et al. 2006; Bombieri et al. 2011).

The clinical presentation of classical CF is distinguished by severe chronic obstructive lung disease with mucus accumulation, persistent bacterial infection and marked inflammation, exocrine pancreatic insufficiency (PI), failure to thrive, elevated sweat Cl^- concentration (~90–110 mmol/L) and male infertility (congenital bilateral absence of the vas deferens, CBAVD) (Knowles and Durie 2002; De Boeck et al. 2006; Bombieri et al. 2011). Other characteristics of classical CF include intestinal blockage (meconium ileus in ~15% of CF neonates and distal intestinal obstructive syndrome (DIOS) in children and adults with CF), liver disease, diabetes (clinically distinct from types 1 and 2 diabetes mellitus), kidney disease, bone and joint disease and increased cancer risk (Knowles and Durie 2002; De Boeck et al. 2006; Bombieri et al. 2011; Ratjen et al. 2015). Without early diagnosis and effective treatment, the progression of classical CF is generally rapid and the prognosis is poor.

People with non-classical CF are distinguished by improved nutritional status because they retain exocrine pancreatic function (termed pancreatic sufficiency, PS) (Knowles and Durie 2002; De Boeck et al. 2006; Bombieri et al. 2011). With their more moderate disease phenotype, people with non-classical CF were diagnosed later prior to the widespread adoption of newborn screening for CF, while their survival is typically longer than people with classical CF (Knowles and Durie 2002; De Boeck et al. 2006; Bombieri et al. 2011). Although people with non-classical CF are PS and their sweat Cl^- concentration is less elevated (~60–90 mmol/L), they are susceptible to pancreatitis and still experience severe chronic obstructive lung disease together with other characteristics of the disease, including liver disease, bone and joint disease and CBAVD in males (Knowles and Durie 2002; De Boeck et al. 2006; Bombieri et al. 2011).

Some individuals with *CFTR* variants do not satisfy the diagnostic criteria for either classical or non-classical CF. They include asymptomatic neonates with a positive newborn screening result, *CFTR* variants and either a normal (<30 mmol/L) or intermediate (30–59 mmol/L) sweat Cl^- concentration. The consensus term CFTR-related metabolic syndrome/CF screen-positive inconclusive diagnosis (CRMS/CFSPID) has been adopted to identify, monitor and care for these neonates (Borowitz et al. 2009; Barben et al. 2021). Although some of these individuals might remain healthy for decades, others develop either CF or CFTR-related disorders, a group of conditions distinguished by the loss of CFTR function in a single organ (Bombieri et al. 2011; Sermet-Gaudelus et al. 2022). The most widely recognised CFTR-related disorder is CBAVD, while others include acute recurrent or chronic pancreatitis, disseminated bronchiectasis and chronic sino-pulmonary disease (Bombieri et al. 2011; Sermet-Gaudelus et al. 2022). Interestingly, using results from a population-based study, Miller et al. (2020) demonstrated that CF carriers are at increased risk of developing conditions associated with CF, including bronchiectasis, pancreatitis, diabetes and constipation. Thus, there is a wide spectrum of disorders associated with pathogenic *CFTR* variants from classical CF to CFTR-related diseases with the clinical phenotype determined by the amount of residual CFTR function (Bombieri et al. 2011).

Environmental factors, not just genetic defects, cause CFTR dysfunction, leading to human disease. The common chronic inflammatory lung disease chronic obstructive pulmonary disease (COPD) shares common pathogenic features with CF, including vicious cycles of lung damage, mucus accumulation, infection and inflammation (Rab et al. 2013). The leading risk factor for COPD, cigarette smoke, causes acquired CFTR dysfunction by reducing CFTR mRNA and protein expression and inhibiting channel gating (Solomon et al. 2017), mechanisms that mirror the impacts of pathogenic *CFTR* variants (see Sect. 4) (Fig. 2). Importantly, the clinically approved CFTR modulator ivacaftor restores channel function to wild-type CFTR inhibited by cigarette smoke (Raju et al. 2017), while a different CFTR potentiator, icenticaftor (QBW251; Novartis Institutes for Biomedical Research, Cambridge, MA, USA), demonstrated clinical efficacy in COPD (Rowe et al. 2020; Le Grand et al. 2021). Together, these studies highlight the therapeutic potential of CFTR modulators beyond the treatment of CF.

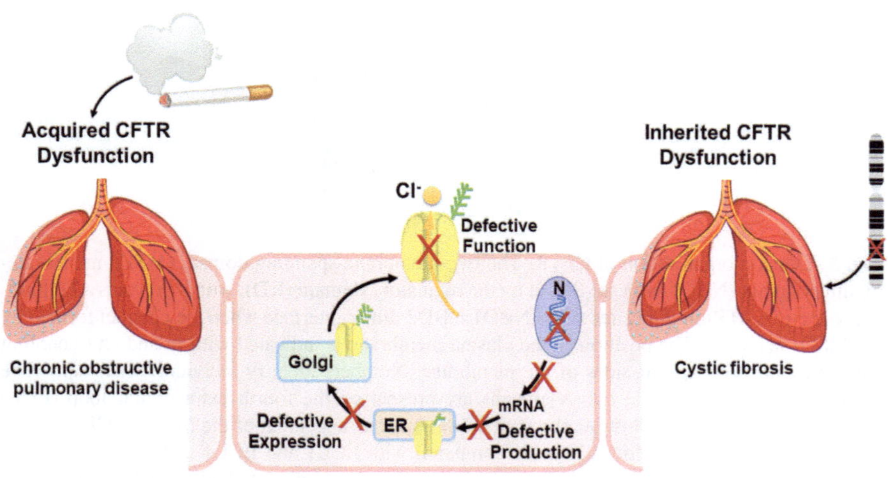

Fig. 2 Acquired and inherited CFTR dysfunction. Genetic defects in CF- and CFTR-related disorders, and cigarette smoke, the leading cause of COPD, disrupt the expression, stability and function of CFTR in airway epithelial cells leading to lung disease. Abbreviations: ER endoplasmic reticulum, N nucleus. The channels decorated with single and multi-branched structures depict the immature, core-glycosylated and mature, fully glycosylated forms of the CFTR protein, respectively. (Figure modified, with permission, from Welsh and Smith (1993))

3 CFTR Structure, Function and Physiological Role

The human *CFTR* gene located on chromosome 7 encodes a 6.5 kb mRNA that specifies the synthesis of the CFTR protein composed of 1480 amino acids assembled into 2 membrane-spanning domain-nucleotide-binding domain (MSD-NBD) units linked by a regulatory domain (RD) (Riordan et al. 1989) (Fig. 3). With two MSD-NBD units, CFTR is an ATP-binding cassette (ABC) transporter (Riordan et al. 1989; Holland et al. 2003; Thomas et al. 2020). While most ABC transporters are ATP-driven pumps that translocate diverse substrates across cellular membranes, CFTR is a unique ABC transporter that functions as an ion channel (Anderson et al. 1991b; Bear et al. 1992; Holland et al. 2003). Chen and Hwang (2008) speculated that CFTR evolved from an ancestral ABC transporter by loss of one of two gates that alternatively control access of the substrate translocation pathway formed by

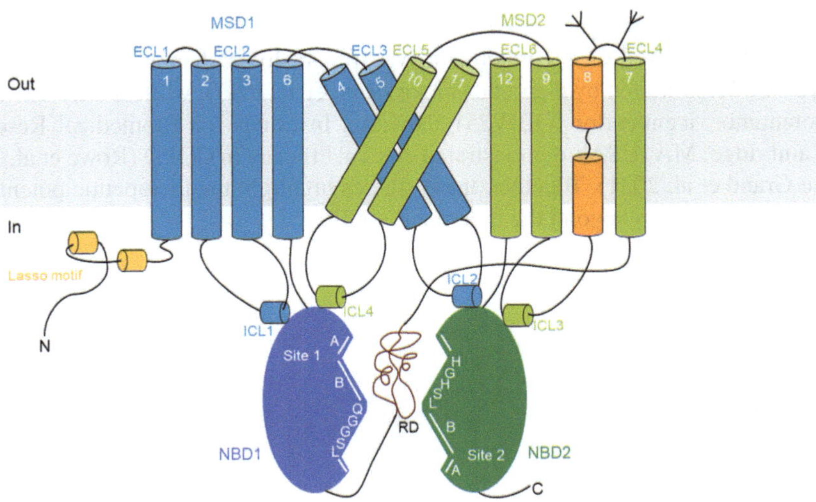

Fig. 3 Domain organisation of CFTR. The two membrane-spanning domain (MSD)-nucleotide-binding domain (NBD) units are linked by the regulatory domain (RD), which prevents ATP binding to the two ATP-binding sites at the NBD1:NBD2 dimer interface when the channel formed by the MSDs is closed. The position of the plasma membrane is indicated with In and Out denoting the intra- and extracellular sides of the membrane. Numbers identify individual transmembrane segments (M1–M12), while two *N*-glycans are present on the fourth extracellular loop (ECL) between M7 and M8. To illustrate the domain-swapped arrangement of the MSDs, ECL2, ECL3, ECL5 and ECL6 are not drawn to scale compared with ECL1 and ECL4 and ICL1–ICL4. The short cylinders in the intracellular loops (ICLs) represent the coupling helices, which mediate communication between the NBDs and the channel pore. The short cylinders in the N-terminus represent the lasso motif, which binds the RD upon channel activation. Abbreviations: A Walker A motif, B Walker B motif, C C-terminus, ECL extracellular loop, ICL intracellular loop, LSGGQ (or LSHGH) ABC signature motif, MSD membrane-spanning domain, N N-terminus, NBD nucleotide-binding domain, Site 1 ATP-binding site 1 (non-canonical), Site 2 ATP-binding site 2 (canonical). See text for further information. (Figure modified with permission of B Kleizen from Kleizen et al. (2020) and Hwang and Sheppard (2009) and Chen et al. (2006); @2006 S. Karger AG, Basel)

the MSDs to either the intra- or extracellular environment. In support of this idea, structural and functional studies revealed that CFTR possesses an external gate located at the narrowest part of the channel pore, but lacks a structure equivalent to the intracellular gate found in other ABC transporters (Gao and Hwang 2015; Zhang and Chen 2016; Liu et al. 2017; Zhang et al. 2017, 2018).

Using cryo-electron microscopy, Zhang and Chen (2016) solved the first atomic structure of CFTR: dephosphorylated zebrafish CFTR in the absence of ATP. Further high-resolution structures rapidly followed, including wild-type human CFTR in dephosphorylated ATP-free and phosphorylated ATP-bound configurations (Fig. 4) as well as wild-type and F508del human CFTR complexed with the clinically approved CFTR modulators ivacaftor, lumacaftor, tezacaftor and elexacaftor (Liu et al. 2017, 2019; Zhang et al. 2018; Fiedorczuk and Chen 2022a, b).[1] These structures verified conclusions from earlier studies, including the monomeric quaternary structure of CFTR, the intrinsically disordered nature of the RD, the organisation of the NBDs as a head-to-tail dimer and the domain-swapped arrangement of the MSDs (Ostedgaard et al. 2000; Lewis et al. 2004; Rosenberg et al. 2004; Mornon et al. 2008; Serohijos et al. 2008). For several reasons, they also provided important new insight into the relationship between CFTR structure and function. First, the N-terminus of CFTR forms a 'lasso'-shaped motif that wraps around transmembrane segments 1 (M1), M2, M10 and M11 just below the level of the plasma membrane, where it binds the RD to regulate channel gating (Liu et al. 2017; Zhang et al. 2018). Second, the RD obstructs NBD dimerisation in the dephosphorylated ATP-free configuration, but is displaced in the phosphorylated, ATP-bound configuration allowing dimerisation and channel opening (Zhang et al. 2018). Third, M8 contains an unstructured region, not found in other ABC transporters, located close to the pore constriction, suggesting that it acts as a hinge that gates the channel pore (Liu et al. 2017; Zhang et al. 2018). For detailed discussion of the relationship between CFTR structure and function, we refer the reader to Hwang et al. (2018) and Csanády et al. (2019).

CFTR is readily distinguished from other anion channels by its physiology and pharmacology. The distinguishing characteristics include:

(i) Regulation by cAMP-dependent phosphorylation and intracellular nucleotides (Anderson et al. 1991a; Berger et al. 1991; Tabcharani et al. 1991)
(ii) Linear or weakly inwardly rectifying current-voltage (I–V) relationship (Berger et al. 1991; Tabcharani et al. 1993; Cai et al. 2003) (Fig. 5)
(iii) Small single-channel conductance (6–10 pS) (Berger et al. 1991; Tabcharani et al. 1997)
(iv) Selectivity for anions over cations (Anderson et al. 1991b; Tabcharani et al. 1997)
(v) Anion permeability sequence $Br^- \geq Cl^- > I^-$ (Anderson et al. 1991b; Tabcharani et al. 1997)

[1] Phosphorylated ATP-bound CFTR structures, including those complexed with CFTR modulators, contain the mutation E1371Q, which eliminates ATP hydrolysis at ATP-binding site 2, stabilising the open-channel configuration (Vergani et al. 2005).

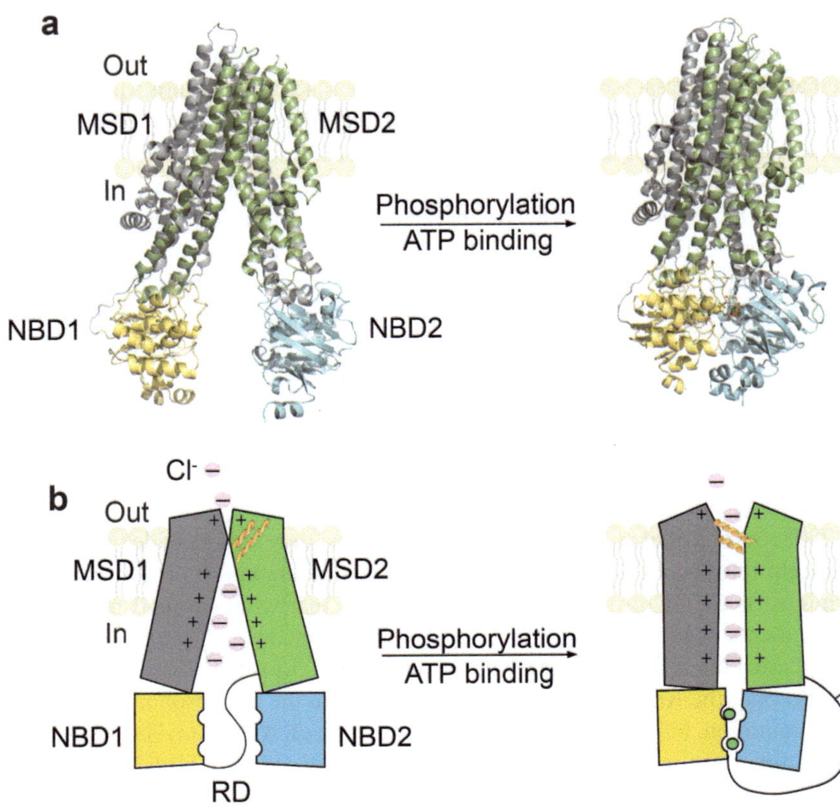

Fig. 4 Conformational changes of human CFTR following PKA-dependent phosphorylation and ATP binding. (**a**) Orthogonal views of molecular dynamics simulations snapshots for the dephosphorylated, ATP-free (PDB id: 5UAK) and phosphorylated, ATP-bound (PDB id: 6MSM) structures of human CFTR and (**b**) schematic representations of these conformations of human CFTR. In (**a**), MSD1 is coloured grey; NBD1, yellow; MSD2, green; and NBD2, blue; the unresolved RD has been omitted from these structures and the phosphorylated, ATP-bound structure lacks the E1371Q mutation. In (**b**), positive charges represent the side chains of positively charged residues in the intra- and extracellular vestibules, which attract Cl^- to the pore constriction; helices depict the ATP-controlled channel gate and the green circles ATP molecules. The lateral tunnel through which anions enter the intracellular vestibule (see Sect. 3.3) is not shown in (b). The lipid bilayers show the position of the plasma membrane with In and Out denoting the intra- and extracellular sides of the membrane. Abbreviations: MSD membrane-spanning domain, NBD nucleotide-binding domain, P phosphorylation of the RD, RD regulatory domain. See text for further information. (Figure modified with permission of J Chen from Zhang et al. (2017) and Zhang et al. (2018))

(vi) Time- and voltage-independent gating (Anderson et al. 1991c; Berger et al. 1991) (Fig. 5)
(vii) Pharmacological profile [e.g. potentiation by ivacaftor (Van Goor et al. 2009) and inhibition by $CFTR_{inh}$-172 (Ma et al. 2002)]

These hallmarks of CFTR are determined by the function of the different domains from which CFTR is assembled.

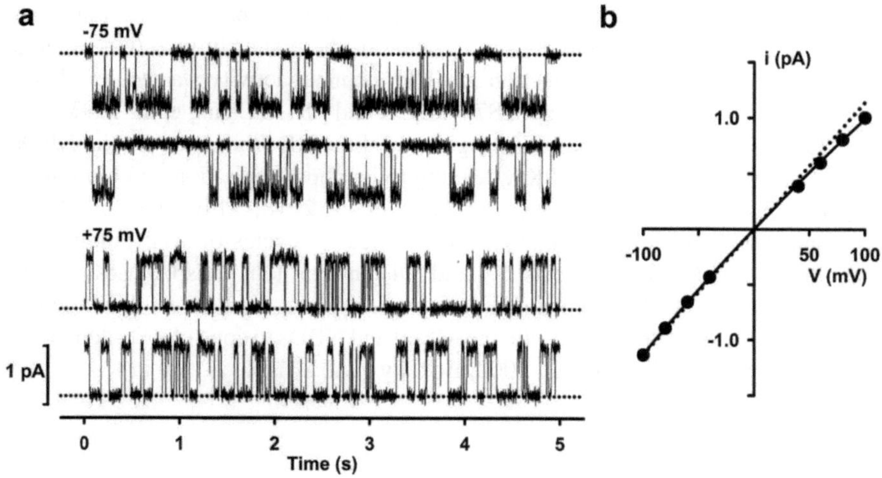

Fig. 5 The single-channel behaviour of wild-type human CFTR. (**a**) Representative recordings of a single CFTR Cl⁻ channel in an excised inside-out membrane patch from a C127 cell acquired at −75 mV (top) and +75 mV (bottom). The membrane patch was bathed in symmetrical 147 mM *N*-methyl-D-glucamine (NMDG)Cl with ATP (1 mM) and PKA (75 nM) continuously present in the intracellular solution; temperature was 37 °C. Dotted lines indicate the closed channel state with downward deflections at −75 mV and upward deflections at +75 mV corresponding to channel openings. (**b**) Single-channel current-voltage (i-V) relationship of wild-type human CFTR. Data are means ± SEM ($n = 10$) at each voltage; error bars are smaller than symbol size. The continuous line is the fit of a second-order regression to the data, while the dotted line shows the i-V relationship of an ion channel with ohmic behaviour. ©2003 Cai Z et al. (Originally published in *The Journal of General Physiology*. https://doi.org/10.1085/jgp.200308921)

3.1 The Regulatory Domain

RD phosphorylation is a prerequisite before intracellular nucleotides regulate CFTR channel gating (Gadsby and Nairn 1999). Thus, the competing actions of protein kinases and phosphatases within cells determine the extent of RD phosphorylation and, hence, the level of channel activity. Although the RD is primarily phosphorylated by protein kinase A (PKA) (Berger et al. 1993), CFTR is activated following RD phosphorylation by protein kinase C (PKC), the type II cGMP-dependent protein kinase and tyrosine kinases with PKC accelerating the rate of channel activation and enhancing the magnitude of channel activity achieved by PKA (Tabcharani et al. 1991; French et al. 1995; Billet et al. 2015). Dephosphorylation of the RD by protein phosphatases is cell-type specific (Gadsby and Nairn 1999). For example, CFTR is dephosphorylated by protein phosphatase 2C (PP2C) in airway and intestinal epithelia, but by PP1 and PP2A in sweat duct epithelia (Reddy and Quinton 1996; Travis et al. 1997).

Elimination of 15 PKA consensus phosphorylation sites is required to render CFTR unresponsive to PKA-dependent phosphorylation (Hegedűs et al. 2009). However, two lines of evidence indicate that the sites at S660, S700, S737, S795 and

S813 are crucial for CFTR regulation by PKA-dependent phosphorylation. First, cAMP agonists stimulated the phosphorylation of these sites in vivo (Cheng et al. 1991). Second, the greatest reduction in CFTR channel activity was achieved with the simultaneous mutation of S660, S737, S795 and S813 (Chang et al. 1993; Rich et al. 1993). Upon spontaneous disengagement of the RD from between the NBDs, the RD becomes accessible to PKA, leading to its phosphorylation and binding by the lasso motif (Mense et al. 2006; Zhang et al. 2018; Mihályi et al. 2020). Subsequently, CFTR channel gating is controlled by cycles of ATP binding and hydrolysis at the NBDs (see Sects. 3.2 and 3.4) until protein phosphatases dephosphorylate the RD, leading to channel deactivation. Thus, the RD is a dynamic hub mediating multisite interactions with the MSDs and NBDs to control ATP-dependent channel gating (Ostedgaard et al. 2000; Bozoky et al. 2013).

3.2 The Nucleotide-Binding Domains

Each NBD is composed of two subdomains, one consisting of the F1-type ATP-binding core and the ABC-specific antiparallel β-region and the other the ABC-specific α-helical region that are distinguished by conserved amino acid sequences (Karpowich et al. 2001; Lewis et al. 2004). The two NBDs assemble as a head-to-tail dimer with the F1-type ATP-binding core of one NBD interacting with the ABC-specific α-helical region of the other NBD (Lewis et al. 2004; Liu et al. 2017; Zhang et al. 2018). The interactions of the subdomains from the opposing NBDs form two ATP-binding sites at the NBD1:NBD2 dimer interface with distinct properties. ATP-binding site 1, formed by the Walker A and B motifs of NBD1 and the ABC signature or LSHGH motif of NBD2, binds ATP stably, but is catalytically inactive (Aleksandrov et al. 2002; Basso et al. 2003; Lewis et al. 2004; Liu et al. 2017; Zhang et al. 2018) (Fig. 3). By contrast, ATP-binding site 2, formed by the Walker A and B motifs of NBD2 and the LSGGQ motif of NBD1, rapidly hydrolyses ATP with a turnover rate of ~1 s^{-1} (Li et al. 1996; Aleksandrov et al. 2002; Basso et al. 2003; Lewis et al. 2004; Liu et al. 2017; Zhang et al. 2018) (Fig. 3). The interaction of ATP with ATP-binding sites 1 and 2 powers dynamic structural rearrangements within the MSDs that gate the channel pore (Vergani et al. 2003, 2005; Liu et al. 2017; Zhang et al. 2018). These structural rearrangements are transduced by the intracellular loops (ICLs), whose coupling helices dock with hydrophobic grooves on the surface of the NBDs (Mornon et al. 2008; Serohijos et al. 2008; Liu et al. 2017; Zhang et al. 2018) (Fig. 3). Thus, ATP binding and hydrolysis at the NBDs gates the CFTR pore, acting as a timing mechanism to control anion flux through the channel (Hwang and Sheppard 2009).

By analysing open-time distributions, Csanády et al. (2010) deduced that ATP binding and hydrolysis at ATP-binding sites 1 and 2 is strictly coupled to channel gating. By contrast, Jih et al. (2012) interpreted the gating behaviour of a pore-lining CFTR mutation to propose that more than one ATP molecule might be hydrolysed during an open-channel burst. This observation led Jih et al. (2012) to propose the energetic coupling model of CFTR channel gating, which predicts that ATP binding and hydrolysis at the NBDs is not strictly coupled to the channel gating cycle. Indeed, using single-molecule fluorescence resonance energy transfer (FRET), Levring et al. (2023) demonstrated that NBD dimerisation precedes, but does not necessarily lead to, gate opening. Monitoring the changes in FRET signals, which reflect the distance between the two NBDs, the authors observed a post-hydrolytic state where ATP had vacated ATP-binding site 2, but not site 1, consistent with previous results (Tsai et al. 2010; Szollosi et al. 2011). Based on this result, Levring et al. (2023) concluded that ATP rebinding to the post-hydrolytic state is possible and, therefore, that multiple gating cycles can occur before complete separation of the NBD1:NBD2 dimer, which initiates gate closure (c.f. Jih et al. 2012).

3.3 The Membrane-Spanning Domains

Functional studies using the substituted cysteine accessibility method (SCAM) identified residues along the length of M6 and M12 that line the pore and showed that regions of M1, M3–M5 and M9–M11 contribute pore-lining residues (for review, see Hwang et al. 2018). Together, the pore-lining segments form an asymmetric hourglass structure with a deep intracellular vestibule and a shallow extracellular vestibule separated by a constriction towards the extracellular end of the pore (for review, see Linsdell 2017). At the extracellular side of the membrane, anions readily enter the channel and are funnelled towards the pore constriction by positively charged residues within the extracellular vestibule (Zhang and Chen 2016; Linsdell 2017; Liu et al. 2017; Zhang et al. 2018). However, the intracellular vestibule is not directly accessible from the intracellular side of the membrane (Zhang and Chen 2016; Linsdell 2017; Liu et al. 2017; Zhang et al. 2018). Instead, anions reach the intracellular vestibule through a lateral tunnel within the ICLs beneath M4 and M6 that runs parallel to the plane of the lipid bilayer (Hoffmann et al. 2018; Li et al. 2018b; Zhang et al. 2018). At the tunnel's entrance and within the intracellular vestibule, positively charged residues direct anion flow towards the pore constriction (Zhang and Chen 2016; Linsdell 2017; Liu et al. 2017; Hwang et al. 2018; Li et al. 2018b; Zhang et al. 2018).

The anion permeability of CFTR is determined by the hydration energy of anions with large, weakly hydrated anions permeating best (Smith et al. 1999; Linsdell et al. 2000). Based on the size of permeant anions, Linsdell and Hanrahan (1998) determined that the pore constriction is ~0.53 nm in diameter, while SCAM data indicate that it is located between L102-I106 in M1 and T338-S341 in M6 (Hwang et al. 2018). Of note, the pore constriction is also the site of an ATP-controlled gate formed by residues from M1, M6, M8 and M12 with the discontinuous region of M8 acting as a flexible hinge that gates the pore (Gao and Hwang 2015; Zhang and Chen 2016; Zhang et al. 2017). When this gate opens, the channel pore is stabilised by salt bridges formed between different transmembrane segments [e.g. R347 (M6)–D924 (M8) and R352 (M6)–D993 (M9); (Cotten and Welsh 1999; Cui et al. 2008)].

3.4 The CFTR Gating Pathway

Opening of the CFTR channel proceeds as a sequence of conformational changes, initiated by formation of the NBD1:NBD2 dimer by ATP binding at ATP-binding sites 1 and 2, which moves towards the MSDs, leading to opening of the channel gate (Csanády et al. 2006; Sorum et al. 2015). By contrast, the hydrolysis of ATP bound at ATP-binding site 2 initiates channel closure (Csanády et al. 2006), which is followed by partial separation of the NBD1:NBD2 dimer (Tsai et al. 2010; Szollosi et al. 2011; Levring et al. 2023) (see Sect. 3.2).

Analysis of the cryo-EM structures of human and zebrafish CFTR provides further insight into the sequence of conformational changes that open the channel pore (Zhang and Chen 2016; Liu et al. 2017; Zhang et al. 2017, 2018). In the dephosphorylated ATP-free structures, the two MSD-NBD units form an inverted V shape with the MSDs locked closed near the extracellular side of the channel and the NBDs held apart by the RD (Zhang and Chen 2016; Liu et al. 2017) (Fig. 4). Following RD displacement and phosphorylation, ATP binding at the two ATP-binding sites drives NBD dimerisation (Vergani et al. 2005; Mense et al. 2006; Zhang et al. 2017, 2018; Mihályi et al. 2020). In turn, conformational changes are transmitted to the MSDs by the ICLs leading to the two MSD-NBD units aligning vertically to each other (Zhang et al. 2017, 2018) (Fig. 4). NBD1 moves with the lasso motif, M1–M3, M6 and M10–M11, while NBD2 is displaced with M4–M5, M7–M9 and M12 (Zhang et al. 2017). The reorientation of the MSDs that occurs packs the transmembrane segments together, screening the CFTR pore from the lipid bilayer (Zhang et al. 2017, 2018). Of note, in both the phosphorylated, ATP-bound structures of human and zebrafish CFTR, the channel is too narrow at the pore constriction for Cl^- flow (Zhang et al. 2017, 2018). Pore occlusion in these CFTR structures might represent a transient state where the NBDs are dimerised, but the channel is closed, such as the brief, flickery closures that interrupt bursts of channel openings (e.g. Cai et al. 2003) (Fig. 5). Thus, a different CFTR structure might better represent the open-channel conformation.

3.5 The Physiological Roles of CFTR in Epithelial Tissues

CFTR is widely expressed in both epithelia lining ducts and tubes throughout the body and various non-epithelial tissues, including cells of the cardiovascular and musculoskeletal systems (e.g. Crawford et al. 1991; Denning et al. 1992b; Nagel et al. 1992; Sprague et al. 1998; Robert et al. 2005; Bonvin et al. 2008). Here, we focus on its physiological role in different epithelial tissues.

In the respiratory airways, CFTR is highly expressed in rare pulmonary ionocytes, whereas it is found at much lower levels in the abundant basal and secretory cells (Montoro et al. 2018; Plasschaert et al. 2018). While it was speculated that the identification of pulmonary ionocytes might lead to a re-evaluation of CFTR function in the respiratory airways (Hawkins and Kotton 2018; Okuda et al. 2021) demonstrated that CFTR-mediated transepithelial ion transport in human airway epithelia is correlated with secretory cells not pulmonary ionocytes. CFTR plays important roles in epithelial ion transport and host defence mechanisms in the respiratory airways (Frizzell and Hanrahan 2012; Ratjen et al. 2015; Stoltz et al. 2015; Saint-Criq and Gray 2017). The balance of CFTR-mediated salt and water secretion and epithelial Na^+ channel (ENaC)-mediated fluid and electrolyte absorption determines the quantity and composition of airway surface liquid (Gaillard et al. 2010; Saint-Criq and Gray 2017). Furthermore, CFTR plays a pivotal role in HCO_3^- secretion by surface airway epithelial cells, both by directly transporting some HCO_3^- and by indirectly facilitating the anion exchanger SLC26A4 (pendrin) to secrete most HCO_3^- (Poulsen et al. 1994; Kim et al. 2019).

CFTR dysfunction compromises host defence mechanisms in the respiratory airways by affecting mucociliary clearance in several ways. First, loss of CFTR-mediated salt and water secretion decreases the volume of airway surface liquid (ASL), hindering the beating of cilia (Ratjen et al. 2015; Stoltz et al. 2015). Second, the changes in ASL alter osmotic forces, concentrating mucus, causing cilia to collapse (Boucher 2007; Ratjen et al. 2015). Third, loss of CFTR-mediated HCO_3^- secretion perturbs the physicochemical properties of mucins, preventing the detachment of mucin strands from submucosal glands, leading to the blockage of respiratory airways by viscous mucus (Quinton 2010; Hoegger et al. 2014; Hansson 2019). CFTR dysfunction further compromises host defence mechanisms in the respiratory airways by inhibiting bacterial killing. Loss of HCO_3^- secretion leads to acidification of ASL by the nongastric H^+/K^+ ATPase (ATP12A), which inactivates antimicrobial peptides, preventing the rapid killing of inhaled bacteria (Pezzulo et al. 2012; Abou Alaiwa et al. 2014; Shah et al. 2016). Similarly, loss of CFTR-dependent SCN^- secretion by surface epithelial cells prevents the production of bactericidal $OSCN^-$ in ASL by the dual oxidase/lactoperoxidase defence system (Pedemonte et al. 2007; Fischer 2009). Thus, loss of host defence mechanisms leaves CF airway epithelia vulnerable to a vicious cycle of infection and inflammation that destroys lung tissue, leading to end-stage lung disease (Ratjen et al. 2015; Stoltz et al. 2015).

In the pancreas, intestine, hepatobiliary system and reproductive tissues, CFTR-driven fluid and electrolyte movements play a key role in lubricating epithelial surfaces to permit the efficient movement of protein-rich cargoes along ducts and tubes (Kopelman et al. 1985; Frizzell and Hanrahan 2012). CFTR dysfunction in these organs therefore leads to the plugging of ducts and tubes, which initiates a pathogenic cascade involving mucus accumulation, mild-to-moderate inflammation and tissue remodelling (Meyerholz et al. 2010). In its severest form, this type of tissue damage causes the destruction and atrophy of the vas deferens in utero and hence, the development of CBAVD (Sun et al. 2019). Thus, the rate of transport of protein-rich cargoes along ducts and tubes is inversely correlated with tissue damage in CF (Kopelman et al. 1985). In the sweat gland, CFTR is predominantly expressed in the sweat duct, a water-impermeable epithelium that absorbs NaCl to generate hypotonic sweat (Quinton 2007). Because the CF sweat duct is structurally unaffected by the tissue damage that afflicts other organs (Kartner et al. 1992), the sweat Cl^- concentration is the gold standard test to diagnose CF and evaluate the response of people with the disorder to CFTR-targeted therapeutics (Knowles and Durie 2002; De Boeck et al. 2006; McCague et al. 2019). Finally, in the kidney, the physiological role of CFTR has long remained enigmatic. CFTR is highly expressed in epithelial cells lining the nephron, yet people with CF experience limited renal symptoms, including episodes of dehydration with metabolic alkalosis and enhanced clearance of some drugs (Jouret and Devuyst 2020). An explanation for the enhanced drug clearance in CF was provided by studies demonstrating that CFTR plays a role in receptor-mediated endocytosis in the proximal tubule by stabilising the expression of the multiligand receptor cubilin in the apical membrane (Jouret et al. 2007). Using a HCO_3^- drinking test, Berg et al. (2020) demonstrated that β-intercalated cell-expressed CFTR plays a key role in urinary HCO_3^- secretion, partially explaining the metabolic alkalosis observed in people with CF. Of note, the authors' data suggest the possibility of using a urine test to evaluate treatment with CFTR-targeted therapeutics.

4 *CFTR* Variants and Their Molecular Mechanisms of CFTR Dysfunction

For several reasons, the abundance, diversity and impact of *CFTR* variants have been, and continue to be, a challenge for the development of CFTR-targeted therapeutics. First, *CFTR* is a highly polymorphic human gene (Cutting 2015). To date, >2100 *CFTR* variants are listed in the Cystic Fibrosis Mutation Database (http://www.genet.sickkids.on.ca/). The largest single group of *CFTR* variants in the database are missense (39%) followed by frameshift (16%), splicing (11%), nonsense (8%), large insertions/deletions (3%), in-frame insertions/deletions (2%) and promoter (1%). A further 13% of the sequence changes in the database are classified as polymorphisms, while the final 8% are unknown variants. Although polymorphisms were originally considered without effect on the CFTR protein, accumulating evidence suggests that some polymorphisms modulate CFTR expression, stability and function with implications for disease treatment using CFTR-targeted therapeutics (Cuppens et al. 1998; Kirchner et al. 2017; Rauscher et al. 2021).

Second, the vast majority of *CFTR* variants are extremely rare. Only 23 variants occur at a frequency greater than 0.1% and 159 at a frequency greater than 0.01% (Watson et al. 2004; Sosnay et al. 2013). The *CFTR* variant F508del is by far the most common cause of CF (Zielenski and Tsui 1995; Bobadilla et al. 2002; Sosnay et al. 2013). Worldwide, approximately 70% of people with CF are homozygous for F508del and carry two copies of the variant, while 90% of people with CF are heterozygous and carry one copy of F508del (Bobadilla et al. 2002). To account for the prevalence of the F508del variant, it has been speculated that one copy of the variant confers a survival advantage possibly against diarrhoeal disease (Gabriel et al. 1994;

Pier et al. 1998) or lactose intolerance (Modiano et al. 2007). Some other *CFTR* variants occur at high frequency in certain populations because of founder effects in religious, ethnic or geographical isolates [e.g. 48% of Ashkenazi Jews with CF carry W1282X (p.Trp1282X; c.3846G>A) (Bobadilla et al. 2002)]. The prevalence of the next most common *CFTR* variants after F508del: G542X (p.Gly542X; c.1624G>T), G551D, N1303K (p.Asn1303Lys; c.3909C>G) and W1282X accounts for just 3.9%, 3.3%, 2.4% and 1.8% of the 89,052 people with CF on the Clinical and Functional Translation of CFTR (CFTR2) website (https://cftr2.org/) (Sosnay et al. 2013), exemplifying the global dominance of F508del. Of these common *CFTR* variants, G542X, N1303K and W1282X currently lack clinically approved CFTR-targeted therapeutics. Thus, while the predominance of F508del assisted the development of elexacaftor-tezacaftor-ivacaftor, particularly by facilitating clinical trials, the scarcity of most other *CFTR* variants hinders the development of CFTR-targeted therapeutics for individual *CFTR* variants.

Third, the complex relationship between *CFTR* variants and disease phenotype, which exhibits tissue-specific consequences for the role of *CFTR* variants, modifier genes and the environment in disease presentation (Cutting 2015). For example, pancreatic insufficiency is predominantly determined by *CFTR* variants, intestinal obstruction by modifier genes and airway obstruction by environmental factors and modifier genes. *CFTR* variants unequivocally identified as CF-causing are listed on the CFTR2 website. As of 7 April, 2023, the 804 annotated *CFTR* variants include 719 CF-causing variants, 49 variants of varying clinical consequence, 25 non-CF-causing and 11 variants of unknown significance (https://cftr2.org/). The clinical impact of the remaining *CFTR* variants listed in the Cystic Fibrosis Mutation Database (http://www.genet.sickkids.on.ca/) is unclear. These *CFTR* variants might cause (i) CF (classical or non-classical), (ii) a CFTR-related disorder, (iii) a varying clinical consequence or (iv) no clinical consequence (Knowles and Durie 2002; De Boeck et al. 2006; Bombieri et al. 2011). As exemplified by the inclusion of non-CF-causing *CFTR* variants in the elexacaftor-tezacaftor-ivacaftor label expansion by the FDA (Salvatore et al. 2022; Raraigh et al. 2022) (see Sect. 8), the correct classification of *CFTR* variants as disease-causing is essential to optimise treatment with CFTR-targeted therapeutics.

Fourth, *CFTR* variants are found throughout the CFTR protein with many affecting functionally important residues. Indeed, early studies of *CFTR* variants demonstrated that the MSDs form the channel pore, the NBDs control channel gating and the ICLs mediate communication between the NBDs and channel pore (e.g. Gregory et al. 1991; Anderson and Welsh 1992; Sheppard et al. 1993; Tabcharani et al. 1993; Cotten et al. 1996; Seibert et al. 1996). These and other studies demonstrated that *CFTR* variants affect the production, intracellular transport and plasma membrane stability of the CFTR protein and its function as an epithelial anion channel (for review, see Wang et al. 2014; De Boeck and Amaral 2016) (Fig. 2). Initial efforts to classify how *CFTR* variants cause CFTR dysfunction identified four classes: defective protein production (class I), defective protein processing (class II), defective channel regulation (class III) and defective channel conduction (class IV) (Welsh and Smith 1993). Two further classes, reduced protein synthesis (class V) and reduced protein stability (class VI), were later introduced (Zielenski and Tsui 1995; Haardt et al. 1999). More recently, De Boeck and Amaral (2016) proposed class VII (no mRNA) to highlight *CFTR* variants unresponsive to CFTR-targeted therapeutics. Time will tell whether further classes are required to discriminate *CFTR* variants. However, within individual classes, different *CFTR* variants cause CFTR dysfunction by distinct mechanisms and varying levels of severity (for review, see Yeh et al. 2022). Thus, the wide spectrum of effects of different *CFTR* variants on the biosynthesis, plasma membrane stability and function of CFTR indicate that there is unlikely to be a single CFTR-targeted therapeutic for all people with CF.

Fifth, reinforcing the argument that CFTR-targeted therapeutics necessitate the use of combinations of small molecules targeting different defects, most *CFTR* variants characterised to date cause CFTR dysfunction by multiple mechanisms (Wang et al. 2014). For example, F508del exhibits a temperature-sensitive folding defect, which prevents correct domain assembly, leading to recognition by the ER quality control mechanism, ubiquitination and destruction by the proteasome (Cheng et al. 1990; Denning et al. 1992a; Ward et al. 1995; Jensen et al. 1995; Du et al. 2005; Fiedorczuk and Chen 2022b). Any F508del-CFTR that escapes the ER and is delivered to the plasma membrane exhibits two further defects: first, reduced protein stability evident from the thermoinstability of F508del-CFTR Cl^- currents and the accelerated endocytosis of F508del-CFTR protein (Lukacs et al. 1993; Swiatecka-Urban et al. 2005). Second, defective channel regulation revealed by aberrant channel activation by PKA-dependent phosphorylation and greatly reduced frequency of channel openings caused by an unstable NBD1:NBD2 dimer (Dalemans et al. 1991; Haws et al. 1996; Wang et al. 2000; Jih et al. 2011; Pasyk et al. 2015; Fiedorczuk and Chen 2022b) (Fig. 6). To accommodate the complex effects of *CFTR* variants on CFTR, Veit et al. (2016) proposed a revised combinatorial classification system derived from the six classes previously proposed, but with classes III (defective channel regulation) and IV (defective channel conduction) combined into a single class. Importantly, this revised classification system provides a framework to classify *CFTR* variants according to their molecular mechanisms of CFTR dysfunction and their sensitivity to different CFTR modulators, facilitating precise personalised therapies for people with CF, termed theratyping (Cutting 2015; Veit et al. 2016; Manfredi et al. 2019).

5 Variant-Specific Therapies for Cystic Fibrosis

The function of CFTR in an epithelium is related to its molecular behaviour by the relationship:

$$I_{\text{CFTR}} = N \times i \times P_o \tag{1}$$

where I_{CFTR} is the CFTR-mediated transepithelial Cl^- current, N is the total number of CFTR Cl^- channels in the apical membrane, i is the current flowing through a

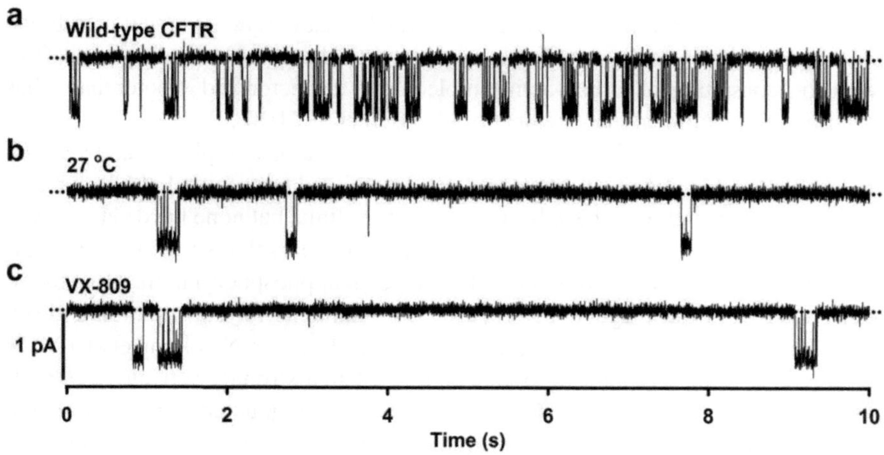

Fig. 6 The F508del *CFTR* variant disrupts CFTR channel gating. (**a–c**) Representative recordings of wild-type and F508del-CFTR Cl^- channels in excised inside-out membrane patches from BHK cells heterologously expressing *CFTR* variants. The plasma membrane expression of F508del-CFTR was rescued by either low temperature incubation (27 °C for 72 – 96 h) (**b**) or treatment with lumacaftor (VX-809; 3 μM) for 24–48 h at 37 °C (**c**). To magnify the size of channel openings, a large Cl^- concentration gradient was imposed across the membrane patch ($[Cl^-]_{internal}$, 147 mM; $[Cl^-]_{external}$, 10 mM) and voltage clamped at −50 mV. ATP (1 mM) and PKA (75 nM) were continuously present in the intracellular solution, and temperature was 37 °C. Dotted lines indicate the closed channel state and downward deflections correspond to channel openings. (Figure modified with permission from Wang et al. (2018))

single open CFTR Cl^- channel and P_o is its open probability. Just as this relationship has been used to explain how *CFTR* variants cause CFTR dysfunction (Sheppard et al. 1993; Sheppard et al. 1995), it can be used to inform the restoration of CFTR-mediated transepithelial Cl^- currents to CF epithelia using small molecules.

Based on the molecular mechanisms of CFTR dysfunction, particularly those of the F508del *CFTR* variant, two types of CFTR-targeted therapeutics have been developed and approved for use in the clinic. First, CFTR correctors, which overcome the F508del-CFTR folding and domain assembly defects, allowing the intracellular transport of the mutant protein to the plasma membrane. These small molecules restore CFTR-mediated transepithelial Cl^- currents by increasing the number of F508del-CFTR Cl^- channels at the apical membrane (Verkman and Galietta 2009; Mijnders et al. 2017). Second, CFTR potentiators, which enhance channel gating following CFTR phosphorylation by PKA. These small molecules restore CFTR-mediated transepithelial Cl^- currents by increasing the P_o of CFTR Cl^- channels at the apical membrane (Verkman and Galietta 2009; Jih et al. 2017). Interestingly, a small number of compounds, including the clinically approved CFTR modulator elexacaftor, have been identified to possess both corrector and potentiator activities (Pedemonte et al. 2011; Phuan et al. 2011; Liu et al. 2018; Laselva et al. 2021a). These small molecules, termed CFTR corrector-potentiators or dual-acting molecules, restore CFTR-mediated transepithelial Cl^- currents by both increasing the number of F508del-CFTR Cl^- channels at the apical membrane and augmenting their P_o. The triple combination therapy elexacaftor-tezacaftor-ivacaftor, consisting of a dual-acting molecule, a corrector and a potentiator, has transformed the clinical management of CF (Bell et al. 2020).

Other classes of CFTR modulators, including activators, amplifiers and stabilisers, have been demonstrated to restore CFTR-mediated transepithelial Cl^- currents in the laboratory and some have been tested in the clinic, but none have yet received regulatory approval. CFTR activators are channel openers that stimulate the CFTR Cl^- channel without its activation by PKA-dependent phosphorylation. Most therapeutic interest in these agents has focused on treatments for constipation and dry eye disease (e.g. Cil et al. 2016; Flores et al. 2016). However, Namkung et al. (2013) identified an amino-carbonitrile-pyrazole, C_{act}-A1 that strongly activates F508del-CFTR-mediated transepithelial Cl^- currents in heterologous and native epithelial cells. Intriguingly, the authors demonstrated that C_{act}-A1 both activates and potentiates F508del-CFTR. Such small molecules warrant further investigation in combination therapies for *CFTR* variants. CFTR amplifiers co-translationally increase CFTR mRNA stability to augment the amount of CFTR protein produced by epithelial cells (Molinski et al. 2017; Dukovski et al. 2020). Because the action of amplifiers is independent of either correctors or potentiators, they might be used by themselves with *CFTR* variants possessing residual function or in combination with correctors and potentiators to enhance the rescue of *CFTR* variants with minimal function (Dukovski et al. 2020). CFTR stabilisers act directly or indirectly to enhance the residency of *CFTR* variants at the plasma membrane. Two approaches that directly target CFTR are suggested by recent studies. First, nanobodies might prevent deleterious protein transitions leading to CFTR misfolding by binding specific CFTR conformations with high affinity (Scholl et al. 2021)[2]. Second, the action of ATP at ATP-binding site 1 (the non-hydrolytic site) suggests that high-affinity ATP analogues acting at this site might stabilise the function of *CFTR* variants (Yeh et al. 2021). When tested in the clinic, the indirectly acting stabiliser cavosonstat, which inhibits S-nitrosoglutathione reductase, did not provide additional benefit to people with CF homozygous for the F508del *CFTR* variant treated with lumacaftor-ivacaftor combination therapy (De Boeck and Davies 2017; Donaldson et al. 2017). For discussion of readthrough agents, which suppress premature termination codons (PTCs) caused by nonsense variants, but which are not selective CFTR modulators, see Sect. 9.

The mechanism of dysfunction of *CFTR* variants that cause defects in channel conduction (class IV) suggests that a further strategy to restore CFTR-mediated transepithelial Cl^- current is to increase i. To date, small molecules that increase current flow through open channels have not been identified or disclosed publicly. Instead, CFTR correctors and potentiators have been used to rescue class IV *CFTR*

[2] Nanobodies are one example of biologics with others including gene therapies and recombinant proteins. Although, to date, no biologics are approved as treatments for CF, different gene therapy strategies have been tested in clinical trials (Alton et al. 2016).

variants, particularly the most common, R117H, which exhibits a marked gating defect that is partially rescued by ivacaftor (Sheppard et al. 1993; Yu et al. 2016). However, the identification of negatively charged activators that open K^+ channels at their selectivity filter, leading to increases in both P_o and i (Schewe et al. 2019), raises the possibility that a similar class of small molecules might be discovered for CFTR.

6 Clinically Approved CFTR Potentiators

Many small molecules, possessing diverse chemical structures, have been demonstrated to act as CFTR potentiators either by hypothesis-driven studies [e.g. phloxine B, (Cai and Sheppard 2002); capsaicin, (Ai et al. 2004)] or by high-throughput screening [e.g. $\Delta F508_{act}$-02, (Yang et al. 2003); ivacaftor, (Van Goor et al. 2009)]. However, few potentiators have been tested in the clinic, and, except for the CFTR corrector-potentiator elexacaftor, ivacaftor remains the only one approved for use by people with CF (Becq et al. 2011; Ramsey et al. 2011; Ramsey and Welsh 2017; Heijerman et al. 2019; Middleton et al. 2019; Laselva et al. 2021a; Veit et al. 2021). At the time of ivacaftor's clinical approval, it was known that the drug enhanced CFTR channel gating and increased CFTR-mediated transepithelial ion transport in primary cultures of CF airway epithelia, leading to improvements in ASL volume and ciliary beat frequency (Van Goor et al. 2009). However, the mechanism of action of ivacaftor was unknown.

Early mechanistic studies of CFTR potentiators, most notably the flavonoid genistein, suggested that these agents enhance CFTR activity by interacting with the NBD1:NBD2 dimer to modify ATP-dependent channel gating (Ai et al. 2004; Moran et al. 2005; Cai et al. 2006). However, the demonstration that once CFTR was phosphorylated by PKA, ivacaftor enhanced both ATP-dependent and ATP-independent channel gating and was without effect on its ATPase activity (Eckford et al. 2012; Jih and Hwang 2013), indicated that it had a different mechanism of action, which did not involve the ATP-binding sites at the NBD1:NBD2 dimer interface. Although the phosphorylation status of CFTR influences ivacaftor potentiation of some *CFTR* variants (Cui et al. 2019), the drug's potentiation of a CFTR construct lacking most of the RD (residues 634–836) excluded its interaction with the RD (Yeh et al. 2015). By contrast, several lines of evidence suggested that ivacaftor might interact with the MSDs: the accumulation of ivacaftor in the inner leaflet of the lipid bilayer; the failure of ivacaftor to potentiate some *CFTR* variants in the MSDs and its additivity with the permeant anion nitrate, which enhances CFTR gating in an identical manner to ivacaftor (Baroni et al. 2014; Van Goor et al. 2014; Yeh et al. 2015; van Willigen et al. 2019). Consistent with this idea, structural and functional studies identified a binding site for ivacaftor at the MSD-lipid interface involving residues from M4, M5 and the unstructured region of M8 (Liu et al. 2019; Yeh et al. 2019) (Fig. 7). Interestingly, the CFTR potentiator GLPG1837 (Galapagos NV, Mechelen, Belgium) binds to the same site as ivacaftor to augment channel

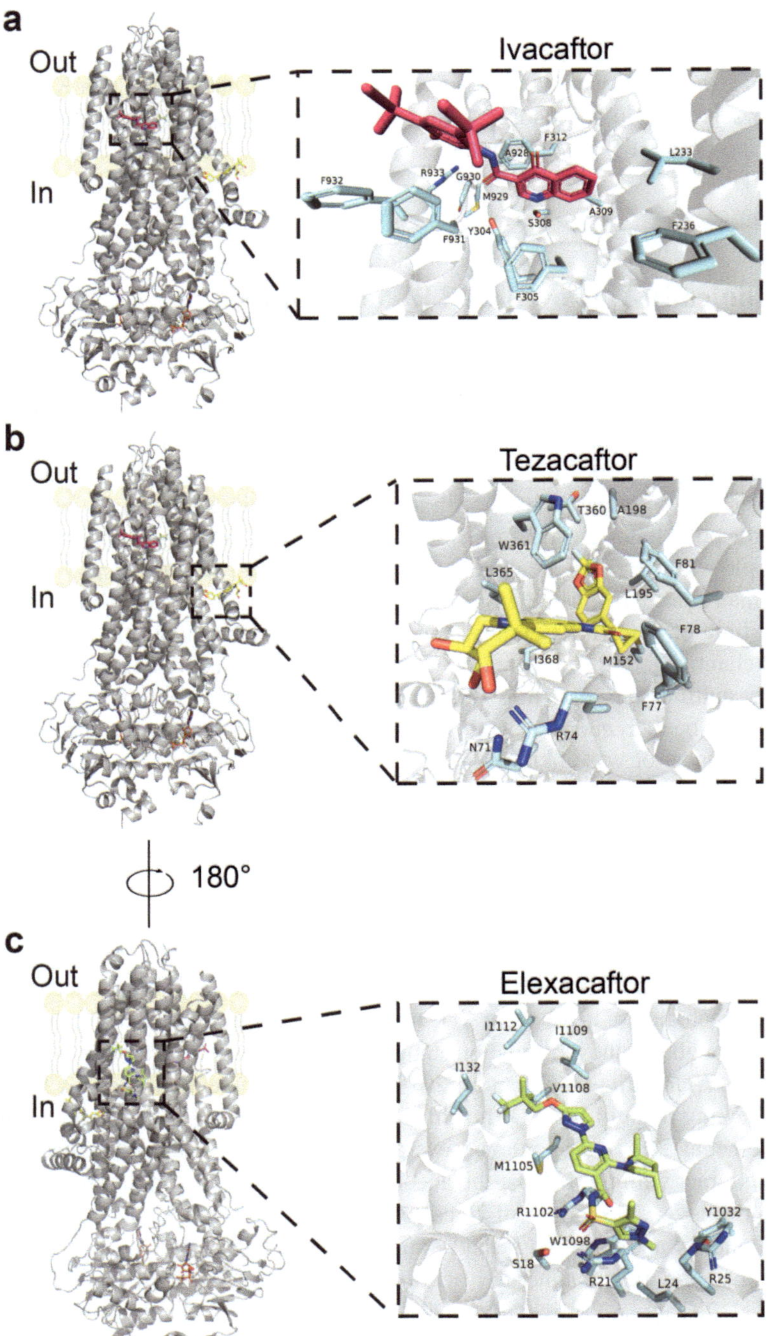

Fig. 7 The binding sites of elexacaftor, tezacaftor and ivacaftor on F508del-CFTR. (**a**–**c**) Orthogonal views of molecular dynamics simulations snapshots of phosphorylated, ATP-bound human F508del-CFTR complexed with ivacaftor (magenta), tezacaftor (yellow) and elexacaftor

gating with increased efficacy by a common mechanism (Yeh et al. 2017, 2019; Van der Plas et al. 2018; Liu et al. 2019).

The location of this potentiator-binding site at the MSD-lipid interface explains why ivacaftor rescues *CFTR* variants throughout the structure of CFTR (Van Goor et al. 2014). It also helps to rationalise how ivacaftor increases the frequency and duration of CFTR channel openings. Evidence from biochemical, structural and functional studies (Cholon et al. 2014; Meng et al. 2017; Byrnes et al. 2018; Wang et al. 2018) suggests that ivacaftor promotes channel openings by destabilising a closed channel configuration. The energetic coupling model of CFTR channel gating (Jih et al. 2012) (see Sect. 3.2) explains how ivacaftor prolongs channel openings by uncoupling the gating cycle from ATP binding and hydrolysis at the NBDs. Alternatively, ivacaftor might impede channel closure by stabilising a pre-hydrolytic open state (Langron et al. 2018) or bind simultaneously to two independent drug-binding sites with equivalent affinities (Csanády and Töröcsik 2019). In support of the latter idea and building on earlier studies (Byrnes et al. 2018), Laselva et al. (2021b) identified an additional binding site for ivacaftor on ICL4 at the interface between the NBDs and MSDs.

Ivacaftor restores wild-type levels of channel activity to some *CFTR* variants (e.g. F508del and G1244E), but not others (e.g. G551D and S549R) (Van Goor et al. 2009; Yu et al. 2012) (Figs. 6 and 8). One potential strategy to restore greater function to *CFTR* variants is co-potentiation, the use of two or more CFTR potentiators that act at different sites on CFTR to additively enhance channel gating (Phuan et al. 2018). To explain the mechanism of co-potentiation, Phuan et al. (2019) identified two classes of CFTR potentiators: class I potentiators (e.g. ivacaftor and GLPG1837), which share a common binding site at the MSD-lipid interface, and class II potentiators (e.g. ASP-11 and apigenin), which likely bind at the NBD1:NBD2 interface in a position distinct from the ATP-binding sites (Moran et al. 2005; Bose et al. 2020). Co-potentiation occurred when a class I potentiator was used with a class II potentiator, but not with two potentiators from the same class (Phuan et al. 2019). Intriguingly, some *CFTR* variants are receptive to co-potentiation (e.g. G551D and S549N), whereas others are not (e.g. A561E and M1101K) (Phuan et al. 2019; Veit et al. 2020a). The identification of CP-628006, a CFTR potentiator with a distinct chemical structure and mechanism of action that co-potentiates G551D-CFTR with ivacaftor (Liu et al. 2022), supports the concept that combinations of CFTR potentiators have additional therapeutic benefit in CF.

Fig. 7 (Continued) (green) based on PDB id: 8EIQ are shown on the left; the unresolved RD has been omitted from these structures. The lipid bilayer depicts the position of the plasma membrane with In and Out identifying the intra- and extracellular sides of the membrane. Residues forming the drug-binding sites are identified in the magnified images of the MSDs shown on the right. The elexacaftor-bound cryo-EM structure (**c**) has been rotated 180° compared to the ivacaftor- and tezacaftor-bound structures in (**a**) and (**b**). (Figure modified with permission of J Chen from Fiedorczuk and Chen (2022b))

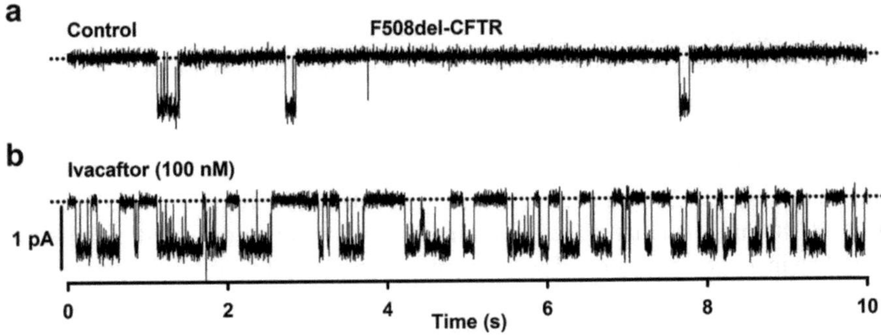

Fig. 8 Ivacaftor potentiates F508del-CFTR channel gating. Representative recordings of F508del-CFTR Cl^- channels in excised inside-out membrane patches from BHK cells heterologously expressing F508del-CFTR made prior to channel deactivation show the effect of ivacaftor (100 nM) on channel gating. Ivacaftor was acutely added to the intracellular solution, which contained ATP (1 mM) and PKA (75 nM); temperature was 37 °C. The plasma membrane expression of F508del-CFTR was rescued by low temperature incubation (27 °C for 72–96 h). To magnify the size of channel openings, a large Cl^- concentration gradient was imposed across the membrane patch ($[Cl^-]_{internal}$, 147 mM; $[Cl^-]_{external}$, 10 mM) and voltage clamped at −50 mV. Dotted lines indicate the closed channel state and downward deflections correspond to channel openings. (Figure modified with permission from Wang et al. (2018))

When compared with new therapeutics for CF that target disease symptoms, the clinical impact of ivacaftor on adults with CF and the G551D *CFTR* variant in its phase III trial was remarkable. The results included a 10% increase in lung function [as measured by percentage of predicted forced expiratory volume in 1 s (FEV_1)] after just 2 weeks in many individuals with irreversible lung disease; the frequency of hospitalisations for the treatment of lung infections, a measure of disease stability, improved by 55%; individuals treated with ivacaftor gained weight (>2.5 kg over 48 weeks) and sweat Cl^- concentration decreased dramatically, falling by almost 50 mmol/L to below the diagnostic level for CF (Ramsey et al. 2011). Clinical trials of ivacaftor in children and infants with CF and *CFTR* gating variants rapidly followed, leading to the drug's approval for use in infants aged ≥4 months (Kleizen et al. 2020; Davies et al. 2021). These and other studies demonstrated that ivacaftor improves lung function by enhancing mucociliary clearance, dampening airway inflammation and decreasing chronic *Pseudomonas aeruginosa* infection, leading to increased microbial diversity (Rowe et al. 2014; Hisert et al. 2017; Gray et al. 2018). Ivacaftor also improved nutrient absorption in the small intestine by markedly alkalinising pH, providing an explanation for the drug-induced weight gain (Rowe et al. 2014). Importantly, analysis of data over a 5-year period from the US and UK CF patient registries, representing approximately half the global CF population, provides real-world verification that ivacaftor slows disease progression (Volkova et al. 2020).

Some treatment benefits of ivacaftor were unexpected. When young children (≤5 years) with *CFTR* gating variants were treated with ivacaftor, there were improvements in faecal elastase-1, a marker of exocrine pancreatic insufficiency

(Davies et al. 2016, 2021; Rosenfeld et al. 2018). This striking result suggests that early treatment with ivacaftor might delay or minimise irreversible damage to the pancreas. Consistent with this idea, using ferrets homozygous for the G551D *CFTR* variant, Sun et al. (2019) demonstrated that in utero and sustained postnatal administration of ivacaftor protects against organ damage, including the highly sensitive male reproductive system. Thus, early pharmacological intervention might prevent disease development in some cases of CF.

7 Clinically Approved CFTR Correctors

Two types of small molecules have been identified that rescue the plasma membrane expression of class II (defective protein processing) *CFTR* variants termed CFTR correctors and proteostasis regulators. CFTR correctors are pharmacological chaperones that interact directly with CFTR to improve protein folding and domain assembly, allowing *CFTR* variants to escape the ER quality control mechanism and traffic to the plasma membrane (Lukacs and Verkman 2012; Fiedorczuk and Chen 2022a, b). By contrast, proteostasis regulators target one or more of the many CFTR-interacting proteins that orchestrate and control the processing of CFTR, its delivery to, and expression at the apical membrane. Thus, proteostasis regulators modulate cellular homoeostasis to control specific cellular functions (Balch et al. 2008). Here, we focus on the CFTR correctors licensed for clinical use: lumacaftor, tezacaftor and elexacaftor (Mall et al. 2020). Lumacaftor and tezacaftor are structural analogues, which share a 1,3-benzodioxol-5-yl-cyclopropane carboxamide (BCC) moiety, whereas elexacaftor has a distinct chemical structure (Fig. 1). For discussion of proteostasis regulators, see Al Salmani et al. (2020).

7.1 Lumacaftor and Tezacaftor

The first CFTR corrector to receive clinical approval was lumacaftor (VX-809; Vertex Pharmaceuticals) (Van Goor et al. 2011). The authors demonstrated that treatment of human bronchial epithelial cells from people with CF (genotype: F508del/F508del) with lumacaftor improved CFTR protein maturation, leading to rescue of CFTR-mediated transepithelial ion transport to 14% of that of non-CF human bronchial epithelial cells (Van Goor et al. 2011). However, by itself, lumacaftor was without clinical benefit to people with CF homozygous for the F508del *CFTR* variant (Clancy et al. 2012), likely due to the variant's multiple effects on the processing and stability of CFTR protein and its channel function (see Sect. 4). Based on these effects of the F508del variant and the greater rescue of F508del-CFTR-mediated transepithelial ion transport by lumacaftor when combined with ivacaftor (Van Goor et al. 2011), clinical trials of the two drugs as a corrector-potentiator combination therapy were conducted in people with CF homozygous for

F508del. These and other clinical trials (see Sect. 8) led to CFTR modulator combination therapy becoming the gold standard treatment for most people with CF.

Mechanistic studies led to different ideas to explain how lumacaftor promotes the plasma membrane expression of F508del-CFTR. Some studies suggested that lumacaftor interacts directly with either MSD1 (Loo et al. 2013; Ren et al. 2013) or NBD1 (Hudson et al. 2017), whereas others suggested that it binds the MSD-NBD interface (Farinha et al. 2013; He et al. 2013; Okiyoneda et al. 2013). Building on these and other studies (Mendoza et al. 2012; Rabeh et al. 2012), Okiyoneda et al. (2013) identified three classes of CFTR correctors: class I (e.g. lumacaftor), which target the MSD-NBD interface; class II [e.g. corr-4a (CF Foundation Therapeutics CFTR Compound Program reference no. C4) (Pedemonte et al. 2005)], which target NBD2; and class III (e.g. the chemical chaperone glycerol (Brown et al. 1996)), which target NBD1. Using this classification system, Veit et al. (2020b) proposed that elexacaftor is a class III corrector based on its synergistic rescue of F508del-CFTR with class I and II correctors (see Sect. 7.2). An alternative classification system based on the necessity to simultaneously target defects in CFTR folding and domain assembly to maximise the rescue of F508del-CFTR (Mendoza et al. 2012; Rabeh et al. 2012) identified two classes of CFTR correctors with complementary mechanisms of action (Li et al. 2018a; Al Salmani et al. 2020). C1-type correctors (e.g. lumacaftor) target early folding events during CFTR biogenesis, whereas C2-type correctors (e.g. elexacaftor) have novel mechanisms of action (Li et al. 2018a; Al Salmani et al. 2020).

Initial attempts to identify the lumacaftor-binding site combined biochemical studies of site-directed mutations with in silico docking of the drug with molecular models of CFTR based on the structure of the ABC transporter Sav1862 (Mornon et al. 2008; Serohijos et al. 2008). These studies identified the interface between NBD1 and ICL1 and ICL4 as a putative lumacaftor-binding site (He et al. 2013; Okiyoneda et al. 2013). By docking lumacaftor and tezacaftor with the dephosphorylated, ATP-free structure of human CFTR (Liu et al. 2017; Molinski et al. 2018) refined this multi-domain binding site to include the lasso motif, ICL1, C-terminus of MSD1, N-terminus of NBD1 and ICL4. By contrast, when Baatallah et al. (2021) docked lumacaftor and tezacaftor with the phosphorylated, ATP-bound structure of human CFTR (Zhang et al. 2018), they identified and verified by site-directed mutation of key residues, a binding site for lumacaftor in MSD1 at the inner leaflet of the lipid bilayer formed by residues from M1, M2, M3 and M6 allosterically coupled to F508 by the K162–E1075 salt bridge.

Support for the lumacaftor/tezacaftor binding site in MSD1 identified by Baatallah et al. (2021) is provided by the cryo-EM structures of lumacaftor and tezacaftor complexed with human CFTR determined by Fiedorczuk and Chen (2022a). In both the dephosphorylated, ATP-free and phosphorylated, ATP-bound conformations of wild-type human CFTR, the lumacaftor binding site is a hydrophobic pocket formed by the intracellular ends of M1, M2, M3 and M6 of MSD1 (Fiedorczuk and Chen 2022a). The cryo-EM structure of tezacaftor complexed with phosphorylated, ATP-bound wild-type human CFTR identified the same hydrophobic pocket as the tezacaftor binding site (Fiedorczuk and Chen 2022a) (Fig. 7).

Comparison of the lumacaftor and tezacaftor complexed structures revealed that the BCC moieties of the two drugs interact with identical residues within the hydrophobic pocket (Fiedorczuk and Chen 2022a). Of note, this pocket creates a cavity within MSD1 that renders the packing of its transmembrane segments thermodynamically unstable during the early stages of CFTR biosynthesis (Fiedorczuk and Chen 2022a). The authors proposed that binding of lumacaftor or tezacaftor within the cavity stabilises the partially assembled MSD1 (i.e. M1, M2, M3 and M6) during folding and inter-domain assembly until it associates with M10 and M11 to form the complete domain-swapped MSD1 unit. Consistent with this idea, Kleizen et al. (2021) demonstrated that lumacaftor improves co-translational folding of MSD1 to enhance the domain's stability. Thus, lumacaftor binding within a cavity that is part of the native conformation of CFTR and not a structural defect caused by CFTR misfolding elegantly explains how the drug promotes the plasma membrane expression of a wide spectrum of *CFTR* variants (Fiedorczuk and Chen 2022a; Han et al. 2018).

7.2 Elexacaftor

When compared with class I correctors, the mechanisms of action of class II and class III correctors are less well understood. Rescue of F508del-CFTR by both classes of correctors is synergistic with lumacaftor, suggesting that they act at distinct binding sites (Van Goor et al. 2011; Okiyoneda et al. 2013; Veit et al. 2020b). In silico docking of the class II corrector corr-4 with molecular models of CFTR suggests that corr-4a binds NBD2 at the domain's interface with either NBD1 or the MSDs (Okiyoneda et al. 2013). However, the binding sites of class II correctors are either unknown or undisclosed, and, to date, no class II correctors are licensed for clinical use.

By contrast, the class III corrector elexacaftor is a component of the highly effective triple combination therapy elexacaftor-tezacaftor-ivacaftor (Heijerman et al. 2019; Middleton et al. 2019) (see Sect. 8). Chronic incubation of human nasal epithelial cells with elexacaftor-tezacaftor-ivacaftor restored substantial function to F508del and some rare *CFTR* variants (e.g. G85E and M1101K) (Veit et al. 2020b). Two lines of evidence led Veit et al. (2020b) to conclude that elexacaftor is a class III corrector that targets NBD1. First, elexacaftor rescue of the plasma membrane expression of F508del-CFTR was synergistic with class I and class II, but not class III correctors (Veit et al. 2020b). Second, direct binding of elexacaftor to isolated NBD1 partially suppressed the loss of domain structure caused by F508del (Veit et al. 2020b). However, using different experimental conditions, which favour domain unfolding, Hillenaar et al. (2022) found that elexacaftor did not improve the stability of F508del-NBD1 but instead enhanced inter-domain assembly. This latter conclusion is supported by cryo-EM structures of phosphorylated, ATP-bound human F508del-CFTR complexed with elexacaftor (elexacaftor; elexacaftor-lumacaftor and elexacaftor-tezacaftor-ivacaftor) (Fiedorczuk and Chen 2022b). In these structures, elexacaftor bound a shallow groove at the MSD-lipid interface, extending from the centre of the lipid bilayer towards the inner leaflet and composed of residues from M11 with additional contributions from M2, M10 and the lasso motif (Fiedorczuk and Chen 2022b) (Fig. 7). In either the absence of CFTR correctors or presence of lumacaftor alone, NBD1 lacked structural stability and appeared amorphous in the cryo-EM structure of phosphorylated, ATP-bound human F508del-CFTR (Fiedorczuk and Chen 2022b). By contrast, elexacaftor stabilised the structure of NBD1: the cryo-EM structure of phosphorylated, ATP-bound human F508del-CFTR complexed with elexacaftor contained dimerised NBDs, but ATP-binding site 1 (the non-hydrolytic site) was solvent exposed, whereas the structures complexed with either elexacaftor-lumacaftor or elexacaftor-tezacaftor-ivacaftor had dimerised NBDs resembling those of phosphorylated, ATP-bound wild-type human CFTR (Fiedorczuk and Chen 2022b). Of note, a recent preprint confirms the binding site of elexacaftor and its synergy with lumacaftor by determining the cryo-EM structures of the *CFTR* gating variant G551D complexed with CFTR modulators (Wang et al. 2022). Taken together, the data demonstrate convincingly that class I and class III correctors act synergistically to correct the structural defects of F508del- and G551D-CFTR. They also suggest that the classification of CFTR correctors should be revised to include a fourth class that indirectly improves NBD1 stability (Hillenaar et al. 2022). Alternatively, the simplified two-class categorisation of CFTR correctors introduced by Li et al. (2018a) should be widely adopted.

Unexpectedly, studies of the rescue of different class II *CFTR* variants by elexacaftor-tezacaftor-ivacaftor demonstrated that elexacaftor is a dual-acting molecule with both CFTR corrector and potentiator activities (Laselva et al. 2021a) (see Sect. 5). These and other studies (Shaughnessy et al. 2021; Veit et al. 2021) revealed that acute addition of elexacaftor enhances CFTR-mediated transepithelial ion transport in native and heterologous epithelia expressing *CFTR* variants. They also showed that elexacaftor acts as a co-potentiator with both class I and class II potentiators (Shaughnessy et al. 2021; Veit et al. 2021) (see Sect. 6), suggesting that elexacaftor might represent a distinct class of CFTR potentiators. Because the mutation R1102A in the elexacaftor binding site abolished both the drug's corrector and potentiator activities, Fiedorczuk and Chen (2022b) proposed that a single binding site is responsible for the dual action of elexacaftor as a CFTR corrector-potentiator. However, the half-maximal effective concentration (EC_{50}) for wild-type and G551D-CFTR potentiation by elexacaftor is >100-fold lower than that for correction of the F508del-CFTR processing defect (correction, EC_{50} ~ 0.28 μM; potentiation, EC_{50} ~ 1 nM) (Veit et al. 2020b, 2021; Shaughnessy et al. 2021), suggesting that there might be different binding sites for the corrector and potentiator activities of elexacaftor. Consistent with this idea, using molecular docking and molecular dynamics simulations, Baatallah et al. (2021) identified two binding sites for elexacaftor distinct from the site identified by Fiedorczuk and Chen (2022b), one in MSD1 and a second in NBD1, shared with lumacaftor. Adding to the complexity of elexacaftor's action, Tomati et al. (2023) demonstrated that rescue of the *CFTR* variant G1244E by elexacaftor-tezacaftor-ivacaftor is affected by the cellular background, while Saint-Criq et al. (2021) found that extracellular phosphate enhances

F508del-CFTR activity following rescue by the triple combination therapy. Thus, while much has been learnt about the mechanisms of action of CFTR correctors, our understanding remains incomplete.

8 Clinically Approved Combination Therapies

As explained in Sect. 7, the rationale for CFTR combination therapy is twofold: first, the complex effects of F508del and many other CFTR variants on the channel (Veit et al. 2016) and, second, the lack of clinical benefit when people with CF homozygous for F508del were treated with either lumacaftor or ivacaftor, by themselves (Clancy et al. 2012; Flume et al. 2012). When tested in people with CF homozygous for F508del, lumacaftor-ivacaftor combination therapy (Orkambi; Vertex Pharmaceuticals) modestly improved lung function, but markedly increased disease stability (Wainwright et al. 2015), leading to regulatory approval in 2015 for use of the drug combination in this group of people with CF. Unfavourable drug-drug interactions (Hanrahan et al. 2017) stimulated the development of tezacaftor, a close analogue of lumacaftor with improved metabolism and pharmacokinetics, but reduced propensity for drug-drug interactions (Kym et al. 2018). Clinical trials demonstrated that combination therapy with tezacaftor and ivacaftor (Symdeko/Symkevi; Vertex Pharmaceuticals) was of greater clinical benefit to people with CF homozygous for F508del than lumacaftor-ivacaftor combination therapy (Taylor-Cousar et al. 2017). They also showed that tezacaftor-ivacaftor combination therapy extended the clinical benefit of CFTR modulators to people with CF carrying F508del and a *CFTR* variant associated with residual CFTR function (Rowe et al. 2017), leading to regulatory approval for this combination therapy in 2018 (Collins 2019).

While dual combination therapies were important milestones in the development of CFTR-targeted therapeutics for CF, only approximately half of all people with CF carried *CFTR* variants that could be rescued with the tezacaftor-ivacaftor combination (Collins 2019). In the search for combination therapies with greater clinical benefit, clinical trials of triple combination therapies with two correctors and one potentiator were conducted (Davies et al. 2018; Keating et al. 2018; Heijerman et al. 2019; Middleton et al. 2019). Elexacaftor-tezacaftor-ivacaftor improved lung function and sweat Cl^- concentration in people with CF homozygous for F508del by similar amounts to the action of ivacaftor on people with CF and the G551D *CFTR* variant (Ramsey et al. 2011; Heijerman et al. 2019). Encouragingly, elexacaftor-tezacaftor-ivacaftor achieved the greatest improvement in lung function (14% increase in percentage of predicted FEV_1 within 2 weeks of starting treatment) in people with CF heterozygous for F508del and a minimal function *CFTR* variant (Middleton et al. 2019). In this group of people with CF, the triple combination therapy also reduced sweat Cl^- concentration by 42 mmol/L to below the diagnostic threshold for CF, improved greatly disease stability (63% reduction in pulmonary exacerbations) and led to weight gain (BMI treatment difference of 1

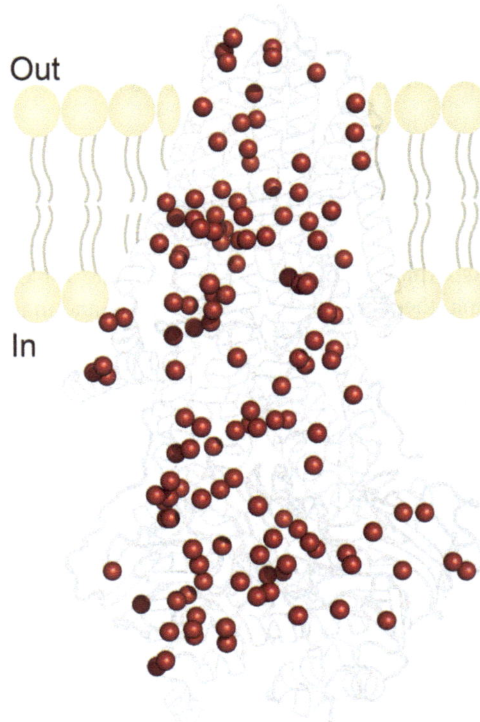

Fig. 9 Location of *CFTR* variants responsive to elexacaftor-tezacaftor-ivacaftor. Orthogonal view of a molecular dynamics simulations snapshot of phosphorylated, ATP-bound human CFTR (PDB id: 6MSM) lacking the E1371Q mutation with red spheres depicting the location of *CFTR* variants responsive to elexacaftor-tezacaftor-ivacaftor (https://www.accessdata.fda.gov/drugsatfda_docs/label/2023/217660s000lbl.pdf). The lipid bilayer shows the position of the plasma membrane with In and Out identifying the intra- and extracellular sides of the membrane. (Figure modified with permission of J Chen from Zhang et al. (2018))

relative to placebo) (Middleton et al. 2019). With the regulatory approval of elexacaftor-tezacaftor-ivacaftor in 2019 for people with CF heterozygous for F508del, ~90% of people with CF aged 12 years and older had a highly effective CFTR-targeted therapeutic (Collins 2019). In 2020, the FDA expanded the elexacaftor-tezacaftor-ivacaftor label to include 177 additional *CFTR* variants (Fig. 9), and in 2023, following clinical trials of children with CF heterozygous for F508del (Zemanick et al. 2021; Goralski et al. 2023), the FDA further extended the use of elexacaftor-tezacaftor-ivacaftor to children aged 2 years and older with these *CFTR* variants. Approval for use in younger infants would further minimise organ damage and, hence, disease development.

Post-regulatory approval studies of elexacaftor-tezacaftor-ivacaftor emphasise the benefit of this triple combination therapy for people with CF (for review, see Tümmler 2023). They reveal that all eligible people with CF profit from the therapy irrespective of their baseline lung function, evidenced by the marked decline in lung transplants and hospitalisations for pulmonary exacerbations (e.g. Burgel et al. 2021; Tümmler 2023). Using lung clearance index and magnetic resonance imaging, Graeber et al. (2022a) demonstrated that elexacaftor-tezacaftor-ivacaftor increases lung ventilation by clearing mucus plugs and reducing airway wall thickening. Encouragingly, the triple combination therapy markedly reduced lung infection by *Staphylococcus aureus* and *Pseudomonas aeruginosa* (Pallenberg et al.

2022; Sheikh et al. 2023), concomitant with reductions in inflammasome activation and the secretion of pro-inflammatory cytokines (Gabillard-Lefort et al. 2022; Sheikh et al. 2023). As a result, microbial communities in the lungs of people with CF treated with elexacaftor-tezacaftor-ivacaftor approached those of healthy lungs (Pallenberg et al. 2022), and there was normalisation of immune cell populations (Sheikh et al. 2023).

In other organ systems affected by CF, similar notable benefits of elexacaftor-tezacaftor-ivacaftor have been observed (Tümmler 2023). Using electrophysiological assays, Graeber et al. (2022b) demonstrated that the combination therapy restored similar, substantial levels of CFTR function to intestinal epithelia as those achieved by airway epithelia. Consistent with these data, two studies using CF-specific patient-reported outcome measures recorded improvements in the gastrointestinal symptoms of CF, albeit stronger effects were observed in one study than the other (Mainz et al. 2022; Schwarzenberg et al. 2023). As observed with infants with *CFTR* gating variants treated with ivacaftor (see Sect. 6), treatment of infants heterozygous for F508del with elexacaftor-tezacaftor-ivacaftor improved exocrine pancreatic function in some children (Goralski et al. 2023). However, in older people with CF, the treatment did not alleviate pancreatic insufficiency, while conflicting results have been observed in studies of glucose homeostasis to evaluate CF-related diabetes (for review, see Tümmler 2023). Of note, elexacaftor-tezacaftor-ivacaftor has markedly improved the reproductive health of women with CF, leading to increased rates of pregnancy (Taylor-Cousar and Jain 2021). For example, the 2021 Annual Data Report of the UK CF Registry records 58 births to women with CF in 2019, but 103 in 2021 following the licensing of elexacaftor-tezacaftor-ivacaftor in the UK (https://www.cysticfibrosis.org.uk/the-work-we-do/uk-cf-registry/reporting-and-resources). Moreover, on at least one occasion, the combination therapy has been used to successfully treat meconium ileus in a foetus with CF in utero by administering the drug to the pregnant mother (Szentpetery et al. 2022). Although the available evidence is encouraging (Taylor-Cousar and Jain 2021), much more is required to develop consensus guidelines for the use of elexacaftor-tezacaftor-ivacaftor during pregnancy and lactation.

Real-world studies demonstrate that elexacaftor-tezacaftor-ivacaftor is well-tolerated, but the frequency of adverse events appears higher, and they include different types to those reported in clinical trials (Dagenais et al. 2020; Tümmler 2023). Three examples are skin rashes, liver injury and altered mental health. Different forms of adverse skin reactions (e.g. drug-induced acne and eruptive melanocytic naevi) have been frequently reported following initiation of elexacaftor-tezacaftor-ivacaftor treatment with most cases resolved with desensitisation protocols (Tümmler 2023). In most people with CF, treatment with elexacaftor-tezacaftor-ivacaftor leads to clinically insignificant rises in liver enzymes even in cases of existing CF-related liver disease (Heijerman et al. 2019; Middleton et al. 2019; Zemanick et al. 2021; Goralski et al. 2023; Tewkesbury et al. 2023). However, in rare cases the combination therapy causes abnormal increases in liver enzymes indicative of drug-induced hepatotoxicity, leading to liver necrosis (Salehi et al. 2021; Tewkesbury et al. 2023). Except for oropharyngeal pain and isolated

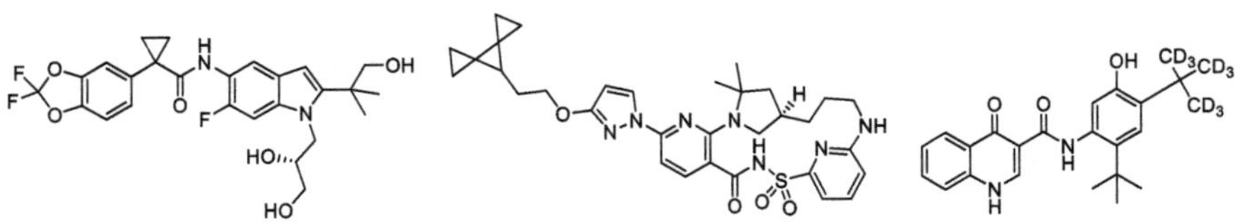

Tezacaftor (VX-661) Vanzacaftor (VX-121) Deutivacaftor (VX-561)

Fig. 10 Chemical structures of the CFTR modulators for once-daily dosing. The CFTR correctors tezacaftor and vanzacaftor are combined with the CFTR potentiator deutivacaftor

cases of abdominal pain and abnormal behaviour, no elexacaftor-tezacaftor-ivacaftor-related neurological adverse events were reported in clinical trials (Heijerman et al. 2019; Middleton et al. 2019; Zemanick et al. 2021; Goralski et al. 2023). By contrast, post-regulatory approval studies have reported a range of neurological and psychological adverse events (Tümmler 2023). For example, Spoletini et al. (2022) described cases of anxiety, low mood, insomnia and 'brain fog' with reduced concentration in 7% of treated adults at a single CF centre. These mental health adverse events were mostly resolved by elexacaftor-tezacaftor-ivacaftor dose adjustment without loss of clinical benefit (Spoletini et al. 2022). Thus, real-world experience reinforces the conclusion that CFTR-targeted combination therapies have unprecedented clinical benefits for people with CF. Optimising adherence to treatment will be crucial to ensure that these benefits have long-term positive impacts on the health of people with CF. Towards this goal, the latest triple combination therapy to be tested in the clinic, vanzacaftor (VX-121)-tezacaftor-deutivacaftor (VX-561) (Vertex Pharmaceuticals) (Uluer et al. 2023) (Fig. 10) has a once-daily dosing regimen.

9 Unmet Need

Around 10% of people with CF carry *CFTR* variants that are unresponsive to CFTR correctors and potentiators, many of whom are individuals from ethnic minority backgrounds (Desai et al. 2022; McGarry et al. 2022). Most of these *CFTR* variants are nonsense variants, which introduce PTCs into the CFTR mRNA (De Boeck and Amaral 2016; Veit et al. 2016). Except for PTCs located in NBD2 and the C-terminus, what little protein that is produced is non-functional because it lacks an intact CFTR pore (Sharma et al. 2018; Yeh and Hwang 2020).

There is a long history of attempts to suppress PTCs in the CFTR mRNA with small molecules (e.g. Howard et al. 1996; Wilschanski et al. 2003; Welch et al. 2007; Kerem et al. 2014; Konstan et al. 2020). However, it is now recognised that these *CFTR* variants are the hardest to tackle with CFTR-targeted therapeutics. There are several reasons why:

First, the efficacy of PTC suppression. Termination of translation at a PTC is tenfold less efficient than that at a naturally occurring stop codon (Keeling et al.

2014). For this and other reasons (for review, see Keeling et al. 2014; Sharma et al. 2020), translation termination at PTCs can be suppressed with small molecules, leading to the production of full-length protein. Building on earlier studies (for review, see Lee and Dougherty 2012), aminoglycoside antibiotics were demonstrated to suppress *CFTR* nonsense variants, leading to the partial restoration of CFTR function in laboratory and clinical studies (Howard et al. 1996; Wilschanski et al. 2003). In the search for readthrough molecules with enhanced efficacy and reduced toxicity (Keeling et al. 2014; Sharma et al. 2020), the orally bioavailable small molecule ataluren (Translarna; PTC 124) was developed by PTC Therapeutics (South Plainfield, NJ, USA) (Welch et al. 2007). Ataluren suppresses PTCs by inserting near-cognate tRNAs at the site of the nonsense codon like those introduced endogenously, but at reduced frequency, by the ribosome (Roy et al. 2016). Despite encouraging data from initial short-term clinical studies (for review, see Lee and Dougherty 2012), ataluren failed to improve either the lung function or disease stability of people with CF bearing *CFTR* nonsense variants in two 48-week randomised placebo-controlled clinical trials (Kerem et al. 2014; Konstan et al. 2020), the second conducted to exclude inhibition of ataluren-mediated PTC suppression by the aminoglycoside antibiotic tobramycin (Roy et al. 2016). A further readthrough molecule, ELX-02 (NB124; Eloxx Pharmaceuticals Inc., Waltham, MA, USA), a synthetic aminoglycoside with reduced toxicity, has been tested in the clinic (Kerem 2020). Although ELX-02 restored some CFTR function to epithelial cells endogenously expressing the *CFTR* nonsense variant G542X (Xue et al. 2014; Crawford et al. 2021), ELX-02 in combination with ivacaftor did not achieve its efficacy endpoints in a phase II clinical trial of people with CF and G542X-CFTR (https://www.clinicaltrials.gov/search?cond=NCT04135495; https://cysticfibrosisnewstoday.com/news/elx-02-plus-kalydeco-fails-reach-goals-cystic-fibrosis-trial/). Taken together, these data highlight the challenge of developing readthrough molecules with clinical efficacy for *CFTR* nonsense variants.

Second, PTCs activate nonsense-mediated decay (NMD), a protective surveillance mechanism, which degrades faulty mRNA transcripts containing PTCs to prevent the production of abnormal proteins (Lykke-Andersen and Jensen 2015; Hug et al. 2016). As a result, few CFTR mRNA transcripts with PTCs are available for protein synthesis. Pharmacological inhibition of NMD with SMG1i (Gopalsamy et al. 2012) in combination with CFTR modulators partially rescued CFTR function in epithelial cells endogenously expressing the *CFTR* variant W1282X, but the additional inclusion of G418 diminished CFTR rescue (Laselva et al. 2020; Venturini et al. 2021), likely because NMD and PTC suppression utilise common regulatory proteins (Venturini et al. 2021). Of note, NMD plays an essential role in the control of gene expression with physiological roles in development, cellular stress and immunity (Lykke-Andersen and Jensen 2015; Hug et al. 2016). While it is therefore unclear whether pharmacological inhibition of NMD is a safe therapeutic strategy to increase the availability of CFTR mRNA transcripts with PTCs (Sharma et al. 2020), cellular regulatory mechanisms might be exploited to achieve this goal. NMD regulates the unfolded protein response (UPR), which is activated when the protein-folding capacity of the ER is overwhelmed, and UPR, in turn, regulates

NMD (Lykke-Andersen and Jensen 2015; Hug et al. 2016). Of note, Oren et al. (2014) exploited these regulatory circuits to downregulate NMD and, hence, increase the availability of CFTR mRNA transcripts with PTCs for PTC suppression using the aminoglycoside G418. Thus, just as lung inflammation appears to augment the action of CFTR modulators under some circumstances (e.g. Gentzsch et al. 2021; Rehman et al. 2021), its induction of UPR (Ribeiro and Hull-Ryde 2022) might be exploited to prevent the degradation of CFTR mRNA transcripts with PTCs.

Third, the identity of the amino acid inserted during PTC suppression. Near-cognate aminoacyl-tRNAs are occasionally accepted by the ribosome allowing the incorporation of an amino acid into the nascent polypeptide at the position of the PTC (Keeling et al. 2014). Using HEK293 cells and a model reporter system, Xue et al. (2017) identified the amino acids inserted at G542X-CFTR and W1282X-CFTR by G418-mediated suppression of the UGA PTC at these *CFTR* variants (amino acid frequency, G542X UGA, cysteine, 44%; tryptophan, 36% and arginine, 20%; W1282X UGA, leucine, 58%; cysteine, 38% and tryptophan, 4%). These data indicate that no full-length CFTR protein generated by G418-mediated suppression of PTCs at residues G542 and W1282 contain the wild-type amino acid residue. They further suggest that the local amino acid sequence surrounding a PTC influences the identity of the amino acid incorporated during readthrough by G418 (Roy et al. 2016; Xue et al. 2017). Excluding leucine, insertion of non-native amino acid residues at positions G542 and W1282 attenuated CFTR-mediated transepithelial Cl^- currents by diminishing CFTR protein maturation with G542W-CFTR generating little or none (Roy et al. 2016; Xue et al. 2017). Thus, depending on the identity of the inserted amino acid, the resulting full-length protein might have none, some or wild-type CFTR function. Nevertheless, in the case of full-length CFTR proteins with some residual function produced by PTC suppression, CFTR modulators augmented the restoration of CFTR function achieved by PTC suppression (e.g. Xue et al. 2017; Sharma et al. 2018; Yeh and Hwang 2020; Venturini et al. 2021).

Taken together, PTC suppression with CFTR-targeted therapeutics requires combination therapy including small molecules to (i) suppress the PTC, (ii) inhibit NMD and (iii) rescue CFTR protein maturation, stability and function. As for other classes of *CFTR* variants (e.g. McDonald et al. 2022), different *CFTR* variants will likely require distinct combinations of CFTR-targeted therapeutics for optimal restoration of CFTR function (Sharma et al. 2018; Yeh and Hwang 2020; Venturini et al. 2021). In the case of *CFTR* variants with PTCs located in NBD2 and the C-terminus that possess an intact CFTR pore, particularly those close to the C-terminus that are insensitive to NMD, PTC suppression combined with CFTR modulators might restore sufficient function (Haggie et al. 2017; Sharma et al. 2018; Yeh and Hwang 2020; Venturini et al. 2021). Future studies should test this possibility.

10 Conclusion

CFTR-targeted therapeutics are now available to most people with CF, transforming their disease from an incurable condition to a manageable chronic illness. The striking clinical success of these innovative medicines provides clear evidence of what is possible and motivates the CF community to try harder to do better for people with CF. However, the CF community arguably now faces its greatest challenge, to develop transformational therapeutics for the remaining people with CF who do not benefit from current CFTR-targeted therapeutics. Achieving this goal, while ensuring equity of access to all, will guarantee that no one with CF is left behind.

Acknowledgements We thank past and present laboratory colleagues and collaborators for stimulating discussions and assistance. Work in the authors' laboratories discussed in this manuscript was supported by the National Institutes of Health, the Cystic Fibrosis Foundation, Ministry of Science and Technology, Taiwan and Veterans General Hospitals and University System of Taiwan Joint Research Program, Taiwan (T-CH) and Cystic Fibrosis Foundation Therapeutics, Cystic Fibrosis Trust and the Medical Research Council (DNS). H-IY is supported by an AbbVie Postdoctoral Fellowship and YMF by the Engineering and Physical Sciences Research Council (DTP 22-24 University of Bristol – EP/W524414/1) and an international Cotutelle Macquarie University Research Excellence Scholarship (Cotutelle 'iMQRES', scholarship allocation no.: 20224678). ASFO acknowledges the Biotechnology and Biological Sciences Research Council (grant no.: BB/W003449/1 and BB/X009831/1) and thanks Oracle for Research for a research fellowship.

Conflict of Interest The authors have no conflicts of interest to declare.

Author Contributions All authors drafted the chapter or revised it critically for important intellectual content. All authors approved the final version of the chapter.

Author's Present Addresses J.D. Bradbury: Department of Pharmacology, University of Oxford, Oxford, OX1 3QT, UK.

References

Abou Alaiwa MH, Reznikov LR, Gansemer ND, et al. pH modulates the activity and synergism of the airway surface liquid antimicrobials β-defensin-3 and LL-37. Proc Natl Acad Sci U S A. 2014;111:18703–8.

Ai T, Bompadre SG, Wang X, et al. Capsaicin potentiates wild-type and mutant cystic fibrosis transmembrane conductance regulator chloride-channel currents. Mol Pharmacol. 2004;65:1415–26.

Al Salmani MK, Sondo E, Balut C, et al. Molecular physiology and pharmacology of the cystic fibrosis transmembrane conductance regulator. In: Hamilton KL, Devor DC, editors. Studies of epithelial transporters and ion channels, physiology in health and disease. 2nd ed. Rockville/Berlin: The American Physiological Society/Springer-Verlag; 2020. p. p605–70.

Aleksandrov L, Aleksandrov AA, Chang X-B, et al. The first nucleotide binding domain of cystic fibrosis transmembrane conductance regulator is a site of stable nucleotide interaction, whereas the second is a site of rapid turnover. J Biol Chem. 2002;277:15419–25.

Alton EWFW, Boyd AC, Davies JC, et al. Genetic medicines for CF: hype versus reality. Pediatr Pulmonol. 2016;51:S5–S17.

Anderson M, Welsh MJ. Regulation by ATP and ADP of CFTR chloride channels that contain mutant nucleotide-binding domains. Science. 1992;257:1701–4.

Anderson MP, Berger HA, Rich DP, et al. Nucleoside triphosphates are required to open the CFTR chloride channel. Cell. 1991a;67:775–84.

Anderson MP, Gregory RJ, Thompson S, et al. Demonstration that CFTR is a chloride channel by alteration of its anion selectivity. Science. 1991b;253:202–5.

Anderson MP, Rich DP, Gregory RJ, et al. Generation of cAMP-activated chloride currents by expression of CFTR. Science. 1991c;251:679–82.

Baatallah N, Elbahnsi A, Mornon J-P, et al. Pharmacological chaperones improve intra-domain stability and inter-domain assembly via distinct binding sites to rescue misfolded CFTR. Cell Mol Life Sci. 2021;78:7813–29.

Balch WE, Morimoto RI, Dillin A, et al. Adapting proteostasis for disease intervention. Science. 2008;319:916–9.

Barben J, Castellani C, Munck A, et al. Updated guidance on the management of children with cystic fibrosis transmembrane conductance regulator-related metabolic syndrome/cystic fibrosis screen positive, inconclusive diagnosis (CRMS/CFSPID). J Cyst Fibros. 2021;20:810–9.

Baroni D, Zegarra-Moran O, Svensson A, et al. Direct interaction of a CFTR potentiator and a CFTR corrector with phospholipid bilayers. Eur Biophys J. 2014;43:341–6.

Basso C, Vergani P, Nairn AC, et al. Prolonged nonhydrolytic interaction of nucleotide with CFTR's NH_2-terminal nucleotide binding domain and its role in channel gating. J Gen Physiol. 2003;122:333–48.

Bear CE, Li C, Kartner N, et al. Purification and functional reconstitution of the cystic fibrosis transmembrane conductance regulator (CFTR). Cell. 1992;68:809–18.

Becq F, Mall MA, Sheppard DN, et al. Pharmacological therapy for cystic fibrosis: from bench to bedside. J Cyst Fibros. 2011;10(Suppl 2):S129–45.

Bell SC, Mall MA, Gutierrez H, et al. The future of cystic fibrosis care: a global perspective. Lancet Respir Med. 2020;8:65–124.

Berg P, Svendsen SL, Sorensen MV, et al. Impaired renal HCO_3^- excretion in cystic fibrosis. J Am Soc Nephrol. 2020;31:1711–27.

Berger HA, Anderson MP, Gregory RJ, et al. Identification and regulation of the cystic fibrosis transmembrane conductance regulator-generated chloride channel. J Clin Invest. 1991;88:1422–31.

Berger HA, Travis SM, Welsh MJ. Regulation of the cystic fibrosis transmembrane conductance regulator Cl^- channel by specific protein kinases and protein phosphatases. J Biol Chem. 1993;268:2037–47.

Billet A, Jia Y, Jensen T, et al. Regulation of the cystic fibrosis transmembrane conductance regulator anion channel by tyrosine phosphorylation. FASEB J. 2015;29:3945–53.

Bobadilla JL, Macek M Jr, Fine JP, et al. Cystic fibrosis: a worldwide analysis of CFTR mutations—correlation with incidence data and application to screening. Hum Mutat. 2002;19:575–606.

Bombieri C, Claustres M, De Boeck K, et al. Recommendations for the classification of diseases as CFTR-related disorders. J Cyst Fibros. 2011;10(Suppl 2):S86–S102.

Bonvin E, Le Rouzic P, Bernaudin J-F, et al. Congenital tracheal malformation in cystic fibrosis transmembrane conductance regulator-deficient mice. J Physiol. 2008;586:3231–43.

Borowitz D, Parad RB, Sharp JK, et al. Cystic Fibrosis Foundation practice guidelines for the management of infants with cystic fibrosis transmembrane conductance regulator-related metabolic syndrome during the first two years of life and beyond. J Pediatr. 2009;155(Suppl 6):S106–16.

Bose SJ, Krainer G, Ng DRS, et al. Towards next generation therapies for cystic fibrosis: folding, function and pharmacology of CFTR. J Cyst Fibros. 2020;19(Suppl 1):S25–32.

Boucher RC. Airway surface dehydration in cystic fibrosis: pathogenesis and therapy. Annu Rev Med. 2007;58:157–70.

Bozoky Z, Krzeminski M, Muhandiram R, et al. Regulatory R region of the CFTR chloride channel is a dynamic integrator of phospho-dependent intra- and intermolecular interactions. Proc Natl Acad Sci U S A. 2013;110:E4427–36.

Brown CR, Hong-Brown LQ, Biwersi J, et al. Chemical chaperones correct the mutant phenotype of the ΔF508 cystic fibrosis transmembrane conductance regulator protein. Cell Stress Chaperones. 1996;1:117–25.

Burgel P-R, Durieu I, Chiron R, et al. Rapid improvement after starting elexacaftor–tezacaftor–ivacaftor in patients with cystic fibrosis and advanced pulmonary disease. Am J Respir Crit Care Med. 2021;204:64–73.

Byrnes LJ, Xu Y, Qiu X, et al. Sites associated with Kalydeco binding on human cystic fibrosis transmembrane conductance regulator revealed by hydrogen/deuterium exchange. Sci Rep. 2018;8:4664.

Cai Z, Sheppard DN. Phloxine B interacts with the cystic fibrosis transmembrane conductance regulator at multiple sites to modulate channel activity. J Biol Chem. 2002;277:19546–53.

Cai Z, Scott-Ward TS, Sheppard DN. Voltage-dependent gating of the cystic fibrosis transmembrane conductance regulator Cl^- channel. J Gen Physiol. 2003;122:605–20.

Cai Z, Taddei A, Sheppard DN. Differential sensitivity of the cystic fibrosis (CF)-associated mutants G551D and G1349D to potentiators of the cystic fibrosis transmembrane conductance regulator (CFTR) Cl^- channel. J Biol Chem. 2006;281:1970–7.

Chang XB, Tabcharani JA, Hou YX, et al. Protein kinase A (PKA) still activates CFTR chloride channel after mutagenesis of all 10 PKA consensus phosphorylation sites. J Biol Chem. 1993;268:11304–11.

Chen T-Y, Hwang T-C. CLC-0 and CFTR: chloride channels evolved from transporters. Physiol Rev. 2008;88:351–87.

Chen J-H, Cai Z, Li H, et al. Function of CFTR protein: ion transport. In: Bush A, Alton EWFW, Davies JC, Griesenbach U, Jaffe A, editors. Cystic fibrosis in the 21st century. Basel: Karger; 2006. p. p38–44.

Cheng SH, Gregory RJ, Marshall J, et al. Defective intracellular transport and processing of CFTR is the molecular basis of most cystic fibrosis. Cell. 1990;63:827–34.

Cheng SH, Rich DP, Marshall J, et al. Phosphorylation of the R domain by cAMP-dependent protein kinase regulates the CFTR chloride channel. Cell. 1991;66:1027–36.

Cholon DM, Quinney NL, Fulcher ML, et al. Potentiator ivacaftor abrogates pharmacological correction of ΔF508 CFTR in cystic fibrosis. Sci Transl Med. 2014;6:246ra96.

Cil O, Phuan P-W, Lee S, et al. CFTR activator increases intestinal fluid secretion and normalizes stool output in a mouse model of constipation. Cell Mol Gastroenterol Hepatol. 2016;2:317–27.

Clancy JP, Rowe SM, Accurso FJ, et al. Results of a phase IIa study of VX-809, an investigational CFTR corrector compound, in subjects with cystic fibrosis homozygous for the *F508del-CFTR* mutation. Thorax. 2012;67:12–8.

Collins FS. Realizing the dream of molecularly targeted therapies for cystic fibrosis. N Engl J Med. 2019;381:1863–5.

Cotten JF, Welsh MJ. Cystic fibrosis-associated mutations at arginine 347 alter the pore architecture of CFTR. J Biol Chem. 1999;274:5429–35.

Cotten JF, Ostedgaard LS, Carson MR, et al. Effect of cystic fibrosis-associated mutations in the fourth intracellular loop of cystic fibrosis transmembrane conductance regulator. J Biol Chem. 1996;271:21279–84.

Crawford I, Maloney PC, Zeitlin PL, et al. Immunocytochemical localization of the cystic fibrosis gene product CFTR. Proc Natl Acad Sci U S A. 1991;88:9262–6.

Crawford DK, Mullenders J, Pott J, et al. Targeting G542X CFTR nonsense alleles with ELX-02 restores CFTR function in human-derived intestinal organoids. J Cyst Fibros. 2021;20:436–42.

Csanády L, Töröcsik B. Cystic fibrosis drug ivacaftor stimulates CFTR channels at picomolar concentrations. The National Library of Medicine abbreviation is Elife. 2019;8:e46450.

Csanády LS, Nairn AC, Gadsby DC. Thermodynamics of CFTR channel gating: a spreading conformational change initiates an irreversible gating cycle. J Gen Physiol. 2006;128:523–33.

Csanády L, Vergani P, Gadsby DC. Strict coupling between CFTR's catalytic cycle and gating of its Cl^- ion pore revealed by distributions of open channel burst durations. Proc Natl Acad Sci U S A. 2010;107:1241–6.

Csanády L, Vergani P, Gadsby DC. Structure, gating, and regulation of the CFTR anion channel. Physiol Rev. 2019;99:707–38.

Cui G, Zhang Z-R, O'Brien ARW, et al. Mutations at arginine 352 alter the pore architecture of CFTR. J Membr Biol. 2008;222:91–106.

Cui G, Stauffer BB, Imhoff BR, et al. VX-770-mediated potentiation of numerous human CFTR disease mutants is influenced by phosphorylation level. Sci Rep. 2019;9:13460.

Cuppens H, Lin W, Jaspers M, et al. Polyvariant mutant cystic fibrosis transmembrane conductance regulator genes. The polymorphic (TG)m locus explains the partial penetrance of the T5 polymorphism as a disease mutation. J Clin Invest. 1998;101:487–96.

Cutting GR. Cystic fibrosis genetics: from molecular understanding to clinical application. Nat Rev Genet. 2015;16:45–56.

Dagenais RVE, Su VCH, Quon BS. Real-world safety of CFTR modulators in the treatment of cystic fibrosis: a systematic review. J Clin Med. 2020;10:23.

Dalemans W, Barbry P, Champigny G, et al. Altered chloride ion channel kinetics associated with the ΔF508 cystic fibrosis mutation. Nature. 1991;354:526–8.

Davies JC, Cunningham S, Harris WT, et al. Safety, pharmacokinetics, and pharmacodynamics of ivacaftor in patients aged 2–5 years with cystic fibrosis and a CFTR gating mutation (KIWI): an open-label, single-arm study. Lancet Respir Med. 2016;4:107–15.

Davies JC, Moskowitz SM, Brown C, et al. VX-659–tezacaftor–ivacaftor in patients with cystic fibrosis and one or two phe508del alleles. N Engl J Med. 2018;379:1599–611.

Davies JC, Wainwright CE, Sawicki GS, et al. Ivacaftor in infants aged 4 to <12 months with cystic fibrosis and a gating mutation. Results of a two-part phase 3 clinical trial. Am J Respir Crit Care Med. 2021;203:585–93.

De Boeck K, Amaral MD. Progress in therapies for cystic fibrosis. Lancet Respir Med. 2016;4:662–74.

De Boeck K, Davies JC. Where are we with transformational therapies for patients with cystic fibrosis? Curr Opin Pharmacol. 2017;34:70–5.

De Boeck KD, Wilschanski M, Castellani C, et al. Cystic fibrosis: terminology and diagnostic algorithms. Thorax. 2006;61:627–35.

Denning GM, Anderson MP, Amara JF, et al. Processing of mutant cystic fibrosis transmembrane conductance regulator is temperature-sensitive. Nature. 1992a;358:761–4.

Denning GM, Ostedgaard LS, Cheng SH, et al. Localization of cystic fibrosis transmembrane conductance regulator in chloride secretory epithelia. J Clin Invest. 1992b;89:339–49.

Desai M, Hine C, Whitehouse JL, et al. Who are the 10%? - non eligibility of cystic fibrosis (CF) patients for highly effective modulator therapies. Respir Med. 2022;199:106878.

Donaldson SH, Solomon GM, Zeitlin PL, et al. Pharmacokinetics and safety of cavosonstat (N91115) in healthy and cystic fibrosis adults homozygous for *F508DEL-CFTR*. J Cyst Fibros. 2017;16:371–9.

Du K, Sharma M, Lukacs GL. The ΔF508 cystic fibrosis mutation impairs domain-domain interactions and arrests post-translational folding of CFTR. Nat Struct Mol Biol. 2005;12:17–25.

Dukovski D, Villella A, Bastos C, et al. Amplifiers co-translationally enhance CFTR biosynthesis via PCBP1-mediated regulation of CFTR mRNA. J Cyst Fibros. 2020;19:733–41.

Eckford PDW, Li C, Ramjeesingh M, et al. Cystic fibrosis transmembrane conductance regulator (CFTR) potentiator VX-770 (ivacaftor) opens the defective channel gate of mutant CFTR in a phosphorylation-dependent but ATP-independent manner. J Biol Chem. 2012;287:36639–49.

Elborn JS. Cystic fibrosis. Lancet. 2016;388:2519–31.

Farinha CM, King-Underwood J, Sousa M, et al. Revertants, low temperature, and correctors reveal the mechanism of F508del-CFTR rescue by VX-809 and suggest multiple agents for full correction. Chem Biol. 2013;20:943–55.

Fiedorczuk K, Chen J. Mechanism of CFTR correction by type I folding correctors. Cell. 2022a;185:158–168.e11.

Fiedorczuk K, Chen J. Molecular structures reveal synergistic rescue of Δ508 CFTR by Trikafta modulators. Science. 2022b;378:284–90.

Fischer H. Mechanisms and function of DUOX in epithelia of the lung. Antioxid Redox Signal. 2009;11:2453–65.

Flores AM, Casey SD, Felix CM, et al. Small-molecule CFTR activators increase tear secretion and prevent experimental dry eye disease. FASEB J. 2016;30:1789–97.

Flume PA, Liou TG, Borowitz DS, et al. Ivacaftor in subjects with cystic fibrosis who are homozygous for the F508del-CFTR mutation. Chest. 2012;142:718–24.

French PJ, Bijman J, Edixhoven M, et al. Isotype-specific activation of cystic fibrosis transmembrane conductance regulator-chloride channels by cGMP-dependent protein kinase II. J Biol Chem. 1995;270:26626–31.

Frizzell RA, Hanrahan JW. Physiology of epithelial chloride and fluid secretion. Cold Spring Harb Perspect Med. 2012;2:a009563.

Gabillard-Lefort C, Casey M, Glasgow AMA, et al. Trikafta rescues CFTR and lowers monocyte P2X7R-induced inflammasome activation in cystic fibrosis. Am J Respir Crit Care Med. 2022;205:783–94.

Gabriel SE, Brigman KN, Koller BH, et al. Cystic fibrosis heterozygote resistance to cholera toxin in the cystic fibrosis mouse model. Science. 1994;266:107–9.

Gadsby DC, Nairn AC. Control of CFTR channel gating by phosphorylation and nucleotide hydrolysis. Physiol Rev. 1999;79(Suppl 1):S77–S107.

Gaillard EA, Kota P, Gentzsch M, et al. Regulation of the epithelial Na^+ channel and airway surface liquid volume by serine proteases. Pflugers Arch. 2010;460:1–17.

Gao X, Hwang T-C. Localizing a gate in CFTR. Proc Natl Acad Sci U S A. 2015;112:2461–6.

Gentzsch M, Cholon DM, Quinney NL, et al. Airway epithelial inflammation in vitro augments the rescue of mutant CFTR by current CFTR modulator therapies. Front Pharmacol. 2021;12:628722.

Gopalsamy A, Bennett EM, Shi M, et al. Identification of pyrimidine derivatives as hSMG-1 inhibitors. Bioorg Med Chem Lett. 2012;22:6636–41.

Goralski JL, Hoppe JE, Mall MA, et al. Phase 3 open-label clinical trial of elexacaftor/tezacaftor/ivacaftor in children aged 2–5 years with cystic fibrosis and at least one F508del allele. Am J Respir Crit Care Med. 2023;208:59–67.

Graeber SY, Renz DM, Stahl M, et al. Effects of elexacaftor/tezacaftor/ivacaftor therapy on lung clearance index and magnetic resonance imaging in patients with cystic fibrosis and one or two F508del alleles. Am J Respir Crit Care Med. 2022a;206:311–20.

Graeber SY, Vitzthum C, Pallenberg ST, et al. Effects of elexacaftor/tezacaftor/ivacaftor therapy on CFTR function in patients with cystic fibrosis and one or two F508del alleles. Am J Respir Crit Care Med. 2022b;205:540–9.

Gray RD, Hardisty G, Regan KH, et al. Delayed neutrophil apoptosis enhances NET formation in cystic fibrosis. Thorax. 2018;73:134–44.

Gregory RJ, Rich DP, Cheng SH, et al. Maturation and function of cystic fibrosis transmembrane conductance regulator variants bearing mutations in putative nucleotide-binding domains 1 and 2. Mol Cell Biol. 1991;11:3886–93.

Haardt M, Benharouga M, Lechardeur D, et al. C-terminal truncations destabilize the cystic fibrosis transmembrane conductance regulator without impairing its biogenesis: a novel class of mutation. J Biol Chem. 1999;274:21873–7.

Haggie PM, Phuan P-W, Tan J-A, et al. Correctors and potentiators rescue function of the truncated W1282X-cystic fibrosis transmembrane regulator (CFTR) translation product. J Biol Chem. 2017;292:771–85.

Han ST, Rab A, Pellicore MJ, et al. Residual function of cystic fibrosis mutants predicts response to small molecule CFTR modulators. JCI Insight. 2018;3:e121159.

Hanrahan JW, Matthes E, Carlile G, et al. Corrector combination therapies for F508del-CFTR. Curr Opin Pharmacol. 2017;34:105–11.

Hansson GC. Mucus and mucins in diseases of the intestinal and respiratory tracts. J Intern Med. 2019;285:479–90.

Harrison PT. CFTR RNA- and DNA-based therapies. Curr Opin Pharmacol. 2022;65:102247.

Hawkins FJ, Kotton DN. Pulmonary ionocytes challenge the paradigm in cystic fibrosis. Trends Pharmacol Sci. 2018;39:852–4.

Haws CM, Nepomuceno IB, Krouse ME, et al. ΔF508-CFTR channels: kinetics, activation by forskolin, and potentiation by xanthines. Am J Physiol Cell Physiol. 1996;270:C1544–55.

He L, Kota P, Aleksandrov AA, et al. Correctors of ΔF508 CFTR restore global conformational maturation without thermally stabilizing the mutant protein. FASEB J. 2013;27:536–45.

Hegedűs T, Aleksandrov A, Mengos A, et al. Role of individual R domain phosphorylation sites in CFTR regulation by protein kinase A. Biochim Biophys Acta Biomembr. 2009;1788:1341–9.

Heijerman HGM, McKone EF, Downey DG, et al. Efficacy and safety of the elexacaftor plus tezacaftor plus ivacaftor combination regimen in people with cystic fibrosis homozygous for the F508del mutation: a double-blind, randomised, phase 3 trial. Lancet. 2019;394:1940–8.

Hillenaar T, Beekman J, van der Sluijs P, et al. Redefining hypo- and hyper-responding phenotypes of CFTR mutants for understanding and therapy. Int J Mol Sci. 2022;23:15170.

Hisert KB, Heltshe SL, Pope C, et al. Restoring cystic fibrosis transmembrane conductance regulator function reduces airway bacteria and inflammation in people with cystic fibrosis and chronic lung infections. Am J Respir Crit Care Med. 2017;195:1617–28.

Hoegger MJ, Fischer AJ, McMenimen JD, et al. Impaired mucus detachment disrupts mucociliary transport in a piglet model of cystic fibrosis. Science. 2014;345:818–22.

Hoffmann B, Elbahnsi A, Lehn P, et al. Combining theoretical and experimental data to decipher CFTR 3D structures and functions. Cell Mol Life Sci. 2018;75:3829–55.

Holland IB, Cole SPC, Kuchler K, et al., editors. ABC proteins. From bacteria to man. London: Academic Press; 2003.

Howard M, Frizzell RA, Bedwell DM. Aminoglycoside antibiotics restore CFTR function by overcoming premature stop mutations. Nat Med. 1996;2:467–9.

Hudson RP, Dawson JE, Chong PA, et al. Direct binding of the corrector VX-809 to human CFTR NBD1: evidence of an allosteric coupling between the binding site and the NBD1:CL4 interface. Mol Pharmacol. 2017;92:124–35.

Hug N, Longman D, Cáceres JF. Mechanism and regulation of the nonsense-mediated decay pathway. Nucleic Acids Res. 2016;44:1483–95.

Hwang T-C, Sheppard DN. Gating of the CFTR Cl^- channel by ATP-driven nucleotide-binding domain dimerisation. J Physiol. 2009;587:2151–61.

Hwang T-C, Yeh J-T, Zhang J, et al. Structural mechanisms of CFTR function and dysfunction. J Gen Physiol. 2018;150:539–70.

Jensen TJ, Loo MA, Pind S, et al. Multiple proteolytic systems, including the proteasome, contribute to CFTR processing. Cell. 1995;83:129–35.

Jih K-Y, Hwang T-C. Vx-770 potentiates CFTR function by promoting decoupling between the gating cycle and ATP hydrolysis cycle. Proc Natl Acad Sci U S A. 2013;110:4404–9.

Jih K-Y, Li M, Hwang T-C, et al. The most common cystic fibrosis-associated mutation destabilizes the dimeric state of the nucleotide-binding domains of CFTR. J Physiol. 2011;589:2719–31.

Jih K-Y, Sohma Y, Hwang T-C. Nonintegral stoichiometry in CFTR gating revealed by a pore-lining mutation. J Gen Physiol. 2012;140:347–59.

Jih K-Y, Lin W-Y, Sohma Y, et al. CFTR potentiators: from bench to bedside. Curr Opin Pharmacol. 2017;34:98–104.

Jouret F, Devuyst O. Targeting chloride transport in autosomal dominant polycystic kidney disease. Cell Signal. 2020;73:109703.

Jouret F, Bernard A, Hermans C, et al. Cystic fibrosis is associated with a defect in apical receptor–mediated endocytosis in mouse and human kidney. J Am Soc Nephrol. 2007;18:707–18.

Karpowich N, Martsinkevich O, Millen L, et al. Crystal structures of the MJ1267 ATP binding cassette reveal an induced-fit effect at the ATPase active site of an ABC transporter. Structure. 2001;9:571–86.

Kartner N, Augustinas O, Jensen TJ, et al. Mislocalization of ΔF508 CFTR in cystic fibrosis sweat gland. Nat Genet. 1992;1:321–7.

Keating D, Marigowda G, Burr L, et al. VX-445–tezacaftor–ivacaftor in patients with cystic fibrosis and one or two phe508del alleles. N Engl J Med. 2018;379:1612–20.
Keeling KM, Xue X, Gunn G, et al. Therapeutics based on stop codon readthrough. Annu Rev Genomics Hum Genet. 2014;15:371–94.
Kerem E. ELX-02: an investigational read-through agent for the treatment of nonsense mutation-related genetic disease. Expert Opin Investig Drugs. 2020;29:1347–54.
Kerem E, Konstan MW, De Boeck K, et al. Ataluren for the treatment of nonsense-mutation cystic fibrosis: a randomised, double-blind, placebo-controlled phase 3 trial. Lancet Respir Med. 2014;2:539–47.
Kim D, Huang J, Billet A, et al. Pendrin mediates bicarbonate secretion and enhances cystic fibrosis transmembrane conductance regulator function in airway surface epithelia. Am J Respir Cell Mol Biol. 2019;60:705–16.
Kirchner S, Cai Z, Rauscher R, et al. Alteration of protein function by a silent polymorphism linked to tRNA abundance. PLoS Biol. 2017;15:e2000779.
Kleizen B, Hunt JF, Callebaut I, et al. CFTR: new insights into structure and function and implications for modulation by small molecules. J Cyst Fibros. 2020;19(Suppl 1):S19–24.
Kleizen B, van Willigen M, Mijnders M, et al. Co-translational folding of the first transmembrane domain of ABC-transporter CFTR is supported by assembly with the first cytosolic domain. J Mol Biol. 2021;433:166955.
Knowles MR, Durie PR. What is cystic fibrosis? N Engl J Med. 2002;347:439–42.
Konstan MW, Vandevanter DR, Rowe SM, et al. Efficacy and safety of ataluren in patients with nonsense-mutation cystic fibrosis not receiving chronic inhaled aminoglycosides: the international, randomized, double-blind, placebo-controlled ataluren confirmatory trial in cystic fibrosis (ACT CF). J Cyst Fibros. 2020;19:595–601.
Kopelman H, Durie P, Gaskin K, et al. Pancreatic fluid secretion and protein hyperconcentration in cystic fibrosis. N Engl J Med. 1985;312:329–34.
Kym PR, Wang X, Pizzonero M, et al. Recent progress in the discovery and development of small-molecule modulators of CFTR. Prog Med Chem. 2018;57:235–76.
Langron E, Prins S, Vergani P. Potentiation of the cystic fibrosis transmembrane conductance regulator by VX-770 involves stabilization of the pre-hydrolytic, O_1 state. Br J Pharmacol. 2018;175:3990–4002.
Laselva O, Eckford PDW, Bartlett C, et al. Functional rescue of c.3846G>A (W1282X) in patient-derived nasal cultures achieved by inhibition of nonsense mediated decay and protein modulators with complementary mechanisms of action. J Cyst Fibros. 2020;19:717–27.
Laselva O, Bartlett C, Gunawardena TNA, et al. Rescue of multiple class II CFTR mutations by elexacaftor+ tezacaftor+ivacaftor mediated in part by the dual activities of elexacaftor as both corrector and potentiator. Eur Respir J. 2021a;57:2002774.
Laselva O, Qureshi Z, Zeng Z-W, et al. Identification of binding sites for ivacaftor on the cystic fibrosis transmembrane conductance regulator. iScience. 2021b;24:102542.
Le Grand D, Gosling M, Baettig U, et al. Discovery of icenticaftor (QBW251), a cystic fibrosis transmembrane conductance regulator potentiator with clinical efficacy in cystic fibrosis and chronic obstructive pulmonary disease. J Med Chem. 2021;64:7241–60.
Lee H-LR, Dougherty JP. Pharmaceutical therapies to recode nonsense mutations in inherited diseases. Pharmacol Ther. 2012;136:227–66.
Levring J, Terry DS, Kilic Z, et al. CFTR function, pathology and pharmacology at single-molecule resolution. Nature. 2023;616:606–14.
Lewis HA, Buchanan SG, Burley SK, et al. Structure of nucleotide-binding domain 1 of the cystic fibrosis transmembrane conductance regulator. EMBO J. 2004;23:282–93.
Li C, Ramjeesingh M, Wang W, et al. ATPase activity of the cystic fibrosis transmembrane conductance regulator. J Biol Chem. 1996;271:28463–8.
Li H, Salomon JJ, Sheppard DN, et al. Bypassing CFTR dysfunction in cystic fibrosis with alternative pathways for anion transport. Curr Opin Pharmacol. 2017;34:91–7.

Li H, Pesce E, Sheppard DN, et al. Therapeutic approaches to CFTR dysfunction: from discovery to drug development. J Cyst Fibros. 2018a;17(Suppl 2):S14–21.

Li M-S, Cowley EA, El Hiani Y, et al. Functional organization of cytoplasmic portals controlling access to the cystic fibrosis transmembrane conductance regulator (CFTR) chloride channel pore. J Biol Chem. 2018b;293:5649–58.

Linsdell P. Architecture and functional properties of the CFTR channel pore. Cell Mol Life Sci. 2017;74:67–83.

Linsdell P, Hanrahan JW. Adenosine triphosphate–dependent asymmetry of anion permeation in the cystic fibrosis transmembrane conductance regulator chloride channel. J Gen Physiol. 1998;111:601–14.

Linsdell P, Evagelidis A, Hanrahan JW. Molecular determinants of anion selectivity in the cystic fibrosis transmembrane conductance regulator chloride channel pore. Biophys J. 2000;78:2973–82.

Liu F, Zhang Z, Csanády L, et al. Molecular structure of the human CFTR ion channel. Cell. 2017;169:85–95.e8.

Liu J, Bihler H, Farinha CM, et al. Partial rescue of F508del-CFTR channel gating with modest improvement of protein processing, but not stability by a dual-acting small molecule. Br J Pharmacol. 2018;175:1017–38.

Liu F, Zhang Z, Levit A, et al. Structural identification of a hotspot on CFTR for potentiation. Science. 2019;364:1184–8.

Liu J, Berg AP, Wang Y, et al. A small molecule CFTR potentiator restores ATP-dependent channel gating to the cystic fibrosis mutant G551D-CFTR. Br J Pharmacol. 2022;179:1319–37.

Loo TW, Bartlett MC, Clarke DM. Corrector VX-809 stabilizes the first transmembrane domain of CFTR. Biochem Pharmacol. 2013;86:612–9.

Lukacs GL, Verkman AS. CFTR: folding, misfolding and correcting the ΔF508 conformational defect. Trends Mol Med. 2012;18:81–91.

Lukacs GL, Chang XB, Bear C, et al. The ΔF508 mutation decreases the stability of cystic fibrosis transmembrane conductance regulator in the plasma membrane. Determination of functional half-lives on transfected cells. J Biol Chem. 1993;268:21592–8.

Lykke-Andersen S, Jensen TH. Nonsense-mediated mRNA decay: an intricate machinery that shapes transcriptomes. Nat Rev Mol Cell Biol. 2015;16:665–77.

Ma T, Thiagarajah JR, Yang H, et al. Thiazolidinone CFTR inhibitor identified by high-throughput screening blocks cholera toxin–induced intestinal fluid secretion. J Clin Invest. 2002;110:1651–8.

Mainz JG, Zagoya C, Polte L, et al. Elexacaftor-tezacaftor-ivacaftor treatment reduces abdominal symptoms in cystic fibrosis-early results obtained with the CF-specific CFAbd-score. Front Pharmacol. 2022;13:877118.

Mall MA, Mayer-Hamblett N, Rowe SM. Cystic fibrosis: emergence of highly effective targeted therapeutics and potential clinical implications. Am J Respir Crit Care Med. 2020;201:1193–208.

Manfredi C, Tindall JM, Hong JS, et al. Making precision medicine personal for cystic fibrosis. Science. 2019;365:220–1.

McCague AF, Raraigh KS, Pellicore MJ, et al. Correlating cystic fibrosis transmembrane conductance regulator function with clinical features to inform precision treatment of cystic fibrosis. Am J Respir Crit Care Med. 2019;199:1116–26.

McDonald EF, Sabusap CMP, Kim M, et al. Distinct proteostasis states drive pharmacologic chaperone susceptibility for cystic fibrosis transmembrane conductance regulator misfolding mutants. Mol Biol Cell. 2022;33:ar62.

McGarry ME, Gibb ER, Oates GR, et al. Left behind: the potential impact of CFTR modulators on racial and ethnic disparities in cystic fibrosis. Paediatr Respir Rev. 2022;42:35–42.

Mendoza JL, Schmidt A, Li Q, et al. Requirements for efficient correction of ΔF508 CFTR revealed by analyses of evolved sequences. Cell. 2012;148:164–74.

Meng X, Wang Y, Wang X, et al. Two small molecules restore stability to a subpopulation of the cystic fibrosis transmembrane conductance regulator with the predominant disease-causing mutation. J Biol Chem. 2017;292:3706–19.

Mense M, Vergani P, White DM, et al. In vivo phosphorylation of CFTR promotes formation of a nucleotide-binding domain heterodimer. EMBO J. 2006;25:4728–39.

Meyerholz DK, Stoltz DA, Pezzulo AA, et al. Pathology of gastrointestinal organs in a porcine model of cystic fibrosis. Am J Pathol. 2010;176:1377–89.

Middleton PG, Mall MA, Dřevínek P, et al. Elexacaftor–tezacaftor–ivacaftor for cystic fibrosis with a single phe508del allele. N Engl J Med. 2019;381:1809–19.

Mihályi C, Iordanov I, Töröcsik B, et al. Simple binding of protein kinase A prior to phosphorylation allows CFTR anion channels to be opened by nucleotides. Proc Natl Acad Sci U S A. 2020;117:21740–6.

Mijnders M, Kleizen B, Braakman I. Correcting CFTR folding defects by small-molecule correctors to cure cystic fibrosis. Curr Opin Pharmacol. 2017;34:83–90.

Miller AC, Comellas AP, Hornick DB, et al. Cystic fibrosis carriers are at increased risk for a wide range of cystic fibrosis-related conditions. Proc Natl Acad Sci U S A. 2020;117:1621–7.

Modiano G, Ciminelli BM, Pignatti PF. Cystic fibrosis: cystic fibrosis and lactase persistence: a possible correlation. Eur J Hum Genet. 2007;15:255–9.

Molinski SV, Ahmadi S, Ip W, et al. Orkambi® and amplifier co-therapy improves function from a rare *CFTR* mutation in gene-edited cells and patient tissue. EMBO Mol Med. 2017;9:1224–43.

Molinski SV, Shahani VM, Subramanian AS, et al. Comprehensive mapping of cystic fibrosis mutations to CFTR protein identifies mutation clusters and molecular docking predicts corrector binding site. Proteins. 2018;86:833–43.

Montoro DT, Haber AL, Biton M, et al. A revised airway epithelial hierarchy includes CFTR-expressing ionocytes. Nature. 2018;560:319–24.

Moran O, Galietta LJV, Zegarra-Moran O. Binding site of activators of the cystic fibrosis transmembrane conductance regulator in the nucleotide binding domains. Cell Mol Life Sci. 2005;62:446–60.

Mornon JP, Lehn P, Callebaut I. Atomic model of human cystic fibrosis transmembrane conductance regulator: membrane-spanning domains and coupling interfaces. Cell Mol Life Sci. 2008;65:2594–612.

Nagel G, Hwang T-C, Nastiuk KL, et al. The protein kinase A-regulated cardiac Cl^- channel resembles the cystic fibrosis transmembrane conductance regulator. Nature. 1992;360:81–4.

Namkung W, Park J, Seo Y, et al. Novel amino-carbonitrile-pyrazole identified in a small molecule screen activates wild-type and ΔF508 cystic fibrosis transmembrane conductance regulator in the absence of a cAMP agonist. Mol Pharmacol. 2013;84:384–92.

Okiyoneda T, Veit G, Dekkers JF, et al. Mechanism-based corrector combination restores ΔF508-CFTR folding and function. Nat Chem Biol. 2013;9:444–54.

Okuda K, Dang H, Kobayashi Y, et al. Secretory cells dominate airway CFTR expression and function in human airway superficial epithelia. Am J Respir Crit Care Med. 2021;203:1275–89.

Oren YS, McClure ML, Rowe SM, et al. The unfolded protein response affects readthrough of premature termination codons. EMBO Mol Med. 2014;6:685–701.

Ostedgaard LS, Baldursson O, Vermeer DW, et al. A functional R domain from cystic fibrosis transmembrane conductance regulator is predominantly unstructured in solution. Proc Natl Acad Sci U S A. 2000;97:5657–62.

Pallenberg ST, Pust MM, Rosenboom I, et al. Impact of elexacaftor/tezacaftor/ivacaftor therapy on the cystic fibrosis airway microbial metagenome. Microbiol Spectr. 2022;10:e0145422.

Pasyk S, Molinski S, Ahmadi S, et al. The major cystic fibrosis causing mutation exhibits defective propensity for phosphorylation. Proteomics. 2015;15:447–61.

Pedemonte N, Lukacs GL, Du K, et al. Small-molecule correctors of defective ΔF508-CFTR cellular processing identified by high-throughput screening. J Clin Invest. 2005;115:2564–71.

Pedemonte N, Caci E, Sondo E, et al. Thiocyanate transport in resting and IL-4-stimulated human bronchial epithelial cells: role of pendrin and anion channels. J Immunol. 2007;178:5144–53.

Pedemonte N, Tomati V, Sondo E, et al. Dual activity of aminoarylthiazoles on the trafficking and gating defects of the cystic fibrosis transmembrane conductance regulator chloride channel caused by cystic fibrosis mutations. J Biol Chem. 2011;286:15215–26.

Pezzulo AA, Tang XX, Hoegger MJ, et al. Reduced airway surface pH impairs bacterial killing in the porcine cystic fibrosis lung. Nature. 2012;487:109–13.

Phuan P-W, Yang B, Knapp JM, et al. Cyanoquinolines with independent corrector and potentiator activities restore ΔPhe508-cystic fibrosis transmembrane conductance regulator chloride channel function in cystic fibrosis. Mol Pharmacol. 2011;80:683–93.

Phuan P-W, Son J-H, Tan J-A, et al. Combination potentiator ('co-potentiator') therapy for CF caused by CFTR mutants, including N1303K, that are poorly responsive to single potentiators. J Cyst Fibros. 2018;17:595–606.

Phuan P-W, Tan J-A, Rivera AA, et al. Nanomolar-potency 'co-potentiator' therapy for cystic fibrosis caused by a defined subset of minimal function CFTR mutants. Sci Rep. 2019;9:17640.

Pier GB, Grout M, Zaidi T, et al. Salmonella typhi uses CFTR to enter intestinal epithelial cells. Nature. 1998;393:79–82.

Plasschaert LW, Žilionis R, Choo-Wing R, et al. A single-cell atlas of the airway epithelium reveals the CFTR-rich pulmonary ionocyte. Nature. 2018;560:377–81.

Poulsen JH, Fischer H, Illek B, et al. Bicarbonate conductance and pH regulatory capability of cystic fibrosis transmembrane conductance regulator. Proc Natl Acad Sci U S A. 1994;91:5340–4.

Quinton PM. Cystic fibrosis: lessons from the sweat gland. Physiology (Bethesda). 2007;22:212–25.

Quinton PM. Role of epithelial HCO_3^- transport in mucin secretion: lessons from cystic fibrosis. Am J Physiol Cell Physiol. 2010;299:C1222–33.

Rab A, Rowe SM, Raju SV, et al. Cigarette smoke and CFTR: implications in the pathogenesis of COPD. Am J Physiol Lung Cell Mol Physiol. 2013;305:L530–41.

Rabeh WM, Bossard F, Xu H, et al. Correction of both NBD1 energetics and domain interface is required to restore ΔF508 CFTR folding and function. Cell. 2012;148:150–63.

Raju SV, Lin VY, Liu L, et al. The cystic fibrosis transmembrane conductance regulator potentiator ivacaftor augments mucociliary clearance abrogating cystic fibrosis transmembrane conductance regulator inhibition by cigarette smoke. Am J Respir Cell Mol Biol. 2017;56:99–108.

Ramsey BW, Welsh MJ. AJRCCM: 100-year anniversary. Progress along the pathway of discovery leading to treatment and cure of cystic fibrosis. Am J Respir Crit Care Med. 2017;195:1092–9.

Ramsey BW, Davies J, McElvaney NG, et al. A CFTR potentiator in patients with cystic fibrosis and the *G551D* mutation. N Engl J Med. 2011;365:1663–72.

Raraigh KS, Lewis MH, Collaco JM, et al. Caution advised in the use of CFTR modulator treatment for individuals harboring specific CFTR variants. J Cyst Fibros. 2022;21:856–60.

Ratjen F, Bell SC, Rowe SM, et al. Cystic fibrosis. Nat Rev Dis Primers. 2015;1:15010.

Rauscher R, Bampi GB, Guevara-Ferrer M, et al. Positive epistasis between disease-causing missense mutations and silent polymorphism with effect on mRNA translation velocity. Proc Natl Acad Sci U S A. 2021;118:e2010612118.

Reddy MM, Quinton PM. Deactivation of CFTR-Cl conductance by endogenous phosphatases in the native sweat duct. Am J Physiol Cell Physiol. 1996;270:C474–80.

Rehman T, Karp PH, Tan P, et al. Inflammatory cytokines TNF-α and IL-17 enhance the efficacy of cystic fibrosis transmembrane conductance regulator modulators. J Clin Invest. 2021;131:e150398.

Ren HY, Grove DE, De La Rosa O, et al. VX-809 corrects folding defects in cystic fibrosis transmembrane conductance regulator protein through action on membrane-spanning domain 1. Mol Biol Cell. 2013;24:3016–24.

Ribeiro CMP, Hull-Ryde EA. Functional role of the ER stress transducer IRE1α in CF airway epithelial inflammation. Curr Opin Pharmacol. 2022;65:102258.

Rich DP, Berger HA, Cheng SH, et al. Regulation of the cystic fibrosis transmembrane conductance regulator Cl^- channel by negative charge in the R domain. J Biol Chem. 1993;268:20259–67.

Riordan JR, Rommens J, Kerem B, et al. Identification of the cystic fibrosis gene: cloning and characterization of complementary DNA. Science. 1989;245:1066–73.

Robert R, Norez C, Becq F. Disruption of CFTR chloride channel alters mechanical properties and cAMP-dependent Cl^- transport of mouse aortic smooth muscle cells. J Physiol. 2005;568:483–95.

Rosenberg MF, Kamis AB, Aleksandrov LA, et al. Purification and crystallization of the cystic fibrosis transmembrane conductance regulator (CFTR). J Biol Chem. 2004;279:39051–7.

Rosenfeld M, Wainwright CE, Higgins M, et al. Ivacaftor treatment of cystic fibrosis in children aged 12 to <24 months and with a CFTR gating mutation (ARRIVAL): a phase 3 single-arm study. Lancet Respir Med. 2018;6:545–53.

Rowe SM, Heltshe SL, Gonska T, et al. Clinical mechanism of the cystic fibrosis transmembrane conductance regulator potentiator ivacaftor in G551D-mediated cystic fibrosis. Am J Respir Crit Care Med. 2014;190:175–84.

Rowe SM, Daines C, Ringshausen FC, et al. Tezacaftor–ivacaftor in residual-function heterozygotes with cystic fibrosis. N Engl J Med. 2017;377:2024–35.

Rowe SM, Jones I, Dransfield MT, et al. Efficacy and safety of the CFTR potentiator icenticaftor (QBW251) in COPD: results from a phase 2 randomized trial. Int J Chron Obstruct Pulmon Dis. 2020;15:2399–409.

Roy B, Friesen WJ, Tomizawa Y, et al. Ataluren stimulates ribosomal selection of near-cognate tRNAs to promote nonsense suppression. Proc Natl Acad Sci U S A. 2016;113:12508–13.

Saint-Criq V, Gray MA. Role of CFTR in epithelial physiology. Cell Mol Life Sci. 2017;74:93–115.

Saint-Criq V, Wang Y, Delpiano L, et al. Extracellular phosphate enhances the function of F508del-CFTR rescued by CFTR correctors. J Cyst Fibros. 2021;20:843–50.

Salehi M, Iqbal M, Dube A, et al. Delayed hepatic necrosis in a cystic fibrosis patient taking elexacaftor/tezacaftor/ivacaftor (Kaftrio). Respir Med Case Rep. 2021;34:101553.

Salvatore D, Pepe A, Carnovale V, et al. Elexacaftor/tezacaftor/ivacaftor for CFTR variants giving rise to diagnostic uncertainty: personalised medicine or over-medicalisation? J Cyst Fibros. 2022;21:544–8.

Schewe M, Sun H, Mert Ü, et al. A pharmacological master key mechanism that unlocks the selectivity filter gate in K^+ channels. Science. 2019;363:875–80.

Scholl D, Sigoillot M, Overtus M, et al. A topological switch in CFTR modulates channel activity and sensitivity to unfolding. Nat Chem Biol. 2021;17:989–97.

Schwarzenberg SJ, Vu PT, Skalland M, et al. Elexacaftor/tezacaftor/ivacaftor and gastrointestinal outcomes in cystic fibrosis: report of promise-GI. J Cyst Fibros. 2023;22:282–9.

Seibert FS, Linsdell P, Loo TW, et al. Disease-associated mutations in the fourth cytoplasmic loop of cystic fibrosis transmembrane conductance regulator compromise biosynthetic processing and chloride channel activity. J Biol Chem. 1996;271:15139–45.

Sermet-Gaudelus I, Girodon E, Vermeulen F, et al. ECFS standards of care on CFTR-related disorders: diagnostic criteria of CFTR dysfunction. J Cyst Fibros. 2022;21:922–36.

Serohijos AWR, Hegedus T, Aleksandrov AA, et al. Phenylalanine-508 mediates a cytoplasmic-membrane domain contact in the CFTR 3D structure crucial to assembly and channel function. Proc Natl Acad Sci U S A. 2008;105:3256–61.

Shah VS, Meyerholz DK, Tang XX, et al. Airway acidification initiates host defense abnormalities in cystic fibrosis mice. Science. 2016;351:503–7.

Sharma N, Evans TA, Pellicore MJ, et al. Capitalizing on the heterogeneous effects of CFTR nonsense and frameshift variants to inform therapeutic strategy for cystic fibrosis. PLoS Genet. 2018;14:e1007723.

Sharma J, Keeling KM, Rowe SM. Pharmacological approaches for targeting cystic fibrosis nonsense mutations. Eur J Med Chem. 2020;200:112436.

Shaughnessy CA, Zeitlin PL, Bratcher PE. Elexacaftor is a CFTR potentiator and acts synergistically with ivacaftor during acute and chronic treatment. Sci Rep. 2021;11:19810.

Sheikh S, Britt RD Jr, Ryan-Wenger NA, et al. Impact of elexacaftor–tezacaftor–ivacaftor on bacterial colonization and inflammatory responses in cystic fibrosis. Pediatr Pulmonol. 2023;58:825–33.

Sheppard DN, Rich DP, Ostedgaard LS, et al. Mutations in CFTR associated with mild-disease-form Cl^- channels with altered pore properties. Nature. 1993;362:160–4.

Sheppard DN, Ostedgaard LS, Winter MC, et al. Mechanism of dysfunction of two nucleotide binding domain mutations in cystic fibrosis transmembrane conductance regulator that are associated with pancreatic sufficiency. EMBO J. 1995;14:876–83.

Sheppard DN, Bear CE, de Jonge HR. Editorial overview: respiratory: transformational therapies for cystic fibrosis. Curr Opin Pharmacol. 2017;34:viii–xi.

Smith SS, Steinle ED, Meyerhoff ME, et al. Cystic fibrosis transmembrane conductance regulator. Physical basis for lyotropic anion selectivity patterns. J Gen Physiol. 1999;114:799–818.

Solomon GM, Fu L, Rowe SM, et al. The therapeutic potential of CFTR modulators for COPD and other airway diseases. Curr Opin Pharmacol. 2017;34:132–9.

Sorum B, Czégé D, Csanády L. Timing of CFTR pore opening and structure of its transition state. Cell. 2015;163:724–33.

Sosnay PR, Siklosi KR, Van Goor F, et al. Defining the disease liability of variants in the cystic fibrosis transmembrane conductance regulator gene. Nat Genet. 2013;45:1160–7.

Spoletini G, Gillgrass L, Pollard K, et al. Dose adjustments of elexacaftor/tezacaftor/ivacaftor in response to mental health side effects in adults with cystic fibrosis. J Cyst Fibros. 2022;21:1061–5.

Sprague RS, Ellsworth ML, Stephenson AH, et al. Deformation-induced ATP release from red blood cells requires CFTR activity. Am J Physiol Heart Circ Physiol. 1998;275:H1726–32.

Stoltz DA, Meyerholz DK, Welsh MJ. Origins of cystic fibrosis lung disease. N Engl J Med. 2015;372:351–62.

Sun X, Yi Y, Yan Z, et al. In utero and postnatal VX-770 administration rescues multiorgan disease in a ferret model of cystic fibrosis. Sci Transl Med. 2019;11:eaau7531.

Swiatecka-Urban A, Brown A, Moreau-Marquis S, et al. The short apical membrane half-life of rescued ΔF508-cystic fibrosis transmembrane conductance regulator (CFTR) results from accelerated endocytosis of ΔF508-CFTR in polarized human airway epithelial cells. J Biol Chem. 2005;280:36762–72.

Szentpetery S, Foil K, Hendrix S, et al. A case report of CFTR modulator administration via carrier mother to treat meconium ileus in a F508del homozygous fetus. J Cyst Fibros. 2022;21:721–4.

Szollosi A, Muallem DR, Csanády L, et al. Mutant cycles at CFTR's non-canonical ATP-binding site support little interface separation during gating. J Gen Physiol. 2011;137:549–62.

Tabcharani JA, Chang X-B, Riordan JR, et al. Phosphorylation-regulated Cl^- channel in CHO cells stably expressing the cystic fibrosis gene. Nature. 1991;352:628–31.

Tabcharani JA, Rommens JM, Hou Y-X, et al. Multi-ion pore behaviour in the CFTR chloride channel. Nature. 1993;366:79–82.

Tabcharani JA, Linsdell P, Hanrahan JW. Halide permeation in wild-type and mutant cystic fibrosis transmembrane conductance regulator chloride channels. J Gen Physiol. 1997;110:341–54.

Taylor-Cousar JL, Jain R. Maternal and fetal outcomes following elexacaftor-tezacaftor-ivacaftor use during pregnancy and lactation. J Cyst Fibros. 2021;20:402–6.

Taylor-Cousar JL, Munck A, McKone EF, et al. Tezacaftor–ivacaftor in patients with cystic fibrosis homozygous for phe508del. N Engl J Med. 2017;377:2013–23.

Tewkesbury DH, Athwal V, Bright-Thomas RJ, et al. Longitudinal effects of elexacaftor/tezacaftor/ivacaftor on liver tests at a large single adult cystic fibrosis centre. J Cyst Fibros. 2023;22:256–62.

Thomas C, Aller SG, Beis K, et al. Structural and functional diversity calls for a new classification of ABC transporters. FEBS Lett. 2020;594:3767–75.

Tomati V, Costa S, Capurro V, et al. Rescue by elexacaftor-tezacaftor-ivacaftor of the G1244E cystic fibrosis mutation's stability and gating defects are dependent on cell background. J Cyst Fibros. 2023;22:525–37.

Travis SM, Berger HA, Welsh MJ. Protein phosphatase 2C dephosphorylates and inactivates cystic fibrosis transmembrane conductance regulator. Proc Natl Acad Sci U S A. 1997;94:11055–60.

Tsai M-F, Li M, Hwang T-C. Stable ATP binding mediated by a partial NBD dimer of the CFTR chloride channel. J Gen Physiol. 2010;135:399–414.

Tümmler B. Post-approval studies with the CFTR modulators elexacaftor-tezacaftor—ivacaftor. Front Pharmacol. 2023;14:1158207.

Uluer AZ, MacGregor G, Azevedo P, et al. Safety and efficacy of vanzacaftor–tezacaftor–deutivacaftor in adults with cystic fibrosis: randomised, double-blind, controlled, phase 2 trials. Lancet Respir Med. 2023;11:550–62.

Van der Plas SE, Kelgtermans H, De Munck T, et al. Discovery of N-(3-carbamoyl-5,5,7,7-tetramethyl-5,7-dihydro-4H-thieno[2,3-c]pyran-2-yl)-lH-pyrazole-5-carboxamide (GLPG1837), a novel potentiator which can open class III mutant cystic fibrosis transmembrane conductance regulator (CFTR) channels to a high extent. J Med Chem. 2018;61:1425–35.

Van Goor F, Hadida S, Grootenhuis PDJ, et al. Rescue of CF airway epithelial cell function in vitro by a CFTR potentiator, VX-770. Proc Natl Acad Sci U S A. 2009;106:18825–30.

Van Goor F, Hadida S, Grootenhuis PDJ, et al. Correction of the F508del-CFTR protein processing defect in vitro by the investigational drug VX-809. Proc Natl Acad Sci U S A. 2011;108:18843–8.

Van Goor F, Yu H, Burton B, et al. Effect of ivacaftor on CFTR forms with missense mutations associated with defects in protein processing or function. J Cyst Fibros. 2014;13:29–36.

van Willigen M, Vonk AM, Yeoh HY, et al. Folding–function relationship of the most common cystic fibrosis–causing CFTR conductance mutants. Life Sci Alliance. 2019;2:e201800172.

Veit G, Avramescu RG, Chiang AN, et al. From CFTR biology toward combinatorial pharmacotherapy: expanded classification of cystic fibrosis mutations. Mol Biol Cell. 2016;27:424–33.

Veit G, Da Fonte DF, Avramescu RG, et al. Mutation-specific dual potentiators maximize rescue of CFTR gating mutants. J Cyst Fibros. 2020a;19:236–44.

Veit G, Roldan A, Hancock MA, et al. Allosteric folding correction of F508del and rare CFTR mutants by elexacaftor-tezacaftor-ivacaftor (Trikafta) combination. JCI Insight. 2020b;5:e139983.

Veit G, Vaccarin C, Lukacs GL. Elexacaftor co-potentiates the activity of F508del and gating mutants of CFTR. J Cyst Fibros. 2021;20:895–8.

Venturini A, Borrelli A, Musante I, et al. Comprehensive analysis of combinatorial pharmacological treatments to correct nonsense mutations in the CFTR gene. Int J Mol Sci. 2021;22:11972.

Vergani P, Nairn AC, Gadsby DC. On the mechanism of MgATP-dependent gating of CFTR Cl^- channels. J Gen Physiol. 2003;121:17–36.

Vergani P, Lockless SW, Nairn AC, et al. CFTR channel opening by ATP-driven tight dimerization of its nucleotide-binding domains. Nature. 2005;433:876–80.

Verkman AS, Galietta LJV. Chloride channels as drug targets. Nat Rev Drug Discov. 2009;8:153–71.

Volkova N, Moy K, Evans J, et al. Disease progression in patients with cystic fibrosis treated with ivacaftor: data from national US and UK registries. J Cyst Fibros. 2020;19:68–79.

Wainwright CE, Elborn JS, Ramsey BW, et al. Lumacaftor–ivacaftor in patients with cystic fibrosis homozygous for phe508del CFTR. N Engl J Med. 2015;373:220–31.

Wang F, Zeltwanger S, Hu S, et al. Deletion of phenylalanine 508 causes attenuated phosphorylation-dependent activation of CFTR chloride channels. J Physiol. 2000;524:637–48.

Wang Y, Wrennall JA, Cai Z, et al. Understanding how cystic fibrosis mutations disrupt CFTR function: from single molecules to animal models. Int J Biochem Cell Biol. 2014;52:47–57.

Wang Y, Cai Z, Gosling M, et al. Potentiation of the cystic fibrosis transmembrane conductance regulator Cl^- channel by ivacaftor is temperature independent. Am J Physiol Lung Cell Mol Physiol. 2018;315:L846–57.

Wang C, Yang Z, Loughlin BJ, et al. Mechanism of dual pharmacological correction and potentiation of human CFTR. bioRxiv. 2022:2022.10.10.510913.

Ward CL, Omura S, Kopito RR. Degradation of CFTR by the ubiquitin-proteasome pathway. Cell. 1995;83:121–7.

Watson MS, Cutting GR, Desnick RJ, et al. Cystic fibrosis population carrier screening: 2004 revision of American College of Medical Genetics mutation panel. Genet Med. 2004;6:387–91.

Welch EM, Barton ER, Zhuo J, et al. PTC124 targets genetic disorders caused by nonsense mutations. Nature. 2007;447:87–91.

Welsh MJ, Smith AE. Molecular mechanisms of CFTR chloride channel dysfunction in cystic fibrosis. Cell. 1993;73:1251–4.

Wilschanski M, Yahav Y, Yaacov Y, et al. Gentamicin-induced correction of CFTR function in patients with cystic fibrosis and *CFTR* stop mutations. N Engl J Med. 2003;349:1433–41.

Xue X, Mutyam V, Tang L, et al. Synthetic aminoglycosides efficiently suppress cystic fibrosis transmembrane conductance regulator nonsense mutations and are enhanced by ivacaftor. Am J Respir Cell Mol Biol. 2014;50:805–16.

Xue X, Mutyam V, Thakerar A, et al. Identification of the amino acids inserted during suppression of CFTR nonsense mutations and determination of their functional consequences. Hum Mol Genet. 2017;26:3116–29.

Yang H, Shelat AA, Guy RK, et al. Nanomolar affinity small molecule correctors of defective ΔF508-CFTR chloride channel gating. J Biol Chem. 2003;278:35079–85.

Yeh J-T, Hwang T-C. Positional effects of premature termination codons on the biochemical and biophysical properties of CFTR. J Physiol. 2020;598:517–41.

Yeh H-I, Yeh J-T, Hwang T-C. Modulation of CFTR gating by permeant ions. J Gen Physiol. 2015;145:47–60.

Yeh H-I, Sohma Y, Conrath K, et al. A common mechanism for CFTR potentiators. J Gen Physiol. 2017;149:1105–18.

Yeh H-I, Qiu L, Sohma Y, et al. Identifying the molecular target sites for CFTR potentiators GLPG1837 and VX-770. J Gen Physiol. 2019;151:912–28.

Yeh H-I, Yu Y-C, Kuo P-L, et al. Functional stability of CFTR depends on tight binding of ATP at its degenerate ATP-binding site. J Physiol. 2021;599:4625–42.

Yeh H-I, Sutcliffe KJ, Sheppard DN, et al. CFTR modulators: from mechanism to targeted therapeutics. Handb Exp Pharmacol. Online ahead of print. 2022

Yu H, Burton B, Huang C-J, et al. Ivacaftor potentiation of multiple CFTR channels with gating mutations. J Cyst Fibros. 2012;11:237–45.

Yu Y-C, Sohma Y, Hwang T-C. On the mechanism of gating defects caused by the R117H mutation in cystic fibrosis transmembrane conductance regulator. J Physiol. 2016;594:3227–44.

Zemanick ET, Taylor-Cousar JL, Davies J, et al. A phase 3 open-label study of elexacaftor/tezacaftor/ivacaftor in children 6 through 11 years of age with cystic fibrosis and at least one *F508del* allele. Am J Respir Crit Care Med. 2021;203:1522–32.

Zhang Z, Chen J. Atomic structure of the cystic fibrosis transmembrane conductance regulator. Cell. 2016;167:1586–1597.e9.

Zhang Z, Liu F, Chen J. Conformational changes of CFTR upon phosphorylation and ATP binding. Cell. 2017;170:483–491.e8.

Zhang Z, Liu F, Chen J. Molecular structure of the ATP-bound, phosphorylated human CFTR. Proc Natl Acad Sci U S A. 2018;115:12757–62.

Zielenski J, Tsui L-C. Cystic fibrosis: genotypic and phenotypic variations. Annu Rev Genet. 1995;29:777–807.

Validation of TMEM16A Modulation as a Therapeutic Approach for the Treatment of Cystic Fibrosis: The Discovery of Novel TMEM16A Potentiators

Henry Danahay and Martin Gosling

Abstract The discovery of transmembrane member 16A (TMEM16A) as the molecular identity of Ca $^{2+}$-activated Cl^- channels (CaCCs) has reignited interest in the modulation of these channels as a therapeutic approach for the treatment of muco-obstructive respiratory diseases, including cystic fibrosis. However, the recent literature is paradoxical with some investigators suggesting that inhibition of TMEM16A is the correct approach, while others advocate increasing (potentiating) the channels activity. The lack of effective pharmacological tools has contributed to this lack of clarity on which modality will deliver benefit. In this chapter we describe the discovery of a novel tool compound (ET002663), which effectively potentiates the activity of TMEM16A, and the studies undertaken to validate the concept of enhancing TMEM16A activity as a novel approach for the treatment of cystic fibrosis.

Keywords TMEM16A · Cystic fibrosis · Potentiator · Bronchial epithelial cells

1 TMEM16A, a Ca^{2+}-Activated Chloride Channel

Ca^{2+}-activated Cl^- channels (CaCCs) activate in response to an increase in intracellular calcium ($[Ca^{2+}]_i$) level and are an important transducer in linking Ca^{2+} signaling to ion homeostasis and cellular electrical activity (Pedemonte and Galietta 2014). CaCCs have widespread expression, encompassing both excitable and non-excitable cells, further highlighting their potential for influencing a variety of vital physiological functions ranging from blood pressure to mucociliary clearance (MCC) (Huang et al. 2012a). Although CaCC currents were first described in 1981 (Cross 1981; Schlichter and Elinson 1981), their molecular identity has been

H. Danahay · M. Gosling (✉)
Enterprise Therapeutics Limited, Sussex Innovation Centre, Sussex, UK
e-mail: henry@enterprisetherapeutics.com; martin@enterprisetherapeutics.com

G. Stephens, E. Stevens (eds.), *Ion Channels as Targets in Drug Discovery*,
https://doi.org/10.1007/978-3-031-52197-3_9

elusive. Several candidates have been proposed including members of the chloride channel accessory family (CLCAs) (Cunningham et al. 1995; Agnel et al. 1999) and bestrophins (Petrukhin et al. 1998; Sun et al. 2002), but these proteins have either (1) failed to recapitulate key CaCC biophysical characteristics (activation by $[Ca^{2+}]_i$ in the sub-micromolar range, outward rectification at low $[Ca^{2+}]_I$ but linear at high, and an ion permeability sequence of $NO_3^- > I^- > Br^- > Cl^- > F^-$), (2) found to be secreted rather than membrane proteins, or (3) their knockdown has not affected native CaCC currents.

The riddle of the CaCC molecular identity was finally solved in 2008 with the identification of transmembrane member 16A (TMEM16A) by three different laboratories (Caputo et al. 2008; Schroeder et al. 2008; Yang et al. 2008) whom all showed that expression of TMEM16A generated currents with classic CaCC biophysical properties and pharmacology. The TMEM16A gene is a member of a family of 10 genes in mammals (Pedemonte and Galietta 2014). Although the 10 paralogs share relatively high sequence homologies, their functions are more diverse. TMEM16A and B form CaCCs, while D, E, F, J, and K function predominantly as lipid scramblases which translocate phospholipids between the lipid layers of cellular membranes; they are also reported to exhibit weak cationic channel activity (reviewed by Kalienkova et al. 2021).

Structurally, TMEM16A is a homodimer, with each subunit contributing an independent pore to the overall channel structure (Paulino et al. 2017). Bioinformatic hydrophobicity analysis of the amino acid sequence suggested that the TMEM16A protein has eight transmembrane domains – this led to the protein also being called "anoctamin-1" (Ano-1) (Yang et al. 2008). However, experimentally determined structures using cryo-EM have clearly shown that the TMEM16A subunit has 10 transmembrane domains, so Ano-1 is arguably an inaccurate and obsolete term (Brunner et al. 2014; Paulino et al. 2017). Cryo-EM has provided significant advances in our understanding of the structure function of TMEM16A including early insight into the binding sites for low molecular weight channel modulators (Cheng et al. 2022; Lam et al. 2022; reviewed Nguyen and Chen 2022).

2 TMEM16A Pharmacology

The pharmacology of CaCCs has been severely hampered by the lack of a molecular identity of the channel. Prior to the discovery of TMEM16A, CaCC pharmacology was characterized by low potency, "generic" non-selective chloride channel blockers exemplified by fenamates (e.g., niflumic acid, flufenamic acid), stilbene sulfonic acids (e.g., 4,4′-diisothoicyano-2,2′-stilbenesulfoinc acid (DIDS)), and 5-nitro-2(3-phenylpropylamino)benzoic acid (NPPB). Efforts to find more potent and selective inhibitors using a screening assay employing cells expressing native functional CaCC yielded $CaCC_{inh}A01$, a low micromolar blocker (De La Fuente et al. 2008). Although originally reported to show some selectivity for CaCC, several reports have shown that $CaCC_{inh}A01$ also inhibits the cystic fibrosis transmembrane conductance regulator (CFTR) in the same concentration range (Gianotti et al. 2016; Miner et al. 2019).

The identification of TMEM16A has enabled high throughput screening versus the channel and catalyzed the discovery of a number of new modulators. This has included novel low molecular weight compounds, repurposed compounds, natural products, and the first reported positive modulators of TMEM16A (see Table 1). However, subsequent characterization of many of these compounds has shown they lack selectivity or modulate the channel only indirectly. Perhaps, the most notable of the latter group is E_{act}, the prototypical TMEM16A activator. In the authors' hands, E_{act} fails to increase TMEM16A currents in cells when $[Ca^{2+}]_i$ is controlled, suggesting its predominant mechanism of action upon the channel is indirect. Recent studies by Genovese et al. (2019) have identified TRPV4, a calcium permeable cation channel, as the target of E_{act} in epithelial cells, as opposed to TMEM16A. Although TMEM16A pharmacology is advancing, care should be taken when interpreting physiological studies with the current batch of widely used compounds. The reports of TMEM16A function now span a broad array of physiological processes, and it is not the purpose of this chapter to review these, but we would rather refer the reader to some excellent reviews (Pedemonte and Galietta 2014; Ji et al. 2019; Al-Hosni et al. 2022). In the context of the discovery of TMEM16A potentiators, we will instead focus specifically on the validation of TMEM16A as a target for the treatment of respiratory diseases.

3 Activation or Inhibition of TMEM16A for the Treatment of Respiratory Diseases?

The hydration of airway mucus is widely accepted as central to the effectiveness of mucus clearance in both health and disease (Batson et al. 2022). Too little fluid and the mucins in the mucus gel become hyperconcentrated, and the altered rheological properties reduce the effectiveness of mucus clearance to a point where static plugs can form, promoting both airway occlusion and microbial colonization. Hyperconcentrated mucins in the mucus gel have been demonstrated in cystic fibrosis (CF) as well chronic obstructive pulmonary disease (COPD) and related diseases (Henderson et al. 2014; Kesimer et al. 2017; Batson et al. 2022). In support of this theory, therapies to promote mucus hydration such as CFTR modulators and inhaled osmotic agents including hypertonic saline have been demonstrated to improve mucus clearance (Rowe et al. 2014; Altes et al. 2017) and to reduced infectious exacerbations of disease and improve lung function (Accurso et al. 2010; Ramsey et al. 2011). To this end, the role of the airway epithelial CaCC has been studied for >30 years as an approach to enhance fluid movement into the mucosa, particularly in CF where CFTR-mediated secretion is attenuated or even absent. The observations that the P2Y-receptor agonists ATP and UTP could stimulate anion secretion in CF airway epithelial cells (Knowles et al. 1991) and enhance mucus clearance

Table 1 Commonly used modulators of TMEM16A

Compound	Structure	Mechanism	Comments[a]	References
$CaCC_{inh}A01$	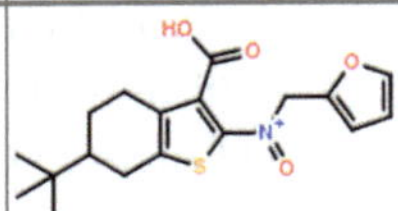	Inhibitor	TMEM16A IC_{50} ~ 5 μM but also blocks CFTR	De La Fuente et al. (2008), Gianotti et al. (2016) and Miner et al. (2019)
$T16_{inh}A01$	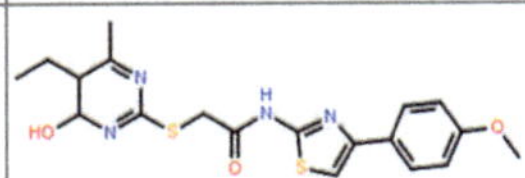	Inhibitor	TMEM16A IC_{50} 1.1 – >30 μM	Namkung et al. (2011a) and Miner et al. (2019)
1PBC; 1-hydroxy-3-(trifluoromethyl) pyrido[1,2-a] benzimidazole-4-carbonitrile	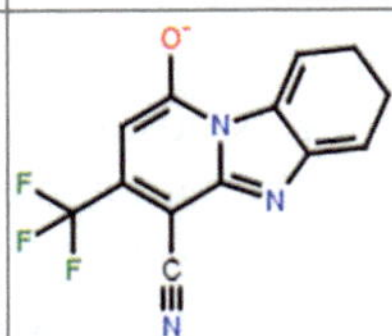	Inhibitor	IC_{50} 2.8–13 μM; pore blocker; also blocks TMEM16B; cryo-EM structure in complex with TMEM16A available	Peters et al. (2015), Miner et al. (2019) and Lam et al. (2022)
NTTP; *N*-(5-nitro-1,3-thiazol-2-yl)-2-thiophenecarboxamide	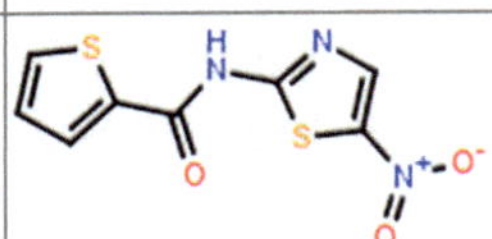	Inhibitor	TMEM16A IC_{50} 4.7–32.9 μM; pore blocker	Peters et al. (2015) and Miner et al. (2019)
MONNA; *N*-((4-methoxy)-2-napthyl)-5-nitroanthranilic acid	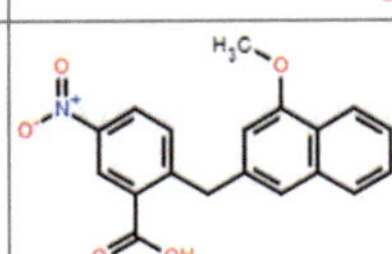	Inhibitor	TMEM16A IC_{50} 0.08–13.6 μM; blocks TMEM16B CFTR IC_{50} 18.99 μM	Oh et al. (2013), Seo et al. (2016) and Miner et al. (2019)
Ani9; 2-(4-chloro-2-methylphenoxy)-N-[(2-methoxyphenyl) methylideneamino]-acetamide	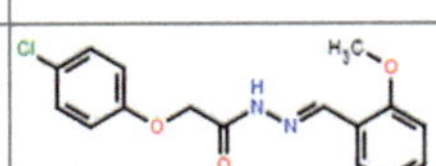	Inhibitor	TMEM16A IC_{50} 0.077 μM; selective over TMEM16B	Seo et al. (2016)
10bm/$TM_{inh}23$	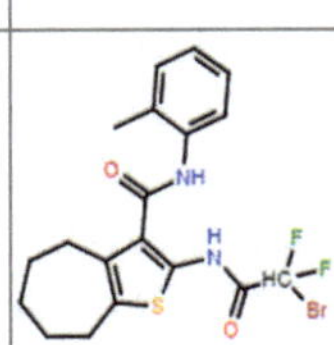	Inhibitor	TMEM16A IC_{50} 0.03 μM; blocks TMEM16B IC_{50} ~ 400 nM; no effect upon CFTR	Truong et al. (2017) and Cil et al. (2019)
Benzbromarone	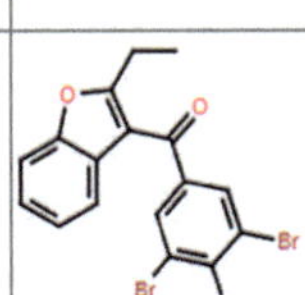	Inhibitor	TMEM16A IC_{50} 4.1–10.4 μM; CFTR IC_{50} 5.2 μM; also blocks TMEM16B	Huang et al. (2012b) and Miner et al. (2019)
Niclosamide	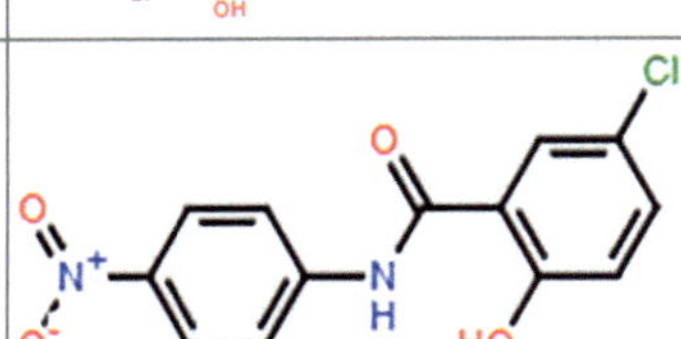	Inhibitor	TMEM16A IC_{50} 2.4 μM;	Miner et al. (2019)

(continued)

Table 1 (continued)

Compound	Structure	Mechanism	Comments[a]	References
Nitazoxanide	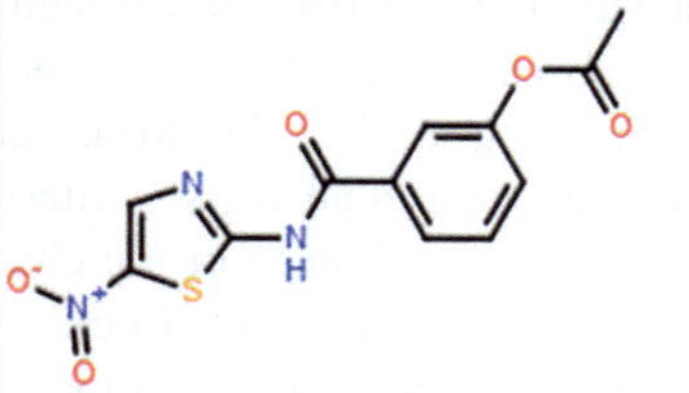	Inhibitor	TMEM16A IC_{50} 1.3 μM;	Miner et al. (2019)
E_{act}	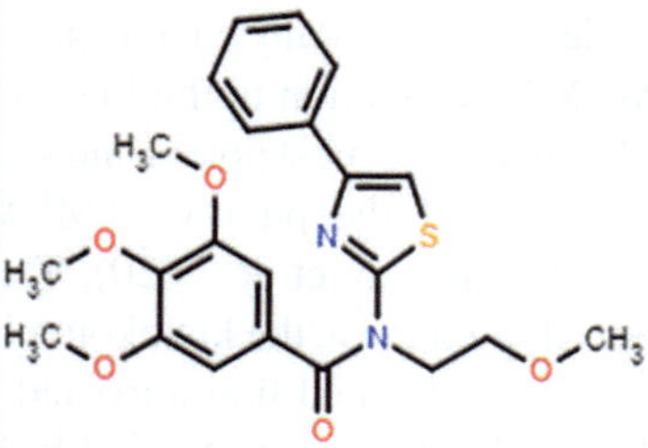	Activator	TMEM16A EC_{50} ~ 3 – > 30 μM	Namkung et al. (2011b), Liu et al. (2016) and Genovese et al. (2019)
F_{act}	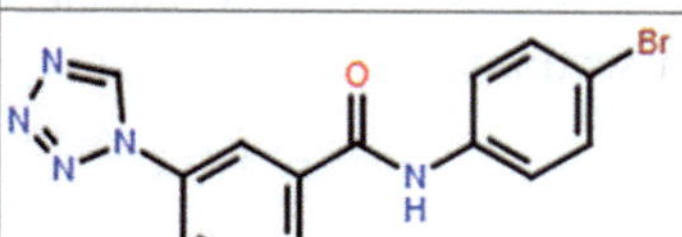	Activator	TMEM16A EC_{50} ~ 6 μM	Namkung et al. (2011b)

The structure and key parameters of TMEM16A modulators routinely used in the scientific literature. In the majority of cases, these molecules have low potency and are poorly selective questioning their utility in studies assessing the physiological role of TMEM16A. Ani9 is the most potent and selective inhibitor with a growing body of data suggesting that the "Act" series of compounds modulate TMEM16A indirectly via increasing intracellular calcium levels

[a]Data cited are either patch clamp or ion transport assay formats

in vivo (Sabater et al. 1999) triggered the development of a longer acting derivative, denufosol (INS-37217). P2Y-receptor stimulation leads to an immediate elevation of $[Ca^{2+}]_i$ that will promote anion secretion through the CaCC (Mason et al. 1991). It is likely that the rapid desensitization of the P2Y-receptors in the airway upon exposure to high concentrations of denufosol, together with a panoply of additional consequences of elevating $[Ca^{2+}]_i$ (including stimulating mucin secretion from goblet cells), contributed to the failure of denufosol in clinical studies (Moss 2013).

With this background, a selective potentiator of TMEM16A would be predicted to amplify any given stimulus to activate TMEM16A and thereby increase anion secretion in the right place at the right time. This profile would be significantly differentiated from denufosol in that the stimulation of TMEM16A and $[Ca^{2+}]_i$ remains under normal physiological regulation. However, recent reports have suggested that this profile of a potentiator might not be so straightforward as additional roles for TMEM16A have been proposed, ranging from the regulation of airway goblet cell function to bronchial smooth muscle contraction (Kunzelmann et al. 2019). Using a combination of mouse genetic models with TMEM16A silenced or TMEM16A inhibitors in human cells, several groups have reported that TMEM16A (1) positively regulates goblet cell metaplasia (Lin et al. 2015; Zhang et al. 2015; Qin et al. 2016; Hara et al. 2019), (2) promotes goblet cell-mediated secretion of airway mucins (Huang et al. 2012b; Benedetto et al. 2019), and (3) facilitates a contractile response in isolated human airways (Huang et al. 2012b; Gallos et al. 2013; Zhang et al. 2013; Danielsson et al. 2015; Wang et al. 2018; Miner et al. 2019). These observations are not however without some controversy. Alternative TMEM16A knockout studies have demonstrated that the absence of the channel increases airway goblet cell numbers and promotes mucus plugging (Rock et al. 2009; He et al. 2020) as a consequence of failed mucus clearance. Furthermore, a growing body of data is highlighting the lack of selectivity of many of the pharmacological tools used in an attempt to regulate TMEM16A function in the literature (see above section). In support of this, recent studies using the most potent and selective TMEM16A blocker Ani9 have failed to show any of the putative TMEM16A phenotypes described above (Simoes et al. 2019; Danahay et al. 2020). That a potentiator of TMEM16A function would necessarily juxtapose the knockout phenotype is also an assumption on the part of much of the published literature and represented a key scientific question that could only be addressed once a suitable TMEM16A potentiator had been identified (see later section).

4 Screening for Novel TMEM16A Potentiators

As the authors' prior experience of high throughput screening of ion channels, and that of others (e.g., Dunlop et al. 2008), has shown that assay format can influence output, Enterprise Therapeutics undertook a multi-format, parallel screening program to identify novel positive modulators (potentiators) of TMEM16A (Fig. 1). Both membrane potential and yellow fluorescent protein (YFP) formats have been used previously to screen Cl^- channels including CFTR and TMEM16A (e.g., Galietta et al. 2001; Caputo et al. 2008; Van Goor et al. 2009). Although these assay formats enable screening of medium-large size compound libraries, they can suffer from a number of assay artifacts, for example, fluorescent compounds (Dunlop et al. 2008). Additionally, and perhaps arguably most importantly for TMEM16A, these assay formats do not have control of $[Ca^{2+}]_i$. As such, compounds which increase $[Ca^{2+}]_i$ (e.g., an agonist of a G_q coupled GPCR) will appear as a positive hit despite not interacting with the TMEM16A channel directly. This limitation of the membrane potential and YFP formats was addressed by inclusion of orthoganol screening in a calcium flux assay and/or a QPatch automated patch clamp assay as part of the hit triage and confirmation process. The QPatch assay (Danahay et al. 2020; Gosling et al. 2021) utilized an intracellular solution with $[Ca^{2+}]_i$ clamped at the EC_{20} for TMEM16A activation along with significant Ca^{2+} buffering enabling the assay to readily remove compounds which had indirectly activated TMEM16A in the fluorescence assays by perturbation of $[Ca^{2+}]_i$. This advantage, combined with the ability to detect weak modulators due to a high signal-to-noise ratio (see Fig. 2), was a key factor in Enterprises' decision to perform two automated patch clamp

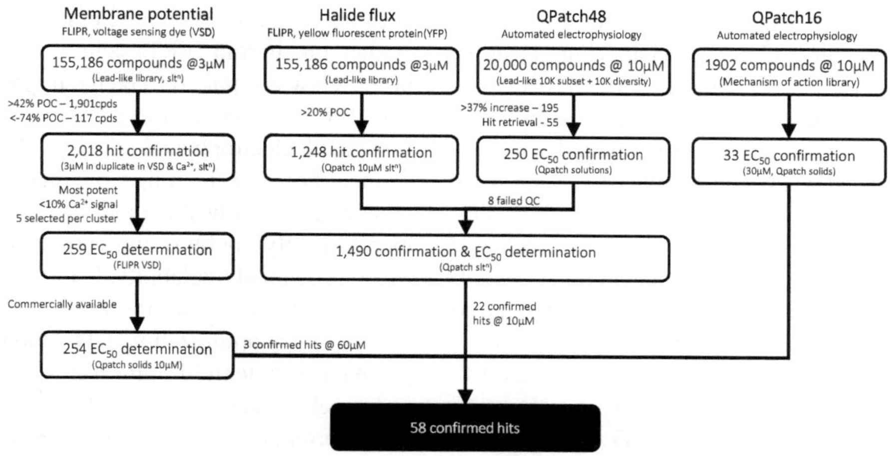

Fig. 1 Screening campaigns to identify TMEM16A potentiators. Enterprise undertook multiple fluorescence and electrophysiology screens in parallel. Membrane potential hits were confirmed prior to testing in a calcium flux assay as an orthogonal screen. Halide flux and electrophysiology screening outputs were triaged using automated patch clamp. All hits were finally confirmed using solid samples in electrophysiology to derive EC_{50} and efficacy values. Abbreviations: *POC* point of center, *sltⁿ* compound solution, *VSD* FLIPR voltage sensing dye, *YFP* halide flux screen using mutant yellow fluorescent protein

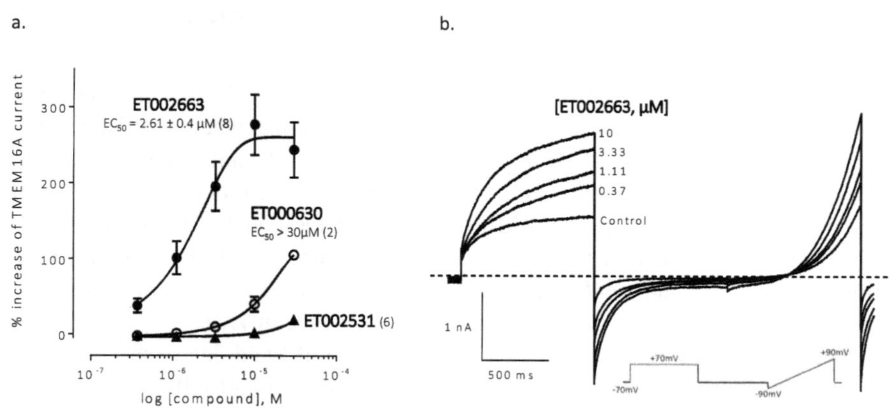

Fig. 2 ET002663 is an effective potentiator of recombinant hTMEM16A. Whole-cell automated patch clamp (QPatch) data for the screening hit ET000630 and associated compounds ET002663 and ET002531. The QPatch assay utilized a Fischer rat thyroid (FRT) cell line stably expressing the human TMEM16A abc variant. Intracellular Ca^{2+} was clamped at the EC_{20} for channel activation, and a step and ramp voltage protocol (as shown in lower inset in **b**) was applied at 0.1 Hz while cells were exposed to increasing concentrations of test compound. ET002663 is a partially optimized derivative of ET000630 with superior efficacy and potency in the QPatch assays (**a**, mean data ± SEM are shown (*n* detailed in parentheses)). ET002531 is a close structural analogue of ET002663 but almost devoid of any TMEM16A potentiator activity. ET002663 preserved the characteristic biophysical profile of TMEM16A (slow time-dependent activation and inactivation kinetics) and increased currents in both the inward and outward directions

campaigns alongside the fluorescence assays. In all cases, the screening assays in the hit finding flowchart (Fig. 1) utilized Fischer rat thyroid (FRT) cells stably expressing the human TMEM16Aabc isoform, which was reported to be the major isoform expressed by human airway epithelial cells (Caputo et al. 2008).

At the end of the compound screening campaign, 58 electrophysiologically validated hits from multiple compound series were identified (Fig. 1). A hit from one of these novel series was ET000630 – this compound gave a robust enhancement of TMEM16A currents in the QPatch assay, approximately doubling the current at 30 μM (Fig. 2a). This feature, along with attractive chemical tractability, led to the ET000630 series being selected to rapidly develop a tool potentiator for target validation studies. After several rounds of medicinal chemistry optimization, ET002663 was identified. ET002663 has an EC_{50} approximately tenfold improved over ET000630 and the ability to potentiate the channel by almost 300% (Fig. 2a). Biophysically, ET002663 potentiates TMEM16A currents, across the entire physiological voltage range, both inward and outward, and preserves the characteristic time-dependent gating parameters of the channel (Fig. 2b). In addition to the identification of ET002663, a close analogue was synthesized (ET002531) which was devoid of TMEM16A potentiation at 10 μM – these two compounds were selected for downstream validation studies.

5 Physiological Validation of TMEM16A Potentiators

Having identified compounds that could potentiate TMEM16A function under patch-clamp conditions, a key next question was to establish whether efficacy could be demonstrated in more physiologically and disease-relevant systems? Would a TMEM16A potentiator enhance Ca^{2+}-dependent anion secretory responses in cell systems expressing endogenous TMEM16A and in the context of an orchestra of ion channels and transporters maintaining epithelial function? What might the magnitude of effect be, and would it translate through to changes in mucosal hydration? Ultimately, could this mechanism demonstrate relevance in vivo?

ET002663, with a patch-clamp EC_{50} value of 2.6 μM and demonstrating 300% efficacy (increasing baseline currents by approximately threefold) together with a relatively benign selectivity profile, was a good candidate with which to address the aforementioned questions. Bronchial epithelial cells (HBE) derived from cadaveric or transplanted lungs provide what is considered to be the gold standard model of human airway mucosal biology when cultured at an air-liquid interface (ALI; Randell et al. 2011). The epithelium that forms is well-differentiated, with abundant secretory and ciliated cells capable of supporting mucociliary transport through the transport of ions and fluid in parallel with the secretion of mucins to form a mucus gel (Fulcher and Randell 2013).

Using the short-circuit current (ISC) technique, two formats of ion transport assay were performed using ALI cultures of HBE derived from CF lungs at transplantation (Danahay et al. 2020). In both formats, the ability of ET002663 to potentiate a Ca^{2+} stimulated anion secretory current was tested. When mounted in Ussing Chambers during the ion transport experiment, CF-HBE are "static," and $[Ca^{2+}]_i$ are at a low resting state. Under these conditions, a TMEM16A potentiator compound would be expected to have a relatively minor effect on anion secretory currents due to the low level of TMEM16A opening which could be enhanced. A true "potentiator" mechanism will effectively "amplify" the response to endogenous stimulation of the channel, as opposed to an "activator" mechanism, which can directly stimulate an effect, that is independent of any physiological signal. To this end, in the first format of ISC experiment, $[Ca^{2+}]_i$ was elevated using an EC_{20} concentration of the SERCA pump inhibitor cyclopiazonic acid (CPA). In CF-HBE, CPA induced a small but stable increase in ISC that was sensitive to the TMEM16A blocker CaCC-A01 (Fig. 3a). The addition of ET002663 to CPA treated CF-HBE induced a further increase in ISC that was also sensitive to CaCC-A01 (Fig. 3b). Further supporting the TMEM16A potentiator mechanism of ET002663 was the observation that ET002531, a structurally related but TMEM16A inactive analogue of ET002663, was without effect in this assay. In the second assay format, CF-HBE were treated with ET002663 prior to stimulating a rise in $[Ca^{2+}]_i$ with exogenous UTP, a P2Y-receptor agonist that is generated by the airway mucosa (Okada et al. 2011). With $[Ca^{2+}]_i$ at resting levels, the addition of ET002663 was without direct effect on the ISC (Fig. 3d); however, the subsequent stimulation with a threshold concentration (EC_{20}) of UTP evoked an increase in ISC that was significantly larger than in the vehicle control (Fig. 3e). Together these studies supported the TMEM16A potentiator mechanism of ET002663, in primary CF cells and under conditions where $[Ca^{2+}]_i$ was unbuffered and in the context of both apical and basolateral channels, pumps, and transporters.

As described in previous sections, the ability for calcium-stimulated secretory mechanisms to drive fluid secretion in the airway has been well described (Tarran et al. 2006). The question as to whether the TMEM16A potentiation observed in the ISC studies could translate through to a fluid secretory response, and if so by how much, remained unknown. To address this question, the effects of ET002663 and ET002531 were evaluated in CF-HBE under conditions of rotational shear stress to mimic some of the mechanical aspects of normal lung physiology. Rotational shear stress, at a force and rate to stimulate tidal breathing, has been demonstrated to stimulate the endogenous release of both ATP and UTP onto the mucosal surface of HBE which in turn drive anion secretory responses and transepithelial fluid movement which can be measured as changes in the height of airway surface liquid (ASL) (Tarran et al. 2005, 2006). ASL height is measured by confocal imaging of the thin film of fluid on the apical surface of the HBE that has been rendered fluorescent by the dissolution of a water soluble, impermeable dye (Texas Red dextran). Under these conditions, a 24 h exposure to rotational shear stress induced an increase in ASL height (Fig. 3f) that was significantly enhanced by ET002663. In contrast, the inactive analogue ET002531 was without effect, further supporting the concept that a TMEM16A potentiator could enhance fluid secretory responses of CF-HBE under conditions whereby $[Ca^{2+}]_i$ was physiologically regulated through the endogenous release of nucleotide triphosphates.

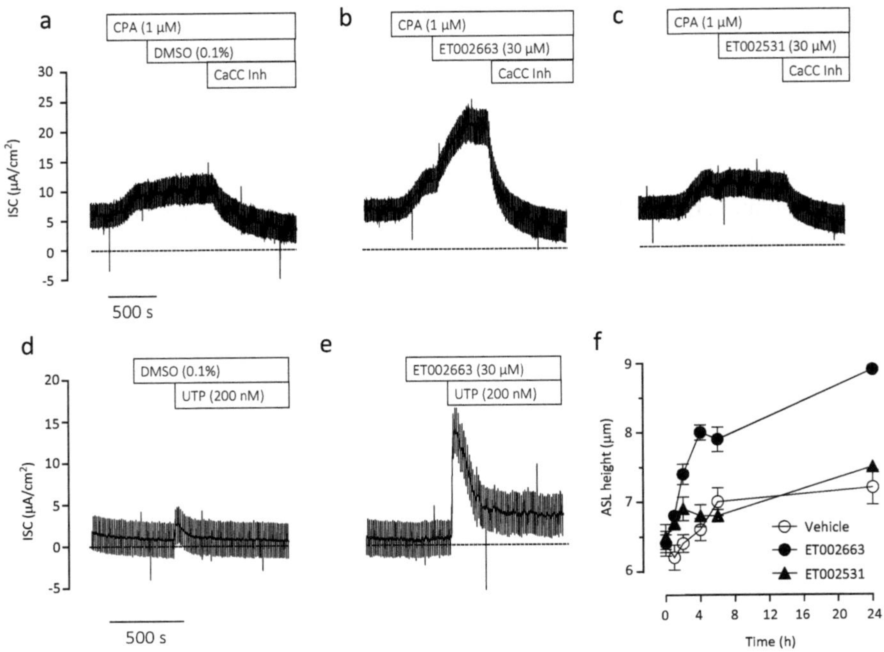

Fig. 3 ET002663 enhances anion and fluid secretion in primary human cells. Human bronchial epithelial cells (HBE) derived from CF lungs were cultured at air-liquid interface to generate a well-differentiated, ion and fluid transporting mucosa. Under short-circuit current (ISC) conditions, and in the presence of amiloride to block ENaC function, Ca^{2+}-dependent anion secretory responses could be stimulated by either the SERCA pump inhibitor CPA (**a–c**) or by the P2Y-receptor agonist UTP (**d**, **e**). Using a sub-maximal concentration of CPA, the increase in ISC was slow to develop and reached a stable plateau response (**a**) that was attenuated by the addition of the TMEM16A blocker, CaCC-InhA01 (CaCC Inh). The subsequent addition of ET002663 induced a further increase in ISC that remained sensitive to CaCC Inh (**b**). The addition of ET002531, a structurally related analogue of ET002663, but that fails to potentiate TMEM16A in patch-clamp assays, likewise had no effect on the CPA-stimulated ISC (**c**). A secretory ISC response of HBE was also stimulated by an EC_{20} concentration of UTP (**d**). When cells had been pre-treated with ET002663 for 5 min prior to the addition of UTP, the ISC response to the P2Y agonist was greatly enhanced both in terms of the peak increase in current and also the duration of the response (**e**). Note that the vertical deflections represent the ISC responses to ±1 mV pulses. Using confocal microscopy to measure the height of fluid (airway surface liquid, ASL) on the mucosal surface of CF HBE, ET002663 stimulated a prosecretory response (**f**). ET002531 had a similar effect to vehicle. Intracellular Ca^{2+} levels were endogenously controlled by the epithelium in the ASL height assay by placing the cells under continual, rotary shear stress, a maneuver to mimic breathing and the oscillatory release of endogenous ATP and UTP. Mean data ± SEM are shown ($n = 6$ individual inserts of cells/group)

To evaluate whether potentiating TMEM16A could promote airway mucociliary clearance in vivo, the effect of inhaled ET002663 on tracheal mucus velocity (TMV) was examined. In this model, radio-opaque disks are applied onto the tracheal surface of conscious, free-breathing sheep, and the rate of movement of the disks is measured (Sabater et al. 1999; Scott et al. 2017). Following an initial baseline

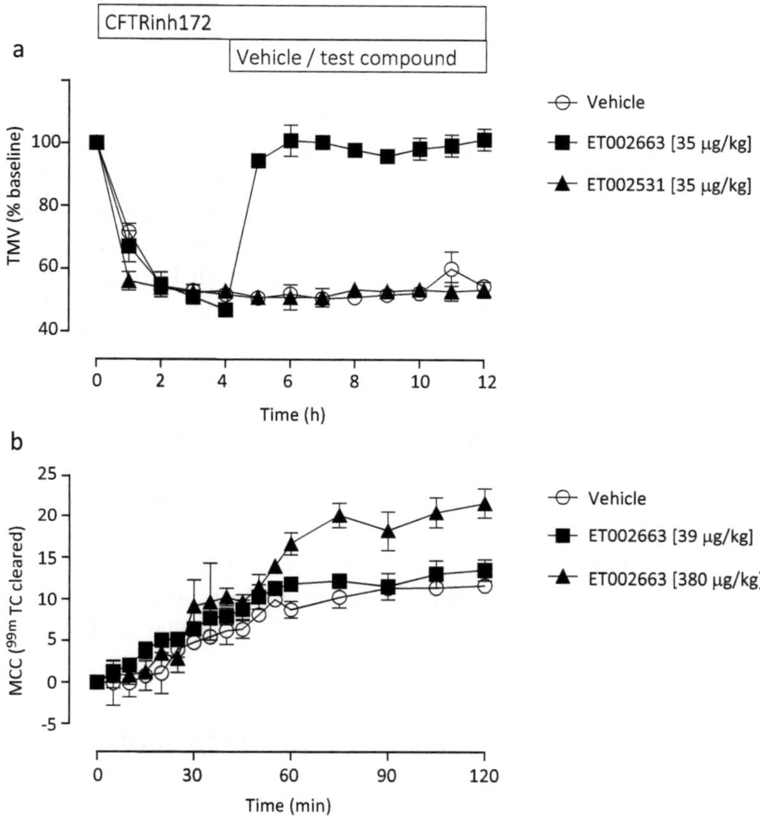

Fig. 4 ET002663 enhances airway mucus clearance in vivo. Inhaled dosing of ET002663 stimulated increases in airway mucociliary clearance in two in vivo models. (**a**) Tracheal mucus velocity (TMV) was measured in healthy conscious sheep. Immediately after measuring the baseline rate (100%), sheep were administered the CFTR inhibitor (CFTR Inh172) which caused a steady decline in TMV to approximately 50% of the baseline rate which remained depressed for at least 12 h. At 4 h, inhaled dosing with ET002663 stimulated a rise in TMV rate to approximately the pre-CFTR Inh172 levels. In contrast, dosing with vehicle or ET002531 was without effect. (**b**) Healthy conscious sheep were administered ET002663 or vehicle by inhalation. At 4 h after dosing, sheep were administered ^{99m}Tc labeled sulfur colloid particles, again by inhalation, and the rate of clearance of ^{99m}Tc was monitored by gamma-scintigraphy for a further 120 min. ET002663 stimulated an increase in the rate of whole lung mucociliary clearance with a duration of action of at least 4 h. Mean data ± SEM are shown (n = 2–3 sheep/group)

measurement of TMV, the CFTR inhibitor Inh-172 was inhaled by the animals following which TMV was slowed over the subsequent 4 h to approximately 50% of the initial rate (Fig. 4a). At this time either test compounds or vehicle were dosed to the sheep, again by inhalation, and TMV was monitored hourly for the next 8 h. Vehicle or ET002531 treated sheep showed no evidence of any change in TMV over this time with the rate remaining at approximately 50% of the starting baseline. In contrast, 60 min after dosing with ET002663, TMV had increased to the initial baseline rate and remained at this level for the duration of the study (Fig. 4a).

A second format of MCC assay in vivo was then utilized to understand whether a TMEM16A potentiator mechanism could (a) enhance whole lung MCC, rather than just in the trachea, (b) could enhance MCC in the presence of functional CFTR, that is, in the absence of a pre-dose with the CFTR inhibitor Inh-172, and (c) show a durable effect. Conscious, free-breathing sheep were again dosed with either ET002663 or vehicle by inhalation and 4 h later inhaled an aerosol of ^{99m}Tc-labeled sulfur colloid particles at which time they were placed under a gamma-camera and serial images of the lungs were obtained (Fig. 4b). Quantification of retained ^{99m}Tc then enabled an assessment of the rate and magnitude of MCC from the whole lung (Danahay et al. 2020). Vehicle-treated sheep cleared approximately 11% of the inhaled ^{99m}TC particles from the lungs over the 120 min assessment of MCC, while the ET002663 (380 μg/kg) group cleared almost double the quantity.

These early, pivotal efficacy studies, which although performed with a sub-optimal compound, provided a first insight into the opportunity for a TMEM16A potentiator to provide the increases in anion and fluid secretion required to promote mucus hydration (human cells) and to enhance mucus clearance (in vivo). On this basis, lead optimization was initiated that resulted in the discovery of ETD002 (ETX001, GDC-6988), a more potent and selective TMEM16A potentiator compound. The findings in both the human cell-based assays as well as the in vivo studies were further validated and extended with ETD002 (Danahay et al. 2020).

6 Exploring Additional Effects of TMEM16A Biology with TMEM16A Potentiators

Having discovered potent and selective TMEM16A potentiators such as ETD002, it was important to begin to evaluate what effect this mechanism might have on alternative areas of TMEM16A biology that had been proposed. Many of these studies had employed TMEM16A inhibitors to explore the various roles of the channel, so it was important to understand whether promoting channel activity may have additional effects that could be important from both an efficacy and safety perspective.

Roles proposed for TMEM16A in the airways beyond anion secretion and mucus hydration include (a) airway smooth muscle function (Huang et al. 2012b; Gallos et al. 2013; Zhang et al. 2013; Danielsson et al. 2015; Wang et al. 2018; Miner et al. 2019), (b) goblet cell formation (Lin et al. 2015; Zhang et al. 2015; Qin et al. 2016; Hara et al. 2019), and (c) mucin secretion (Huang et al. 2012b; Benedetto et al. 2019). In studies using freshly isolated human bronchial smooth muscle, ETD002 was without effect on either constrictor or relaxant responses, and in vivo there was no effect of ETD002 on lung function in rats when instilled directly into the airways (Danahay et al. 2020). When primary HBE were exposed to ETD002 for 4 consecutive days, there was no change in the density of mucin producing goblet cells, whereas the positive control treatment of IL-13 stimulated a profound elevation of MUC5AC producing cells. Similarly, ETD002 had no effect on goblet cell density when co-administered with IL-13. Exposure of either normal or IL-13 pre-treated HBE to ETD002 also failed to show any evidence of a stimulation of mucin release or goblet cell degranulation, when assessed both biochemically and using quantitative imaging of electron microscopy images. To evaluate goblet cell exocytosis in vivo, a goblet cell metaplasia was induced mice by repeated IL-13 exposure (Zhu et al. 2015). Subsequent exposure to inhaled ATP induced a rapid emptying of the goblet cells. Administration of ETX004, a TMEM16A potentiator that is structurally related to ETD002, shows high potency on murine TMEM16A and has extended systemic exposure failed to induce any emptying of the goblet cells at levels that would provide full target coverage.

In addition to functions in airway tissues, TMEM16A has also been reported to influence vascular smooth muscle function (Dam et al. 2014; Hubner et al. 2015; Leblanc et al. 2015). As an inhaled therapeutic, tissues in the lung will be exposed to higher concentrations than those in the periphery, so the pulmonary vasculature might be reasonably considered to be the vascular smooth muscle bed at greatest risk of a TMEM16A-mediated effect. Furthermore, a role for TMEM16A in the pulmonary artery with a putative role in pulmonary hypertension has been proposed (Papp et al. 2019). Using wire myography to detect any contractile or relaxant activity, ETD002 was without any effect on freshly isolated human pulmonary artery smooth muscle preparations. When dosed to telemetered rats by intra-tracheal instillation, there was likewise no effect on heart rate or blood pressure.

The lack of effects of TMEM16A potentiators on many of these endpoints is perhaps surprising in view of the literature; however, it should be considered that a potentiator of any given target will not necessarily generate the opposite phenotype to that established in a knockout mouse model. Furthermore, as discussed earlier in the chapter, many of these literature reports supported genetic models with the use of TMEM16A inhibitors which at best are non-selective and in some cases are not even direct blockers (Danahay et al. 2022), but indirectly affect the channel by influencing $[Ca^{2+}]_i$. In the human airway smooth muscle studies described above, niclosamide did relax the preparations, which is consistent with the published data (Miner et al. 2019) but through a mechanism that is unlikely to be TMEM16A-mediated, as the potent and selective blocker Ani9 was without effect. In goblet cell formation studies, Ani9 is without effect, whereas both niclosamide and benzbromarone appear to reduce goblet cell numbers. Niclosamide and benzbromarone also attenuated ciliated cell numbers in parallel, increased LDH levels in cell culture media and reduced the transepithelial resistance of primary HBE at ALI, effects consistent with cellular toxicity with these compounds rather than any TMEM16A pharmacology (Danahay et al. 2022).

7 Concluding Remarks

Understanding the pharmacology of any novel compounds is central to building confidence in the physiological process being regulated. In the case in hand, a willingness to initiate drug discovery programs to identify positive modulators of

TMEM16A may well have been hampered by the contrasting reports of physiological roles that could be considered serious safety risks and thus temper enthusiasm to understand the significance of any putative efficacy. Through the combined efforts to profile historical TMEM16A modulators as well as the early de-risking of newly discovered TMEM16A potentiators, it was possible to balance the risks of target engagement in the airways. To this end, the efficacy profiles of TMEM16A potentiators, together with some disarming of the potential literature-based side effect risk and a clean bill of health in 35-day safety studies in both rat and dog, have culminated in the progression of ETD002 toward clinical studies. Data from the now completed Phase 1 and upcoming Phase 2 programs are anticipated to understand whether this therapy could offer benefit to people with CF and especially those not genetically suited to the established CFTR modulators.

Acknowledgments The authors would like to thank the following colleagues and collaborators for their contributions to the work outlined in this chapter: Drs Sarah Lilley, Roy Fox, Holly Charlton (University of Sussex, UK); Drs Gary Clark, Emma Richardson (Charles River, UK); Dr. Brian Button (University of North Carolina Chapel Hill, US); Prof Bill Abraham, Dr. Juan Sabater (University of Miami, US).

References

Accurso FJ, Rowe SM, Clancy JP, Boyle MP, Dunitz JM, Durie PR, Sagel SD, Hornick DB, Konstan MW, Donaldson SH, Moss RB, Pilewski JM, Rubenstein RC, Uluer AZ, Aitken ML, Freedman SD, Rose LM, Mayer-Hamblett N, Dong Q, Zha J, Stone AJ, Olsen ER, Ordonez CL, Campbell PW, Ashlock MA, Ramsey BW. Effect of VX-770 in persons with cystic fibrosis and the G551D-CFTR mutation. N Engl J Med. 2010;363(21):1991–2003.

Agnel M, Vermat T, Culouscou JM. Identification of three novel members of the calcium-dependent chloride channel (CaCC) family predominantly expressed in the digestive tract and trachea. FEBS Lett. 1999;455:295–301.

Al-Hosni R, Ilkan Z, Agostinelli E, Tammaro P. The pharmacology of the TMEM16A channel: therapeutic opportunities. Trends in Pharmacological Sciences 2022;43(9):712–725.

Altes TA, Johnson M, Fidler M, Botfield M, Tustison NJ, Leiva-Salinas C, de Lange EE, Froh D, Mugler JP 3rd. Use of hyperpolarized helium-3 MRI to assess response to ivacaftor treatment in patients with cystic fibrosis. J Cyst Fibros. 2017 16(2):267–274.

Batson BD, Zorn BT, Radicioni G, Livengood SS, Kumagi T, Dang H, Ceppe A, Clapp PW, Tunney M, Elborn JS, McElvaney NG, Muhlebach MS, Boucher RC, Tiemeyer M, Wolfgang MC, Kesimer M. Cystic fibrosis airway mucus hyperconcentration prodcues a vicious cycle of mucin, pathogen, and inflammatory interactions that promotes disease persistence. Am J Respir Cell Mol Biol. 2022;67(2):253–65.

Benedetto R, Cabrita I, Schreiber R, Kunzelmann K. TMEM16A is indispensable for basal mucus secretion in airways and intestine. FASEB J. 2019;33(3):4502–12.

Brunner JD, Lim NK, Schenck S, Duerst A, Dutzler R. X-ray structure of a calcium-activated TMEM16 lipid scramblase. Nature. 2014;516:207–12.

Caputo A, Caci E, Ferrera L, Pedemonte N, Barsanti C, Sondo E, Pfeffer U, Ravazzolo R, Zegarra-Moran O, Galietta LJ. TMEM16A, a membrane protein associated with calcium-dependent chloride channel activity. Science. 2008;322:590–4.

Cheng Y, Feng S, Puchades C, Ko J, Figueroa E, Chen Y, Wu H, Gu S, Han T, Junrui L, Ho B, Shoichet B, Jan YN, Jan L. Identification of a conserved drug binding pocket in TMEM16 proteins. Res Sq. 2022;10:rs.3.rs-1296933.

Cil O, Anderson MO, Yen R, Kelleher B, Huymh TL, Seo Y, Nilsen SP, Turner JR, Verkman AS. Slowed gastric emptying and improved oral glucose tolerance produced by a nanomolar-potency inhibitor of calcium-activated chloride channel TMEM16A. FASEB J. 2019;33(10):11247–57.

Cross NL. Initiation of the activation potential by an increase in intracellular calcium in eggs of the frog, Rana pipiens. Dev Biol. 1981;85(2):380–4.

Cunningham SA, Awayda MS, Bubien JK, Ismailov II, Arrate MP, Berdiev BK, Benos DJ, Fuller CM. Cloning of an epithelial chloride channel from bovine trachea. J Biol Chem. 1995;270:31016–26.

Dam VS, Boedtkjer DMB, Aalkjaer C, Matchkov V. The bestrophin- and TMEM16A-associated Ca^{2+} activated Cl^- channels in vascular smooth muscle. Channels (Austin). 2014;8(4):361–9.

Danahay HL, Lilley S, Fox R, Charlton H, Sabater J, Button B, McCarthy C, Collingwood SP, Gosling M. TMEM16A potentiation, a novel therapeutic approach for the treatment of cystic fibrosis. Am J Respir Crit Care Med. 2020;201(8):946–54.

Danahay H, Lilley S, Adley K, Charlton H, Fox R, Gosling M. Niclosamide is not a blocker of the calcium activated chloride channel, TMEM16A. Front Pharmacol. 2022; Submitted

Danielsson J, Perez-Zoghbi J, Benrstein K, Barajas MB, Zhang Y, Kumar S, Sharma PK, Gallos G, Emala CW. Antagonists of the TMEM16A calcium-activated chloride channel modulate airway smooth muscle tone and intracellular calcium. Anesthesiology. 2015;123(3):569–81.

De La Fuente R, Namkung W, Mills A, Verkman AS. Small-molecule screen identifies inhibitors of a human intestinal calcium-activated chloride channel. Mol Pharmacol. 2008;73(3):758–68.

Dunlop J, Bowlbly M, Peri R, Tawa G, LaRocque J, Soloveva V, Morin J. Ion channel screening. Comb Chem High Throughput Screen. 2008;11(7):514–22.

Fulcher ML, Randell SH. Human nasal and tracheo-bronchial respiratory epithelial cell culture. Methods Mol Biol. 2013;945:109–21.

Galietta LV, Jayaraman S, Verkman AS. Cell-based assay for high-throughput quantitative screening of CFTR chloride transport agonists. Am J Phys. 2001;281(5):C1734–42.

Gallos G, Remy KE, Danielsson J, Funayama H, Fu XW, Chang H-YS, Yim P, Xu D, Emala CW. Functional expression of the TMEM16 family of calcium-activated chloride channels in airway smooth muscle. Am J Phys Lung Cell Mol Phys. 2013;305(9):L625–34.

Genovese M, Borrelli VA, Guidone D, Caci E, Viscido G, Gambardella G, di Bernardo D, Scudieri P, Galietta LJV. TRPV4 and purinergic signalling pathways are separately linked in airway epithelia to CFTR and TMEM16A chloride channels. J Physiol. 2019;597(24):5859–78.

Gianotti A, Ferrera L, Philip AR, Caci E, Zegarra-Moran O, Galietta LJV, Flores CA. Pharmacological analysis of epithelial chloride secretion mechanisms in adult murine airways. Eur J Pharmacol. 2016;781:100–8.

Gosling M, Lilley S, Binzer T, Bell D. Automated patch clamp in cystic fibrosis drug discovery. 2021.

Hara K, Kondo M, Tsuji M, Takeyama K, Tamaoki J. Clarithromycin suppresses IL-13-dependent goblet cell metaplasia via the TMEM16A-dependent pathway in guinea pig airway epithelial cells. Respir Investig. 2019;57(1):79–88.

He M, Wu B, Ye W, Le DD, Sinclair AW, Padovano V, Chen Y, Li K-X, Sit R, Tan M, Caplan MJ, Neff N, Jan YN, Darmanis S, Jan LY. Chloride channels regulate differentiation and barriers functions of the mammalian airway. elife. 2020;9:e53085.

Henderson AG, Ehre C, Button B, Abdullah LH, Cai L-H, Leigh MW, DeMaria GC, Matsui H, Donaldson SH, Davis CW, Sheehan JK, Boucher RC, Kesimer M. Cystic fibrosis airway secretions exhibit mucin hyperconcentration and increased osmotic pressure. J Clin Invest. 2014;124(7):3047–60.

Huang F, Wong X, Jan LY. International union of basic and clinical pharmacology. LXXXV: calcium-activated chloride channels. Pharmacol Rev. 2012a;64(1):1–15.

Huang F, Zhang H, Wu M, Yang H, Kudo M, Peters CJ, Woodruff PG, Solberg OD, Donne ML, Huang X, Sheppard D, Fahy JV, Wolters PJ, Hogan BL, Finkbeiner WE, Li M, Jan YN, Jan LY, Rock JR. Calcium-activated chloride channel TMEM16A modulates mucin secretion and airway smooth muscle contraction. Proc Natl Acad Sci U S A. 2012b;109(40):16354–9.

Hubner CA, Schroeder BC, Ehmke H. Regulation of vascular tone and arterial blood pressure: role of chloride transport in vascular smooth muscle. Pflugers Arch. 2015;467(3):605–14.

Ji Q, Guo S, Wang X, Pang C, Zhan Y, Chen Y, An H. Recent advances in TMEM16A: structure, function, and disease. J Cell Physiol. 2019;234(6):7856–73.

Kalienkova V, Mosina VC, Paulino C. The groovy TMEM16 family: molecular mechanisms of lipid scrambling and ion conduction. J Mol Biol. 2021;433(16):166941.

Kesimer M, Ford AA, Ceppe A, Radicioni G, Cao R, Davis CW, Doerschuk CM, Alexis NE, Anderson WH, Henderson AG, Barr RG, Bleecker ER, Christensen SA, Cooper CB, Han MK, Hansel NN, Hastie AT, Hoffman EA, Kanner RE, Martinez F, Paine R, Woodruff PG, O'Neal WK, Boucher RC. Airway mucin concentration as a marker of chronic bronchitis. N Engl J Med. 2017;377:911–22.

Knowles MR, Clarke LL, Boucher RC. Activation by extracellular nucleotides of chloride secretion in the airway epithelia of patients with cystic fibrosis. N Engl J Med. 1991;325(8):533–8.

Lam AKM, Rutz S, Dutzler R. Inhibition mechanism of the chloride channel TMEM16A by pore blocker 1PBC. Nat Commun. 2022;13(1):2798.

Leblanc N, Forrest AS, Ayon RJ, Wiwchar M, Angermann JE, Pritchard HA, Singer CA, Valencik ML, Britton F, Greenwood IA. Molecular and functional significance of Ca(2+)-activated Cl(-) channels in pulmonary arterial smooth muscle. Pulm Circ. 2015;5(2):244–268.

Lin J, Jiang Y, Liu Y, Tang H, Jiang D. TMEM16A mediates the hypersecretion of mucus induced by interleukin-13. Exp Cell Res. 2015;334(2):260–9.

Liu S, Feng J, Luo J, Yang P, Brett TJ, Hu H. Eact, a small molecule activator of TMEM16A activates TRPV1 and elicits pain- and itch-related behaviours. Br J Pharmacol. 2016;173(7):1208–18.

Mason SJ, Paradiso AM, Boucher RC. Regulation of transepithelial ion transport and intracellular calcium by extracellular ATP in human normal and cystic fibrosis airway epithelium. Br J Pharmacol. 1991;103(3):1649–56.

Miner K, Labitzke K, Liu B, Wang P, Henckels K, Gaida K, Elliott R, Chen JJ, Lui L, Leith A, Trublood E, Hensley K, Xia X-Z, Homann O, Bennett B, Fiorino M, Whoriskey J, Yu G, Escobar S, Wong M, Born TL, Budelsky A, Comeau M, Smith D, Phillips J, Johnston JA, McGivern JG, Weikl K, Powers D, Kunzelmann K, Mohn D, Hochheimer A, Sullivan JK. Drug repurposing: the antihelmintics niclosamide and nitazoxanide are potent TMEM16A antagonists that fully bronchodilate airways. Front Pharmacol. 2019;10:51.

Moss RB. Pitfalls of drug development: lessons learned from the trials of denufosol in cystic fibrosis. J Pediatr. 2013;162(4):676–80.

Namkung W, Phuan P-W, Verkman AS. TMEM16A inhibitors reveal TMEM16A as a minor component of calcium-activated chloride channel conductance in airway and intestinal epithelial cells. J Biol Chem. 2011a;286(3):2365–74.

Namkung W, Yao Z, Finkbeiner WE, Verkman AS. Small-molecule activators of TMEM16A, a calcium-activated chloride channel, stimulate epithelial chloride secretion and intestinal contraction. FASEB J. 2011b;25(11):4048–62.

Nguyen DN, Chen T-Y. Structure and function of calcium-activated chloride channels and phospholipid scramblases in the TMEM16 family. In: Handbook of experimental pharmacolology. Berlin/Heidelberg: Springer; 2022.

Oh S-J, Hwang SJ, Jung J, Yu K, Kim J, Choi YJ, Hartzell HC, Roh EJ, Lee CJ. MONNA, a potent and selective blocker for transmembrane protein with unknown function 16/anoctamin-1. Mol Pharmacol. 2013;84(5):726–35.

Okada SF, Zhang L, Kreda SM, Abdullah LH, Davis CW, Pickles RJ, Lazarowski ER, Boucher RC. Coupled nucleotide and mucin hypersecretion from goblet-cell metaplastic human airway epithelium. Am J Respir Cell Mol Biol. 2011;45(2):253–60.

Papp R, Nagaraj C, Zabini D, Nagy BM, Lengyel M, Skofic Maurer D, Sharma N, Egemnazarov B, Kovacs G, Kwapiszewska G, Marsh LM, Hrzenjak A, Höfler G, Didiasova M, Wygrecka

M, Sievers LK, Szucs P, Enyedi P, Ghanim B, Klepetko W, Olschewski H, Olschewski A. Targeting TMEM16A to reverse vasoconstriction and remodelling in idiopathic pulmonary arterial hypertension. Eur Res. 2019;53(6):1800965.

Paulino C, Neldner Y, Lam AKM, Kalienkova V, Brunner JD, Schenck S, Dutzler R. Structural basis for anion conduction in the calcium-activated chloride channel TMEM16A. elife. 2017;6:e26232.

Pedemonte N, Galietta LJ. Structure and function of TMEM16 proteins (anoctamins). Physiol Rev. 2014;94(2):419–59.

Peters CJ, Yu H, Tien J, Jan YN, Li M, Jan LY. Four basic residues critical for the ion selectivity and pored blocker sensitivity of TMEM16A calcium-activated chloride channels. Proc Natl Acad Sci U S A. 2015;112(11):3547–52.

Petrukhin K, Koisti MJ, Bakall B, Li W, Xie G, Marknell T, Sandgren O, Forsman K, Holmgren G, Andreasson S, Vujic M, Bergen AA, McCarty-Dugan V, Figueroa D, Austin CP, Metzker ML, Caskey CT, Wadelius C. Identification of the gene responsible for best macular dystrophy. Nat Genet. 1998;19(3):383–9.

Qin Y, Jiang Y, Sheikh AS, Shen S, Liu J, Jiang D. Interleukin-13 stimulates MUC5AC expression via a STAT6-TMEM16A-ERK1/2 pathway in human airway epithelial cells. Int Immunopharmacol. 2016;40:106–14.

Ramsey BW, Davies J, McElvaney NG, Tullis E, Bell SC, Drevinek P, Griese M, McKone EF, Wainwright CE, Konstan MW, Moss R, Ratjen F, Sermet-Gaudelus I, Rowe SM, Dong Q, Rodriguez S, Yen K, Ordonez C, Elborn JS. A CFTR potentiator in patients with cystic fibrosis and the G551D mutation. N Engl J Med. 2011;365(18):1663–72.

Randell SH, Fulcher ML, O'Neal W, Olsen JC. Primary epithelial cell models for cystic fibrosis research. Methods Mol Biol. 2011;742:285–310.

Rock JR, O'Neal WK, Gabriel SE, Randell SH, Harfe BD, Boucher RC, Grubb BR. Transmembrane protein 16A (TMEM16A) is a Ca^{2+}-regulated Cl^- secretory channel in mouse airways. J Biol Chem. 2009;284(22):14875–80.

Rowe SM, Heltshe SL, Gonska T, Donaldson SH, Borowitz D, Gelfond D, Segal SD, Khan U, Mayer-Hamblett N, Van Dalfsen JM, Joseloff E, Ramsey BW. Clinical mechanism of the cystic fibrosis transmembrane conductance regulator potentiator ivacaftor in G551D-mediated cystic fibrosis. Am J Respir Crit Care Med. 2014;190(2):175–84.

Sabater JR, Mao YM, Shaffer C, James MK, O'Riordan TG, Abraham WM. Aerosolization of P2Y(2)-receptor agonists enhances mucociliary clearance in sheep. J Appl Physiol. 1999;87(6):2191–6.

Schlichter LC, Elinson RP. Electrical responses of immature and mature Rana pipiens oocytes to sperm and other activating stimuli. Dev Biol. 1981;83(1):33–41.

Schroeder BC, Cheng T, Jan YN, Jan LY. Expression cloning of TMEM16A as a calcium-activated chloride channel subunit. Cell. 2008;134:1019–29.

Scott DW, Walker MP, Sesma J, Wu B, Stuhlmiller TJ, Sabater JR, Abraham WM, Crowder TM, Christensen DJ, Tarran R. SPX-101 is a novel epithelial sodium channel-targeted therapeutic for cystic fibrosis that restores mucus transport. Am J Respir Crit Care Med. 2017;196(6):734–44.

Seo Y, Lee HK, Park J, Jeon D-K, Jo S, Jo M, Namkung W. Ani9, a novel potent small-molecule ANO1 inhibitor with negligible effect on ANO2. PLoS One. 2016;11(5):e0155771.

Simoes FB, Quaresma MC, Clarke LA, Silva IA, Pankonien, Railean V, Kmit A, Amaral MD. TMEM16A chloride channel does not drive mucus production. Life Sci Alliance. 2019;2(6):e201900462.

Sun H, Tsunenari T, Yau KW, Nathan J. The vitelliform macular dystrophy protein defines a new family of chloride channels. Proc Natl Acad Sci U S A. 2002;99:4008–13.

Tarran R, Button B, Picher M, Paradiso AM, Ribeiro CM, Lazarowski ER, Zhang L, Collins PL, Pickles RJ, Fredberg JJ, Boucher RC. Normal and cystic fibrosis airway surface liquid homeostasis. The effects of phasic shear stress and viral infections. J Biol Chem. 2005;280(42):35751–9.

Tarran R, Button B, Boucher RC. Regulation of normal and cystic fibrosis airway surface liquid volume by phasic sheer stress. Annu Rev Physiol. 2006;68:543–61.

Truong EC, Phuan PW, Reggi AL, Ferrera L, Galietta LJV, Levy SE, Moises AC, Cil O, Diez-Cecilia E, Lee S, Verkman AS, Anderson MO. Substituted 2-acylaminocycloalkylthiohene-3-carboxylic acid arylamides as inhibitors of the calcium-activated chloride channel transmembrane protein 16A (TMEM16A). J Med Chem. 2017;60(11):4626–35.

Van Goor F, Hadida S, Grootenhuis PDJ, Burton B, Cao D, Neuberger T, Turnbull A, Singh A, Joubran J, Hazelwood A, Zhou J, McCartney J, Arumguan V, Decker C, Yang J, Young C, Olson ER, Winbe JJ, Frizzell RA, Ashlock M, Negulescu P. Rescue of CF airway epithelial function in vitro by a CFTR potentiator, VX-770. Proc Natl Acad Sci U S A. 2009;106(44):18825–30.

Wang P, Zhao W, Sun J, Tao T, Chen X, Zheng Y-Y, Zhang C-H, Chen Z, Gao Y-Q, She F, Li Y-Q, Wi L-S, Lu P, Chen C-P, Zhou J, Wang D-Q, Chen L, Shi X-H, Deng L, ZhuGe R, Chen H-Q, Zhu M-S. Inflammatory mediators mediate airway smooth muscle contraction through a G-protein-coupled receptor-transmembrane protein 16A-voltage-dependent Ca^{2+} channel axis and contribute to bronchial hyperresponsiveness in asthma. J Allergy Clin Immunol. 2018;141(4):1259–68.

Yang YD, Cho H, Koo JY, Tak MH, Cho Y, Shim W-S, Park SP, Lee J, Lee B, Kim B-M, Raouf R, Shin YK, Oh U. TMEM16A confers receptor0activated calcium-dependent chloride conductance. Nature. 2008;455(7217):1210–5.

Zhang C-H, Li Y, Zhao W, Lifshitz LM, Li H, Harfe BD, Zhu M-S, ZhuGe R. The transmembrane protein 16A Ca^{2+}-activated Cl^- channel in airway smooth muscle contributes to airway hyperresponsiveness. Am J Respir Crit Care Med. 2013;187(4):374–81.

Zhang Y, Wang X, Wang H, Jiao J, Li Y, Fan E, Zhang L, Bachert C. TMEM16A-mediated mucin secretion in IL-13-induced nasal epithelial cells from chronic rhinosinusitis patients. Allergy, Asthma Immunol Res. 2015;7(4):367–75.

Zhu Y, Abdullah LH, Doyle SP, Nguyen K, Ribeiro CM, Vasquez PA, Forest MG, Lethem MI, Dickey BF, Davis CW. Baseline goblet cell mucin secretion in the airways exceeds stimulated secretion over extended time periods, and is sensitive to shear stress and intracellular mucin stores. PLoS One. 2015;10(5):e0127267.

(2017) Use of hyperpolarized helium-3 MRI to assess response to ivacaftor treatment in patients with cystic fibrosis Journal of Cystic Fibrosis 16(2) 267–274.

(2019) TMEM16A in Cystic Fibrosis: Activating or Inhibiting? Frontiers in Pharmacology.

Targeting Acid-Sensing Ion Channels in Disease

Victor Senescall, Elena Budusan, Adriana Knezic, Maleeha Waqar, Neville Butcher, Nemat Khan, Claudia McCarthy, Robert Widopp, Aneesh Karatt-Vellat, and Lachlan D. Rash

Abstract Acidosis is a central feature of many pathological conditions, and while the activity of many ion channels is suppressed under acidic conditions, acid-sensing ion channels (ASICs) are exquisitely sensitive to decreases in pH and activated by protons. The expression and function of ASICs is enhanced under conditions of ischemia, inflammation, and trauma. Here, we will focus on this family of ion channels whose main role appears to be the detection of perturbations to local tissue pH, and review the evidence for their role, and whether they represent viable druggable targets in a multitude of currently poorly treated conditions. This chapter provides an overview of the key evidence for the role of ASICs in pathological states, a commentary on their potential as therapeutic targets in humans identifying several major challenges we face in developing ASIC targeting therapies.

Keywords Acid-sensing ion channel · Pharmacology · Acidosis · Ischemia-related disease · Pain

---

Authors Victor Senescall, Lachlan D Rash, Maleeha Waqar, Elena Budusan, Adriana Knezic, Nemat Khan, and Neville Butcher wrote the chapter. All authors contributed to content design and editing the text.

V. Senescall · E. Budusan · M. Waqar · N. Butcher · N. Khan · L. D. Rash (✉)
School of Biomedical Sciences, The University of Queensland, St. Lucia, QLD, Australia
e-mail: victor.senescall@uqconnect.edu.au; e.budusan@uq.edu.au; m.waqar@uq.edu.au; n.butcher@uq.edu.au; n.khan2@uq.edu.au; L.Rash@uq.edu.au

A. Knezic · C. McCarthy · R. Widopp
Department of Pharmacology, Monash University, Clayton, VIC, Australia
e-mail: Adriana.Knezic@monash.edu; Claudia.mccarthy@monash.edu; robert.widdop@monash.edu

A. Karatt-Vellat
Maxion Therapeutics Ltd., Cambridge, UK
e-mail: akv@maxion.co.uk

G. Stephens, E. Stevens (eds.), *Ion Channels as Targets in Drug Discovery*,
https://doi.org/10.1007/978-3-031-52197-3_10

1 Acidosis and Acid-Sensing Ion Channels (ASICs)

Homeostasis is the maintenance of a stable internal physical and chemical environment within an organism. It is crucial for survival as it provides optimal conditions for cellular, tissue, and organ function. Perturbations to homeostasis, such as changes in temperature, pH, and ionic balance (concentrations or gradients), can result in dysfunction or significant damage to cells, organs, or the whole organism. Severe and/or prolonged dyshomeostasis leads to death.

Protons/hydrogen ions (H^+) are released by acids into surrounding solutions and, as they are ever-present in biological processes, are important both functionally and as highly sensitive indicators of perturbations to homeostasis in biological systems. pH is defined as the negative base 10 log of proton (H^+) concentration, such that as H^+ concentration increases, pH is reduced. The normal physiological blood and extracellular pH of ~7.4 corresponds to an approximate proton concentration of 40 nM. Changes in pH can modulate physiological processes by (1) changing the ionization states of weak acids and bases and proteins modulating the bound vs free, biologically active ion concentrations, (2) regulating the function of enzymes, transporters, and ion channels, and (3) changing proton gradients that are used to power various transporters and processes. This is exploited by tissues and subcellular compartments to carry out a range of tasks crucial to survival. For example, the very low pH of the stomach (~2) provides protection from ingested pathogens and aids digestion, neurotransmitter vesicles in neurons are loaded by virtue of a strong proton gradient driving transporters, and acidic pH is essential for powering the degradation and recycling endosomal/lysosomal system. Finally, carbonic acid increases as CO_2 is retained with the lowering of blood pH driving the respiratory center in the brain keeping us alive. It is not surprising then that pH is strictly regulated or that perturbations to pH are often, if not always, associated with pathological conditions.

Acidosis occurs in ischemia, trauma, tumors, and infection and is a hallmark of inflammation during which the extracellular pH can reach as low as 5.5 (Deval and Lingueglia 2015). Different cell types exposed to prolonged acidosis respond in different ways: sensory neurons and immune cells are sensitized/activated while CNS neurons can be suppressed or fatally over-activated (Xiong et al. 2004). With the discovery of several families of acidosis-sensitive receptors (Zha et al. 2022), how particular types of cell sense and respond to acidosis is becoming much better understood.

1.1 Acid-Sensing Ion Channels (ASICs)

Acid-sensing ion channels (ASICs) are primary pH sensors in mammals and are expressed widely in neuronal and non-neuronal cells (Deval and Lingueglia 2015; Foster et al. 2021). They are part of the epithelial sodium channel/degenerin ((ENaC)/DEG) family of ion channels. While they are mainly permeable to Na^+, their activation results in depolarization and cell excitation/activation and often calcium influx. In the context of inhibitory nerves, ASIC expression and activation will result in decreased activity of surrounding neurons. There are six main ASIC isoforms in mammals, ASIC1a/b, ASIC2a/b, ASIC3, and ASIC4, encoded by genes ASIC1, ASIC2, ASIC3, and ASIC4, with ASIC1 and ASIC2 being alternatively spliced into the isoforms: 1a/1b, 2a/2b (Kellenberger and Schild 2015). In humans ASIC3 also has multiple isoforms (3a, 3b, and 3c); however, little is known about the function of isoforms b and c (Delaunay et al. 2012). ASICs are trimeric, voltage-independent, ligand-gated ion channels activated by protonation, whereby a decrease in pH leads to channel opening with thresholds for activation ranging from pH 7 (ASIC1a and ASIC3) to 6 (ASIC2a) and pH_{50} values ranging from 6 to 5 (Verkest et al. 2022). Prolonged exposure to acidosis leads to varying degrees of channel inactivation, depending on the subtype and extracellular conditions (e.g., ASIC1a fully desensitizes whereas ASIC1b and ASIC3 have sustained currents that increase in the presence of lactate). Mild sustained decreases in pH lead to steady-state desensitization (SSD) with the pH_{50} of this process likely to most accurately reflect the proton sensitivity of a given subtype (Gründer 2020). As homomeric channels, ASIC3 and ASIC1a are the most pH-sensitive isoforms followed by ASIC1b, then ASIC2a, with heteromeric channels increasing the dynamic pH sensing range (see Verkest et al. (2022) for an excellent summary table of rodent vs human ASIC pH sensitivity). How these aspects of ASIC function relate to different pathological states will be elaborated in the preamble to each disease class below.

1.1.1 ASIC Structure and Function

The first structure determined for the ENaC/Deg family was that of a truncated version of chicken ASIC1a (with ~90% sequence identity to rat and human ASIC1a) in a presumed desensitized state. This prototypical structure revealed that ASICs are trimeric channels (Jasti et al. 2007). Twenty-two structures of ASICs have now been reported covering resting, open, and desensitized functional states (note there are likely to be multiple variations of these states) and with a range of bound ligands (reviewed in Rook et al. 2021). Recently, three structures of human ASIC1a have been reported in an apo-resting state and two ligand-bound resting states (Sun et al. 2020; Wu et al. 2021). All structures indicate that three subunits are required to form a functional channel (Fig. 1), and together they have revealed much about the function and pharmacology of this channel family. Each ASIC subunit contains short cytoplasmic N- and C-termini and two small hydrophobic transmembrane (TM1 and TM2) domains which are separated by a large extracellular domain (ECD) of ~370 amino acids (Fig. 1). The structural arrangement of the ECD of a single subunit was described as resembling a clenched hand holding a ball with defined domains named the wrist, palm, finger, knuckle, thumb, and β-ball (Fig. 1a). The central feature of each channel subunit is the palm domain formed by several β-sheets that are connected to all extracellular subdomains and the transmembrane helices TM1/TM2 of the channel by short loops that form the wrist. The thumb is

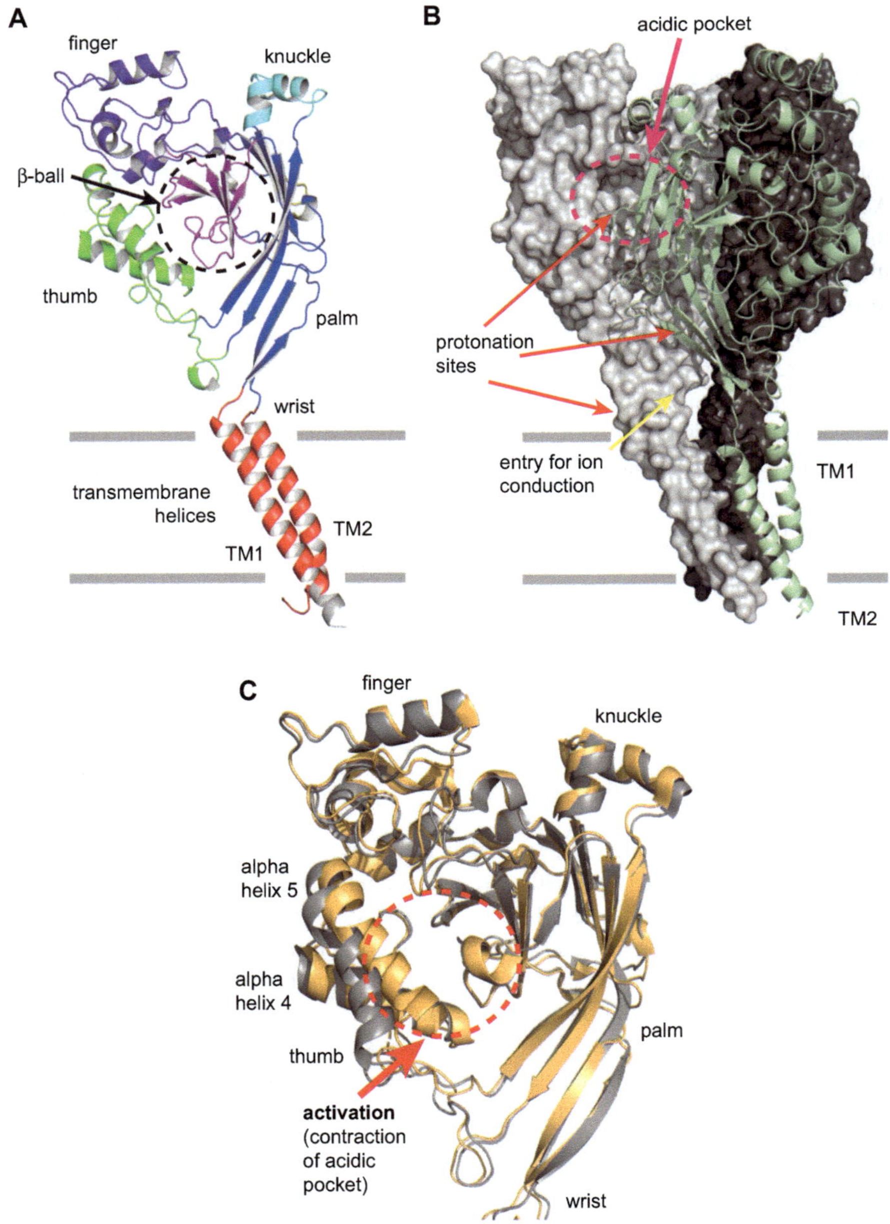

Fig. 1 Structure of ASICs. (**a**) Single ASIC subunit [PDB 2QTS] showing the relative positions of the structural features described as the wrist, palm, knuckle, finger, thumb, and beta ball. (**b**) Combined ribbon and surface structure showing the trimeric arrangement of ASICs [PDB 4FZ0]. (**c**) Overlay of the resting (gray: PDB 5WKV) and open (orange; PDB 4FZ0) structures of a single subunit (cASIC1a) showing movement of helix 5 in the thumb domain that occurs upon protonation and activation of the channel. (Figure made using PyMol 2.5.2)

formed by two α-helices and stabilized by seven disulfide bonds (Jasti et al. 2007; Yoder et al. 2018). The interface of the thumb, finger, and palm domain forms an "acidic pocket," as it contains many acidic residues and is negatively charged at physiological pH. Within this region there are three pairs of carboxyl-carboxylate interactions, in short distance to each other (~2.8 Å) when one residue per pair is protonated in low pH conditions (pH 5–6). Due to the presence of so many negatively charged residues, the acidic pocket was predicted to be the proton sensor which determines the activation and desensitization of acid-sensing ion channels (Jasti et al. 2007). This idea has been partly confirmed by studies showing that mutations within the acidic pocket lead to a decrease in pH sensitivity of ASICs (Jasti et al. 2007; Sherwood et al. 2009).

The current model of ASIC gating suggests that it is highly complex with multiple protonatable sites away from the acidic pocket contributing to the process (Rook et al. 2021). In the resting or closed state of the channel, the acidic pocket and the thumb domain are expanded, and ion conduction is not possible (Yoder et al. 2018; Sun et al. 2020). Upon channel exposure to extracellular acidification, at least one of the acidic residues of each pair of carboxyl-carboxylate interactions is thought to be protonated. This leads to the collapse of the acidic pocket and to closure of the thumb domain toward the threefold axis (Fig. 1c). These conformational changes are transmitted via the lower palm and thumb domains to the transmembrane domains resulting in gate opening which allows passage of ions through the pore (Yoder et al. 2018). After activation, desensitization of the channel occurs resulting in the rearrangement of the β1–β2 and β11–β12 linkers which connect the palm domain with the transmembrane domains. The central and extracellular vestibules contract slightly, closing the pore (Rook et al. 2020). Interestingly, by preventing the movement of the β11–β12 linker by trapping it in the resting state, the process of desensitization can be suppressed. This indicates that the rearrangement of the β11–β12 linker is not a consequence of, but a necessity for, channel desensitization (Rook et al. 2020). The resting state of the channel is achieved by deprotonation of the residues in the acidic pocket and the rearrangement of the carboxyl-carboxylate pairs, causing the acidic pocket to expand again (Rook et al. 2021). Consistent with this model, a recent study using voltage-clamp fluorometry combined with fluorophore-quencher pairing showed that conformational changes that lead to ASIC activation occur simultaneously in the acidic pocket, finger, palm, thumb-palm loops, and the wrist. Slow conformational changes of the channel associated with desensitization tend to occur in the palm domain, whereas fast changes associated with activation occur in the thumb-palm loops (Vullo et al. 2021).

1.1.2 ASIC Expression and Physiological Roles

In both rodents and humans, ASICs are most highly expressed in the central nervous system (CNS), then the peripheral nervous system (PNS). So, it is not surprising that ASICs are most often described as neuronal acidosis sensors. Nevertheless,

ASICs are also widely distributed throughout the body in non-neuronal cells. Below we provide an overview of ASIC expression and physiological roles. The reader is directed to these papers/resources for more detailed information on the expression and evidence for expression in rodents vs humans (Deval and Lingueglia 2015; Gründer 2020; Foster et al. 2021) and the Human Protein Altlas.org (Sjostedt et al. 2020; Karlsson et al. 2021).

ASIC1a is the dominant subtype expressed in the spinal cord and throughout the brain (forebrain, midbrain, brainstem, pons, cerebellum, hippocampus, cortex) of both rodents and humans (Alvarez de la Rosa et al. 2003; Li et al. 2010; Delaunay et al. 2012). ASIC1a has been shown to have a role in CNS pathologies and is an emerging target of interest for these diseases (see below). ASIC1b has not been characterized in human CNS yet but has been shown in rat CNS in spinal cord and brain (Alvarez de la Rosa et al. 2003; Gründer 2020). Oligodendrocyte-lineage cells (OLC), particularly in the white matter, have been shown to express ASICs, with subtypes 1a, 2a, and 4 being detected in oligodendrocyte progenitor cells and weak expression of ASIC3 in differentiating OLCs only (Feldman et al. 2008). ASIC 2a and 2b are widely expressed in spinal cord and brain, ASIC3 is expressed in spinal cord with very low expression in the brain, and ASIC4 is expressed in the brain but not in spinal cord (Price et al. 1996; Feldman et al. 2008; Delaunay et al. 2012). ASIC1a and ASIC2a have been shown to localize in the synaptic area of excitatory neurons suggesting their role in synaptic plasticity (Wemmie et al. 2002, 2003). ASIC1a is abundant in amygdala and plays a role in fear conditioning due to its presence of fear circuits. ASIC1 null mice show deficits in fear conditioning cues (Wemmie et al. 2003). In patients with panic disorder, CO_2 inhalation induced panic attacks suggesting the role of ASIC1a in chemosensing, as the CO_2 causes pH drop and activates the channel (Ziemann et al. 2009). CNS resident immune cells such as microglia have shown to express ASIC1 and ASIC2a which are upregulated upon treatment with lipopolysaccharide (LPS) (Yu et al. 2015). ASIC1, ASIC2, and ASIC3 have been detected in astrocytes with ASIC1a being the most prevalent subtype in human astrocytes (Huang et al. 2010).

Within the rodent PNS, almost all subtypes are expressed in nociceptive sensory neurons and play a role in pain perception, with ASIC3 and ASIC1b being the primary acidosis-related pain sensors (Deval et al. 2008; Hoagland et al. 2010). In contrast, ASIC1 seems to be more highly expressed than ASIC3 in human sensory neurons (at least at the mRNA level (see Fig. S1 in Delaunay et al. 2012)) and is likely to be dominated by ASIC1a rather than ASIC1b as in rodent dorsal root ganglion (DRG) neurons (Hoagland et al. 2010). This has obvious clinical translation implications for ASICs as analgesic targets as discussed below. ASIC2 and ASIC3, and to a lesser degree ASIC1, are expressed in mechanoreceptors that innervate various rodent tissues and organs including the skin, the cardiovascular system, and muscles, while little is known about ASIC expression in human mechanosensing neurons or the exact mechanisms underlying how ASICs are mechanically gated (Cheng et al. 2018).

A growing area in ASIC research is investigating the widespread expression of ASICs in many immune cell types and the potential role they play in immune cell function (recently reviewed in Foster et al. 2021). Acid-induced (pH 6) currents in dendritic cells (expressing ASICs 1, 2, and 3) were increased following 24-h stimulation with LPS and resulted in increased cell activation, an effect prevented by ASIC inhibitors, demonstrating functional surface expression of ASICs (Tong et al. 2011). Similar observations have been made for mouse bone marrow-derived macrophages that express ASIC1 and ASIC3, whereby exposure to acidic pH increased their phagocytotic activity and upregulated key cell surface markers related to maturation of macrophages, effects blocked by the application of amiloride (Kong et al. 2013). In addition to being the main isoform expressed on CNS neurons (Li et al. 2010; Zha 2013; Lin et al. 2018), ASIC1a is the dominant subtype and functionally expressed in microglia (Yu et al. 2015; Lin et al. 2018), astrocytes (Huang et al. 2010; Yu et al. 2015; Lin et al. 2018) and oligodendrocytes (both rodent and human Feldman et al. 2008; Vergo et al. 2011; Arun et al. 2013). Surface expression of ASIC1a is increased under both acidosis and inflammation suggesting that functional expression of ASIC1a on glial/immune cells may play an important role in how local tissue acidosis enhances the inflammatory response (Foster et al. 2021).

The expression and function of ASICs in the vasculature is another emerging and therapeutically relevant field, particularly in light of the pathological role of ASIC1 in ischemia-induced damage in a variety of tissues (discussed below). ASIC1 is expressed in vascular smooth muscle cells from pulmonary, cerebral, and mesenteric resistance arteries (Lin et al. 2014; Nitta et al. 2014; Garcia et al. 2020) and in endothelial cells (Garcia et al. 2023). Activation of ASIC1 in pulmonary vessels results in vasoconstriction, whereas its activation on mesenteric artery endothelial cells leads to vasodilation; thus, the role of ASICs in regulating local blood flow appears to be complex and is likely to be tissue dependent (summarized in Garcia et al. 2023). Importantly, the distribution of ASIC1 in various cerebral arteries was the same in both rats and humans (Lin et al. 2014), suggesting that their contribution to regulating blood flow (at least in the brain) is likely conserved in mammals.

In the heart, ASIC3 is expressed in sensory nerves and plays a role in ischemia sensing (Immke and McCleskey 2001; Sutherland et al. 2001). ASIC1a has been detected, albeit at low levels, in human-induced pluripotent stem cell (iPSC)-derived cardiomyocytes (Redd et al. 2021). In kidneys, ASIC1a, ASIC2, and ASIC3 expression is detected in proximal tubular cells in rodents (Song et al. 2019) and in humans (mRNA expression) (Sjostedt et al. 2020; Karlsson et al. 2021), but their physiological role is not known. ASICs are expressed in the retina of many species and appear to be involved in both physiological and pathological conditions (Ettaiche et al. 2006; Vina et al. 2015). While ASIC1a overactivation contributes to cell damage during ischemia or trauma (Dibas et al. 2018; Stankowska et al. 2018), ASIC2 plays a role in the maintenance of retinal integrity (Ettaiche et al. 2004).

In addition to the tissues discussed above, ASICs have been detected in taste receptor cells (ASIC1–3), cochlear hair cells (ASIC1b), various cells in the testis (hASIC3), the pituitary gland (ASIC4), lung epithelial cells (ASIC1a and ASIC3), adipose cells (ASIC3), and cells from bone and joints (ASIC1–3) (Xu and Chen

2020; Kellenberger and Rash 2021). For more comprehensive reviews covering the proposed physiological roles of ASICs, see Gründer (2020) and Storozhuk et al. (2021).

2 ASIC Pharmacology

In this section we describe the main pharmacological tools that have been used to investigate the role of ASICs in in vivo studies. The pharmacology is summarized in Fig. 2, while the details of how these agents have been used to study ASICs in various disease models are provided in each relevant section and the associated summary tables. For more in-depth information and a historical perspective on ASIC pharmacology, we direct the reader to (Baron and Lingueglia 2015; Rash 2017; Kellenberger and Rash 2021).

2.1 *Small Molecule Inhibitors*

Amiloride and Benzamil: Amiloride is a clinically used diuretic due to its moderately potent inhibition (IC_{50} 100 nM) of epithelial Na channels (ENaCs) in the kidney. It is non-selective, inhibiting all ASIC homomeric channels via open pore block

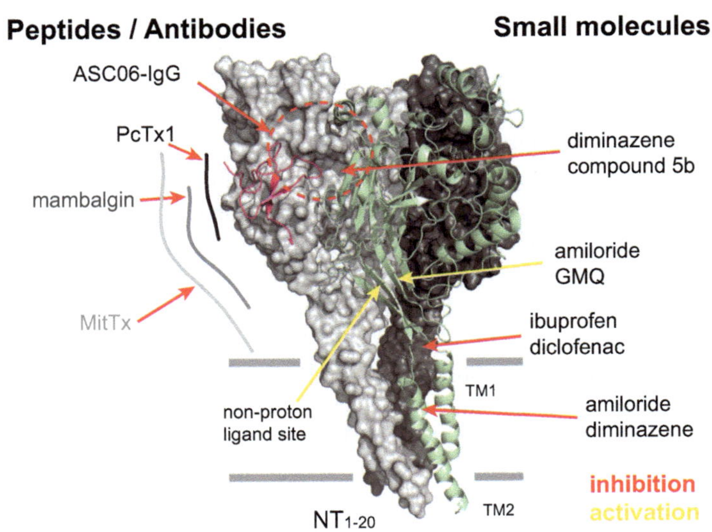

Fig. 2 Binding sites of ASIC pharmacological tools discussed in this chapter. The structure shown is PcTx1 bound to an open state of chicken ASIC1a (PDB 4FZ0). (Figure made using PyMol 2.5.2)

(IC_{50} 10–30 μM), and works in a voltage-dependent manner (Adams et al. 1999). Amiloride is a promiscuous drug with a wide variety of non-ASIC/non-ENaC targets (including sodium/hydrogen antiporters, sodium/calcium exchangers, and GPCRs) (Rash 2017). Nevertheless, amiloride has been used extensively as a research tool to study ASICs in vitro, and in vivo and in several clinical trials to assess ASICs as clinically relevant targets in several diseases, mostly with negative outcomes. Benzamil (benzyl amiloride) is ~4–5 times more potent as an ASIC inhibitor than amiloride (Leng et al. 2016) but retains potent effects on ENaCs and many other ion channels and transporters (Castaneda et al. 2019).

Diminazene is a commercially available, veterinary anti-parasitic drug and potent (IC_{50} ~ 300 nM), non-selective ASIC blocker. It seems to have two binding sites, with an open-channel pore-blocking mechanism that is voltage-dependent similar to amiloride (Schmidt et al. 2017; Lee et al. 2018) but may also interact with the acidic pocket to inhibit activation (Gornati et al. 2021). Despite being more potent and less promiscuous than amiloride, and clinically used as a veterinary drug, diminazene has been used in few in vivo studies of ASIC function. A recent structure activity study on diminazene resulted in the development of an analogue (3b) which has a trend toward ASIC1a selectivity and appears to be a useful in vivo research tool (Gornati et al. 2021); see section on stroke below.

Compound 5b (C5b) is a small molecule inhibitor of ASIC1a, designed through molecular modeling based on the interaction of the triple Arg motif of PcTx1 with the acidic pocket. Although very potent for a small molecule ASIC inhibitor (IC_{50} 27 nM with pH stimulus 6.7), the activity of C5b is highly pH-dependent with potency dropping when more acidic stimuli are used (IC_{50} 10 μM at pH 5) (Buta et al. 2015). C5b has recently been used in vivo for target validation of ASIC1a in mouse model of stroke when given peripherally (Qi et al. 2022).

A-317567 was developed by Abbott laboratories as an analgesic lead molecule targeting ASICs. It initially appeared to be relatively non-selective among subtypes of ASIC-like currents in rat DRG neurons (IC_{50}s: ASIC1-like 2 μM, ASIC2-like 30 μM, ASIC3-like 10 μM) (Dube et al., 2005). It was subsequently shown to be more potent at ASIC3 and, along with a more potent analogue (10b), was used as a tool to study the role of this channel in pain. However, both compounds had substantial off-target activity and were analgesic in both wild-type and ASIC3 KO mice (Kuduk et al. 2010).

Ibuprofen and flurbiprofen were two of several non-steroidal anti-inflammatory drugs (NSAIDs) found to directly modulate ASIC function. They are quite selective for ASIC1a, and although not particularly potent, their IC_{50} of ~350 μM (Voilley et al. 2001) falls in the plasma concentrations observed clinically for doses >600 mg (Davies 1998). A structure-activity study carried out by the Pless group revealed that ibuprofen binds to the wrist domain to decouple activation-associated conformational changes in the extracellular domain from transferring to the transmembrane domains and thus oppose channel opening (Lynagh et al. 2017).

2.2 Peptides

Psalmotoxin 1 (PcTx1) is a 40-amino acid disulfide-rich peptide that was isolated from the venom of the Trinidad Chevron tarantula, *Psalmopoeus cambridgei*. It was the first highly selective and potent ASIC1a inhibitor discovered (IC_{50} ~ 1 nM) (Escoubas et al. 2000) and has been used extensively in both in vivo and in vitro studies. PcTx1 binds to the acidic pocket of ASIC1a increasing its apparent affinity to H^+ thereby stabilizing its desensitized state (Chen et al. 2005). Although PcTx1 has played an important role in helping to validate ASICs as therapeutic targets in many disease models, there may be limitations for its translational potential (Cristofori-Armstrong et al. 2019). PcTx1 is not ideally selective, potentiating rodent and human ASIC1b and has variable effects on ASIC1a/2a heteromultimers showing both inhibition and potentiation depending on the surrounding pH conditions (Joeres et al. 2016; Liu et al. 2018a). It is also not as biologically stable as one might expect for a venom peptide (Er et al. 2017). Nevertheless, PcTx1 does seem to work very well as a potent ASIC inhibitor in vivo in multiple animal models as discussed below.

APETx2 is a potent peptide inhibitor of ASIC3 and ASIC3-containing heteromers (IC_{50} 37–2000 nM) isolated from the sea anemone *Anthopleura elegantissima* (Diochot et al. 2004). The binding site and mechanism of action of APETx2 are still poorly understood, partly because APETx2 is not easy to produce, making SAR studies rather challenging (Jensen et al. 2009; Anangi et al. 2012). Unlike many cysteine-rich peptides (Mobli et al. 2017), APETx2 is not particularly stable in mammalian biological fluids, and attempts to improve its stability unfortunately resulted in substantial decreases in potency (Jensen et al. 2012). Although being widely used in vitro and in vivo as a selective ASIC3 inhibitor, APETx2 also inhibits several voltage-gated channels with only slightly lower potency ($Na_V1.8$, and hERG) (Blanchard et al. 2012; Jensen et al. 2014), or possibly equivalent potency (IC_{50} 114 nM at rat $Na_V1.2$, IC_{50} 55 nM at rat $Na_V1.8$) as seen in an independent study (Peigneur et al. 2012). Thus, caution is required when interpreting results using this peptide, particularly in in vivo studies.

MitTx (isolated from the venom of the Texas coral snake, *Micrurus tener*) is a noncovalent heterodimeric peptide and potent activator of ASIC1a (IC_{50} 9 nM) and ASIC1b (IC_{50} 23 nM), with 88- to 35-fold less activity at ASIC3 (IC_{50} 800 nM) (Bohlen et al. 2011). The individual components (a Kunitz-like α-subunit and phospholipase A2-like β-subunit) are inactive on their own, but when applied together directly activate ASICs involved in nociception, consistent with bites by the Texas coral snake being very painful. MitTx is of great value as an ASIC research tool as it allows the assessment of selective ASIC activation in the absence of acidosis which, in itself, can have many effects on protein function.

Pi-HXTX-Hi1a (Hi1a), a funnel-web spider venom peptide (8.7 kDa), is highly potent and selective against ASIC1a (IC_{50} 0.5 nM) (Chassagnon et al. 2017). Hi1a is comprised of two tandem PcTx1-like sequences joined by a short linker and, in contrast to the mechanism of PcTx1, seems to stabilize the resting state of ASIC1a.

Furthermore, Hi1a is a non-competitive inhibitor, is less pH-dependent, and has slower reversibility compared to PcTx1. Hi1a is not commercially available; thus, published studies on this peptide are limited. Nevertheless, it shows promising potential for validating ASIC1a as a target in vivo (Chassagnon et al. 2017; Redd et al. 2021) and is patented as a novel drug lead for ischemia-related diseases (King et al. 2021).

Mambalgins are a family of 57-residue long three-finger toxins isolated from mamba snake venoms (*Dendroaspis polylepis polylepis* and *Dendroaspis angusticeps*). The sequences differ by a single point mutation, and their pharmacological activity is essentially identical. Mambalgins are potent inhibitors of rodent ASIC1a and ASIC1b and appear to stabilize the resting/closed state of ASICs (Salinas et al. 2014; Mourier et al. 2015; Sun et al. 2020). Although mambalgins potently inhibit human ASIC1a in a pH-independent manner (as for rat ASIC1a *and* 1b), they have pH-dependent effects on human ASIC1b, being a potentiator under mild pH stimulus (pH 7–6, as would be encountered in vivo), but an inhibitor if the channel is stimulated under more extreme acidosis (pH < 6) (Cristofori-Armstrong et al. 2021). The mambalgins have played an important role in identifying the contribution of ASIC1b to nociception in rodents (Diochot et al. 2012, 2016; Chang et al. 2019) and investigating the role of ASIC1a in glioma (Bychkov et al. 2020, 2021).

NT_{1-20} is a synthetic membrane-penetrating peptide made up of the first 20 N-terminal residues of ASIC1a preceded by a cell-penetrating TAT tag. This peptide was designed to enter cells and interact with the positively charged, intracellular C-terminal region of ASIC1a that has been shown to recruit and activate RIPK1, leading to activation of the necroptosis pathway (Wang et al. 2015, 2020). This peptide represents the first successful attempt to modulate ASIC1a signaling by interrupting intracellular protein-protein interaction.

2.3 Antibodies

In addition to the pharmacological tools outlined above, several anti-ASIC antibodies and an inhibitory nanobody-PcTx1 fusion protein have been developed. Regeneron Pharmaceuticals Inc. were the first to develop functional anti-ASIC antibodies and showed efficacy for two of these (H1M6718N or H4H6721N2) in multiple pain models (MacDonald et al. 2015; Santos et al. 2015). The Regeneron antibodies are described as being anti-ASIC1 and targeting exclusively human ASIC1, or ASIC1 across multiple species. H1M6781N is highly potent as an inhibitor of human ASIC1 (IC_{50} 860 pM), but there is no available information on its effects on the ASIC1 splice variant ASIC1b, and the binding epitope is not disclosed (MacDonald et al. 2015). Anti-ASIC1a antibody ASC06-IgG1 is the result of using structurally intact, truncated, human ASIC1a in nanodiscs to pan a human scFv combinatorial antibody library using phage display technology and converting the best binding fragment into a full-length IgG1 form (Qiang et al. 2018). This antibody binds to the structurally intact channel at the subunit interface in the upper part of the extracellular domain with substantial contact on the finger domain. The finger domain is before the splice site in ASIC1, consistent with the very high selectivity of ASC06-IgG1 for ASIC1a over all other subtypes. ASC06-IgG1 prevents hASIC1a activation with an IC_{50} of 85 nM for current inhibition in a manual patch clamp assay and 2.9 nM in an acidosis-induced calcium influx assay (Qiang et al. 2018). In addition, this antibody also showed in vivo efficacy in the mouse model of middle cerebral artery occlusion (MCAO), thus representing a very promising research tool and therapeutic lead.

Using a different approach to anti-ASIC ligand discovery, Cecilia Canessa's group raised alpaca-derived nanobodies against hASIC1a and selected for a lead that had high affinity but did not modify channel function, with the goal of using the nanobody to aid hASIC1a protein production for structural studies (Wu et al. 2021). The resulting lead, Nb.C1, binds to the lower thumb domain and overlaps the binding site of MitTx and mambalgins, but not PcTx1. As the thumb is identical between ASIC1a and ASIC1b, Nb.C1 will likely also bind to ASIC1b. Fusion of Nb.C1 to PcTx1 with either a rigid or flexible linker resulted in a potent (>90% inhibition at 10 nM) and slowly reversible ASIC1a inhibitor that adds another potent and valuable tool to the ASIC1 research toolbox (Wu et al. 2021). Nb.C1 and its PcTx1 fusion do not appear to have been used in vivo as yet, but the authors suggested they could be used to treat the pain of Texas coral snake bite or in other ASIC1a-related pathologies.

3 The Role of ASICs in Disease

3.1 *ASICs in Ischemic Disease*

During ischemia, the lack of blood supply to the affected tissue causes cells to undergo anaerobic respiration leading to an increase in the production of lactic acid (Zupping et al. 1971; Rehncrona et al. 1981). The resulting acidosis can lower pH to 7–6.4 (Kelmanson et al. 2021) and even down to transient lows of ~5.5 under hyperglycemia (Chopp et al. 1987), more than sufficient to cause the activation of ASICs. In addition to accumulation of protons, many other mediators are released at various stages of the ischemic cascade, including lactate (Zupping et al. 1971; Rehncrona et al. 1981), spermine (Fage et al. 1992), reactive oxygen species (ROS) (Lakhan et al. 2009), and proteases (Gingrich and Traynelis 2000). Strikingly, all of these mediators are associated with enhancing ischemia-induced damage, and all have actions that increase ASIC1a function (Immke and McCleskey 2001; Poirot et al. 2004; Cadiou et al. 2007; Duan et al. 2011; Wu et al. 2020). Together this provides ideal conditions for the overactivation of ASICs during sustained tissue ischemia as occurs in stroke, myocardial infarction, kidney injury, and in donor organs awaiting transplant.

3.1.1 Brain/Stroke

During ischemic stroke, which represents ~71% of all strokes (Campbell et al. 2019), there is a loss of blood supply to the brain resulting in varying levels and durations of lactic acidosis. The concentration of lactate and protons that accumulate is directly correlated to the severity of the tissue damage (Kalimo et al. 1981; Rehncrona et al. 1981; Paljarvi et al. 1983; Paljarvi 1984; Chopp et al. 1987). Zhi-Gang Xiong and colleagues were the first to demonstrate a role for ASIC1a in mediating ischemic brain injury and suggested that ASIC1a-mediated calcium influx was the primary mechanism of acidosis-induced cell death (Xiong et al. 2004). In addition to causing sodium and calcium influx, the activation of ASIC1a has been shown to trigger necroptosis via activation of receptor-interacting protein kinase 1 (RIPK1) (Wang et al. 2015), a pro-inflammatory mechanism of cell death that may be common to many pathological conditions (Degterev et al. 2019; Yuan et al. 2019). There is now a very substantial body of evidence (covering many different interventions, in vitro and in vivo models and species) that the inhibition of ASIC1a is beneficial in ischemic stroke. The key in vivo evidence to support a role for ASIC1a in ischemic stroke is summarized in Table 1 and discussed below.

In terms of genetic level evidence for a role of ASIC1a in stroke, four studies have reported the effects of global (whole body) genetic deletion of ASIC1a in transient MCAO models of stroke using the mouse line developed in Michael Welsh's group (Table 1 and Wemmie et al. 2002). The seminal study in this area showed that heterozygous mice (ASIC1a$^{+/-}$) exposed to transient (100 min) MCAO had 32.7% smaller infarct volume than wild-type mice (WT) while full knockout mice (ASIC1a$^{-/-}$) had 61% smaller infarcts as assessed at 24 h post ischemia (Xiong et al. 2004). The same group later showed that ASIC1a activity is enhanced by the endogenous polyamine spermine and that the spermine-enhanced damage during MCAO is more than halved in ASIC1a KO mice (Duan et al. 2011). The neuroprotective effects of ASIC1a KO have been confirmed in two subsequent studies conducted by independent groups (Wang et al. 2020; Qi et al. 2022) but using the same ASIC1a$^{-/-}$ mouse line. In addition to decreased infarct volume, the most recent of these studies also reported improved motor function and reduced mortality in ASIC$^{-/-}$ mice (Qi et al. 2022). These results from engineered ASIC KO mice are nicely supported by an example of evolutionary adaptation to extreme environments. Cortical and hippocampal neurons from naked mole rats, which are famously resistant to hypoxia, have functional ASIC1a expression levels that are ~sevenfold lower than equivalent mouse neurons, and are essentially resistant to pH 5-induced cell death (low pH applied for 2 h (Husson and Smith 2018)) and oxygen-nutrient deprivation for up to 24 h (Nathaniel et al. 2013).

Central administration of amiloride, a non-selective ASIC inhibitor, before and after MCAO significantly reduced infarct volume by ~32% (Xiong et al. 2004). However, amiloride is not selective for ASIC1a and has also been shown to inhibit multiple other proteins involved in sodium transport including the Ca^{2+}/Na^{+} exchanger, which results in intracellular Ca^{2+} overload. Indeed, prolonged administration of amiloride resulted in neuronal injury in cultured mouse cortical neurons

Table 1 ASICs in ischemia-related conditions

ASIC subtype	Intervention	Model/tissue (inc. spp.)	Outcome/therapeutic effect	References
Ischemia of brain				
1a	KO	Mouse cultured cortical neurons Mouse transient MCAO model of IS	ASIC1a$^{-/-}$ cultured cortical neurons protected from acidosis and oxygen-glucose deprivation-induced cell death ASIC1a$^{+/-}$ mice had ~33% reduction in infarct volume while ASIC1a$^{-/-}$ mice showed a 50–60% reduction in infarct volume ASIC1a$^{-/-}$ mice had >50% reduction in infarct volume in spermine (2.5 mM) enhanced MCAO mouse model	Xiong et al. (2004), Duan et al. (2011), Wang et al. (2020) and Qi et al. (2022)
1a	Decreased expression	Cortical and hippocampal neurons in culture (naked mole rats (NMR) and C56/bl6 mice)	NMR CNS neurons have sevenfold lower ASIC current density than mouse CNS neurons and are resistant to acidosis-induced cell death	Husson and Smith (2018) and Nathaniel et al. (2013)
1a	Amiloride and benzamil	Mouse cultured cortical neurons Amiloride: Rat transient MCAO model of IS Benzamil: Mouse transient MCAO model of IS	Amiloride reduced LDH release from acidosis exposed cortical neurons in vitro and decreased infarct volume (~32%) when given i.c.v. in mice Benzamil (12 μM, i.c.v.) reduced infarct volume when given 30 min prior (56%), or 3 h after (31%) MCAO in mice	Xiong et al. (2004) and Leng et al. (2016)
1a	Diminazene analogue 3b	Human neurons (differentiated progenitor cells) Rat cultured cortical neurons Mouse transient MCAO model of IS	3b (10 μM) inhibits ASIC1a currents in human neurons; 0.4 μM 3b protected rat cultured cortical neurons from OGD induced cell death; 3b at 0.65 mg/kg, given i.p. 1, 3, and 6 h post MCAO reduced infarct volume by ~40% in mice	Gornati et al. (2021)
1a	Compound 5b	Mouse cultured cortical neurons Mouse transient MCAO model of IS	10–30 μM 5b protected cultured neurons from pH 6-induced death but 100 μM was cytotoxic; 5b decreased infarct volume when given either i.c.v. (1 μl of 10 mM) or i.v. (5 mg/kg). 5b did not decrease infarct volume in ASIC1-KO mice, confirming ASIC1a as the target	Qi et al. (2022)

(continued)

Table 1 (continued)

ASIC subtype	Intervention	Model/tissue (inc. spp.)	Outcome/therapeutic effect	References
1a	PcTx1	Mouse cultured cortical neurons Mouse transient MCAO model of IS Rat conscious (endothethin-1) MCAO model of IS	PcTx1 containing venom protected cultured cortical neurons from acidosis and OGD induced cell death and reduced infarct volume by up to 60% when given either i.c.v. or i.n. PcTx1 venom given i.c.v. was protective when administered up to 5 h post MCAO in mice PcTx1 (pure peptide @ 1 ng/kg) was neuroprotective when given i.c.v. in a conscious rat model of IS at 2 h post MCAO (~75% reduction in infarct volume) and at 4.6 ng total dose i.c.v. in a mouse model of IS exacerbated by the presence of 2.5 mM spermine	Xiong et al. (2004), Pignataro et al. (2007), Duan et al. (2011), McCarthy et al. (2015), Qiang et al. (2018)
1a	Hi1a	Rat mixed cortical culture Rat conscious (endothethin-1) MCAO model of IS	Hi1a (2 ng/kg i.c.v.) was neuroprotective in conscious rat MCAO model, up to 8 h post-stroke onset: reduced infarct size by up to 80% and improved neurological and motor function	Chassagnon et al. (2017)
1a	Peptide NT_{1-20}	Mouse cultured cortical neurons Mouse transient MCAO model of IS	NT_{1-20} protected mouse cultured cortical neurons against acidosis (pH 6) induced necroptosis; decreased infarct volume by ~43% in mice	Wang et al. (2020)
1a	ASC06-IgG	CHO-K1 cells expressing human ASIC1a Mouse transient (60 min) MCAO model of IS	ASC06-IgG1 (up to 1 μM) provided mild protection of hASIC1a/CHO cells exposed to pH 5.5, when given at 0.5 mg/kg i.c.v. ASC06-IgG1 decreased infarct volume by ~50% in mice	Qiang et al. (2018)
Ischemia of heart				
1a	KO, Hi1a, PcTx1	Human stem-cell (hiPSC) derived cardiomyocytes Langendorff mouse heart, IR injury (25′/45′) Mouse model of myocardial IR injury	ASIC1a KO, Hi1a (10 nM) and PcTx1 (10 nM) protected hiPSC derived cardiomyocytes acute IRI in vitro; ASIC1a KO, Hi1a (10 nM) and PcTx1 (10 nM) decreased IR injury in ex vivo, perfused mouse heart preparation; Hi1a (1 mg/kg, i.v.) decreased adverse chamber remodeling and improves cardiac function after IRI in vivo	Redd et al. (2021)

(continued)

Table 1 (continued)

ASIC subtype	Intervention	Model/tissue (inc. spp.)	Outcome/therapeutic effect	References
Ischemia of kidney				
1a	PcTx1	Human HK-2 cells (in culture) Mouse model of renal IR injury	ASIC1a inhibition decreased hypoxia-induced calcium influx and apoptosis in HK-2 cells; inhibiting 1a with PcTx1 (2–10 nmol/kg i.v. prior to ischemia) protected mice against renal IR injury	Song et al. (2019)
Organ transplant				
1a	Hi1a	Ex vivo donor heart models (rat) direct transplant (18 min ischemia/60 min reperfusion) 8 h hypothermic storage model	Hi1a (10 nM) improved the functional measures of hearts (aortic flow, cardiac output, and pulse pressure), demonstrating cardio protection in models mimicking clinical heart transplant	Redd et al. (2021)

Table abbreviations: *BBB* blood-brain barrier, *IR* ischemia/reperfusion, *i.c.v.* intracerebroventricular, *i.p.* intraperitoneally, *i.v.* intravenously, *i.n.* intra-nasally, *KO* knockout, *IS* ischemic stroke, *PcTx1* psalmotoxin, *NMR* naked mole rat, *MCAO* middle cerebral artery occlusion, *OGD* oxygen-glucose deprivation, route of administrations

highlighting the need for more selective inhibitors of ASIC1a (Xiong et al. 2004). In addition to amiloride, its benzyl analogue has also been observed to inhibit ASIC currents in cortical neurons, with improved activity at reducing infarct size in a mouse model of MCAO when administered i.c.v., even reducing infarct volume by 31% when administration was delayed to 3 h post-stroke (Leng et al. 2016). In these studies, benzamil was effective at 24- and 12-fold lower doses than amiloride, suggesting that this clinically available drug is a better candidate for studying ASICs in vivo. However, it should be noted that benzamil is still more potent at ENaCs and other channels than ASICs (Leng et al. 2016; Castaneda et al. 2019).

Diminazene analogues have also been developed which have been shown to inhibit ASIC1a currents in human neurons. These analogues decreased cell death in an in vitro model of ischemia, using rat primary cortical neurons exposed to oxygen and glucose deprivation followed by reoxygenation (Gornati et al. 2021). Additionally, these analogues were administered to mice i.p. immediately after reperfusion with additional doses at 3 and 6 h following MCAO, with Compound 3b significantly reducing infarct size and improving neurological deficits following stroke (Gornati et al. 2021). Given that Compound 3b is neuroprotective even when administered peripherally following stroke at 0.65 mg/kg, it is a promising candidate for future studies targeting ASICs in vivo.

C5b is a small molecule mimic of the positively charged loop of PcTx1. C5b reduced cell death of mouse cortical neurons following acid-induced cell death, with C5b having no effect in ASIC1a KO neurons, confirming ASIC1a inhibition as its mechanism of reducing cell death (Qi et al. 2022). In the context of stroke, central administration of C5b 15 min prior to stroke improved motor coordination and infarct volume in a mouse model of MCAO (Qi et al. 2022). In the same study, C5b also showed a similar level of protection when administered systemically compared to i.c.v. administration, indicating that changing to a more clinically relevant route of administration did not decrease efficacy, although this was still given as a pre-treatment prior to stroke. Intravenous administration of C5b had no effect on infarct size or behavioral outcomes in ASIC1a KO mice, confirming that the improvements in stroke outcome following C5b administration are ASIC1a-dependent (Qi et al. 2022). These results look promising; however, for such a potent inhibitor of ASIC1a in vitro (27 nM IC_{50}), a rather high dose was required in vivo to see protective effects – this may be due to the strongly pH-dependent nature of the inhibition. Furthermore, C5b was cytotoxic in in vitro assays at three to tenfold higher concentrations than its most protective concentrations. How valuable this molecule is for in vivo target validation of ASIC1a remains to be confirmed.

PcTx1-venom inhibits proton-gated currents in mouse cerebellar granule cells (Escoubas et al. 2000), and reduced LDH release in neurons subjected to acidosis-induced injury (Xiong et al. 2004). Interestingly, PcTx1 did not reduce LDH release in neurons from ASIC1 KO mice, confirming an ASIC1-dependent mechanism of neuroprotection (Xiong et al. 2004). Additionally, in a model of acidosis-induced injury in combination with oxygen-glucose deprivation, PcTx1 treatment also reduced LDH release (Xiong et al. 2004). The therapeutic effect of PcTx1 has also been tested in several models of ischemic stroke. Central administration of whole PcTx1-containing venom 30 min prior to and after transient MCAO reduced infarct volume by 60% in mice (Xiong et al. 2004), while pure PcTx1 (also i.c.v.) provided protection against spermine enhanced damage in a mouse MCAO model (Duan et al. 2011). Extending these initial findings, pure PcTx1 was administered i.c.v. 2 h after ET-1 induced MCAO in rats, and animals receiving treatment had improved stroke outcome, such that infarct volume was reduced by 70% (McCarthy et al. 2015). The neuroprotective effects of PcTx1 have also been observed following systemic drug administration. Specifically, intranasal administration of PcTx1 at 1- and 4-h post-stroke in mice subjected to transient MCAO reduced infarct volume by 50% (Pignataro et al. 2007). Despite these exciting initial findings, when central administration of PcTx1 was delayed to 6 h post-stroke, the neuroprotective effects were lost (Pignataro et al. 2007). For PcTx1 to be a clinically viable therapeutic for stroke, it would need to have a longer therapeutic time window than 6 h and be effective when delivered intravenously. Another caveat with the use of PcTx1 as a therapeutic lead is that it potentiates both rodent and human ASIC1b, which is involved in peripheral nociception, at least in rodents (see below). Thus, more selective ASIC1a inhibitors with an extended therapeutic window need to be examined.

The Australian funnel-web spider peptide Hi1a has improved ASIC1a selectivity compared to PcTx1 and was shown to increase cell viability in co-cultured primary neurons and astrocytes undergoing oxygen and glucose deprivation to a greater extent than cells treated with PcTx1 (Chassagnon et al. 2017). In vivo, a single dose of Hi1a injected directly into the brain up to 8 h after ET-1-induced MCAO dramatically decreased infarct volume and behavioral deficits in rats (Chassagnon et al. 2017). However, the neuroprotective effects of Hi1a have only been tested when this drug was given centrally, which is invasive and not a clinically viable route of administration. Therefore, further studies are needed to establish the effectiveness of Hi1a using more clinically relevant routes of administration.

Only one ASIC1a-blocking antibody has been examined in the context of stroke. The antibody ASC06-IgG increases cell viability in acidosis-induced cell death (Qiang et al. 2018). Additionally, i.c.v. administration of ASC06-IgG 3 h after MCAO in mice resulted in reduced infarct volume (Qiang et al. 2018). ASC06-IgG is potent, the most selective ASIC1a inhibitor known, and likely has the favorable pharmacokinetics of antibodies drugs, so it will be of great interest to see if it can gain sufficient access to infarcted tissue when given by i.v. injection or infusion.

Unlike all other ASIC modulators used to date, the ASIC modulatory peptide NT_{1-20} acts intracellularly to target the interaction between ASIC1a and RIPK1, decreasing acidosis-mediated activation of necroptosis. This approach to inhibiting ASIC1a signaling reduced cortical neuron acidosis-induced cell death in vitro, and central administration of NT_{1-20} 30 min prior to MCAO in mice-reduced infarct size, but its inactive version $NT_{1-20}^{E/A}$ did not influence infarct size (Wang et al. 2020). This is a very novel approach to targeting ASIC1a signaling that may selectively decrease the pathological outcomes of overactivation via inhibiting RIPK1-phosphorylation, while leaving the ion conduction function of the channel intact. It remains to be seen if NT_{1-20} is effective when given peripherally, or post-stroke, rather than prior to ischemic insult.

3.1.2 Myocardial Infarct and Organ Transplant

Cardiac ischemia occurs during myocardial infarction and in cases of insufficient blood supply to meet metabolic demand. Within the heart, acid-evoked currents occur in cardiac afferents, resulting in angina during ischemia (Sutherland et al. 2001). ASIC3 mimics the functional features of the channel which underlies these currents (Sutherland et al. 2001); therefore, it has been suggested that it is the sensor of myocardial acidity responsible for cardiac pain (Sutherland et al. 2001). However, sensing cardiac pain due to ischemia is not the only role of ASICs within cardiac ischemia, with ASIC1a recently being implicated in cardiac cell death.

In a comprehensive study on the role of ASICs in ischemia-induced tissue damage in the heart, Redd et al. showed impressive levels of protection via decreased ASIC1a function (Table 1). ASIC1a KO in human iPSC-derived cardiomyocytes increased cardiomyocyte viability compared to WT cells after acidosis-induced cell injury, and hearts excised from ASIC1a KO mice had higher left ventricular developed pressure and improved coronary flow recovery compared to WT mice following ischemia-reperfusion injury (IRI), as assessed using an ex vivo Langendorff apparatus (Redd et al. 2021). Consistent with the results observed using ASIC1a KO, the ASIC1a inhibitors Hi1a or PcTx1 provided similar levels of cell/tissue protection (Redd et al. 2021). The protective effect of Hi1a in the Langendorff IRI model was lost in hearts from ASIC1a KO mice, confirming ASIC1a as the therapeutic target (Redd et al. 2021). In an in vivo model of myocardial infarction, where mice underwent cardiac IRI, mice pre-treated with 1 mg/kg of Hi1a showed protection against ischemia-induced injury with an increase in ejection fraction, and reduced cardiac fibrosis compared to untreated animals (Redd et al. 2021). Finally, the authors assessed Hi1a as a conditioning agent for improving viability of hearts during organ transplant, using a rodent model of donation after circulatory death (DCD) (Redd et al. 2021). Hi1a treatment resulted in improved viability, aortic flow, and cardiac output when combined with clinically used supplements, compared to treatment with these supplements or Hi1a alone (Redd et al. 2021). In another model of DCD, where hearts were also exposed to 8 h of cold storage, combination treatment of Hi1a with the supplement Celsior improved aortic flow and cardiac output compared to Celsior alone (Redd et al. 2021). Overall, these data all suggest a promising cardioprotective effect of inhibiting ASIC1a.

3.1.3 Kidney

Acidosis and oxidative stress play a critical role in acute kidney injury following ischemia. ASICs are expressed in renal proximal tubule cells in mice and humans and are activated during kidney ischemia and inflammation (Song et al. 2019). Pretreatment of human kidney HK-2 cells with PcTx1 prior to hypoxia resulted in decreased Ca^{2+} overload and apoptosis and, in an in vivo model of renal IRI, injection of PcTx1 30 min prior to the ischemic insult reduced renal injury and apoptosis (Table 1 and Song et al. 2019).

3.2 *ASICs in Multiple Sclerosis*

As acidosis is common to pathological conditions that attract and activate immune cells, it is not surprising that low pH affects their behavior (Erra Diaz et al. 2018), or that immune cells express ASICs (Foster et al. 2021). Acidosis and inflammatory mediators can increase ASIC expression, current density, and the number of ASIC-expressing cells (Mamet et al. 2002; Duan et al. 2007; Tong et al. 2011). Furthermore, ASIC function is enhanced by many endogenous inflammatory mediators. For example, lactate, serotonin, spermine, arachidonic acid, lysophosphatidylcholine (LPC), nitric oxide, prokineticin2, and adenosine triphosphate (ATP) all potentiate the function of ASICs either directly (via decreasing desensitization or increasing activation) or indirectly (via phosphorylation, Qiu et al. 2012; Baron and Lingueglia 2015; Marra et al. 2016). Together, this supports a direct role for ASICs in the development and maintenance of inflammation, as well as contributing to acidosis-induced nerve and cardiac cell death as discussed above. This idea is well illustrated when looking at the role of ASICs in the autoimmune/neurodegenerative disease, multiple sclerosis (studies are summarized in Table 2).

Table 2 ASICs in multiple sclerosis

ASIC subtype	Intervention	Model/tissue (inc. spp.)	Outcome/therapeutic effect	References
Multiple sclerosis				
1	ASIC1a knockout	ASIC2$^{-/-}$ vs WT and ASIC1$^{-/-}$ mice in EAE model of MS Human brain tissue of 62 cases of secondary progressive MS vs 32 control ASIC1a$^{-/-}$ and ASIC2$^{-/-}$ vs WT mice models in EAE	ASIC2$^{-/-}$ mice observed to have reduced clinical score in EAE mouse model (however, less than ASIC1 reductions, and for less time); ASIC2 mRNA increased in human MS samples 1a involved in EAE-induced motor deficits, but no change in neuropathic pain development	Friese et al. (2007), Vergo et al. (2011), Wang et al. (2017) and Fazia et al. (2019)
1	Amiloride	Mice oligodendrocytes and axonal cells; human spinal cord/optic nerve tissue cells; ASIC1 KO vs WT mice in EAE model with amiloride administered	ASIC1 upregulated in axons and oligodendrocytes in mouse and human MS lesions Blockade of ASIC1 with amiloride (10 mg/kg) protected myelin and neurons from damage in mouse EAE model	Friese et al. (2007) and Vergo et al. (2011)
1a	Amiloride (human trials)	Progressive MS patients: open label pilot study Secondary progressive MS patients: double blind clinical trial (MS-SMART trial)	Amiloride reduced brain atrophy in 9 out of 14 patients with primary progressive MS in a pilot study with no control group Amiloride (10 mg/day) failed to show neuroprotection in patients in the MS-SMART trial	Arun et al. (2013) and Chataway et al. (2020)

Table abbreviations: *EAE* Experimental autoimmune encephalitis, *MS* multiple sclerosis.

Multiple sclerosis (MS) is a chronic disease of the central nervous system. Local acidosis in MS lesions occurs due to hypoxia (Halder and Milner 2021), with strong contributions from highly activated immune cells such as macrophages and microglia shifting their metabolism to glycolysis (Lauro and Limatola 2020), and in the CNS from widespread synaptic release. Furthermore, oxidative bursts from activated microglia also release protons (Ritzel et al. 2021), as does the resulting energy failure of highly stressed neurons (Friese et al. 2007; Dendrou et al. 2015). As a result, inflamed spinal cords of experimental autoimmune encephalomyelitis (EAE) mice can reach a pH of 6.5, and lactate levels are increased in the lesions (Bitsch et al. 1999), cerebrospinal fluid (Albanese et al. 2016), and serum of MS patients and correlates strongly with disease severity (Amorini et al. 2014). Being the main isoform expressed in CNS neurons, microglia, astrocytes, and oligodendrocytes, ASIC1a is perfectly positioned to be an effector and sensor of sustained CNS immune cell activation, contributing to the vicious cycle of CNS inflammation and to neuronal and oligodendrocyte dysfunction and death in MS. Indeed, increased expression of ASIC1a has been associated with demyelination and axonal injury in post-mortem spinal cord tissues from mice with EAE and from patients with MS (Vergo et al. 2011). Moreover, inhibition of ASIC1a with amiloride (Friese et al. 2007; Vergo et al. 2011) or ASIC1a genetic knockout significantly decreased clinical deficits in EAE mice (Friese et al. 2007; Wang et al. 2017; Fazia et al. 2019). Thus, there is good evidence that ASIC1a inhibition protects both neurons and oligodendrocytes in MS, making centrally expressed ASIC1a a promising candidate as a neuroprotective drug target in MS as well as stroke. A limitation in this field is that amiloride is the only pharmacological agent used in MS models so far. The use of more selective ASIC1a inhibitors as has been reported in stroke studies is required.

ASIC2a is the second most common subtype expressed in the CNS, and two studies using ASIC2 knockout mice suggest it may also contribute to the pathology of EAE; however, the results were conflicting with one study showing a trend toward worsened clinical deficits (Wang et al. 2017), while Fazia et al. saw a mild reduction in clinical deficits over a narrow time window near the peak of symptom development (Fazia et al. 2019). ASIC2a is less sensitive to acidosis than ASIC1a, and its increased expression (increasing abundance of ASIC1a/2a heteromers that are less pH sensitive than ASIC1a homomers) is postulated to be protective in the CNS. This idea is consistent with a trend for worsening of disease in ASIC2 KO mice as seen by Wang et al. with increased ASIC1a driven response to acidosis, and ASIC1a is the driver of acidosis-induced cell death in CNS hypoxia/inflammation. Currently, the exact role of ASIC2a in EAE/MS remains unclear but should be considered when investigating the effects of ASIC inhibitors in models of MS and neurological conditions in general.

3.3 ASICs in Pain

Chronic pain conditions affect up to 50% of the population at some stage in their life and inflict a huge burden on society both for quality of life and economically (Mills et al. 2019). Upon the first discovery of acidosis-activated currents in trigeminal neurons (Krishtal and Pidoplichko 1981) and the identification of ASIC1a as the ion channel underlying these currents, ASICs were predicted to play a role in nociception (Waldmann et al. 1997a, b). Pain is often associated with tissue acidosis and the release of inflammatory mediators (Dube et al. 2009), conditions that can lead to sustained ASIC activation (Pattison et al. 2019). ASICs are widely expressed in critical regions involved in pain sensing at all levels of the nervous system, consistent with a key role in nociception and pain perception (Waldmann et al. 1997a, b; Pattison et al. 2019; Verkest et al. 2022). Pain-related plasticity and signal transmission occur in the dorsal spinal cord where ASIC1a and ASIC2a are expressed (Wu et al. 2004). Expression of ASIC1a is also observed in the ventral and dorsal regions of the periaqueductal gray matter, regions that are important for central pain perception and modulation (Sluka et al. 2009). ASIC1b and ASIC3 are predominantly expressed in sensory neurons in the PNS, where they play a role in local inflammation and pain sensation (Walder et al. 2010). The expression of ASICs is upregulated during inflammation, and channel function is increased directly by inflammatory mediators such as lactate and arachidonic acid (Molliver et al. 2005; Smith et al. 2007; Marra et al. 2016). Indeed, ASICs play an important role in central sensitization in nociceptive signaling in the spinal cord as revealed by elegant in vivo electrophysiological experiments (Duan et al. 2007; Chafai et al. 2023). The use of knockout mice alone to study the role of ASICs in nociception/pain has resulted in varied and often conflicting results (reviewed in Lin et al. 2015). To date the most compelling evidence that ASICs are involved in pain has been provided by studies combining genetic knockout in mice with pharmacological inhibition or activation of ASICs (summarized in Table 3).

The venom peptide APETx2 inhibits ASIC3-containing homomers and heteromers (Diochot et al. 2004). Peripheral administration of APETx2 in mice and rats in three independent studies showed decreased pain-related behaviors with potent analgesic and antihyperalgesic effects in acid-evoked and inflammatory pain (Deval et al. 2008; Karczewski et al. 2010; Lee et al. 2018). Furthermore, the role of ASIC3 in acid-induced bone pain was demonstrated by the injection of APETx2 in rodent models of arthritis. APETx2 inhibited the upregulation of ASIC3 expression in knee joints and prevented cartilage damage in rats (Izumi et al. 2012) and prevented the development of mechanical hypersensitivity in mice injected with B02/B09 monoclonal antibodies (Jurczak et al. 2022). The non-proton ligand, 2-guanidine-4-methylquinazoline (GMQ), selectively modulates ASIC3 gating so that the channel stays open at neutral pH. Peripheral injection of GMQ into the paw of wild-type mice triggered acute pain responses, an effect that was not observed in ASIC3 knockout animals (Yu et al. 2010), providing compelling evidence for a role of ASIC3 in peripheral pain, consistent with it being the dominant ASIC subtype in rodent sensory neurons.

The snake toxin MitTx is an agonist of ASIC1a, ASIC1b, and ASIC3 at neutral pH, and potentiator of low-pH-induced currents of ASIC2a (Bohlen et al. 2011). MitTx induced robust pain responses when injected into the hind paw of rodents, a response that was ablated in ASIC1 knockout mice. However, the MitTx-evoked pain response persisted in ASIC3-knockout animals, suggesting that ASIC1a and/or ASIC1b are also involved in peripheral pain sensation. As MitTx activates both ASIC1a and ASIC1b subtypes, it was unclear which ASIC1 subtype was responsible for peripheral pain (Bohlen et al. 2011). Mambalgin peptides were discovered to be the first potent inhibitors of rodent ASIC 1a *and* ASIC1b (Diochot et al. 2012). Mamabalgin-1 is analgesic upon central and peripheral injection. siRNA-mediated gene silencing of ASIC1b in nociceptors of ASIC1a knockout mice caused sustained analgesia, revealing the importance of ASIC1b in peripheral pain sensing. In contrast, central injection of mambalgin-1 in ASIC1a knockout mice resulted in a loss of analgesia. Furthermore, central injection of PcTx1 is analgesic in rodent models of pain, highlighting the involvement of ASIC1a in central pain (Duan et al. 2007; Mazzuca et al. 2007). It has since been demonstrated that mamabalgin-1 produces peripheral analgesia through ASIC1b inhibition in several animal pain

Table 3 ASICs in pain and migraine

ASIC subtype	Intervention	Model/tissue (inc. spp.)	Outcome/therapeutic effect	References
Acute and surgical pain				
1a	PcTx1	Acute pain with various tests (tail/paw immersion test, hot plate test, etc.) in mice	PcTx1 i.t. and i.c.v. injection produced blockade of 1a and resulted in reduced nociception via activation of the endogenous enkephalin pathway	Mazzuca et al. (2007)
1	MitTx	Injection of MitTx into WT mice hindpaws	Nocifensive response produced by injection of MitTx, which diminished in ASIC1-deficienct mice but not ASIC3 KO	Bohlen et al. (2011)
3	KO GMQ (agonist)	ASIC3 KO vs WT mouse with GMQ	GMQ-activation of peripheral ASIC3 caused acute pain response which was not observed in ASIC3 KO mouse	Yu et al. (2010)
1a, 1b, 1a/2a heteromers	KO Mambalgin	Multiple in vivo acute pain readouts, 1a and 2a in central neurons; 1b in nociceptors	Mambalgin blockade of 1a and 1a/2a heteromers in central neurons and 1b in nociceptors produced analgesia Paw-flick and tail-flick test in mouse models; thermal pain mouse models	Diochot et al. (2012)
3	APETx2	Rat model of postoperative pain (hindpaw skin/muscle incision)	ASIC3 inhibition with PcTx1 and ASIC3 KO reduced postoperative pain behaviors	Deval et al. (2011)
Inflammatory and neuropathic pain				
1a, 1b	Mambalgin	Mambalgin injection in Carrageenan injection mouse model of inflammatory pain; unilateral peripheral mononeuropathy induced mouse model	Mambalgin i.v. reduced inflammatory pain via 1a-independent effect; intrathecal injection reduced pain via 1a-dependent partially naloxone-sensitive effect	Diochot et al. (2016)
3	APETx2 Diminazene	Mouse and rat Freund's complete adjuvant inflammatory pain models	APETx2 provided dose-dependent anti-hyperalgesia in rodents in three independent studies. Non-selective ASIC inhibitor diminazene was shown for first time to have potent anti-hyperalgesic effects in rat pain model	Deval et al. (2008), Karczewski et al. (2010) and Lee et al. (2018)

(continued)

Table 3 (continued)

ASIC subtype	Intervention	Model/tissue (inc. spp.)	Outcome/therapeutic effect	References
Chronic muscle pain/fibromyalgia				
1b	ASIC1b KO Mambalgin	Dorsal root ganglion neurons from mice; acid IM injection-induced fibromyalgia pain model in ASIC1b$^{-/-}$	Deletion of 1b or blockade by mambalgin reduced acid-induced acute and chronic hyperalgesia; 1a and 3 blockers had mild/no effect on pain in this model	Chang et al. (2019)
3	Acid injection	Dorsal root ganglion neurons; acid IM injection-induced fibromyalgia pain model in ASIC3$^{-/-}$ mice	ASIC3 activation led to increased, long-lasting hyperalgesia via plasticity changes/priming of tetrodotoxin-sensitive and resistant voltage-gated sodium channels in muscle nociceptors	Chen et al. (2014)
Bone pain and erosion				
3	APETx2	Collagen antibody induced arthritis in mouse model of joint inflammation Mono-iodoacetate-induced rat model of osteoarthritis (OA)	B02/B09-induced mechanical hypersensitivity inhibited by ASIC3 block with APETx2 ASIC3 upregulated in knee joint afferents and APETX2 inhibited both pain and disease progression in rat OA model	Izumi et al. (2012) and Jurczak et al. (2022)
Migraine				
1b	Mambalgin	Rat model of acute and chronic migraine induced by i.p. injections of isosorbide dinitrate	ASIC1 inhibition reduced cutaneous mechanical allodynia; 1a not involved in this effect, as 1a KO mice did not show allodynic effects with mambalgin-1 administration	Verkest et al. (2018)
3	APETx2	NO-mediated trigeminal nociceptive response and cortical spreading depression models in rats; nitroglycerin-evoked periorbital mechanosensitivity model in mice	APETx2 blockade of ASIC3 inhibits: Acid-induced CGRP release from trigeminal neurons Durovascular-evoked trigeminal nociceptive processing and reduces trigeminal sensitization induced by NO, acidosis and acidosis with inflammatory mediators, in vivo and in vitro	Yan et al. (2011, 2013), Durham and Masterson (2013) and Holton et al. (2020)

(continued)

Table 3 (continued)

ASIC subtype	Intervention	Model/tissue (inc. spp.)	Outcome/therapeutic effect	References
1a	Amiloride	Cortical spreading depression induced in WT vs ASIC1$^{-/-}$ mice Small human clinical trial	Amiloride blockade reduced CSD/trigeminal activation in mice via ASIC1 pathway Amiloride (10–15 mg) decreased symptoms in 4 of 7 patients with intractable migraine with aura	Holland et al. (2012)

Table abbreviations: *GMQ* 2-guani-dine-4-methylquinazoline, *NO* nitric oxide, *NP* Nucleus pulposus

models related to acidosis, such as in fibromyalgia, inflammatory pain, and neuropathic pain (Diochot et al. 2012, 2016; Mourier et al. 2015; Chang et al. 2019). The above studies together have demonstrated roles for ASIC1a in central pain processing and ASIC1b and ASIC3 in peripheral nociception in rodents.

In humans, the direct infusion or iontophoresis of acidic solution (pH 6.0) into skin triggered local sensation of pain (Jones et al. 2004). The application of NSAIDs such as ibuprofen and diclofenac (inhibitors of ASIC1a and ASIC3) after iontophoresis of acid reduced acid-evoked pain. However, skin desensitized by repeated application of capsaicin, an agonist of transient receptor potential vanilloid receptor-1 (TRPV1) did not significantly reduce the acid-evoked pain. These results suggest that cutaneous acid-induced pain is primarily associated with ASICs, whereas TRPV1 only plays a minor role (Jones et al. 2004). The acid-induced pain response could also be alleviated by subcutaneous injection of amiloride, supporting the involvement of ASICs in human peripheral pain pathways (Ugawa et al. 2002; Jones et al. 2004). As amiloride is a non-selective ASIC inhibitor, it remains to be determined which ASIC subtypes are most important for peripheral pain in humans. Further studies using selective inhibitors with well-characterized activity at human ASIC subtypes will be essential for determining the function of ASICs in pain sensing and validating them as potential analgesic targets in humans.

More than 15% of the world's population suffer from the debilitating and recurring severe headache, nausea, and sensory symptoms of migraine, making it one of the most common pain and neurological conditions (Ferrari et al. 2022). The exact pathophysiology underlying migraines is still not clear; however, three main events seem to take place. Cortical spreading depolarization is proposed to activate trigeminovascular nociceptors (rich in pro-inflammatory and pain enhancing calcitonin gene-related peptide CGRP) and, under prolonged migraine, lead to peripheral and central sensitization of this pain pathway (Ferrari et al. 2022). Approximately 80% of the trigeminal nociceptors that innervate the dura in rats display acid-induced currents that are blocked by amiloride. Furthermore, direct application of pH 5 solution to the dura in conscious animals resulted in migraine-like behavior that was unaffected by a TRPV1 antagonist but inhibited by amiloride, providing a solid link between ASIC activation and migraine pathology (Yan et al. 2011).

Shortly after, it was discovered that ASIC3 underlies acidosis-induced CGRP release from rat trigeminal neurons in vitro (Durham and Masterson 2013). ASIC3 was then shown to be functionally expressed in the majority of dural afferents and is the dominant ASIC subtype underlying trigeminal nociceptor activation. This translated to in vivo studies where both amiloride and APETx2 prevented low pH-induced migraine behavior to the same extent (Yan et al. 2013). In the same study, it was observed in vitro that simulated inflammatory conditions (mimicking mast cell degranulation) substantially shifted low pH-induced dural afferent activation to more neutral values. This was also observed in vivo whereby a milder acidosis was required to induce migraine pain when combined with simulated inflammation (Yan et al. 2013). A key role for ASIC3 in the pathology of migraine as suggested in these studies was recently supported by an independent group again using APETx2, showing that it could reverse durovascular and NO-evoked trigeminal nociceptive responses in rats and a model of cortical spreading depolarization in mice (Holton et al. 2020). All these studies used APETx2 as the ASIC3 inhibitor, which is still the most potent and selective ASIC3 inhibitor available; however, it is not selective for ASIC3 and has also been reported to inhibit rat $Na_V1.2$ channels in the same concentration range. Given the frequent lack of selectivity of venom derived Na_V channel inhibitors particularly for CNS Na_V and Ca_V isoforms (Peigneur et al. 2012; Klint et al. 2014; Wingerd et al. 2017; Finol-Urdaneta et al. 2022), and the role of voltage-gated ion channels in migraine (Ferrari et al. 2022), it would be of great value to have the role of ASIC3 confirmed by other means. The effects of APETx2 were comparable to those of amiloride; however, amiloride is also a highly promiscuous drug, particularly against other sodium and calcium transporters and ion channels (Kleyman and Cragoe Jr. 1988).

A role for ASIC1b in a rat model of migraine-related cutaneous allodynia has also been suggested. Systemic administration of the ASIC1 inhibiting mambalgin-1 inhibited the development of cephalic and extra-cephalic acute cutaneous mechanical allodynia and delayed the establishment of chronic allodynia. Amiloride and mambalgin-1 also reversed established chronic allodynia in this model, effects that were still present in ASIC1a$^{-/-}$ mice confirming the role of ASIC1b (Verkest et al. 2018). Overall, targeting ASICs in pain and migraine shows substantial promise for the development of novel and hopefully well-tolerated therapeutics. Which subtype of ASIC1 is most appropriate for human pain is further discussed in Sect. 4.

3.4 ASICs in Neurological Conditions and Trauma

Growing evidence from animal models and humans has demonstrated that neurodegenerative diseases, characterized by a persistent inflammatory state, are associated with a significant reduction in brain pH (Tyrtyshnaia et al. 2016; Decker et al. 2021). ASIC1a, predominantly expressed in the CNS (particularly hippocampus, cortex, striatum, and amygdala), provides a non-voltage-gated mechanism for increasing intracellular Ca^{2+} (either directly or indirectly), thereby modulating calcium-related processes such as synaptic plasticity, learning and memory, as well as neuronal death during neurological disorders (Mango and Nistico 2021). We have covered the compelling data that supports a key role for ASIC1a in ischemia- and inflammation-induced CNS tissue injury in stroke and MS above. Here we summarize what is known about the potential role of ASICs in primary neurodegenerative diseases, acute trauma of the CNS, and epilepsy (see also Table 4).

Table 4 ASICs in neurological disease

ASIC subtype	Intervention	Model/tissue (inc. spp.)	Outcome/therapeutic effect	References
Alzheimer's disease				
1a	PcTx1	CA1 pyramidal neurons; APP transgenic mouse model of AD	1a modulates Aβ-mediated increase of CA1 pyramidal neuron excitability; 1a blockade by PcTx1 rescued DHPG-LTD facilitation, suggesting 1a may contribute to synaptic alterations in AD PcTx1 reduced Aβ-mediated death of mouse primary cortical neurons	Liu et al. (2018b and Mango and Nistico (2018)
Parkinson's disease				
1a	Amiloride PxTc1	Dopaminergic neurons in substantia nigra; Mouse MPTP model of PD	Amiloride administration protected neurons from degeneration PcTx1 (as whole venom) showed partial neuroprotection of dopaminergic neurons	Arias et al. (2008)
1a	KO	Dopaminergic neurons in substantia nigra; Mouse MPTP model of PD	ASIC1a KO mice did not show any protection of dopaminergic neurons	Komnig et al. (2016)
Huntington's disease				
1a	Benzamil	R6/2 mouse model of HD	Blocking 1a with benzamil (amiloride derivative) markedly reduced the huntingtin-polyglutamine (htt-polyQ) aggregation	Wong et al. (2008)
Traumatic brain and spinal cord injury				
1a	KO	TBI mouse model comparing ASIC1a−/− vs WT	ASIC1a KO mice showed reduced lesion volume and hippocampal damage, reduced edema and less brain damage, neuroinflammation; showing role of 1a chronic neurodegeneration after TBI ASIC1a KO mice showed reduced neuronal cell death at 24 h and partially improved spatial memory deficits 4 days after the insult	Yin et al. (2013) and Cheng et al. (2021)

(continued)

Table 4 (continued)

ASIC subtype	Intervention	Model/tissue (inc. spp.)	Outcome/therapeutic effect	References
1a	Amiloride PcTx1 siRNA KD	SCI rat model	ASIC expression and function increased at injury site, intrathecal PcTx1, amiloride or siRNA decreased SCI and functional impairments ASIC1a inhibition with s.c. infusion of PcTx1 decreased white matter damage and improved locomotor function	Hu et al. (2011) and Koehn et al. (2016)
Epilepsy				
1a	KO vs overexpression	Kainate and PTZ-induced seizures in 1a KO vs WT mice; interneurons in 1a KO vs WT mice; CO2 inhalation model of seizures	1a inhibition increased severity and duration of seizures; overexpression had opposite effect; suggests ASIC1a is involved in the termination of seizures induced by acidosis	Ziemann et al. (2008)
1a	Gene expression study	Assessing ASIC1a polymorphism in temporal lobe epilepsy	*ASIC1a* gene as a possible temporal lobe epilepsy (TLE) susceptibility gene in humans and suggested that ASIC1a activation plays a role in the generation and maintenance of epileptic seizures	Lv et al. (2011)
1a	AAV vectors to to suppress or restore 1a	Temporal lobe epilepsy (TLE) model	Inhibition of astrocytic ASIC1a expression in the TLE mouse model resulted in reduced spontaneous seizures	Yang et al. (2016)
1a, 3	Gene expression	Pilocarpine induced status epilepticus model in rats	Amiloride reduced gene expression of ASIC1a and ASIC3 to suppress epileptic seizures	N'Gouemo (2008) and Liang et al. (2015)
1a	Amiloride	Pentylenetetrazol (PTZ)-induced seizures in mice	Amiloride showed antiepileptic effects	Ali et al. (2004, 2006)

Table abbreviations: *APP* Amyloid precursor protein, *AD* Alzheimer's disease, *MPTP* 1-methyl-4-henyl-1,2,3,6-tetra hydropyridine, *PD* Parkinson's disease, *PTZ* pentylenetetrazole

Alzheimer's disease (AD) is characterized by a progressive decline in cognitive function that is associated with formation of amyloid plaques (AP) and neurofibrillary tangles leading to neuronal loss in various brain regions, particularly in the hippocampus (DeTure and Dickson 2019). Despite being the most common neurodegenerative disease and with a known link to CNS acidosis, only two studies have so far looked directly at a possible role for ASICs in AD pathology. The first used hippocampal slices from either Tg2576 mice or WT mice treated with neurotoxic Aβ oligomers and showed that in both preparations inhibition of ASIC1a with PcTx-1 reduced metabotropic glutamate receptor (mGluR)-mediated excitability of CA1 pyramidal neurons (Mango and Nistico 2018). The author also demonstrated that long-term depression (LTD) induced by 3,5-dihydroxyphenylglyine (DHPG), a group I mGlu receptor agonist, in genetic and non-genetic models of AD, was rescued by PcTx-1, suggesting functional coupling between ASIC1a and mGlu receptors (Mango and Nistico 2018). A second report in the same year went a step further and showed that PcTx1 at 200 nM could prevent the Aβ oligomer-induced death of mouse cortical neurons (Liu et al. 2018b).

Parkinson's disease (PD) is the second most common neurodegenerative disorder after AD. PD is characterized by progressive degeneration of dopaminergic neurons in the substantia nigra pars compacta and cytoplasmic inclusions of misfolded alpha-synuclein, termed Lewy bodies (Aarsland et al. 2021). The neurodegeneration typically causes involuntary movements including tremor, rigidity, and bradykinesia, in addition to cognitive impairment (Aarsland et al. 2021). Non-selective ASIC inhibition using amiloride had substantial neuroprotective effects in a mouse model of PD induced with the neurotoxin MPTP (1-methyl-4-phenyl-1,2,3,6-tetrahydropyridine). Interestingly, more selective ASIC1a inhibition with PcTx-1 showed more modest neuroprotection, reducing deficits only in striatal dopamine transporter (DAT) binding and dopamine (DA) levels in the MPTP model (Arias et al. 2008). In stark contrast, ASIC1a KO mice did not show any protection of dopaminergic neurons in the substantia nigra following induction of MPTP toxicity (Komnig et al. 2016). Further studies using more selective pharmacological tools with well-characterized CNS distribution would seem warranted to confirm a role for ASIC1a in PD.

Huntington's disease (HD) is a rare hereditary neurodegenerative disease caused by a dominantly inherited CAG trinucleotide repeat expansion in the "huntingtin gene" on chromosome 4 (Tabrizi et al. 2019). This progressive brain disorder leads to involuntary movements of the fingers, limbs, or facial muscles (called chorea), in addition to decline in cognition (Tabrizi et al. 2019). Currently, there is a scarcity of reports to suggest the direct pathobiological role of ASICs in HD. However, one study demonstrated that benzamil, the more potent amiloride analogue, could rescue ASIC-dependent acidotoxicity in R6/2 animal model of HD. Notably, benzamil markedly reduced huntingtin-polyglutamine (htt-polyQ) aggregation, a main hallmark of HD, in the striatum of R6/2 mice (Wong et al. 2008).

Tissue damage after spinal cord injury (SCI) and traumatic brain injury (TBI) occurs by mechanical, ischemic, and inflammatory mechanisms. Vascular disruption in the central gray matter is a prominent feature of human and animal SCIs, leading to marked reductions in spinal cord blood flow (hypoperfusion) that extend beyond the initial injury site (Rivlin and Tator 1978; Tei et al. 2005). The ensuing local hypoxia and inflammation following TBI (Ritzel et al. 2021) or SCI (Li et al. 2021) leads to pronounced tissue acidosis at the injury site and penumbra. Thus, the pathology of TBI and SCI share many features with the CNS damage of stroke. It is perhaps not surprising that inhibition of ASICs is reported to be neuroprotective in both SCI and TBI. ASIC expression and function was increased in the penumbra of SCI in a rat model and contributed to neuronal injury. Decreasing ASIC function via intrathecal administration of PcTx1-venom, amiloride, or anti-ASIC1a oligonucleotides decreased SCI-induced tissue damage and preserved neurological function (Hu et al. 2011). A subsequent study showed that preventing ASIC1a activation using a 3-day infusion of pure PcTx1 (total dose 25 μg/kg) via subcutaneous minipump reduced white matter loss and improved locomotor function in adult Sprague Dawley rats following thoracic SCI (Koehn et al. 2016). Notably, this study showed efficacy using a small total dose of PcTx1 over 3 days delivered peripherally. Similar promising results have been replicated in animal models of TBI. ASIC1a KO mice showed reduced neuronal cell death at 24 h and partially improved spatial memory deficits 4 days after the insult (Yin et al. 2013). A more recent study showed that ASIC1a KO mice exhibited reduced brain edema at 24 h after TBI and improved cognitive function, reduced depression-like behavior, and reduced neuroinflammation (particularly microglia activation) at 6 months after TBI compared to control mice (Cheng et al. 2021).

Epilepsy is a common chronic neurological disorder characterized by spontaneous recurrence of unprovoked seizures. Despite several groups investigating a role of ASICs in epilepsy, their exact role in seizure generation, propagation, and termination remains controversial; for instance, one study reported that activation of ASIC1a leads to termination of seizures induced by acidosis (Ziemann et al. 2008), whereas others have reported the *ASIC1* gene as a possible temporal lobe epilepsy (TLE) susceptibility gene in humans and suggested that ASIC1a activation plays a role in the generation and maintenance of epileptic seizures (Lv et al. 2011). In agreement with the latter notion, others have shown that inhibition of astrocytic ASIC1a expression in the TLE mouse model resulted in reduced spontaneous seizures (Yang et al. 2016). Additionally, amiloride was shown to reduce gene expression of ASIC1a and ASIC3 to suppress epileptic seizures in a pilocarpine animal model of status epilepticus in rats (N'Gouemo 2008; Liang et al. 2015). A number of other studies have also confirmed the antiepileptic effects of amiloride in pentylenetetrazol (PTZ)-induced seizures or in mice (Ali et al. 2004, 2006). Given that the models used in these studies are diverse and epilepsy is a very heterogeneous condition, more work with better pharmacological tools is required to see if there is role for ASIC1a as a potential therapeutic target.

3.5 ASICs in Psychological Disorders

Based on the abundant expression of ASIC1a in brain structures associated with emotive behaviors, such as the amygdala, bed nucleus of the stria terminalis (BNST), periaqueductal gray, dorsal striatum, and nucleus accumbens, numerous studies have linked the role of ASICs – particularly ASIC1a – with psychological behaviors such as fear/anxiety and depression (Wemmie et al. 2003; Price et al. 2014; Taugher et al. 2014, 2017) (summarized in Table 5).

Detailed studies of ASIC1a expression in the brain, supported by experiments employing genetic deletion or pharmacological inhibition of ASIC1a, showed altered synaptic plasticity and reduced fear response in mice (Wemmie et al. 2003; Coryell et al. 2007). In contrast, overexpressing human ASIC1a in the CNS of mice enhanced fear conditioning (Wemmie et al. 2004). In humans, genetic variants of the *ASIC1* gene have been linked with anxiety and panic disorder (Smoller et al. 2014; Gugliandolo et al. 2016). The BNST is a key structure in the brain that was linked with CO_2-evoked fear-related behavior for the first time in studies showing that CO_2-evoked freezing behavior was mediated by ASIC1a (Taugher et al. 2014). Further investigation by Taugher et al. confirmed the neuronal site of ASIC1a relevant to fear conditioning using SynAsic1a KO mice, where Cre-mediated disruption of ASIC1a was restricted to neurons (Taugher et al. 2017). The localization and functional role of ASIC2 is not as well studied as ASIC1; however, a comparison of ASIC2 and ASIC1 KO mice suggests that both subtypes can mediate fear-like responses (Price et al. 2014). ASIC3 is predominately distributed in the peripheral nervous system, and ASIC3 KO in mice did not alter CNS synaptic plasticity or motor function. Nevertheless, loss of ASIC3 did reduce anxiety and aggression in mice (Wu et al. 2010). The authors suggested that a decrease in ASIC3 mediated sensory input could lead to the emotional and behavioral deficits observed.

ASIC1a expression in the amygdala has also been implicated in depression, with animal studies showing pharmacological inhibition of ASIC1a by PcTx1 and A-317567 producing antidepressant-like effects (Coryell et al. 2009). These results were supported by showing that genetic disruption of ASIC1a in mice reduced depressant-like effects during forced swim test and tail suspension test that was reversed by restoring ASIC1a to the amygdala of ASIC1a KO mice (Coryell et al. 2009). More recently, amiloride was shown to reduce depression-related behaviors in a chronic unpredictable mild stress rat model by reducing stress-induced elevation of ACTH and corticosteroid levels, suggesting that ASICs play a role in regulating the hypothalamus-pituitary-adrenal axis (Zhou et al. 2019).

3.6 ASICs in Cancer

The extracellular tumor microenvironment plays a crucial role in regulating cell proliferation, epithelial to mesenchymal transition (EMT), and cell migration (Thews and Riemann 2019). An acidic extracellular environment enhances all these properties of cancer cells, increasing tumor growth and metastasis. The decreased pH of the microenvironment is a result of a switch from oxidative phosphorylation to glycolysis (Warburg effect), poor blood supply (hypoxia), and metabolic waste accumulation, in particular lactate. This environment also results in cancer cells

Table 5 ASICs in psychological disorders

ASIC subtype	Intervention	Model/tissue (inc. spp.)	Outcome/therapeutic effect	References
Anxiety				
1a	KO	Brain tissues and cultured amygdala neurons from ASIC1$^{+/+}$ and $^{-/-}$ mice	ASIC1 was predominantly expressed at glomerulus of the olfactory bulb, whisker barrel cortex, cingulate cortex, striatum, nucleus accumbens, amygdala, and cerebellar cortex. H+-gated currents in cultured amygdala neurons from ASIC1$^{+/+}$ mice, with no current observed in neurons from ASIC1$^{-/-}$ mice following reducing extracellular pH to 5.0. ASIC1$^{-/-}$ mice had decreased fear condition responses	Wemmie et al. (2003)
1a	KO PcTx1	Open-field, acoustic startle, and predator odor fear tests	Loss of ASIC1a decreased fear responses in all three tests. Intracerebroventricular administered PcTx1 reduced odor-evoked freezing in ASIC1$^{+/+}$ mice but not ASIC1$^{-/-}$ mice	Coryell et al. (2007)
1a	Human gene expression	Humans with panic disorder compared to healthy subjects	Participants were screened for *ASIC1* single nucleotide polymorphisms. Identified single nucleotide polymorphisms (SNPs) human *ASIC1a* gene that were associated with panic disorder	Smoller et al. (2014) and Gugliandolo et al. (2016)
1a	KO	AAV vectors to express ASIC1a in ASIC1a$^{-/-}$ mice at BNST SynAsic1a selective KO in neurons	Disrupting ASIC1a in the BNST reduced TMT-evoked freezing SynAsic1a KO mice showed significant deficits during multiple fear-related behaviors including Pavlovian fear conditioning to cue and context, predator odor-evoked freezing and freezing responses to carbon dioxide inhalation	Taugher et al. (2014, 2017)
1, 2	KO	Compared ASIC1$^{-/-}$ and ASIC2$^{-/-}$ mice for conditioned and innate fear assays	ASIC1 and ASIC2 were expressed as co-localized in the brain regions associated with fear including basolateral nucleus (BLA), amygdala and the bed nucleus of the stria terminalis (BST); ASIC1$^{-/-}$ and ASIC2$^{-/-}$ mice showed reduced freezing in conditioned and innate fear assays in addition to increased mobility in the forced swim assay	Price et al. (2014)

(continued)

Table 5 (continued)

ASIC subtype	Intervention	Model/tissue (inc. spp.)	Outcome/therapeutic effect	References
3	KO	ASIC3$^{-/-}$ compared with WT for anxiety-related tests	ASIC3 KO mice had altered anxiety behaviors, with no effect on synaptic plasticity of hippocampus and displayed less aggressive behavior A second study showed ASIC3KO decreased long-term potentiation (LTP) in corticostriatal circuits accounting for altered behavioral responses	Wu et al. (2010, 2019)
Depression				
1a	KO PcTx1 A-317567 AAV vectors to restore ASIC1a	Compared ASIC1a$^{+/+}$ and ASIC1a$^{-/-}$ mice in depression-related assays (forced swim and tail suspension tests)	Loss of ASIC1a reduces depression-related behavior. Pharmacological inhibition of ASIC1a by PcTx1 and A-317567 producing antidepressant-like effects Restoring ASIC1a in the amygdala with AAV vectors eliminated the antidepressant-like phenotype in ASIC1aKO mice	Coryell et al. (2009)
1, 2	Amiloride	Rat chronic stress model of depression	Amiloride reduced chronic stress-elevation of ACTH and corticosteroid levels, reversing depression-related behaviors, possibly by reducing hyperactivity of HPA axis	Zhou et al. (2019)

often having substantially higher intracellular sodium levels than non-tumor cells (Leslie et al. 2019). Sustained exposure to these conditions is lethal to normal cells yet allows cancer cells to thrive. ASIC1a and ASIC3 are expressed in a variety of different cancers and may play an important role in linking the acidic microenvironment to enhanced cancer cell growth, migration, EMT, and metastasis. Over the last decade, there have been only about 20 published studies that have investigated the potential role of ASICs in cancer development and progression, and they are listed and briefly described in Table 6. Many studies have focused on ASIC1 in glioma, but a growing body of evidence suggests that ASICs, in particular ASIC1a, play an important role in a range of cancers and have potential as novel therapeutic targets and/or biomarkers of disease outcome.

Elevated ASIC1 and/or 3 mRNA and/or protein expression in patient cancer samples compared with adjacent non-cancerous tissues have generally been associated with a poorer prognosis. Patients with metastatic melanoma (stage III) and high ASIC1 expression had a poorer overall survival (Bychkov et al. 2021), as did patients with liver cancer (Jin et al. 2017), lung cancer (stage II–IV) (Sudarikova et al. 2022), and breast cancer (Gupta et al. 2016). In patients with prostate cancer, up-regulated expression of ASIC1 and 3 was associated with poor tumor differentiation, advanced tumor stage, and lymph node and distant metastasis (Zhu et al. 2017). The exception was glioma, where high expression of either ASIC1a or 3 was

Table 6 ASICs in cancer

ASIC subtype	Intervention	Model/tissue (inc. spp.)	Outcome/therapeutic effect	References
Glioma				
1a	Mambalgin-2	U251-MG and A172 glioma cells and normal astrocytes; patch clamp and cell culture studies in human cells	Mambalgin-2 inhibition of ASIC1a induced cell cycle arrest and caused apoptosis in glioma cells	Bychkov et al. (2020)
1a, 3	PcTx1, APETx2	GSC lines R54 and R8; whole-cell patch clamping, microarray	Expression of either ASIC1a or 3 was associated with improved patient survival	Tian et al. (2017)
1	PcTx1, RNAi KD	U251-MG glioma cell line; whole-cell patch clamp, transwell migration assay	PcTx1 and ASIC1 KD both attenuated the constitutively activated amiloride-sensitive current in U251-MG cells. CaMKII physically associated with and activated ASIC1, and was required for cell migration	Sun et al. (2013)
1	Acidosis, PcTx1, siRNA KD	U87MG and A172 glioma cells; whole-cell patch clamp and transwell migration and wound-healing assays	Weak acidosis (pH 7.0) increased migration of glioma cells. PcTx1 and ASIC1 KD abolished the acid-induced migration	Sheng et al. (2021)
1a	Acidosis, PcTx1, MitTx1, KO	GSC lines R54 and R8; spheroid formation assay	Reduced spheroid growth at acidic pH was ASIC1a-dependent and involved RIP1, but not MLKL. ASIC1a KO rendered cells resistant to acid-induced reduction in spheroid growth. MitTx1 activation of ASIC1a reduced spheroid formation at pH 7.4	Clusmann et al. (2022)
1a, 2, 3	Acidosis	GSC line R54; 2.5D spheroid formation assay	No evidence of ASIC involvement in acid-dependent migration of glioma cells	Cortes Franco et al. (2023)
1	Amiloride, PcTx1, DN KD	D54-MG human glioma cells and normal human astrocytes; whole-cell patch clamp, migration assay	KD of ASIC1a reduced whole-cell patch clamp current in glioma cells and reduced cell migration	Kapoor et al. (2009)

(continued)

Table 6 (continued)

ASIC subtype	Intervention	Model/tissue (inc. spp.)	Outcome/therapeutic effect	References
Breast cancer				
1a	PcTx1, siRNA KD	MCF-7 and MDA-MB-231 cells; breast cancer tissue; immunohistochemistry, transwell migration/invasion assay, in vivo xenograft nude mouse model	ASIC1a was upregulated in breast cancer tissue compared to adjacent normal tissue. Down-regulation by siRNA or inhibition by PcTx1 reduced proliferation, migration, and invasion of cancer cell lines in vitro and reduced cancer cell proliferation in vivo	Yang et al. (2020)
1	Acidosis, PcTx1, amiloride, siRNA KD	Human breast cancer cell lines MCF-7, MDA-MB-231, LM-4142; cell culture studies with human cell lines, in vitro invasion assay, in vivo metastasis/tumor growth xenograft nude mouse model	Acidosis-induced ROS production and in vitro invasion were mediated by ASIC1, and were blocked by inhibitors, siRNA and/or KO. ASIC1 was required for tumor growth and lung metastasis in a xenografted nude mouse model. ASIC1 mRNA expression in breast cancer samples were 3.7-fold higher than control tissue. Analysis of cDNA microarray dataset showed elevated ASIC1 was associated with poor prognosis	Gupta et al. (2016)
1	Acidosis, siRNA KD	Human breast cancer cell line MDA-MB-231; transwell migration/invasion assays, wound-healing assays	Acidic conditions increased breast cancer cell migration and invasion and was caused by increased IL-8 expression which was mediated by ASIC1. ASIC1 siRNA reduced IL-8 levels	Nakanishi et al. (2021)

(continued)

Table 6 (continued)

ASIC subtype	Intervention	Model/tissue (inc. spp.)	Outcome/therapeutic effect	References
Skin cancer				
1a	Mambalgin-2	Human metastatic melanoma mel P cells and human immortalized oral keratinocyte Het-1A cells; transwell migration/invasion assay, wound-healing assay, TCGA database analysis	Acidification upregulated cell surface ASIC1a and promoted metastatic cell proliferation, migration, and invasion. Mambalgin-2 inhibition of ASIC1a induced apoptosis of metastatic melanoma cells and inhibited the expression/activation of some melanoma progression markers. Patients with metastatic melanoma at disease stage III with lower ACCN2 expression had a better survival prognosis. There was no correlation for primary melanoma	Bychkov et al. (2021)
Colorectal cancer				
2	Acidosis, shRNA KD	Human CRC cell lines SW480, SW620, HT-29, HCT116, and human CRC tissues; transwell migration/invasion assay, colony formation assay, immunohistochemistry, in vivo metastasis/tumor growth xenograft nude mouse model	ASIC2 promoted acidosis-induced CRC cell proliferation and invasion in vitro that was prevented by ASIC2 KD. ASIC2 promoted CRC cell proliferation and liver metastases in vivo that was inhibited by ASIC2 KD. ASIC2 expression was increased in CRC samples compared to adjacent tissue, and correlated with tumor stage, distant metastasis, and recurrence. High expression of ASIC2 was associated with shorter overall survival rates of CRC patients	Zhou et al. (2017)

(continued)

Table 6 (continued)

ASIC subtype	Intervention	Model/tissue (inc. spp.)	Outcome/therapeutic effect	References
Pancreatic cancer				
1, 3	Acidosis, PcTx1, amiloride, siRNA KD, OE	Human PC cell lines PANC-1, BxPC-3, SW1990, AsPc-1, and normal pancreatic ductal cells, and human tumor tissues; transwell invasion assay, wound-healing assay, in vivo metastasis/tumor growth xenograft nude mouse model	ASIC1 and 3 were expressed at higher levels in PC cell compared to normal cells and contributed to acidosis-induced EMT, migration, and invasion. KD or inhibition of ASIC1 or 3 suppressed migration and invasion in vitro, while OE had the opposite effect. KD of ASIC1 or 3 inhibited liver and lung metastases in vivo but did not affect tumor growth. Higher expression of both ASIC1 and 3 in human PC tissue compared to adjacent tissue was correlated with poor tumor differentiation, advanced tumor stage, lymph node and distant metastasis	Zhu et al. (2017)
Liver cancer				
1a	CRISPR/Cas9 KO	SMMC-7721 and Huh-7 cells; HCC tissues; immunohistochemistry, colony formation assay, in vivo xenograft nude mouse model	ASIC1a KO reduced cancer cell growth in vitro and in vivo. ASIC1a KO induced cell cycle arrest and apoptosis. ASIC1a was up-regulated in liver cancer tissue and was associated with poor prognosis	Jin et al. (2017)
1a	Acidosis, PcTx1, shRNA KD, OE	HepG2 (H^R) and Bel7402 (B^R) cells; HCC tissues; colony formation assay, wound-healing assay	ASIC1a was over-expressed in HCC tissue compared to adjacent non-cancerous tissue, and mediated EMT and chemosensitivity in drug-resistant cell lines. Inhibition or KD of ASIC1a suppressed EMT and enhanced chemosensitivity, while OE had the opposite effect	Zhang et al. (2022)

(continued)

Table 6 (continued)

ASIC subtype	Intervention	Model/tissue (inc. spp.)	Outcome/therapeutic effect	References
1a	RNAi KD, OE	HepG2, HepG2/FU, Bel7402 and Bel7402/ADR HCC cells, L-02 human liver cells, and human HCC tissues; MTT viability assay, calcium imaging	ASIC1a was highly expressed in human HCC tissue and up-regulated in drug-resistant HCC cells. ASIC1a KD enhanced the chemosensitivity of drug-resistant HCC cells, while OE contributed to drug-resistance in HCC cells. ASIC1a-mediated calcium influx activated PI3K/AKT signaling and increased resistance	Zhang et al. (2017a)
Lung cancer				
1a	Mambalgin-2, siRNA KD	A549 lung cancer cells and WI-38 lung fibroblasts; whole-cell patch clamp, spheroid formation assay, wound-healing assay, TCGA database analysis	Mambalgin-2 inhibited acidosis-mediated cancer cell growth and migration and induced cell cycle arrest and apoptosis. Elevated ASIC1a expression correlated with poor survival prognosis for Stage II–IV cancers	Sudarikova et al. (2022)
1a	Acidosis, amiloride, PcTx1, siRNA KD, OE	A549 lung cancer cells; whole-cell patch clamp, wound-healing assay	ASIC1a mediates acidosis-induced cell proliferation and migration and increased [Ca^{2+}]. Inhibition or KD of ASIC1a reduced the effects of acidosis, while OE did the opposite	Wu et al. (2017)
Gastric cancer				
1a	Acidosis, PcTx1, shRNA KD	AGS GC cells and human GC tissues; immunohistochemistry, transwell invasion assay, wound-healing assay, in vivo tumor growth xenograft nude mouse model	ASIC1 expression was up-regulated in GC tissues showing metastasis compared to non-metastasis and adjacent non-tumor tissues. ASIC1a KD suppressed acidosis-induced cell migration and invasion in vitro and reduced tumor growth and extended survival in vivo	Chen et al. (2018)

(continued)

Table 6 (continued)

ASIC subtype	Intervention	Model/tissue (inc. spp.)	Outcome/therapeutic effect	References
1	shRNA KD	SGC-7901 GC cells, GES-1 cells, and human GC tissues; RNA-seq, in vivo xenograft nude mouse model	ASIC1a was up-regulated in human GC tissue compared to normal gastric tissue. ASIC1 KD inhibited GC cell growth in vitro by activating autophagy, and reduced tumor growth and extended survival in vivo	Zhang et al. (2017b)
Prostate cancer				
1	Acidosis, PcTx1, CRISPR/Cas9 KO	22Rv1, DU-145, LNCaP, and PC-3 prostate cancer cells; transwell invasion assay, in vivo xenograft nude mouse model	ASIC1 promotes acidosis-induced invasion in vitro and increased EMT mediators, which were both prevented by ASIC1 KO. Tumor growth in vivo and castration resistance were reduced by ASIC1a KO	Chen et al. (2016)

Table abbreviations: *CRC* colorectal cancer, *DN* dominant negative mutation, *EMT* epithelial to mesenchymal transition, *GC* gastric cancer, *GSC* glioblastoma stem cell, *HCC* hepatocellular cancer, *KD* knockdown, *KO* knockout, *OE* over-expression, *PC* pancreatic cancer, *PcTx1* psalmotoxin

associated with improved overall patient survival (Tian et al. 2017). The reason for this difference between glioma and the other cancers discussed above is currently unknown. There also has been one study reporting that elevated ASIC2 expression in colorectal cancer was correlated with tumor stage, distant metastasis, and disease recurrence, and was associated with poor overall survival (Zhou et al. 2017). While ASIC1 appears to be the main ASIC involved in cancer development and progression, it is clear that ASIC2 and ASIC3 are also important in certain cancer types.

The clinical data showing the importance of ASICs in cancer development and progression have been corroborated by independent groups using different in vitro and in vivo models. The in vivo studies used a nude mouse xenograft model to assess tumor growth and metastasis of injected cancer cell lines with and without ASIC CRISPR/Cas9 knockout, shRNA knockdown, or PcTx1 inhibition (see Table 6). All but one of the eight published studies that performed in vivo experiments reported that a decrease in ASIC expression/activity resulted in reduced tumor growth (Chen et al. 2016, 2018; Gupta et al. 2016; Jin et al. 2017; Zhang et al. 2017a; Zhou et al. 2017; Yang et al. 2020). In addition, reduced ASIC expression/activity decreased liver and lung metastases (Gupta et al. 2016; Zhou et al. 2017; Zhu et al. 2017), extended survival rates (Zhang et al. 2017a; Chen et al. 2018), and reduced castration resistance for prostate cancer cells (Zhang et al. 2017a). The study that failed to show a change in tumor growth did report reduced liver and lung metastases when ASIC1 or 3 were knocked down, and this was consistent with their in vitro data showing knockdown or inhibition of ASIC1 or 3 suppressed cell migration and invasion (Zhu et al. 2017).

The results from in vitro experiments using various cancer cell lines where ASICs had been manipulated by CRISPR/Cas9 knockout, shRNA/siRNA/RNAi knockdown, over-expression vectors, or inhibitors/activators were robust and consistent with the in vivo and clinical data. Using techniques such as transwell migration/invasion assays, wound-healing assays, colony formation assays, and various cell proliferation assays, it was demonstrated that acid-induced cancer cell growth, migration, and invasion in vitro could be suppressed by decreasing ASIC expression/activity in breast (Gupta et al. 2016; Yang et al. 2020; Nakanishi et al. 2021), skin (Bychkov et al. 2021), colorectal (Zhou et al. 2017), pancreatic (Zhu et al. 2017), liver (Jin et al. 2017), lung (Wu et al. 2017; Sudarikova et al. 2022), gastric (Zhang et al. 2017a; Chen et al. 2018), and prostate (Chen et al. 2016) cancer cells. Inhibition or loss of ASIC1a also resulted in cell-cycle arrest and apoptosis, with subsequent reduced cell proliferation in several different cancer cell lines (Jin et al. 2017; Bychkov et al. 2020, 2021; Sudarikova et al. 2022). In addition, some studies investigated the role of ASICs in the acidosis-induced EMT of cancer cells which promotes cell migration and invasion. Loss of ASIC expression/activity suppressed EMT in liver (Zhang et al. 2022), pancreatic (Zhu et al. 2017), and prostate (Chen et al. 2016) cancer cells, while ASIC over-expression promoted EMT (Zhu et al. 2017). Although glioma cells are the most studied with respect to ASICs and cancer, the involvement of ASICs remains equivocal. There are three studies that used glioma cell lines and showed that knockdown or inhibition of ASIC1a reduced acidosis-induced cell migration (Kapoor et al. 2009; Sheng et al. 2021) and induced cell cycle arrest and caused apoptosis (Bychkov et al. 2020). However, studies using glioblastoma stem cells report conflicting results, with one demonstrating ASIC1a involvement in acidosis-induced spheroid formation (Clusmann et al. 2022), while the other reported no evidence of ASIC involvement (Tian et al. 2017). In drug-resistant liver cancer cells, inhibition, or knockdown of ASIC1a enhanced chemosensitivity, while ASIC1a over-expression contributed to drug-resistance of the cells (Zhang et al. 2017a, 2022). This suggests that ASIC1a inhibition could re-sensitize drug-resistant liver cancer to conventional chemotherapy.

While there is compelling evidence for ASIC involvement in cancer development and progression, the molecular mechanisms that link the acidic microenvironment to signaling pathways in cancer cells remains to be fully elucidated and is likely different for different cancer types, sub-types, and tumor microenvironments.

4 Challenges and Barriers in Targeting ASICs in Disease

From the information presented in the sections above, it seems clear that ASICs play a role in a wide range of pathological conditions, and being predominantly cell surface proteins represent attractive druggable targets for preclinical development and translation to the clinic. However, there are still many challenges and unknowns on this journey.

A major issue with any novel, promising drug target is whether the pathological role of the target and the pharmacology observed in rodent models translates to humans. In the case of ASIC1a as a mediator of ischemic tissue damage, this role seems to be conserved between rodents and humans, and many of the pharmacological inhibitors available work on both the rodent and human forms of the channel. In support of this, several studies have shown effective ASIC1a inhibition in assays using human cell lines (see cancer studies in Table 6), human iPSC derived myocytes (Redd et al. 2021), and human primary cortical neurons (Li et al. 2010). The case for ASICs as promising targets for new analgesics is a little less clear. ASIC3 and ASIC1b are the dominant isoforms in rodent sensory neurons; however, ASIC1a rather than ASIC1b is likely to be the dominant isoform in human sensory neurons (Hoagland et al. 2010). This may cause some issues for translating preclinical findings from rodent models to humans. However, on the bright side, if ASIC1a is indeed the dominant human pain ASIC isoform, the drug candidates that show promise in ischemia-related indications should also modulate pain, and there are many more ASIC1a inhibitors than selective ASIC1b ligands.

A substantial factor that needs to be considered is the consequence of inhibiting ASICs in the long term, as would be required to treat chronic diseases such as MS, rheumatoid arthritis, cancer, and neurodegenerative diseases. What we know about the physiological role of ASICs comes mostly from studies on rodents, and there is no guarantee that they are the same in humans. Assuming they are, long-term ASIC inhibition with molecules that can access the CNS may have detrimental effects on learning and memory, but on the plus side might also decrease anxiety, depression, and inflammation. Until we know more about the developmental and maintenance roles of ASICs in humans, it might be wise to focus efforts on ASIC targeting therapeutics for acute medical needs, such as stroke and myocardial infarct. As we have seen from ASIC KO animals, all subtypes can be deleted globally and result in reproductively viable mice with no overt deleterious phenotypes – thus, acute ASIC inhibition in humans over several days may be well tolerated.

Another barrier to breach relates to this question. *If ASIC inhibition is so promising in preclinical studies of MS, ischemic stroke, migraine,* etc., *why is the clinically available ASIC inhibitor amiloride not being used already?* This is a great and pertinent question. Based on promising antimigraine effects of amiloride (10 mg/kg) in rat and mouse studies, a very small, open label trial of amiloride was carried out in seven patients suffering migraine with intractable aura (Holland et al. 2012). Four of the seven patients showed reduced symptoms while taking 10–15 mg of amiloride per day; however, no control group was included. A larger, placebo-controlled trial was approved in France to take place in 2019; however, this does not appear to have been completed, and the current status is listed as "unknown" (NCT04063540). Around the same time as the migraine trial, an initial small scale clinical trial of amiloride (10 mg/day for 1 year) in 14 patients with primary progressive MS showed a promising decrease in the rate of brain volume loss compared to pre-treatment rates in nine of the fourteen patients, again with no control group (Arun et al. 2013).

Based on the promising results amiloride was tested in a larger randomized, placebo-controlled trial (again at 10 mg/day in the MS-SMART trial); however, this time it failed to show any difference from the placebo control group at reducing brain volume loss (Chataway et al. 2020). This result is disappointing for many reasons, but considering the pharmacology of amiloride, its failure in human trials so far is perhaps not surprising. In most rodent studies investigating the role of ASICs in disease models to date, researchers have used 10 mg/kg, whereas *10 mg total* (~0.15 mg/kg) was used in the human trials mentioned above. It is almost certain that the systemic doses of amiloride used in clinical studies to inhibit ASICs are way too low to affect ASIC function. In order to gain useful clinical data for the therapeutic value of ASIC inhibition, we need clinically suitable, selective inhibitors. With luck the most promising peptide and antibody inhibitors will provide these needed tools.

5 Conclusions

ASICs were first discovered in sensory neurons (Krishtal and Pidoplichko 1981), and a role in nociception was predicted. 25 years of research, plus the fact that all endogenous mediators of ASICs found to date are potentiators and related to pain, seem to support Krishtal and Pidoplichko's original hypothesis. The years of research have also revealed other neuronal roles in sensory transduction, synaptic plasticity, and neuronal excitability. ASICs are also expressed in non-neuronal tissues where they likely participate in regulating breathing and regional blood flow and inflammation. There is also compelling evidence for ASICs as novel drug targets in a wide range of currently poorly managed diseases.

Of the many potential therapeutic applications of ASIC inhibitors, the prevention of ischemia-induced tissue damage seems to be the most well supported – not only from the observed effect of ASIC inhibitors in in vitro and in vivo experiments, but also from what we know from nature. Naked mole rats are well known for their very long life span and resistance to cancer and hypoxia. Human CNS neurons are more susceptible to hypoxia-induced death than those from mice, while naked mole rat neurons are essentially hypoxia resistant. One way in which naked mole rats achieve this is by massively downregulating the expression of ASIC1a in CNS neurons. We hope through further well-planned and executed research efforts, we will be able to pharmacologically replicate in humans what nature has achieved in the naked mole rat, resistance to ischemia-related disease and longer healthy lives.

Conflict of Interest LDR is an inventor on the patent for use of Hi1a as a neuroprotective agent, AK-V works for a biotech company (Maxion Therapeutics Ltd.) who are targeting ASICs.

References

Aarsland D, Batzu L, Halliday GM, Geurtsen GJ, Ballard C, Ray Chaudhuri K, et al. Parkinson disease-associated cognitive impairment. Nat Rev Dis Primers. 2021;7(1):47.

Adams CM, Price MP, Snyder PM, Welsh MJ. Tetraethylammonium block of the BNC1 channel. Biophys J. 1999;76(3):1377–83.

Albanese M, Zagaglia S, Landi D, Boffa L, Nicoletti CG, Marciani MG, et al. Cerebrospinal fluid lactate is associated with multiple sclerosis disease progression. J Neuroinflammation. 2016;13:36.

Ali A, Ahmad FJ, Pillai KK, Vohora D. Evidence of the antiepileptic potential of amiloride with neuropharmacological benefits in rodent models of epilepsy and behavior. Epilepsy Behav. 2004;5(3):322–8.

Ali A, Pillai KP, Ahmad FJ, Dua Y, Vohora D. Anticonvulsant effect of amiloride in pentetrazole-induced status epilepticus in mice. Pharmacol Rep. 2006;58(2):242–5.

Alvarez de la Rosa D, Krueger SR, Kolar A, Shao D, Fitzsimonds RM, Canessa CM. Distribution, subcellular localization and ontogeny of ASIC1 in the mammalian central nervous system. J Physiol. 2003;546(Pt 1):77–87.

Amorini AM, Nociti V, Petzold A, Gasperini C, Quartuccio E, Lazzarino G, et al. Serum lactate as a novel potential biomarker in multiple sclerosis. Biochim Biophys Acta. 2014;1842(7):1137–43.

Anangi R, Rash LD, Mobli M, King GF. Functional expression in *Escherichia coli* of the disulfide-rich sea anemone peptide APETx2, a potent blocker of acid-sensing ion channel 3. Mar Drugs. 2012;10(7):1605–18.

Arias RL, Sung ML, Vasylyev D, Zhang MY, Albinson K, Kubek K, et al. Amiloride is neuroprotective in an MPTP model of Parkinson's disease. Neurobiol Dis. 2008;31(3):334–41.

Arun T, Tomassini V, Sbardella E, de Ruiter MB, Matthews L, Leite MI, et al. Targeting ASIC1 in primary progressive multiple sclerosis: evidence of neuroprotection with amiloride. Brain. 2013;136(Pt 1):106–15.

Baron A, Lingueglia E. Pharmacology of acid-sensing ion channels – physiological and therapeutical perspectives. Neuropharmacology. 2015;94:19–35.

Bitsch A, Bruhn H, Vougioukas V, Stringaris A, Lassmann H, Frahm J, et al. Inflammatory CNS demyelination: histopathologic correlation with in vivo quantitative proton MR spectroscopy. AJNR Am J Neuroradiol. 1999;20(9):1619–27.

Blanchard MG, Rash LD, Kellenberger S. Inhibition of voltage-gated Na^+ currents in sensory neurones by the sea anemone toxin APETx2. Br J Pharmacol. 2012;165(7):2167–77.

Bohlen CJ, Chesler AT, Sharif-Naeini R, Medzihradszky KF, Zhou S, King D, et al. A heteromeric Texas coral snake toxin targets acid-sensing ion channels to produce pain. Nature. 2011;479(7373):410–4.

Buta A, Maximyuk O, Kovalskyy D, Sukach V, Vovk M, Ievglevskyi O, et al. Novel potent orthosteric antagonist of ASIC1a prevents NMDAR-dependent LTP induction. J Med Chem. 2015;58(11):4449–61.

Bychkov M, Shulepko M, Osmakov D, Andreev Y, Sudarikova A, Vasileva V, et al. Mambalgin-2 induces cell cycle arrest and apoptosis in glioma cells via interaction with ASIC1a. Cancers (Basel). 2020;12(7):1837.

Bychkov ML, Kirichenko AV, Shulepko MA, Mikhaylova IN, Kirpichnikov MP, Lyukmanova EN. Mambalgin-2 inhibits growth, migration, and invasion of metastatic melanoma cells by targeting the channels containing an ASIC1a subunit whose up-regulation correlates with poor survival prognosis. Biomedicine. 2021;9(10):1324.

Cadiou H, Studer M, Jones NG, Smith ES, Ballard A, McMahon SB, et al. Modulation of acid-sensing ion channel activity by nitric oxide. J Neurosci. 2007;27(48):13251–60.

Campbell BCV, De Silva DA, Macleod MR, Coutts SB, Schwamm LH, Davis SM, et al. Ischaemic stroke. Nat Rev Dis Primers. 2019;5(1):70.

Castaneda MS, Tonini R, Richards CD, Stocker M, Pedarzani P. Benzamil inhibits neuronal and heterologously expressed small conductance Ca(2+)-activated K(+) channels. Neuropharmacology. 2019;158:107738.

Chafai M, Delrocq A, Inquimbert P, Pidoux L, Delanoe K, Toft M, et al. Dual contribution of ASIC1a channels in the spinal processing of pain information by deep projection neurons revealed by computational modeling. PLoS Comput Biol. 2023;19(4):e1010993..

Chang CT, Fong SW, Lee CH, Chuang YC, Lin SH, Chen CC. Involvement of acid-sensing ion channel 1b in the development of acid-induced chronic muscle pain. Front Neurosci. 2019;13:1247.

Chassagnon IR, McCarthy CA, Chin YK, Pineda SS, Keramidas A, Mobli M, et al. Potent neuroprotection after stroke afforded by a double-knot spider-venom peptide that inhibits acid-sensing ion channel 1a. Proc Natl Acad Sci U S A. 2017;114(14):3750–5.

Chataway J, De Angelis F, Connick P, Parker RA, Plantone D, Doshi A, et al. Efficacy of three neuroprotective drugs in secondary progressive multiple sclerosis (MS-SMART): a phase 2b, multiarm, double-blind, randomised placebo-controlled trial. Lancet Neurol. 2020;19(3):214–25.

Chen X, Kalbacher H, Gründer S. The tarantula toxin psalmotoxin 1 inhibits acid-sensing ion channel (ASIC) 1a by increasing its apparent H^+ affinity. J Gen Physiol. 2005;126(1):71–9.

Chen WN, Lee CH, Lin SH, Wong CW, Sun WH, Wood JN, et al. Roles of ASIC3, TRPV1, and NaV1.8 in the transition from acute to chronic pain in a mouse model of fibromyalgia. Mol Pain. 2014;10:40.

Chen B, Liu J, Ho TT, Ding X, Mo YY. ERK-mediated NF-kappaB activation through ASIC1 in response to acidosis. Oncogene. 2016;5(12):e279.

Chen X, Sun X, Wang Z, Zhou X, Xu L, Li F, et al. Involvement of acid-sensing ion channel 1a in gastric carcinoma cell migration and invasion. Acta Biochim Biophys Sin Shanghai. 2018;50(5):440–6.

Cheng YR, Jiang BY, Chen CC. Acid-sensing ion channels: dual function proteins for chemo-sensing and mechano-sensing. J Biomed Sci. 2018;25(1):46.

Cheng S, Mao X, Lin X, Wehn A, Hu S, Mamrak U, et al. Acid-ion sensing channel 1a deletion reduces chronic brain damage and neurological deficits after experimental traumatic brain injury. J Neurotrauma. 2021;38(11):1572–84.

Chopp M, Frinak S, Walton DR, Smith MB, Welch KM. Intracellular acidosis during and after cerebral ischemia: in vivo nuclear magnetic resonance study of hyperglycemia in cats. Stroke. 1987;18(5):919–23.

Clusmann J, Franco KC, Suarez DAC, Katona I, Minguez MG, Boersch N, et al. Acidosis induces RIPK1-dependent death of glioblastoma stem cells via acid-sensing ion channel 1a. Cell Death Dis. 2022;13(8):702.

Cortes Franco KD, Brakmann IC, Feoktistova M, Panayotova-Dimitrova D, Grunder S, Tian Y. Aggressive migration in acidic pH of a glioblastoma cancer stem cell line in vitro is independent of ASIC and K(Ca)3.1 ion channels, but involves phosphoinositide 3-kinase. Pflugers Arch. 2023;475(3):405–16.

Coryell MW, Ziemann AE, Westmoreland PJ, Haenfler JM, Kurjakovic Z, Zha XM, et al. Targeting ASIC1a reduces innate fear and alters neuronal activity in the fear circuit. Biol Psychiatry. 2007;62(10):1140–8.

Coryell MW, Wunsch AM, Haenfler JM, Allen JE, Schnizler M, Ziemann AE, et al. Acid-sensing ion channel-1a in the amygdala, a novel therapeutic target in depression-related behavior. J Neurosci. 2009;29(17):5381–8.

Cristofori-Armstrong B, Saez NJ, Chassagnon IR, King GF, Rash LD. The modulation of acid-sensing ion channel 1 by PcTx1 is pH-, subtype- and species-dependent: importance of interactions at the channel subunit interface and potential for engineering selective analogues. Biochem Pharmacol. 2019;163:381–90.

Cristofori-Armstrong B, Budusan E, Rash LD. Mambalgin-3 potentiates human acid-sensing ion channel 1b under mild to moderate acidosis: implications as an analgesic lead. Proc Natl Acad Sci U S A. 2021;118(8):e2021581118.

Davies NM. Clinical pharmacokinetics of ibuprofen. The first 30 years. Clin Pharmacokinet. 1998;34(2):101–54.

Decker Y, Nemeth E, Schomburg R, Chemla A, Fulop L, Menger MD, et al. Decreased pH in the aging brain and Alzheimer's disease. Neurobiol Aging. 2021;101:40–9.

Degterev A, Ofengeim D, Yuan J. Targeting RIPK1 for the treatment of human diseases. Proc Natl Acad Sci U S A. 2019;116(20):9714–22.

Delaunay A, Gasull X, Salinas M, Noel J, Friend V, Lingueglia E, et al. Human ASIC3 channel dynamically adapts its activity to sense the extracellular pH in both acidic and alkaline directions. Proc Natl Acad Sci U S A. 2012;109(32):13124–9.

Dendrou CA, Fugger L, Friese MA. Immunopathology of multiple sclerosis. Nat Rev Immunol. 2015;15(9):545–58.

DeTure MA, Dickson DW. The neuropathological diagnosis of Alzheimer's disease. Mol Neurodegener. 2019;14(1):32.

Deval E, Lingueglia E. Acid-sensing ion channels and nociception in the peripheral and central nervous systems. Neuropharmacology. 2015;94:49–57.

Deval E, Noel J, Lay N, Alloui A, Diochot S, Friend V, et al. ASIC3, a sensor of acidic and primary inflammatory pain. EMBO J. 2008;27(22):3047–55.

Deval E, Noel J, Gasull X, Delaunay A, Alloui A, Friend V, et al. Acid-sensing ion channels in postoperative pain. J Neurosci. 2011;31(16):6059–66.

Dibas A, Millar C, Al-Farra A, Yorio T. Neuroprotective effects of psalmotoxin-1, an acid-sensing ion channel (ASIC) inhibitor, in ischemia reperfusion in mouse eyes. Curr Eye Res. 2018;43(7):921–33.

Diochot S, Baron A, Rash LD, Deval E, Escoubas P, Scarzello S, et al. A new sea anemone peptide, APETx2, inhibits ASIC3, a major acid-sensitive channel in sensory neurons. EMBO J. 2004;23(7):1516–25.

Diochot S, Baron A, Salinas M, Douguet D, Scarzello S, Dabert-Gay AS, et al. Black mamba venom peptides target acid-sensing ion channels to abolish pain. Nature. 2012;490(7421):552–5.

Diochot S, Alloui A, Rodrigues P, Dauvois M, Friend V, Aissouni Y, et al. Analgesic effects of mambalgin peptide inhibitors of acid-sensing ion channels in inflammatory and neuropathic pain. Pain. 2016;157(3):552–9.

Duan B, Wu LJ, Yu YQ, Ding Y, Jing L, Xu L, et al. Upregulation of acid-sensing ion channel ASIC1a in spinal dorsal horn neurons contributes to inflammatory pain hypersensitivity. J Neurosci. 2007;27(41):11139–48.

Duan B, Wang YZ, Yang T, Chu XP, Yu Y, Huang Y, et al. Extracellular spermine exacerbates ischemic neuronal injury through sensitization of ASIC1a channels to extracellular acidosis. J Neurosci. 2011;31(6):2101–12.

Dube GR, Elagoz A, Mangat H. Acid sensing ion channels and acid nociception. Curr Pharm Des. 2009;15(15):1750–66.

Durham PL, Masterson CG. Two mechanisms involved in trigeminal CGRP release: implications for migraine treatment. Headache. 2013;53(1):67–80.

Er SY, Cristofori-Armstrong B, Escoubas P, Rash LD. Discovery and molecular interaction studies of a highly stable, tarantula peptide modulator of acid-sensing ion channel 1. Neuropharmacology. 2017;127:185.

Erra Diaz F, Dantas E, Geffner J. Unravelling the interplay between extracellular acidosis and immune cells. Mediat Inflamm. 2018;2018:1218297.

Escoubas P, De Weille JR, Lecoq A, Diochot S, Waldmann R, Champigny G, et al. Isolation of a tarantula toxin specific for a class of proton-gated Na^+ channels. J Biol Chem. 2000;275(33):25116–21.

Ettaiche M, Guy N, Hofman P, Lazdunski M, Waldmann R. Acid-sensing ion channel 2 is important for retinal function and protects against light-induced retinal degeneration. J Neurosci. 2004;24(5):1005–12.

Ettaiche M, Deval E, Cougnon M, Lazdunski M, Voilley N. Silencing acid-sensing ion channel 1a alters cone-mediated retinal function. J Neurosci. 2006;26(21):5800–9.

Fage D, Voltz C, Scatton B, Carter C. Selective release of spermine and spermidine from the rat striatum by N-methyl-D-aspartate receptor activation in vivo. J Neurochem. 1992;58(6):2170–5.

Fazia T, Pastorino R, Notartomaso S, Busceti C, Imbriglio T, Cannella M, et al. Acid sensing ion channel 2: a new potential player in the pathophysiology of multiple sclerosis. Eur J Neurosci. 2019;49(10):1233–43.

Feldman DH, Horiuchi M, Keachie K, McCauley E, Bannerman P, Itoh A, et al. Characterization of acid-sensing ion channel expression in oligodendrocyte-lineage cells. Glia. 2008;56(11):1238–49.

Ferrari MD, Goadsby PJ, Burstein R, Kurth T, Ayata C, Charles A, et al. Migraine. Nat Rev Dis Primers. 2022;8(1):2.

Finol-Urdaneta RK, Ziegman R, Dekan Z, McArthur JR, Heitmann S, Luna-Ramirez K, et al. Multitarget nociceptor sensitization by a promiscuous peptide from the venom of the King Baboon spider. Proc Natl Acad Sci U S A. 2022;119(5):e2110932119.

Foster VS, Rash LD, King GF, Rank MM. Acid-sensing ion channels: expression and function in resident and infiltrating immune cells in the central nervous system. Front Cell Neurosci. 2021;15:738043.

Friese MA, Craner MJ, Etzensperger R, Vergo S, Wemmie JA, Welsh MJ, et al. Acid-sensing ion channel-1 contributes to axonal degeneration in autoimmune inflammation of the central nervous system. Nat Med. 2007;13(12):1483–9.

Garcia SM, Herbert LM, Walker BR, Resta TC, Jernigan NL. Coupling of store-operated calcium entry to vasoconstriction is acid-sensing ion channel 1a dependent in pulmonary but not mesenteric arteries. PLoS One. 2020;15(7):e0236288.

Garcia SM, Naik JS, Resta TC, Jernigan NL. Acid-sensing ion channel 1a activates IKCa/SKCa channels and contributes to endothelium-dependent dilation. J Gen Physiol. 2023;155(2):e202213173.

Gingrich MB, Traynelis SF. Serine proteases and brain damage – is there a link? Trends Neurosci. 2000;23(9):399–407.

Gornati D, Ciccone R, Vinciguerra A, Ippati S, Pannaccione A, Petrozziello T, et al. Synthesis and characterization of novel mono- and bis-guanyl hydrazones as potent and selective ASIC1 inhibitors able to reduce brain ischemic insult. J Med Chem. 2021;64(12):8333–53.

Gründer S. Acid-sensing ion channels. In: Bhattacharjee A, editor. The Oxford handbook of neuronal ion channels. Oxford: Oxford University Press; 2020.

Gugliandolo A, Gangemi C, Caccamo D, Curro M, Pandolfo G, Quattrone D, et al. The RS685012 polymorphism of ACCN2, the human ortholog of murine acid-sensing ion channel (ASIC1) gene, is highly represented in patients with panic disorder. NeuroMolecular Med. 2016;18(1):91–8.

Gupta SC, Singh R, Asters M, Liu J, Zhang X, Pabbidi MR, et al. Regulation of breast tumorigenesis through acid sensors. Oncogene. 2016;35(31):4102–11.

Halder SK, Milner R. Hypoxia in multiple sclerosis; is it the chicken or the egg? Brain. 2021;144(2):402–10.

Hoagland EN, Sherwood TW, Lee KG, Walker CJ, Askwith CC. Identification of a calcium permeable human acid-sensing ion channel 1 transcript variant. J Biol Chem. 2010;285(53):41852–62.

Holland PR, Akerman S, Andreou AP, Karsan N, Wemmie JA, Goadsby PJ. Acid-sensing ion channel 1: a novel therapeutic target for migraine with aura. Ann Neurol. 2012;72(4):559–63.

Holton CM, Strother LC, Dripps I, Pradhan AA, Goadsby PJ, Holland PR. Acid-sensing ion channel 3 blockade inhibits durovascular and nitric oxide-mediated trigeminal pain. Br J Pharmacol. 2020;177(11):2478–86.

Hu R, Duan B, Wang D, Yu Y, Li W, Luo H, et al. Role of acid-sensing ion channel 1a in the secondary damage of traumatic spinal cord injury. Ann Surg. 2011;254(2):353–62.

Huang C, Hu ZL, Wu WN, Yu DF, Xiong QJ, Song JR, et al. Existence and distinction of acid-evoked currents in rat astrocytes. Glia. 2010;58(12):1415–24.

Husson Z, Smith ESJ. Naked mole-rat cortical neurons are resistant to acid-induced cell death. Mol Brain. 2018;11(1):26.

Immke DC, McCleskey EW. Lactate enhances the acid-sensing Na^+ channel on ischemia-sensing neurons. Nat Neurosci. 2001;4(9):869–70.

Izumi M, Ikeuchi M, Ji Q, Tani T. Local ASIC3 modulates pain and disease progression in a rat model of osteoarthritis. J Biomed Sci. 2012;19:77.

Jasti J, Furukawa H, Gonzales EB, Gouaux E. Structure of acid-sensing ion channel 1 at 1.9 Å resolution and low pH. Nature. 2007;449(7160):316–23.

Jensen JE, Durek T, Alewood PF, Adams DJ, King GF, Rash LD. Chemical synthesis and folding of APETx2, a potent and selective inhibitor of acid sensing ion channel 3. Toxicon. 2009;54(1):56–61.

Jensen JE, Mobli M, Brust A, Alewood PF, King GF, Rash LD. Cyclisation increases the stability of the sea anemone peptide APETx2 but decreases its activity at acid-sensing ion channel 3. Mar Drugs. 2012;10(7):1511–27.

Jensen JE, Cristofori-Armstrong B, Anangi R, Rosengren KJ, Lau CH, Mobli M, et al. Understanding the molecular basis of toxin promiscuity: the analgesic sea anemone peptide APETx2 interacts with acid-sensing ion channel 3 and hERG channels via overlapping pharmacophores. J Med Chem. 2014;57(21):9195–203.

Jin C, Yuan FL, Gu YL, Li X, Liu MF, Shen XM, et al. Over-expression of ASIC1a promotes proliferation via activation of the beta-catenin/LEF-TCF axis and is associated with disease outcome in liver cancer. Oncotarget. 2017;8(16):25977–88.

Joeres N, Augustinowski K, Neuhof A, Assmann M, Grunder S. Functional and pharmacological characterization of two different ASIC1a/2a heteromers reveals their sensitivity to the spider toxin PcTx1. Sci Rep. 2016;6:27647.

Jones NG, Slater R, Cadiou H, McNaughton P, McMahon SB. Acid-induced pain and its modulation in humans. J Neurosci. 2004;24(48):10974–9.

Jurczak A, Delay L, Barbier J, Simon N, Krock E, Sandor K, et al. Antibody-induced pain-like behavior and bone erosion: links to subclinical inflammation, osteoclast activity, and acid-sensing ion channel 3-dependent sensitization. Pain. 2022;163(8):1542–59.

Kalimo H, Rehncrona S, Soderfeldt B, Olsson Y, Siesjo BK. Brain lactic acidosis and ischemic cell damage: 2. Histopathology. J Cereb Blood Flow Metab. 1981;1(3):313–27.

Kapoor N, Bartoszewski R, Qadri YJ, Bebok Z, Bubien JK, Fuller CM, et al. Knockdown of ASIC1 and epithelial sodium channel subunits inhibits glioblastoma whole cell current and cell migration. J Biol Chem. 2009;284(36):24526–41.

Karczewski J, Spencer RH, Garsky VM, Liang A, Leitl MD, Cato MJ, et al. Reversal of acid-induced and inflammatory pain by the selective ASIC3 inhibitor, APETx2. Br J Pharmacol. 2010;161(4):950–60.

Karlsson M, Zhang C, Mear L, Zhong W, Digre A, Katona B, et al. A single-cell type transcriptomics map of human tissues. Sci Adv. 2021;7(31):eabh2169.

Kellenberger S, Rash LD. Acid-sensing (proton-gated) ion channels (ASICs) in GtoPdb v.2021.3. Br J Pharmacol. 2021;158(Suppl 1):S124–5.

Kellenberger S, Schild L. International Union of Basic and Clinical Pharmacology. XCI. Structure, function, and pharmacology of acid-sensing ion channels and the epithelial Na+ channel. Pharmacol Rev. 2015;67(1):1–35.

Kelmanson IV, Shokhina AG, Kotova DA, Pochechuev MS, Ivanova AD, Kostyuk AI, et al. In vivo dynamics of acidosis and oxidative stress in the acute phase of an ischemic stroke in a rodent model. Redox Biol. 2021;48:102178.

King GF, Rash LD, Chassagnon I, Pineda SS, Widdop RE, Ireland CA. Neuroprotective agents derived from spider venom peptides. University of Queensland, Monash University; 2021.

Kleyman TR, Cragoe EJ Jr. Amiloride and its analogs as tools in the study of ion transport. J Membr Biol. 1988;105(1):1–21.

Klint JK, Berecki G, Durek T, Mobli M, Knapp O, King GF, et al. Isolation, synthesis and characterization of omega-TRTX-Cc1a, a novel tarantula venom peptide that selectively targets L-type Cav channels. Biochem Pharmacol. 2014;89(2):276–86.

Koehn LM, Noor NM, Dong Q, Er SY, Rash LD, King GF, et al. Selective inhibition of ASIC1a confers functional and morphological neuroprotection following traumatic spinal cord injury. F1000Res. 2016;5:1822.

Komnig D, Imgrund S, Reich A, Grunder S, Falkenburger BH. ASIC1a deficient mice show unaltered neurodegeneration in the subacute MPTP model of Parkinson disease. PLoS One. 2016;11(11):e0165235.

Kong X, Tang X, Du W, Tong J, Yan Y, Zheng F, et al. Extracellular acidosis modulates the endocytosis and maturation of macrophages. Cell Immunol. 2013;281(1):44–50.

Krishtal OA, Pidoplichko VI. A "receptor" for protons in small neurons of trigeminal ganglia: possible role in nociception. Neurosci Lett. 1981;24(3):243–6.

Kuduk SD, Di Marco CN, Bodmer-Narkevitch V, Cook SP, Cato MJ, Jovanovska A, et al. Synthesis, structure-activity relationship, and pharmacological profile of analogs of the ASIC-3 inhibitor A-317567. ACS Chem Neurosci. 2010;1(1):19–24.

Lakhan SE, Kirchgessner A, Hofer M. Inflammatory mechanisms in ischemic stroke: therapeutic approaches. J Transl Med. 2009;7:97.

Lauro C, Limatola C. Metabolic reprograming of microglia in the regulation of the innate inflammatory response. Front Immunol. 2020;11:493.

Lee JYP, Saez NJ, Cristofori-Armstrong B, Anangi R, King GF, Smith MT, et al. Inhibition of acid-sensing ion channels by diminazene and APETx2 evoke partial and highly variable antihyperalgesia in a rat model of inflammatory pain. Br J Pharmacol. 2018;175(12):2204–18.

Leng TD, Si HF, Li J, Yang T, Zhu M, Wang B, et al. Amiloride analogs as ASIC1a inhibitors. CNS Neurosci Ther. 2016;22(6):468–76.

Leslie TK, James AD, Zaccagna F, Grist JT, Deen S, Kennerley A, et al. Sodium homeostasis in the tumour microenvironment. Biochim Biophys Acta Rev Cancer. 2019;1872(2):188304.

Li M, Inoue K, Branigan D, Kratzer E, Hansen JC, Chen JW, et al. Acid-sensing ion channels in acidosis-induced injury of human brain neurons. J Cereb Blood Flow Metab. 2010;30(6):1247–60.

Li Y, Ritzel RM, He J, Cao T, Sabirzhanov B, Li H, et al. The voltage-gated proton channel Hv1 plays a detrimental role in contusion spinal cord injury via extracellular acidosis-mediated neuroinflammation. Brain Behav Immun. 2021;91:267–83.

Liang JJ, Huang LF, Chen XM, Pan SQ, Lu ZN, Xiao ZM. Amiloride suppresses pilocarpine-induced seizures via ASICs other than NHE in rats. Int J Clin Exp Pathol. 2015;8(11):14507–13.

Lin LH, Jin J, Nashelsky MB, Talman WT. Acid-sensing ion channel 1 and nitric oxide synthase are in adjacent layers in the wall of rat and human cerebral arteries. J Chem Neuroanat. 2014;61-62:161–8.

Lin SH, Sun WH, Chen CC. Genetic exploration of the role of acid-sensing ion channels. Neuropharmacology. 2015;94:99–118.

Lin LH, Jones S, Talman WT. Cellular localization of acid-sensing ion channel 1 in rat nucleus tractus solitarii. Cell Mol Neurobiol. 2018;38(1):219–32.

Liu Y, Hagan R, Schoellerman J. Dual actions of Psalmotoxin at ASIC1a and ASIC2a heteromeric channels (ASIC1a/2a). Sci Rep. 2018a;8(1):7179.

Liu Z, Pei H, Zhang L, Tian Y. Mitochondria-targeted DNA nanoprobe for real-time imaging and simultaneous quantification of Ca(2+) and pH in neurons. ACS Nano. 2018b;12(12):12357–68.

Lv R-j, He J-s, Fu Y-h, Zhang Y-q, Shao X-q, Wu L-w, et al. ASIC1a polymorphism is associated with temporal lobe epilepsy. Epilepsy Res. 2011;96(1):74–80.

Lynagh T, Romero-Rojo JL, Lund C, Pless SA. Molecular basis for allosteric inhibition of acid-sensing ion channel 1a by ibuprofen. J Med Chem. 2017;60(19):8192–200.

MacDonald L, Gao M, Morra MR, Nicole M, Alessandri-Haber NM, LaCroix-Fralish ML. Anti-ASIC1 antibodies and uses thereof. Regeneron Pharmaceuticals, Inc; 2015.

Mamet J, Baron A, Lazdunski M, Voilley N. Proinflammatory mediators, stimulators of sensory neuron excitability via the expression of acid-sensing ion channels. J Neurosci. 2002;22(24):10662–70.

Mango D, Nistico R. Role of ASIC1a in Abeta-induced synaptic alterations in the hippocampus. Pharmacol Res. 2018;131:61–5.

Mango D, Nistico R. Neurodegenerative disease: what potential therapeutic role of acid-sensing ion channels? Front Cell Neurosci. 2021;15:730641.

Marra S, Ferru-Clement R, Breuil V, Delaunay A, Christin M, Friend V, et al. Non-acidic activation of pain-related acid-sensing ion channel 3 by lipids. EMBO J. 2016;35(4):414–28.

Mazzuca M, Heurteaux C, Alloui A, Diochot S, Baron A, Voilley N, et al. A tarantula peptide against pain via ASIC1a channels and opioid mechanisms. Nat Neurosci. 2007;10(8):943–5.

McCarthy CA, Rash LD, Chassagnon IR, King GF, Widdop RE. PcTx1 affords neuroprotection in a conscious model of stroke in hypertensive rats via selective inhibition of ASIC1a. Neuropharmacology. 2015;99:650–7.

Mills SEE, Nicolson KP, Smith BH. Chronic pain: a review of its epidemiology and associated factors in population-based studies. Br J Anaesth. 2019;123(2):e273–e83.

Mobli M, Undheim EAB, Rash LD. Modulation of ion channels by cysteine-rich peptides: from sequence to structure. Adv Pharmacol. 2017;79:199–223.

Molliver DC, Immke DC, Fierro L, Pare M, Rice FL, McCleskey EW. ASIC3, an acid-sensing ion channel, is expressed in metaboreceptive sensory neurons. Mol Pain. 2005;1:35.

Mourier G, Salinas M, Kessler P, Stura EA, Leblanc M, Tepshi L, et al. Mambalgin-1 pain-relieving peptide: stepwise solid-phase synthesis, crystal structure and functional domain for acid-sensing ion channel 1a inhibition. J Biol Chem. 2015;291:2616.

Nakanishi M, Korechika A, Yamakawa H, Kawabe N, Nakai K, Muragaki Y. Acidic microenvironment induction of interleukin-8 expression and matrix metalloproteinase-2/–9 activation via acid-sensing ion channel 1 promotes breast cancer cell progression. Oncol Rep. 2021;45(3):1284–94.

Nathaniel TI, Otukonyong EE, Okon M, Chaves J, Cochran T, Nathaniel AI. Metabolic regulatory clues from the naked mole rat: toward brain regulatory functions during stroke. Brain Res Bull. 2013;98:44–52.

N'Gouemo P. Amiloride delays the onset of pilocarpine-induced seizures in rats. Brain Res. 2008;1222:230–2.

Nitta CH, Osmond DA, Herbert LM, Beasley BF, Resta TC, Walker BR, et al. Role of ASIC1 in the development of chronic hypoxia-induced pulmonary hypertension. Am J Physiol Heart Circ Physiol. 2014;306(1):H41–52.

Paljarvi L. Brain lactic acidosis and ischemic cell damage: a topographic study with high-resolution light microscopy of early recovery in a rat model of severe incomplete ischemia. Acta Neuropathol. 1984;64(2):89–98.

Paljarvi L, Rehncrona S, Soderfeldt B, Olsson Y, Kalimo H. Brain lactic acidosis and ischemic cell damage: quantitative ultrastructural changes in capillaries of rat cerebral cortex. Acta Neuropathol. 1983;60(3–4):232–40.

Pattison LA, Callejo G, St John Smith E. Evolution of acid nociception: ion channels and receptors for detecting acid. Philos Trans R Soc Lond Ser B Biol Sci. 2019;374(1785):20190291.

Peigneur S, Beress L, Moller C, Mari F, Forssmann WG, Tytgat J. A natural point mutation changes both target selectivity and mechanism of action of sea anemone toxins. FASEB J. 2012;26(12):5141–51.

Pignataro G, Simon RP, Xiong ZG. Prolonged activation of ASIC1a and the time window for neuroprotection in cerebral ischaemia. Brain. 2007;130(Pt 1):151–8.

Poirot O, Vukicevic M, Boesch A, Kellenberger S. Selective regulation of acid-sensing ion channel 1 by serine proteases. J Biol Chem. 2004;279(37):38448–57.

Price MP, Snyder PM, Welsh MJ. Cloning and expression of a novel human brain Na^+ channel. J Biol Chem. 1996;271(14):7879–82.

Price MP, Gong H, Parsons MG, Kundert JR, Reznikov LR, Bernardinelli L, et al. Localization and behaviors in null mice suggest that ASIC1 and ASIC2 modulate responses to aversive stimuli. Genes Brain Behav. 2014;13(2):179–94.

Qi X, Lu JF, Huang ZY, Liu YJ, Cai LB, Wen XL, et al. Pharmacological validation of ASIC1a as a druggable target for neuroprotection in cerebral ischemia using an intravenously available small molecule inhibitor. Front Pharmacol. 2022;13:849498.

Qiang M, Dong X, Zha Z, Zuo X-K, Song X-L, Zhao L, et al. Selection of an ASIC1a-blocking combinatorial antibody that protects cells from ischemic death. Proc Natl Acad Sci U S A. 2018;115(32):E7469–E77.

Qiu CY, Liu YQ, Qiu F, Wu J, Zhou QY, Hu WP. Prokineticin 2 potentiates acid-sensing ion channel activity in rat dorsal root ganglion neurons. J Neuroinflammation. 2012;9:108–19.

Rash LD. Acid-sensing ion channel pharmacology, past, present, and future .... In: Geraghty DP, Rash LD, editors. Advances in pharmacology, ion channels down under. Academic; 2017. p. 35–66.

Redd MA, Scheuer SE, Saez NJ, Yoshikawa Y, Chiu HS, Gao L, et al. Therapeutic inhibition of acid-sensing ion channel 1a recovers heart function after ischemia-reperfusion injury. Circulation. 2021;144(12):947–60.

Rehncrona S, Rosen I, Siesjo BK. Brain lactic acidosis and ischemic cell damage: 1. Biochemistry and neurophysiology. J Cereb Blood Flow Metab. 1981;1(3):297–311.

Ritzel RM, He J, Li Y, Cao T, Khan N, Shim B, et al. Proton extrusion during oxidative burst in microglia exacerbates pathological acidosis following traumatic brain injury. Glia. 2021;69(3):746–64.

Rivlin AS, Tator CH. Regional spinal cord blood flow in rats after severe cord trauma. J Neurosurg. 1978;49(6):844–53.

Rook ML, Williamson A, Lueck JD, Musgaard M, Maclean DM. beta11-12 linker isomerization governs acid-sensing ion channel desensitization and recovery. elife. 2020;9:9.

Rook ML, Musgaard M, MacLean DM. Coupling structure with function in acid-sensing ion channels: challenges in pursuit of proton sensors. J Physiol. 2021;599(2):417–30.

Salinas M, Besson T, Delettre Q, Diochot S, Boulakirba S, Douguet D, et al. Binding site and inhibitory mechanism of the mambalgin-2 pain-relieving peptide on acid-sensing ion channel 1a. J Biol Chem. 2014;289(19):13363–73.

Santos PL, Guimaraes AG, Barreto RS, Serafini MR, Quintans JS, Quintans-Junior LJ. A review of recent patents on the ASICs as a key drug target. Recent Pat Biotechnol. 2015;9(1):30–41.

Schmidt A, Rossetti G, Joussen S, Grunder S. Diminazene is a slow pore blocker of acid-sensing ion channel 1a (ASIC1a). Mol Pharmacol. 2017;92:665.

Sheng Y, Wu B, Leng T, Zhu L, Xiong Z. Acid-sensing ion channel 1 (ASIC1) mediates weak acid-induced migration of human malignant glioma cells. Am J Cancer Res. 2021;11(3):997–1008.

Sherwood T, Franke R, Conneely S, Joyner J, Arumugan P, Askwith C. Identification of protein domains that control proton and calcium sensitivity of ASIC1a. J Biol Chem. 2009;284(41):27899–907.

Sjostedt E, Zhong W, Fagerberg L, Karlsson M, Mitsios N, Adori C, et al. An atlas of the protein-coding genes in the human, pig, and mouse brain. Science. 2020;367(6482):eaay5947.

Sluka KA, Winter OC, Wemmie JA. Acid-sensing ion channels: a new target for pain and CNS diseases. Curr Opin Drug Discov Devel. 2009;12(5):693–704.

Smith ES, Cadiou H, McNaughton PA. Arachidonic acid potentiates acid-sensing ion channels in rat sensory neurons by a direct action. Neuroscience. 2007;145(2):686–98.

Smoller JW, Gallagher PJ, Duncan LE, McGrath LM, Haddad SA, Holmes AJ, et al. The human ortholog of acid-sensing ion channel gene ASIC1a is associated with panic disorder and amygdala structure and function. Biol Psychiatry. 2014;76(11):902–10.

Song N, Lu Z, Zhang J, Shi Y, Ning Y, Chen J, et al. Acid-sensing ion channel 1a is involved in ischaemia/reperfusion induced kidney injury by increasing renal epithelia cell apoptosis. J Cell Mol Med. 2019;23(5):3429–40.

Stankowska DL, Mueller BH 2nd, Oku H, Ikeda T, Dibas A. Neuroprotective effects of inhibitors of acid-sensing ion channels (ASICs) in optic nerve crush model in rodents. Curr Eye Res. 2018;43(1):84–95.

Storozhuk M, Cherninskyi A, Maximyuk O, Isaev D, Krishtal O. Acid-sensing ion channels: focus on physiological and some pathological roles in the brain. Curr Neuropharmacol. 2021;19(9):1570–89.

Sudarikova AV, Bychkov ML, Kulbatskii DS, Chubinskiy-Nadezhdin VI, Shlepova OV, Shulepko MA, et al. Mambalgin-2 inhibits lung adenocarcinoma growth and migration by selective interaction with ASIC1/alpha-ENaC/gamma-ENaC heterotrimer. Front Oncol. 2022;12:904742.

Sun X, Zhao D, Li YL, Sun Y, Lei XH, Zhang JN, et al. Regulation of ASIC1 by Ca2+/calmodulin-dependent protein kinase II in human glioblastoma multiforme. Oncol Rep. 2013;30(6):2852–8.

Sun D, Liu S, Li S, Zhang M, Yang F, Wen M, et al. Structural insights into human acid-sensing ion channel 1a inhibition by snake toxin mambalgin1. elife. 2020;9:e57096.

Sutherland SP, Benson CJ, Adelman JP, McCleskey EW. Acid-sensing ion channel 3 matches the acid-gated current in cardiac ischemia-sensing neurons. Proc Natl Acad Sci U S A. 2001;98(2):711–6.

Tabrizi SJ, Ghosh R, Leavitt BR. Huntingtin lowering strategies for disease modification in Huntington's disease. Neuron. 2019;101(5):801–19.

Taugher RJ, Lu Y, Wang Y, Kreple CJ, Ghobbeh A, Fan R, et al. The bed nucleus of the stria terminalis is critical for anxiety-related behavior evoked by CO2 and acidosis. J Neurosci. 2014;34(31):10247–55.

Taugher RJ, Lu Y, Fan R, Ghobbeh A, Kreple CJ, Faraci FM, et al. ASIC1A in neurons is critical for fear-related behaviors. Genes Brain Behav. 2017;16(8):745–55.

Tei R, Kaido T, Nakase H, Sakaki T. Secondary spinal cord hypoperfusion of circumscribed areas after injury in rats. Neurol Res. 2005;27(4):403–8.

Thews O, Riemann A. Tumor pH and metastasis: a malignant process beyond hypoxia. Cancer Metastasis Rev. 2019;38(1–2):113–29.

Tian Y, Bresenitz P, Reska A, El Moussaoui L, Beier CP, Gründer S. Glioblastoma cancer stem cell lines express functional acid sensing ion channels ASIC1a and ASIC3. Sci Rep. 2017;7(1):13674–14.

Tong J, Wu WN, Kong X, Wu PF, Tian L, Du W, et al. Acid-sensing ion channels contribute to the effect of acidosis on the function of dendritic cells. J Immunol. 2011;186(6):3686–92.

Tyrtyshnaia AA, Lysenko LV, Madamba F, Manzhulo IV, Khotimchenko MY, Kleschevnikov AM. Acute neuroinflammation provokes intracellular acidification in mouse hippocampus. J Neuroinflammation. 2016;13(1):283.

Ugawa S, Ueda T, Ishida Y, Nishigaki M, Shibata Y, Shimada S. Amiloride-blockable acid-sensing ion channels are leading acid sensors expressed in human nociceptors. J Clin Invest. 2002;110(8):1185–90.

Vergo S, Craner MJ, Etzensperger R, Attfield K, Friese MA, Newcombe J, et al. Acid-sensing ion channel 1 is involved in both axonal injury and demyelination in multiple sclerosis and its animal model. Brain. 2011;134(Pt 2):571–84.

Verkest C, Piquet E, Diochot S, Dauvois M, Lanteri-Minet M, Lingueglia E, et al. Effects of systemic inhibitors of acid-sensing ion channels 1 (ASIC1) against acute and chronic mechanical allodynia in a rodent model of migraine. Br J Pharmacol. 2018;175(21):4154–66.

Verkest C, Salinas M, Diochot S, Deval E, Lingueglia E, Baron A. Mechanisms of action of the peptide toxins targeting human and rodent acid-sensing ion channels and relevance to their in vivo analgesic effects. Toxins. 2022;14(10):709.

Vina E, Parisi V, Sanchez-Ramos C, Cabo R, Guerrera MC, Quiros LM, et al. Acid-sensing ion channels (ASICs) 2 and 4.2 are expressed in the retina of the adult zebrafish. Cell Tissue Res. 2015;360(2):223–31.

Voilley N, de Weille J, Mamet J, Lazdunski M. Nonsteroid anti-inflammatory drugs inhibit both the activity and the inflammation-induced expression of acid-sensing ion channels in nociceptors. J Neurosci. 2001;21(20):8026–33.

Vullo S, Ambrosio N, Kucera JP, Bignucolo O, Kellenberger S. Kinetic analysis of ASIC1a delineates conformational signaling from proton-sensing domains to the channel gate. elife. 2021;10:e66488.

Walder RY, Rasmussen LA, Rainier JD, Light AR, Wemmie JA, Sluka KA. ASIC1 and ASIC3 play different roles in the development of hyperalgesia after inflammatory muscle injury. J Pain. 2010;11(3):210–8.

Waldmann R, Bassilana F, de Weille J, Champigny G, Heurteaux C, Lazdunski M. Molecular cloning of a non-inactivating proton-gated Na^+ channel specific for sensory neurons. J Biol Chem. 1997a;272(34):20975–8.

Waldmann R, Champigny G, Bassilana F, Heurteaux C, Lazdunski M. A proton-gated cation channel involved in acid-sensing. Nature. 1997b;386(6621):173–7.

Wang YZ, Wang JJ, Huang Y, Liu F, Zeng WZ, Li Y, et al. Tissue acidosis induces neuronal necroptosis via ASIC1a channel independent of its ionic conduction. elife. 2015;4:4.

Wang IC, Chung CY, Liao F, Chen CC, Lee CH. Peripheral sensory neuron injury contributes to neuropathic pain in experimental autoimmune encephalomyelitis. Sci Rep. 2017;7:42304.

Wang JJ, Liu F, Yang F, Wang YZ, Qi X, Li Y, et al. Disruption of auto-inhibition underlies conformational signaling of ASIC1a to induce neuronal necroptosis. Nat Commun. 2020;11(1):475.

Wemmie JA, Chen J, Askwith CC, Hruska-Hageman AM, Price MP, Nolan BC, et al. The acid-activated ion channel ASIC contributes to synaptic plasticity, learning, and memory. Neuron. 2002;34(3):463–77.

Wemmie JA, Askwith CC, Lamani E, Cassell MD, Freeman JH Jr, Welsh MJ. Acid-sensing ion channel 1 is localized in brain regions with high synaptic density and contributes to fear conditioning. J Neurosci. 2003;23(13):5496–502.

Wemmie JA, Coryell MW, Askwith CC, Lamani E, Leonard AS, Sigmund CD, et al. Overexpression of acid-sensing ion channel 1a in transgenic mice increases acquired fear-related behavior. Proc Natl Acad Sci U S A. 2004;101(10):3621–6.

Wingerd JS, Mozar CA, Ussing CA, Murali SS, Chin YK, Cristofori-Armstrong B, et al. The tarantula toxin beta/delta-TRTX-Pre1a highlights the importance of the S1-S2 voltage-sensor region for sodium channel subtype selectivity. Sci Rep. 2017;7(1):974.

Wong HK, Bauer PO, Kurosawa M, Goswami A, Washizu C, Machida Y, et al. Blocking acid-sensing ion channel 1 alleviates Huntington's disease pathology via an ubiquitin-proteasome system-dependent mechanism. Hum Mol Genet. 2008;17(20):3223–35.

Wu LJ, Duan B, Mei YD, Gao J, Chen JG, Zhuo M, et al. Characterization of acid-sensing ion channels in dorsal horn neurons of rat spinal cord. J Biol Chem. 2004;279(42):43716–24.

Wu WL, Lin YW, Min MY, Chen CC. Mice lacking Asic3 show reduced anxiety-like behavior on the elevated plus maze and reduced aggression. Genes Brain Behav. 2010;9(6):603–14.

Wu Y, Gao B, Xiong QJ, Wang YC, Huang DK, Wu WN. Acid-sensing ion channels contribute to the effect of extracellular acidosis on proliferation and migration of A549 cells. Tumour Biol. 2017;39(6):1010428317705750.

Wu WL, Cheng SJ, Lin SH, Chuang YC, Huang EY, Chen CC. The effect of ASIC3 knockout on corticostriatal circuit and mouse self-grooming behavior. Front Cell Neurosci. 2019;13:86.

Wu BM, Bargaineer J, Zhang L, Yang T, Xiong ZG, Leng TD. Upregulation of acid sensing ion channel 1a (ASIC1a) by hydrogen peroxide through the JNK pathway. Acta Pharmacol Sin. 2020;42:1248.

Wu Y, Chen Z, Sigworth FJ, Canessa CM. Structure and analysis of nanobody binding to the human ASIC1a ion channel. elife. 2021;10:e67115.

Xiong ZG, Zhu XM, Chu XP, Minami M, Hey J, Wei WL, et al. Neuroprotection in ischemia: blocking calcium-permeable acid-sensing ion channels. Cell. 2004;118(6):687–98.

Xu Y, Chen F. Acid-sensing ion channel-1a in articular chondrocytes and synovial fibroblasts: a novel therapeutic target for rheumatoid arthritis. Front Immunol. 2020;11:580936.

Yan J, Edelmayer RM, Wei X, De Felice M, Porreca F, Dussor G. Dural afferents express acid-sensing ion channels: a role for decreased meningeal pH in migraine headache. Pain. 2011;152(1):106–13.

Yan J, Wei X, Bischoff C, Edelmayer RM, Dussor G. pH-evoked dural afferent signalling is mediated by ASIC3 and is sensitized by mast cell mediators. Headache. 2013;53(8):1250–61.

Yang F, Sun X, Ding Y, Ma H, Yang TO, Ma Y, et al. Astrocytic acid-sensing ion channel 1a contributes to the development of chronic epileptogenesis. Sci Rep. 2016;6:31581.

Yang C, Zhu Z, Ouyang X, Yu R, Wang J, Ding G, et al. Overexpression of acid-sensing ion channel 1a (ASIC1a) promotes breast cancer cell proliferation, migration and invasion. Transl Cancer Res. 2020;9(12):7519–30.

Yin T, Lindley TE, Albert GW, Ahmed R, Schmeiser PB, Grady MS, et al. Loss of acid sensing ion channel-1a and bicarbonate administration attenuate the severity of traumatic brain injury. PLoS One. 2013;8(8):e72379.

Yoder N, Yoshioka C, Gouaux E. Gating mechanisms of acid-sensing ion channels. Nature. 2018;555(7696):397–401.

Yu Y, Chen Z, Li WG, Cao H, Feng EG, Yu F, et al. A nonproton ligand sensor in the acid-sensing ion channel. Neuron. 2010;68(1):61–72.

Yu XW, Hu ZL, Ni M, Fang P, Zhang PW, Shu Q, et al. Acid-sensing ion channels promote the inflammation and migration of cultured rat microglia. Glia. 2015;63(3):483–96.

Yuan J, Amin P, Ofengeim D. Necroptosis and RIPK1-mediated neuroinflammation in CNS diseases. Nat Rev Neurosci. 2019;20(1):19–33.

Zha XM. Acid-sensing ion channels: trafficking and synaptic function. Mol Brain. 2013;6:1.

Zha XM, Xiong ZG, Simon RP. pH and proton-sensitive receptors in brain ischemia. J Cereb Blood Flow Metab. 2022;42(8):1349–63.

Zhang Y, Zhang T, Wu C, Xia Q, Xu D. ASIC1a mediates the drug resistance of human hepatocellular carcinoma via the Ca(2+)/PI3-kinase/AKT signaling pathway. Lab Investig. 2017a;97(1):53–69.

Zhang Q, Wu S, Zhu J, Chai D, Gan H. Down-regulation of ASIC1 suppressed gastric cancer via inhibiting autophagy. Gene. 2017b;608:79–85.

Zhang Y, Cao N, Gao J, Liang J, Liang Y, Xie Y, et al. ASIC1a stimulates the resistance of human hepatocellular carcinoma by promoting EMT via the AKT/GSK3beta/Snail pathway driven by TGFbeta/Smad signals. J Cell Mol Med. 2022;26(10):2777–92.

Zhou ZH, Song JW, Li W, Liu X, Cao L, Wan LM, et al. The acid-sensing ion channel, ASIC2, promotes invasion and metastasis of colorectal cancer under acidosis by activating the calcineurin/NFAT1 axis. J Exp Clin Cancer Res. 2017;36(1):130.

Zhou W, Ye S, Luo R, Wu LM, Wang W. Inhibition of acid-sensing ion channels reduces the hypothalamus-pituitary-adrenal axis activity and ameliorates depression-like behavior in rats. RSC Adv. 2019;9(16):8707–13.

Zhu S, Zhou HY, Deng SC, Deng SJ, He C, Li X, et al. ASIC1 and ASIC3 contribute to acidity-induced EMT of pancreatic cancer through activating ca(2+)/RhoA pathway. Cell Death Dis. 2017;8(5):e2806.

Ziemann AE, Schnizler MK, Albert GW, Severson MA, Howard MA 3rd, Welsh MJ, et al. Seizure termination by acidosis depends on ASIC1a. Nat Neurosci. 2008;11(7):816–22.

Ziemann AE, Allen JE, Dahdaleh NS, Drebot II, Coryell MW, Wunsch AM, et al. The amygdala is a chemosensor that detects carbon dioxide and acidosis to elicit fear behavior. Cell. 2009;139(5):1012–21.

Zupping R, Kaasik AE, Raudam E. Cerebrospinal fluid metabolic acidosis and brain oxygen supply. Studies in patients with brain infarction. Arch Neurol. 1971;25(1):33–8.

Transient Receptor Potential Canonical (TRPC) as a Therapeutic Drug Target

Hussein N. Rubaiy

Abstract Canonical or classical transient receptor potential (TRPC) proteins are nonselective cationic channels and a subset of TRP superfamily which are ubiquitously expressed in mammalian cells. Activation of TRPC channels permits the inflow of calcium (Ca^{2+}) and other monovalent alkali cations such as Na^{+} into the cytosol resulting in cell depolarization and an increase in intracellular Ca^{2+} concentration. Hence, TRPC channels are multifunctional signaling protein complexes that play crucial roles in a vast range of human physiological and pathological processes such as cancer, neurological disorders, cardiovascular diseases, pain, and chronic kidney disease. Therefore, they serve as potential drug target for several human diseases and disorders. In this chapter, the TRPC structure, function, regulation, and recently identified potent and selective modulators of these channels will be discussed.

Functional tetrameric TRPC channels assemble as homo- or heterotetrameric complexes. To date, seven TRPC subfamily members have been identified (TRPC1–TRPC7), and TRPC2 is a pseudogene in humans. Recent TRPC structural studies have offered new insights into channel architecture and assembly. High-resolution structures of TRPC3, TRPC4, TRPC5, and TRPC6 homotetramers have been reported. Together, potentially compound profiling will lead to discovering and characterizing novel drug candidates that are intended to generate innovative therapeutic therapies.

Keywords Ion channel · TRP channelss · TRPC · Calcium signaling · Small-molecule modulators · Drug discovery · Therapeutic

H. N. Rubaiy (✉)
Department of Laboratory Medicine, Division of Clinical Pharmacology, Karolinska Institute & Karolinska University Hospital, Stockholm, Sweden
e-mail: hussein.rubaiy@ki.se

G. Stephens, E. Stevens (eds.), *Ion Channels as Targets in Drug Discovery*,
https://doi.org/10.1007/978-3-031-52197-3_11

1 Introduction

As a background on TRP channels, Cosens and Manning discovered a spontaneous mutation in *Drosophila melanogaster* in 1969, which was revealed by an electroretinogram defect in which the steady electrical response normally evoked by a sustained light pulse becomes transient (Cosens and Manning 1969). This impairment occurred in the photoreceptor cells' receptor potential. As a result, the mutant was named Transient Receptor Potential, or TRP. The gene was cloned by Montell and Rubin in 1989, and they hypothesized that it encoded a new ion channel. Several laboratories later verified this (Montell and Rubin 1989; Wong et al. 1989; Hardie and Minke 1992; Phillips et al. 1992). The first mammalian TRP channel sequence was discovered in 1995 (Wes et al. 1995; Zhu et al. 1995). This sequence is now known as TRPC1, and it is one of the seven members of the canonical or classical TRP (TRPC) channel subfamily (Minard et al. 2018).

It is worth mentioning that three Nobel Prizes (1963, 1991, and 2003) have been awarded in recognition of the vital importance of ion channel research by the scientific community (Rubaiy 2017; The official website of the Nobel Prize 2022). Moreover, last year (2021), the scientific community awarded the fourth Nobel Prize to David Julius and Ardem Patapoutian for their work on TRP and Piezo channels (Reeh and Fischer 2022). In addition, several Nobel Prizes were awarded related to the ion channel field (1932, 1936, 1944, 1970, and 2000) (The official website of the Nobel Prize 2022).

Today, it is well established that there are 28 different mammalian members that make up the family of TRP channels (Rubaiy 2019; Bon et al. 2022). Each of these members is an individual cellular sensor that reacts to a wide variety of different external and internal signals. TRP channels are engaged in a wide variety of physiological processes and are expressed by most mammalian cells (Diver et al. 2022). The vast majority of TRP channels are found in the plasma membrane which depolarize cells, leading to an increase in the concentrations of sodium (Na^+) and calcium (Ca^{2+}) ions inside the cell (Clapham 2003; Nishida et al. 2006; Vangeel and Voets 2019). Additionally, it is known that intracellular Ca^{2+} has the ability to modulate all TRP channels and can activate some mammalian TRP channels on its own (Clapham 2003; Owsianik et al. 2006; Vangeel and Voets 2019).

Seven subfamilies comprise the TRP superfamily: TRPC (canonical or classical, seven members), TRPV (vanilloid, six members), TRPM (melastatin, eight members), TRPN (name derived from the fly mutant no mechanoreceptor potential C, one member), TRPA (ankyrin, one member), TRPP (polycystic, three members), and TRPML (mucophilin, three members) (Vangeel and Voets 2019). These channels variously sense photo-, chemo-, thermo-, mechano-, osmoreception, pain, and itching (Moore et al. 2017).

The classical or canonical TRP channels, also known as TRPCs, are a class of nonselective cationic channels that have a stronger preference for the cation Ca^{2+} (Beech 2013). This family consists of seven members (TRPC1-7), each of which contains a TRP box in the carboxyl terminus and 3–4 ankyrin repeats in the amino terminal. The members of this family are referred to as TRPCs (Rubaiy 2019). The TRPC channels are involved in a wide variety of signaling pathways and exhibit a variety of gating mechanisms; however, they do have significant similarities in their characteristics (Gees et al. 2012). The purpose of this chapter is to provide up-to-date discussions regarding TRPC channels as a pharmacological target in the research and development of new drugs.

2 Structure of TRPC Channel

Recent structural investigations on TRPC ion channels have yielded unique insights into the architecture and assembly of TRPC channels (Guo and Chen 2019). Up to the present, high-resolution structures of TRPC3, TRPC4, TRPC5, and TRPC6 homotetramers have been reported; the structures of the cytoplasmic domains of TRPC3 and TRPC6 have also been described (Rubaiy 2017; Azumaya et al. 2018; Fan et al. 2018; Sierra-Valdez et al. 2018; Tang et al. 2018; Vinayagam et al. 2018; Duan et al. 2018a; Duan et al. 2019; Guo and Chen 2019; Kim et al. 2019; Li et al. 2019). TRPC channels share a two-layer design, the cytoplasmic domain layer (CTD) and the transmembrane domain layer (TMD). N- and C-terminal residues fold the CTD. N-terminal residues contain a flexible loop, linker helices, and ankyrin repeats (Fig. 1). C-terminal residues contain a TRP helix, a flexible linker, two helices, and flexible residues (Guo and Chen 2019). The transmembrane domain structure is comparable to other TRP channel members, with the voltage sensor-like domain (VSLD) joined to the pore domain by an S4-S5 linker. In addition, TRPC's TMD structure most closely mimics that of TRPM and TRPN, which both feature the pre-S1 elbow (Jin et al. 2017; Moore et al. 2017; Winkler et al. 2017; Autzen et al. 2018; Tang et al. 2018; Vinayagam et al. 2018; Duan et al. 2018b).

The TMD of TRPC channels, like many other TRP channels, has the ability to bind a variety of ligands or lipids and is the primary target of synthetic compounds. The TRPC4/TRPC5 contains ion-binding sites in the voltage sensor-like domain. Moreover, in TRPC4, the putative ion lies in the pocket between S2 and S3, suggesting it might correspond to Na^{+}, whereas in TRPC5, it could be Ca^{2+} (Vinayagam et al. 2018; Duan et al. 2018c; Guo and Chen 2019).

Membrane proteins frequently engage in crucial and particular interactions with membrane lipids, which can serve stabilizing or regulatory functions. Lipids have a role in the control of TRPC channel activity; hence, the detection of lipids in some TRPC cryogenic electron microscopy (cryo-EM) is noteworthy. There are two major locations containing lipid-like molecules. Lipid site 1 is located in the inner leaflet of the VSLD, while lipid site 2 is detected in the outer leaflet, coupled to the P loop and S6 helix of nearby subunits. In the absence of small-molecule modulators, both TRPC5:C5 and TRPC4:C4 possess a well-defined lipid in site 2 (Vinayagam et al. 2018; Duan et al. 2018c; Duan et al. 2019; Wright et al. 2020; Bon et al. 2022).

As discussed earlier, recent structural investigations of several groups on TRPC ion channels have yielded unique insights into the architecture and assembly of

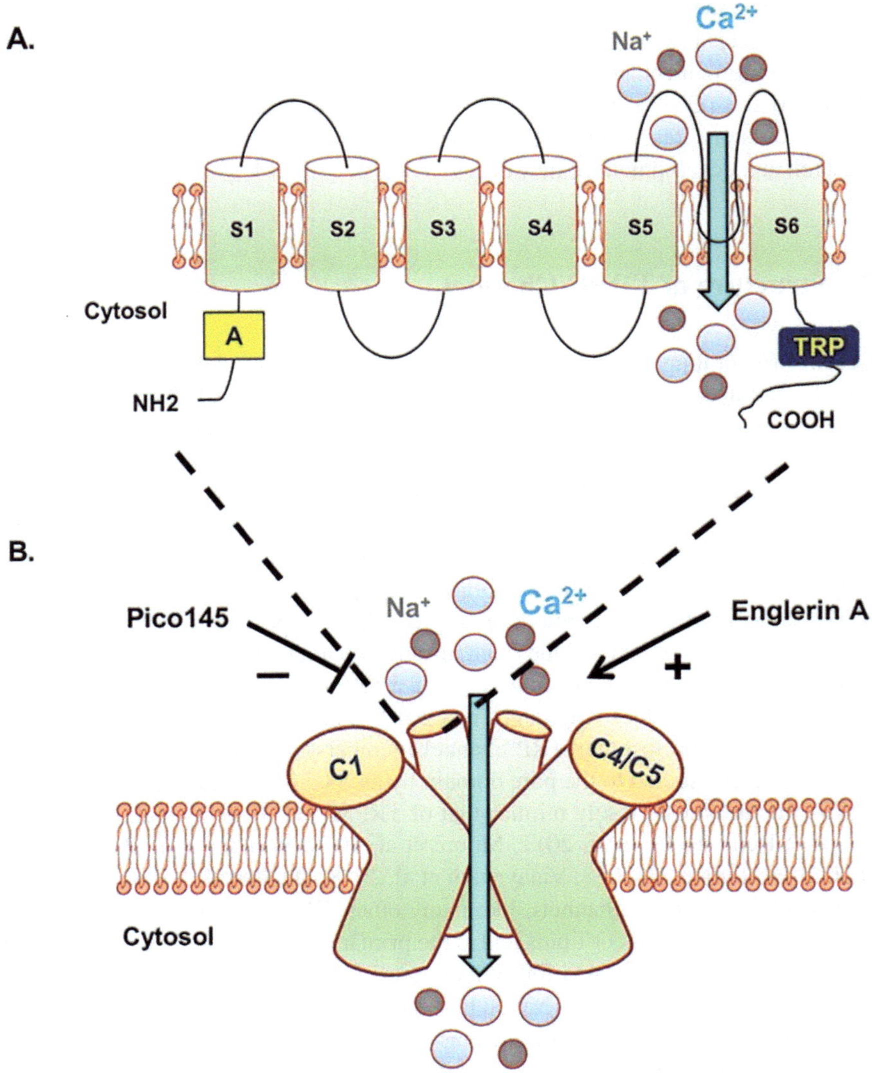

Fig. 1 Schematic membrane topology structure of the TRPC channels. (**a**) The proposed structural topology of monomeric TRPC channels consists of six membrane-spanning domains (S1–S6), with A representing the ankyrin-like repeat domain and TRP representing the TRP-like domain. (**b**) Illustrates the functional nonselective cationic TRPC1/4/5 channels that assemble as homo-tetrameric or hetero-tetrameric channels with high calcium permeability. The novel, selective, and most potent agonists (EA) and antagonists (Pico145) are demonstrated in (B)

TRPC channels and, importantly, reported the small-molecule binding sites of these channels and the binding modes. As drug discovery is a time-consuming and resource-intensive procedure, this significant development in TRPC channels might bring significant benefits. Notably, the structures of TRPC channels in the open state remain enigmatic, despite the fact that activators and inhibitors have been used to solve the structures. For unknown stoichiometries TRPC1/4/5 channels, heteromeric states may be present in numerous tissues, suggesting that the natural state of many TRPC channels is not in fact a homotetramer (Bon et al. 2022).

Remarkably, structures of TRPC5:C5 obtained in the presence of Pico145 (a picomolar, selective, and subtype-specific antagonist) indicated that Pico145 binds to lipid site 2 and substitutes the phospholipid that was detected in this position in structures of TRPC4/5 obtained in the complete absence of small-molecule modulators (Bon et al. 2022). Interestingly, similar substitution of lipid 2 was seen in the structure of TRPC6:C6 once the TRPC6 agonist AM-0883 was present (Bai et al. 2020). Hopefully, these breakthroughs in the field of TRPC channels, which have excessive pharmaceutical promise, could open the gates for structure-based drug design targeting diereses and disorders which are related to these proteins.

3 Activation and Regulation of TRPC Channels

Mammalian TRPC channels are triggered by phospholipase C (PLC)-signaling receptors, involving Gq/11-coupled receptors and receptor tyrosine kinases (Trebak et al. 2007; Chen et al. 2020). The PLC pathway entails a number of stages, among them hydrolysis of phosphatidylinositol 4,5-bisphosphate (PIP_2) at the plasma membrane, inositol 1,4,5-trisphosphate (IP_3) and diacylglycerol (DAG) production, IP_3 receptor activation at the endoplasmic reticulum membrane, Ca^{2+} store depletion, and $[Ca^{2+}]_i$ increase (Trebak et al. 2003; James and Putney 2007; Chen et al. 2020).

PIP_2 plays an important role for TRPC channel gating; nevertheless, the effects of PIP_2 on TRPC activation are complex. At basal levels of PIP_2, and its predecessors phosphatidylinositol 4-phosphate (PIP) and phosphatidylinositol (PI) are found on the plasma membrane (Beech 2013; Hille et al. 2015); PLC breaks down PIP_2 which at the same time decreases PIP and PI levels due to rapid PI to PIP and PIP to PIP_2 conversion by kinases, which eases the inhibition and allow channel activation. This is while some levels of PIP_2 might be necessary to maintain TRPC channel activity, as DAG, IP_3, and H^+ are dependent on PIP_2; however, this is not fully understood. Studies have shown that PIP_2 plays a vital role in supporting TRPC3, TRPC6, and TRPC7 function, while exhibiting differences in the sensitivity to PIP_2 (Venkatachalam et al. 2003; Chen et al. 2020). It has been reported that TRPC5 activation was augmented by hydrolyzing PIP_2 with PLC or impeding the kinase which phosphorylates PIP to PIP_2, but inhibited by dephosphorylating PIP_2 to PIP via a phosphatase (Ningoo et al. 2021). In 2009, Trebak et al. showed that direct application of PIP_2 to the cytoplasmic side of inside-out patches increased TRPC5 function (Trebak et al. 2009). Several studies have shown that TRPC4 activation is PIP_2-dependent (Kim et al. 2013; Thakur et al. 2016; Thakur et al. 2020). Kim et al. demonstrated that lower levels of endogenous PIP_2 in the presence of overexpression of a constitutively active Gq protein, or phosphatase that dephosphorylated

PIP_2 to PIP, fully inhibited TRPC4 activation (Jeon et al. 2012). Later, the same group proposed that while PIP_2 is not an activator of TRPC4, it is still required for TRPC4 regulation (Kim et al. 2013). In regard to TRPC4 and both its isoforms α and β, their activation is dependent on PLCδ1 and $G_{i/o}$ protein signaling (Thakur et al. 2016).

DAG and PIP_2 play an important role in the activation of TRPC6 and TRPC7 channels, while channel inactivation is associated with the loss of PIP_2 but not with DAG (Itsuki et al. 2014). In 1999, Hofmann et al. reported that natural 1-stearoyl-2-arachidonyl-sn-glycerol (SAG) and synthetic DAG analog 1-oleoyl-2-acetyl-sn-glycerol (OAG) activated TRPC3 and TRPC6 when applied from the extracellular or cytoplasmic sides (Hofmann et al. 1999). Moreover, in separate studies, it was shown that DAG activated TRPC2 and TRPC7 (Okada et al. 1999; Lucas et al. 2003; Beech 2012). It was later revealed that TRPC4/5 are also regulated by DAG; however, the lipid sensitivity was masked by the PDZ domain protein, Na^+/H^+ exchanger regulatory factor (NHERF). This regulatory factor binds to the ends of C-terminal for TRPC4/5 (Tang et al. 2000; Storch et al. 2017). Following this finding, it was shown that DAGs appear to directly activate all mammalian TRPC channels (Chen et al. 2020; Wang et al. 2020a).

When comparing receptor activation in electrophysiological experiments, the synthetic DAG analogue OAG, which is often preferred as it has easier access and better solubility in aqueous solutions than naturally occurring DAGs, appears not to recapitulate all features of TRPC channel activation (Wang et al. 2020a). The recordings show that the magnitude of OAG-evoked current seems to be smaller and the kinetics of its activation are slower than TRPC currents activated by receptor stimulation. Another difference between TRPC channel activation by OAG and receptor agonists is solubility and site of lipid application; OAG is usually applied extracellularly, while natural DAGs appear initially in the inner leaflet of the plasma membrane following the hydrolysis of PIP_2 (Wang et al. 2020a).

Of the different TRPC subgroups 1–7, it was first thought that TRPC6 was activated in a receptor-operated manner; however, it was later shown that TRPC6 along with TRPC3 were activated by agonists of G-protein coupled receptors (GPCRs) in a PLC-dependent but store-depletion independent manner (Hofmann et al. 1999; Zitt et al. 1997). Later, it was reported that TRPC3, TRPC6, and TRPC7 can be directly activated by DAG or OAG in a PKC-independent manner (Hofmann et al. 1999; Okada et al. 1999; Trebak et al. 2003). Trebak et al. discovered that PIP_2 degradation is vital for TRPC3, TRPC6, and TRPC7 channel activation by DAG, as well as for the closing of these three channels. DAG is also involved in the activation of TRPC2 and TRPC7 (Trebak et al. 2009).

Regarding TRPC1, TRPC4, and TRPC5, which are considered DAG-insensitive, discussions have been held whether they are activated via the PLC-dependent mechanism without store-deletion. Some groups presented supporting evidence for the store-independent activation, while others claimed that the store-operated mechanism is the major route for the activation (Chen et al. 2020). It has been shown that TRPC4 and TRPC5 are activated via GPCRs or receptor tyrosine kinases downstream of PLC; however, they function independently of store-depletion (Okada et al. 1998; Schaefer et al. 2000). Furthermore, it has been reported that the TRPC1 channel is activated by store-depletion (Zitt et al. 1996; Subedi et al. 2017). It has also been reported that TPRCs can be activated in a PLC-independent manner. TRPC5 and TRPC6 can be activated by lysophospholipids (So et al. 2005; Flemming et al. 2006), whereas sphingosine-1-phosphate can activate TRPC5 and TRPC5-TRPC1 heteromeric channels while regulating smooth muscle cell motility (Xu et al. 2006). TRPC5 can also be activated by perfusion of extracellular hypoosmotic solution or pressure-induced membrane stretch. There is also evidence that TRPC6 can be activated by H_2O_2 via a cysteine oxidation-dependent pathway, while also sensitizing TRPC6 to DAG-activation and stimulating its trafficking to the plasma membrane (Chen et al. 2020). Furthermore, a group led by Spassova suggested that TRPC6 might function as a mechanosensitive, stretch-activated channel and presented a TRPC6 activation model involving changes in plasma membrane geometry. Based on Spassova's model, TRPC6 opens by stretch-induced plasma membrane thinning or by PLC-dependent PIP_2 breakdown to DAG leading a significant change in local membrane curvature (Spassova et al. 2006).

It was determined that activation of all TRPC channels cause increased intracellular Ca^{2+} concentration and/or depolarization of membrane potential (Chen et al. 2020). However, the two subgroups, TRPC3/6/7 and TRPC1/4/5, express a number of dissimilar properties such as how they potentiate and inhibit. TRPC4 and TRPC5 are potentiated by acidic pH, while TRPC6 is blocked by protons (Semtner et al. 2007). TRPC4 and TRPC5 can also be potentiated by ions such as lanthanide (La^{3+}) and gadolinium (Gd^{3+}), whereas TRPC6 is inhibited by lanthanides (Rubaiy et al. 2003, Rubaiy 2019).

4 TRPC in Human Health and Diseases

TRPC channels are involved in many physiological and pathophysiological conditions. Studies involving TRPC channels have shown that its activity or excessive activation are involved in diseases and disorders within the vital systems such as cardiovascular and metabolic and central nervous system as well as cancer (Fig. 2) (Rubaiy 2019; Chen et al. 2020; Wang et al. 2020a; Bon et al. 2022). Despite being linked to such important systems, a study in mice has shown that TRPC is not essential for life as genetic knockout of the channel did not lead to death (Bernal et al. 2021). Therefore, TRPC inhibitors may play a key role in the treatment of some forms of illness without causing severe or possibly adverse effects.

5 Cardiovascular and Metabolic Diseases

In cardiac cells, TRPC channels may act as receptor- and store-operated Ca^{2+} entry channels, and it is suggested that they are involved in stretch-activated Ca^{2+} entry (Ward et al. 2008; Freichel et al. 2017). Studies involving gene knockout and

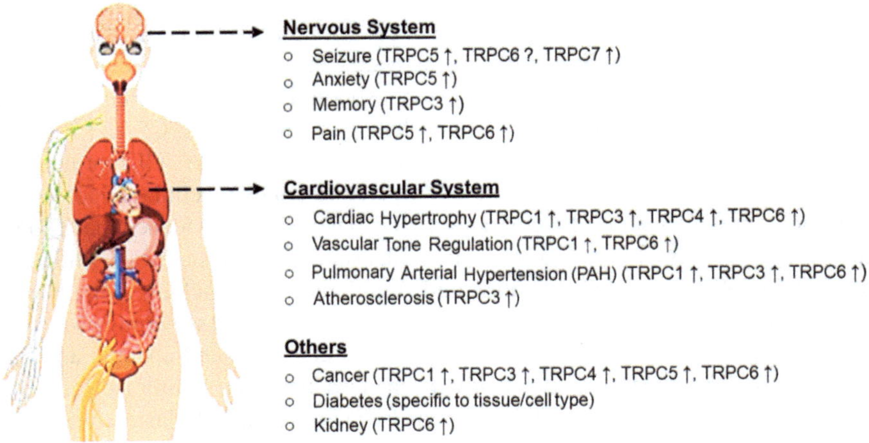

Fig. 2 TRPC channels in human diseases and disorders

dominant-negative expression reported both in vitro and murine data indicating stimulation of GPCRs by hypertrophic agents and mechanical stress causes activation of TRPC3/6/7 which leads to cardiac hypertrophy, that is, thickening of the heart (Camacho Londoño et al. 2015). Reduction of background Ca^{2+} entry into cardiac myocytes in TRPC1/4 double-knockout mice enhanced cardiac hypertrophy triggered by neurohormonal and mechanical stimulation, while TRPC3/6 had no role (Camacho Londoño et al. 2015). In separate studies, Seo et al. revealed that TRPC3/6 inhibition reduces cardiac hypertrophy, and a study by Kiyonaka et al. revealed that Pyr3, a TRPC3 inhibitor, blocked hypertrophic grown in rat neonatal cardiomyocytes and in pressure overload-induced cardiac hypertrophy in mice (Kiyonaka et al. 2009; Seo et al. 2014). This is while a TRPC3/6/7 inhibitor, BI749327, reduced interstitial fibrosis in a pressure overload mouse model, which led to an improved left heart function (Lin et al. 2019).

It has been shown an upregulation of TRPC1/3/4/6 channels in post-myocardial infarction heart and TRPC3/6/7 knockout reduced the infarct size and tissue damage (He et al. 2017). Furthermore, TRPC3/6 are able to physically interact and activate reactive oxygen species (ROS)-producing nicotinamide adenine dinucleotide phosphate (NADPH) oxidase (Nox) enzymes which causes an induction of oxidative stress in cardiomyocytes and cardiac fibroblasts, resulting in cardiac remodeling and fibrosis (Kitajima et al. 2016; Oda et al. 2017; Numaga-Tomita and Nishida 2020). Other studies have shown augmented TRPC5 and TRPC6 expression in failing human hearts (Bush et al. 2006; Kuwahara et al. 2006). This evidence suggests that TRPC channels could be a target for heart failure. Sabourin et al. suggest that TRPC channels might be involved in electromechanical conduction in developing hearts (Sabourin et al. 2011). Moreover, Nikolova-Krstevski proposed that TRPC6 are linked to the development of myocardial calcium transients and arrhythmia, as it is implied that stretch-dependent modulation of TRPC6 expression in atrial endocardium regulates endothelin-1 (ET-1) production (Nikolova-Krstevski et al. 2017).

Studies in regard to atherosclerosis and vascular inflammation have shown that TRPC3 could play a role in atherosclerosis management, as Ca^{2+} influx through TRPC3 channels is vital for cell adhesion-molecule expression and endothelial cell activity and that Pyr10, a TRPC3 inhibitor, lessened endoplasmic reticulum stress-induced apoptosis in endothelial cells (Smedlund et al. 2010; Schleifer et al. 2012; Ampem et al. 2016). It has also been reported that TRPC3 activity in macrophages can play a protective role against apoptosis, which might augment atherosclerotic lesion progression (Ampem et al. 2016). In regard to TRPC1 and its role in atherosclerosis and vascular inflammation, it has been shown that blocking TRPC1 reduced neointima formation in pig models of vascular injury and in vitro human vein culture models, and that TRPC1 expression was boosted in the same models. TRPC1 also induced smooth muscle migration and proliferation (Kumar et al. 2006).

TRPC channels play a role in primary systemic hypertension (Zhu et al. 2016; Xiao et al. 2017). Zhu et al. reported defective Ca^{2+} homeostasis and increased TRPC1, TRPC3, and TRPC5 expression in the vasculature and peripheral blood cells of hypertensive humans and animals (Zhu et al. 2016). In a separate study, TRPC6 knockout mice suffered from hypertension, which could be due to a compensatory augmentation in TRPC3 expression (Dietrich et al. 2005). This is while mesenteric vessels of 11-deoxycorticosterone acetate-treated hypertensive rats displayed increased expression and activity of TRPC6 channels (Bae et al. 2007). With this said, it is thought that the role of specific TRPC channels in contraction and resultant blood pressure regulation is both complex and dependent on the partner channels. Moreover, the pathophysiological context proposes TRPC channels as promising targets for the treatment of pulmonary arterial hypertension (PAH), as hypoxic pulmonary vasoconstriction is reduced when treated with a β-carboline derivative or larixyl acetate through TRPC6 channel inhibition (Urban et al. 2012; Reyes et al. 2018). A separate study by Maier et al. showed SAR7334, a TRPC3/6/7 inhibitor subdued acute hypoxic pulmonary vasoconstriction in mice, while in rats with spontaneously hypertensive the mean arterial pressure remained unaltered (Maier et al. 2015).

TRPC5 plays a vital role in the cardiovascular system including in atherosclerosis. TPRC5 is involved in vascular smooth muscle cell migration and proliferation, endothelial cell injury and dysfunction, lipid deposition, and cardiac hypertrophy through controlling intracellular Ca^{2+} homeostasis, oxidative stress, and apoptosis. By blocking TRPC5 channels, atherosclerotic plaque formation can be prevented, cardiac hypertrophy associated gene expression can be reduced, and cardiac electrical activity can be stably maintained (Du et al. 2021).

TRPC channels have been shown to be involved in diabetic complications including development of nephropathy, neuropathy, retinopathy, and vasculopathy (Smani et al. 2015; Liu et al. 2017; Dryer et al. 2019). In regard to homeostasis, TRPC3 channels in hypothalamic neurons play a vital role in sensing glucose, insulin secretion, and glucose regulation (Chrétien et al. 2017). In pancreatic cells, ATP-sensitive potassium (K_{ATP}) channels regulate glucose-induced insulin release (Rubaiy 2016, 2017).

Activation of leptin-induced TRPC4 causes K_{ATP} channels to traffic to the plasma membrane of pancreatic β cells during fasting, which leads to reduced insulin secretion (Park et al. 2013). As diabetes increases TRPC channel expression in endothelial cells and vascular smooth muscle cells, it is suggested that TRPC-mediated pathological signaling, which causes vascular disorders, could be augmented in diabetes (Graham et al. 2012). Recent studies have underscored the complex roles of TRPC channels in different tissues and conditions,including knockout studies involving TRPC6 which have shown that its signaling may have various effects on diabetic nephropathy (Wang et al. 2019). Another study showed no hyperglycemia-induced retinal changes in quadruple-knockout (TRPC1/4/5/6) mice through the preservation of Muller and microglial functions (Sachdeva et al. 2018). This is while it has been reported that TRPC6 plays a role in peripheral neuropathy in the streptozotocin-induced rate model of diabetes (Roa-Coria et al. 2019).

TRPC channels are also linked to adiposity (also known as obesity). A study using global TRPC1 knockout mice showed that this channel plays a vital role in cellular energy metabolism regulation and that inhibiting TRPC1 channels leads to increased fat mass and insulin resistance in the high-fat feeding mice compared to wild-type (Krout et al. 2017). TRPC1 regulation of adiposity occurs through its contrasting impacts on autophagy and apoptosis of adipocytes. It is suggested that TRPC1:C5 inhibition could be a therapeutic strategy in metabolic disorders, as well as boosting positive effects of exercising. This is while TRPC1 disruption leads to weight gain and caused negative effects on metabolic profile by decreasing thermogenic gene expression in brown adipose tissue (Wolfrum et al. 2018). It has also been suggested that TRPC5 plays a vital role in leptin regulation of hunger/satiety and energy homeostasis as a neuronal and proopiomelanocortin-specific TRPC5 knockout model leads to weight gain in knockout animals (Gao et al. 2017).

6 Nervous System Diseases and Disorders

TRPC channels are ubiquitously expressed throughout the brain and play a vital role in the control of physiological activities as well as in the pathophysiology of human disorders associated with ROS (TRP Channels in the Brain 2022). TRPC1 is involved in either onset/progression of Parkinson's disease, and restoration of TRPC1 levels could restrict neuronal degeneration in 1-methyl-4-phenyl-1,2,3,6-tetrahyrdro-pyridine (MPTP) mediated Parkinson's disease (Selvaraj et al. 2010). This was shown when either silencing TRPC1 or applying TRPC1 channel antagonists that lowered mitochondrial membrane potential. This is while TRPC1 activation reestablished mitochondrial membrane potential; moreover, it was revealed that overexpression of TRPC1 blocked Bax translocation to the mitochondria preventing cytochrome C release and mitochondrial-mediated cell death cells (Selvaraj et al. 2010).

Neuner et al. reported that overexpression of TRPC3 in hippocampus of aging mice is involved in mediating enhanced afterhyperpolarization (AHP) and TRPC3 play a role in regulating hippocampal neuron excitability associated with memory function (Neuner et al. 2015). The group suggested that therapeutics which reduce TRPC3 expression or impair its function may work as neurocognitive enhancers (Neuner et al. 2015). TRPC3 plays a role in the function of microglia including cytokines and nitric oxide (NO) release (Mizoguchi et al. 2014). It is also suggested that Ca^{2+} influx and TRPC3-mediated concentration regulation are key players in improving neurodegenerative diseases (Mizoguchi et al. 2014). It has also been suggested that TRPC3 is involved in neurodegeneration, as it is directly activated by oxidative stress. TRPC3 affects Ca^{2+} influx which is linked to neuronal cell death (Selvaraj et al. 2010). A separate study reported that TRPC3-mediated Ca^{2+} influx is associated with Na^+/Ca^{2+} exchange (Rosker et al. 2004). It is suggested that with an increase of Na^+ influx via TRPC3 amid oxidative stress could lower Ca^{2+} influx, which may be beneficial in the treatment of Parkinson's disease. A study by Zhou et al. showed TRPC3 inhibition depolarized GABA neurons in the substantia nigra pars reticulate which is linked to parkinsonism (Zhou et al. 2008a).

Studies have also been performed on TRPC3 and its involvement in seizure disorders such as epilepsy. Data retrieved through genetic strategies and pharmacological approaches in various animal models show that TRPC3 is a novel antiseizure target (Yu et al. 2022).

TRPC5 channels, along with TRPC1, have high expression levels in substantia nigra, and they play a vital role in dopaminergic neurons (De March et al. 2006). It is also suggested that TRPC5 regulates neuronal growth cone morphology and is involved in development of the nervous system. In the downstream processes in which semaphorin 3A is involved, growth cone collapse is triggered by cleavage and activation of TRPC5 using calpain (Kaczmarek et al. 2012). It has also been reported that reduced TRPC5 activity by ML204-specific block boosted striatal neurons' survival and behavioral motor symptoms (Hong et al. 2020). There is also evidence that TRPC5 plays a role in the control of fear-related behaviors: acquired (learned) fear and innate fear (anxiety). Blinded behavioral studies have revealed that mice with inhibited TRCPC5 currents showed substantially less anxiety in response to naturally aversive stimuli than their wild-type littermates (Riccio et al. 2009).

The role of TRPC6 in the nervous system has been investigated, and it has been reported that it upregulates the cAMP-response element-binding protein pathway which is a key player in dendritic growth, and that TRPC6 promotes synapse and dendritic spine formation, spatial memory, as well as learning (Tai et al. 2008; Zhou et al. 2008b). Furthermore, a study has shown that TRPC6 acts as a negative regulator which inhibits NMDA-induced Ca^{2+} influx in hippocampal neurons (Shen et al. 2013).

Thus, TRPC6 modulation might have a protective impact on neurons against excitotoxicity in the primary stage of ischemic brain damage. Despite TRPC6 activation having beneficial effects in the brain, TRPC6 activation causes toxic effects on other cellular systems, such as the immune system, and it is linked to breast cancer cells (Guilbert et al. 2008; Ramirez et al. 2018). In regard to TRPC channels and their role in pain, it has been revealed that they are expressed in dorsal root ganglion (DRG) neurons, and it is suggested that they play a vital role in sensory physiology with TRPC1, TRPC3, and TRPC6 being the most predominant, TRPC2, TRPC4, and TRPC5 at low levels, and TRPC7 at very low levels (Sun et al. 2020).

It has been reported that TRPC1, TRPC3, TRPC4, and TRPC6 are expressed at mRNA and protein levels in the intervertebral disc. In degenerated intervertebral discs, TRPC6 exhibited enhanced gene and protein expressions compared to nongenerated intervertebral disc, associating TRPC6 with development of low back pain in humans (Sadowska et al. 2019). However, further studies are required to investigate TRPC expression and its involvement in pain pathway in humans. It has also been reported that TRPC3 and TRPC6 play a vital role in normal mechanotransduction in subsets of sensory neurons and cochlear hair cells (Quick et al. 2012). Double knockout TRPC3/TRPC6 mice have deficits in light touch, and aound half of small-diameter sensory neurons that display mechanically-activated, rapidly-adapting mechanosensitive currents were silenced. This is while mice with single knockout TRPC3 and TRPC6, respectively, showed no behavioral phenotype, indicating that these TRPCs possibly show some compensation in function, in line with other observed functional redundancy (Dietrich et al. 2005). Furthermore, it has been reported that TRPC3/TRPC6 knockout mice displayed hearing impairment, vestibular deficits, and impaired auditory brain stem responses to high frequency sounds (Sun et al. 2020). There has been progress made in our knowledge of TRPC1-7 in chronic pain; nevertheless, further research is required to better our comprehension of the physiological and pathological activities of TRPC channels in pain and sensitization.

7 TRPC Channels in Cancer

TRPC1 as a Ca^{2+} channel is involved in pancreatic cancer cell proliferation and its progression, acting possibly as a downstream effector of TGFβ signaling (Dong et al. 2010). It was discovered that levels of TRPC1, TRPC4, and TRPC6 were high in pancreatic cancer cells, suggesting that they function as TGFβ mediators for Ca^{2+} entry (Dong et al. 2010). This group reported that TRPC1 knockdown by siRNA or treatment with the inhibitor 2′-aminoethoxydiphenyl borate (2-APB), significantly blocked pancreatic cancer cell motility triggered by TGFβ. It is worth mentioning that the same type of knockdown of TRPC4 and TRPC6 had no effect on the motility (Dong et al. 2010).

TRPC subtypes have also been reported in breast cancer. It is suggested that TRPC1 expression might have a role in facilitating epithelial–mesenchymal transition (EMT) and proliferation in breast cancer (Elzamzamy et al. 2020). Zhang et al. showed that TRPC1 overexpression blocks proliferation and migration of estrogen receptor-positive (ER +) breast cancer, contributing to a better prognosis by inhibiting activation of PI3K/AKT signaling pathway. The group suggested that TRPC1 could be an independent prognostic predictor of breast cancer (Zhang et al. 2020). It was also reported that overexpression of TRPC1 was linked to expression and increased proliferative and invasive capacity of small grade I breast cancer tumors (Dhennin-Duthille et al. 2011).

TRPC6 overexpression has been detected in breast cancer cells; however, TRPC6 levels did not have any effect on the tumor grade, expression of estrogen receptor, or lymph node positive tumors (Guilbert et al. 2008). TRPC is also found in the aggressive cancer type glioblastoma which occur in the brain or spinal cord. TRPC1 channels play a role in the proliferation of human malignant gliomas (Zhong et al. 2022). Knockdown of TRPC1 in D54MG glioma cells via shRNA or inhibition by SKF96365 deters cell proliferation and incomplete cell division. In vivo results suggested that suppression of TPRC1 via shRNA decreased the size of flank tumours (Bomben and Sontheimer 2010). It was also reported that caveolar lipid rafts mark the TRPC1 channels mediating glioma chemotactic metastasis, suggesting a drug target for TRPC channel inhibitors (Maklad et al. 2019).

In lung cancer, TRPC1, TRPC3, TRPC4, and TRPC6 are overexpressed in comparison to other TRPC channels. It was also reported that overexpressed TRPC1 and TRPC6 in A549 non-small-cell lung carcinoma (NSCLC) cell line causes increased proliferation, while inhibiting TRPC channels via 2-APB or TRPC1 antibody T1E3, blocked proliferation of the A549 cells (Jiang et al. 2013). TRPC1 is unregulated in NSCLC tissue, and overexpression of the channel is linked to lymph node metastasis and worsened disease-free survival in NSCLC patients (Ke and Long 2022). TRPC channels are not only involved in the mentioned cancers; they also play a role in other types such as colorectal, ovarian, prostate, gastric, and medulloblastoma (Cai et al. 2009; Yang et al. 2009; Yue et al. 2009; Wei et al. 2017; Sun et al. 2021).

As a conclusion for the role of TRPC in human health and diseases, to fully understand the physiological significance of TRPC channels in native cells and in vivo animal models it is helpful to use the pharmacological tools of these channels, which will be detailed in the next section. Importantly, utilizing these compounds, disorders mediated by these channels may be investigated, potentially launching drug development efforts to develop novel therapeutic treatments.

Up arrows (↑) represent the high expression of TRPC isoforms subtypes. It is worth mentioning that inhibition of some isoforms pharmacologically or by knockdown inhibited the diseases or disorders. As an example, TRPC5 channels are extensively expressed in human sensory neurons, and it is suggested that TRPC5 inhibitors could be non-opioid-based therapies for tactile and spontaneous pain. Combination of TRPC1/TRPC4 knockdown prevented cardiac hypertrophy. This is a brief overview of the expression of TRPC isoform linked to disease and disorders (Nilius et al. 2005; Morrell et al. 2009; Riccio et al. 2009; Graham et al. 2012; Kim and Kang 2015; Neuner et al. 2015; Fels et al. 2018; Martín-Bórnez et al. 2020; Numaga-Tomita and Nishida 2020; Sadler et al. 2021).

8 The Pharmacology of TRPC Channel

8.1 TRPC Channel Activators

Recently, nonselective and weak pharmacological tools such as ions (La^{3+}, Gd^{3+}), GPCR agonists, carbachol (CCH), 2-APB, and OAG were utilized to study the physiological and pathological role of TRPC channels. However, in this section recent novel, potent, and selective agonists and antagonists of TRPC proteins will be discussed (Table 1, 2, 3, and 4).

8.2 TRPC1/4/5 Channel Activators

8.2.1 (-)-Englerin A (EA)

Of all the TRPC1/4/5 channel activators, the most potent and selective is the natural product (-)- (EA) as it activates its currents at low nanomolar concentrations (Akbulut et al. 2015a; Ludlow et al. 2017). EA is effective against certain human cancer cells, and its cytotoxic effects target A498 renal cancer cells and SW982 synovial carcinoma cells resulting from Na^{+} influx mediated by heteromeric TRPC1:C4 channels (Ludlow et al. 2017; Muraki et al. 2017). EA activates homomeric TRPC5 at a EC_{50} value of 7.6 nM, TRPC4 at 123 nM, TRPC5-TRPC1

Table 1 TRPC1/4/5 channels agonists

Name	EC_{50} (Channel)	Chemical structure	Antagonized by	Reference
(–)-EA	11.2 nM (TRPC4) 7.6 nM (TRPC5) 10 nM (TRPC1/ TRPC4)	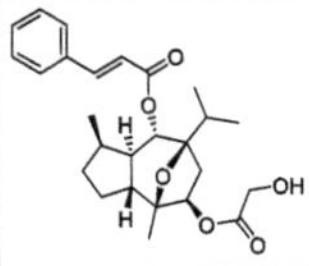	Pico145	(Akbulut et al. 2015b)
Tonantzitlolone	123 nM (TRPC4) 83 nM (TRPC5) 140 nM (TRPC4-TRPC1, concatemers) 61 nM (TRPC5-TRPC1, concatemers)	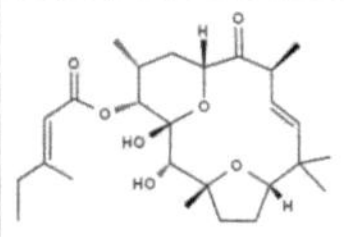	Pico145	(Rubaiy et al. 2018b)
BTD	1.4 μM (TRPC5)	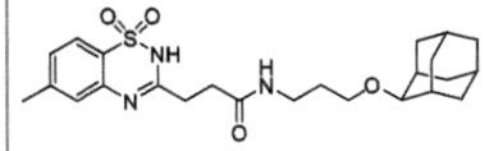	Clemizole	(Beckmann et al. 2017)
Riluzole	9.2 μM (TRPC5)	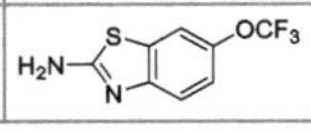	2-APB	(Richter et al. 2014a)

Table 2 Selected TRPC3/6/7 channels agonists

Name	EC_{50} (channel)	Chemical structure	Antagonized by	Reference
GSK1702934A	TRPC3 (80 nM) TRPC6 (440 nM)	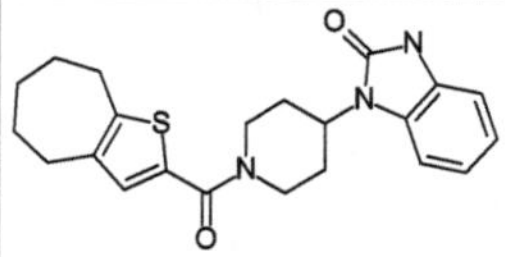	GSK417651A	(Xu et al. 2013)
Pyrazolopyrimidines 4n	TRPC3 (19 nM) TRPC6 (1.4 μM) TRPC7 (90 nM)	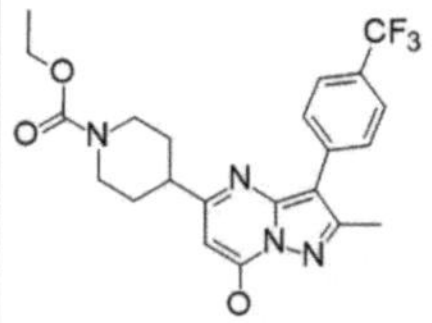	Extracellular Ca^{2+} 2-APB	(Qu et al. 2017)

Table 3 Selected TRPC1/4/5 channels antagonists

Name	IC_{50} (channel)	Chemical structure	Channel activator	Reference
Pico145	49 pM (TRPC1/TRPC4, endogenous) 33pM (TRPC4-TRPC1, concatemers) 199 pM (TPRC5-TRPC1, concatemers) 63 pM (TRPC4) 1.3 nM (TRPC5)	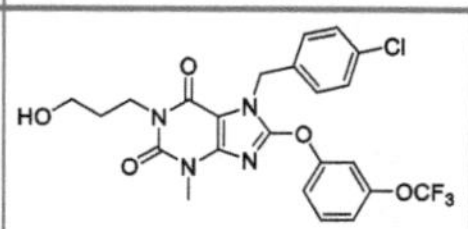	EA	(Rubaiy et al. 2017a)
Clemizole hydrochloride	6.4 μM (TRPC4) 1.1 μM (TRPC5) 9.1 μM (TRPC3) 11.3 μM (TRPC6) 26.5 μM (TRPC7)	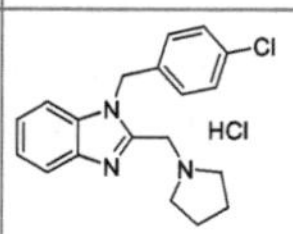	TRPC4 – CCh TRPC5 – Riluzole TRPC3, 6, and 7 – Amix: ATP, CCh, and thrombin	(Richter et al. 2014b)
ML204	0.99 μM (TRPC4) 9.2 μM (TRPC5) 18.4 μM (TRPC6) 58 μM (TRPC4/TRPC1)	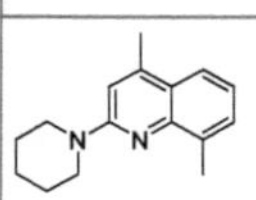	TRPC4 – CCh + M2R/μOR TRPC5 – DAMGO TRPC4/TRPC1 – EA TRPC6 – Ach	(Miller et al. 2011; Zhou et al. 2017)
AM-12	0.28 μM (TRPC5)	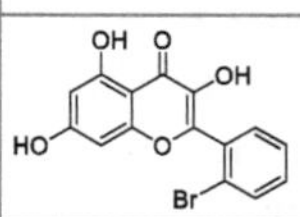	Gd^{3+}	(Naylor et al. 2016)

Table 4 Selected TRPC3/6/7 channels antagonists

Name	IC_{50} (channel)	Chemical structure	Channel activator	Reference
Pyr3	0.8 μM (TRPC3)	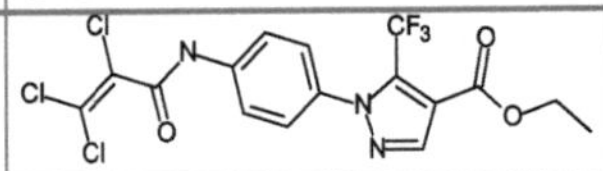	OAG	(Kiyonaka et al. 2009)
GSK2833503A	21 nM (TRPC3) 3 nM (TRPC6)	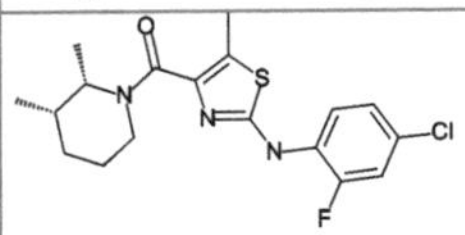	CCh	(Seo et al. 2014)
SAR7334	282 nM (TRPC3) 7.9–9.5 nM (TRPC6) 226 nM (TRPC7)	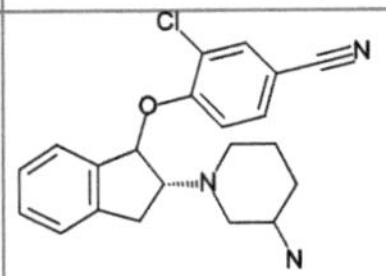	OAG	(Maier et al. 2015)
DS88790512	11 nM (TRPC6)	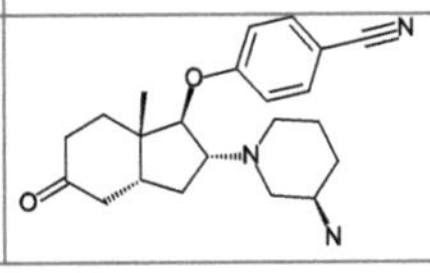	OAG	(Motoyama et al. 2018)

concatemers at 61 nM, and TRPC4-TRPC1 concatemers at 140 nM (Table 1). Renal cell carcinoma (RCC) is among the top ten most common forms of cancer and is the most common malignancy of the kidney. Clear cell renal carcinoma (cc-RCC), the most common type of RCC, is one of the most refractory cancers with an incidence that is on the rise. Screening of plant extracts in search of new anti-cancer agents resulted in the discovery of EA, a guaiane sesquiterpene with potent cytotoxicity against renal cancer cells and a small subset of other cancer cells. Though a few cellular targets have been identified for EA, it is still not clear what mechanisms account for the cytotoxicity of EA in RCC, which occurs at concentrations well below those used to engage the targets previously identified (Beutler et al. 2019). Unlike any prior study, the current study used a systems biology approach to explore the mechanism(s) of action of EA. Metabolomics analyses indicated that EA profoundly altered lipid metabolism by 24 h in cc-RCC cell lines and generated significant levels of ceramides that were highly toxic to these cells. Microarray analyses determined that EA induced ER stress signaling and an acute inflammatory response, which was confirmed by quantitative PCR and Western blot analyses. Additionally, fluorescence confocal microscopy revealed that EA at 25 nM disrupted the morphology of the ER, confirming the deleterious effect of EA on the ER. Collectively, our findings suggest that cc-RCC is highly sensitive to disruptions in lipid metabolism and ER stress and that these vulnerabilities can be targeted for the treatment of cc-RCC and possibly other lipid storing cancers. Furthermore, our results suggest that ceramides may be a mediator of some of the actions of EA. Lastly, the acute inflammatory response induced by EA may mediate anti-tumor immunity (Batova et al. 2017). EA induces an acute inflammatory response and reveals lipid metabolism and ER stress as targetable vulnerabilities in renal cell carcinoma, as revealed through using metabolomics analyses that show EA significantly changes metabolism by 24 h in clear cell renal carcinoma (cc-RCC) cell lines and produced significant levels of ceramides, which are highly toxic to these cell lines (Batova et al. 2017). In addition, the toxicity of EA in mice was on-target, and blocked by Pico145, which is an antagonist of TRPC1/4/5 channels (Cheung et al. 2018). Notably, an analog of EA, A54, was identified as a competitive antagonist which strongly blocked EA-mediated activation of TRPC4/TRPC1 or TRPC5 and weakly inhibited TRPC4 activation. It is suggested that A54 could possibly be used an antidote to EA poisoning (Rubaiy et al. 2018a).

8.2.2 Tonantzitlolone (TZL)

TZL, a diterpene ester, has similar effect to EA, and it is a natural product that has cytotoxic effect toward certain types of cancer cell such as renal cell carcinoma cells (Rubaiy et al. 2018b). TZL has been shown to be a novel potent agonist for TRPC1/4/5 channels. The results from the study revealed that TZL activated channels at EC_{50} levels of 123 nM (TRPC4), 83 nM (TRPC5), 140 nM (TRPC4-TRPC1), and 61 nM (TRPC5-TRPC1) (Table 1). Furthermore, the study showed that TZL effects could be antagonized by TRPC1/4/5 inhibitor Pico145, while TZL did not activate TRPC3 channels in HEK293 cells (Rubaiy et al. 2018b). TZL could be beneficial for evaluating the functionality of this type of ion channel and determining how TRPC1/4/5 agonists achieve selective cytotoxicity against specific cancer cell types.

8.2.3 Benzothiadiazine Derivative (BTD)

BTD was discovered by a screening of a ChemBioNet compound library, and it is a novel TRPC5 agonist with an EC_{50} of 1.4 μM in Ca^{2+} assays and EC_{50} of 1.3 μM (Table 1) in electrophysiological whole-cell patch-clamp recordings (Beckmann et al. 2017). BTD is selective for TRPC5; however, it is unable to activate other homomeric TRPC channels such as TRPC3, TRPC6, and TRPC7. However, BTD can trigger TRPC1:C5 and TRPC4:C5 channels, but not TRPC4:C4 and TRPC1:C4 channels. The effects of BTD are long-lasting and reversible, and it is sensitive to the TRPC5 antagonist clemizole (Beckmann et al. 2017).

8.2.4 Riluzole

Riluzole, a glutamate release inhibitor, is the only globally licensed drug treatment for the progressive neurodegenerative disease, amyotrophic lateral sclerosis (ALS), and it is also clinically administered to treat psychiatric disorders. Riluzole effects are modest, prolonging ALS patients lives by 2–3 months (Blyufer et al. 2021). This inhibitor is a derivative of benzothiazole and has also been identified as a TRPC5 channel activator with an EC_{50} of 9.2 μM (Richter et al. 2014a), and it's effects are antagonized by 2-APB (Table 1). Riluzole's mechanism of action as a TRPC5 activator is independent of G protein signaling and PLC activity, and its activation is reversible upon washout. Excised inside-out patch recordings have shown that riluzole has a relatively direct effect on homomeric TRPC5 channels (Richter et al. 2014a). Interestingly, riluzole does not activate any other TRPC subfamily, suggesting that it could be a useful pharmacological tool to detect drug targets for TRPC channel modulation (Rubaiy 2019).

8.3 TRPC3/6/7 Channel Activators

8.3.1 1-Oleoyl-2-Acetyl-sn-Glycerol (OAG)

OAG is described as the first known modulator of TRPC activity. This cell permeable DAG analog was first identified as a direct activator of TRPC3 and TRPC6 channels, acting in a PKC-independent manner (Hofmann et al. 1999). There are some differences between OAG-activated and receptor-activated TRPC currents, such as the magnitude of OAG-triggered currents tends to be smaller, and the activation kinetics are slower than receptor-induced TRPC currents. Also, OAG is commonly applied extracellularly, while the endogenously produced DAGs are seen in the inner leaflet of plasma membrane. This could be a contributing factor to the differences (Wang et al. 2020a).

8.3.2 GSK1702934A

GSK1702934A, from GlaxoSmithKline, selectively activates TRPC3/6, and studies with murine Langendorff hearts showed that the activator enhanced contractility and evoked arrhythmia (Xu et al. 2013; Doleschal et al. 2015). This activator is a small 1,3-dihydro-2H-benzo[d]midazole-2-one-based potent agonist. GSK1702934A is most potent toward TRPC3 with an EC_{50} of 0.08 mM in comparison to TRPC6 with an EC_{50} of 0.44 mM (Table 2). Electrophysical whole-cell patch-clamp recordings have shown that this potent activator has the ability to trigger TRPC3 and TRPC6 currents in HEK293 cells (de la Cruz et al. 2017). The effects of activating TRPC6 by GSK1702934A increased the systolic and diastolic blood pressure variability, and it was dose-dependently exacerbated after the activation. However, the TRPC6 activation induced by this agonist were dose-dependently diminished after TRPC6 is blocked by SAR7334 (Wang et al. 2020b). This is while activation of TRPC3 by GSK1702934A plays a vital role in cardiac contractility and excitability (Doleschal et al. 2015).

8.3.3 Pyrazolopyrimidines

Studies on structure-activity relationships have revealed pyrazolopyrimidines activate TRPC3/6/7 but not TRPC4, TRPC5, or several other TRP channels (Qu et al. 2017; Bon et al. 2022). Among pyrazolopyrimidine compound, 4n has remarkable properties, showing more than sevenfold increase in potency on TRPC3 channels compared to other compounds and more than 70-fold increased on TRPC6 (Qu et al. 2017). The highly potent pyrazolopyrimidine 4n activates TRPC6 at an EC_{50} of 1.4 nM, TPRC7 at 90 nM, and TRPC3 at 19 nM (Table 2). This potent compound should be investigated further to study TRPC channels' roles in the physiology and pathogenesis of diseases including glomerulosclerosis and cancer (Qu et al. 2017).

8.3.4 AM-0883

A novel TRPC6 agonist, AM-0883, was discovered from a small molecule in vitro high throughput screening at Amgen. AM-0883 activates TRPC6 channels at a EC_{50} value of 46 nM. This agonist induced roughly the same maximum activity level as OAG; however, it was about 70 times more potent (Bai et al. 2020). AM-0883 contains a chloro-indole, a piperidine, and a benzodioxine, and it binds to TRPC6 at transmembrane domain. Cryo-EM showed that AM-0883 occupies at a site between S6 of one subunit and the pore helix of the adjoining subunit revealed by (Bai et al. 2020).

9 TRPC Channel Blockers

9.1 TRPC1/4/5 Blockers

9.1.1 Pico145

Pico145 is a remarkably selective and subtype-specific small-molecule antagonist of TRPC1/4/5 channels, and its potency varied from 9 to 1300 pM (Table 3), dependent upon the TRPC1/4/5 subtype and activation mechanism (Yu et al. 2022). Notably, Pico145 was selective with respect to TRPC3, TRPC6, TRPV1, TRPV4, TRPA1, TRPM2, TRPM8, and the ORAI calcium channels (Rubaiy et al. 2017a; Rubaiy et al. 2017b). Given its characteristics, pico145 is an excellent pharmacological tool for further exploration into TRPC1/4/5 channels.

9.1.2 Clemizole Hydrochloride

Clemizole hydrochloride (Table 3) is a histamine H1-receptor agonist which also potently inhibits homomeric TRPC5 channels (Richter et al. 2014b). Homomeric TRPC5 channels are inhibited by clemizole at an IC_{50} value of 1.1 μM. TRPC5 is sixfold more sensitive to this antagonist compared to TRPC4 which inhibits at 6.4 μM, TRPC3 at 9.1 μM, TRPC6 at 11.3 μM, and TRPC7 at 26.5 μM. It has also been shown that clemizole blocks homomeric TRPC5 currents regardless of activation mode such as GPCRs stimulation or by riluzole, a direct activator. Cytosolic components are not vital for its inhibitory effects on TRPC5 channels (Richter et al. 2014b).

9.1.3 GFB-8438

Goldfinch Bio has reported that GFB-8438 is a potent inhibitor of TRPC4:C4 and TRPC5:C5 channels in rat and humans. GFB-8483 has beneficial physicochemical properties and good selectivity against TRPC3/6/7; however, it has yet to be determined how effective GFB-8483 is against TRPC1/4/5 (Yu et al. 2019; Vinayagam et al. 2020). GFB-8483 is most potent toward TRPC5 with a EC_{50} of 0.18–0.28 μM and then TRPC4 at 0.29 μM and less potent toward hERG at 8.7 μM (Vinayagam et al. 2020). Further studies need to be performed to determine any data for heteromeric channels. However, it has been reported that GFB-8483 seems to have efficacy in focal segmental glomerulosclerosis, scarring on kidneys which leads to permanent kidney damage and even kidney failure (Yu et al. 2019).

9.1.4 ML204

Small molecule ML204 inhibits homomeric TRPC4 and homomeric TRPC5 at low micromolar levels, while having high selectivity toward other channels, receptors, and transporters. This antagonist was discovered through a fluorescent screen evaluating compounds from the Molecular Libraries Small Molecule Repository library (Miller et al. 2011). ML204 inhibits TRPC4 at an IC_{50} of 0.99 μM, TRPC5 at 9.2 μM, and TRPC6 at 18.4 μM (Table 3). Heteromeric TRPC4/TRPC1 (concatemers) is less sensitive to ML204 inhibiting at an IC_{50} of 58 μM (Rubaiy 2019).

Studies on the pharmacological effects of ML204 have shown homomeric TRPC4 currents induced by either μ receptor stimulation or intracellular dialysis of GTPγS were inhibited by ML204, and there was no block of the signal transduction pathways, meaning that homomeric TRPC4 inhibition is independent of GPCR activation (Miller et al. 2011).

9.1.5 Galangin and AM-12

Through screening of various natural products, flavonol galangin, a member of the ginger family, was found to inhibit human overexpressed homomeric TRPC5 channels in HEK293 cells (Naylor et al. 2016). Galagin inhibited lathanide-evoked Ca^{2+} entry at an IC_{50} of 0.45 μM. Further structure-activity relationship studies of 48 natural and synthetic flavanols identified a more potent analogue, AM-12 (Table 3), which inhibited homomeric TRPC5 channels at an IC_{50} of 0.28 μM (Rubaiy 2019). This is while studies showed that both galangin and AM-12 were effective in excised electrophysiological outside-out membrane patches, suggesting these two compounds might have a direct effect on homomeric TRPC5 channels. It is yet not determined if galangin and AM-12 directly bind to the homomeric TRPC5 channels (Naylor et al. 2016).

9.2 TRPC3/6/7 Channel Inhibitors

9.2.1 Pyr3

In 2009, Kiyonaka et al. reported the pyrazole compound Pyr3 (Table 4) as the first submicromolar (IC_{50} = 0,8 μM) antagonist of TRPC3. In addition, Pyr3 was able to inhibit the activation of nuclear factor of activated T cells, which is a Ca^{2+}-dependent transcription factor, likewise hypertrophic growth in rat neonatal cardiomyocytes, suggesting that Pyr3 may help treat TRPC3-related diseases such cardiac hypertrophy. Pyr3 showed selectivity toward TRPC5 channels and might bind directly to TRPC3 based on photoaffinity labeling revealed that Pyr3 directly bind to TRPC3 (Kiyonaka et al. 2009).

9.2.2 GSK2833503A

In 2014, Seo et al. revealed that GSK2833503A (Table 4) is a highly selective TRPC3 (IC_{50} = 21–100 nM) and TRPC6 (IC_{50} = 3–16 nM) channels blocker (Seo et al. 2014). Also, this compound was able to block pathological hypertrophic signaling in a variety of cell types, including adult cardiac myocytes. The enantiomer of GSK2833503 is tenfold less potent against TRPC3 and 100-fold less potent against TRPC6, enabling it to be used possibly as a control compound with well-matched physicochemical properties. The chemical structure of GSK2833503A is shown in Table 4.

9.2.3 SAR7334

It was demonstrated that the indane derivative, SAR7334, inhibits TRPC3/6/7 channels at nanomolar concentrations (TRPC3: IC_{50} = 282 nM), TRPC6: IC_{50} = 9.5 nM), TRPC7: IC_{50} = 226 nM) while preserving selectivity for TRPC4 and TRPC5 channels. Furthermore, pharmacokinetic data for SAR7334 indicated that this compound was suitable for chronic oral administration (orally bioavailable); also it inhibits acute pulmonary vasoconstriction in mice (Maier et al. 2015). The chemical structure for SAR7334 is illustrated in Table 4.

9.2.4 DS88790512

DS88790512 (Table 4) is a bicyclo[4.3.0]nonane derivative that blocked TRPC6 with an $IC_{50} = 11$ nM, and markedly, it demonstrated superb selectivity over hERG ($IC_{50} > 100$ μM) and hNaV1.5 ($IC_{50} > 300$ μM) channels (Motoyama et al. 2018). In addition, a pharmacokinetic parameters analysis of DS88790512 in mice demonstrated that this compound was determined to be orally available with a fraction equal to 50%.

9.2.5 BTDM

BTDM was utilized in cryo-EM studies to identify the small-molecule binding site on the TRPC6 channels. Pharmacological studies revealed that BTDM is a potent blocker of TRPC3 ($IC_{50} = 11$ nM) and TRPC6 ($IC_{50} = 10$ nM) (Tang et al. 2018).

9.2.6 SH045

Häfner et al. by employing high-throughput Ca^{2+} FLIPR screening assays and electrophysiological patch-clamp recordings identified a derivative of the natural product (+)-larixol as a potent, high affinity, selective agent against TRPC6 (Häfner et al. 2018). The OAG-Ca^{2+} evoked entry through TRPC3/6/7 channels was blocked by SH045 as follows: TRPC3 (IC_{50} = 440–634 nM), TRPC6 (IC_{50} = 5.2–5.8 nM), and TRPC7 (IC_{50} = 18–22 nM). Remarkably, SH045 is selective with regard to TRPC4, TRPC5, and other TRP channels and also resulted in a reduction in the development of edema in explanted mouse lungs.

9.2.7 BI749327

An orally bioavailable, potent, selective TRPC6 antagonist, BI 749327, was described by Lin et al. with $IC_{50} = 13$ nM against mouse TRPC6 (Lin et al. 2019). BI 749327 exhibited 85- and 42-fold selectivity over the channels that are most closely related to TRPC6, namely TRPC3 and TRPC7. This TRPC6 blocker improved heart function and reduces chamber dilation and fibrosis in hemodynamic stress. Moreover, BI 749327 showed less sensitivity to other channels (TRPC5, TRPM8, TRPV1, TRPA1, Nav1.5, and hERG) (Lin et al. 2019).

9.2.8 AM-1473

AM-1473, which currently is the most potent TRPC6 inhibitor reported with $IC_{50} = 0.13$ nM (activation by OAG), is a chemically related analogue to SAR7334. Cryo-EM was used in combination with AM-1473 for the purpose of identifying an indane binding site within the TRPC6:C6 channel (Bai et al. 2020).

10 Summary and Future Directions

The year 1969 marked the discovery of classical or canonical TRPC proteins, and in 1995 the first TRPC channel (TRPC1) was cloned which expanded the scope of scientific endeavors (Cosens and Manning 1969; Hardie and Minke 1992; Wes et al. 1995; Zhu et al. 1995). The subfamily of TRPC are nonselective cationic channels with high calcium permeability that play important roles in a variety of physiological processes. In a nutshell, the TRPC channels are thought of as the founding members of the mammalian TRP channel superfamily. Additionally, these channels are the homologs that are the most similar to the *Drosophila* TRP channel, which is involved in photo transduction. TRPC1 and TRPC3-7 are the six TRPC channels found in humans, and these channels play essential roles in cellular signaling, regulation, and function.

To identify the physiological and pathological role of TRPC channels, various genetic approaches such as knockout or transgenic overexpression mouse models and pharmacological tools were utilized (Nilius and Szallasi 2014; Minard et al. 2018; Rubaiy 2019; Bon et al. 2022). For almost three decades, research in TRPC channels has revealed that these proteins complexes are involved and play essential role in human diseases and disorders; therefore, they serve as potential drug target for several human diseases and disorders (cardiovascular, nervous system, cancer, and others) (Wang et al. 2017; Chen et al. 2020; Wen et al. 2020). Recent advances in cryo-EM techniques have resulted in various high-resolution TRP channel structures. Likewise, in the field of TRPC, cryo-EM techniques has been used to resolve the structures of various members of these ion channel family, at resolutions varying from 2.8 Å to 5.8 Å (Guo and Chen 2019; Kim et al. 2019; Li et al. 2019).

In conclusion, in the field of drug discovery and development, targeting TRPCs has been limited due to a lack of understanding of its molecular activation processes, structural features, and physiological circumstances, and a lack of subtype-specific compounds. However, as a final point, by employing potent, selective, and subtype-specific pharmacological tools, it is now possible to explore illnesses that are mediated by these channels, which potentially will lead to discovering and characterizing novel drug candidates to generate innovative therapeutic therapies.

References

Akbulut Y, Gaunt HJ, Muraki K, Ludlow MJ, Amer MS, Bruns A, et al. (−)-Englerin a is a potent and selective activator of TRPC4 and TRPC5 calcium channels. Angew Chem Int Ed. 2015a;54(12):3787–91.

Akbulut Y, Gaunt HJ, Muraki K, Ludlow MJ, Amer MS, Bruns A, et al. (−)-Englerin a is a potent and selective activator of TRPC4 and TRPC5 calcium channels. Angew Chem. 2015b;127(12):3858–62.

Ampem PT, Smedlund K, Vazquez G. Pharmacological evidence for a role of the transient receptor potential canonical 3 (TRPC3) channel in endoplasmic reticulum stress-induced apoptosis of human coronary artery endothelial cells. Vasc Pharmacol. 2016;76:42–52.

Autzen HE, Myasnikov AG, Campbell MG, Asarnow D, Julius D, Cheng Y. Structure of the human TRPM4 ion channel in a lipid nanodisc. Science. 2018;359(6372):228–32.

Azumaya CM, Sierra-Valdez F, Cordero-Morales JF, Nakagawa T. Cryo-EM structure of the cytoplasmic domain of murine transient receptor potential cation channel subfamily C member 6 (TRPC6). J Biol Chem. 2018;293(26):10381–91.

Bae YM, Kim A, Lee YJ, Lim W, Noh YH, Kim EJ, et al. Enhancement of receptor-operated cation current and TRPC6 expression in arterial smooth muscle cells of deoxycorticosterone acetate-salt hypertensive rats. J Hypertens. 2007;25(4):809–17.

Bai Y, Yu X, Chen H, Horne D, White R, Wu X, et al. Structural basis for pharmacological modulation of the TRPC6 channel. Csanády L, Aldrich RW, Moiseenkova-Bell VY, Csanády L, editors. eLife. 2020;9:e53311.

Batova A, Altomare D, Creek KE, Naviaux RK, Wang L, Li K, et al. Englerin A induces an acute inflammatory response and reveals lipid metabolism and ER stress as targetable vulnerabilities in renal cell carcinoma. PLoS One. 2017;12(3):e0172632.

Beckmann H, Richter J, Hill K, Urban N, Lemoine H, Schaefer M. A benzothiadiazine derivative and methylprednisolone are novel and selective activators of transient receptor potential canonical 5 (TRPC5) channels. Cell Calcium. 2017;66:10–8.

Beech DJ. Integration of transient receptor potential canonical channels with lipids. Acta Physiol Oxf Engl. 2012;204(2):227–37.

Beech DJ. Characteristics of transient receptor potential canonical calcium-permeable channels and their relevance to vascular physiology and disease. Circ J Off J Jpn Circ Soc. 2013;77(3):570–9.

Bernal L, Sotelo-Hitschfeld P, König C, Sinica V, Wyatt A, Winter Z, et al. Odontoblast TRPC5 channels signal cold pain in teeth. Sci Adv. 2021;7(13):eabf5567.

Beutler JA, Echavarren A, Chain W, Beech DJ, Wu Z, Suppo JS, Bravo F, Rubaiy HN (2019). Englerin derivatives for treatment of cancer. (US. Patent No. WO2019010298A1). U.S. Patent Trademark Office. https://patents.google.com/patent/WO2019010298A1/en

Blyufer A, Lhamo S, Tam C, Tariq I, Thavornwatanayong T, Mahajan SS. Riluzole: A neuroprotective drug with potential as a novel anti-cancer agent (review). Int J Oncol. 2021;59(5):95.

Bomben VC, Sontheimer H. Disruption of transient receptor potential canonical channel 1 causes incomplete cytokinesis and slows the growth of human malignant gliomas. Glia. 2010;58(10):1145–56.

Bon RS, Wright DJ, Beech DJ, Sukumar P. Pharmacology of TRPC channels and its potential in cardiovascular and metabolic medicine. Annu Rev Pharmacol Toxicol. 2022;62:427–46.

Bush EW, Hood DB, Papst PJ, Chapo JA, Minobe W, Bristow MR, et al. Canonical transient receptor potential channels promote cardiomyocyte hypertrophy through activation of calcineurin signaling. J Biol Chem. 2006;281(44):33487–96.

Cai R, Ding X, Zhou K, Shi Y, Ge R, Ren G, et al. Blockade of TRPC6 channels induced G2/M phase arrest and suppressed growth in human gastric cancer cells. Int J Cancer. 2009;125(10):2281–7.

Camacho Londoño JE, Tian Q, Hammer K, Schröder L, Camacho Londoño J, Reil JC, et al. A background Ca2+ entry pathway mediated by TRPC1/TRPC4 is critical for development of pathological cardiac remodelling. Eur Heart J. 2015;36(33):2257–66.

Chen X, Sooch G, Demaree IS, White FA, Obukhov AG. Transient receptor potential canonical (TRPC) channels: then and now. Cell. 2020;9(9):1983.

Cheung SY, Henrot M, Al-Saad M, Baumann M, Muller H, Unger A, et al. TRPC4/TRPC5 channels mediate adverse reaction to the cancer cell cytotoxic agent (-)-Englerin A. Oncotarget. 2018;9(51):29634–43.

Chrétien C, Fenech C, Liénard F, Grall S, Chevalier C, Chaudy S, et al. Transient receptor potential canonical 3 (TRPC3) channels are required for hypothalamic glucose detection and energy homeostasis. Diabetes. 2017;66(2):314–24.

Clapham DE. TRP channels as cellular sensors. Nature. 2003;426(6966):517–24.

Cosens DJ, Manning A. Abnormal electroretinogram from a Drosophila mutant. Nature. 1969;224(5216):285–7.

de la Cruz GG, Svobodova B, Lichtenegger M, Tiapko O, Groschner K, Glasnov T. Intensified microwave-assisted N-acylation procedure – synthesis and activity evaluation of TRPC3 channel agonists with a 1,3-dihydro-2H-benzo[d]imidazol-2-one core. Synlett Acc Rapid Commun Synth Org Chem. 2017;28(6):695–700.

De March Z, Giampà C, Patassini S, Bernardi G, Fusco FR. Cellular localization of TRPC5 in the substantia nigra of rat. Neurosci Lett. 2006;402(1):35–9.

Dhennin-Duthille I, Gautier M, Faouzi M, Guilbert A, Brevet M, Vaudry D, et al. High expression of transient receptor potential channels in human breast cancer epithelial cells and tissues: correlation with pathological parameters. Cell Physiol Biochem Int J Exp Cell Physiol Biochem Pharmacol. 2011;28(5):813–22.

Dietrich A, Mederos y Schnitzler M, Gollasch M, Gross V, Storch U, Dubrovska G, et al. Increased vascular smooth muscle contractility in TRPC6−/− mice. Mol Cell Biol. 2005;25(16):6980–9.

Diver MM, Lin King JV, Julius D, Cheng Y. Sensory TRP channels in three dimensions. Annu Rev Biochem. 2022;91:629–49.

Doleschal B, Primessnig U, Wölkart G, Wolf S, Schernthaner M, Lichtenegger M, et al. TRPC3 contributes to regulation of cardiac contractility and arrhythmogenesis by dynamic interaction with NCX1. Cardiovasc Res. 2015;106(1):163–73.

Dong H, Shim KN, Li JMJ, Estrema C, Ornelas TA, Nguyen F, et al. Molecular mechanisms underlying Ca2+-mediated motility of human pancreatic duct cells. Am J Physiol Cell Physiol. 2010;299(6):C1493–503.

Dryer SE, Roshanravan H, Kim EY. TRPC channels: regulation, dysregulation and contributions to chronic kidney disease. Biochim Biophys Acta (BBA) Mol Basis Dis. 2019;1865(6):1041–66.

Du SL, Jia ZQ, Zhong JC, Wang LF. TRPC5 in cardiovascular diseases. Rev Cardiovasc Med. 2021;22(1):127–35.

Duan J, Li J, Zeng B, Chen GL, Peng X, Zhang Y, et al. Structure of the mouse TRPC4 ion channel. Nat Commun. 2018a;9(1):1–10.

Duan J, Li Z, Li J, Hulse RE, Santa-Cruz A, Valinsky WC, et al. Structure of the mammalian TRPM7, a magnesium channel required during embryonic development. Proc Natl Acad Sci. 2018b;115(35):E8201–10.

Duan J, Li J, Zeng B, Chen GL, Peng X, Zhang Y, et al. Structure of the mouse TRPC4 ion channel. Nat Commun. 2018c;9(1):3102.

Duan J, Li J, Chen GL, Ge Y, Liu J, Xie K, et al. Cryo-EM structure of TRPC5 at 2.8-Å resolution reveals unique and conserved structural elements essential for channel function. Sci Adv. 2019;5(7):eaaw7935.

Elzamzamy OM, Penner R, Hazlehurst LA. The role of TRPC1 in modulating cancer progression. Cell. 2020;9(2):388.

Fan C, Choi W, Sun W, Du J, Lü W. Structure of the human lipid-gated cation channel TRPC3. elife. 2018;7:e36852.

Fels B, Bulk E, Pethő Z, Schwab A. The role of TRP channels in the metastatic cascade. Pharmaceuticals. 2018;11:48.

Flemming PK, Dedman AM, Xu SZ, Li J, Zeng F, Naylor J, et al. Sensing of lysophospholipids by TRPC5 calcium channel. J Biol Chem. 2006;281(8):4977–82.

Freichel M, Berlin M, Schürger A, Mathar I, Bacmeister L, Medert R, et al. TRP channels in the heart. In: Emir TLR, editor. Neurobiology of TRP Channels [Internet]. Boca Raton: CRC Press/Taylor & Francis; 2017 [cited 2022 Oct 15]. (Frontiers in Neuroscience). Available from: http://www.ncbi.nlm.nih.gov/books/NBK476106/.

Gao Y, Yao T, Deng Z, Sohn JW, Sun J, Huang Y, et al. TrpC5 mediates acute leptin and serotonin effects via pomc neurons. Cell Rep. 2017;18(3):583–92.

Gees M, Owsianik G, Nilius B, Voets T. TRP Channels. In: Comprehensive physiology [Internet]. John Wiley & Sons, Ltd; 2012 [cited 2022 Oct 7]. p. 563–608. Available from: https://onlinelibrary.wiley.com/doi/abs/10.1002/cphy.c110026.

Graham S, Yuan JP, Ma R. Canonical transient receptor potential (TRPC) channels in diabetes. Exp Biol Med Maywood. 2012;237(2):111–8.

Guilbert A, Dhennin-Duthille I, Hiani YE, Haren N, Khorsi H, Sevestre H, et al. Expression of TRPC6 channels in human epithelial breast cancer cells. BMC Cancer. 2008;8:125.

Guo W, Chen L. Recent progress in structural studies on canonical TRP ion channels. Cell Calcium. 2019;83:102075.

Häfner S, Burg F, Kannler M, Urban N, Mayer P, Dietrich A, et al. A (+)-Larixol congener with high affinity and subtype selectivity toward TRPC6. ChemMedChem. 2018;13(10):1028–35.

Hardie RC, Minke B. The trp gene is essential for a light-activated Ca2+ channel in Drosophila photoreceptors. Neuron. 1992;8(4):643–51.

He X, Li S, Liu B, Susperreguy S, Formoso K, Yao J, et al. Major contribution of the 3/6/7 class of TRPC channels to myocardial ischemia/reperfusion and cellular hypoxia/reoxygenation injuries. Proc Natl Acad Sci U S A. 2017;114(23):E4582–91.

Hille B, Dickson EJ, Kruse M, Vivas O, Suh BC. Phosphoinositides regulate ion channels. Biochim Biophys Acta. 2015;1851(6):844–56.

Hofmann T, Obukhov AG, Schaefer M, Harteneck C, Gudermann T, Schultz G. Direct activation of human TRPC6 and TRPC3 channels by diacylglycerol. Nature. 1999;397(6716):259–63.

Hong C, Choi SH, Kwak M, Jeong B, Ko J, Park HJ, et al. TRPC5 channel instability induced by depalmitoylation protects striatal neurons against oxidative stress in Huntington's disease. Biochim Biophys Acta BBA Mol Cell Res. 2020;1867(2):118620.

Itsuki K, Imai Y, Hase H, Okamura Y, Inoue R, Mori MX. PLC-mediated PI(4,5)P2 hydrolysis regulates activation and inactivation of TRPC6/7 channels. J Gen Physiol. 2014;143(2):183–201.

James W, Putney J. Multiple mechanisms of TRPC activation [Internet]. TRP ion channel function in sensory transduction and cellular signaling cascades. CRC Press/Taylor & Francis; 2007 [cited 2022 Oct 11]. Available from: https://www.ncbi.nlm.nih.gov/books/NBK5263/.

Jeon JP, Hong C, Park EJ, Jeon JH, Cho NH, Kim IG, et al. Selective Gαi subunits as novel direct activators of transient receptor potential canonical (TRPC)4 and TRPC5 channels. J Biol Chem. 2012;287(21):17029–39.

Jiang HN, Zeng B, Zhang Y, Daskoulidou N, Fan H, Qu JM, et al. Involvement of TRPC channels in lung cancer cell differentiation and the correlation analysis in human non-small cell lung cancer. PLoS One. 2013;8(6):e67637.

Jin P, Bulkley D, Guo Y, Zhang W, Guo Z, Huynh W, et al. Electron cryo-microscopy structure of the mechanotransduction channel NOMPC. Nature. 2017;547(7661):118–22.

Jung S, Mühle A, Schaefer M, Strotmann R, Schultz G, Plant TD. Lanthanides potentiate TRPC5 currents by an action at extracellular sites close to the pore mouth. J Biol Chem. 2003;278(6):3562–71.

Kaczmarek JS, Riccio A, Clapham DE. Calpain cleaves and activates the TRPC5 channel to participate in semaphorin 3A-induced neuronal growth cone collapse. Proc Natl Acad Sci. 2012;109(20):7888–92.

Ke C, Long S. Dysregulated transient receptor potential channel 1 expression and its correlation with clinical features and survival profile in surgical non-small-cell lung cancer patients. J Clin Lab Anal. 2022;36(3):e24229.

Kim YJ, Kang TC. The role of TRPC6 in seizure susceptibility and seizure-related neuronal damage in the rat dentate gyrus. Neuroscience. 2015;307:215–30.

Kim H, Jeon JP, Hong C, Kim J, Myeong J, Jeon JH, et al. An essential role of PI(4,5)P2 for maintaining the activity of the transient receptor potential canonical (TRPC)4β. Pflüg Arch Eur J Physiol. 2013;465(7):1011–21.

Kim J, Ko J, Hong C, So I. Structure–function relationship and physiological roles of transient receptor potential canonical (TRPC) 4 and 5 channels. Cell. 2019;9(1):73.

Kitajima N, Numaga-Tomita T, Watanabe M, Kuroda T, Nishimura A, Miyano K, et al. TRPC3 positively regulates reactive oxygen species driving maladaptive cardiac remodeling. Sci Rep. 2016;6:37001.

Kiyonaka S, Kato K, Nishida M, Mio K, Numaga T, Sawaguchi Y, et al. Selective and direct inhibition of TRPC3 channels underlies biological activities of a pyrazole compound. Proc Natl Acad Sci. 2009;106(13):5400–5.

Krout D, Schaar A, Sun Y, Sukumaran P, Roemmich JN, Singh BB, et al. The TRPC1 Ca2+-permeable channel inhibits exercise-induced protection against high-fat diet-induced obesity and type II diabetes. J Biol Chem. 2017;292(50):20799–807.

Kumar B, Dreja K, Shah SS, Cheong A, Xu SZ, Sukumar P, et al. Upregulated TRPC1 channel in vascular injury in vivo and its role in human neointimal hyperplasia. Circ Res. 2006;98(4):557–63.

Kuwahara K, Wang Y, McAnally J, Richardson JA, Bassel-Duby R, Hill JA, et al. TRPC6 fulfills a calcineurin signaling circuit during pathologic cardiac remodeling. J Clin Invest. 2006;116(12):3114–26.

Li J, Zhang X, Song X, Liu R, Zhang J, Li Z. The structure of TRPC ion channels. Cell Calcium. 2019;80:25–8.

Lin BL, Matera D, Doerner JF, Zheng N, del Camino D, Mishra S, et al. In vivo selective inhibition of TRPC6 by antagonist BI 749327 ameliorates fibrosis and dysfunction in cardiac and renal disease. Proc Natl Acad Sci U S A. 2019;116(20):10156–61.

Liu B, He X, Li S, Xu B, Birnbaumer L, Liao Y. Deletion of diacylglycerol-responsive TRPC genes attenuates diabetic nephropathy by inhibiting activation of the TGFβ1 signaling pathway. Am J Transl Res. 2017;9(12):5619–30.

Lucas P, Ukhanov K, Leinders-Zufall T, Zufall F. A diacylglycerol-gated cation channel in vomeronasal neuron dendrites is impaired in TRPC2 mutant mice: mechanism of pheromone transduction. Neuron. 2003;40(3):551–61.

Ludlow MJ, Gaunt HJ, Rubaiy HN, Musialowski KE, Blythe NM, Vasudev NS, et al. (−)-Englerin A-evoked cytotoxicity is mediated by Na+ influx and counteracted by Na+/K+-ATPase *. J Biol Chem. 2017;292(2):723–31.

Maier T, Follmann M, Hessler G, Kleemann HW, Hachtel S, Fuchs B, et al. Discovery and pharmacological characterization of a novel potent inhibitor of diacylglycerol-sensitive TRPC cation channels. Br J Pharmacol. 2015;172(14):3650–60.

Maklad A, Sharma A, Azimi I. Calcium signaling in brain cancers: roles and therapeutic targeting. Cancers. 2019;11(2):E145.

Martín-Bórnez M, Galeano-Otero I, del Toro R, Smani T. TRPC and TRPV channels' role in vascular remodeling and disease. Int J Mol Sci. 2020;21(17):6125.

Miller M, Shi J, Zhu Y, Kustov M, Tian JB, Stevens A, et al. Identification of ML204, a novel potent antagonist that selectively modulates native TRPC4/C5 ion channels. J Biol Chem. 2011;286(38):33436–46.

Minard A, Bauer CC, Wright DJ, Rubaiy HN, Muraki K, Beech DJ, et al. Remarkable progress with small-molecule modulation of TRPC1/4/5 channels: implications for understanding the channels in health and disease. Cell. 2018;7(6):E52.

Mizoguchi Y, Kato TA, Seki Y, Ohgidani M, Sagata N, Horikawa H, et al. Brain-derived neurotrophic factor (BDNF) induces sustained intracellular Ca2+ elevation through the up-regulation of surface transient receptor potential 3 (TRPC3) channels in rodent microglia*. J Biol Chem. 2014;289(26):18549–55.

Montell C, Rubin GM. Molecular characterization of the Drosophila trp locus: a putative integral membrane protein required for phototransduction. Neuron. 1989;2(4):1313–23.

Moore C, Gupta R, Jordt SE, Chen Y, Liedtke WB. Regulation of pain and itch by TRP channels. Neurosci Bull. 2017;34(1):120–42.

Morrell NW, Adnot S, Archer SL, Dupuis J, Jones PL, MacLean MR, et al. Cellular and molecular basis of pulmonary arterial hypertension. J Am Coll Cardiol. 2009;54(1 Suppl):S20–31.

Motoyama K, Nagata T, Kobayashi J, Nakamura A, Miyoshi N, Kazui M, et al. Discovery of a bicyclo[4.3.0]nonane derivative DS88790512 as a potent, selective, and orally bioavailable blocker of transient receptor potential canonical 6 (TRPC6). Bioorg Med Chem Lett. 2018;28(12):2222–7.

Muraki K, Ohnishi K, Takezawa A, Suzuki H, Hatano N, Muraki Y, Hamzah N, Foster R, Waldmann H, Nussbaumer P, Christmann M, Bon RS, Beech DJ. Na+ entry through heteromeric TRPC4/C1 channels mediates (-)Englerin A-induced cytotoxicity in synovial sarcoma cells. Sci Rep. 2017 Dec 5;7(1):16988.

Naylor J, Minard A, Gaunt HJ, Amer MS, Wilson LA, Migliore M, et al. Natural and synthetic flavonoid modulation of TRPC5 channels. Br J Pharmacol. 2016;173(3):562–74.
Neuner SM, Wilmott LA, Hope KA, Hoffmann B, Chong JA, Abramowitz J, et al. TRPC3 channels critically regulate hippocampal excitability and contextual fear memory. Behav Brain Res. 2015;281:69–77.
Nikolova-Krstevski V, Wagner S, Yu ZY, Cox CD, Cvetkovska J, Hill AP, et al. Endocardial TRPC-6 channels act as atrial mechanosensors and load-dependent modulators of endocardial/myocardial cross-talk. JACC Basic Transl Sci. 2017;2(5):575–90.
Nilius B, Szallasi A. Transient receptor potential channels as drug targets: from the science of basic research to the art of medicine. Pharmacol Rev. 2014;66(3):676–814.
Nilius B, Voets T, Peters J. TRP channels in disease. Sci STKE. 2005;2005(295):re8.
Ningoo M, Plant LD, Greka A, Logothetis DE. PIP2 regulation of TRPC5 channel activation and desensitization. J Biol Chem. 2021;296:100726.
Nishida M, Hara Y, Yoshida T, Inoue R, Mori Y. TRP channels: molecular diversity and physiological function. Microcirculation. 2006;13(7):535–50.
Numaga-Tomita T, Nishida M. TRPC channels in cardiac plasticity. Cell. 2020;9(2):454.
Oda S, Numaga-Tomita T, Kitajima N, Toyama T, Harada E, Shimauchi T, et al. TRPC6 counteracts TRPC3-Nox2 protein complex leading to attenuation of hyperglycemia-induced heart failure in mice. Sci Rep. 2017;7(1):7511.
Okada T, Shimizu S, Wakamori M, Maeda A, Kurosaki T, Takada N, et al. Molecular cloning and functional characterization of a novel receptor-activated TRP Ca2+ channel from mouse brain. J Biol Chem. 1998;273(17):10279–87.
Okada T, Inoue R, Yamazaki K, Maeda A, Kurosaki T, Yamakuni T, et al. Molecular and functional characterization of a novel mouse transient receptor potential protein homologue TRP7. Ca(2+)-permeable cation channel that is constitutively activated and enhanced by stimulation of G protein-coupled receptor. J Biol Chem. 1999;274(39):27359–70.
Owsianik G, D'hoedt D, Voets T, Nilius B. Structure-function relationship of the TRP channel superfamily. Rev Physiol Biochem Pharmacol. 2006;156:61–90.
Park SH, Ryu SY, Yu WJ, Han YE, Ji YS, Oh K, et al. Leptin promotes K_{ATP} channel trafficking by AMPK signaling in pancreatic β-cells. Proc Natl Acad Sci. 2013;110(31):12673–8.
Phillips AM, Bull A, Kelly LE. Identification of a Drosophila gene encoding a calmodulin-binding protein with homology to the trp phototransduction gene. Neuron. 1992;8(4):631–42.
Qu C, Ding M, Zhu Y, Lu Y, Du J, Miller M, et al. Pyrazolopyrimidines as potent stimulators for transient receptor potential canonical 3/6/7 channels. J Med Chem. 2017;60(11):4680–92.
Quick K, Zhao J, Eijkelkamp N, Linley JE, Rugiero F, Cox JJ, et al. TRPC3 and TRPC6 are essential for normal mechanotransduction in subsets of sensory neurons and cochlear hair cells. Open Biol. 2012;2(5):120068.
Ramirez GA, Coletto LA, Sciorati C, Bozzolo EP, Manunta P, Rovere-Querini P, et al. Ion channels and transporters in inflammation: special focus on TRP channels and TRPC6. Cell. 2018;7(7):70.
Reeh PW, Fischer MJM. Nobel somatosensations and pain. Pflugers Arch. 2022;474(4):405–20.
Reyes RV, Castillo-Galán S, Hernandez I, Herrera EA, Ebensperger G, Llanos AJ. Revisiting the Role of TRP, Orai, and ASIC Channels in the Pulmonary Arterial Response to Hypoxia. Front Physiol. 2018 May 7;9:486.
Riccio A, Li Y, Moon J, Kim KS, Smith KS, Rudolph U, et al. Essential role for TRPC5 in amygdala function and fear-related behavior. Cell. 2009;137(4):761–72.
Richter JM, Schaefer M, Hill K. Riluzole activates TRPC5 channels independently of PLC activity. Br J Pharmacol. 2014a;171(1):158–70.
Richter JM, Schaefer M, Hill K. Clemizole hydrochloride is a novel and potent inhibitor of transient receptor potential channel TRPC5. Mol Pharmacol. 2014b;86(5):514–21.
Roa-Coria JE, Pineda-Farias JB, Barragán-Iglesias P, Quiñonez-Bastidas GN, Zúñiga-Romero Á, Huerta-Cruz JC, et al. Possible involvement of peripheral TRP channels in the hydrogen sulfide-induced hyperalgesia in diabetic rats. BMC Neurosci. 2019;20(1):1.

Rosker C, Graziani A, Lukas M, Eder P, Zhu MX, Romanin C, et al. Ca2+ Signaling by TRPC3 involves Na+ entry and local coupling to the Na+/Ca2+ exchanger*. J Biol Chem. 2004;279(14):13696–704.

Rubaiy HN. The influence of hyperglycemia on cardiovascular diseases; 2017. p. 2–27.

Rubaiy HN. The therapeutic agents that target ATP-sensitive potassium channels. Acta Pharma. 2016;66(1):23–34.

Rubaiy HN. A short guide to electrophysiology and ion channels. J Pharm Pharm Sci. 2017;20:48–67.

Rubaiy HN. Treasure troves of pharmacological tools to study transient receptor potential canonical 1/4/5 channels. Br J Pharmacol. 2019;176(7):832–46.

Rubaiy HN, Ludlow MJ, Henrot M, Gaunt HJ, Miteva K, Cheung SY, et al. Picomolar, selective, and subtype-specific small-molecule inhibition of TRPC1/4/5 channels. J Biol Chem. 2017a;292(20):8158–73.

Rubaiy HN, Ludlow MJ, Bon RS, Beech DJ. Pico145 – powerful new tool for TRPC1/4/5 channels. Channels. 2017b;11(5):362–4.

Rubaiy HN, Seitz T, Hahn S, Choidas A, Habenberger P, Klebl B, et al. Identification of an (−)-englerin A analogue, which antagonizes (−)-englerin A at TRPC1/4/5 channels. Br J Pharmacol. 2018a;175(5):830–9.

Rubaiy HN, Ludlow MJ, Siems K, Norman K, Foster R, Wolf D, et al. Tonantzitlolone is a nanomolar potency activator of transient receptor potential canonical 1/4/5 channels. Br J Pharmacol. 2018b;175(16):3361–8.

Sabourin J, Robin E, Raddatz E. A key role of TRPC channels in the regulation of electromechanical activity of the developing heart. Cardiovasc Res. 2011;92(2):226–36.

Sachdeva R, Schlotterer A, Schumacher D, Matka C, Mathar I, Dietrich N, et al. TRPC proteins contribute to development of diabetic retinopathy and regulate glyoxalase 1 activity and methylglyoxal accumulation. Mol Metab. 2018;9:156–67.

Sadler KE, Moehring F, Shiers SI, Laskowski LJ, Mikesell AR, Plautz ZR, et al. Transient receptor potential canonical 5 mediates inflammatory mechanical and spontaneous pain in mice. Sci Transl Med. 2021;13(595):eabd7702.

Sadowska A, Hitzl W, Karol A, Jaszczuk P, Cherif H, Haglund L, et al. Differential regulation of TRP channel gene and protein expression by intervertebral disc degeneration and back pain. Sci Rep. 2019;9(1):18889.

Schaefer M, Plant TD, Obukhov AG, Hofmann T, Gudermann T, Schultz G. Receptor-mediated regulation of the nonselective cation channels TRPC4 and TRPC5. J Biol Chem. 2000;275(23):17517–26.

Schleifer H, Doleschal B, Lichtenegger M, Oppenrieder R, Derler I, Frischauf I, et al. Novel pyrazole compounds for pharmacological discrimination between receptor-operated and store-operated Ca(2+) entry pathways. Br J Pharmacol. 2012;167(8):1712–22.

Selvaraj S, Sun Y, Singh BB. TRPC channels and their implications for neurological diseases. CNS Neurol Disord Drug Targets. 2010;9(1):94–104.

Semtner M, Schaefer M, Pinkenburg O, Plant TD. Potentiation of TRPC5 by protons. J Biol Chem. 2007;282(46):33868–78.

Seo K, Rainer PP, Shalkey Hahn V, Lee DI, Jo SH, Andersen A, et al. Combined TRPC3 and TRPC6 blockade by selective small-molecule or genetic deletion inhibits pathological cardiac hypertrophy. Proc Natl Acad Sci U S A. 2014;111(4):1551–6.

Shen H, Pan J, Pan L, Zhang N. TRPC6 inhibited NMDA current in cultured hippocampal neurons. NeuroMolecular Med. 2013;15(2):389–95.

Sierra-Valdez F, Azumaya CM, Romero LO, Nakagawa T, Cordero-Morales JF. Structure–function analyses of the ion channel TRPC3 reveal that its cytoplasmic domain allosterically modulates channel gating. J Biol Chem. 2018;293(41):16102–14.

Smani T, Shapovalov G, Skryma R, Prevarskaya N, Rosado JA. Functional and physiopathological implications of TRP channels. Biochim Biophys Acta. 2015;1853(8):1772–82.

Smedlund K, Tano JY, Vazquez G. The constitutive function of native TRPC3 channels modulates vascular cell adhesion molecule-1 expression in coronary endothelial cells through nuclear factor kappaB signaling. Circ Res. 2010;106(9):1479–88.

So I, Chae MR, Kim SJ, Lee SW. Lysophosphatidylcholine, a component of atherogenic lipoproteins, induces the change of calcium mobilization via TRPC ion channels in cultured human corporal smooth muscle cells. Int J Impot Res. 2005;17(6):475–83.

Spassova MA, Hewavitharana T, Xu W, Soboloff J, Gill DL. A common mechanism underlies stretch activation and receptor activation of TRPC6 channels. Proc Natl Acad Sci. 2006;103(44):16586–91.

Storch U, Forst AL, Pardatscher F, Erdogmus S, Philipp M, Gregoritza M, et al. Dynamic NHERF interaction with TRPC4/5 proteins is required for channel gating by diacylglycerol. Proc Natl Acad Sci U S A. 2017;114(1):E37–46.

Subedi KP, Ong HL, Ambudkar IS. Assembly of ER-PM junctions: a critical determinant in the regulation of SOCE and TRPC1. In: Krebs J, editor. Membrane dynamics and calcium signaling [Internet]. Cham: Springer International Publishing; 2017 [cited 2022 Oct 15]. p. 253–76. (Advances in Experimental Medicine and Biology). Available from: https://doi.org/10.1007/978-3-319-55858-5_11.

Sun ZC, Ma SB, Chu WG, Jia D, Luo C. Canonical transient receptor potential (TRPC) channels in nociception and pathological pain. Neural Plast. 2020;2020:3764193.

Sun Y, Ye C, Tian W, Ye W, Gao YY, Feng YD, et al. TRPC1 promotes the genesis and progression of colorectal cancer via activating CaM-mediated PI3K/AKT signaling axis. Oncogenesis. 2021;10(10):1–13.

Tai Y, Feng S, Ge R, Du W, Zhang X, He Z, et al. TRPC6 channels promote dendritic growth via the CaMKIV-CREB pathway. J Cell Sci. 2008;121(14):2301–7.

Tang Y, Tang J, Chen Z, Trost C, Flockerzi V, Li M, et al. Association of mammalian trp4 and phospholipase C isozymes with a PDZ domain-containing protein, NHERF J Biol Chem. 2000;275(48):37559–64.

Tang Q, Guo W, Zheng L, Wu JX, Liu M, Zhou X, et al. Structure of the receptor-activated human TRPC6 and TRPC3 ion channels. Cell Res. 2018;28(7):746–55.

Thakur DP, Tian JB, Jeon J, Xiong J, Huang Y, Flockerzi V, et al. Critical roles of $G_{i/o}$ proteins and phospholipase C-δ1 in the activation of receptor-operated TRPC4 channels. Proc Natl Acad Sci. 2016;113(4):1092–7.

Thakur DP, Wang Q, Jeon J, Tian JB, Zhu MX. Intracellular acidification facilitates receptor-operated TRPC4 activation through PLCδ1 in a Ca2+-dependent manner. J Physiol. 2020;598(13):2651–67.

The official website of the Nobel Prize [Internet]. NobelPrize.org. [cited 2022 Nov 22]. Available from: https://www.nobelprize.org/.

Trebak M, Bird GSJ, McKay RR, Birnbaumer L, Putney JW. Signaling mechanism for receptor-activated canonical transient receptor potential 3 (TRPC3) channels. J Biol Chem. 2003;278(18):16244–52.

Trebak M, Lemonnier L, Smyth JT, Vazquez G, Putney JW. Phospholipase C-coupled receptors and activation of TRPC channels. Handb Exp Pharmacol. 2007;179:593–614.

Trebak M, Lemonnier L, DeHaven WI, Wedel BJ, Bird GS, Putney JW. Complex functions of phosphatidylinositol 4,5-bisphosphate in regulation of TRPC5 cation channels. Pflüg Arch Eur J Physiol. 2009;457(4):757–69.

TRP Channels in the Brain – Neurobiology of TRP channels – NCBI bookshelf [Internet]. [cited 2022 Oct 20]. Available from: https://www.ncbi.nlm.nih.gov/books/NBK476104/.

Urban N, Hill K, Wang L, Kuebler WM, Schaefer M. Novel pharmacological TRPC inhibitors block hypoxia-induced vasoconstriction. Cell Calcium. 2012;51(2):194–206.

Vangeel L, Voets T. Transient receptor potential channels and calcium signaling. Cold Spring Harb Perspect Biol. 2019;11(6):a035048.

Venkatachalam K, Zheng F, Gill DL. Regulation of canonical transient receptor potential (TRPC) channel function by diacylglycerol and protein kinase C*. J Biol Chem. 2003;278(31):29031–40.

Vinayagam D, Mager T, Apelbaum A, Bothe A, Merino F, Hofnagel O, et al. Electron cryo-microscopy structure of the canonical TRPC4 ion channel. elife. 2018;7:e36615.

Vinayagam D, Quentin D, Yu-Strzelczyk J, Sitsel O, Merino F, Stabrin M, et al. Structural basis of TRPC4 regulation by calmodulin and pharmacological agents. elife. 2020;9:e60603.

Wang Y, Bu J, Shen H, Li H, Wang Z, Chen G. Targeting transient receptor potential canonical channels for diseases of the nervous system. Curr Drug Targets. 2017;18(12):1460–5.

Wang L, Chang JH, Buckley AF, Spurney RF. Knockout of TRPC6 promotes insulin resistance and exacerbates glomerular injury in Akita mice. Kidney Int. 2019;95(2):321–32.

Wang H, Cheng X, Tian J, Xiao Y, Tian T, Xu F, et al. TRPC channels: structure, function, regulation and recent advances in small molecular probes. Pharmacol Ther. 2020a;209:107497.

Wang Y, Liu L, Tao H, Wen L, Qin S. TRPC6 participates in the development of blood pressure variability increase in sino-aortic denervated rats. Heart Vessel. 2020b;35(12):1755–65.

Ward ML, Williams IA, Chu Y, Cooper PJ, Ju YK, Allen DG. Stretch-activated channels in the heart: contributions to length-dependence and to cardiomyopathy. Prog Biophys Mol Biol. 2008;97(2–3):232–49.

Wei W, Huang W, Lin Y, Becker EBE, Ansorge O, Flockerzi V, et al. Functional expression of calcium-permeable canonical transient receptor potential 4-containing channels promotes migration of medulloblastoma cells. J Physiol. 2017;595(16):5525–44.

Wen H, Gwathmey JK, Xie LH. Role of transient receptor potential canonical channels in heart physiology and pathophysiology. Front Cardiovasc Med. 2020;7:24.

Wes PD, Chevesich J, Jeromin A, Rosenberg C, Stetten G, Montell C. TRPC1, a human homolog of a Drosophila store-operated channel. Proc Natl Acad Sci. 1995;92(21):9652–6.

Winkler PA, Huang Y, Sun W, Du J, Lü W. Electron cryo-microscopy structure of a human TRPM4 channel. Nature. 2017;552(7684):200–4.

Wolfrum C, Kiehlmann E, Pelczar P. TRPC1 regulates brown adipose tissue activity in a PPARγ-dependent manner. Am J Physiol Endocrinol Metab. 2018;315(5):E825–32.

Wong F, Schaefer EL, Roop BC, LaMendola JN, Johnson-Seaton D, Shao D. Proper function of the Drosophila trp gene product during pupal development is important for normal visual transduction in the adult. Neuron. 1989;3(1):81–94.

Wright DJ, Simmons KJ, Johnson RM, Beech DJ, Muench SP, Bon RS. Human TRPC5 structures reveal interaction of a xanthine-based TRPC1/4/5 inhibitor with a conserved lipid binding site. Commun Biol. 2020;3(1):1–11.

Xiao X, Liu HX, Shen K, Cao W, Li XQ. Canonical transient receptor potential channels and their link with cardio/cerebro-vascular diseases. Biomol Ther. 2017;25(5):471–81.

Xu SZ, Muraki K, Zeng F, Li J, Sukumar P, Shah S, et al. A Sphingosine-1–phosphate-activated calcium channel controlling vascular smooth muscle cell motility. Circ Res. 2006;98(11):1381–9.

Xu X, Lozinskaya I, Costell M, Lin Z, Ball JA, Bernard R, et al. Characterization of small molecule TRPC3 and TRPC6 agonist and antagonists. Biophys J. 2013;104(2):454a.

Yang SL, Cao Q, Zhou KC, Feng YJ, Wang YZ. Transient receptor potential channel C3 contributes to the progression of human ovarian cancer. Oncogene. 2009;28(10):1320–8.

Yu M, Ledeboer MW, Daniels M, Malojcic G, Tibbitts TT, Coeffet-Le Gal M, et al. Discovery of a potent and selective TRPC5 inhibitor, efficacious in a focal segmental glomerulosclerosis model. ACS Med Chem Lett. 2019;10(11):1579–85.

Yu Y, Li W, Jiang J. TRPC channels as emerging targets for seizure disorders. Trends Pharmacol Sci. 2022;43(9):787–98.

Yue D, Wang Y, Xiao JY, Wang P, Ren CS. Expression of TRPC6 in benign and malignant human prostate tissues. Asian J Androl. 2009;11(5):541–7.

Zhang LY, Zhang YQ, Zeng YZ, Zhu JL, Chen H, Wei XL, et al. TRPC1 inhibits the proliferation and migration of estrogen receptor-positive breast cancer and gives a better prognosis by inhibiting the PI3K/AKT pathway. Breast Cancer Res Treat. 2020;182(1):21–33.

Zhong T, Zhang W, Guo H, Pan X, Chen X, He Q, et al. The regulatory and modulatory roles of TRP family channels in malignant tumors and relevant therapeutic strategies. Acta Pharm Sin B. 2022;12(4):1761–80.

Zhou FW, Matta SG, Zhou FM. Constitutively active TRPC3 channels regulate basal ganglia output neurons. J Neurosci. 2008a;28(2):473–82.

Zhou J, Du W, Zhou K, Tai Y, Yao H, Jia Y, et al. Critical role of TRPC6 channels in the formation of excitatory synapses. Nat Neurosci. 2008b;11(7):741–3.

Zhou Y, Castonguay P, Sidhom EH, Clark AR, Dvela-Levitt M, Kim S, et al. A small-molecule inhibitor of TRPC5 ion channels suppresses progressive kidney disease in animal models. Science. 2017;358(6368):1332–6.

Zhu X, Chu PB, Peyton M, Birnbaumer L. Molecular cloning of a widely expressed human homologue for the Drosophila trp gene. FEBS Lett. 1995;373(3):193–8.

Zhu Z, Xiong S, Li Q. The role of transient receptor potential channels in hypertension and metabolic vascular damage. Exp Physiol. 2016;101(11):1338–44.

Zitt C, Zobel A, Obukhov AG, Harteneck C, Kalkbrenner F, Lückhoff A, et al. Cloning and functional expression of a human Ca2+-permeable cation channel activated by calcium store depletion. Neuron. 1996;16(6):1189–96.

Zitt C, Obukhov AG, Strübing C, Zobel A, Kalkbrenner F, Lückhoff A, et al. Expression of TRPC3 in Chinese hamster ovary cells results in calcium-activated cation currents not related to store depletion. J Cell Biol. 1997;138(6):1333–41.

Nicotinic Receptors of the Neuronal and Non-neuronal Cholinergic Systems as Therapeutic Targets: Opportunities and Limitations

Sue Wonnacott

Abstract Nicotinic acetylcholine receptors (nAChRs) are widely distributed throughtout the body. Classically, nAChRs were considered to mediate fast synaptic transmission, typified by muscle and ganglionic nAChRs. Now there is a vast diversity of subtypes of "neuronal" nAChRs, expressed in both neurons and non-neuronal cells, where they exhibit a remarkable range of cellular and subcellular locales, signaling mechanisms, and cellular functions, with a more modulatory or integrative role predominating in the central nervous system and non-neuronal cholinergic system. Thus, nAChRs contribute to many brain functions and physiological processes, and are implicated in a host of associated disease states. This prompted substantial interest from the pharmaceutical industry in the quest for novel therapeutic drugs. While this activity has generated many new compounds that are valuable tools for better understanding nAChR function and regulation, the translation of promising preclinical candidates to new therapeutics has been almost totally disappointing. After surveying the diversity of the nAChR family and the opportunities that they present for drug discovery, factors that have limited the successful therapeutic exploitation of nAChRs are discussed.

Keywords Orthosteric and Non-canonical binding · Pain · Positive allosteric modulator · Schizophrenia · Skin diseases · Nicotine dependence · Smoking cessation

1 Introduction

Nicotinic acetylcholine receptors (nAChRs) are arguably the best characterized ligand-gated ion channels (LGICs). Their inference by Langley in 1905 led to the concept of neurotransmitter receptors as transduction systems for intercellular

S. Wonnacott (✉)
Department of Life Sciences, University of Bath, Bath, UK
e-mail: bsssw@bath.ac.uk

G. Stephens, E. Stevens (eds.), *Ion Channels as Targets in Drug Discovery*,
https://doi.org/10.1007/978-3-031-52197-3_12

communication, and the study of their roles in neuromuscular and ganglionic transmission in the mid-twentieth century resulted in a level of understanding of receptor function that was transformative. This was followed by pioneering biochemical approaches that resulted in the nAChR being the first receptor to be purified and its molecular structure determined (Bouzat and Sine 2018; Changeux 2020). All these studies were facilitated by an abundance of natural products targeted at nAChRs. As with other LGIC receptors, advances in molecular biology identified a multiplicity of subunit genes. nAChR subunits have distinct expression patterns in muscle and neurons and, unexpectedly, are also present in other cell types. Research has progressed to give a sophisticated understanding of receptor expression, localization, regulation, and turnover, as well as atomic resolution structural information, biophysical properties of channel function, and network consequences of receptor activation and desensitization (Poorhuis and Mansvelder 2013; Crespi et al. 2018a; Delgado-Velez et al. 2021; Letsinger et al. 2022). The generic LGIC structure offers key sites for drug action: orthosteric binding sites for agonists and competitive antagonists, the ion channel for non-competitive inhibitors and functional modulators, and distinct sites for allosteric modulators (Fig. 1). In addition, recognition of associated proteins and other molecules offers scope for targeting more regulatory aspects of receptor function, such as assembly, localization, membrane expression, and turnover (Araud et al. 2010).

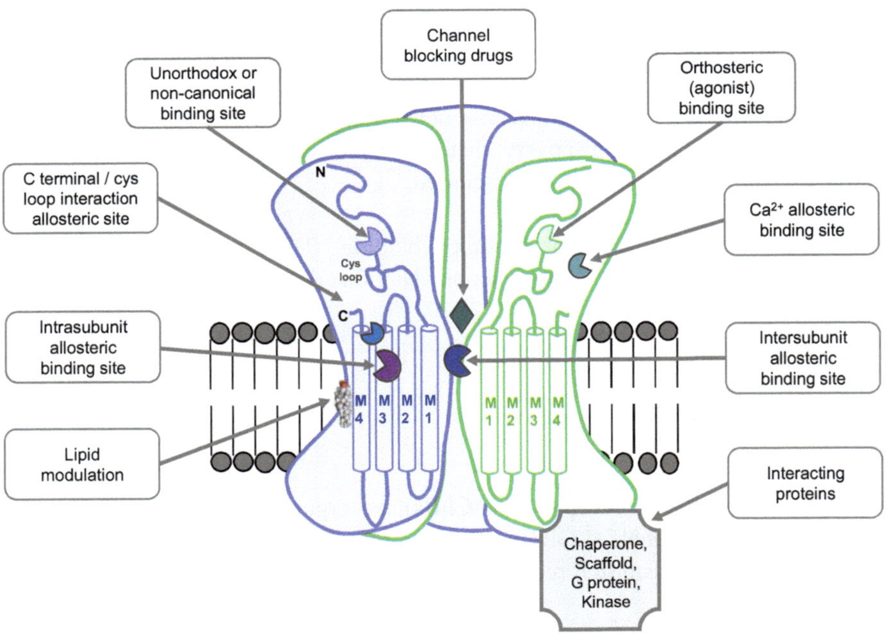

Fig. 1 Binding sites associated with nAChRs. A schematic heteromeric nAChR illustrating the diversity of sites for interacting drugs and other molecules that may modulate receptor function. Only one site of each type is shown for clarity, but multiple sites may occur (e.g., two orthosteric binding sites per heteromeric nAChR). Interacting proteins are not only intracellular but may be extracellular, membrane tethered, or transmembrane proteins. (Adapted from Wonnacott 2020)

The diversity of receptor subtypes adds another dimension to the challenge and opportunity for drug discovery. For pentameric nAChRs, the assembly of subunits generates a vast array of receptor subtypes. While the assembly of muscle nAChRs is constrained to two subtypes, the subunit composition of neuronal nAChRs is much more diverse (Zoli et al. 2018). Their selective distribution and modulatory behavior in the CNS and more recently recognised in non-neuronal cells and tissues, has nurtured interest in finding selective drugs to tap into their therapeutic potential for a wide range of clinical conditions, from smoking cessation to Alzheimer's disease and pain. This aspiration has been encouraged by the selectivity and potency of many natural products. To date, however, there have been very few examples of drugs targeting nAChRs reaching the clinic. In this chapter the structural and functional diversity of nAChRs will be summarized and potential target sites for modifying their actions identified. While including a brief discourse of old successes in drug discovery, targeting muscle and ganglionic nAChRs, the focus will be on neuronal nAChRs, with their clinical repertoire now extended by the expression of neuronal nAChR subunits in non-neuronal cells and tissues. The pros and cons of targeting different classes of functional site will be appraised, with consideration of why no drug candidate has so far succeeded in clinical trials, with the singular exception of varenicline for smoking cessation. While many excellent recent reviews provide in-depth coverage of particular aspects, this chapter aims to summarize current understanding of the breadth of nAChR contributions to physiology and disease, the potential for therapeutic intervention, and lessons to be learned from the dismal success in translating drug discovery into approved therapeutics.

2 The nAChR Family

2.1 Subunit Diversity and Stoichiometry of nAChRs

nAChRs and their constituent subunits conform to a general structural pattern that is also shared by other members of the "cys-loop" family of LGIC receptors that includes $5HT_3$, $GABA_A$, and glycine receptors. These pentameric receptors are comprised of five membrane-spanning subunits that form a central ion channel (Fig. 2). Each subunit has a long extracellular N-terminal domain, containing the eponymous 13 amino acid cysteine-bonded loop, and traverses the membrane 4 times (M1–4) ending in a short extracellular C-terminal tail. The second transmembrane span, M2, from each subunit lines the ion channel and influences channel properties including ion selectivity and permeability (Karlin 2002). nAChRs are cation channels that allow the entry of Na^+ and, to varying degrees, Ca^{2+} into cells and exit of K^+, producing depolarizing currents under standard physiological conditions.

Agonist binding sites reside in the N-terminal extracellular domain, at the interface between adjacent subunits. In nAChRs, the *principal* face of these orthosteric binding sites is found on α subunits that are distinguished by having a pair of vicinal cysteine residues close to the binding site. The adjacent subunit contributes the *complementary* face of the binding site (Fig. 2; see Sect. 4.1). Agonist binding results in channel opening and ion flux, which is terminated by channel closure and the receptor entering a desensitized state in which it is refractory to activation (agonist may remain bound in the desensitized state; several desensitized states have been defined). The precise mechanism that translates agonist binding to channel opening is still unresolved (Auerbach 2015; Changeux 2018).

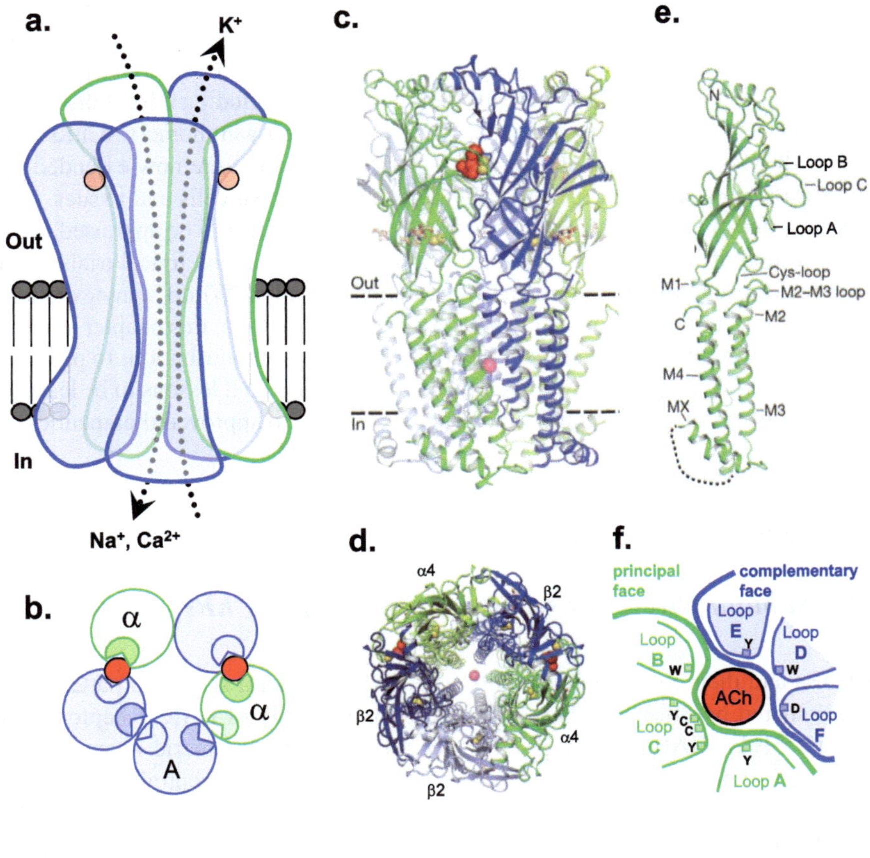

Fig. 2 Structural aspects of nAChRs. (**a**) A schematic heteromeric nAChR comprised of five membrane spanning subunits creating a central ion channel permeable to cations. Two ACh molecules (red spheres) are bound to the orthosteric sites. (**b**) Top-down view of (**a**) illustrating the location of the orthosteric sites between an α subunit (green) and its neighbor. The fifth "accessory" subunit (A) has two putative binding sites. (**c**) Crystal structure of human $(\alpha4)_2(\beta2_3)$ receptor, with nicotine (red) bound to the orthosteric site visible at the front. A Na^+ ion is located in the channel (pink sphere). (**d**) Top-down view of (**c**). (**e**) A single α4 subunit with key features identified. (**f**) A schematic illustrating the six conserved loops that coordinate ACh in the orthosteric binding site. Key conserved amino acids are indicated. (**a**, **b**, and **f** adapted from Wonnacott (2020); **b**, **c**, and **e** reproduced from Morales-Perez et al. (2016) with permission from Springer Nature)

As the first nAChR to be purified, the muscle nAChR revealed a heteromeric composition (Changeux 2020). Purification was facilitated by the very high density of nAChRs in electric organs of *Torpedo* and *Electrophorus* species. These electric tissues consist of stacks of electrocytes that generate powerful electrical discharges for defence and predation. This electromotor system resembles the neuromuscular junction but is devoid of muscle tissue. N-terminal sequencing of purified subunits enabled cloning of the corresponding cDNA sequences and subsequent cloning of related subunits. Thus, five distinct subunits forming two subtypes of nAChR were identified in mammalian muscle: a fetal form (also found at extrajunctional locations), comprised of $(\alpha)_2\beta\gamma\delta$ subunits, and a junctional nAChR in which the γ subunit is replaced by an ε subunit. *Torpedo* electrocytes express the fetal $(\alpha)_2\beta\gamma\delta$ nAChR, reflecting their origin in embryonic muscle cells. The γ and ε subunits confer subtle differences in nAChR function and localization (Cetin et al. 2020). The stoichiometry of subunits identified two copies of the α subunit per pentamer and hence two orthosteric binding sites. These sites act cooperatively to increase the probability of channel opening. They are non-identical as they differ in their complementary face, which is provided by γ/ε and δ subunits, respectively. This distinction can have functional consequences, with $\alpha\gamma$ having higher affinity for the antagonist d-tubocurarine, for example, compared with $\alpha\delta$ (Sine 1993). The β subunit does not contribute to the orthosteric sites of muscle nAChRs (in contrast to the subunit nomenclature of neuronal nAChRs).

Neuronal and muscle nAChRs were first distinguished pharmacologically in the classic studies of Paton and Zaimis (1949) that differentiated C10 polymethylene bistrimethyl-ammonium compounds, such as decamethonium, as neuromuscular blocking agents from C6 analogues, like hexamethonium, that are ganglionic blockers. It took some 30 years for the molecular basis of this distinction to be clarified with the cloning of a distinct family of nAChR subunits expressed in neurons.

The portfolio of mammalian neuronal nAChR subunits is now established (Zoli et al. 2018). Six α subunits are expressed in the CNS (α2-α7), with α9 and α10 localized to cochlear hair cells and some other sensory neurons (α1 is expressed in muscle; α8 occurs in avian neurons and is not found in mammalian cells). Subunits lacking the vicinal cysteines (Cys 192, 193, *Torpedo* α1 numbering) that define an α subunit are designated β, with three subunits, β2-β4, discovered (β1 is in muscle nAChRs). Uniquely, humans also have an additional gene (*CHRFAM7A*) coding for a truncated version of the α7 subunit (dupα7). This arose from an evolutionarily recent partial gene duplication and lacks the first 146 amino acids from the N-terminal domain (Sinkus et al. 2015).

The subunits combine in various stoichiometries to form functional nAChRs (Zoli et al. 2018). The rules of assembly require at least two α subunits, to give two orthosteric binding sites. In this regard, neuronal α5 subunits are excluded: despite having the definitive vicinal cysteines, a conserved tyrosine located in the principal face of the binding site in all other α subunits (Tyr190, *Torpedo* numbering) is replaced by an aspartate in the α5 subunit (Karlin 2002). α7 and α9 subunits are exceptional in their capacity to form homomeric receptors. Although $\alpha7\beta2$ nAChRs have been identified in basal forebrain neurons (Wu et al. 2016), native α7-containing nAChRs are predominantly homomeric in the mammalian brain. However, evidence from heterologous expression systems shows that human α7 and dupα7 subunits can form hetero-pentamers in which the duplicated gene product acts as a dominant negative regulator (Sinkus et al. 2015). In contrast to α7 nAChRs, and despite the ability of α9 subunits to form robust homomeric nAChRs in heterologous expression systems, α9α10 heteromeric nAChRs are the functional receptors in mechanosensory hair cells (Lipovsek et al. 2021). These subunits can combine in two stoichiometries: $(\alpha9)_2(\alpha10)_3$ or $(\alpha9)_3(\alpha10)_2$. Like muscle nAChRs, α7 and α9* nAChRs bind α-bungarotoxin with high affinity; this snake toxin is a potent antagonist and selective radioligand for these nAChR subtypes. However, α9 and α10 subunits are the most structurally divergent, conferring a distinctive, mixed nicotinic-muscarinic pharmacology, with nicotine and some other agonists acting as antagonists. They are also blocked by antagonists of other cys-loop LGICs (Elgoyhen 2022).

The other α subunits (α2, α3, α4, α6) combine with β subunits to form obligate heteromers. β2 and β4 can pair with these α subunits to create two orthosteric binding sites (hence, β2 and β4 are analogous to γ, δ, and ε muscle nAChR subunits); the β3 subunit is unable to fulfill this role (and is the true equivalent of the muscle β1 subunit). Thus, α5 and β3 subunits are relegated to occupying the fifth position in a pentamer between two αβ pairs that form the orthosteric binding sites; they are referred to as *accessory subunits* (Kuryatov et al. 2008; Fig. 1b; Sect. 4.3). Interest in α5 and β3 subunits has been enhanced by the identification of single nucleotide polymorphisms (SNPs) in their genes as risk factors for nicotine addiction and lung cancer (Bierut 2010). The common non-synonymous SNP in the human *CHRNA5 gene* changes aspartate (D) to asparagine (N) in position 398 of the protein. Located at the intracellular face of the nAChR, this change has a negative impact on receptor function (Maskos 2020). α5 and β3 are not the only subunits able to reside in the fifth position: for example, the common α4β2 nAChR can have the stoichiometry $(\alpha4)_3(\beta2)_2$ or $(\alpha4)_2(\beta2)_3$ with α4 and β2, respectively, in the accessory role. This is not without functional implications for drug discovery (Sect. 4.3). The fifth subunit can also influence nAChR trafficking and efficiency of export to the plasma membrane (Crespi et al. 2018b). When the precise subunit composition of a nAChR subtype is not resolved, typically for the fifth position, an asterisk is used to indicate the possibility of additional subunits being present (e.g., α4β2*, α3β4* nAChR; Lukas et al. 1999).

α4β2* nAChRs are characterized by high affinity binding of orthosteric agonists, enabling nicotine, cytisine, and epibatidine to be employed as radioligands. Together with null mutant mice lacking a particular nAChR subunit, these tools have helped define nAChR populations. $[^3H]$nicotine and $[^3H]$cytisine only label α4β2* nAChRs, whereas $[^3H]$epibatidine binding is more extensive. Cytisine-resistant $[^3H]$epibatidine binding sites correspond to several subtypes of nAChR that lack α4, including minor populations containing the β4 subunit (Baddick and Marks 2011; Meyers et al. 2015). α-Conotoxins, secreted by cone snails, are peptide antagonists of nAChRs that exhibit exquisite subtype selectivity (Azam and McIntosh 2009). For example, α-conotoxin MII from *Conus magus* is selective for α3β2* and α6β2* nAChR subtypes and has served as a definitive radioligand (Baddick and Marks 2011).

2.2 *nAChR Distribution and Localization*

2.2.1 Muscle

Expression of muscle nAChR subunits is limited to skeletal muscle cells, with a developmental switch from γ to ε subunits, giving rise to two subtypes: junctional ($(\alpha1)_2\beta1\varepsilon\delta$) and extrajunctional/fetal ($(\alpha1)_2\beta1\gamma\delta$) nAChRs. Many factors, including electrical activity and release of trophic substances from motor nerve endings, notably agrin, orchestrate the developmental transition in gene expression from γ to ε subunits and the change in localization of nAChRs from a diffuse extrajunctional distribution to clustering at the end plate (Cetin et al. 2020).

In contrast, the formation of neuronal nAChRs from the array of α and β subunits is more complex and less well understood. The reasons for this include the complexity and diversity of nervous tissue, the potential to express multiple nAChRs in a single neuron, and more promiscuous subunit options for nAChR assembly.

2.2.2 Autonomic Neurons

nAChRs are responsible for fast excitatory transmission in autonomic ganglia. Autonomic neurons receive multiple inputs and have an integrative function (Wang et al. 2002). The predominant nAChR subunits expressed in ganglionic neurons are α3, α5, α7, β2, and β4 that can give rise to homomeric α7 nAChRs and multiple heteromeric nAChRs: α3β2, α3β4, α3β2β4, α3α5β2, α3α5β4, and α3α5β2β4 (Wang et al. 2002; Urriola & Adelstein, 2022). Transgenic approaches have also been instrumental in deciphering the physiological significance of different subunits. Knockout mice lacking the α3 subunit do not survive long after birth and display many autonomic dysfunctions (Xu et al. 1999a). This is consistent with the ubiquitous expression of the α3 subunit in autonomic neurons and its essential role in nAChRs mediating ganglionic transmission. The α3 knockout phenotype also aligns with a rare human autonomic recessive condition megacystis-microcolon-intestinal hypoperistalsis syndrome (MMIHS). In contrast, mice lacking any of the other subunits (α5, α7, β2, β4) exhibit no gross physiological abnormalities under normal conditions, but a β2−/−β4−/−double knockout recapitulates the α3−/− phenotype (Xu et al. 1999b). α3β2 and α3β4 nAChRs differ in channel properties and pharmacological profiles. As the β4 subunit is more abundant than β2 in autonomic neurons, and confers the more ganglionic-like properties, it can be inferred that α3β4* nAChRs predominate as the postsynaptic receptor mediating synaptic transmission. The β2 subunit may additionally be present in a minority of receptors and can effectively substitute for β4 in the absence of the latter (David et al. 2010). Incorporation of the α5 subunit will further modulate receptor properties. Recent functional data have identified α3α5β4 nAChRs as the predominant subtype in mouse superior cervical ganglia, accounting for about 70% of the functional nAChRs expressed (Simeone et al. 2019). It is striking that the genes encoding these three subunits constitute a gene cluster (*CHRNA5-CHRNA3-CHRNB4)* on human chromosome 15, suggesting coordinated transcriptional regulation (Liu et al. 1999). It is likely that the proportions of the various heteromeric nAChR subtypes will vary between sympathetic, parasympathetic, and enteric ganglia and different development stages or physiological conditions. The α3 nAChR subunit is necessary and sufficient for the postsynaptic localization of ganglionic nAChRs.

In contrast, α7 nAChRs are localized perisynaptically at autonomic synapses, suggesting a modulatory influence on excitability rather than a role in fast ganglionic transmission (Wang et al. 2002). Interpretation of expression patterns is also complicated by the presynaptic location of nAChRs constituted from the same pool of expressed subunits.

2.2.3 Sensory Neurons

Less well characterized are populations of sensory neurons that relay information from the external or internal environment to the CNS. Best studied are dorsal root ganglion (DRG) nociceptive neurons, including those of human DRG, that express many nAChR subunits, including α9 and α10, but show marked species differences (Hone et al. 2018; Zhang et al. 2019).

2.2.4 CNS

Extensive studies of rodent brain from wild-type and genetically engineered animals have provided a detailed and complex picture of nAChR subunit and subtype expression, with all neuronal subunits (α2-α7, β2-β4) detected, in varying amounts and with distinct but overlapping distributions (Zoli et al. 2018). α4, β2, and α7 nAChR subunits are the most abundant: α4 and β2 subunits are coordinately expressed throughout the brain, consistent with the formation of α4β2* heteromers, while α7 nAChR subunits have a distinct distribution. Highest expression of α4 and β2 is found in the thalamus, cortex, and striatum, whereas α7 nAChRs predominate in the hippocampus, as well as cortex, hypothalamus, and brain stem nuclei. Sparse α7β2 heteromers are found in regions associated with learning and memory innervated by basal forebrain cholinergic neurons (Wu et al. 2016). α7β2 nAChRs display subtly altered properties; increasing the number of β2 subunits (from 0 to 3) progressively increases channel open times, reflecting decreased desensitization (Nielsen et al. 2018).

Other nAChR subunits have more localized distributions. α6 and β3 subunits show comparable expression patterns, with a restricted distribution in both rodents and primates suggesting assembly into a α6β3* the same heterooligomer (except in the medial habenula where β3 predominates) (Le Novère and Changeux 1996). The highest expression of α6 and β3 nAChR subunits is found in catecholaminergic neurons, and several α6* nAChR subtypes, with and without β3, have been characterized in midbrain dopaminergic neurons (Quik and Wonnacott 2011).

The "ganglionic" trio of subunits α3, α5, and β4 are also found together in the CNS, with a limited distribution and highest expression in the habenula-interpeduncular nucleus (IPN) pathway (Zoli et al. 2018). nAChR subunit expression in the medial habenula and IPN is very diverse; immunoprecipitation and ligand binding experiments indicate the assembly of multiple nAChR subtypes (Grady et al. 2009; Scholze et al. 2012; Wills et al. 2022).

The α2 nAChR subunit is sparsely expressed in rodent brain, with transcripts localized to populations of GABAergic interneurons in the IPN, hippocampus, striatum, and some other areas, including Renshaw cells of the spinal cord (Ishii et al. 2005). There is functional evidence for α2* nAChRs in some of these cells, facilitated by a *Chrna2*-Cre mouse (expressing Cre under the control of the *Chrna*2 promoter; Hilscher et al. 2019; Tokarska and Silberberg 2022). Deletion of the gene produces only subtle motor effects but generates some altered nicotine-induced behaviors (Lotfipour et al. 2013). In primates a much greater abundance and wider distribution of α2 nAChR subunit expression, comparable to that of α4 and β2 subunits, has been reported (Han et al. 2000). This is in striking contrast to the rather similar expression patterns of the other neuronal nAChR subunits between rodents and primates. α2* nAChRs should not be overlooked as a potential drug target for human brain disorders.

In contrast to skeletal muscle and autonomic neurons, where nAChRs predominantly occupy a postsynaptic location and mediate synaptic transmission, in the CNS nAChRs are more commonly found at presynaptic and extra- or perisynaptic sites (Wonnacott 1997; Lendvai and Vizi 2008). In these locations nAChRs exert a modulatory influence. nAChRs are Na^+ and Ca^{2+} permeable and can trigger Ca^{2+}-dependent processes in addition to electrical signals (Dajas-Bailador and Wonnacott 2004). Ca^{2+} permeability is nAChR subtype-dependent and influenced by the subunit composition (Fucile 2004). For example, inclusion of an α5 subunit increases Ca^{2+} flux (Gerzanich et al. 1998; Sciaccaluga et al. 2015), whereas α7 homomers and α9* nAChRs display the highest Ca^{2+} permeabilities. One determinant is a conserved glutamate residue at the extracellular mouth of the ion channel: this is present in most α subunits but absent from β2 and β4 subunits, making heteromeric nAChRs markedly less Ca^{2+} permeable than their homomeric counterparts (Fucile 2004). Mutagenesis has confirmed the importance of this glutamate residue for the differential Ca^{2+} permeabilities of the two stoichiometries of α4β2 nAChRs (Tapia et al. 2007). This can also explain the enhancing effect of incorporating an α5 subunit.

2.2.5 Non-neuronal Cholinergic System

Non-neuronal acetylcholine in many tissues and organs is recognized to contribute to diverse physiological and pathological processes (Kawashima and Fujii 2008). Increasing appreciation of the non-neuronal cholinergic system has been matched by evidence for its receptors in non-neuronal cells. mRNAs coding for all the neuronal nAChR subunits, including α9 and α10, have been variously reported in immune, epithelial, endothelial and skin cells, glia, and, surprisingly, in skeletal muscle fibers (Zoli et al. 2018). However, mRNA transcripts should not be equated with translation into protein and assembly of functional receptors. Low levels of expression, concerns over the validity of some immunodetection assays, lack of selective ligands, and unconventional modes of operation have impeded the characterization of nAChRs in non-neuronal cells. Nevertheless, the body of reports to have emerged in recent years supports the widespread distribution of nAChRs of pathophysiological significance in these systems. Hence, nAChRs in many non-neuronal situations are attracting attention in the context of various diseases, while the normal physiological significance of such nAChRs is less well understood (Sect. 3.6).

2.2.6 Mitochondrial nAChRs

Adding to the complexity, neuronal nAChRs containing α3, α4, α7, or α9 subunits have been reported to occur in the outer membrane of mitochondria from a variety of tissues (Skok 2022). The possible association of choline acetyltransferase with mitochondria could provide a local source of agonist. Mitochondrial nAChRs are suggested to influence cell survival by regulating the release of pro-apoptotic Bcl-2 proteins from mitochondria.

3 Physiological and Clinical Significance of nAChRs

The wide reach of the cholinergic system, with its neuronal and non-neuronal arms, is mirrored by the distribution of its cognate receptors. Thus, nAChRs contribute to a host of physiological processes and their associated diseases, for which nAChRs may contribute causal or protective influences (Fig. 3). Their roles in the peripheral nervous system are relatively well defined in contrast to the CNS where the diversity of neuronal nAChRs, their modulatory functions, and the complexity of neural systems confound understanding. Moreover, nAChRs in non-neuronal cells remain generally ill-defined but are increasingly recognized in healthy and disease settings.

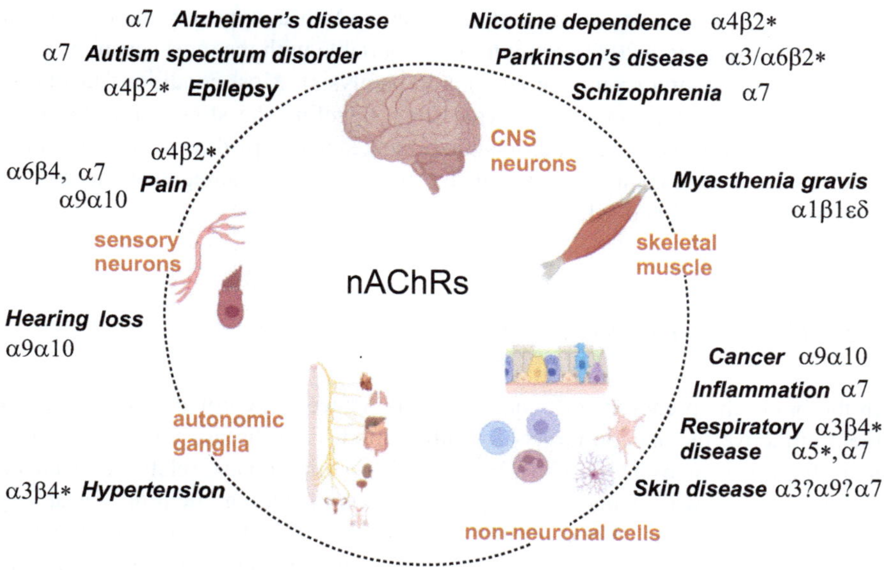

Fig. 3 nAChRs and disease. Summary of the tissue distribution of nAChRs and conditions discussed in the text, in which various nAChR subtypes contribute to the disease process or serve as potential therapeutic targets. (Illustrated tissues are courtesy of Biorender)

3.1 Muscle nAChRs

Muscle $(\alpha)_2\beta\varepsilon\delta$ nAChRs mediate neuromuscular transmission and muscle contraction. Myasthenia gravis, most commonly due to an autoimmune attack on muscle nAChRs, is the most notable disease concerning these receptors and involves muscle weakness that varies in extent and severity (Cetin et al. 2020). A congenital myasthenic syndrome can arise from mutations in any of the muscle nAChR subunits, including the fetal γ subunit but most commonly in the ε subunit, resulting in a range of phenotypes that vary in severity (Engel et al. 2015). Long-lived open channel blockers (e.g., fluoxetine and quinidine) are used to treat slow-channel forms of congenital myasthenia gravis, with the aim of preventing depolarization block and desensitization.

3.2 Autonomic Ganglia nAChRs

The autonomic nervous system, comprising the sympathetic, parasympathetic, and enteric branches, controls the function of diverse organs and tissues through complex mechanisms (Wang et al. 2002). nAChRs mediate fast synaptic transmission at autonomic synapses. Autoimmune autonomic ganglionopathy is a rare condition caused by autoantibodies directed at ganglionic nAChRs, typically the α3 subunit (Urriola and Adelstein 2022). Symptoms are variable with the most severe producing pandysautonomia (complete autonomic paralysis). Most conditions involving the autonomic system are treated with drugs targeting the slower, metabotropic components. However, one condition that provoked a therapy directed at ganglionic nAChRs is hypertension, treated with the noncompetitive antagonist mecamylamine (see Sects. 4.2 and 5.1).

3.3 CNS Heteromeric nAChRs

In the brain, the predominantly modulatory influence of neuronal nAChRs makes them attractive therapeutic targets for subtly modifying disease processes. A primary focus was on α4β2* nAChRs whose wide distribution and relative abundance associates them with many physiological systems, including attention, rewarding behaviors, appetite, and mood (Picciotto and Kenny 2021). However, their modulatory roles are complex and multifaceted, confounding the ambition to generate rational nicotinic therapies. nAChRs are activated by acetylcholine released tonically or phasically from varicosities or cholinergic interneurons or, less commonly, by synaptic contacts (Lendvai and Vizi 2008). nAChRs typically produce local depolarization, shaping neuronal excitability. Simplistically, nAChRs can enhance, prolong, or increase the likelihood of neurotransmitter release presynaptically, and increase the probability of action potential generation or influence synaptic plasticity postsynaptically. For example, postsynaptic heteromeric nAChRs on cell bodies and dendrites of midbrain dopamine neurons receive synaptic inputs from cholinergic projection neurons of the pedunculopontine and laterodorsal tegmental nuclei and modulate the firing patterns of the dopamine cells (Maskos 2008), while presynaptic heteromeric nAChRs on dopamine terminals respond to tonic acetylcholine release from local interneurons to shape striatal dopamine transmission (Exley and Cragg 2008). Both α4β2* and α7 nAChRs are also associated with GABAergic interneurons, and can thus promote inhibition or disinhibition, contributing to network regulation in many brain regions (Poorhuis and Mansvelder 2013; Letsinger et al. 2022).

3.3.1 Nicotine Dependence

Tobacco addiction and the rewarding properties of nicotine first stimulated the quest to find brain nAChRs. The well-documented upregulation of α4β2 nAChRs by chronic nicotine in animals and smokers (Wonnacott 1990) and their association with the mesolimbic dopamine system that is important for reinforcing behaviors implicated this nAChR subtype in nicotine dependence (Wittenberg et al. 2020). This is supported by the absence of several behaviors associated with nicotine reinforcement, including nicotine self-administration, in mice lacking either the α4 or β2 subunit (Picciotto and Kenny 2021; Wills et al. 2022). However, the story has become more complex. α6 subunits are also necessary for systemic nicotine self-administration, reflecting the heterogeneity of β2* nAChRs and relative abundance of α6β2* nAChRs in the terminal fields of ascending dopamine neurons (Exley et al. 2011; Wills et al. 2022).

The genome-wide association studies linking in the α3α5β4 gene cluster with nicotine dependence (Bierut 2010) has further complicated the picture. The preponderance of these subunits in the habenular-IPN pathway has identified an aversive circuit that limits nicotine intake, with both α4*β2*α5* and α3*β4** nAChRs proposed to contribute to nicotine aversion (Wills et al. 2022). The D397N SNP in the α5 subunit confers a partial loss of function that weakens this aversive brake, increasing nicotine intake and exacerbating the adverse health consequences. Localization of the α2 subunit in the IPN and altered acquisition of self-administered nicotine in mice lacking this subunit (Lotfipour et al. 2013), together with much higher prevalence of the α2 subunit in primate brains, also implicates α2* *nAChRs* in regulating nicotine intake in smokers (Wills et al. 2022). Thus, β2* nAChRs in the "reward" circuitry would seem to drive nicotine reinforcement, whereas the degree of nicotine intake is shaped by nAChRs in the habenular-IPN circuit.

3.3.2 Parkinson's Disease (PD)

Epidemiological data showing a reduced incidence of PD in smokers, together with the prevalence of α4β2* and α6β2* nAChRs in dopamine systems including the nigrostriatal pathway, have highlighted a potential protective nicotinic mechanism with respect to PD (Quik and Wonnacott 2011) and movement disorders more generally (Quik et al. 2019). Initial focus on nicotine's ability to enhance dopamine release as a symptomatic treatment has been largely superseded by interest in nAChR-induced neuroprotective and anti-inflammatory mechanisms. Intriguingly, a heteromer comprised of a β2* nAChR coupled to a dopamine D_3 receptor and expressed in dopamine neurons has been ascribed neurotrophic and neuroprotective mechanisms not shown by either component independently (Mutti et al. 2022). Another therapeutic focus is the treatment of L-DOPA-induced dyskinesia: preclinical studies indicate that nicotine or other nAChR agonists can reduce the severity of this common side effect of parkinsonian medication. This is attributed to desensitization of striatal nAChRs (Quik et al. 2019).

3.3.3 Autosomal Dominant Sleep-Related Hypermotor Epilepsy (ADSHE)

Mutations in α4, β2, and α2 nAChR subunit genes give rise to this rare familial epilepsy (formerly referred to as autosomal dominant nocturnal frontal lobe epilepsy, ADNFLE) that is characterized by hyperkinetic focal seizures arising in the neocortex during non-rapid eye movement (NREM) sleep (Becchetti et al. 2020). Mutations are commonly located in the M2 and M3 transmembrane domains, with a few located in the cytoplasmic loops (Weltzin et al. 2021). They typically increase agonist responses when tested in heterologous expression systems to mimic the heterozygous genetics, which could reflect altered subunit stoichiometry or channel kinetics, for example. In contrast, α2 nAChR subunit mutations produce a loss of function phenotype (Becchetti et al. 2020). More elusive is how these changes trigger focal seizures in the neocortex during non-REM sleep; recapitulation of the mutations in animal models suggests altered GABAergic inhibition (Steinlein 2010; Becchetti et al. 2020). There are some reports that nicotine patches can be beneficial in controlling seizures, raising interest in a precision medicine approach to treating patients with selective agonists to target particular nAChR subtypes that contain subunits bearing mutations (Becchetti et al. 2020).

3.3.4 Cognition and Alzheimer's Disease (AD)

Anecdotal reports that some smokers regulate nicotine intake to boost their concentration on demanding tasks, and the preponderance of α4β2* nAChRs in frontal cortical regions encouraged an interest in nicotinic mechanisms relating to cognition. Subsequent preclinical and human studies showed that nicotine and novel nicotinic agonists and allosteric modulators can enhance several aspects of cognition including attention, working memory and executive function (Demeter and Sarter 2013; Terry and Callaghan 2019). This evidence initially encouraged targeting α4β2* nAChRs for conditions with cognitive dysfunction. For AD, direct activation of nAChRs offered an appealing means of bypassing the loss of acetycholine due to the degeneration of the basal forebrain cholinergic system. In addition to this potential for symptomatic improvement, neuroprotective capabilities ascribed to nicotine and novel nicotinic ligands enhanced the therapeutic credibility of nAChRs. However, the focus has shifted to α7 nAChRs (see below), and interest in α4β2* nAChRs as drug targets for cognitive conditions has declined.

3.4 *CNS α7 nAChRs*

The high Ca^{2+} permeability and propensity of α7 nAChRs to desensitize, coupled with precise timing of signals, can generate complex modulatory effects, exemplified by α7 nAChRs in hippocampal circuits (Letsinger et al. 2022). In humans, complexity may be further complicated by the dupα7 subunit (Sinkus et al. 2015) and by engaging a metabotropic mode of signaling, first described in non-neuronal cells and subsequently reported in the CNS (Kabbani and Nichols 2018; see Sect. 3.6).

3.4.1 Schizophrenia

The high density of α7 nAChRs in the cortex and hippocampus, their prevalence on GABA and glutamate neurons, and their ability to promote synaptic plasticity support a role in learning and memory mechanisms (Yakel 2014). Cognitive deficits, including impaired working memory and attentional processing, seen in α7 null mutant mice (Levin 2012), have been likened to cognitive impairments in schizophrenia, and there has been sustained research into the relationship between α7 nAChRs and this disease. Key clinical observations include the high incidence of tobacco smoking among schizophrenics, considered to reflect an attempt at self-medication to relieve negative symptoms; decreased levels of α7 nAChR binding sites in the postmortem brains of schizophrenic patients; genetic association studies that link chromosome 15q13.3 containing the genes *CHRNA7* and *CHRFAM7A* (encoding α7 and dupα7 subunits, respectively) to schizophrenia; and sensory gating deficits associated with polymorphisms in the *CHRNA7* promoter region (Sinkus et al. 2015). The substantial history of drug discovery targeted at α7 nAChRs for schizophrenia is summarized in Sinkus et al. (2015), with subsequent studies reviewed by Antonio-Tolentino and Hopkins (2020).

3.4.2 Autism Spectrum Disorder (ASD)

Micro-deletions and duplications in the unstable chromosome 15 locus mentioned in relation to schizophrenia are also associated with a number of other neurodevelopmental disorders, including bipolar disease, ASD, and ADHD (Deutsch et al. 2015; Sinkus et al. 2015). Focusing here on ASD, genetic evidence of a link to α7 nAChRs is accruing (Deutsch and Burket 2020; Budisteanu et al. 2021). The gene dosage effects of genomic changes are predicted to disrupt the level of signaling through α7 nAChRs. This could impact on a developmental role for these receptors in neuronal circuit formation (Lozada et al. 2012) and/or regulation of GABAergic interneurons important for synchronous oscillatory outputs (Deutsch and Burket 2020). Promoting α7 nAChR activity in individuals with ASD is proposed as a potential beneficial intervention.

3.4.3 Alzheimer's Disease (AD)

Polymorphisms in *CHRNA7* are also risk factors for AD (Sinkus et al. 2015). α7 nAChRs have a complex relationship with the disease. β-Amyloid (Aβ) peptide binds with high affinity to α7 nAChRs, and either potentiates or inhibits receptor function, in a concentration-dependent manner (Parri et al. 2011; Oz et al. 2013). α7 nAChRs are accused of mediating adverse effects of soluble Aβ peptides (Talantova et al. 2013), but genetic deletion of α7 enhances the development of AD-like pathology and memory deficits in mouse models with (Hernandez et al. 2010) or without altered genes linked to familial AD (Tropea et al. 2021). Such seemingly paradoxical results may reflect physiological roles of low levels of Aβ peptide in the healthy brain, in part enacted through α7 nAChRs to promote synaptic plasticity, versus toxic roles of high peptide concentrations or aberrant forms that may compromise α7 nAChR functions. In addition to any direct engagement with Aβ peptides, α7 nAChRs may also exert protective effects. Current focus is particularly on anti-inflammatory actions that may be mediated by α7 nAChRs residing on astrocytes and microglia (Mizrachi et al. 2021) (see Sect. 3.6).

3.5 Sensory Neurons nAChRs

3.5.1 Dorsal Root Ganglia and Pain

Pain is a complex perception stimulated by diverse external or endogenous noxious stimuli that activate various nociceptive sensory receptors; the progression of pain signals through the DRG to the spinal cord and brain is modulated at all levels to generate a very individualized pain response. Initially α4β2 nAChRs dominated interest in nicotinic targets for pain, due to the impressive analgesic effects of the high affinity agonist epibatidine (Badio and Daly 1994). While α4β2 nAChRs may modulate pain signals centrally, it now seems that they have only minor roles in pain transmission through the DRG. Expression of the anatomically more restricted α6 subunit in subsets of nociceptive neurons of the DRG has attracted interest (Hone et al. 2018). A negative correlation between *Chrna6* mRNA expression in DRG neurons and allodynia in different mouse strains, as well as increased pain sensitivity in a small sample of people homozygous for a SNP in the human *CHRNA6* promoter, has made α6β4* nAChRs candidates of interest for anti-nociceptive drugs (Wieskopf et al. 2015).

Functional α7 nAChRs are expressed in DRG nociceptive neurons, and anterograde transport to the dorsal horn of the spinal cord, where DRG afferents synapse, has been observed (Shelukhina et al. 2015). In the spinal cord α7 nAChRs were reported to promote glutamate transmission (Genzen and McGehee 2003), although α7 nAChR-evoked glutamate release could not be demonstrated (Shelukhina et al. 2015). However, current interest in α7 nAChRs in relation to pain is dominated by their anti-inflammatory roles mediated by immune cells (Hone et al. 2018; see Sect. 3.6).

Identification of α9α10-selective α-conotoxins showing some efficacy in models of neuropathic pain has encouraged interest in α9α10 nAChRs as candidate therapeutic targets for pain (Hone et al. 2018). While expression of the α10 subunit is consistently observed in DRG neurons, it is not matched by comparable expression of α9, calling into question these sensory neurons as the locus of influence on pain signals (see Sect. 5.2). As with α7 nAChRs, there is more compelling evidence that α9α10 nAChRs in immune cells account for their antinociceptive actions (Hone et al. 2018).

3.5.2 Other Sensory Neurons

It is unclear if patterns of nAChR expression and function observed in DRG neurons also occur in other, less well-studied, sensory neurons. Nicotine, in smoking cessation products as well as in tobacco smoke, is an irritant that activates trigeminal and vagus nerve sensory nerve endings in the skin, buccal mucosa, trachea, and lungs (Carstens and Carstens 2022). It is presumed that nicotine's effect is mediated by nAChRs. A report that nicotine directly activates chemosensory TRPA1 channels heterologously expressed in HEK cells (Talavera et al. 2009) is not supported for TRPA1 channels in rat and human DRG (Zhang et al. 2019). Gene expression of α3, β3, and β4 nAChR subunits is reported to be upregulated in nociceptive trigeminal neurons in response to inflammatory stimulants (Albers et al. 2014).

3.5.3 Auditory Neurons and Hearing Loss

Outer hair cells are sensory receptors of the cochlear and vestibular systems that amplify auditory signals transduced by inner hair cells for transmission to the brain via the cochlear nerve. Outer hair cells receive efferent cholinergic innervation from the superior olivary complex. α9α10 nAChRs mediate this cholinergic transmission that results in hyperpolarization of these cells (Fuchs 2002; Lipovsek et al. 2021). A function of the inhibitory response is to modulate auditory signals by suppressing distracting irrelevant sounds. It is argued that enhancing this filtering of signals by boosting cholinergic inhibition could be beneficial for treating age-related hearing loss (Alvarez-Munoz et al. 2022; Elgoyhen 2022).

3.6 nAChRs in Non-neuronal Cells

The autocrine/paracrine cholinergic system pervades multiple, diverse tissues and organs with implications for many physiological processes. A cholinergic niche supporting stem cell proliferation and differentiation mediated by muscarinic receptors and nAChRs, notably in the brain, skin, and gut, has also been proposed (Takahashi 2021). As yet, there are limited examples of well-characterized cellular mechanisms of defined nAChR subtypes in the physiological processes of non-neuronal cells, with α7 nAChRs being the best studied, assisted by the availability of definitive ligands (Fig. 3).

In the CNS, functional α7 nAChRs in astrocytes (Sharma and Vijayaraghaven 2001; Shen and Yakel 2012; Wang et al. 2013) and microglia (Brawek et al. 2021) shape Ca^{2+} signals and may contribute to the tripartite synapse and immune surveillance/inflammation, respectively. As mediators of the cholinergic anti-inflammatory system, α7 nAChRs came to prominence in macrophages, as modulators of the Jak-STAT pathway (Wang et al. 2003; see below), although more complex mechanisms involving contributions from other immune cells expressing additional nAChR subtypes (Van der Zanden et al. 2009; Mizrachi et al. 2021) and dominant negative regulation by the human dupα7 isoform (Maldifassi et al. 2018) have emerged.

The mechanism of α7 nAChR-mediated signaling in some non-neuronal cells, including macrophages and T cells, presented a conundrum as α7 nAChR agonists failed to elicit detectable currents (Razani-Boroujerdi et al. 2007). This has been resolved by the discovery that α7 nAChRs can adopt a metabotropic mode of signaling (Kabbani and Nicholls 2018). Heterotrimeric GTP binding protein (G protein) can bind to the intracellular loop between M3 and M4 of the α7 nAChR and is activated when the receptor enters the desensitized configuration. The mechanistic details of G protein dissociation are unclear, but release of $G_{\alpha q}$ activates phospholipase C and promotes Ca^{2+} release from intracellular stores. This response does not appear to require α7 nAChRs to pass through the open channel state as so-called silent agonists are effective in eliciting a metabotropic response (Papke et al. 2015). Metabotropic signaling has now been reported for other LGICs including α4β2 and α3β4 nAChRs (Dore et al. 2017; Acharya et al. 2020; Kumar et al. 2020; Falcon-Moya and Rodríguez-Moreno 2021).

3.6.1 Inflammation

Non-neuronal acetylcholine is implicated in a wide range of immune and inflammatory conditions (Cox et al. 2020). Some 20 years ago, a cholinergic anti-inflammatory mechanism in macrophages was shown to be promoted by α7 nAChRs, following activation by acetylcholine released from the vagus nerve (Wang et al. 2003). The resultant decrease in the production of inflammatory cytokines, notably TNF-α, is mediated by activation of the JAK2/STAT3 pathway while also inhibiting the pro-inflammatory nuclear translocation of NF-κB (Mizrachi et al. 2021). In addition to biochemical and pharmacological evidence for the specificity for α7 nAChRs in this role in macrophages in vitro, serum TNF-α in response to endotoxin was shown to be significantly elevated in α7 null mutant mice compared with wild-type controls (Wang et al. 2003). Non-neuronal acetylcholine has since been shown to be released from T cells (Cox et al. 2020), and expression of α7 nAChRs mediating anti-inflammatory responses in other immune cells has also been proposed (Mizrachi et al. 2021), together with alternative cellular pathways for the α7 nAChR-mediated anti-inflammatory responses (Zhu et al. 2021). There are some more twists to this story, with the metabotropic mode of action of α7 nAChRs in immune cells (Papke et al. 2015), as described above. Furthermore, the chaperone RIC-3 (see Sect. 4.4) is also present in immune cells, and its expression is regulated by pro-inflammatory signals (Ben-David et al. 2020); thus, levels of anti-inflammatory α7 nAChRs may be tuned to demand. The cholinergic anti-inflammatory response is relevant to many inflammatory conditions, including arthritis, inflammatory bowel disease, and sepsis (Cox et al. 2020).

Microglia are the resident macrophages in the brain, and α7 nAChRs have an inhibitory effect on the production of pro-inflammatory molecules in activated microglia, as in peripheral macrophages. However, it is not clear if the mechanisms are identical. α7 nAChRs may additionally upregulate glutamate transporters in microglia to enhance removal of cytotoxic glutamate (Morioka et al. 2018).

3.6.2 Respiratory System and Lung Disease

The association of tobacco smoking with lung disease has long prompted an interest in the effects of nicotine on the airways. The association of genomic variants of the *CHRNA5-CHRNA3-CHRNB4* gene cluster, encoding α5, α3, and β4 subunits, with lung disease as well as nicotine dependence has reinforced this interest (Bierut 2010). In the lung (and other mucosal membranes), non-neuronal acetylcholine is produced by chemosensory bush cells that express the synthetic enzyme choline acetyltransferase and line the trachea (Krasteva et al. 2011); acetylcholine is also released from local sensory nerve endings. It acts as an autocrine and paracrine growth factor for bronchial epithelial cells (Friedman et al. 2019), and α7 and α5* nAChRs contribute to this effect by promoting proliferation of epithelial and smooth muscle cells, including during lung development.

Airway epithelial cells express almost all nAChR subunits (Hollenhorst and Krasteva-Christ 2021). Metabotropic signaling via α3β4 nAChRs in tracheal epithelial cells has been implicated in mucociliary clearance in the airways, a mechanism for the elimination of inhaled pathogens (Kumar et al. 2020). This functional study with pharmacological tools and null mutant mice is persuasive evidence for α3β4 nAChRs; it is not yet clear if they signal via direct interaction with G proteins or via another mediator such as β-arrestin (Hollenhorst and Krasteva-Christ 2021).

Tobacco smoking is the major cause of chronic obstructive pulmonary disorder (COPD). Activation of the α7 nAChR-mediated anti-inflammatory process has been mooted as a potential therapy, with the caveat that the proliferative effects of α7 nAChRs could generate undesirable side effects (Hollenhorst and Krasteva-Christ 2021). Indeed, α7 nAChRs have been credited with mediating proliferative, pro-angiogenic, and pro-metastatic actions attributed to nicotine in lung cancer (Friedman et al. 2019). α5* nAChRs have also been implicated in such actions in non-small cell lung cancer (NSCLC) (Sun et al. 2017), offering an alternative target for inhibition if selective drugs become available (Friedman et al. 2019).

3.6.3 Cancer

Numerous nAChR subunits appear to be upregulated in many different types of cancer cells; nAChRs promote proliferation, angiogenesis, and migration and are anti-apoptotic, as indicated above for lung cancer (Friedman et al. 2019). Thus, actions that may be considered physiological during development, or neuroprotective for neurodegenerative conditions, promote tumorigenesis and tumor progression. The prevalence of α9 and α10 subunit expression in cancer cells has been highlighted (Pucci et al. 2022). Given the limited expression of these subunits in the body, α9α10 nAChRs could serve as biomarkers for cancer cells, as well as offering a more selective nicotinic therapeutic target.

3.6.4 Keratinocytes and Skin Conditions

Non-neuronal acetylcholine is produced in keratinocytes and regulates epidermal differentiation, proliferation, and cell adhesion (Kurzen et al. 2007). Experimental blockade of nAChRs in skin disrupts adhesion in the basal layers of the epidermis, producing acantholysis. This is a characteristic of pemphigus vulgaris, an autoimmune skin disease in which the disrupted epidermis produces blisters. While most autoantibodies in this condition appear to be directed against adhesion molecules, autoantibodies to nAChRs, notably α3 and α9 subunits, have also been detected (Grando 2012), suggesting that nicotinic agonists could be useful therapeutically. This is supported by the negative association between pemphigus and tobacco smoking, attributed to beneficial effects of nicotine (Lai et al. 2018). Immune actions effected by nAChRs, including the cholinergic anti-inflammatory response mediated by α7 nAChRs, may contribute to the presumed beneficial effects of nicotine. α7 nAChRs in keratinocytes are also proposed to regulate epidermal homeostasis, revealed in the rare skin condition Mal de Meleda (Sect. 4.4), and α7 nAChRs have been proposed as a therapeutic target for both inflammatory and fibrotic skin conditions (Stegemann and Bohm 2020).

4 nAChRs as Potential Drug Targets: Sites for Ligand Interaction

Historically, LGIC receptors offered two modes of drug action: agonism at the orthosteric site and inhibition, either competitive at the orthosteric site or non-competitive by blocking ion conduction. As detailed knowledge of nAChR structure and function increases, many new sites for ligand interaction have emerged, their characterization fueled by the desire to exploit new therapeutic opportunities (Fig. 1). An aspiration for drug discovery is rational drug design, with the assumption that understanding the molecular basis of ligand binding and its functional consequences will advance this ambition. This is extended by the range of associated molecules that present novel sites for modulating nAChR expression and function. In this section current details of the different classes of target site will be briefly summarized.

4.1 Orthosteric Binding Site

The agonist or orthosteric binding site resides at the interface between an α subunit (principal face) and the adjacent subunit (complementary face; Fig. 2). Thus, heteromeric nAChRs have two orthosteric sites that display cooperative behavior with respect to channel opening. Homomeric nAChRs have five agonist binding sites, but occupancy of just one site is sufficient for channel activation (Andersen et al. 2013; Bouzat et al. 2018): this is speculated to facilitate uniform responses to low and variable acetylcholine concentrations that prevail with volume transmission. Atomic resolution structures have provided molecular insight into ligand recognition and receptor operation. Pioneering EM structures of *Torpedo* nAChR (Unwin 2013) have been complemented more recently by X-ray crystallographic and cryo-EM structures of muscle and neuronal nAChRs, notably α4β2, α3β4, and α7 nAChRs (Morales-Perez et al. 2016; Delgado-Velez et al. 2021) (Fig. 2). A particular contribution to characterization of the orthosteric site has come from studies of an acetylcholine binding protein (AChBP) secreted by certain molluscs. The AChBP is a homologue of the extracellular domain of a nAChR, a homopentamer most closely related to the α7 nAChR (Brejc et al. 2001). The high-resolution structures confirm that the orthosteric site is highly conserved, not only among nAChR subtypes and across species, but more generally within the larger family of cys-loop receptors and prokaryotic prototypes (Sauguet et al. 2015). The site is determined by 6 non-contiguous loops in the protein structure, three from each both the principal and complementary subunits, labeled A-F and bearing conserved, predominantly aromatic residues (Fig. 2). Agonists (e.g., acetylcholine, nicotine) stabilize the mobile loop C of the principal binding subunit in close apposition to the adjacent (complementary) subunit; loops A, B, C, and D form an "aromatic box" with the tryptophan residue of loop B coordinating the quaternary nitrogen of acetylcholine (Purohit and Auerbach 2013; Post et al. 2017). Agonists with a protonated secondary nitrogen (e.g., varenicline) make an additional cation-π interaction with a tyrosine in loop C. Understanding differences in binding modes between agonists offers clues for designing potential subtype-selective agonists.

Partial agonists activate receptors with low efficacy, and this can be a subtype-specific feature. For example, the smoking cessation drugs cytisine and varenicline are high affinity partial agonists at α4β2 nAChRs and full agonists (but with lower potency) at α7 nAChRs (Mihalak et al. 2006; Gotti and Clementi 2021). NS6740 is a selective but very low efficacy agonist at α7 nAChRs, referred to as a "silent agonist" because it produces negligible discernible channel activation. The agonist behavior of such compounds is revealed in the presence of allosteric modulators that overcome desensitization (Papke et al. 2015). It has been demonstrated that NS7640 does indeed bind to the orthosteric binding site of the α7 nAChR (Blunt and Dougherty 2019; Pismataro et al. 2021). Interestingly, NS6740 is effective in promoting metabotropic signaling through α7 nAChRs (Papke et al. 2015; Kabbani and Nichols 2018).

Sazetidine-A has the appearance of a classical partial agonist at α4β2 nAChRs, but its actions are more complex. Sazetidine-A interacts differentially with the two stoichiometries of α4β2 nAChRs (Eaton et al. 2014). It is a highly efficacious agonist at $(\alpha4)_2(\beta2)_3$ nAChRs, similar to acetylcholine, but has approximately 12% of the efficacy of ACh at $(\alpha4)_3(\beta2)_2$ nAChRs. The explanation of this discrepancy concerns the additional non-canonical acetylcholine binding site (see Sect. 4.3). Comprehensive understanding of the molecular basis of different modes of partial agonism, which has some desirable therapeutic qualities, would inform drug design (Rucktooa et al. 2012; Blom et al. 2019; Minguez-Viñas et al. 2021). Sazetidine-A has received some interest in several preclinical models (Rahman et al. 2015).

Competitive antagonists (e.g., the natural products and important research tools DHβE for heteromeric neuronal nAChRs and MLA for α7 nAChRs) also bind at or close to the orthosteric binding site but stabilize loop C in an extended conformation that precludes channel opening (Hansen et al. 2005). There has been limited interest in such molecules for therapeutic indications, reflecting the generally favorable effects of neuronal nAChR activation.

4.2 Ion Channel Sites

Structurally diverse channel blocking drugs act as non-competitive antagonists by occluding the lumen of the ion channel. Photoaffinity labeling of Torpedo nAChR with the neuroleptic [^{3}H]chlorpromazine (that blocks nAChRs at non-clinical concentrations) identified binding sites at the cytoplasmic end of the M2 transmembrane helix (Chiara et al. 2009). The clinically approved drug mecamylamine is an open channel blocker of nAChRs (see Sect. 5.1). Although it blocks all nAChRs, there are differences in the duration of inhibition between subtypes, with muscle and α7 nAChRs showing transient inhibition compared with a sustained block of heteromeric neuronal nAChRs (Papke et al. 2001). Mecamylamine is racemic, and comparison of the S-(+) and R-(-) enantiomers has indicated some subtle differences in their interaction with nAChRs, prompting interest in new therapeutic possibilities (Nickell et al. 2013).

4.3 Allosteric Binding Sites

nAChRs possess a variety of sites for drug interaction or ion binding, distinct from the orthosteric sites (Chatzadiki and Millar 2015; Changeux 2018; Fig. 1). The effect of molecules acting at these sites is allosteric as they modulate the transduction process indirectly. Consequently, such compounds are classified as either positive (PAMs) or negative allosteric modulators (NAMs). Type I PAMs increase peak responses, whereas type II PAMs also enhance the duration of channel opening by delaying desensitization, making them more efficacious (Bouzat et al. 2018). Type II PAMs are distinguished by their ability to reactivate desensitized nAChRs.

4.3.1 Accessory Subunit Site(s)

The evolution of heteromeric nAChRs from a homomeric ancestor introduced a new potential binding site. This site resides between the principal face of the "accessory" subunit in the fifth position and the complementary face of the adjacent α subunit (Fig. 2), making it a homologue of the orthosteric site, and is coined the *non-canonical* site (Wang and Lindstrom 2018). Characterization of this site is an active research area, with the diversity of subunit permutations yielding a plethora of data and an evolving landscape of putative interactions. In some cases this site can accommodate acetylcholine, as well as other agonists, making it an *unorthodox* agonist binding site. For binding acetylcholine it was generally accepted that the accessory subunit should be an α subunit (other than α5) in order to provide an orthodox principal face (Wang and Lindstrom 2018). This site has been best characterized for the α4β2 nAChR with the stoichiometry $(\alpha4)_3(\beta2)_2$ that yields an α4/α4 unorthodox binding site, capable of binding acetylcholine with lower affinity and with a distinct pharmacology from the α4/β2 orthosteric site (Mazzaferro et al. 2011; Eaton et al. 2014; Wang and Lindstrom 2018). NS9283 is a selective ligand for the α4/α4 binding site and enhances $(\alpha4)_3(\beta2)_2$ nAChR responses to classical agonists. NS9283 was initially considered to be an allosteric modulator but is now classified as an agonist that selectively binds at this unorthodox acetylcholine binding site (Olsen et al. 2014). For the purposes of this chapter, the precise name ascribed to this site is a moot point as the orthosteric sites dominate agonist responses, with binding to the unorthodox agonist binding site modulating their response (Weltzin et al. 2019), hence inclusion in this section on allosteric binding sites.

Whereas NS9283 is a selective agonist for the α4/α4 binding site and does not bind to the orthosteric sites, sazetidine-A is an agonist at the orthosteric sites that is unable to bind to the non-canonical site (Wang and Lindstrom 2018). Hence, acetylcholine at high concentrations can act via the α4/α4 non-canonical binding site of $(\alpha4)_3(\beta2)_2$ nAChRs to increase function per receptor several-fold, compared to that produced by acetylcholine occupying the two α4/β2 orthosteric sites only, whereas sazetidine-A is unable to reproduce this enhancement in efficacy and this contributes to its characterization as a partial agonist (Mazzaferro et al. 2014; Sect. 4.1). In populations of native α4β2 nAChRs, the proportions of the two stoichiometries are generally not known, but NS9283 and sazetidine-A serve as useful probes (DeDominicis et al. 2017; Lamotte-d'Incamps et al. 2018).

Surprisingly, there is some evidence that the α5 or β3 nAChR subunit in the accessory position in a α4β2* nAChR can also constitute an unorthodox ACh binding site with the adjacent α4 subunit, despite the absence of one or more key residues from loops A, B, and C (Jain et al. 2016). These sites can bind acetylcholine but cannot accommodate, NS9283, indicating a distinct pharmacology compared with the α4/α4 non-canonical binding site. In addition, the α5 accessory subunit has been proposed to confer sensitivity to galantamine, the acetylcholinesterase inhibitor, and Alzheimer's disease drug that also potentiates nicotinic responses by acting as a type I PAM (Kuryatov et al. 2008). However, the involvement of α5 subunits in the nAChR sensitivity to galantamine has recently been challenged (Prevost et al. 2021).

In contrast, β2 and β4 accessory subunits generate true allosteric non-canonical sites, unable to bind acetylcholine, at the subunit interface. For example, the anthelmintic drug morantel acts as a PAM at $(\alpha3)_2(\beta2)_3$ nAChRs by binding to the β2/α3 site, whereas its analogue oxantel acts as a NAM via the β2/α4 site of $(\alpha4)_2(\beta2)_3$ nAChRs (Wang and Lindstrom 2018). These examples encourage interest in the generation of subtype-selective allosteric ligands for β/α non-canonical sites.

4.3.2 C-terminal/cys-Loop Interactions

The PAMs β-estradiol, desformylflustrabromine, and Br-PBTC interact with the cys-loop to influence channel gating (Alcaino et al. 2017; Norleans et al. 2019). Homology modeling and mutagenesis studies implicate an intramembrane binding site close to the C terminal, enabling communication with the Cys-loop to facilitate potentiation of channel gating. Br-PBTC is selective for α2 and α4 (but not α3) subunits (Norleans et al. 2019), whereas desformylflustrabromine is selective for β2* nAChRs and is attracting considerable interest in a variety of preclinical models (Pandya 2016).

4.3.3 Ca^{2+} Binding Sites

Most nAChRs are potentiated by Ca^{2+} acting at sites in the extracellular domain, below the level of the orthosteric binding site (Changeux 2018). α7 nAChRs show the highest potentiation, reflecting the presence of five Ca^{2+} binding sites (Le Novere et al. 2002), which is particularly striking given the high Ca^{2+} permeability of α7 nAChRs and their propensity to modulate intracellular Ca^{2+} signaling pathways.

4.3.4 Transmembrane Sites

A variety of agents that act as PAMs or NAMs interact with the transmembrane region to modulate nAChRs. Several distinct sites have been defined between or within subunits (Fig. 1), based on evidence from chimeric and mutated receptors, and molecular simulation studies (Alcano et al. 2017; Bouzat et al. 2018). A well-studied example of an intra-subunit PAM is PNU-120596, an α7 nAChR-selective type II PAM predicted to bind in a cavity between the four TM spans of an α7 nAChR subunit; this PNU-120596 binding site overlaps a cholesterol binding site (see below; Chatzidaki and Millar 2015; Targowska-Duda et al. 2019; Zhuang et al. 2022). Thus, five putative binding sites are predicted in a homomer, and occupancy of at least four sites is reported to be required for positive modulation of channel function by PNU-120596 (da Costa and Sine 2013). Other type II PAMs and the type I PAM NS-1728 may also bind at or near this intramembrane site (Bouzat et al. 2018). It has been recently suggested that these α7 nAChR-selective PAMs can also modulate $GABA_A$ receptors by acting at the transmembrane inter-subunit anesthetic site (Arias et al. 2022). This introduces a note of caution with respect to their specificity, although their effects at $GABA_A$ receptors is very modest compared with the potentiation of α7 nAChRs.

An inter-subunit site has also been defined for ivermectin, an anthelmintic drug that acts as a PAM at various cys-loop anionic LGICs and also potentiates α7 nAChRs (but not other nAChR subtypes or $5HT_3$ receptors). Collins and Millar (2010) deduced that ivermectin binds in close proximity to a predicted inter-subunit membrane cavity in α7 nAChRs, which is consistent with crystal structures of ivermectin complexed with glycine receptors (Huang et al. 2017). This site may also bind the type I PAM 5-hydroxyindole (Targowska-Duda et al. 2019) although additional interactions in the extracellular domain are predicted for type I PAMs (Bouzat et al. 2018).

4.3.5 Lipids and the Membrane Interface

A somewhat neglected topic is the influence of the membrane environment on nAChRs. It is well established that membrane lipids enhance receptor function, and this is proposed to result from stabilizing the nAChR in an agonist-activatable conformation (Baenziger et al. 2015). Experimental evidence, from mutagenesis studies to single particle cryo-EM technology and molecular dynamics, supports a cholesterol binding site between M4 (the transmembrane αhelix most exposed to the lipid bilayer), M3 and M1 of the adjacent subunit (Baenziger et al. 2015; Delgado-Velez et al. 2021; Zhuang et al. 2022). In the α7 nAChR, the PAM PNU-120596 is proposed to displace cholesterol that occupies this site in the desensitized conformation. This reverses desensitisation to sustain the activated state, while agonist remains bound and displaced cholesterol contacts the intracellular MX helix between M3 and M4 (Figure 2e; Zhuang et al. 2022).

General anesthetics also interact with nAChRs at the lipid-protein interface to allosterically modulate nAChR function, as well as acting at other transmembrane and extracellular allosteric sites (Foreman et al. 2015). The distinct lipid characteristics of neuronal and muscle plasma membranes, as well as those of the diverse tissues modulated by the non-neuronal cholinergic system, will impose differential influences on nAChRs. Moreover, the lipid composition of local membrane domains where nAChRs are localized (e.g., α7 nAChRs in cholesterol-rich lipid rafts) further complicates the picture. The local lipid environment can also influence receptor folding, assembly, trafficking, and clustering (Baenziger et al. 2015). Better understanding of these environments and how they affect nAChR behavior could suggest new therapeutic interventions (Schneider et al. 2017; Fabiani and Antollini 2019).

4.4 Interacting Proteins

Progress in uncovering the players in the cell biology of nAChR assembly, trafficking, localization, stabilization, regulation, and turnover is revealing new prospects for potential drug targets (Matta et al. 2021). The life cycle of $(\alpha)_2\beta\varepsilon\delta$ nAChRs at the mammalian muscle endplate is the best characterized, with a plethora of components involved in the different stages (Li et al. 2018; Rudolf and Straka 2019). The ability of associated proteins to contribute to disease is illustrated by congenital myasthenic syndrome produced by defects in the genes that encode the five proteins that directly interact with muscle nAChRs to stabilize their end-plate localization (agrin, LRP4, MuSK, Dok-7, and rapsyn) although this syndrome is more commonly due to mutations in the muscle nAChR subunit genes themselves (Engel et al. 2015; Sect. 3.1). The regulation of neuronal nAChR subtypes is more complex, diverse, and less well understood. Proteomic analysis of proteins associated with α7 nAChRs in mouse brain has produced a daunting interactome of more than 120 partners associated with many cellular functions (Mulcahy et al. 2018). Some examples of the better characterized interacting proteins in key regulatory or signaling roles that might offer therapeutic opportunities are given below.

4.4.1 Chaperones

The efficiency of heterologous expression of neuronal nAChRs varies with subtype and cell type. In contrast to α4β2 nAChRs that express readily, functional heterologous expression of α7 nAChRs in the plasma membrane has proven difficult to achieve, and this prompted a quest for factors that facilitate expression. Two proteins, RIC-3 and NACHO, overcome this difficulty by promoting the correct folding and assembly of α7 nAChRs (Crespi et al. 2018a). RIC-3, first discovered in *C. elegans* for its involvement in nematode nAChR maturation, is an ER-resident transmembrane protein. The influence of RIC-3 is limited to nAChRs and $5HT_3$ receptors, with complex effects dependent on isoform and ratio to receptor subunits. Whereas RIC-3 promotes the membrane expression of α7 nAChRs, it can be a negative regulator of heteromeric nAChRs (Treinin 2008). Of therapeutic interest, RIC-3 is emerging as a modulator of α7 nAChR expression in inflammatory conditions (Mizrachi et al. 2021).

The requirement for RIC-3 for α7 nAChR expression is not absolute but cell-type dependent. A second ER-resident chaperone, NACHO, was subsequently identified. Evidence from transgenic mice lacking NACHO indicates that, in contrast to RIC-3, NACHO is essential for α7 nAChR assembly and also has a profound effect on other nAChRs expressed in the CNS, but not other types of LGIC or ganglionic nAChRs (Matta et al. 2021). Null mutant mice lacking NACHO show deficits in locomotor and cognitive functions consistent with loss of brain nAChRs. NACHO is proposed to act upstream of RIC-3, via the N-linked glycosylation machinery and another ER chaperone, calnexin. Models for the concerted actions of NACHO and RIC-3 on the folding and assembly of α7 nAChRs have recently been scrutinized (Loring 2022).

The diverse locations of nAChRs in neurons (postsynaptic, extrasynaptic, and presynaptic) also demand precise trafficking mechanisms. Insight into these is provided by α6-containing nAChRs (Matta et al. 2021). α6β2β3 nAChRs are expressed in the striatal terminals of dopamine neurons and require at least four proteins: NACHO for assembly, LAMP5 and SULT2B1 for trafficking and surface expression, and BARP for channel activation (Gu et al. 2019). A different sequence is proposed for α6β4 nAChRs in sensory neurons: NACHO is not required, IRE1α promotes folding and assembly, and then BARP facilitates surface expression (Knowland et al. 2020).

Many more endogenous factors conspire to regulate nAChR assembly (Crespi et al. 2018a). Exogenous substances can also influence this process. The upregulation of α4β2 nAChRs by nicotine is attributed to its ability to promote the assembly and surface expression of high affinity $(\alpha4)_2(\beta2)_3$ nAChRs (Fox-Loe et al. 2017). It has been suggested that this reflects stabilization of the desensitized conformation by bound nicotine, with the M4 transmembrane helix more favorably aligned with the lipid bilayer (Baenzinger et al. 2015). An equivalent physiological role has been proposed for acetylcholine in the expression of α9α10 nAChRs in cochlear hair cells, thus timing nAChR expression with cholinergic innervation (Gu et al. 2020). Menthol, a cigarette additive, is asserted to upregulate nAChRs, notably in dopamine neurons, via an allosteric interaction (Matta et al. 2021). Upregulation of brain nAChRs by exogenous substances presents a potential opportunity for therapeutic intervention but must also be considered when such agents are administered for other indications.

4.4.2 Scaffold Proteins

The synaptic or perisynaptic localization of nAChRs is organized via interacting proteins that stabilize their location through connection to the underlying cytoskeleton; this intricate arrangement can also orchestrate interactions with regulatory mechanisms and downstream targets. The molecular determinants of neuromuscular junction formation, with rearrangement of the postjunctional membrane and clustering of nAChRs, provide an insight into the complexity of the process (Li et al. 2018). Stabilization of clustered muscle nAChRs occurs through multiple interactions of nAChR subunits, notably β, with rapsyn dimers to form a stable lattice (Zuber and Unwin 2013). Rapsyn, a 43 kDa cytoplasmic protein tethered to the muscle membrane through myristoylation, is essential for nAChR clustering. It binds to sites within the intracellular domains of muscle nAChR subunits and restrains nAChRs at the endplate, as rapsyn's multiple other binding partners include the actin cytoskeleton and β-dystroglycan (Ferns 2021).

α3β4* nAChRs are concentrated postsynaptically at ganglionic synapses by phosphorylation-dependent binding to the 14-3-3 adaptor protein. 14-3-3 in turn connects to the postsynaptic scaffold by interacting with the adenomatous polyposis coli (APC) protein which has a rapsyn-like role in organizing a multimolecular postsynaptic complex (Rosenberg et al. 2008). The α3 subunit is the major determinant of the postsynaptic localization of this nAChR subtype, and this is speculated to occur via a conserved 14-3-3 binding motif in the large intracellular loop between M3 and M4 of α3 subunits (Ferns 2021). In contrast, α7 nAChRs are directed to a perisynaptic location at ganglionic synapses via lipid rafts and are also associated with lipid rafts in the CNS where they can occupy a similar perisynaptic locale, as well as being present at presynaptic sites. Their association with lipid rafts has implicated α7 nAChRs in Alzheimer's disease because of a spatial relationship with Aβ peptide processing and partitioning (Fabiani and Antollini 2019).

While glutamate and GABA synapses provide insight into the dynamic organization of LGIC receptors at postsynaptic and extrasynaptic sites (Boeckers 2006; Lorenz-Guertin and Jacob 2018), much remains to be learned about the organization and stabilization of neuronal nAChRs in the plasma membrane of both neurons and non-neuronal cells. The submembrane scaffold orchestrates interactions with multiple partners. These include mediators of posttranslational modifications, such as kinases and phosphatases, that regulate receptor function, turnover, and plasticity. This interactome comprises potential targets for modulating receptor activity. One example is the PDZ scaffold protein PICK1, best characterized for binding AMPA receptors and critically considered as a therapeutic target for treating chronic pain (Sorensen et al. 2022); interestingly, PICK1 has also been reported to interact with α7 nAChRs to negatively affect their clustering (Baer et al. 2007).

4.4.3 Ly6/UPAR Family Proteins

This large family of prototoxins is characterized by the presence of a three-finger fold also found in elapid snake toxins, including α-bungarotoxin. Indeed, several members of the Ly6 branch of the family interact with nAChRs. These comprise GPI-anchored lynx1 and lynx2, as well as some less well-characterized Ly proteins and two secreted proteins, SLURP-1 and -2; sequence comparisons reflect their evolutionary relationship and structural similarity with snake αneurotoxins (Miwa et al. 2019; Tsetlin et al. 2021). These proteins interact with the extracellular domain of nAChRs to have a generally inhibitory effect.

Lynx1 was the first to be discovered and is the best characterized (Miwa 2021). It occurs throughout the brain, notably in the hippocampus, cortex, and cerebellum, in association with both α7 and α4β2 nAChRs (Miwa et al. 2019). It also interacts with most other neuronal nAChR subtypes. Among its many effects, lynx1 alters agonist affinity, rate and recovery of desensitization, and single channel kinetics. While these effects are compatible with the extracellular binding of lynx1 to nAChRs at the plasma membrane, it can also influence the assembly of α4β2 nAChRs at the level of the ER in favor of the low affinity $(\alpha4)_3(\beta2)_2$ isoform (Nichols et al. 2014). This is consistent with molecular dynamics simulations of lynx1 binding at the non-canonical (unorthodox) α4/α4 interface in close proximity to loop C (Miwa 2021). Both altered stoichiometry and impaired movement of loop C could account for some of the changes in nAChR properties such as agonist sensitivity and desensitization kinetics. Lynx1 can also influence the expression of other nAChR subtypes (Miwa 2021).

Behavioral studies, including the use of lynx1 null mutant mice, revealed increased synaptic plasticity and associative learning in the absence of lynx1, but this was accompanied by enhanced neurodegeneration with aging (Miwa et al. 2006; Morishita et al. 2010). Thus, it is proposed that lynx1 acts a "molecular brake" to balance plasticity with survival. Modulating lynx1 to manipulate this balance has theoretical therapeutic potential for certain conditions (Miwa 2021). In a similar vein, a recent report describes lynx1 in skeletal muscle where its prime function seems to relate to the maintenance of adult and aging neuromuscular junctions (Doss et al. 2022).

Lynx2 displays a complementary and non-overlapping distribution in the brain compared with lynx1 (Miwa et al. 2019). Its presence in the amygdala and prefrontal cortex, regions associated with anxiety, and the increase in anxiety-like behaviors in lynx2 null mutant mice (Tekinay et al. 2009) indicate distinctive roles for lynx2 compared with lynx1. A recent characterization of molecular interactions, using atomic force microscopy, revealed marked differences in the nAChR subtype preferences of the two lynx proteins: lynx1 showed only a mild preference for α7 over α3β4 nAChRs, whereas lynx2 has two orders of magnitude preference for α3β4 nAChRs (Pisapati et al. 2021).

SLURP-1 and -2 are secreted soluble forms of this group of nAChR-associated protein, found in peripheral tissues in conjunction with the non-neuronal cholinergic system (Miwa et al. 2019; Tsetlin et al. 2021). Conflicting results for their precise effects (positive or negative) have been attributed to differences between recombinant, synthetic, and native forms of these proteins. One report that SLURP-2 also acts on muscarinic receptors (Lyukmanova et al. 2016) complicates interpretation of its actions. There is an emerging consensus that SLURP-1 acts allosterically to inhibit α7 and heteromeric nAChRs including α9α10 (Durek et al. 2017). In addition to their extracellular actions, SLURP-1 and -2 are also located in the cytoplasm where they regulate growth and cell cycle activity. Null mutant mice lacking either SLURP-1 or SLURP-2 develop palmoplantar keratoderma, an abnormal thickening of the skin on the soles of the feet and palms of the hands, reflecting the loss of control of epidermal growth. Palmoplantar keratoderma is characteristic of the human condition Mal de Meleda, a rare autosomal recessive condition caused by mutations in the *SLURP-1* gene (Shulepko et al. 2021). This illustrates the important influence of nAChRs in skin (Sect. 3.6) notably the control of growth, differentiation, and apoptosis by α7 nAChRs that is restrained by SLURP-1. SLURP-1 and -2 are also proposed to inhibit the proliferation of cancer cells (Miwa et al. 2019): SLURP-1 and lynx-1 have been found in lung epithelium where they may have protective effects with respect to lung cancer, as both reduce proliferation of lung adenocarcinoma cells in vitro (Bychkov et al. 2021; Hollenhorst and Krasteva-Christ 2021). SLURP-1 is downregulated in inflammatory conditions such as asthma (Friedman et al. 2019), and peptide mimics could be useful for treating inflammatory lung disease and lung cancer (Bychkov et al. 2021; Hollenhorst and Krasteva-Christ 2021).

4.4.4 Metabotropic Partners

Other interacting partners include signaling molecules such as nitric oxide synthase (Jayaker et al. 2014), G proteins (Kabbani and Nichols 2018), and growth factors (Bychkov et al. 2021) that introduce the concept of nicotinic signaling independent of channel activation, as discussed in Sect. 3.6. Spatially apposed or dimeric receptors, where a nAChR physically interacts with another receptor to produce distinctive properties, for example P_2X (Khakh et al. 2005; Wieskopf et al. 2015) or D_3 dopamine receptor (Sect. 3.3; Mutti et al. 2022), have also been described (Bono et al. 2021). These examples expand the repertoire of nicotinic signaling modes and the opportunities for intervention.

5 Drug Discovery: Unmet Aspirations and New Prospects

The previous sections document the myriad of neuronal nAChR subtypes with different subunit compositions, stoichiometries, and binding partners, their widespread distribution reflecting the neuronal and non-neuronal cholinergic systems, and their association with a diversity of physiological processes and clinical conditions. In the early 1990s, this prompted tremendous interest from pharmaceutical companies in neuronal nAChRs as drug targets, with the generation of a multitude of novel compounds. But, with the single exception of varenicline, none has yet made it to the clinic. Now this lack of success has led most companies to abandon their nAChR drug discovery programs. Some of the reasons underlying this failure to deliver novel nicotinic drugs have recently been cogently discussed by Bertrand and Terry (2018); this section will summarize some of the key factors and consider the future. But first it is worth noting some precedents for the development of successful nicotinic drugs.

5.1 Examples of Clinically Approved Nicotinic Receptor Drugs

5.1.1 Muscle Relaxants

Neuromuscular blocking agents have been used as muscle relaxants in surgery for 80 years and provide an example of rational drug development. Synthetic drugs based on the competitive antagonist (+)-tubocurarine have been optimized to promote desirable properties, notably fast, non-depolarizing blockade of muscle nAChRs and more rapid reversal, while minimizing side effects. For example, the aminosteroids vecuronium and rocuronium, currently used as adjuncts in anesthesia, for tracheal intubation and in mechanical ventilation, incorporate acetylcholine moieties to enhance their affinity at the muscle nAChR (Bowman 2006).

5.1.2 Antihypertensive Autonomic Ganglia Blockers

In the 1950s several nAChR antagonists were developed as ganglion blockers. The most successful of these was the secondary amine mecamylamine (see Sect. 4.2), which was the first orally available drug approved for treating hypertension. Thirty years later its use was terminated (except for severe essential hypertension and malignant hypertension) due to side effects attributed to inhibition of other parasympathetic functions, now known to be mediated by $\alpha3\beta4^*$ nAChRs. Mecamylamine readily crosses the blood-brain barrier and centrally mediated nAChR blockade by mecamylamine, and its individual stereoisomers have been considered for a number of indications, including Tourette's syndrome, autism, and mood disorders (Sanberg et al. 2012; Nickell et al. 2013). At low doses, which have shown some efficacy in preliminary clinical studies of these conditions, parasympathetic side effects are minimized.

5.1.3 Smoking Cessation Agents

Nicotine is a pioneer for drug discovery endeavors. Recognition that nicotine is largely responsible for tobacco addiction but does not itself cause the major health issues associated with smoking paved the way for the development of nicotine products (nicotine replacement therapies, NRTs) for smoking cessation (Stead et al. 2012). The clinical approval NRTs inspired a search for new, more efficacious drugs to counter tobacco addiction. Another natural product that was also inspirational is cytisine, which has traditionally been used for smoking cessation in Eastern Europe where it is marketed as Tabex (Tutka et al. 2019). Like nicotine, cytisine is a high affinity agonist at $\alpha4\beta2^*$ and $\alpha6\beta2^*$ nAChRs (i.e., nAChRs associated with the mesolimbic dopamine system; see Sect. 3.3) but cytisine activates these nAChRs with lower efficacy than nicotine. The partial agonist profile of cytisine is replicated in the synthetic drug varenicline, developed by Pfizer in 1997 and licensed by the FDA as Champix for smoking cessation in 2006 (Jordan and Xi 2019). Partial agonists are appealing for moderating nicotinic responses where mild activation of $\beta2^*$ nAChRs (to prevent withdrawal symptoms) together with a degree of inhibition (to avoid reinforcement) is deemed desirable (Rollema and Hurst 2018). Varenicline is superior to NRT in sustained quit rates; a recent random controlled trial of cytisine versus varenicline reported that cytisine is "non-inferior" (Courtney et al. 2021). It should be noted that different pharmacokinetics, dosing regimes, and the timepoints selected for assessing quit rates can confound meaningful comparisons. A further five novel compounds are known to have been developed for smoking cessation, but all were discontinued due to inadequate clinical efficacy (Rollema and Hurst 2018). Mecamylamine has also been considered, as an adjunct therapy to NRT, but not pursued (Nickell et al. 2013).

5.2 *What Kind of Nicotinic Drugs for Which Clinical Conditions?*

5.2.1 Agonist Versus Antagonist

The rationale for the type of nicotinic ligand most suitable for treating a particular clinical condition is often limited, reflecting an imprecise understanding of the role of the cholinergic system, and nAChRs in particular, in neurological and neuropsychiatric conditions and in non-neuronal tissues and associated conditions. The examples in Sect. 5.1 provide a precedent for the clinical acceptability, in principle, of competitive antagonists, channel blocking drugs, and agonists. Drug discovery programs have focused on developing nicotinic agonists, with a view to emulating endogenous cholinergic mechanisms. However, choosing between agonist and antagonist characteristics is not straightforward; the propensity of neuronal nAChRs to rapidly desensitize in the continued presence of agonist confers mixed properties to drugs that acutely serve to activate nAChRs (Bertrand and Terry 2018). This is a built-in safety mechanism to avoid over-stimulation; but desensitization caused by the lingering presence of a drug can also prevent endogenous acetylcholine from eliciting its physiological responses.

5.2.2 Partial Agonists

The balance between nAChR activation and desensitization is further shifted in partial agonists as they produce less activation than an equipotent concentration of acetylcholine. This property is advantageous for smoking cessation (Rollema and Hurst 2018). The efficacy of partial agonists differs between drugs and nAChR subtypes, reflecting different mechanistic interactions, and is exemplified by the low efficacy of sazetidine-A and NS6740 at $\alpha4\beta2$ and $\alpha7$ nAChRs, respectively (see Sect. 4.1).

Following on from their success for smoking cessation, there has been interest in $\beta2^*$ nAChR partial agonists, including varenicline, cytisine, and sazetidine-A, for alcohol and drug dependence, with promising preclinical data. There are only limited outcomes, to date, from preliminary human studies (Rahman et al. 2015). Recent data suggest that sazetidine-A promotes alcohol aversion in mice by acting through non-$\alpha4$ nAChRs (Touchette et al. 2021).

NS6740, a "silent agonist" at 7 nAChRs, is antinociceptive in some preclinical pain models (Papke et al. 2015). Its ability to engage metabotropic signaling processes in non-neuronal cells, rather than nAChR channel activation, is credited as important for its nociceptive actions. Another α7 nAChR partial agonist, EVP-6124 (encenicline), reached phase III clinical trials for cognitive improvement in schizophrenia and AD. But the schizophrenia trial failed due to lack of statistical difference from placebo (Antonio-Tolentino and Hopkins 2020), whereas the AD trial was halted because of adverse gastrointestinal effects (Alzforum 2015). Possible reasons for these failures, including overlap between α7 and $5HT_3$ receptors, are discussed by Bertrand and Terry (2018).

5.2.3 Allosteric Modulators

PAMs (Sect. 4.3) offer a means of avoiding the desensitization problem. Having no effect alone, they act to amplify the responses to endogenous agonist, that is, acetylcholine (and possibly choline in the case of α7 and α9α10 nAChRs). Thus, PAMs offer an attractive therapeutic strategy to augment endogenous receptor activation by acetylcholine that, in principle, would confer greater specificity and temporal patterning than achieved with an exogenous agonist (Uteshev 2014; Bouzat et al. 2018; Wilkerson et al. 2020). However, despite encouraging preclinical data, results from clinical trials with α7 nAChR PAMs to date have been disappointing, with no efficacy shown in phase 1b clinical trials for JNJ-3939406 or AVL-3288 in schizophrenia (Winterer et al. 2013; Antonio-Tolentino and Hopkins 2020), or JNJ-3939406 in unipolar depression (Davidson et al. 2021). PAMs selective for α4β2 nAChRs have been described, but none have progressed to clinical trials (Grupe et al. 2015; Wilkerson et al. 2020). However, the acetylcholinesterase inhibitor galantamine that is prescribed for mild to moderate AD is credited with also being an α4β2 nAChR PAM; whether this allosteric action contributes to the therapeutic efficacy of galantamine is debated (Wilkerson et al. 2020). An intriguing possibility is "combination therapy" of PAM with a low dose of agonist or acetylcholinesterase inhibitor, offering an improved agonist safety profile (Grupe et al. 2015).

5.2.4 Antagonists

There has been less focus on nAChR antagonists for therapeutic intervention although mecamylamine and its enantiomers have sparked some interest, as noted in Sect. 5.1. In light of the suggestion that mecamylamine, like nicotine, has antidepressant properties, a small molecule α4β2 nAChR antagonist, ropanicant, has been developed. Following preclinical characterization, it has so far been reported to be safe and well tolerated in two phase 1 clinical trials in healthy human subjects (Nirogi et al. 2022).

α-Conotoxins are highly subtype-selective nAChR antagonists that have contributed enormously to basic research and preclinical studies. One α-conotoxin, Vc1.1 (ACVI), characterized as a potent antagonist of rat α9α10 nAChRs and an effective analgesic in rat pain models (Hone et al. 2018), progressed to a phase 2 clinical trial for neuropathic pain. However, despite promising results, the trial was subsequently terminated when it emerged that the human α9α10 nAChR is 100 times less sensitive to Vc1.1 than the corresponding rat nAChR, making the drug uneconomic (Mohammadi and Christie 2015). A further limitation of these peptide toxins is their restricted bioavailability, requiring intrathecal or intramuscular administration. However, peptide toxins can provide a template for small molecule mimics.

A novel application of α-conotoxin ImI, an antagonist of α7 nAChRs, was recently proposed for tumor targeting. Taking advantage of the overexpression of α7 nAChRs in some tumor cells such as NSCLC, micelles loaded with an anti-cancer drug and bearing α-conotoxin ImI on their surface would be directed to the α7 nAChR-bearing tumor cells for a more targeted drug delivery (Mei et al. 2018).

5.2.5 New Prospects

Increasing knowledge of interacting partners and their influence on nAChRs (Sect. 4.4) raises the possibility of circumventing some of the issues encountered with nAChR ligands by targeting interacting molecules to regulate nAChR expression and function. Directly targeting such partners is an emerging concept, with the modulatory potential of lynx1 advocated as a stimulus to revive therapeutic development targeted at nAChRs (Miwa 2021). Another example attracting interest is the intimate relationship in immune cells between RIC-3 and α7 nAChRs that underpins the cholinergic anti-inflammatory response (Mizrachi et al. 2021).

Interacting proteins also provide a practical service to drug discovery: chaperones facilitate expression of nAChR subtypes that are otherwise refractory to heterologous expression (Knowland et al. 2020; Matta et al. 2021) and auxiliary proteins like lynx1 generate nAChR complexes whose subunit stoichiometry and behavior may better match that of native nAChRs (Miwa 2021).

5.3 Which nAChR Subtype? Is Subtype Specificity Always Desirable?

A dogma in drug discovery is that high specificity and subtype selectivity is paramount. Thus, in addition to specificity for nAChRs, enhancing the discrimination between subtypes has been a priority. This is warranted for avoiding peripheral side effects at muscle and ganglionic nAChRs and minimizing potential off target effects in the non-neuronal cholinergic system. However, the relative contributions of particular neuronal nAChR subtypes to complex disease phenotypes or underlying mechanisms are usually far from clear. This problem is not limited to nAChRs: it has been estimated that around 40% of clinical trials fail due to lack of clear drug target linkage to the disease (Bespalov et al. 2016). Neurons can express multiple nAChR subtypes, and targeting one subtype can prompt compensatory changes in others. A case for PAMs is that they avoid these issues, relying only on endogenous agonist for nAChR activation, but most of the PAMs documented show a subtype preference for α4β2 or α7 nAChRs (Grupe et al. 2015). Noting the broad specificity of nicotine in relation to its cognate receptors, together with its clinically useful properties, it is questionable whether a highly selective ligand is always the best option. Furthermore, the cholinergic system acts through both nAChRs and muscarinic receptors, but there has been limited interest in the consequences of dual targeting.

As members of the cys-loop LGIC family, nAChRs share structural and functional homology with other receptors in this class with consequences for drug specificity. A striking example derives from the similarities between α7 and $5HT_3$ receptors: tropisetron, a selective competitive antagonist of the $5HT_3$ receptor, is a selective partial agonist at α7 nAChRs (Macor et al. 2001). Tropisetron has shown some efficacy in preclinical models of schizophrenia, with blockade by MLA implicating α7 nAChRs in at least some of its effects (Hashimoto 2015). The improvement in cognitive and sensory deficits in a small, randomized control trial of non-smoking schizophrenic patients treated with tropisetron is attributed to its agonism at α7 nAChRs (Zhang et al. 2012). However, other $5HT_3$ receptor antagonists with little or no effect at α7 nAChRs have also been credited with improving sensory gating (Tsitsipa et al. 2022). It remains to be established whether activation of α7 nAChRs and inhibition of $5HT_3$ receptors achieve equivalent effects (with respect to sensory deficits at least) or if there is any benefit, such as greater efficacy or lower drug concentrations, from dual targeting.

Most of the compounds discussed in Sect. 5.2 show a degree of selectivity for α4β2 or α7 nAChRs, the two most prevalent subtypes in the mammalian brain. As such, they contribute to the cholinergic modulation of multiple brain systems, and this may compromise their therapeutic focus. The more sparsely distributed nAChR subtypes could offer better strategic targets. α6β2* nAChRs have been highlighted for Parkinson's disease (Quik and Wonnacott 2011), but finding ligands that can discriminate between the numerous β2* nAChR subtypes and stoichiometries is challenging. The relatively sparsely distributed α5 subunit has attracted attention as an accessory subunit that influences receptor function (Sect. 2.1), reinforced by the high frequency, in humans, of its D398N SNP that is functionally relevant to the regulation of nicotine intake (Wills et al. 2022; Sect. 3.3). Selective PAMs have been advocated as a potential means of restoring normal function to individuals homozygous for this SNP (Maskos 2020). α6β4* nAChRs have emerged as candidates for analgesia (Wieskopf et al. 2015), as well as the α9α10 nAChR. With its restricted localization to sensory and immune cells (Sect. 3.5) and its distinctive pharmacology (Sect. 2.1), the α9α10 nAChR is advocated as an appealing target for treating pain (Hone et al. 2018), hearing loss (Elgoyhen 2022), and cancer (Pucci et al. 2022).

5.4 Which Diseases to Target?

It is striking that many of the brain conditions covered in Sect. 3 (Fig. 3) as targets for nicotinic therapeutic intervention are broad, ill-defined, and/or poorly understood, for example, Alzheimer's disease, schizophrenia, and mood disorders. These conditions are typically heterogeneous, symptoms are variable, diagnosis imprecise, and biomarkers limited. There are arguments for better stratification of heterogeneous conditions to provide less disparate patient groups for clinical trials (Bertrand and Terry 2018). The great pressure to find better medications for these prevalent conditions with currently inadequate and/or suboptimal treatment options has driven drug discovery programs, but the case for a nicotinic cholinergic component varies in its robustness, in some cases supported by genetic links or an epidemiological relationship with tobacco smoking. Moreover, the rationale evolves as knowledge accrues and new concepts arise, exemplified in the shift of focus away from cognitive enhancers (initially aimed at α4β2 nAChRs) to anti-inflammatory strategies (mediated by α7 nAChRs) for treating neurodegenerative conditions (Mizrachi et al. 2021).

Outside of the CNS, more precise conditions with better characterized nicotinic mechanisms should, theoretically, be more tractable. An example is neuropathic pain, where trials have focused on pain associated with a particular clinical cause, such as diabetic neuropathy or sciatic pain. But even here, the target role of α9α10 nAChRs has shifted from DRG to immune cells, and an alternative target has also been mooted to explain the preclinical efficacy of the candidate nicotinic antagonist Vc1.1: $GABA_B$ receptors governing N-type Ca^{2+} channel inhibition (Mohammedi and Christie 2015; Hone et al. 2018).

Targeting nAChRs for peripheral conditions influenced by the non-neuronal cholinergic system holds some promise. An example is the case made for α7 nAChR agonists to treat both inflammatory (e.g., psoriasis) and non-inflammatory (fibrosis) skin conditions (Stegemann and Bohm 2020). Nicotinic ligands for topical application and variants that do not penetrate the brain would be desirable, perhaps presenting new opportunities for drugs rejected from early stages of CNS drug development because of poor bioavailability. In general, a better understanding of the associated nAChR subtypes and their functions in peripheral tissues is still needed; the unconventional modes of signaling emerging for nAChRs in such systems could provide additional tractable targets.

5.5 Lost in Translation: The Reality Gap of Preclinical Models and Clinical Trials

The vast majority of drugs with efficacy in preclinical animal models fail in clinical trials, typically due to lack of efficacy or safety (toxicity or unacceptable side effects). This is particularly true for drugs aimed at brain disorders (Gribkoff and

Kaczmarak 2017) and accounts for the almost total failure of all new nAChR ligands. Why is translation of drugs from animal to human so unsuccessful? There are many limitations of both the preclinical and clinical arms, and these have been comprehensively discussed in recent reviews (Cook et al. 2014; Bespalov et al. 2016; Bertrand and Terry 2018). Some key points are summarized below.

5.5.1 Animal Models

Preclinical testing in animal models is intended to provide "proof of concept" for the safety, efficacy, and target of drug candidates (Singh and Seed 2021). Mice and rats comprise the vast majority of animals used in preclinical research. Rodent cohorts are homogeneous (genetically and with respect to age, diet, lifestyle, etc.), compared with even the most refined human cohorts. Genomic differences between animals and humans should not be overlooked. For nAChRs the much greater prevalence of α2 subunit expression in the brains of primates, including humans (Han et al. 2000), the unique presence of the Dupα7 gene in humans (Sinkus et al. 2015), and the species-limited occurrence of SNPs, for example, the α5 D398N SNP that is common in humans and some other primates (Maskos 2020), provide examples of differences that can have functional consequences.

Producing animal models of complex and heterogeneous conditions whose causes are largely unknown is challenging and often leads to treating symptoms rather than causes, as is the case for AD, PD, and pain, for example (Gribkoff and Kaczmarak 2017). For modelling neurodegenerative conditions, there has been reliance on genetic defects responsible for a very small proportion of cases that show Mendelian inheritance (Dawson et al. 2018). Familial forms of AD and PD may not be fully representative of sporadic disease which arises from unknown causal factors. The emphasis on these mouse models generating protein aggregates has also been criticized, as they generally fail to reproduce the full gambit of biochemical and behavioral changes associated with the human conditions (Gribkoff and Kaczmarak 2017). The short lifespan of rodents has also been flagged as a limitation in this respect (MacDougall et al. 2021). It is very costly to investigate diseases of old age in aged animals, so these conditions are often modeled in young healthy animals with rapid development of symptoms or markers promoted to facilitate shorter study times, potentially compromising model validity.

Neuropsychiatric conditions lack clear biomarkers, and animal models are based on providing behavioral analogues of human symptoms, bearing in mind that some symptoms, such as paranoid delusions, hallucinations, and suicidal thoughts, are uniquely human. Schizophrenia is a heterogeneous neurodevelopmental disorder without a clear etiology. Developmental, drug-induced, and genetic models have been generated, and multiple behavioral paradigms defined as correlates of the negative, positive, and cognitive symptoms of schizophrenia (Ang et al. 2021). While no model satisfies all the defining behaviors, it is argued that these models further understanding of the mechanistic basis of schizophrenia that, in turn, stimulates an iterative improvement of the models (Canetta and Kallendonk 2018). Whether they are sufficiently representative to be predictive for novel therapeutics remains to be demonstrated.

Confounding factors that are overlooked in animal models include comorbidities that typically afflict some clinical populations (e.g., obesity, diabetes) as well as smoking history and prescribed drug treatments. The selective efficacy of an α7 nAChR-specific ligand only in schizophrenics who were non-smokers (Bertrand and Terry 2018) emphasizes the relevance of nicotine exposure/smoking status in both the preclinical and clinical appraisal of nicotinic ligands. The rigor with which preclinical studies are designed and executed (including cohort sizes, blinding, randomization, dose selection, treatment timecourse, and replication) has also been criticized as a factor in their poor predictive value, as judged by the failure of drug candidates in subsequent clinical trials (Bespalov et al. 2016; Singh and Seed 2021).

Are there alternatives to animal models? Human-induced pluripotent stem cells (hiPSCs) have been advocated for capturing the inherent variation in the clinical population (MacDougall et al. 2021). They are presently most useful for cellular studies, for example, the use of hiPSC-derived dopamine neurons to characterize the impact of the α5 nAChR D398N SNP (Maskos 2020). The generation of co-cultures, organoids, and organ-on chip technology could offer more sophisticated models to address more complex questions. Such in vitro preparations cannot replace animal models for behavioral assessments of key phenotypic markers of modeled brain diseases, but they could provide valuable data supporting drug target mechanisms. Sophisticated in vitro human tissue models could help close the translation gap for diseases associated with the non-neuronal cholinergic system. Examples that could benefit from better defined target mechanisms and biomarkers from using such in vitro models are skin diseases (Khurana et al. 2021) and lung conditions (Pun et al. 2021).

5.5.2 Clinical Trials

The structure and execution of clinical trials is not without criticism (Bertrand and Terry 2018). Poor stratification of patients and placebo effects can result in group sizes being insufficiently powered or compromised by having multiple study sites that increase variability. Trial design with respect to aspects such as dose range and timecourse examined is not always optimal. Indeed, it has been argued that clinical trials for neurodegenerative disorders should focus on earlier timepoints in disease progression, including presymptomatic individuals selected on the basis of refined biomarkers, and treatment timecourses should be greatly extended (Gribkoff and Kaczmarak 2017).

Given the expense of running clinical trials, especially phase 3 trials, only well-validated candidates from robust preclinical studies should proceed to clinical trials. Those human trials should be optimized to maximize producing clearly defined outcomes. Perhaps then there would be a lower attrition rate and renewed faith in the cornucopia of nicotinic drugs delivering some valuable new therapeutics.

6 Concluding Remarks

I have watched with wonder the incredible expansion of knowledge about neuronal nAChRs over the last 40 years. Back before the first neuronal nAChR subunit (α3) was cloned (Boulter et al. 1986), who would have predicted the fascinating α7 nAChR, the preponderance of presynaptic nAChRs, nicotinic modulation of synaptic plasticity and network function, and now their involvement beyond the nervous system in tissues throughout the body. Neuronal nAChRs should offer many therapeutic possibilities. Yet the rise and fall of pharmaceutical company interest in this target, with only varenicline making the grade, is disappointing evidence of the arduous climb from drug discovery to clinical approval, with all but the fittest falling by the wayside. With some refinement of tractable disease targets, preclinical models and clinical evaluation strategies, their day will surely come.

Acknowledgments I have aimed at a broad survey of a broad topic. Inevitably, there will be many omissions – of aspects, findings, and opinions – and I apologize to all those who are left out. I hope that the remarkable structural and functional diversity of neuronal nAChRs and their great (as yet unrealized) potential comes across.

I am grateful to Eloise Kuijer and Paul Whiteaker for their critical comments.

References

Acharya S, Kundu D, Choi HJ, Kim KM. Metabotropic signalling cascade involved in a4ß2 nicotinic acetylcholine receptor-mediated PKCßII activation. Biochim Biophys Acta, Mol Cell Res. 2020;1867:118721.

Albers KM, Zhang XL, Diges CM, Schwartz ES, Yang CI, Davis BM, Gold MS. Artemin growth factor increases nicotinic cholinergic receptor subunt expression and activity in nociceptive sensory neurons. Mol Pain. 2014;10:31.

Alcaino C, Musgaard M, Minguez T, Mazzaferro S, Faundez M, Iturriaga-Vasquez P, et al. Role of the cys loop and transmembrane domain in the allosteric modulation of a4ß2 nicotinic acetylcholine receptors. J Biol Chem. 2017;292:551–62.

Alvarez-Munoz H, Vicencio-Jimenez S, Jorratt P, Delano PH, Terreros G. Corticofugal and brainstem functions associated with medial olivocochlear cholinergic transmission. Front Neurosci. 2022;16:866161.

Alzforum. 2015. https://www.alzforum.org/news/research-news/rare-severe-side-effects-sideline-some-phase-3-encenicline-trials

Andersen N, Corradi J, Sine SM, Bouzat C. Stoichiometry for activation of neuronal a7 nicotinic receptors. Proc Natl Acad Sci U S A. 2013;110:20819–24.

Ang MJ, Lee S, Kim JC, Kim SH, Moon C. Behavioural tasks evaluating schizophrenia-like symptoms in animal models: a recent update. Curr Neuropharmacol. 2021;19:641–64.

Antonio-Tolentino K, Hopkins CR. Selective a7 nicotinic receptor agonists and positive allosteric modulators for the treatment of schizophrenia. Expert Opin Investig Drugs. 2020;29:603–10.

Arias HR, Germann SL, Pierce SR, Sakamoto S, Marcelo MO, Hamachi I, et al. Modulation of the mammalian $GABA_A$ receptor by type I and type II positive allosteric modulators of the a7 nicotinic acetylcholine receptor. Br J Pharmacol. 2022; 179(24):5323–5337.

Araud T, Wonnacott S, Bertrand D. Associated proteins: the universal toolbox controlling ligand gated ion channel function. Biochem Pharmacol. 2010;80:160–9.

Auerbach A. Agonist activation of a nicotinic acetylcholine receptor. Neuropharmacology. 2015;96:150–6.

Azam L, McIntosh JM. Alpha-conotoxins as pharmacological probes of niocotinic acetylcholine receptors. Acta Pharmacol Sin. 2009;30:771–83.

Baddick CG, Marks MJ. An autoradiographic survey of mouse brain nicotinic acetylcholine receptors defined by null mutants. Biochem Pharmacol. 2011;82:828–41.

Badio B, Daly JW. Epibatidine, a potent analgesic and nicotinic agonist. Mol Pharmacol. 1994;45:563–9.

Baenziger JE, Hénault CM, Therien JP, Sun J. Nicotinic acetylcholine receptor-lipid interactions: mechanistic insight and biological function. Biochim Biophys Acta. 2015;1848:1806–17.

Baer K, Bürli T, Huh KH, Wiesner A, Erb-Vögtli S, Göckeritz-Dujmovic D, et al. PICK1 interacts with a7 neuronal nicotinic acetylcholine receptors and controls their clustrering. Mol Cell Neurosci. 2007;35:339–55.

Becchetti A, Grandi LC, Colombo G, Meneghini S, Amadeo A. Nicotinic receptors in sleep-related hypermotor epilepsy: pathophysiology and pharmacology. Brain Sci. 2020;10:907.

Ben-David Y, Kagan S, Cohen Ben-Ami H, Rostami J, Mizrahi T, Kulkarni AR, et al. RIC3, the cholinergic anti-inflammatory pathway, and neuroinflammation. Int Immunopharmacol. 2020;83:106381.

Bertrand D, Terry AV Jr. The wonderland of neuronal nicotinic acetylcholine receptors. Biochem Pharmacol. 2018;151:214–25.

Bespalov A, Steckler T, Altevogt B, Koustova E, Skolnick P, Deaver D, et al. Failed trials for central nervous system disorders do not necessarily invalidate preclinical models and drug targets. Nat Rev Drug Discov. 2016;15:516.

Bierut LJ. Convergence of genetic findings for nicotine dependence and smoking related diseases with chromosome 15q24-25. Trends Pharmacol Sci. 2010;31:46–51.

Blom AEM, Campello HR, Lester HA, Gallagher T, Dougherty DA. Probing binding interactions of cytisine derivatives to the a4ß2 nicotinic acetylcholine receptor. J Am Chem Soc. 2019;141:15840–9.

Blunt CEW, Dougherty DA. Binding interactions of NS9740, a silent agonist of the a7 nicotinic acetylcholine receptor. Mol Pharmacol. 2019;96:212–8.

Boeckers TM. The postsynaptic density. Cell Tissue Res. 2006;326:409–22.

Bono F, Mutti V, Fiorentini C, Missale C. Dopamine D3 receptor heteromerization: implications for neuroplasticity and neuroprotection. Biomol Ther. 2021;10:1016.

Boulter J, Evans K, Goldman D, Martin G, Treco D, Heinemann S, Patrick J. Isolation of a cDNA clone coding for a possible neural nicotinic acetylcholine receptor alpha-subunit. Nature. 1986;319:368–74.

Bouzat C, Sine SM. Nicotinic acetylcholine receptors at the single-channel level. Br J Pharmacol. 2018;175:1789–804.

Bouzat C, Lasala M, Nielsen BE, Corradi J, Esandi MDC. Molecular function of a7 nicotinic receptors as drug targets. J Physiol. 2018;596:1847–61.

Bowman WC. Neuromuscular block. Br J Pharmacol. 2006;147:S277–86.

Brawek B, Skok M, Garaschuk O. Changing functional signatures of microglia along the axis of brain aging. Int J Mol Sci. 2021;223:1091.

Brejc K, van Dijk WJ, Klaassen RV, Schuurmans M, van Der Oost J, Smit AB, Sixma TK. Crystal structure of an ACh-binding protein reveals the ligand binding domain of nicotinic receptors. Nature. 2001;411:269–76.

Budisteanu M, Papuc SM, Streata I, Cucu M, Pirvu A, et al. The phenotypic spectrum of 15q13.3 region duplications: report of 5 patients. Genes (Basel). 2021;12:1025.

Bychkov ML, Shulepko MA, Shlepova OV, Kulbatskii DS, Chulina IA, et al. SLURP-1 controls growth and migration of lung adenocarcinoma cells, forming a complex with a7-nAChR and PDGFR/EGFR heterodimer. Front Cell Dev Biol. 2021;9:739391.

Canetta S, Kellendonk C. Can we use mice to study schizophrenia? Philos Trans R Soc Lond Ser B Biol Sci. 2018;373:20170032.

Carstens E, Carstens MI. Sensory effects of nicotine and tobacco. Nicotine Tob Res. 2022;24:306–15.

Cetin H, Beeson D, Vincent A, Webster R. The structure, function, and physiology of the fetal and adult acetylcholine receptor in muscle. Front Mol Neurosci. 2020;13:581097.

Changeux J-P. The nicotinic acetylcholine receptor: a typical 'allosteric machine'. Philos Trans R Soc Lond Ser B Biol Sci. 2018;373:20170174.

Changeux J-P. Discovery of the first neurotransmitter receptor: the acetylcholine nicotinic receptor. Biomol Ther. 2020;10:547.

Chatzidaki A, Millar NS. Allosteric modulation of nicotinic acetylcholine receptors. Biochem Pharmacol. 2015;97:408–17.

Chiara DC, Hamouda AK, Ziebell MR, Mejia LA, Garcia G 3rd, Cohen JB. [^{3}H]chlorpromazine photolabeling of the torpedo nicotinic acetylcholine receptor identifies two state-dependent binding sites in the ion channel. Biochemistry. 2009;48:10066–77.

Collins T, Millar NS. Nicotinic acetylcholine receptor transmembrane mutations convert ivermectin from a positive to a negative allosteric modulator. Mol Pharmacol. 2010;78:198–204.

Cook D, Brown D, Alexander R, March R, Morgan P, Satterthwaite G, Pangalos MN. Lessons learned from the fate of AstraZeneca's drug pipeline: a five-dimensional framework. Nat Rev Drug Discov. 2014;13:419–31.

Courtney RJ, McRobbie H, Tutka P, Weaver NA, Petrie D, Mendelsohn CP, et al. Effect of cytisine vs varenicline on smoking cessation: a randomised clinical trial. JAMA. 2021;326:56–64.

Cox MA, Bassi C, Saunders ME, Nechanitzky R, Morgado-Palacin I, Zheng C, Mak TW. Beyond neurotransmission: acetylcholine in immunity and inflammation. J Intern Med. 2020;287:120–33.

Crespi A, Colombo SF, Gotti C. Proteins and chemical chaperones involved in neuronal nicotinic receptor expression and function: an update. Br J Pharmacol. 2018a;175:1869–79.

Crespi A, Plutino S, Sciaccaluga M, Righi M, Borgese N, Fucile S, et al. The fifth subunit in a3ß4 nicotinic receptor is more than an accessory subunit. FASEB J. 2018b;32:4190–202.

DaCosta CJ, Sine SM. Stoichiometry for drug potentiation of a pentameric ion channel. Proc Natl Acad Sci U S A. 2013;110:6595–600.

Dajas-Bailador F, Wonnacott S. Nicotinic acetylcholine receptors and the regulation of neuronal signalling. Trends Pharmacol Sci. 2004;25:317–24.

David R, Ciuraszkiewicz A, Simeone X, Orr-Urtreger A, Papke RL, McIntosh JM, et al. Biochemical and functional properties of distinct nicotinic acetylcholine receptors in the superior cervical ganglion of mice with targeted deletions of nAChR subunit genes. Eur J Neurosci. 2010;31:978–93.

Davidson M, Levi L, Park J, Nastas I, Ford L, Rassnick S, et al. The effects of JNJ-39393406 a positive allosteric nicotine modulator on mood and cognition in patients with unipolar depression: a double-blind, add-on, placebo-controlled trial. Eur Neuropsychopharmacol. 2021;51:33–42.

Dawson TM, Golde TE, Lagier-Tourenne C. Animal models of neurodegenerative disease. Nat Neurosci. 2018;21:1370–9.

DeDominicis KE, Sahibzada N, Olson TT, Xiao Y, Wolfe BB, Kellar KJ, Yasuda RP. The $(a4)_3(ß2)_2$ stoichiometry of the nicotinic acetylcholine receptor predominates in the rat mpotor cortex. Mol Pharmacol. 2017;92:327–37.

Delgado-Vélez M, Quesada O, Villalobos-Santos JC, Maldonado-Hernández R, Asmar-Rovira G, Stevens RC, Lasalde-Dominicci JA. Pursuing high-resolution structures of nicotinic acetylcholine receptors: lessons learned from five decades. Molecules. 2021;26:5753.

Demeter E, Sarter M. Leveraging the cortical cholinergic system to enhance attention. Neuropharmacology. 2013;64:294–304.

Deutsch SI, Burket JA. An evolving therapeutic rationale for targeting the a7 nicotinic acetylcholine receptor in autism spectrum disorder. Curr Top Behav Neurosci. 2020;45:167–208.

Deutsch SI, Burket JA, Urbano MR, Benson AD. The a7 nicotinic acetylcholine receptor: a mediator of pathogenesis and therapeutic target in autism spectrum disorders and Down syndrome. Biochem Pharmacol. 2015;97:363–77.

Dore K, Stein IS, Brock JA, Castillo PE, Zito K, Sjöström PJ. Unconventional NMDA receptor signalling. J Neurosci. 2017;37:10800–7.

Doss SV, Barbat-Artigas S, Lopes M, Pradhan BS, Prószynski TJ, Robitaille R, Valdez G. Expression and roles of Lynx1, a modulator of cholinergic transmission, in skeletal muscles and neuromuscular junctions in mice. Front Cell Dev Biol. 2022;10:838612.

Durek T, Shelukhina IV, Tae HS, Thongyoo P, Spirova EN, Kudryavtsev DS, et al. Interaction of synthetic human SLURP-1 with the nicotinic acetylcholine receptors. Sci Rep. 2017;7:16606.

Eaton JB, Lucero LM, Stratton H, Chang Y, Cooper JF, Lindstrom JM, et al. The unique a4+/-a4 agonist binding site in $(a4)_3(ß2)_2$ subtype nicotinic acetylcholine receptors per mits differential agonist desensitization pharmacology versus the $(a4)_2(ß2)_3$ subtype. Pharmacol Exp Ther. 2014;348:46–58.

Elgoyhen AB. The a9a10 nicotinic acetylcholine receptor: a compelling drug target for hearing loss? Expert Opin Ther Targets. 2022;26:291–302.

Engel AG, Shen XM, Selcen D, Sine SM. Congenital myasthenic syndromes: pathogenesis, diagnosis, and treatment. Lancet Neurol. 2015;14:461.

Exley R, Cragg SJ. Presynaptic nicotinic receptors: a dynamic and diverse cholinergic filter of striatal dopamine neurotransmission. Br J Pharmacol. 2008;153:S283–97.

Exley R, Maubourguet N, David V, Eddine R, Evrard A, Pons S, et al. Distinct contributions of nicotinic acetylcholine receptor subunit a4 and subunit a6 to the reinforcing effects of nicotine. Proc Natl Acad Sci U S A. 2011;108:7577–82.

Fabiani C, Antollini SS. Alzheimer's disease as a membrane disorder: spatial cross-talk among beta-amyloid peptides, nicotinic acetylcholine receptors and lipid rafts. Front Cell Neurosci. 2019;13:309.

Falcón-Moya R, Rodríguez-Moreno A. Metabotropic actions of kainate receptors modulating glutamate release. Neuropharmacology. 2021;197:108696.

Ferns M. An inside job: molecular determinants for postsynaptic localisation of nicotinic acetylcholine receptors. Molecules. 2021;26:3065.

Forman SA, Chiara DC, Miller KW. Anesthetics target interfacial transmembrane sites in nicotinic actylcholine receptors. Neuropharmacology. 2015;96:169–77.

Fox-Loe AM, Moonschi FH, Richards CI. Organelle-specific single-molecule imaging of a4ß2 nicotinic receptors reveals the effects of nicotine on receptor assembly and cell-surface trafficking. J Biol Chem. 2017;292:21159–69.

Friedman JR, Richbart SD, Merritt JC, Brown KC, Nolan NA, Akers AT, et al. Acetylcholine signalling system in progression of lung cancers. Pharmacol Ther. 2019;194:222–54.

Fuchs P. The synaptic physiology of cochlear hair cells. Audiol Neurootol. 2002;7:40–4.

Fucile S. Ca^{2+} permeability of nicotinic acetylcholine receptors. Cell Calcium. 2004;35:1–8.

Genzen JR, McGehee DS. Short- and long-term enhancement of excitatory transmission in the spinal cord dorsal horn by nicotinic acetylcholine receptors. Proc Natl Acad Sci U S A. 2003;100:6807–12.

Gerzanich V, Wang F, Kuryatov A, Lindstrom J. a5 Subunit alters desensitization, pharmacology, Ca^{++} permeability and Ca^{++} modulation of human neuronal a3 nicotinic receptors. J Pharmacol Exp Ther. 1998;286:311–20.

Gotti C, Clementi F. Cytisine and cytisine derivatives: more than smoking cessation aids. Pharmacol Res. 2021;170:105700.

Grady SR, Moretti M, Zoli M, Marks MJ, Zanardi A, Pucci L, et al. Rodent habenulo-interpeduncular pathway expresses a large variety of uncommon nAChR subtypes, but only the a3ß4 and a3ß3ß4 subtypes mediate acetylcholine release. J Neurosci. 2009;29:2272–82.

Grando SA. Pemphigus autoimmunity: hypotheses and realities. Autoimmunity. 2012;45:7–35.

Gribkoff VK, Kaczmarek LK. The need for new approaches in CNS drug discovery: why drugs have failed and what can be done to improve outcomes. Neuropharmacology. 2017;120:11–9.

Grupe M, Grunnet M, Bastlund JF, Jensen AA. Targeting α4β2 nicotinic acetylcholine receptors in central nervous system disorders: perspectives on positive allosteric modulation as a therapeutic approach. Basic Clin Pharmacol Toxicol. 2015;116(3):187–200. Epub 2014 Dec 29. PMID: 25441336.

Gu S, Matta JA, Davini WB, Dawe GB, Lord B, Bredt DS. a6-Containing nicotinic acetylcholine receptor reconstitution involves mechanistically distinct accessory components. Cell Rep. 2019;26:866–874.e3.

Gu S, Knowland D, Matta JA, O'Carroll ML, Davini WB, Dhara M, et al. Hair cell a9a10 nicotinic acetylcholine receptor functional expression regulated by ligand binding and deafness gene products. Proc Natl Acad Sci U S A. 2020;117:24534–44.

Han ZY, Le Novère N, Zoli M, Hill JA Jr, Champtiaux N, Changeux J-P. Localization of nAChR subunit mRNAs in the brain of Macaca mulatta. Eur J Neurosci. 2000;12:3664–74.

Hansen SB, Sulzenbacher G, Huxford T, Marchot P, Taylor P, Bourne Y. Structures of aplysia AChBP complexes with nicotinic agonists and antagonists reveal distinctive binding interfaces and conformations. EMBO J. 2005;24:3635–46.

Hashimoto K. Targeting of a7 nicotinic acetylcholine receptors in the treatment of schizophrenia and the use of auditory sensory gating as a translational biomarker. Curr Pharm Des. 2015;21:3797–806.

Hernandez CM, Kayed R, Zheng H, Sweatt JD, Dineley KT. Loss of a7 nicotinic receptors enhances ß-amyloid oligomer accumulation, exacerbating early-stage cognitive decline and septo-hippocampal pathology in a mouse model of Alzheimer's disease. J Neurosci. 2010;30:2442–53.

Hilscher MM, Nogueira I, Mikulovic S, Kullander K, Leão RN, Leão KE. Chrna2-OLM interneurons display different membrane properties and h-current magnitude depending on dorsoventral location. Hippocampus. 2019;29:1224–37.

Hollenhorst MI, Krasteva-Christ G. Nicotinic acetylcholine receptors in the respiratory tract. Molecules. 2021;26:6097.

Hone AJ, Servant D, McIntosh JM. a9-containing nicotinic acetylcholine receptors and the modulation of pain. Br J Pharmacol. 2018;175:1915–27.

Huang X, Chen H, Shaffer PL. Crystal structures of human GlyRa3 bound to ivermectin. Structure. 2017;25:945–950.e2.

Ishii K, Wong JK, Sumikawa K. Comparison of alpha2 nicotinic acetylcholine receptor subunit mRNA expression in the central nervous system of rats and mice. J Comp Neurol. 2005;493:241–60.

Jain A, Kuryatov A, Wang J, Kamenecka TM, Lindstrom J. Unorthodox acetylcholine binding sites formed by a5 and ß3 accessory subunits in a4ß2* nicotinic acetylcholine receptors. J Biol Chem. 2016;291:23452–63.

Jayakar SS, Pugh PC, Dale Z, Starr ER, Cole S, Margiotta JF. PACAP induces plasticity at autonomic synapses by nAChR-dependent NOS1 activation and AKAP-mediated PKA targeting. Mol Cell Neurosci. 2014;63:1–12.

Jordan CJ, Xi ZX. Discovery and development of varenicline for smoking cessation. Expert Opin Drug Discovery. 2019;13:671–83.

Kabbani N, Nichols RA. Beyond the channel: metabotropic signalling by nicotinic receptors. Trends Pharmacol Sci. 2018;39:354–66.

Karlin A. Emerging structure of the nicotinic acetylcholine receptors. Nat Rev Neurosci. 2002;3:102–14.

Kawashima K, Fujii T. Basic and clinical aspects of non-neuronal acetylcholine: overview of non-neuronal cholinergic systems and their biological significance. J Pharmacol Sci. 2008;106:167–73.

Khakh BS, Fisher JA, Nashmi R, Bowser DN, Lester HA. An angstrom scale interaction between plasma membrane ATP-Gated P2X 2 and a 4 ß 2 nicotinic channels measured with fluorescence resonance energy transfer and total internal reflection fluorescence microscopy. J Neurosci. 2005;25(29):6911–6920.

Khurana P, Kolundzic N, Flohr C, Ilic D. Human pluripotent stem cells: an alternative for 3D in vitro modelling of skin disease. Exp Dermatol. 2021;30:1572–87.

Knowland D, Gu S, Eckert WA 3rd, Dawe GB, Matta JA, Limberis J, et al. Functional a6ß4 acetylcholine receptor expression enables pharmacological testing of nicotinc agaonists with analgesic properties. J Clin Invest. 2020;130:6158–70.

Krasteva G, Canning BJ, Hartmann P, Veres TZ, Papadakis T, Mühlfeld C, et al. Cholinergic chemosensory cells in the trachea regulate breathing. Proc Natl Acad Sci U S A. 2011;108:9478–83.

Kumar P, Scholze P, Fronius M, Krasteva-Christ G, Hollenhorst MI. Nicotine stimulates ion transport via metabotropic ß4 subunit containing nicotinic ACh receptors. Br J Pharmacol. 2020;177:5595–608.

Kuryatov A, Onksen J, Lindstrom J. Roles of accessory subunits in alpha4beta2(*) nicotinic receptors. Mol Pharmacol. 2008;74:132–43.

Kurzen H, Wessler I, Kirkpatrick CJ, Kawashima K, Grando SA. The non-neuronal cholinergic system of human skin. Horm Metab Res. 2007;39:125–35.

Lai O, Recke A, Zillikens D, Kasperkiewicz M. Influence of cigarette smoking on pemphigus – a systematic review and pooled analysis of the literature. J Eur Acad Dermatol Venereol. 2018;32:1256–62.

Lamotte d'Incamps B, Zorbaz T, Dingova D, Krejci E, Ascher P. Stoichiometry of the heteromeric nicotinic receptors of the Renshaw cell. J Neurosci. 2018;38:4943–56.

Le Novère N, Changeux J-P. Neuronal nicotinic receptor a6 subunit mRNA is selectively concentrated in catecholaminergic nuclei of the rat brain. Eur J Neurosci. 1996;8:2428–39.

Le Novère N, Grutter T, Changeux J-P. Models of the extracellular domain of the nicotinic receptors and of agonist- and Ca^{2+}-binding sites. Proc Natl Acad Sci U S A. 2002;99:3210–5.

Lendvai B, Vizi ES. Nonsynaptic chemical transmission through nicotinic acetylcholine receptors. Physiol Rev. 2008;88:333–49.

Letsinger AC, Gu L, Yakel JL. a7 nicotinic acetylcholine receptors in the hippocampal circuit: taming complexity. Trends Neurosci. 2022;45:145–57.

Levin ED. a7-Nicotinic receptors and cognition. Curr Drug Targets. 2012;13:602–6.

Li L, Xiong WC, Mei L. Neuromuscular junction n formation, aging, and disorders. Annu Rev Physiol. 2018;80:159–88.

Lipovsek M, Marcovich I, Elgoyhen AB. The hair cell a9a10 nicotinic acetylcholine receptor: odd cousin in an old family. Front Cell Neurosci. 2021;15:785265.

Liu Q, Melnikova IN, Hu M, Gardner PD. Cell type-specific activation of neuronal nicotinic acetylcholine receptor subunit genes by Sox10. J Neurosci. 1999;19:9747–55.

Lorenz-Guertin JM, Jacob TC. GABA type A receptor trafficking and the architecture of synaptic inhibition. Dev Neurobiol. 2018;78:238–70.

Loring RH. Speculation on how RIC-3 and other chaperones facilitate a7 nicotinic receptor folding andassembly. Molecules. 2022;27:4527.

Lotfipour S, Byun JS, Leach P, Fowler CD, Murphy NP, Kenny PJ, et al. Targeted deletion of the mouse a2 nicotinic acetylcholine receptor subunit gene (Chrna2) potentiates nicotine-modulated behaviors. J Neurosci. 2013;33:7728–41.

Lozada AF, Wang X, Gounko NV, Massey KA, Duan J, Liu Z, Berg DK. Glutamatergic synapse formation is promoted by a7-containing nicotinic acetylcholine receptors. J Neurosci. 2012;32:7651–61.

Lukas RJ, Changeux JP, Le Novère N, Albuquerque EX, Balfour DJ, Berg DK, et al. International Union of Pharmacology. XX. Current status of the nomenclature for nicotinic acetylcholine receptors and their subunits. Pharmacol Rev. 1999;51:397–401.

Lyukmanova EN, Shulepko MA, Shenkarev ZO, Bychkov ML, Paramonov AS, Chugunov AO, et al. Secreted isoform of human lynx1 (SLURP-2): spatial structure and pharmacology of interactions with different types of acetylcholine receptors. Sci Rep. 2016;6:30698.

MacDougall G, Brown LY, Kantor B, Chiba-Falek O. The path to progress preclinical studies of age-related neurodegenerative diseases: a perspective on rodent and hiPSC-derived models. Mol Ther. 2021;29:949–72.

Macor JE, Gurley D, Lanthorn T, Loch J, Mack RA, Mullen G, et al. The 5-HT3 antagonist tropisetron (ICS 205-930) is a potent and selective a7 nicotinic receptor partial agonist. Bioorg Med Chem Lett. 2001;11:319–21.

Maldifassi MC, Martín-Sánchez C, Atienza G, Cedillo JL, Arnalich F, Bordas A, et al. Interaction of the a7-nicotinic subunit with its human-specific duplicated dupa7 in mammalian cells. J Biol Chem. 2018;293:13874–13913 isoform888.

Maskos U. The cholinergic mesopontine tegmentum is a relatively neglected nicotinic master modulator of the dopaminergic system: relevance to drugs of abuse and pathology. Br J Pharmacol. 2008;153:S438–45.

Maskos U. The nicotinic receptor alpha5 coding polymorphism rs16969968 as a major target in disease: functional dissection and remaining challenges. J Neurochem. 2020;154:241–50.

Matta JA, Gu S, Davini WB, Bredt DS. Nicotinic acetylcholine receptor redux: discovery of accessories opens therapeutic vistas. Science. 2021;373:eabg6539.

Mazzaferro S, Benallegue N, Carbone A, Gasparri F, Vijayan R, Biggin PC, Moroni M, Bermudez I. Additional acetylcholine (ACh) binding site at alpha4/alpha4 interface of (alpha4beta2)2alpha4 nicotinic receptor influences agonist sensitivity. J Biol Chem. 2011;286(35):31043–31054.

Mazzaferro S, Gasparri F, New K, Alcaino C, Faundez M, Iturriaga Vasquez P, et al. Non-equivalent ligand selectivity of agonist sites in $(a4ß2)_2a4$ nicotinic acetylcholine receptors: a key determinant pf agonist efficacy. J Biol Chem. 2014;289:21795–806.

Mei D, Zhao L, Chen B, Zhang X, Wang X, Yu Z, et al. a-Conotoxin IMI-modified polymeric micelles as potential nanocarriers for targeted docetaxel delivery to a7-nAChR overexpressed non-small cell lung cancer. Drug Deliv. 2018;25:493–503.

Meyers EE, Loetz EC, Marks MJ. Differential expression of the beta4 neuronal nicotinic receptor subunit affects tolerance development and nicotinic binding sites following chronic nicotine treatment. Pharmacol Biochem Behav. 2015;130:1–8.

Mihalak KB, Carroll FI, Luetje CW. Varenicline is a partial agonist at a4ß2 and a full agonist at a7 neuronal nicotinic receptors. Mol Pharmacol. 2006;70:801–5.

Minguez-Viñas T, Nielsen BE, Shoemark DK, Gotti C, Sessions RB, Mulholland AJ, et al. A conserved arginine with non-conserved function is a key determinant of agonist selectivity in a7 nicotinic acetylcholine receptors. Br J Pharmacol. 2021;178:1651–68.

Miwa JM. Lynx1 prototoxins: critical accessory proteins of neuronal nicotinic acetylcholine receptors. Curr Opin Pharmacol. 2021;56:46–51.

Miwa JM, Stevens TR, King SL, Caldarone BJ, Ibanez-Tallon I, Xiao C, et al. The prototoxin lynx1 acts on nicotinic acetylcholine receptors to balance neuronal activity and survival in vivo. Neuron. 2006;51:587–600.

Miwa JM, Anderson KR, Hoffman KM. Lynx prototoxins: roles of endogenous mammalian neurotoxin-like proteins in modulating nicotinic acetylcholine receptor function to influence complex biological processes. Front Pharmacol. 2019;10:343.

Mizrachi T, Vaknin-Dembinsky A, Brenner T, Treinin M. Neuroinflammation modulation via a7 nicotinic acetylcholine receptor and its chaperone RIC-3. Molecules. 2021;26:6139.

Mohammadi SA, Christie MJ. Conotoxin interactions with a9a10-nAChRs: is the a9a10-nicotinic acetylcholine receptor an important therapeutic target for pain management? Toxins (Basel). 2015;7:3916–32.

Morales-Perez CL, Noviello CM, Hibbs RE. X-ray structure of the human a4ß2 nicotinic receptor. Nature. 2016;538:411–5.

Morioka N, Hisaoka-Nakashima K, Nakata Y. Regulation by nicotinic acetylcholine receptors of microglial glutamate transporters: role of microglia in neuroprotection. In: Akaike A, Shimohama S, Misu Y, editors. Nicotinic acetylcholine receptor signaling in neuroprotection. Singapore: Springer; 2018.

Morishita H, Miwa JM, Heintz N, Hensch TK. Lynx1, a cholinergic brake, limits plasticity in adult visual cortex. Science. 2010;330:1238–40.

Mulcahy MJ, Paulo JA, Hawrot E. Proteomic investigation of murine neuronal a7-micotinic acetylcholine receptor interacting proteins. J Proteome Res. 2018;17:3959–75.

Mutti V, Bono F, Tomasoni Z, Bontempi L, Guglielmi A, Bolognin S, et al. Structural plasticity of dopaminergic neurons requires the activation of the D3R-nAChR heteromer and the PI3K-ERK1/2/Akt-induced expression of c-FOS and p70S6K signalling pathway. Mol Neurobiol. 2022;59:2129–49.

Nichols WA, Henderson BJ, Yu C, Parker RL, Richards CI, Lester HA, Miwa JM. Lynx1 shifts a4ß2 nicotinic receptor subunit stoichiometry by affecting assembly in the endoplasmic reticulum. J Biol Chem. 2014;289:31423–32.

Nickell JR, Grinevich VP, Siripurapu KB, Smith AM, Dwoskin LP. Potential therapeutic uses of mecamylamine and its stereoisomers. Pharmacol Biochem Behav. 2013;108:28–43.

Nielsen BE, Minguez T, Bermudez I, Bouzat C. Molecular function of the novel a7ß2 nicotinic receptor. Cell Mol Life Sci. 2018;75:2457–247.

Nirogi R, Benade V, Goyal VK, Pandey SK, Mohammed AR, Shinde A, et al. Safety, tolerability, and pharmacokinetics of ropanicant (SUVN-911), a novel a4ß2 noctinic acetylcholine receptor antagonist, in healthy adult elderly subjects. Clin Drug Investig. 2022;42:747–62.

Norleans J, Wang J, Kuryatov A, Leffler A, Doebelin C, Kamenecka TM, Lindstrom J. Discovery of an intrasubunit nicotinic acetylcholine receptor-binding site for the positive allosteric modulator Br-PBTC. J Biol Chem. 2019;294:12132–45.

Olsen JA, Ahring PK, Kastrup JS, Gajhede M, Balle T. Structural and functional studies of the modulator NS9283 reveal agonist-like mechanism of action at a4ß2 nicotinic acetylcholine receptors. Biol Chem. 2014;289:24911–21.

Oz M, Lorke DE, Yang KH, Petroianu G. On the interaction of ß-amyloid peptides and a7-nicotinic acetylcholine receptors in Alzheimer's disease. Curr Alzheimer Res. 2013;10:618–30.

Pandya AA. Desformylflustrabromine: a novel positive allosteric modulator for ß2 subunit caontaining nicotinic receptor sub-types. Curr Pharm Des. 2016;22:2057–63.

Papke RL, Sanberg PR, Shytle RD. Analysis of mecamylamine stereoisomers on human nicotinic receptor subtypes. J Pharmacol Exp Ther. 2001;297:646–56.

Papke RL, Bagdas D, Kulkarni AR, Gould T, AlSharari SD, Thakur GA, Damaj MI. The analgesic-like properties of the a7 nAChR silent agonist NS6740 is associated with non-conducting conformations of the receptor. Neuropharmacology. 2015;91:34–42.

Parri HR, Hernandez CM, Dineley KT. Research update: Alpha7 nicotinic acetylcholine receptor mechanisms in Alzheimer's disease. Biochem Pharmacol. 2011;82:931–42.

Paton WDM, Zaimis EJ. The pharmacological actions of polymethylene bistrimethyl-ammonium salts. Br J Pharmacol Chemother. 1949;4:381–400.

Picciotto MR, Kenny PJ. Mechanisms of nicotine addiction. Cold Spring Harb Perspect Med. 2021;11:a039610.

Pisapati AV, Cao W, Anderson KR, Jones G, Holick KH, Whiteaker P, et al. Biophysical characterization of lynx-nicotinic receptor interactions using atomic force microscopy. FASEB Bioadv. 2021;3:1034–42.

Pismataro MC, Horenstein NA, Stokes C, Dallanoce C, Thakur GA, Papke RL. Stable desensitization of a7 nicotinic acetylcholine receptors by NS6740 requires interaction with S36 in the orthosteric agonist binding site. Eur J Pharmacol. 2021;905:174179.

Poorhuis RB, Mansvelder HD. Nicotinic acetylcholine receptors controlling attention: behavior, circuits and sensitivity to disruption by nicotine. Biochem Pharmacol. 2013;86:1089–98.

Post MR, Tender GS, Lester HA, Dougherty DA. Secondary ammonium agonists make dual cation-p interactions in a4ß2 nicotinic receptors. eNeuro. 2017;4:ENEURO.0032-17.2017.

Prevost MS, Bouchenaki H, Barilone N, Gielen M, Corringer P-J. Concatemers to re-investigate the role of a5 in a4ß2 nicotinic receptors Cellular and Molecular Life Sciences 2021;78(3):1051–1064.

Pucci S, Zoli M, Clementi F, Gotti C. a9-Containing nicotinic receptors in cancer. Front Cell Neurosci. 2022;15:805123.

Pun S, Haney LC, Barrile R. Modelling human physiology on-chip: historical perspectives and future directions. Micromachines (Basel). 2021;12:1250.

Purohit P, Auerbach A. Loop C and the mechanism of acetylcholine receptor-channel gating. J Gen Physiol. 2013;141:467–78.

Quik M, Wonnacott S. a6ß2* and a4ß2* nicotinic acetylcholine receptors as drug targets for Parkinson's disease. Pharmacol Rev. 2011;63:938–66.

Quik M, Boyd JT, Bordia T, Perez X. Potential therapeutic application for nicotinic receptor drugs in movement disorders. Nicotine Tob Res. 2019;21(3):357–69.

Rahman S, Engleman EA, Bell RL. Nicotinic receptor modulation to treat alcohol and drug dependence. Front Neurosci. 2015;8:426.

Razani-Boroujerdi S, Boyd RT, Dávila-García MI, Nandi JS, Mishra NC, Singh SP, et al. T cells express a7-nicotinic acetylcholine receptor subunits that require a functional TCR and leukocyte- specific protein kinase for nicotine-induced Ca^{2+} response. J Immunol. 2007;179:2889–98.

Rollema H, Hurst RS. The contribution of agonist and antagonist activities of a4ß2* ligands to smoking cessation efficacy: a quantitative analysis of literature data. Psychopharmacology. 2018;235:2479–505.

Rosenberg MM, Yang F, Giovanni M, Mohn JL, Temburni MK, Jacob MH. Adenomatous polyposis coli plays a key role, in vivo, in coordinating assembly of the neuronal nicotinic postsynaptic complex. Mol Cell Neurosci. 2008;38:138–52.

Rucktooa P, Haseler CA, van Elk R, Smit AB, Gallagher T, Sixma TK. Structural characterization of binding mode of smoking cessation drugs to nicotinic acetylcholine receptors through study of ligand complexes with acetylcholine binding protein. J Biol Chem. 2012;287:23283–93.

Rudolf R, Straka T. Nicotinic acetylcholine receptor at vertebrate motor endplates: endocytosis, recycling, and degradation. Neurosci Lett. 2019;711:134434.

Sanberg PR, Vindrola-Padros C, Shytle RD. Translating laboratory discovery to the clinic: from nicotine and mecamylamine to Tourette's: depression, and beyond. Physiol Behav. 2012;107:801–8.

Sauguet L, Shahsavar A, Delarue M. Crystallographic studies of pharmacological sites in pentameric ligand-gated ion channels. Biochim Biophys Acta. 2015;1850:511–23.

Schneider M, Levant B, Reichel M, Gulbins E, Kornhuber J, Müller CP. Lipids in psychiatric disorders and preventive medicine. Neurosci Biobehav Rev. 2017;76:336–62.

Scholze P, Koth G, Orr-Urtreger A, Huck S. Subunit composition of a5-containing nicotinic receptors in the rodent habenula. J Neurochem. 2012;121:551–60.

Sciaccaluga M, Moriconi C, Martinello K, Catalano M, Bermudez I, Stitzel J, et al. Crucial role of nicotinic a5 subunit variants for Ca^{2+} fluxes in ventral midbrain neurons. FASEB J. 2015;29:3389–98.

Sharma G, Vijayaraghavan S. Nicotinic cholinergic signalling in hippocampal astrocytes involves calcium-induced calcium release from intracellular stores. Proc Natl Acad Sci U S A. 2001;98:4148–53.

Shelukhina I, Paddenberg R, Kummer W, Tsetlin V. Functional expression and axonal transport of a7 nAChRs by peptidergic nociceptors of rat dorsal root ganglion. Brain Struct Funct. 2015;220:1885–99.

Shen JX, Yakel JL. Functional a7 nicotinic ACh receptors on astrocytes in rat hippocampal CA1 slices. J Mol Neurosci. 2012;48:14–21.

Shulepko MA, Bychkov ML, Shenkarev ZO, Kulbatskii DS, Makhonin AM, Paramonov AS, et al. Biochemical basis of skin disease Mal de Meleda: SLURP-1 mutants differently affect keratinocyte proliferation and apoptosis. J Invest Dermatol. 2021;141:2229–37.

Simeone X, Karch R, Ciuraszkiewicz A, Orr-Urtreger A, Lemmens-Gruber R, Scholze P, Huck S. The role of the nAChR subunits a5, ß2, and ß4 on synaptic transmission in the mouse superior cervical ganglion. Phys Rep. 2019;7:e14023.

Sine SM. Molecular dissection of subunit interfaces in the acetylcholine receptor: identification of residues that determine curare selectivity. Proc Natl Acad Sci U S A. 1993;90:9436–40.

Singh VK, Seed TM. How necessary are animal models for modern drug discovery? Expert Opin Drug Discovery. 2021;16:1391–7.

Sinkus ML, Graw S, Freedman R, Ross RG, Lester HA, Leonard S. The human CHRNA7 and CHRFAM7A genes: a review of the genetics, regulation, and function. Neuropharmacology. 2015;96:274–88.

Skok M. Mitochondrial nicotinic acetylcholine receptors: mechanisms of functioning and biological significance. Int J Biochem Cell Biol. 2022;143:106138.

Sørensen AT, Rombach J, Gether U, Madsen KL. The scaffold protein PICK1 as a target in chronic pain. Cell. 2022;11:1255.

Stead LF, Perera R, Bullen C, Mant D, Hartmann-Boyce J, Cahill K, Lancaster T. Nicotine replacement therapy for smoking cessation. Cochrane Database Syst Rev. 2012;11:CD000146.

Stegemann A, Böhm M. Targeting a7 nicotinic acetylcholine receptors – a novel road towards the future treatment of skin diseases. Exp Dermatol. 2020;29:924–31.

Steinlein OK. Animal models for autosomal dominant frontal lobe epilepsy: on the origin of seizures. Expert Rev Neurother. 2010;10:1859–67.

Sun HJ, Jia YF, Ma XL. a5 nicotinic acetylcholine receptor contributes to nicotine-induced lung cancer development and progression. Front Pharmacol. 2017;8:573.

Takahashi T. Multiple roles for cholinergic signalling from the perspective of stem cell function. Int J Mol Sci. 2021;22:666.

Talantova M, Sanz-Blasco S, Zhang X, Xia P, Akhtar MW, Okamoto S, et al. Aß induces astrocytic glutamate release, extrasynaptic NMDA receptor activation, and synaptic loss. Proc Natl Acad Sci U S A. 2013;110:E2518–27.

Talavera K, Gees M, Karashima Y, Meseguer VM, Vanoirbeek JA, Damann N, et al. Nicotine activates the chemosensory cation channel TRPA1. Nat Neurosci. 2009;12:1293–9.

Tapia L, Kuryatov A, Lindstrom J. Ca^{2+} permeability of the $(a4)_3(ß2)_2$ stoichiometry greatly exceeds that of $(a4)_2(ß2)_3$ human. Mol Pharmacol. 2007;71:769–76.

Targowska-Duda KM, Kaczor AA, Jozwiak K, Arias HR. Molecular interactions of type I and type II positive allosteric modulators with the human a7 nicotinic acetylcholine receptor: an in silico study. J Biomol Struct Dyn. 2019;37:411–39.

Tekinay AB, Nong Y, Miwa JM, Lieberam I, Ibanez-Tallon I, Greengard P, Heintz N. A role for lynx2 in anxiety-related behaviour. Proc Natl Acad Sci U S A. 2009;106:4477–82.

Terry AV, Callahan PM. Nicotinic acetylcholine receptor ligands, cognitive function, and preclinical approaches to drug discovery. Nicotine Tob Res. 2019;21:383–94.

Tokarska A, Silberberg G. GABAergic interneurons expressing the a2 nicotinic receptor subunit are functionally integrated in the striatal microcircuit. Cell Rep. 2022;39:110842.

Touchette JC, Moen JK, Robinson JM, Lee AM. Enhancement of alcohol aversion by the nicotinic acetylcholine receptor drug sazetidine-A. Addict Biol. 2021;26:e12908.

Treinin M. RIC-3 and nicotinic acetylcholine receptors: biogenesis, properties, and diversity. Biotechnol J. 2008;3:1539–47.

Tropea MR, Li Puma DD, Melone M, Gulisano W, Arancio O, Grassi C, Conti F, Puzzo D. Genetic deletion of nicotinic acetylcholine receptors induces an age-dependent Alzheimer's disease-like pathology. Prog Neurobiol. 2021;206:102154.

Tsetlin VI, Kasheverov IE, Utkin YN. Three-finger proteins from snakes and humans acting on nicotinic receptors: old and new. J Neurochem. 2021;158:1223–35.

Tsitsipa E, Rogers J, Casalotti S, Belessiotis-Richards C, Zubko O, Weil RS, et al. Selective 5HT3 antagonists and sensory processing: a systematic review. Neuropsychopharmacology. 2022;47:880–90.

Tutka P, Vinnikov D, Courtney RJ, Benowitz NL. Cytisine for nicotine addiction treatment: a review of pharmacology, therapeutics and an update of clinical trial evidence for smoking cessation. Addiction. 2019;114:1951–69.

Unwin N. Nicotinic acetylcholine receptor and the structural basis of neuromuscular transmission: insights from Torpedo postsynaptic membranes. Q Rev Biophys. 2013;46:283–322.

Urriola N, Adelstein S. Autoimmua7ne autonomic ganglionopathy: ganglionic acetylcholine receptor autoantibodies. Autoimmun Rev. 2022;21:102988.

Uteshev VV. The therapeutic promise of positive allosteric modulation of nicotinic receptors. Eur J Pharmacol. 2014;727:181–5.

Van Der Zanden EP, Boeckxstaens GE, de Jonge WJ. The vagus nerve as a modulator of intestinal inflammation. Neurogastroenterol Motil. 2009;21:6–17.

Wang J, Lindstrom J. Orthosteric and allosteric potentiation of heteromeric neuronal nicotinic acetylcholine receptors. Br J Pharmacol. 2018;175:1805–21.

Wang N, Orr-Urtreger A, Korczyn AD. The role of neuronal nicotinic acetylcholine receptor subunits in autonomic ganglia: lessons from knockout mice. Prog Neurobiol. 2002;68:341–60.

Wang H, Yu M, Ochani M, Amella CA, Tanovic M, Susarla S, et al. Nicotinic acetylcholine receptor a7 subunit is an essential regulator of inflammation. Nature. 2003;421:384–8.

Wang X, Lippi G, Carlson DM, Berg DK. Activation of a7-containing nicotinic receptors on astrocytes triggers ANMPA receptor recruitment to glutamatergic synapses. J Neurochem. 2013;127:632–43.

Weltzin MM, George AA, Lukas RJ, Whiteaker P. Distinctive single-channel properties of a4ß2 nicotinic acetylcholine receptor isoforms. PLoS One. 2019;14:e0213143.

Weltzin MM, George AA, Lukas RJ, Whiteaker P. Sleep-related hypermotor epilepsy associated mutations uncover important kinetic roles of a4ß2 nicotinic acetylcholine receptor intracellular structures. PLoS One. 2021;16:e0247825.

Wieskopf JS, Mathur J, Limapichat W, Post MR, Al-Qazzaz M, Sorge RE, et al. The nicotinic a6 subunit gene determines variability in chronic pain sensitivity via cross-inhibition of P2X2/3 receptors. Sci Transl Med. 2015;7:287ra72.

Wilkerson JL, Deba F, Crowley ML, Hamouda AK, McMahon LR. Advances in the in vitro and in vivo pharmacology of a4ß2 nicotinic receptor positive allosteric modulators. Neuropharmacology. 2020;168:108008.

Wills L, Ables JL, Braunscheidel KM, Caligiuri SPB, Elayouby KS, Fillinger C, et al. Neurobiological mechanisms of nicotine reward and aversion. Pharmacol Rev. 2022;74:271–310.

Winterer G, Gallinat J, Brinkmeyer J, Musso F, Kornhuber J, Thuerauf N, et al. Allosteric alpha-7 nicotinic receptor modulation and P50 sensory gating in schizophrenia: a proof of mechanism study. Neuropharmacology. 2013;64:197–204.

Wittenberg RE, Wolfman SL, De Biasi M, Dani JA. Nicotinic acetylcholine receptors and nicotine addiction: a brief introduction. Neuropharmacology. 2020;177:108256.

Wonnacott S. The paradox of nicotinic acetylcholine receptor upregulation by nicotine. Trends Pharmacol Sci. 1990;11:216–9.

Wonnacott S. Presynaptic nicotinic ACh receptors. Trends Neurosci. 1997;20:92–8.

Wonnacott S. Nicotinic receptors. In: Offermanns S, Rosenthal W, editors. Encyclopedia of molecular pharmacology. Cham: Springer; 2020.

Wu J, Liu Q, Tang P, Mikkelsen JD, Shen J, Whiteaker P, Yakel JL. Heteromeric a7ß2 nicotinic acetylcholine receptors in the brain. Trends Pharmacol Sci. 2016;37:562–74.

Xu W, Gelber S, Orr-Urtreger A, Armstrong D, Lewis RA, Ou CN, Patrick J, et al. Megacystis, mydriasis, and ion channel defect in mice lacking the a3 neuronal nicotinic acetylcholine receptor. Proc Natl Acad Sci U S A. 1999a;96:5746–51.

Xu W, Orr-Urtreger A, Nigro F, Gelber S, Sutcliffe CB, Armstrong D, et al. Multiorgan autonomic dysfunction in mice lacking the ß2 and the ß4 subunits of neuronal nicotinic acetylcholine receptors. J Neurosci. 1999b;19:9298–305.

Yakel JL. Nicotinic ACh receptors in the hippocampal circuit; Functional expression and role in synaptic plasticity. J Physiol. 2014;592:4147–53.

Zhang XY, Liu L, Liu S, Hong X, Chen DC, Xiu MH, et al. Short-term tropisetron treatment and cognitive and P50 auditory gating deficits in schizophrenia. Am J Psychiatry. 2012;169:974–81.

Zhang X, Hartung JE, Friedman RL, Koerber HR, Belfer I, Gold MS. Nicotine evoked currents in human primary sensory neurons. J Pain. 2019;20:810–8.

Zhu S, Huang S, Xia G, Wu J, Shen Y, Wang Y, et al. Anti-inflammatory effects of a7-nicotinic ACh receptors are exerted through interactions with adenylyl cyclase-6. Br J Pharmacol. 2021;178:2324–38.

Zhuang Y, Noviello CM, Hibbs RE, Howard RJ, Lindahl E. Differential interactions of resting, activated, and desensitized states of the a7 nicotinic acetylcholine receptor with lipidic modulators. Proc Natl Acad Sci U S A. 2022;119:e2208081119.

Zoli M, Pucci S, Vilella A, Gotti C. Neuronal and extraneuronal nicotinic acetylcholine receptors. Curr Neuropharmacol. 2018;16:338–49.

Zuber B, Unwin N. Structure and superorganization of acetylcholine receptor-rapsyn complexes. Proc Natl Acad Sci U S A. 2013;110:10622–7.

Purinergic P_2X Receptors as Therapeutic Targets

Leanne Stokes, Stefan Bidula, and Ronald Sluyter

Abstract Purinergic P_2X receptors are ATP-gated ion channels involved in cation flux across membranes. P_2X receptors are assembled as trimers from three identical subunits (homomeric) or from a combination of subunits (heteromeric) and display different biophysical and pharmacological properties. This chapter provides an overview of the P_2X family and their roles in various disease states as documented over the last 20 years. We highlight the potential of multiple P_2X receptors as therapeutic targets for inflammatory disorders, psychiatric disorders, pain disorders, chronic cough and cancer.

Keywords Purinergic receptors · P_2X · Inflammatory disorders · Psychiatric disorders · Pain disorders · Chronic cough · Cancer

1 Introduction

P_2X receptors ($P_2X_{1\text{-}7}$) are a family of trimeric ion channels activated by extracellular adenosine 5′-triphosphate (ATP). The late Geoffrey Burnstock was instrumental in discovering membrane receptors activated by extracellular nucleotides which were later subdivided into P1 and P2 receptors. P2 receptors were further divided into a P_2Y type which were G-protein coupled receptors and P_2X type which were ligand-gated ion channels (Burnstock and Kennedy 1985). Measurements of native responses to ATP and various ATP analogues revealed differences in purinergic ion channel features such as desensitisation and sensitivity to agonists suggesting

L. Stokes (✉) · S. Bidula
School of Pharmacy, University of East Anglia, Norwich, UK
e-mail: l.stokes@uea.ac.uk; s.bidula@uea.ac.uk

R. Sluyter
Molecular Horizons and School of Chemistry and Molecular Bioscience, University of Wollongong, Wollongong, NSW, Australia
e-mail: rsluyter@uow.edu.au

G. Stephens, E. Stevens (eds.), *Ion Channels as Targets in Drug Discovery*,
https://doi.org/10.1007/978-3-031-52197-3_13

heterogeneity of P_2X receptors in cells and tissues. P_2X subunit RNA transcripts were cloned throughout the 1990s as reviewed in (North and Surprenant 2000), and this began intensive investigations into the properties and pharmacology of P_2X ion channels in various recombinant expression systems (Table 1). Some properties of homomeric recombinant P_2X channels matched native responses, but this approach also allowed the identification of native heteromeric channels such as $P_2X2/3$ which displays unique pharmacological properties with regard to sensitivity to the ATP analogue αβ-methylene-ATP (Spelta et al. 2003). Electrophysiology experiments detailed the ion selectivity of P_2X ion channels, all of which are cation-selective permitting entry of Na^+ and Ca^{2+} and efflux of K^+ (North and Surprenant 2000). Activation and deactivation (or desensitisation) rates vary between the homomeric subtypes with P_2X_1 and showing the fastest activation and deactivation; P_2X_4 and P_2X_2 activate rapidly but exhibit differing rates of desensitisation; P_2X_4 has a moderate desensitisation rate, whereas P_2X2 does not desensitise. P_2X_7 is activated more

Table 1 Properties of human homomeric P_2X receptors

	Gene location	Protein length	Splice variants (protein length)	Agonists	Desensitisation	Potential disease links
P_2X_1	17p13.2	399	nd	ATP	Fast	Sepsis
				αβMe-ATP		
				BzATP		
P_2X_2	12q24.33	471	P_2X_2b (404)	ATP	None	Deafness
			P_2X_2c (447)			Intestinal motility disorders
			P_2X_2d (497)			Depression
P_2X_3	11q12.1	397	nd	ATP	Fast	Chronic cough (RCC)
				αβMe-ATP		Pain
				BzATP		
P_2X_4	12q24.31	388	nd	ATP	Moderate	Neuropathic pain
						Inflammatory pain
						Sepsis
P_2X_5	17p13.2	444[a]	P_2X_5A	ATP	–	Infection
			P_2X_5B (397)			
			P_2X_5C (178)			
P_2X_6	22q11.21	441	nd	ATP	–	
P_2X_7	12q24.32	595	P_2X_7B (364)	ATP	None	Inflammation
			P_2X_7E (275)	BzATP		Psychiatric disorders
			P_2X_7J (258)			Cancer
			P_2X_7L (506)			
			nfP_2X_7 (nd)			

[a]Non-functional in most races
nd not determined

slowly (concentration dependent manner) and typically does not desensitise throughout the agonist application (North and Surprenant 2000).

The first high-resolution crystal structure of a P_2X receptor (zebrafish P_2X_4) was published in 2009 (Kawate et al. 2009) and confirmed that each P_2X subunit comprised of two transmembrane (TM) domains, a large extracellular loop and intracellular N- and C-termini as predicted from the original sequence data (Valera et al. 1994). This crystal structure gave the first insight into the closed-state structure of zebrafish P_2X_4 and was followed by an ATP-bound crystal structure in 2012 (Hattori and Gouaux 2012). This allowed the molecular mechanism involved in P_2X ion channel opening and key determinants for agonist action to be deduced. Since 2012, multiple structures have been reported for human P_2X_3 (Mansoor et al. 2016), P_2X_7 in complex with various antagonists (Karasawa and Kawate 2016) and, more recently, a full-length cryo-electron microscopy structure of rat P_2X_7 showing the structure of the intracellular C-terminus (Fig. 1) (McCarthy et al. 2019). Other studies yielded important advances in understanding ion access to the pore of P_2X ion channel. Samways et al. suggested that the lateral portals – openings on the side of the ion channel complex (see Fig. 1) – were important for cations gaining access to the channel pore rather than travelling through the centre of the complex (Samways et al. 2011). Li et al. demonstrated that an ion selectivity filter is located within the

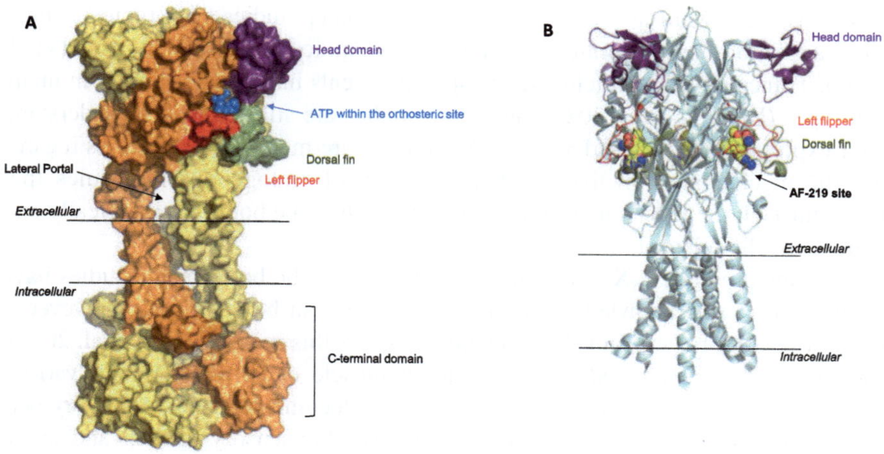

Fig. 1 3D structures of P_2X ion channels with chemical entities bound. (**a**) full length structure of trimeric rat P_2X_7 with ATP bound (PDB: 6U9V). P_2X_7 subunits within the trimer are coloured yellow and orange to allow clear identification. Three domains are indicated in different colours – head domain (purple), left flipper (red) and dorsal fin (green) using terminology From the original structural articles (Kawate et al. 2009; Hattori and Gouaux 2012). ATP is shown as spheres (blue) within the defined orthosteric site. The lateral portal is indicated as the cation entry point for the ion channel. (**b**) Structure of human P_2X_3 (PDB: 5YVE) in complex with AF-219 (Gefapixant) as a cartoon visualisation. Three domains are indicated in different colours; head domain (purple), left flipper (red) and dorsal fin (green). AF-219 is shown as spheres in an allosteric site located between the left flipper and dorsal fin underneath the ATP site

TM domains with the narrowest point being around the middle of TM2 (Li et al. 2008).

Early studies suggested the movement of large organic molecules through some P_2X ion channels in addition to charged ions. This well-studied yet somewhat controversial phenomenon is associated with P_2X_7 receptors as reviewed in (Ugur and Ugur 2019). Often referred to as secondary pore formation or a macropore, it is thought that the open P_2X_7 ion channel is large enough to allow organic molecules to directly pass rather than a gradual dilation over time. Nanometre-sized dyes are able to permeate the P_2X_7 pore as shown by large methanesulphonate compounds reacting with cysteines introduced into the TM2 region (Browne et al. 2013). Single-channel recordings have not found evidence for altered channel conductance over time (Pippel et al. 2017), and high-resolution structures have not revealed a second dilated state (McCarthy et al. 2019). One hypothesis for the P_2X_7-dependent uptake of organic molecules was the regulation of a distinct interacting protein. One candidate protein, the hemichannel pannexin-1, was discovered in 2006, and some evidence suggested it may be involved (Pelegrin and Surprenant 2006). However, several studies have since ruled out the likelihood of a pannexin being involved. Incorporation of a purified P_2X_7 ion channel into liposomes was sufficient for fluorescent dye movement across a lipid bilayer in cell-free experiments (Karasawa et al. 2017), and this was modulated by the lipid environment, demonstrating a role for cholesterol in regulating organic molecule movement (Karasawa et al. 2017). Furthermore, the dye movement was also observed when a C-terminal truncated P_2X_7 was used (Karasawa et al. 2017), whereas other studies have typically indicated a role for the C-terminus in regulating large molecule movement (Smart et al. 2003). Truncated P_2X_7 proteins due to splicing events have been reported in many cell types (De Salis et al. 2022), and more studies are still required to understand their expression profiles and activity. While there are many open questions regarding the mechanisms involved and the identity of endogenous molecules that move through the P_2X_7 ion channel pore, it is likely to be both physiologically and therapeutically relevant.

In terms of where P_2X ion channels are located in the body, many studies have contributed to this knowledge, and not all articles can be cited here; however, a recent IUPHAR review on P_2X ion channels covers this in detail (Illes et al. 2021). In brief, P_2X_1 is expressed in most smooth muscle cells, platelets and various immune cells; P_2X_2 is expressed in neurons with documented roles in sensory perception such as taste and hearing; P_2X_3 is expressed in sensory neurons and spinal cord afferent neurons (nociceptive); P_2X_4 is expressed in neurons, glia, immune cells, cardiac muscle, smooth muscle and endothelial cells; P_2X_7 is expressed in immune cells, glia, endothelial and epithelial cells and potentially in some neurons (Illes et al. 2021). P_2X_5 is predominantly non-functional in humans (King 2022), and P_2X_6 does not form homomeric channels (Ormond et al. 2006), but is found in endothelial cells together with P_2X_4 (Glass et al. 2002).

The rest of this chapter will describe the evidence for various P_2X receptors as therapeutic targets for several different disorders summarising the pre-clinical/clinical progress in each area.

2 Inflammatory Disorders

P_2X_7 has received widespread interest as a therapeutic target due to its involvement in the pathophysiology of inflammatory disorders of the lung (pulmonary hypertension, asthma, chronic obstructive pulmonary disease, acute lung injury, pulmonary fibrosis), gut (inflammatory bowel disease, colitis, Crohn's), liver (fibrosis), kidneys (glomerulonephritis, diabetic nephropathy, tubulointerstitial nephritis, hypertensive nephropathy, acute/chronic kidney injury), cardiovascular system (thrombosis, endothelial dysfunction), neuroinflammation (traumatic brain injury, Parkinson's disease, Alzheimer's disease, ischemia, epilepsy, Huntington's disease, multiple sclerosis) and in diabetes. In general, P_2X_7 exacerbates inflammation due to its links with pro-inflammatory signalling in many cell types. This includes the rapid release of pro-inflammatory cytokines and chemokines, with the activation of the NLRP3-caspase1 inflammasome and release of interleukin-1 family cytokines the best described mechanism to date (Di Virgilio et al. 2017; Savio et al. 2018). There has been much interest in blocking P_2X_7 channels with antagonists in inflammatory diseases, and many candidate molecules have been synthesised and characterised over the last 15 years (Dane et al. 2022).

P_2X_7 antagonists were the first to undergo clinical trials for inflammatory disorders. In 2008, Pfizer conducted a phase 2a clinical trial (NCT00628095) to assess whether the P_2X_7 antagonist CE-224,535 could be used to control the symptoms of rheumatoid arthritis. Although CE-224,535 was shown to be a safe and tolerable drug to administer, it failed to display any benefit when compared with the placebo (Stock et al. 2012). AstraZeneca conducted phase 2a and 2b clinical trials for P_2X_7 antagonist AZD9056 in the treatment of rheumatoid arthritis (NCT00520572), but AZD9056 was not efficacious (Keystone et al. 2012). GlaxoSmithKline conducted a clinical trial to test the efficacy of the P_2X_7 antagonist GSK1482160 in the treatment of inflammatory pain, but the results of this trial have not yet been released (NCT00849134). These trial results contrasted with pre-clinical data where P_2X_7 antagonism was shown to reduce ankle swelling, decrease the pathological and radiological scores and inhibit mechanical hyperalgesia in rat models of arthritis. Moreover, in ex vivo models, AZD9056 could prevent the production of the inflammatory cytokines IL-1β and IL-18 in LPS-primed monocytes (McInnes et al. 2014). Perhaps certain features of the clinical trials and patient recruitment affected the lack of efficacy (e.g. patients were refractory to conventional treatment). AstraZeneca has also conducted phase 2a trials with AZD9056 to assess its efficacy and safety profile in the treatment of Crohn's disease (Eser et al. 2015). Here AZD9056 was shown to increase the clinical remission, decrease disease activity and prevent pain-but did not decrease inflammatory markers (Eser et al. 2015), with the authors suggesting that P_2X_7 antagonism might be more useful to treat chronic abdominal pain. Additional clinical trials have been conducted exploring the potential involvement of P_2X_7 in osteoporosis (NCT00697983/NCT00293189) and cardiac allograft vasculopathy (NCT02082821), but the results of these studies have not yet been released.

Much of the research into inflammatory diseases has focussed on P_2X_7, but there is growing evidence supporting the involvement of P_2X_1, P_2X_2, P_2X_4 and P_2X_5 in inflammation. P_2X_1 can be found on neutrophils where this ion channel may have multiple roles in inflammation. In a mouse model of acute colitis, loss of P_2X_1 appeared to promote recruitment of neutrophils to the site of injury and exacerbate thrombosis and intestinal bleeding (Wéra et al. 2020). Some studies have identified that knockout of P_2X_1 or P_2X_1 antagonism is protective in sepsis (Lecut et al. 2012; Maître et al. 2015; Skals et al. 2019). P_2X_2 has recently been shown to promote cytokine release from enteric glial cells and drive intestinal inflammation (enteric gliosis). Treatment with ambroxol (Fig. 2), a P_2X_2 antagonist, prevented ATP-induced enteric gliosis, inflammation and dysmotility in mice, while reducing enteric gliosis in the human intestine following surgical trauma (Schneider et al. 2021). Thus, P_2X_2 blockade represents a novel therapeutic target to treat immune-driven intestinal motility disorders.

The most notable role for P_2X_4 in inflammation is its involvement in inflammatory pain(see later); however deleterious roles have also been indicated in cardiovascular, respiratory and gastrointestinal inflammation. Low-density lipoprotein

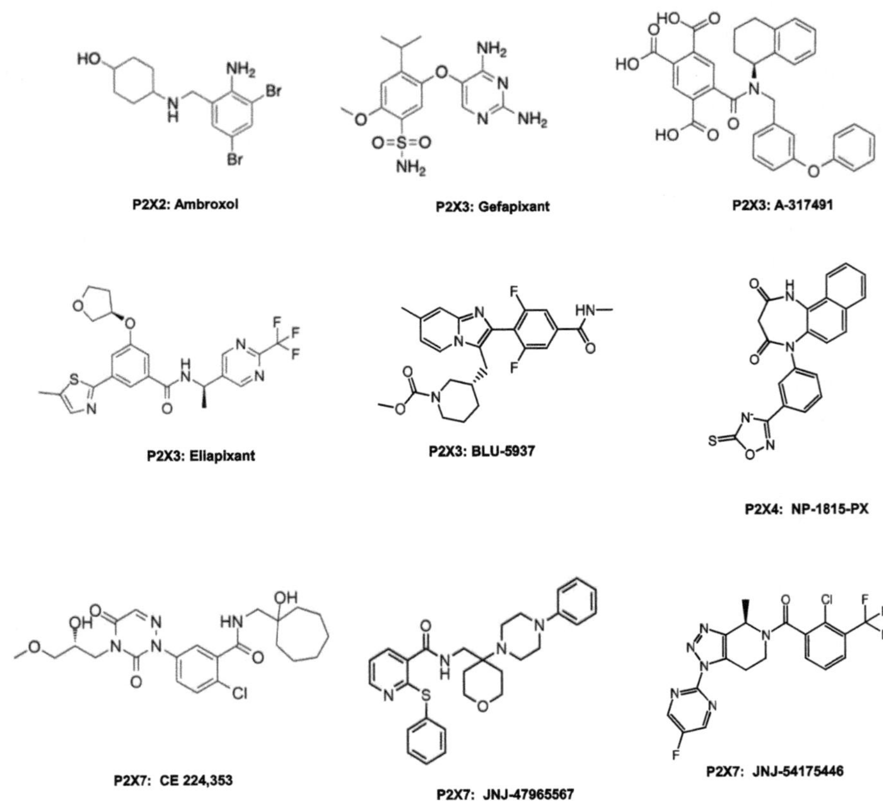

Fig. 2 Chemical structures of antagonists targeting key P_2X ion channels

(LDL)-deficient mice fed a high-cholesterol diet were found to have greater expression of P_2X_4 in atherosclerotic aortic arches (Peikert et al. 2022). When *P2rx4*-knockout (KO) mice were crossed with LDL-deficient mice and fed the same diet, *P2rx4*-KO mice displayed much smaller atherosclerotic lesions (Schneider et al. 2021). This was accompanied by reduced RNA expression of several cytokines (IL-6, TNF-α), chemokines (CCL2, CXCL1, CXCL2) and inflammasome activation (Peikert et al. 2022). P_2X_4 blockade by two novel and selective P_2X_4 antagonists, NC-2600 and NP-1815-PX (Fig. 2), was shown to prevent weight loss, colonic tissue damage and IL-1β production in a 2,4-dinitrobenzene sulphonic acid (DNBS)-induced colitis mouse models (D'Antongiovanni et al. 2022). Further investigations in vitro identified that P_2X_4 was driving inflammasome activation and concluded that the P_2X_4 inflammasome axis could be a potential approach to manage bowel inflammation. There is also growing evidence for the roles of P_2X_4 in airway inflammation. In a murine experimental asthma model, P_2X_4 expression in the inflamed lung was increased, and P_2X_4 antagonism with 5-BDBD could attenuate airway inflammation and contribute to airway remodelling (Chen et al. 2016). Similarly, P_2X_4 mRNA expression was found to be increased in the bronchoalveolar lavage and peripheral mononuclear cells of chronic obstructive pulmonary disease patients. Furthermore, cigarette smoke was shown to elevate P_2X_4 expression in mice, and blocking P_2X_4 could prevent airway inflammation and lung tissue damage (Schneider et al. 2022). Finally, the use of *P2rx4*-KO and *P2rx4* reporter mice have demonstrated a beneficial role for macrophage P_2X_4 in bacterial killing and protection from sepsis (Csóka et al. 2018). Administration of the P_2X_4 positive modulator ivermectin could protect mice from mortality in sepsis (Csóka et al. 2018).

Very little is known about the physiological roles of P_2X_5 in general, but current research has highlighted an emerging role in inflammasome activation. Mice lacking P_2X_5 displayed defective inflammasome activation following infection with the bacterial pathogen *Listeria monocytogenes* (Jeong et al. 2020). P_2X_7 was not involved in inflammasome activation in these infections and macrophages taken from P_2X_5-deficient mice exhibited a poor ability to kill the bacteria (Jeong et al. 2020). Additionally, P_2X_5 was found to be required for inflammasome activation in osteoclasts and promoted osteoclast-mediated inflammatory bone loss (Kim et al. 2017). Thus, the P_2X_5-inflammasome axis may be a novel therapeutic target in infection and immunity. However, as explained in (King 2022), full-length P_2X_5 appears to be absent in most human races except for those of black origin, and in these people P_2X_5 may be a useful future therapeutic target.

3 Psychiatric Disorders

Bipolar disorder is an affective disorder characterised by cyclic episodes of mania and depression (Martinowich et al. 2009). Early genetic association studies identified a locus on chromosome 12 implicated in bipolar disorder heritability close to where *P2RX7* and *P2RX4* are located on 12q24.32 and near *P2RX2* at 12q24.33.

There have been many investigations into single-nucleotide polymorphisms (SNPs) in the human *P2RX7* gene highlighting one non-synonymous SNP – rs2230912 – encoding Q460 > R in the C-terminus (Barden et al. 2006; Lucae et al. 2006; McQuillin et al. 2009) as potentially associated with bipolar disorder. This SNP is not inherited in isolation in many individuals but rather inherited as a haplotype block with rs1718119 encoding A348 > T and most often with rs208294 encoding H155 > Y. The net effect of this combination of mutations is a P_2X_7 receptor with improved responses (or "gain of function") (Stokes et al. 2010). This proposed genetic association has not been repeated in larger genome-wide association studies, e.g. (Wellcome Trust Case Control, C 2007), and several meta-analyses have since been published to attempt a better understanding of whether this association holds up (Feng et al. 2014; Czamara et al. 2018). One suggestion is that the association lies with heterozygote carriers in a heterozygote disadvantage model (Barden et al. 2006; McQuillin et al. 2009). A humanised P_2X_7 mouse model incorporating the rs2230912 (Q460 > R) genetic change was made, and mouse behaviour was assessed (Metzger et al. 2017). The main significant finding was that heterozygosity for rs2230912 was associated with increased fragmented sleep, a risk factor for developing mood disorders. In contrast, there was no difference in behavioural responses between genotypes in response to a chronic social defeat stress paradigm to assess depressive-like changes (Metzger et al. 2017). To understand the involvement of P_2X_7 in pathophysiology of bipolar disorder requires robust animal models to explore either gene deletion or the impact of pharmacological modulators. There are few such animal models with none as yet able to mimic the cyclical nature of bipolar disorder (Beyer and Freund 2017). The amphetamine-induced hyperactivity model is commonly used for assessing mania, and P_2X_7 gene deletion has a significant effect (Csölle et al. 2013a). The non-selective P_2X_7 antagonist brilliant blue G (BBG) and the P_2X_7-selective antagonist A438079 were effective in reducing hyperactivity in mice (Csölle et al. 2013a; Gubert et al. 2016, 2020), and newer central nervous system (CNS)-penetrant P_2X_7 antagonists such as JNJ-47965567 have shown efficacy in rats (Bhattacharya et al. 2013). Mechanistically it is suggested that P_2X_7 is linked to neuroinflammation and dysregulated neurotransmitter release, as reviewed in (Gonçalves et al. 2022).

Around the same time as the genome-wide association study discussed above, several groups investigated the behavioural effect of *P2rx7* gene KO in mice. Consistently these studies found that *P2rx7*-KO mice displayed antidepressive-like behaviour (Basso et al. 2009; Boucher et al. 2011; Csölle et al. 2013a; Csölle et al. 2013b) particularly in response to stress paradigms. This has been confirmed in subsequent studies using restraint stress, chronic unpredictable mild stress and sleep deprivation (Otrokocsi et al. 2017; Yue et al. 2017; Xia et al. 2020; Troubat et al. 2021). Taking P_2X_7 forward as a therapeutic target for depression requires strong pre-clinical evidence gathered from using selective antagonists such as A-804598 and A-438079 (Iwata et al. 2016; Yue et al. 2017). Janssen has demonstrated several of its new CNS-penetrant P_2X_7 antagonists show efficacy in animal models of depression, for example, JNJ-55308942 (Bhattacharya et al. 2018). From their successful pre-clinical studies, two drug candidates, JNJ-54175446 (Fig. 2) and

JNJ-55308942, are currently in clinical trials (NCT04116606, NCT03151486). Mechanistically, the links between inflammation and depression have long been described with well-supported evidence for similarities between sickness behaviour and depressive behaviour (Miller and Raison 2016; Lucido et al. 2021). Acute restraint stress can increase brain ATP, glutamate and cytokine levels as measured by microdialysis experiments and inflammasome activation was also seen (Iwata et al. 2016). Elevated extracellular ATP concentrations in the hippocampus have been confirmed (Yue et al. 2017), and injections of ATP or BzATP into the hippocampus induce depressive-like behaviours in rats over a 3-week interval (Yue et al. 2017). In a learned helplessness (inescapable foot-shock) model, structural abnormalities can be observed with a decrease in spine synapse number in the dentate gyrus of the hippocampus which was not seen in *P2rx7*-KO mice (Otrokocsi et al. 2017). It is proposed that P_2X_7 plays a role in microglial activation and that CNS penetrant P_2X_7 antagonists can reduce neuroinflammation and depressive-like behaviour (Bhattacharya et al. 2018), with some evidence for a reduction in microglial number following chronic unpredictable mild stress (Troubat et al. 2021). Other links between P_2X_7 and changes in the brain include involvement in corticosterone-induced high-mobility group box 1 (HMGB1) release from astrocytes (Hisaoka-Nakashima et al. 2020) and alterations in brain-derived neurotrophic factor (BDNF) signalling pathways (Ribeiro et al. 2019). There is also emerging evidence for P_2X receptor involvement in schizophrenia. In a phencyclidine (PCP)-induced schizophrenia model, *P2rx7*-KO was protective against certain behavioural changes (hyperlocomotion, stereotype behaviour) (Koványi et al. 2016). Pharmacological blockade of P_2X_7 mimicked this abolishing the effects of low-dose PCP (Koványi et al. 2016). A 2021 study showed JNJ-47965567 could reverse PCP-induced behavioural changes (Huang et al. 2021).

Much of the focus in psychiatric disorders has been on P_2X_7; however, a 2013 study suggested that P_2X_2 expressed in the medial prefrontal cortex (mPFC) may be involved in the regulation of depressive-like behaviour (Cao et al. 2013). This study investigated chronic social defeat stress (CSDS) and highlighted a reduction in extracellular ATP levels in the prefrontal cortex and hippocampus regions. Injecting ATP into the mPFC had an antidepressant effect, and this was dependent on activation of P_2X_2 in this region (Cao et al. 2013). A more recent study showed P_2X_2 mRNA expression increased in the mPFC following CSDS (10 days) and that P_2X_2 conditional KO mice (pyramidal neurons) were resilient to depressive-like behaviour (Kuang et al. 2022). More work is needed to elucidate the potential for P_2X_2 targeting in psychiatric disorders.

4 Pain

Chronic pain is a major health issue globally with millions of people suffering with this condition (Cohen et al. 2021). Neuropathic pain disorders result from lesions or diseases of the somatosensory nervous system (Finnerup et al. 2021); other categories include nociceptive pain from tissue injury and nociplastic pain due to a sensitisation of the nervous system (Cohen et al. 2021). Chronic pain has contributing factors including many psychological and biological influences (Cohen et al. 2021), and understanding the pathophysiology associated with the development of chronic pain states is crucial for identification of potential therapeutic targets. One such target for neuropathic pain is P_2X_4, which was first highlighted in a 2005 study demonstrating that pharmacological antagonism of P_2X_4, or antisense oligonucleotidesreducing P_2X_4 expression, reversed tactile allodynia (Tsuda et al. 2003). This was followed by a detailed investigation into the underlying mechanism showing that ATP activation of P_2X_4 on spinal microglia regulates the secretion of BDNF which alters the anion reversal potential in spinal lamina I neurons (Coull et al. 2005). Further evidence came from the use of *P2rx4*-KO mice showing that a lack of mechanical hypersensitivity in peripheral nerve injury models correlated with impaired BDNF release (Ulmann et al. 2008; Tsuda et al. 2009). Several studies determined that cytokines and chemokines, e.g. CCL21, interferon (IFN)-γ and CCL2, can upregulate P_2X_4 in spinal microglia, providing other potential therapeutic targets for modulating the initiation of pain states, as reviewed in (Beggs and Salter 2013). However, this P_2X_4-BDNF microglial-dependent pathway is restricted only to male mice (Mapplebeck et al. 2016, 2018). In addition to neuropathic pain, P_2X_4 is also implicated in musculoskeletal pain (Oliveira-Fusaro et al. 2020) and in inflammatory pain, where *P2rx4*-KO mice displayed reduced pain following intraplantar CFA injection (Tsuda et al. 2009) and in the formalin hind paw model (Ulmann et al. 2010).

High-affinity selective antagonists are required for good pharmacological targeting of P_2X_4 for pre-clinical and progression to clinical studies. Progress with selective and potent P_2X_4 antagonists was initially slow but is now picking up momentum. Early studies utilised the non-selective antagonist 2′,3′-trinitrophenyl (TNP)-ATP (Tsuda et al. 2003), and then paroxetine and duloxetine, both better known as selective serotonin reuptake inhibitors (SSRIs), were shown to block P_2X_4 with efficacy in neuropathic pain models (Nagata et al. 2009). In 2016, NP-1815-PX (Fig. 2) demonstrated efficacy in a model of herpetic pain (Matsumura et al. 2016), and in 2019, a series of P_2X4 antagonists developed in the Müller lab, e.g. PSB-15417, demonstrated efficacy in reversing diabetes-induced neuropathic hyperalgesia (Teixeira et al. 2019). Bayer revealed a new class of P_2X_4 antagonists with characterisation of BAY-1797 and demonstration of activity in an inflammatory pain model (Werner et al. 2019). BX430 is another selective P_2X_4 antagonist (Ase et al. 2015), but its activity is restricted to human P_2X_4, and therefore it is unable to be tested in animal models of pain. In addition to small molecules, antibodies to the head domain of P_2X_4 have been developed and shown to reverse neuropathic pain in mice (Williams et al. 2019). The anti-P_2X_4 antibody IgG#191 was delivered via intrathecal administration and provided a long-lasting reversal of neuropathic pain (Williams et al. 2019). Small single-chain antibodies targeting P_2X_4 been shown to permanently reverse chronic neuropathic pain in male mice (Westlund et al. 2021). These studies highlight an alternative approach for inhibiting P_2X_4, opening up more opportunities for peptides and antibody or nanobody development (Bergmann et al. 2019).

P_2X_4 is not the only purinergic receptor implicated in pathophysiology of chronic pain: both P_2X3 and P_2X_7 receptors are also implicated (Bernier et al. 2018; Inoue and Tsuda 2021). Knockout mouse studies highlighted a role for P_2X_7 in neuropathic pain (Chessell et al. 2005), where both male and female *P2rx7*-KO mice were protected against thermal and mechanical hypersensitivity following partial nerve ligation and complete Freund's adjuvant (CFA) inflammatory pain model (Chessell et al. 2005). Abbott Laboratories developed multiple P_2X_7 antagonists which were analysed in neuropathic pain models such as A-740003 (Honore et al. 2006), A-839977 (Honore et al. 2009) and a series of cyanoguanidine antagonists (Perez-Medrano et al. 2009). Other companies followed suit, such as GSK with their pyroglutamide series of P_2X_7 antagonists (Abdi et al. 2010) and Janssen with their CNS penetrant antagonists, e.g. JNJ-47965567 (Fig. 2) (Bhattacharya et al. 2013). Intriguingly, some of the high-affinity P_2X_7 antagonists developed by Janssen were found to show no efficacy in neuropathic or inflammatory pain models such as JNJ-42253432 (Lord et al. 2014) and a series of carboxamide antagonists (Rech et al. 2016). Mechanistically, it seems that microglial P_2X_7 is involved, and it is proposed that the regulation of pro-inflammatory cytokine secretion is causative (Chessell et al. 2005; Honore et al. 2009). In other types of pain, such as bone cancer-induced pain states, blockade of P_2X_7 has no effect (Hansen et al. 2011). Pre-clinical development of P_2X_7 antagonists for neuropathic pain has slowed, and to date, no antagonists have yet been pushed forward into clinical trials for pain. An alternative approach by Yamashita et al. screened thousands of clinically approved drugs for activity at P_2X_7, revealing cilnidipine as a calcium channel blocker with antagonist action at rat P_2X_7 and reversal of nerve injury-induced neuropathic pain in rats (Yamashita et al. 2021).

P_2X_3 is expressed on sensory neurons and nociceptive fibres, and agonists such as ATP or αβ-methylene-ATP can induce peripheral sensitisation and pain states (Hamilton et al. 1999). Following nerve injury, such as in the chronic constriction injury (CCI) model or diabetic neuropathy, P_2X_3 is upregulated (Novakovic et al. 1999; Xu et al. 2011). Notably these receptors on small and medium diameter dorsal root ganglion (DRG) neurons are sensitised (Chen et al. 2005). *P2rx3*-KO mice show altered pain behaviour in response to ATP injection, plus pain responses to the formalin test were reduced (Cockayne et al. 2000). Antisense oligonucleotide studies also support a role for P_2X_3 in rat pain models (Barclay et al. 2002; Honore et al. 2002). Development of the P_2X_3-selective antagonist A-317491 (Fig. 2) by Abbott Laboratories provided promising pharmacological evidence for P_2X_3 (or $P_2X2/3$ heteromers) as a therapeutic target for chronic inflammatory and neuropathic pain states (Jarvis et al. 2002) although this antagonist was poorly CNS penetrant and not a good in vivo tool (Sharp et al. 2006). This was followed by development of diaminopyridines such as AF-353 (RO-4) (Carter et al. 2009) and RO-51 (Jahangir et al. 2009), and both were effective in an inflammatory pain model (Ballini et al. 2011). AF-353 had more favourable properties for in vivo studies including oral bioavailability (Gever et al. 2010) and was thought to have an allosteric mechanism of action rather than being a competitive antagonist. Oral administration of AF-353 was also shown to attenuate bone cancer-induced pain in rats (Kaan et al. 2010), while A-317491 was similarly effective in mice (Hansen et al. 2012). Around this same time, a peptide derived from spider venom, purotoxin-1, was discovered to have an inhibitory effect on P_2X_3 (Hansen et al. 2012). Purotoxin-1 is thought to reduce P_2X_3 ion channel recovery from desensitisation and had an antinociceptive effect in an inflammatory pain model (Grishin et al. 2010). A fantastic insight into the convincing case for P_2X_3 antagonists as therapeutics for pain circa 2011 is a recommended read (Ford 2012). AstraZeneca described a series of P_2X_3-selective antagonists (Cantin et al. 2012), and in 2016, monoclonal antibodies specific for P_2X_3 developed by Pfizer were tested in visceral hypersensitivity models with good efficacy (Shcherbatko et al. 2016). AF-219, also known as MK-7264 and gefapixant (after Geoffrey Burnstock, the pioneer of the purinergic field) (Richards et al. 2019), progressed into clinical trials for chronic cough rather than pain (see section below). A high-resolution crystal structure of human P_2X_3 in complex with AF-219 revealed the location of the first negative allosteric binding pocket (Wang et al. 2018) close to the ATP binding pocket where antagonists are thought to interfere with the movement of the left flipper, lower body and the dorsal fin domains (Fig. 1) (Wang et al. 2018). Other P_2X_3 antagonists such as dioxotriazine derivatives (Kai et al. 2021a) have been described recently with analgesic effects in neuropathic pain in rats, with sivopixant put forward as a clinical candidate from Shionogi and Co Ltd. (Kai et al. 2021b). This was closely followed by eliapixant (Fig. 2) (BAY1817080) with pre-clinical evidence from in vivo models for reducing neurogenic inflammation associated with endometriosis (Davenport et al. 2021). Eliapixant was in clinical trials for endometriosis (NCT04614246) and for diabetic neuropathic pain (NCT04641273), but development has been discontinued due to a risk of hepatotoxicity (Zhang et al. 2022).

5 Chronic Cough

Chronic cough conditions affect approximately 12–16% of the general population (Satia et al. 2016), and persistent coughing without identifiable causes is classed as refractory chronic cough (RCC). P_2X_3 was highlighted as a new therapeutic target-for chronic cough due to its role in sensitisation of primary afferent sensory neurons and exaggeration of autonomic reflexes in the respiratory tract (Ford and Undem 2013). Animal studies for cough are complicated by significant species differences to the sensitising effect of ATP and cough behaviour being difficult to measure in smaller rodents (Ford et al. 2021; Zhang et al. 2022). This presented a challenge for testing P_2X_3 compounds in pre-clinical models and many P_2X_3 antagonists have been tested in a guinea pig cough reflex model (Zhang et al. 2022). Much of the testing work for discovery and development of P_2X_3 compounds was performed on recombinant receptors in 1321N1 astrocytoma cells using FLIPR and electrophysiology assays (Ford et al. 2021). Roche pursued patient studies with their P_2X_3

antagonists for chronic cough with phase 1 trials in 2007 switching their therapeutic program to this over pain, as reviewed in (Ford et al. 2021). In 2015, a phase 2 trial by Afferent Pharmaceuticals showed reduced cough frequency with AF-219 (gefapixant) (Abdulqawi et al. 2015), and as of 2022, this is approved for clinical use in Japan (Markham 2022) but is still awaiting approval in US and European sectors following phase 3 trials for RCC (Markham 2022; McGarvey et al. 2022). One of the issues with gefapixant is an adverse effect on taste due to antagonism of $P_2X2/3$ receptors in taste cells causing a loss of taste. Trials with eliapixant also documented taste-related side effects (in 24% of patients) (Zhang et al. 2022); however, such taste-related side effects are less frequent with BLU-5937 (Fig. 2), a P_2X_3 antagonist in development by BELLUS Health (Garceau and Chauret 2019). Preclinical studies demonstrated BLU-5937 reduced frequency of histamine-induced or citric acid-induced coughs in guinea pigs and did not affect taste perception in rats (Garceau and Chauret 2019). Sivopixant demonstrated promising antitussive activity in a phase 2a trial (Niimi et al. 2022) but did not demonstrate efficacy in a phase 2b trial (Zhang et al. 2022). The use of P_2X_3 antagonists in the treatment of chronic cough makes them the most successful clinical candidates in targeting a P_2X ion channel. Future studies will hopefully improve selectivity for homomeric P_2X_3 alleviating taste-related side effects, and further optimisation may enhance efficacy.

6 Cancer

Early studies on P_2X receptors in cancer began with chronic lymphocytic leukaemia (CLL) and observations of higher P_2X_7 responses in ex vivo lymphocytes from patients (Adinolfi et al. 2002). Inheritance of a SNP in the C-terminus (rs3751143) of P_2X_7 was higher in CLL patients (Wiley et al. 2002), and this offered a potential explanation for resistance to apoptosis, as P_2X_7 receptors containing this mutation did not undergo cell death (Gu et al. 2001). Many studies have since been published documenting P_2X_7 expression in a number of cancers including melanoma (White et al. 2005), glioma (Morrone et al. 2005), neuroblastoma (Raffaghello et al. 2006) and ovarian carcinoma (Vázquez-Cuevas et al. 2014), and all indicate a role for P_2X_7 in supporting proliferation rather than induction of cell death. In 2006 a splice variant called P_2X_7 J was discovered, a non-functional P_2X_7 ion channel lacking both TM2 and the C-terminus and acting as a dominant negative subunit (Feng et al. 2006); whether this is related to a variant known as nf-P_2X_7 is still unclear. Nf-P_2X_7 was first reported in 2004 in prostate cancer biopsies but was not found in normal prostate epithelium (Slater et al. 2004). An anti-human P_2X_7 antibody (BIL010t) recognises a non-functional form of P_2X_7 containing a typically hidden epitope around amino acids 200–216 and has now been further studied in a number of other cancers (Gilbert et al. 2019); the antibody is currently in clinical trials for basal cell carcinoma (Gilbert et al. 2017). Around 2005, the discovery of additional splice variants of P_2X_7 revealed a C-terminal truncated version termed P_2X_7-B

(Cheewatrakoolpong et al. 2005), and Adinolfi et al. showed this variant was correlated with proliferative activity (Adinolfi et al. 2010) and linked to enhanced mitochondrial calcium signalling (Adinolfi et al. 2005). In an in vivo xenograft model, P_2X_7-transfected cells exhibit enhanced tumourigenesis (Adinolfi et al. 2012), and in neuroblastoma, P_2X_7 is linked to enhanced proliferation via the PI3K/VEGF pathway (Amoroso et al. 2015). Currently not enough is known about expression of all splice variants in different cancers. The discovery of an additional variant P_2X_7-L missing exon 7–8 brings the total of known splice variants up to eight (De Salis et al. 2022; Skarratt et al. 2020). There is good therapeutic potential for P_2X_7 antagonists in cancer to limit tumour growth, as reviewed in (Roger et al. 2015).

P_2X_7 is involved in the immune system where it may enhance the immune response to tumour cells. This was revealed in a study using P_2X_7-knockout mice where tumour progression was accelerated (Adinolfi et al. 2015). Host immune cell expression of P_2X_7 was critical for the anti-tumour immune response; therefore, this brings into question the value of systemic pharmacological P_2X_7 blockade in cancer. Some studies indicate that blocking P_2X_7 may enhance T cell cancer immunotherapy, as P_2X_7-deficient $CD8^+$ T cells performed better in anti-tumour responses (Romagnani et al. 2020). Similarly, P_2X_7 is linked to NK cell anergy, and activation may impair NK anti-tumour activity (Baroja-Mazo et al. 2023). However, others suggest that CD8 T cell P_2X_7 activation by ATP is beneficial for tumour control (Wanhainen et al. 2022). In an alternative approach, another study showed that using a positive allosteric modulator of P_2X_7 to promote dendritic cell anti-tumour immune responses was effective in combination with anti-PD-1 immunotherapy in a lung cancer model (Douguet et al. 2021). In addition to small molecules, biologics such as nanobodies may be useful for targeting P_2X_7 either by blocking or potentiating P_2X_7, and some have been evaluated in tumour models (Demeules et al. 2022). Potentiation of P_2X_7 on cancer cells may increase tumour cell death, and this is thought to be important for inducing anti-tumour immune responses (Grassi and De Ponte Conti 2021). The tumour microenvironment contains high levels of extracellular nucleotides and can affect both tumour growth and the immune cell function (Di Virgilio et al. 2018).

More recently, evidence is emerging for P_2X_4 being involved in various types of cancer. P_2X_4 is highly expressed in prostate cancer cell lines, and blocking P_2X_4 with 5-BDBD or PSB-12062 could reduce growth and mobility (He et al. 2020). Furthermore, tissue microarray analysis of prostate cancer cases showed a significantly elevated expression of P_2X_4 over benign cases (Maynard et al. 2022). Knockdown of P_2X_4 in prostate cancer cell lines reduced proliferation, migration and invasion in in vitro assays and reduced allograft growth in a mouse model (Maynard et al. 2022). A similar picture is emerging in breast cancer with upregulated P_2X_4 expression found in biopsies and demonstration of pro-malignant activities in vitro (Chadet et al. 2022). P_2X_4 is suggested to regulate autophagy and to promote an aggressive phenotype through lysosome-autophagosome fusion and lysosome exocytosis under challenging metabolic conditions (Chadet et al. 2022). This may be linked to the predominant localisation of P_2X_4 within the lysosomal compartment of cells (Murrell-Lagnado and Frick 2019). Activation of P_2X_4 has also been linked to mTOR-dependent signalling promoting tumour growth in colon tumour organoids (Schmitt et al. 2022). Collectively these studies recommend inhibition of P_2X_4 as a therapeutic strategy for cancer; however, an alternative point of view is suggested by Draganov et al. where using the P_2X4/P_2X_7 positive allosteric modulator ivermectin could induce cancer cell death in vitro (Draganov et al. 2015) and synergises with anti-PD1 antibody in vivo to reduce tumour growth (Draganov et al. 2021).

7 Conclusions

Almost 50 years on from the first discoveries of ionotropic receptors for extracellular nucleotides and their impact on physiology, the first medicine has been approved targeting a purinergic ion channel for the treatment of chronic cough. This success story will hopefully be followed by other candidates already in clinical trials or pre-clinical development for chronic pain, inflammation, cancer and psychiatric disorders as discussed here. Other future areas to watch include cardiovascular diseases, autism-spectrum disorders, neurodegeneration and in infectious diseases. In this latter case, an alternative strategy of boosting ATP-gated ion channel responses may prove to be effective in pathogen clearance.

References

Abdi MH, et al. Discovery and structure-activity relationships of a series of pyroglutamic acid amide antagonists of the P2X7 receptor. Bioorg Med Chem Lett. 2010;20(17):5080–4.

Abdulqawi R, et al. P2X3 receptor antagonist (AF-219) in refractory chronic cough: a randomised, double-blind, placebo-controlled phase 2 study. Lancet. 2015;385(9974):1198–205.

Adinolfi E, et al. P2X7 receptor expression in evolutive and indolent forms of chronic B lymphocytic leukemia. Blood. 2002;99(2):706–8.

Adinolfi E, et al. Basal activation of the P2X7 ATP receptor elevates mitochondrial calcium and potential, increases cellular ATP levels, and promotes serum-independent growth. Mol Biol Cell. 2005;16(7):3260–72.

Adinolfi E, et al. Trophic activity of a naturally occurring truncated isoform of the P2X7 receptor. FASEB J. 2010;24(9):3393–404.

Adinolfi E, et al. Expression of P2X7 receptor increases in vivo tumor growth. Cancer Res. 2012;72(12):2957–69.

Adinolfi E, et al. Accelerated tumor progression in mice lacking the ATP receptor P2X7. Cancer Res. 2015;75(4):635–44.

Amoroso F, et al. The P2X7 receptor is a key modulator of the PI3K/GSK3beta/VEGF signaling network: evidence in experimental neuroblastoma. Oncogene. 2015;34(41):5240–51.

Ase AR, et al. Identification and characterization of a selective allosteric antagonist of human P2X4 receptor channels. Mol Pharmacol. 2015;87(4):606–16.

Ballini E, et al. Characterization of three diaminopyrimidines as potent and selective antagonists of P2X3 and P2X2/3 receptors with in vivo efficacy in a pain model. Br J Pharmacol. 2011;163(6):1315–25.

Barclay J, et al. Functional downregulation of P2X3 receptor subunit in rat sensory neurons reveals a significant role in chronic neuropathic and inflammatory pain. J Neurosci. 2002;22(18):8139–8147.

Barden N, et al. Analysis of single nucleotide polymorphisms in genes in the chromosome 12Q24.31 region points to P2RX7 as a susceptibility gene to bipolar affective disorder. Am J Med Genet B Neuropsychiatr Genet. 2006;141B(4):374–82.

Baroja-Mazo A, et al. P2X7 receptor activation impairs anti-tumour activity of natural killer cells. Br J Pharmacol. 2023;180(1):11–128.

Basso AM, et al. Behavioral profile of P2X7 receptor knockout mice in animal models of depression and anxiety: relevance for neuropsychiatric disorders. Behav Brain Res. 2009;198(1):83–90.

Beggs S, Salter MW. The known knowns of microglia–neuronal signalling in neuropathic pain. Neurosci Lett. 2013;557:37–42.

Bergmann P, et al. Generation and characterization of specific monoclonal antibodies and nanobodies directed against the ATP-gated channel P2X4. Front Cell Neurosci. 2019;13:498.

Bernier LP, Ase AR, Seguela P. P2X receptor channels in chronic pain pathways. Br J Pharmacol. 2018;175(12):2219–30.

Beyer DKE, Freund N. Animal models for bipolar disorder: from bedside to the cage. Int J Bipolar Disord. 2017;5(1):35.

Bhattacharya A, et al. Pharmacological characterization of a novel centrally permeable P2X7 receptor antagonist: JNJ-47965567. Br J Pharmacol. 2013;170(3):624–40.

Bhattacharya A, et al. Neuropsychopharmacology of JNJ-55308942: evaluation of a clinical candidate targeting P2X7 ion channels in animal models of neuroinflammation and anhedonia. Neuropsychopharmacology. 2018;43(13):2586–96.

Boucher AA, et al. Resilience and reduced c-Fos expression in P2X7 receptor knockout mice exposed to repeated forced swim test. Neuroscience. 2011;189:170–7.

Browne LE, et al. P2X7 receptor channels allow direct permeation of nanometer-sized dyes. J Neurosci. 2013;33(8):3557.

Burnstock G, Kennedy C. Is there a basis for distinguishing two types of P2-purinoceptor? Gen Pharmacol Vasc S. 1985;16(5):433–40.

Cantin L-D, et al. Discovery of P2X3 selective antagonists for the treatment of chronic pain. Bioorg Med Chem Lett. 2012;22(7):2565–71.

Cao X, et al. Astrocyte-derived ATP modulates depressive-like behaviors. Nat Med. 2013;19(6):773–7.

Carter DS, et al. Identification and SAR of novel diaminopyrimidines. Part 1: The discovery of RO-4, a dual P2X3/P2X2/3 antagonist for the treatment of pain. Bioorg Med Chem Lett. 2009;19(6):1628–31.

Chadet S, et al. P2X4 receptor promotes mammary cancer progression by sustaining autophagy and associated mesenchymal transition. Oncogene. 2022;41(21):2920–31.

Cheewatrakoolpong B, et al. Identification and characterization of splice variants of the human P2X7 ATP channel. Biochem Biophys Res Commun. 2005;332(1):17–27.

Chen Y, et al. Mechanisms underlying enhanced P2X receptor-mediated responses in the neuropathic pain state. Pain. 2005;119(1–3).

Chen H, et al. Effect of P2X4R on airway inflammation and airway remodeling in allergic airway challenge in mice. Mol Med Rep. 2016;13(1):697–704.

Chessell IP, et al. Disruption of the P2X7 purinoceptor gene abolishes chronic inflammatory and neuropathic pain. Pain. 2005;114(3).

Cockayne DA, et al. Urinary bladder hyporeflexia and reduced pain-related behaviour in P2X3-deficient mice. Nature. 2000;407(6807):1011–5.

Cohen SP, Vase L, Hooten WM. Chronic pain: an update on burden, best practices, and new advances. Lancet. 2021;397(10289):2082–97.

Coull JAM, et al. BDNF from microglia causes the shift in neuronal anion gradient underlying neuropathic pain. Nature. 2005;438(7070):1017–21.

Csóka B, et al. Macrophage P2X4 receptors augment bacterial killing and protect against sepsis. JCI Insight. 2018;3(11):e99431.

Csölle, C., et al., The absence of P2X7 receptors (P2rx7) on non-haematopoietic cells leads to selective alteration in mood-related behaviour with dysregulated gene expression and stress reactivity in mice. Int. J. Neuropsychopharmacol. 2013 16(1):213–233.

Csölle C, et al. Neurochemical changes in the mouse hippocampus underlying the antidepressant effect of genetic deletion of P2X7 receptors. PLoS One. 2013b;8(6):e66547.

Czamara, D., B. Müller-Myhsok, and S. Lucae, The P2RX7 polymorphism rs2230912 is associated with depression: a meta-analysis. Prog. Neuropsychopharmacol Biol Psychiatry 2018;82:272–277.

D'Antongiovanni V, et al. Anti-inflammatory effects of novel P2X4 receptor antagonists, NC-2600 and NP-1815-PX, in a murine model of colitis. Inflammation. 2022;45(4):1829–47.

Dane C, Stokes L, Jorgensen WT. P2X receptor antagonists and their potential as therapeutics: a patent review (2010–2021). Expert Opin Ther Pat. 2022;32(7):769–90.

Davenport AJ, et al. Eliapixant is a selective P2X3 receptor antagonist for the treatment of disorders associated with hypersensitive nerve fibers. Sci Rep. 2021;11(1):19877.

De Salis SKF, et al. Alternatively spliced isoforms of the P2X7 receptor: structure, function and disease associations. Int J Mol Sci. 2022;23.

Demeules M, et al. Evaluation of nanobody-based biologics targeting purinergic checkpoints in tumor models in vivo. Front Immunol. 2022;13.

Di Virgilio F, et al. The P2X7 receptor in infection and inflammation. Immunity. 2017;47(1):15–31.

Di Virgilio F, et al. Extracellular ATP and P2 purinergic signalling in the tumour microenvironment. Nat Rev Cancer. 2018;18(10):601–18.

Douguet L, et al. A small-molecule P2RX7 activator promotes anti-tumor immune responses and sensitizes lung tumor to immunotherapy. Nat Commun. 2021;12(1):653.

Draganov D, et al. Modulation of P2X4/P2X7/Pannexin-1 sensitivity to extracellular ATP via Ivermectin induces a non-apoptotic and inflammatory form of cancer cell death. Sci Rep. 2015;5:16222.

Draganov D, et al. Ivermectin converts cold tumors hot and synergizes with immune checkpoint blockade for treatment of breast cancer. NPJ Breast Cancer. 2021;7(1):22.

Eser A, et al. Safety and efficacy of an oral inhibitor of the purinergic receptor P2X7 in adult patients with moderately to severely active Crohn's disease: a randomized placebo-controlled, double-blind, phase IIa study. Inflamm Bowel Dis. 2015;21(10):2247–53.

Feng Y-H, et al. A truncated P2X7 receptor variant (P2X7-j) endogenously expressed in cervical cancer cells antagonizes the full-length P2X7 receptor through hetero-oligomerization*. J Biol Chem. 2006;281(25):17228–37.

Feng W-P, et al. Lack of association of P2RX7 gene rs2230912 polymorphism with mood disorders: a meta-analysis. PLoS One. 2014;9(2):e88575.

Finnerup NB, Kuner R, Jensen TS. Neuropathic pain: from mechanisms to treatment. Physiol Rev. 2021;101(1):259–301.

Ford AP. In pursuit of P2X3 antagonists: novel therapeutics for chronic pain and afferent sensitization. Purinergic Signal. 2012;8(1):3–26.

Ford A, Undem B. The therapeutic promise of ATP antagonism at P2X3 receptors in respiratory and urological disorders. Front Cell Neurosci. 2013;7.

Ford AP, et al. The discovery and development of gefapixant. Auton Neurosci. 2021;235:102859.

Garceau D, Chauret N. BLU-5937: a selective P2X3 antagonist with potent anti-tussive effect and no taste alteration. Pulm Pharmacol Ther. 2019;56:56–62.

Gever JR, et al. AF-353, a novel, potent and orally bioavailable P2X3/P2X2/3 receptor antagonist. Br J Pharmacol. 2010;160(6):1387–98.

Gilbert SM, et al. A phase I clinical trial demonstrates that nfP2X7-targeted antibodies provide a novel, safe and tolerable topical therapy for basal cell carcinoma. Br J Dermatol. 2017;177(1):117–24.

Gilbert SM, et al. ATP in the tumour microenvironment drives expression of nfP2X7, a key mediator of cancer cell survival. Oncogene. 2019;38(2):194–208.

Glass R, et al. P2X4 and P2X6 receptors associate with VE-cadherin in human endothelial cells. Cell Mol Life Sci. 2002;59(5):870–81.
Gonçalves MCB, Andrejew R, Gubert C. The purinergic system as a target for the development of treatments for bipolar disorder. CNS Drugs. 2022;36(8):787–801.
Grassi F, De Ponte Conti B. The P2X7 receptor in tumor immunity. Front Cell Dev Biol. 2021:9.
Grishin EV, et al. Novel peptide from spider venom inhibits P2X3 receptors and inflammatory pain. Ann Neurol. 2010;67(5):680–3.
Gu BJ, et al. A Glu-496 to ala polymorphism leads to loss of function of the human P2X7 receptor. J Biol Chem. 2001;276(14):11135–42.
Gubert, C., et al., Role of P2X7 receptor in an animal model of mania induced by D-amphetamine. Mol. Neurobiol. 2016 53(1):611–620.
Gubert C, et al. P2X7 purinergic receptor is involved in the pathophysiology of mania: a preclinical study. Mol Neurobiol. 2020;57(3):1347–60.
Hamilton SG, Wade A, McMahon SB. The effects of inflammation and inflammatory mediators on nociceptive behaviour induced by ATP analogues in the rat. Br J Pharmacol. 1999;126(1):326–32.
Hansen RR, et al. P2X7 receptor-deficient mice are susceptible to bone cancer pain. Pain. 2011;152(8):1766–1776.
Hansen RR, et al. Chronic administration of the selective P2X3, P2X2/3 receptor antagonist, A-317491, transiently attenuates cancer-induced bone pain in mice. Eur J Pharmacol. 2012;688(1):27–34.
Hattori M, Gouaux E. Molecular mechanism of ATP binding and ion channel activation in P2X receptors. Nature. 2012;485(7397):207–12.
He J, et al. Inhibiting the P2X4 receptor suppresses prostate cancer growth in vitro and in vivo, suggesting a potential clinical target. Cells 2020;9(11):2511.
Hisaoka-Nakashima K, et al. Corticosterone induces HMGB1 release in primary cultured rat cortical astrocytes: involvement of pannexin-1 and P2X7 receptor-dependent mechanisms. Cell. 2020;9(5):1068.
Honore P, et al. Analgesic profile of intrathecal P2X3 antisense oligonucleotide treatment in chronic inflammatory and neuropathic pain states in rats. Pain. 2002;99(1-2):1–19.
Honore P, et al. A-740003 [N-(1-{[(cyanoimino)(5-quinolinylamino) methyl]amino}-2,2-dimethylpropyl)-2-(3,4-dimethoxyphenyl)acetamide], a novel and selective P2X7 receptor antagonist, dose-dependently reduces neuropathic pain in the rat. J Pharmacol Exp Ther. 2006;319(3):1376.
Honore P, et al. The antihyperalgesic activity of a selective P2X7 receptor antagonist, A-839977, is lost in IL-1αβ knockout mice. Behav Brain Res. 2009;204(1):77–81.
Huang H, et al. The potential of the P2X7 receptor as a therapeutic target in a sub-chronic PCP-induced rodent model of schizophrenia. J Chem Neuroanat. 2021;116:101993.
Illes P, et al. Update of P2X receptor properties and their pharmacology: IUPHAR review 30. Br J Pharmacol. 2021;178(3):489–514.
Inoue K, Tsuda M. Nociceptive signaling mediated by P2X3, P2X4 and P2X7 receptors. Biochem Pharmacol. 2021;187:114309.
Iwata M, et al. Psychological stress activates the Inflammasome via release of adenosine triphosphate and stimulation of the purinergic type 2X7 receptor. Biol Psychiatry. 2016;80(1):12–22.
Jahangir A, et al. Identification and SAR of novel diaminopyrimidines. Part 2: The discovery of RO-51, a potent and selective, dual P2X3/P2X2/3 antagonist for the treatment of pain. Bioorg Med Chem Lett. 2009;19(6):1632–5.
Jarvis MF, et al. A-317491, a novel potent and selective non-nucleotide antagonist of P2X3 and P2X2/3 receptors, reduces chronic inflammatory and neuropathic pain in the rat. Proc Natl Acad Sci. 2002;99(26):17179–84.
Jeong YH, et al. Mice lacking the purinergic receptor P2X5 exhibit defective Inflammasome activation and early susceptibility to listeria monocytogenes. J Immunol. 2020;205(3):760–6.
Kaan TKY, et al. Systemic blockade of P2X3 and P2X2/3 receptors attenuates bone cancer pain behaviour in rats. Brain. 2010;133(9):2549–64.

Kai H, et al. Dioxotriazine derivatives as a new class of P2X3 receptor antagonists: identification of a lead and initial SAR studies. Bioorg Med Chem Lett. 2021a;37:127833.
Kai H, et al. Discovery of clinical candidate Sivopixant (S-600918): lead optimization of dioxotriazine derivatives as selective P2X3 receptor antagonists. Bioorg Med Chem Lett. 2021b;52:128384.
Karasawa A, Kawate T. Structural basis for subtype-specific inhibition of the P2X7 receptor. elife. 2016;5:e22153.
Karasawa A, et al. The P2X7 receptor forms a dye-permeable pore independent of its intracellular domain but dependent on membrane lipid composition. Elife. 2017:6.
Kawate T, et al. Crystal structure of the ATP-gated P2X4 ion channel in the closed state. Nature. 2009;460(7255):592–8.
Keystone EC, et al. Clinical evaluation of the efficacy of the P2X7 purinergic receptor antagonist AZD9056 on the signs and symptoms of rheumatoid arthritis in patients with active disease despite treatment with methotrexate or sulphasalazine. Ann Rheum Dis. 2012;71(10):1630.
Kim H, et al. The purinergic receptor P2X5 regulates inflammasome activity and hyper-multinucleation of murine osteoclasts. Sci Rep. 2017;7(1):196.
King BF. Rehabilitation of the P2X5 receptor: a re-evaluation of structure and function. Purinergic Signalling 2022;19(2):421–439.
Koványi B, et al. The role of P2X7 receptors in a rodent PCP-induced schizophrenia model. Sci Rep. 2016;6(1):36680.
Kuang X-J, et al. P2X2 receptors in pyramidal neurons are critical for regulating vulnerability to chronic stress. Theranostics. 2022;12(8):3703–18.
Lecut C, et al. ATP-gated P2X1 ion channels protect against endotoxemia by dampening neutrophil activation. J Thromb Haemost. 2012;10(3):453–65.
Li M, et al. Gating the pore of P2X receptor channels. Nat Neurosci. 2008;11(8):883–7.
Lord B, et al. Pharmacology of a novel central nervous system–penetrant P2X7 antagonist JNJ-42253432. J Pharmacol Exp Ther. 2014;351(3):628.
Lucae S, et al. P2RX7, a gene coding for a purinergic ligand-gated ion channel, is associated with major depressive disorder. Hum Mol Genet. 2006;15(16):2438–45.
Lucido MJ, et al. Aiding and abetting anhedonia: impact of inflammation on the brain and pharmacological implications. Pharmacol Rev. 2021;73(3):1084–1117.
Maître B, et al. The P2X1 receptor is required for neutrophil extravasation during lipopolysaccharide-induced lethal endotoxemia in mice. J Immunol. 2015;194(2):739–49.
Mansoor SE, et al. X-ray structures define human P2X(3) receptor gating cycle and antagonist action. Nature. 2016;538(7623):66–71.
Mapplebeck JC, Beggs S, Salter MW. Sex differences in pain: a tale of two immune cells. Pain. 2016;157(Suppl 1):S2–6.
Mapplebeck JCS, et al. Microglial P2X4R-evoked pain hypersensitivity is sexually dimorphic in rats. Pain. 2018;159(9).
Markham A. Gefapixant: first approval. Drugs. 2022;82(6):691–5.
Martinowich K, Schloesser RJ, Manji HK. Bipolar disorder: from genes to behavior pathways. J Clin Invest. 2009;119(4):726–36.
Matsumura Y, et al. A novel P2X4 receptor-selective antagonist produces anti-allodynic effect in a mouse model of herpetic pain. Sci Rep. 2016;6:32461.
Maynard JP, et al. P2X4 purinergic receptors offer a therapeutic target for aggressive prostate cancer. J Pathol. 2022;256(2):149–63.
McCarthy AE, Yoshioka C, Mansoor SE. Full-length P2X7 structures reveal how palmitoylation prevents channel desensitization. Cell. 2019;179(3):659–670 e13.
McGarvey LP, et al. Efficacy and safety of gefapixant, a P2X3 receptor antagonist, in refractory chronic cough and unexplained chronic cough (COUGH-1 and COUGH-2): results from two double-blind, randomised, parallel-group, placebo-controlled, phase 3 trials. Lancet. 2022;399(10328):909–23.
McInnes, I.B., et al., Targeting the P2X7 receptor in rheumatoid arthritis: biological rationale for P2X7 antagonism. Clin Exp Rheumatol. 2014;32(6):878–882.

McQuillin A, et al. Case-control studies show that a non-conservative amino-acid change from a glutamine to arginine in the P2RX7 purinergic receptor protein is associated with both bipolar- and unipolar-affective disorders. Mol Psychiatry. 2009;14(6):614–20.
Metzger MW, et al. Heterozygosity for the mood disorder-associated variant Gln460Arg alters P2X7 receptor function and sleep quality. J Neurosci. 2017;37(48):11688–700.
Miller AH, Raison CL. The role of inflammation in depression: from evolutionary imperative to modern treatment target. Nat Rev Immunol. 2016;16(1):22–34.
Morrone FB, et al. Increased resistance of glioma cell lines to extracellular ATP cytotoxicity. J Neuro-Oncol. 2005;71(2):135–40.
Murrell-Lagnado RD, Frick M. P2X4 and lysosome fusion. Curr Opin Pharmacol. 2019;47:126–32.
Nagata K, et al. Antidepressants inhibit P2X4 receptor function: a possible involvement in neuropathic pain relief. Mol Pain. 2009;5:20.
Niimi A, et al. Randomised trial of the P2X3 receptor antagonist sivopixant for refractory chronic cough. Eur Respir J. 2022;59(6):2100725.
North RA, Surprenant A. Pharmacology of cloned P2X receptors. Annu Rev Pharmacol Toxicol. 2000;40(1):563–80.
Novakovic SD, et al. Immunocytochemical localization of P2X3 purinoceptors in sensory neurons in naive rats and following neuropathic injury. Pain. 1999;80(1):273–282.
Oliveira-Fusaro MC, et al. P2X4 receptors on muscle macrophages are required for development of hyperalgesia in an animal model of activity-induced muscle pain. Mol Neurobiol. 2020;57(4):1917–29.
Ormond SJ, et al. An uncharged region within the N terminus of the P2X6 receptor inhibits its assembly and exit from the endoplasmic reticulum. Mol Pharmacol. 2006;69(5):1692.
Otrokocsi L, Kittel A, Sperlagh B. P2X7 receptors drive spine synapse plasticity in the learned helplessness model of depression. Int J Neuropsychopharmacol. 2017;20(10):813–22.
Peikert A, et al. P2X4 deficiency reduces atherosclerosis and plaque inflammation in mice. Sci Rep. 2022;12(1):2801.
Pelegrin P, Surprenant A. Pannexin-1 mediates large pore formation and interleukin-1beta release by the ATP-gated P2X7 receptor. EMBO J. 2006;25(21):5071–82.
Perez-Medrano A, et al. Discovery and biological evaluation of novel cyanoguanidine P2X7 antagonists with analgesic activity in a rat model of neuropathic pain. J Med Chem. 2009;52(10):3366–76.
Pippel A, et al. Localization of the gate and selectivity filter of the full-length P2X7 receptor. Proc Natl Acad Sci. 2017;114(11):E2156–65.
Raffaghello L, et al. The P2X7 receptor sustains the growth of human neuroblastoma cells through a substance P–dependent mechanism. Cancer Res. 2006;66(2):907–14.
Rech JC, et al. The discovery and preclinical characterization of 6-chloro-N-(2-(4,4-difluoropiperidin-1-yl)-2-(2-(trifluoromethyl)pyrimidin-5-yl)ethyl)quinoline-5-carboxamide based P2X7 antagonists. Bioorg Med Chem Lett. 2016;26(19):4781–4.
Ribeiro DE, et al. Antidepressant-like effect induced by P2X7 receptor blockade in FSL rats is associated with BDNF signalling activation. J Psychopharmacol. 2019;33(11):1436–46.
Richards D, et al. Action of MK-7264 (gefapixant) at human P2X3 and P2X2/3 receptors and in vivo efficacy in models of sensitisation. Br J Pharmacol. 2019;176(13):2279–91.
Roger S, et al. Understanding the roles of the P2X7 receptor in solid tumour progression and therapeutic perspectives. Biochim Biophys Acta. 2015;1848(10 Pt B):2584–602.
Romagnani A, et al. P2X7 receptor activity limits accumulation of T cells within tumors. Cancer Res. 2020;80(18):3906–19.
Samways DSK, et al. Preferential use of unobstructed lateral portals as the access route to the pore of human ATP-gated ion channels (P2X receptors). Proc Natl Acad Sci. 2011;108(33):13800–5.
Satia I, et al. Towards understanding and managing chronic cough. Clin Med. 2016;16(Suppl 6):s92.
Savio LEB, et al. The P2X7 receptor in inflammatory diseases: angel or demon? Front Pharmacol. 2018;9:52.
Schmitt M, et al. Colon tumour cell death causes mTOR dependence by paracrine P2X4 stimulation. Nature. 2022;612(7939):347–353.

Schneider R, et al. A novel P2X2-dependent purinergic mechanism of enteric gliosis in intestinal inflammation. EMBO Mol Med. 2021;13(1):e12724.

Schneider S, et al. Blocking P2X purinoceptor 4 signalling alleviates cigarette smoke induced pulmonary inflammation. Respir Res. 2022;23(1):148.

Sharp CJ, et al. Investigation into the role of P2X3/P2X2/3 receptors in neuropathic pain following chronic constriction injury in the rat: an electrophysiological study. Br J Pharmacol. 2006;148(6):845–52.

Shcherbatko A, et al. Modulation of P2X3 and P2X2/3 receptors by monoclonal antibodies. J Biol Chem. 2016;291(23):12254–70.

Skals M, et al. P2X1 receptor blockers reduce the number of circulating thrombocytes and the overall survival of urosepsis with haemolysin-producing Escherichia coli. Purinergic Signal. 2019;15(2):265–76.

Skarratt KK, et al. A P2RX7 single nucleotide polymorphism haplotype promotes exon 7 and 8 skipping and disrupts receptor function. FASEB J. 2020;34(3):3884–901.

Slater M, et al. Early prostate cancer detected using expression of non-functional cytolytic P2X7 receptors. Histopathology. 2004;44(3):206–15.

Smart ML, et al. P2X7 receptor cell surface expression and cytolytic pore formation are regulated by a distal C-terminal region. J Biol Chem. 2003;278(10):8853–60.

Spelta V, et al. ATP analogues with modified phosphate chains and their selectivity for rat P2X2 and P2X2/3 receptors. Br J Pharmacol. 2003;140(6):1027–34.

Stock TC, et al. Efficacy and safety of CE-224,535, an antagonist of P2X7 receptor, in treatment of patients with rheumatoid arthritis inadequately controlled by methotrexate. J Rheumatol. 2012;39(4):720.

Stokes, L., et al. Two haplotypes of the $P_2X(7)$ receptor containing the Ala-348 to Thr polymorphism exhibit a gain-of-function effect and enhanced interleukin-1beta secretion. FASEB J. 2010;24(8):2916–27.

Teixeira JM, et al. Diabetes-induced neuropathic mechanical hyperalgesia depends on P2X4 receptor activation in dorsal root ganglia. Neuroscience. 2019;398:158–70.

Troubat R, et al. Brain immune cells characterization in UCMS exposed P2X7 knock-out mouse. Brain Behav Immun. 2021;94:159–74.

Tsuda M, et al. P2X4 receptors induced in spinal microglia gate tactile allodynia after nerve injury. Nature. 2003;424(6950):778–83.

Tsuda M, et al. Behavioral phenotypes of mice lacking purinergic P2X4 receptors in acute and chronic pain assays. Mol Pain. 2009;5:28.

Ugur M, Ugur Ö. A mechanism-based approach to P2X7 receptor action. Mol Pharmacol. 2019;95(4):442–450.

Ulmann L, et al. Up-regulation of P2X4 receptors in spinal microglia after peripheral nerve injury mediates BDNF release and neuropathic pain. J Neurosci. 2008;28(44):11263–8.

Ulmann L, Hirbec H, Rassendren F. P2X4 receptors mediate PGE_2 release by tissue-resident macrophages and initiate inflammatory pain. EMBO J. 2010;29(14):2290–300.

Valera S, et al. A new class of ligand-gated ion channel defined by P2X receptor for extracellular ATP. Nature. 1994;371(6497):516–9.

Vázquez-Cuevas FG, et al. Paracrine stimulation of P2X7 receptor by ATP activates a proliferative pathway in ovarian carcinoma cells. J Cell Biochem. 2014;115(11):1955–66.

Wang J, et al. Druggable negative allosteric site of P2X3 receptors. Proc Natl Acad Sci. 2018;115(19):4939–44.

Wanhainen KM, et al. P2RX7 enhances tumor Control by CD8+ T cells in adoptive cell therapy. Cancer Immunol Res. 2022;10(7):871–84.

Wellcome Trust Case Control Consortium. Genome-wide association study of 14,000 cases of seven common diseases and 3,000 shared controls. Nature. 2007;447(7145):661–78.

Wéra O, et al. P2X1 ion channel deficiency causes massive bleeding in inflamed intestine and increases thrombosis. J Thromb Haemost. 2020;18(1):44–56.

Werner S, et al. Discovery and characterization of the potent and selective P2X4 inhibitor N-[4-(3-chlorophenoxy)-3-sulfamoylphenyl]-2-phenylacetamide (BAY-1797) and structure-guided amelioration of its CYP3A4 induction profile. J Med Chem. 2019;62(24):11194–217.

Westlund KN, et al. Single-dose P2 X4R single-chain fragment variable antibody permanently reverses chronic pain in male mice. Int J Mol Sci. 2021;22(24).

White N, Butler PEM, Burnstock G. Human melanomas express functional P2X7 receptors. Cell Tissue Res. 2005;321(3):411–8.

Wiley JS, et al. A loss-of-function polymorphic mutation in the cytolytic P2X7 receptor gene and chronic lymphocytic leukaemia: a molecular study. Lancet. 2002;359(9312):1114–9.

Williams WA, et al. Antibodies binding the head domain of P2X4 inhibit channel function and reverse neuropathic pain. Pain. 2019;160(9):1989–2003.

Xia M, et al. Sleep deprivation selectively down-regulates astrocytic 5-HT2B receptors and triggers depressive-like behaviors via stimulating P2X7 receptors in mice. Neurosci Bull. 2020;36(11):1259–70.

Xu G-Y, et al. Mechanisms underlying purinergic P2X3 receptor-mediated mechanical allodynia induced in diabetic rats. Mol Pain. 2011;7:60.

Yamashita T, et al. New inhibitory effects of Cilnidipine on microglial P2X7 receptors and IL-1beta release: an involvement in its alleviating effect on neuropathic pain. Cells 2021;10(2):434.

Yue N, et al. Activation of P2X7 receptor and NLRP3 inflammasome assembly in hippocampal glial cells mediates chronic stress-induced depressive-like behaviors. J Neuroinflammation. 2017;14(1):102.

Zhang M, et al. ATP, an attractive target for the treatment of refractory chronic cough. Purinergic Signal. 2022;18(3):289–305.

Immune Cell Ion Channels as Therapeutic Targets

Anna Selezneva, Alasdair J. Gibb, and Dean Willis

Abstract The immune system and its components represent the body's defensive apparatus against microbes, toxins, trauma, and neoplasms. Consisting of a complex network of cellular and tissue barriers, soluble extracellular regulatory and effector molecules, and various tissue resident or highly mobile cell types, the system can be divided into the innate and adaptive immune system. Conceptually the innate system can be considered a pre-programmed response, while the adaptive system exhibits immunological memory. While fundamental for the continued well-being of an individual, the number of diseases in which the immune system has been implicated in has greatly expanded over the past 30 years. These now include not only the classical inflammatory diseases such as rheumatoid arthritis and asthma but also metabolic and cardiovascular diseases, tissue ischemic reperfusion injury, neurodegenerative diseases, and neoplastic development. It is therefore not surprising that intense research is focused on developing therapeutics that can suppress, enhance, or modulate components of the immune the system.

Ion channels have been extensively researched as regulators of cellular function in electrically excitable cells, and this has led to the development of several important drug classes. In comparison the role of ion channels in immune cells has received relatively little attention. In this review we report on the roles of ion channels in individual cell types of the innate and adaptive immune system and highlight recent research which could indicate the potential of targeting ion channels in these cells to treat a variety of diseases.

Keywords Immune system · Innate immune system · Adaptive immune system · Ion channels · Lymphocytes · Leukocytes · Inflammation

A. Selezneva · A. J. Gibb · D. Willis (✉)
Department of Neuroscience, Physiology, and Pharmacology,
University College London, London, UK
e-mail: dean.willis@ucl.ac.uk

G. Stephens, E. Stevens (eds.), *Ion Channels as Targets in Drug Discovery*,
https://doi.org/10.1007/978-3-031-52197-3_14

1 Introduction

A correctly functioning immune system is absolutely critical for our survival and continued well-being. The immune system must efficiently recognise deleterious agents, exogenous or endogenous, and initiate an appropriate defence against these agents and/or their source; this usually involves killing, removing, diluting, or walling-off the agents and finally facilitating a return to tissue/cellular homeostasis usually via wound-healing/tissue regeneration mechanisms. Dysfunction of the immune system, be it an insufficient response or an inappropriate/excessive activation, can lead to significant pathologies and contributes to the high burden of chronic diseases seen throughout the globe (Schmidt et al. 2017).

At the conceptual level, the immune system must recognise a signal which is associated with a possible damage/disease-causing agent/environment, differentiate this from a normal environment or innoxious signals, elicit an appropriate molecular/cellular/tissue response, and, finally, provide mechanisms of learning/memory to increase the efficiency of the system. Importantly, these mechanisms must be responsive to dynamic changes in signals in order to maintain an optimal activity, while not resulting in an overshoot of the immune response, which could lead to tissue damage. To accomplish this, many functions of the immune system involve a myriad of signalling pathways and mediators (Delves et al. 2017).

Ion channels regulate ion flux and membrane potential of all cell membranes, both plasma and intracellular membranes, such as mitochondrial, lysosomal, endoplasmic reticulum membranes, and nuclear membranes. Ion channels are known to be important cellular sensors and participate in cell signalling events at or local to membranes in numerous cell types, and immune cells are of no exception (Feske et al. 2015; Selezneva et al. 2022a). In this chapter we provide an up-to-date overview of how ion channels may regulate important functions of immune cells and discuss the potential of immune cell ion channels as therapeutic targets.

2 Immune System: A Brief Overview

A comprehensive description of the immune system is beyond the scope of this chapter; however, Table 1 provides an overview of immune cell types together with their main effector functions and a list of functional ion channels found in these cells. Readers are referred to the classical immunological textbook *Roitt's Essential Immunology* (Delves et al. 2017) for further details of these individual cellular functions.

The immune system is traditionally divided into two branches – the innate immune system and the adaptive or acquired immune system. The innate immune system provides a general defence against noxious agents and microbes and contains both humoral (e.g. complement coagulation cascades) and cell components, the main being myeloid cells. This mechanism is based on the recognition by

Table 1 Ion channels regulating effector functions of immune system cells (Feske et al. 2015; Delves et al. 2017)

Cells of the immune system			Main effector functions	Ion channels
Lymphoid cells	T lymphocytes	Helper	Assist antibody production by B lymphocytes Enhance macrophage and natural killer cell-mediated killing Promote granulocyte recruitment Promote wound healing	CRAC, IP3R, Ca_v, P_2X receptors, TRPM2/M4/M7/C5, K_{2P}, $K_v1.3$, $K_{Ca}3.1$, $Na_v1.5$, $GABA_A$, NMDAR, VRAC, Piezo1
		Cytotoxic	Killing of virus-infected cells Anti-tumour immunity Cytokine production	
		Regulatory	Regulation of T cell-mediated immunity	
		Memory	Provide the immune system with “memory” against previously encountered pathogens	
	B lymphocytes		Antibody production	CRAC, TRPM7, $K_v1.3$, $K_{Ca}3.1$, VRAC, H_V1
	Natural killer cells		Antibody-dependent cellular cytotoxicity Killing of virus-infected cells Anti-tumour immunity	CRAC, Ca_v, TRPM2, K_{2P}, $K_v1.3$, $K_{Ca}3.1$
Myeloid cells	Dendritic cells		Antigen processing and presentation T cell differentiation Cytokine production	CRAC, Ca_v, P_2X receptors, TRPM4/V1, $K_v1.3$, $K_v1.5$, $K_{Ca}3.1$, $Na_v1.7$, Piezo1, Ryanodine receptors

(continued)

Table 1 (continued)

<table>
<tr><th colspan="3">Cells of the immune system</th><th>Main effector functions</th><th>Ion channels</th></tr>
<tr><td>-</td><td colspan="2">Macrophages</td><td>Multiple homestatic functions
Phagocytosis
Cytokine production
Antigen presentation</td><td>CRAC, IP3R, Ca_v, P_2X receptors, TRP, K_{2P}, K_v, K_{Ca}, K_{ir}, K_{ATP}, TMEM175, TMEM16A, Na_v, $GABA_A$, nAChR, 5-HT3R, VRAC, VDAC, CLIC, Piezo1, ASIC, TPC, H_V1, Gap junction hemichannel</td></tr>
<tr><td rowspan="4"></td><td rowspan="3">Granulocytes</td><td>Neutrophils</td><td>Phagocytosis
NET formation
Cytokine production</td><td rowspan="4">CRAC, P_2X receptors, TRP, $K_v1.3$, $K_{Ca}3.1$, H_V1</td></tr>
<tr><td>Eosinophils</td><td>Degranulation/ extracellular parasite killing
Histamine production
Cytokine production</td></tr>
<tr><td>Basophils</td><td>Degranulation/extracellular killing
Histamine production
Cytokine production</td></tr>
<tr><td colspan="2">Mast cells</td><td>Histamin, acute mediator production
Production of chemotactic factors
Cytokine production</td></tr>
</table>

Abbreviations: *ASIC* acid-sensing ion channel, *CRAC* Ca^{2+} release-activated Ca^{2+} channel, *IP3R* inositol 1,4,5-trisphosphate receptor, Ca_v voltage-gated Ca^{2+} channel, Na_v voltage-gated Na^+ channel, K_v voltage-gated K^+ channel, K_{Ca} Ca^{2+}-dependent K^+ channel, K_{2P} two-pore domain K^+ channel, *NMDAR* N-methyl-d-aspartate receptor, *nACh* nicotinic acetylcholine, *5-HT3* 5-hydroxytryptamine3, *NET* neutrophil extracellular trap *TMEM* transmembrane protein, K_{ir} inward rectifier K^+ channel, K_{ATP} ATP-sensitive K^+ channel, *VRAC* Cl^--permeable volume-regulated anion channel, *VDAC* voltage-dependent anion channel, *CLIC* chloride intracellular channel, *TRP* transient receptor potential channel, H_V1 voltage-gated proton channel, *TPC* two-pore channel

pattern recognition receptors (PRR) of a limited profile of chemical motifs associated with microbial infection and damaged tissue, pathogen-associated molecular patterns (PAMPs), and damage-associated molecular patterns (DAMPs), respectively. Myeloid cells provide the heavy lifting of the immune response by engaging and killing microbes, clearing dead cells and cellular debris, and producing an array of inflammatory and bioactive mediators/cytokines. It is now becoming apparent that the actions of myeloid cells are not limited to "activated" immune responses, with these cells also having roles in normal tissue homeostasis (Kulkarni et al.

2016). Indeed, dysfunction of myeloid "physiological homeostatic" functions may contribute to the development of diseases.

The adaptive immune system is evolutionary younger than the innate system but like the innate system contains both humoral and cellular components. This branch of the immune system relies on the generation of a large repertoire of lymphocyte cells which can recognise specific peptide sequences/antigens, which when presented to them in conjunction with correct co-regulatory receptors and mediators can generate an adaptive immune response against that antigen. This provides a specific immune response against pathogens and also establishes immune memory which allows a more efficient response against the pathogen if subsequently encountered. Immune tolerance is the mechanism by which endogenous antigens, i.e. our own peptides, do not initiate an adaptive immune response, as described further below.

In vertebrates, the two branches of the immune system work together facilitating, supporting, and influencing each other's actions to provide an integrated immune response. An example of this is antigen presentation, where antigen-presenting cells (APCs), macrophages and/or dendritic cells present antigen via major histocompatibility complex (MHC) to T cell receptors (TCRs) found on T lymphocytes. Co-regulatory receptors (e.g. CD4/8, CD28, B7-1 & B7-2) present on the APCs and lymphocytes are required for full activation of lymphocytes, and the presence of various cytokines in the local environment contributes to the final T lymphocyte phenotype. T lymphocytes can also produce a range of cytokines themselves, which can have profound effects on innate cell activation.

Finally, another key characteristic of many immune cells essential for their function is their mobile nature, e.g. neutrophils moving to a site of infection to kill invading bacteria. Immune cells are constantly subjected to dynamic changes, biochemical and mechanical stimuli, all of which could impact on their ion channel biology.

3 T Lymphocytes

The importance of maturation and development of lymphocyte repertoire cannot be underestimated. This process leads to the host not responding to its own peptides, while maintaining a significant population of lymphocytes which respond to foreign antigens (de Greef et al. 2020). Before exiting the thymus, naïve double-positive (CD4/CD8) T lymphocytes undergo positive selection, where lymphocytes expressing TCRs with good affinity for MHCs are saved. At this point depending on which MHC class molecules the T lymphocytes have affinity for, they either become CD4 (affinity for MHC class II molecule) or CD8 (affinity for MHC class I molecule) signal-positive cells. This is followed by negative selection, where lymphocytes expressing TCRs with high affinity for self-antigen-MHCs undergo apoptosis or develop into T-regulatory cells. This ensures that only naïve T lymphocytes which leave the thymus express either CD4 or CD8 and a single TCR variant that does not have a high reactivity against self-peptide, but with potential to detect foreign antigens presented by the individual's own MHCs.

To date, several ion channels have been demonstrated to shape lymphocyte development and effector functions. Loss of inositol 1,4,5-trisphosphate receptor (IP3R)-dependent Ca^{2+} signalling results in abnormal thymocyte development and acute lymphoblastic leukaemia (Ouyang et al. 2014). The presence of L-type voltage-gated Ca^{2+} channel (Ca_v)1.4 subunits was shown to be critical for naïve T lymphocyte homeostasis, where the absence of a β3 regulatory subunit of the channel leads to defective survival of naïve $CD8^+$ T lymphocytes (Jha et al. 2009; Omilusik et al. 2011). Recently, Erdogmus et al. suggested that Ca_v channels both in naïve and activated T lymphocytes have a novel non-channel function mediated by an accessory β1 subunit to limit T lymphocyte apoptosis after viral infection (Erdogmus et al. 2022). P_2X receptors (Freedman et al. 1999), transient receptor potential (TRP)M7 channels, and Cl^--permeable volume-regulated anion channel (VRAC) subunits LRRC8 appear no less significant for T lymphocyte development and selection. T lymphocyte-specific deletion of TRPM7 disrupts thymopoiesis (Jin et al. 2008), while lack of LRRC8A is associated with reduced AKT phosphorylation in the thymus, decreased proliferation, and increased apoptosis of thymocytes and impaired peripheral T lymphocyte function (Kumar et al. 2014). On the other hand, a recent study showed another subunit of VRAC, LRRC8C, to suppress T lymphocyte functions via regulating STING-p35 signalling (Concepcion et al. 2022). Voltage-gated Na^+ channels (Na_v)1.5 amplify TCR-induced Ca^{2+} signalling and are essential for the positive selection of $CD4^+$ T lymphocytes (Lo et al. 2012).

In the secondary lymphoid tissue, lymph nodes and spleen, the differentiation pathways and effector functions of T lymphocytes are shaped by the cytokine environment accompanying antigen presentation process by APCs such as dendritic cells. Recently, the mechanosensitive ion channel, Piezo1, was discovered to function as a mechanotransducer at immunological synapses, optimising TCR activation (Liu et al. 2018). Following the activation of naïve T lymphocytes, they undergo clonal proliferation, which is known to depend on various cytokines e.g. IL-2, IL-4, IL-6, IL-12, and IL-23, as well as ion channel activity, including that of GABA-A receptors (Sparrow et al. 2021) and differentiation to effectors. There are indications that expression patterns of ion channels change with T lymphocyte activation and differentiation; for example, naïve T lymphocytes predominantly express $K_v1.3$, but $K_{Ca}3.1$ expression is upregulated upon activation (Ghanshani et al. 2000). The effector T lymphocytes can be grouped into cytotoxic (Tc), regulatory (Treg), memory (Tm), and helper (Th) (Table 1), which can be further subdivided into four main subsets, Th1, Th2, Th17, and follicular helper (Tfh), based on the cytokine combinations they secrete (Delves et al. 2017). $CD4^+$ naïve T lymphocytes give rise mostly to Treg and Th that recognise antigen-bound MHCII, while $CD8^+$ naïve T lymphocytes give rise to Tc, recognising antigen-bound MHCI. Th1 mainly produces IFN-γ, IL-2, and TNF, which makes them adept at activating macrophages and fighting against intracellular pathogens. Th2 produces IL-4, IL-5, IL-6, and IL-13, assisting B lymphocyte proliferation and differentiation and eosinophil production and coordinating responses to extracellular pathogens. Both Th1 and Th2 produce IL-3 and

GM-CSF to induce the production of neutrophils and macrophages. IL-17A-producing Th17 lymphocytes promote acute inflammatory responses to extracellular bacterial and fungal infections and recruit neutrophils, while Tfh assists B lymphocytes by triggering the formation and maintenance of germinal centres within B lymphocyte follicles (Delves et al. 2017).

An important concept in adaptive immunity is the balance between the different T lymphocyte subsets. Shifts in this balance can lead to various pathologies including autoimmunity and cancer. The N-methyl-D-aspartate receptor (NMDAR) is upregulated upon T lymphocyte activation and influences the balance of Th1-type versus Th2-type immunity (Orihara et al. 2018). TRPM4 channel has also been reported to differentially regulate Th1 and Th2 functions and cytokine production by altering Ca^{2+} signalling and nuclear localisation of NFAT (Weber et al. 2010). A role for another member of TRPM channels, the TRPM2 channel, in T lymphocyte proliferation, proinflammatory cytokine production, and autoimmune central nervous system inflammation has been suggested (Melzer et al. 2012), while its importance in the innate immune system branch is better documented (see below). $Ca_v1.2$ and $Ca_v1.3$ ion channel knock-down in Th2 cells was shown to result in impaired cytokine production and reduced airway inflammation in a model of asthma (Cabral et al. 2010). However, as already mentioned above, Erdogmus and colleagues suggested that Ca_v channels in T lymphocytes have a non-channel function mediated by an accessory subunit (Erdogmus et al. 2022). Dysfunctional Ca^{2+} signalling and transcription factor activation in response to TCR activation upon antigen presentation leads to inappropriate T lymphocyte proliferation and differentiation and, consequently, dysfunctional adaptive immune response. The calcium release-activated calcium (CRAC) channel is believed to be critical for the functions of both $CD4^+$ and $CD8^+$ T lymphocytes. CRAC, encoded by ORAI1 gene, interacts with endoplasmic reticulum (ER) protein STIM1/2 upon depletion of ER Ca^{2+} stores, which results in store-operated Ca^{2+} entry (SOCE). T lymphocyte homing to lymph nodes, chemotaxis, migration towards the site of infection up a concentration gradient of chemokines/cytokines, and the spatiotemporal regulation of T lymphocyte-APC interaction in lymph nodes have all been reported to depend on CRAC-mediated SOCE (Negulescu et al. 1996; Lioudyno et al. 2008; Greenberg et al. 2013). Another essential function of Ca^{2+} influx through CRAC in lymphocytes is the regulation of cytokine gene expression, primarily via nuclear factor of activated T lymphocyte (NFAT) pathway (Macian 2005). CRAC dysfunction leads to profound defects in the production of key T lymphocyte cytokines including IL-2, IL-4, IL-10, IL-17, IFN-γ, and TNF (Feske et al. 2001). While people with ORAI1/CRAC dysfunction suffer from severe immunodeficiencies (McCarl et al. 2009; Feske 2010), T lymphocyte-specific deletion of STIM1/2 is able to protect mice from a model of multiple sclerosis, an autoimmune disease, by impairing effector functions of Th1 and Th17 (Ma et al. 2010; Schuhmann et al. 2010). This comparison highlights the importance and the difficulty of maintaining a healthy balance between pro- and anti-inflammatory immune responses. Th17 has been reported to predominantly express and depend on $K_v1.3$ channel for activation and IL-17 production, in contrast to Th1, Th2, and Treg subsets predominantly expressing $K_{Ca}3.1$. Interestingly, while Th1 and Th2 effector functions depend on $K_{Ca}3.1$ activity, Treg cell functions seem not to (Di et al. 2010). A third type of K^+-permeable ion channels known to regulate T lymphocyte proliferation, cytokine production, and regulatory volume decrease is the two-pore domain K^+ channel (K_{2P}) (Meuth et al. 2008; Bittner et al. 2010; Bobak et al. 2011).

Treg lymphocytes are responsible for "policing" the immune responses and regulating them when and where required through the release of immunosuppressive cytokines such as IL-10, TGFβ, IL-35, and wound-healing cytokines such as amphiregulin and through Treg-mediated killing of APCs and potentially autoreactive T lymphocytes. The suppressive functions of Treg were found to depend heavily on ion channel activity, where a notable lymphoproliferative phenotype and a decrease in Treg numbers were observed upon T lymphocyte-specific ablation of STIM proteins (Oh-hora et al. 2008). Furthermore, P_2X_7 receptor activity in Tregs has been reported to inhibit Treg immunosuppressive effects and induce Treg conversion into pro-inflammatory Th17 following the loss of a Treg signature transcription factor Foxp3 (Schenk et al. 2011). Another study showed that Treg autoimmune suppression mediated by cross-linking of GM1 ganglioside in effector T lymphocytes by galectin-1 in Treg is dependent on TRPC5 channel activity in the effector T lymphocytes (Wang et al. 2009). Inhibition of another TRP channel, TRPM7, has recently been reported to potentiate the thymic development of Tregs by promoting IL-2 signalling (Mendu et al. 2020). Last but not least, a novel channel Piezo1 (already mentioned above) was found to selectively restrain Tregs, while preserving other effector T lymphocyte responses. Importantly, this recent study reported that removing Piezo1 from T lymphocytes significantly attenuates multiple sclerosis in an animal model, while preserving immune defence against other infection (Jairaman et al. 2021).

Tc lymphocytes are crucial for the killing of virally infected cells either via Fas/Fas ligand pathway or via perforin/granzyme pathway (Delves et al. 2017) and are thought to contribute to the protection against cancer cells. Feske and colleagues report the importance of CRAC-mediated SOCE in Tc lymphocytes for the anti-tumour immunity (Weidinger et al. 2013). Tc effector functions, including degranulation and target cell lysis, and the maintenance of virus-specific memory $CD8^+$ T lymphocytes have been demonstrated to depend on a number of ion channels including CRAC (Maul-Pavicic et al. 2011; Shaw et al. 2014). Since Ca^{2+} signalling established itself as a universal second messenger in eukaryotic cells, its significance for lymphocyte development and function is not surprising and is orchestrated by CRAC, NMDAR, IP3R, P_2XR, Ca_V, and TRP channels as described above (Nakayama et al. 1992; Aifantis et al. 2001; Bhakta et al. 2005; Erdogmus et al. 2022).

Tm lymphocytes are a long-lived subset which enables a faster and more efficient secondary immune response upon re-exposure to an antigen previously already encountered. While the memory cell generation pathway is still being debated, IL-7 and IL-15 have emerged as key regulators of the Tm pool. Several studies demonstrate the importance of $K_V1.3$ and $K_{Ca}3.1$ channel activity for Tm contact with APCs, differentiation, proliferation, and cytokine production (Hu et al. 2012). Targeting of these K^+ channels in Tm holds therapeutic promise for multiple sclerosis (Wulff et al. 2003; Rangaraju et al. 2009) and other T lymphocyte-mediated autoimmune diseases (Beeton et al. 2006; Azam et al. 2007).

4 B Lymphocytes

B lymphocytes' major function is antibody/immunoglobulin (Ig) production. Following B cell receptor (BCR) binding to antigen presented on follicular dendritic cell, B lymphocytes undergo proliferation and affinity maturation, involving somatic hypermutation, Ig class switching, and clonal selection, in lymphoid germinal centres overseen by Tfh. These steps are crucial for the subsequent generation of antibodies with increased affinity for antigen. LRRC8A subunit of VRAC ion channel has been reported to be essential for the normal development of B lymphocytes (Sawada et al. 2003; Kumar et al. 2014). Eventually, B lymphocytes differentiate either into memory B lymphocytes or plasma cells depending on cytokine and other signals received. Plasma cells then produce large amounts of soluble antibodies specific for this particular antigen and secrete these high-affinity antibodies into the local environment, where the antibodies recruit other elements of the immune response via Fc receptor (FcR) binding (Delves et al. 2017). BCR stimulation results in the CRAC-mediated SOCE, found to be important for B lymphocyte proliferation, IL-10 production, and regulatory function required to limit autoimmunity (Matsumoto et al. 2011). Inhibition of $K_v1.3$ or $K_{Ca}3.1$ channels in B lymphocytes affects their proliferation (Wulff et al. 2004). Interestingly, $K_v1.3$ and $K_{Ca}3.1$ are differentially expressed in B lymphocyte subsets, potentially allowing selective manipulation of B lymphocyte subsets for the treatment of immune disorders. TRPM7 has also been reported to regulate B lymphocyte proliferation and viability (Schmitz et al. 2003; Sahni et al. 2010). The voltage-gated proton channel (H_V1) was found to be essential for BCR signalling via generation of reactive oxygen species and for healthy antibody responses (Capasso et al. 2010). Finally, any ion channel activity influencing Tfh (see Sect. 3) and dendritic cell (see Sect. 6) generation and cytokine production will indirectly influence B lymphocyte development and functions.

5 Natural Killer (NK) Cells

NK cells are lymphoid cells of the innate immune system branch, in contrast to T and B lymphocytes forming the adaptive branch. NK cells can rapidly kill virally infected cells by recognising abnormal MHC receptor profiles. NK cells are also able to display IgG antibody receptors (FC receptors) on their surface; thus they seek out and kill infected cells in a process termed antibody-dependent cellular cytotoxicity. Importantly, NK cells together with Tc lymphocytes have long been known for their tumour cell killing abilities (Panyi et al. 2014). Cytokines produced in response to tumour cells, such as IL-2, IL-12, IL-15, IL-18, and NK cell stimulatory factor (NKSF), modulate NK cell activity. While IL-15 enhances NK cell cytotoxicity and tumour cell killing, certain cytokines produced by tumour microenvironments, including IL-1α, IL-6, and TGFβ, are immunosuppressive and impair NK cell functions. Similar to Tc lymphocytes, NK cells utilise the granzyme/perforin degranulation process or Fas ligand pathway for target cell lysis. Activated NK cells also produce IFN-γ, enhancing microbicidal activity of macrophages and antigen presentation function of dendritic cells (Delves et al. 2017). The importance of calcium signalling for NK cell functions has been well-documented. Studies report that defective CRAC-mediated SOCE results in severely defective cytotoxic granule exocytosis, impaired target cell lysis, and reduced cytokine production by NK cells. Interestingly, the quantitative SOCE requirements for NK and Tc cytotoxicity seem to differ, with moderate SOCE being sufficient for Tc function but not for NK (Maul-Pavicic et al. 2011). Lack of other Ca^{2+} ion channel activity, TRPM2 or Ca_v, in NK cells also reduces cytolytic granule polarisation and degranulation and impairs NK cytotoxicity towards tumour cells (Solovera et al. 1987; Poggi and Zocchi 2006; Rah et al. 2015). Blocking K^+-permeable ion channels, $K_V1.3$, inhibits proliferation and degranulation of non-adherent NK cells, with minimal effects on adherent type. By contrast, $K_{Ca}3.1$ block increases proliferation, degranulation, and tumour/target cell killing of adherent NK cells, as well as proliferation of a non-adherent type (Koshy et al. 2013). Another K^+-permeable ion channel known to regulate NK cell proliferation, degranulation, and tumour cell targeting is K_{2P} (Schulte-Mecklenbeck et al. 2015). All these ion channels may potentially serve as therapeutic targets to reactivate vital NK cell functions lost in cancer (see Sect. 9). Other innate lymphoid cells are type 1, type 2, type 3, and lymphoid tissue-inducer cells, but little is known about ion channel regulation in these cells (Vivier et al. 2018).

6 Dendritic Cells

The main function of dendritic cells is antigen presentation, linking innate and adaptive immune system branches. Following the stimulation of dendritic cells in the periphery by PAMPs/DAMPs and by cytokines from other innate immune cells, dendritic cells undergo a maturation process, which is characterised by a decrease in their phagocytic/antigen capture activity, a dramatic increase in MHC expression, as well as upregulation of certain chemokine receptors, facilitating dendritic cell migration. This maturation process is regulated by a number of ion channels including CRAC (Félix et al. 2013), ryanodine receptors (Vukcevic et al. 2008), and TRP channels (Tóth et al. 2009; Szöllosi et al. 2013). Activation of ryanodine receptors in immature dendritic cells via $Ca_v1.2$ was reported to cause a rapid increase in the MHC expression on the cell surface (Vukcevic et al. 2008). On the other hand, TRPV1 signalling was found to inhibit phagocytosis, activation, and pro-inflammatory cytokine secretion in human dendritic cells (Tóth et al. 2009). Other ion channels appear more influential at the migration stage. $Na_v1.7$ has been found to maintain the negative membrane potential and regulate the chemokine-induced dendritic cell migration (Kis-Toth et al. 2011). Activity of $K_{Ca}3.1$ (Shao et al. 2011; Crottès et al. 2016) and TRPM4 (Barbet et al. 2008) was reported to enhance human dendritic cell migration through controlling Ca^{2+} homeostasis.

Mature dendritic cells migrate to the lymph nodes where they present antigens via MHC to the cells of the adaptive immune system branch. The unique infection-specific combinations of PAMPs/DAMPs and activated PRRs on innate immune cells trigger particular cytokine combinations being released and, hence, dictate the type of adaptive immune response generated, for example, if response is dominated by antibody-producing B lymphocytes or Tc lymphocytes targeting virus-infected or tumour cells. As discussed in the Sect. 3, the cytokine environment accompanying antigen presentation process by APCs plays critical role in shaping T lymphocyte differentiation. For example, IL-12 and IFN-γ promote Th1 generation, while inhibiting Th2 responses. CRAC-mediated SOCE (Félix et al. 2013), $K_v1.3$, $K_v1.5$ (Mullen et al. 2006), and P_2X_7R (Ghiringhelli et al. 2009) have all been implicated in the regulation of cytokine secretion from dendritic cells. P_2X_7R activation by ATP release from dying tumour cells triggers NLRP3 inflammasome activation and IL-1β release, critical for Tc lymphocyte production and, hence, anti-tumour immunity (Ghiringhelli et al. 2009). On the other hand, ATP-induced P_2X_7R activation on dendritic cells was reported to enhance graft-versus-host disease with implications for transplantation medicine (Wilhelm et al. 2010). $K_v1.3$ and $K_v1.5$ channels, regulating dendritic cell chemotaxis and IL-12 production, were found on dendritic cells from human multiple sclerosis brain tissue, but not healthy brain tissue, leading to the suggestion of these ion channels as therapeutic targets (Mullen et al. 2006). Finally, a recent study reports Piezo1 in dendritic cells, where it is suggested to direct Th1 and Treg differentiation in cancer. Genetic deletion of Piezo1 increases TGFβ and decreases IL-12 secretion in dendritic cells, thus inhibiting the generation of Th1 lymphocytes, upregulating Treg development, and promoting cancer growth (Wang et al. 2022). Therapeutic potential of targeting ion channels in dendritic and other immune cells for cancer and a range of different pathologies will be discussed in more detail in the Sect. 9.

7 Monocytes and Macrophages

Monocytes and macrophages represent important cells of the innate immune systems. In most cases (Locati et al. 2020), macrophage populations are derived via the differentiation of monocytes, with monocytes derived from the embryonic yolk sac contributing to tissue resident macrophage populations and their renewal, e.g. microglia (see further below) and monocytes derived from stem cells in the bone marrow responsible for inflammatory macrophages. Traditionally it was believed that the main role of monocytes was as the highly migratory precursor of the macrophage in inflammation. This view of monocytes is now changing with the recognition that the cell contributes to inflammatory mechanisms such as phagocytosis and cytokine production and shows different functional subsets akin to macrophages (Shapouri-Moghaddam et al. 2018); see below. However, the vast majority of ion channels research in this area concentrates on their role in macrophages.

Tissue resident macrophages are distributed throughout the body where they contribute to tissue/organ homeostasis mechanisms such as the removal of dead cells which occurs under normal physiological conditions. Importantly, the local tissue environment significantly influences resident macrophages, aligning local physiological/pathophysiological requirement to cell functionality (Blériot et al. 2020). These resident macrophages also play a significant role in initiating inflammation, by recognising microbial and tissue damage associated chemical motifs, PAMPs & DAMPs, and releasing pro-inflammatory mediators, which cause, among other things, the large influx of leukocytes into the inflammatory lesion (Davies et al. 2013). Neutrophils are usually the first cells to arrive at the site of inflammation; however an influx of monocytes often follows. Within the lesion the monocytes differentiate into inflammatory macrophages, resulting in the inflammatory macrophage becoming the dominate type of macrophage present (Davies et al. 2013). Within the lesion inflammatory macrophages have diverse functions including killing, elimination, and removal of microbes and noxious agents, antigen presentation, and tissue remodelling. However, inappropriate activation/control of macrophage function can lead to tissue destruction and pathology. The ability of macrophages to successfully implement the required tissue function, while avoiding/minimising self-inflected tissue damage, is predicated on the cell utilising sensor-rich signalling cascades which ensure that the multiple environment cues encountered by the macrophage are integrated to produce a timely and appropriate response. At the cellular level, the ability of macrophages to align the local tissue conditions to cell functionality is reflected in the diverse number of macrophage phenotypes which have been characterised in different tissues and disease setting.

The classification of macrophage phenotypes is an active area of research and debate in the immunological community; however, a convenient way of classifying macrophages is to regard them as either a M1-type macrophage which has a pro-inflammatory phenotype or a M2-type macrophage which has an anti-inflammatory phenotype, although M2 subtypes can also have pro-inflammatory effects. While this is undoubtedly a simplification of macrophage heterogeneity in vivo, this classification is often encountered in the literature and for convenience will be used here. However, it is important to note that how macrophage diversity arises, i.e. via macrophage population heterogeneity and/or plasticity, and the extent to which macrophage functionality exists as a continuum or as distinct phenotypic populations requires further research (Guilliams and Svedberg 2021; Zwicker et al. 2021).

A number of reviews have highlighted the importance of various ion channels in macrophage biology and the shaping of macrophage functionality including macrophage polarisation, phagocytosis and endosome maturation, inflammatory mediator production and inflammasome activation, migration, and cell survival (Selezneva et al. 2022b). Below we highlight some recent findings concerning the role of ion channels in macrophage biology.

Differentiation of human monocytes to macrophages increases the expression of functional $K_{Ca}3.1$ channels, and while the channel is present in both M1- and M2-type macrophages, $K_{Ca}3.1$ inhibition only appears to suppress M1 macrophage functions as measured by the release of pro-inflammatory cytokines (Xu et al. 2017). ATP-sensitive K^+ channels (K_{ATP}), which contain the inward rectifier K^+ (K_{ir}) $K_{ir}6.x$-type subunits, also appear to upregulate in M1 macrophages with the channel's subunits inward rectifier K^+ (K_{ir}) $K_{ir}6.1$, SUR1 and SUR2, being expressed at higher levels than M2 macrophages. In this same study, inhibition of the K_{ATP} channels with glibenclamide inhibited markers of M1 macrophages, while enhancing the expression of M2 macrophage markers in cells which had already been induced to differentiate down this pathway; interestingly, the K_{ATP} channel opener pinacidil appeared to have the opposite effect of inducing M1 macrophage markers (Vicente et al. 2006; Villalonga et al. 2010; Kan et al. 2016; Li et al. 2016). These studies would suggest that $K_{Ca}3.1$ and K_{ATP} have a more significant role in M1 macrophages than M2. However, it is important to consider the multiple signalling molecules that macrophages engage in situ and the local tissue environment when suggesting that some macrophage phenotypes depend on particular ion channels. This can readily be illustrated by looking at recent studies on $K_{ir}2.1$.

Kir2.1 has been reported to be important in the polarisation of macrophages towards an M1 phenotype, with inhibition of the channel promoting an M2 phenotype (Chen et al. 2022a). However, Kir2.1 has also been demonstrated to be important in tumour-associated macrophage (TAM) function, which has a functionality similar to M2 macrophages. Inhibition of $K_{ir}2.1$ in the TAMs drove the cell phenotype towards macrophages demonstrating an anti-tumour phenotype, which is more in line with a M1 macrophage phenotype (Chen et al. 2022b). A question that therefore arises is how can these two observations on the role of $K_{ir}2.1$ be reconciled? Both studies referred to the effects of high extracellular K^+ on macrophage function. 70 mM extracellular K^+ (the normal concentration is ~5 mM) resulted in an inhibition of Ca^{2+} influx in M1 macrophages, which coincided with a reduction in activity of the CaMKII-ERK-NFκB signalling pathway, while promoting the expression of p-STAT6, a marker of M2 macrophages (Chen et al. 2022a). In comparison, high extracellular K^+ (~ 15 to 28 mM) encountered in tumours was associated with a decrease in the M1-like phenotype in the TAM population in tumours and activation of M2 phenotypes (Chen et al. 2022b). In both these situations, increasing extracellular K^+ would depolarise the membrane and reduce Ca^{2+} entry into the cell via store-operated Ca^{2+} channels (but not voltage-gated Ca^{2+} channels); indeed, there is good evidence of functional coupling between K^+ channels and Ca^{2+} channels in a number of cells, including macrophages, and that this relates to functionality/type of the cells, i.e. neuron and smooth muscle cells. Together, these results for $K_{ir}2.1$ in macrophages suggest that these ion channels act as flexible sensors of the local tissue environment and are not rigorously linked to the M1 or M2 phenotypes (see "Therapeutic Opportunities").

A number of other K^+ channels have been implicated in macrophage activation; these include $K_v1.3$ and $K_v1.5$ (Vicente et al. 2006; Villalonga et al. 2010; Kan et al. 2016), $K_v11.1$, TWIK2 (Di et al. 2018; Wang et al. 2023b), TMEM175, and $K_{Ca}1.1$

(BK) channels (Selezneva et al. 2021). It is interesting to note that the role of these channels in macrophage activation may not be limited to plasma membrane-located ion channels but may also involve ion channels in intracellular membranes (Selezneva et al. 2022b), and this requires further investigation.

As expected, intracellular calcium concentrations and oscillations regulate a number of important macrophage functions (Vaeth et al. 2015; Zumerle et al. 2019; Liu et al. 2021), and it is in this area that ion channels may have their most significant effect on macrophage biology. In naïve/non-stimulated macrophages, sometimes referred to as the M0 macrophage phenotype, the ORAI1-encoded CRAC channel contributes to increases in intracellular Ca^{2+}; however, upon activation of the cell resulting in a M1 macrophage phenotype, TRPC1 becomes the dominant channel with regard to Ca^{2+} influx and contributes to the increased expression of pro-inflammatory proteins (Chauhan et al. 2018). Further investigation demonstrated that the M2 macrophage phenotypes were dependent on ORAI1-mediated Ca^{2+} influx, while TRPC1 was not required (Nascimento Da Conceicao et al. 2021). Therefore, M1 macrophages were associated with TRPC1 channels, while ORAI1 appeared important in M2 macrophages. Importantly, while both these channels allow Ca^{2+} entry into the cytoplasm, the spatial and temporal dynamics of the Ca^{2+} signal will be different, and this can lead to differential gene expression (Moccia et al. 2023).

In addition to TRPC1, other TRP channels have been implicated in macrophage polarisation. These include TRPA1, TRPM7, TRPM8, and TRPV1 which appear to facilitate M2 macrophage development or inhibit M1 macrophages, while TRPC3 and TRPM2 promote M1 polarisation, reviewed (Selezneva et al. 2022b). In comparison, members of the TRP mucolipin family, TRPML1, TRPML2, and TRPML3, appear to be important in phagocytosis and phagosome biology (Spix et al. 2020). In fact, a number of TRP channels have been reported to have a role in regulating macrophage phagocytosis; these include TRPV2 (Link et al. 2010; Lévêque et al. 2018), TRPC6 (Riazanski et al. 2015; Ramirez et al. 2018), TRPM4 which is permeable to monovalent cations (Serafini et al. 2012), and TRPV4 (Scheraga et al. 2015; Acharya et al. 2022). It is interesting to note that the TRPV4 channel responds to stretch, and the role of mechanosensitive ion channels in macrophage represents a rapidly developing area of research.

As indicated above, macrophages "read" their local environment to aid in the execution of the appropriate cellular response. To achieve this, macrophages, as well as other immune cells, rely not only on local chemical signals but also on mechanical signals such as stretch, shear stress, compression, and changes in cell volume which can occur during phagocytosis. These signals are detected via a variety of different molecules (Lim et al. 2018), of which ion channels are an important class (Gu and Gu 2014; Kefauver et al. 2020). Importantly, these "force" signals, together with the mechanotransduction pathways that they initiate (Tschumperlin 2011), are increasingly being recognised as regulators of immune cell function (Huse 2017; Du et al. 2022; Lee et al. 2022; Selezneva et al. 2022a). In macrophages, a particular focus has been on the force-sensing ion channels TRPV4 and Piezo1. Both these channels regulate Ca^{2+} influx, and their role(s) in immune cells has recently been reviewed (Michalick and Kuebler 2020; Orsini et al. 2021; Du et al. 2022; Selezneva et al. 2022b). The initial impression from studies on the role of Piezo1 in macrophages is that its activation, via changes in growth substrate stiffness (kPa) or via channel openers, facilitates the activation of the macrophage, drives them towards the M1 phenotype, and regulates iron metabolism (Solis et al. 2019; Atcha et al. 2021; Geng et al. 2021; Ma et al. 2021; He et al. 2022; Leng et al. 2022). However, it has also been reported that Piezo1 is also associated with M2 macrophage phenotypes (Zhang et al. 2022b; Cai et al. 2023; Wang et al. 2023a). Interestingly, TRPV4 has also been implicated in both M1 pro-inflammatory macrophage phenotype (Dutta et al. 2020; Sun et al. 2022; Chen et al. 2023) and M2 anti-inflammatory phenotype (Scheraga et al. 2015; Atsumi et al. 2023). TRPV4 is also involved in the formation of foreign body giant cells from macrophages (Goswami et al. 2019, 2021; Arya et al. 2021), a mechanism which contributes to granuloma formation in response to infection and biomaterials (Pagán and Ramakrishnan 2018; Ahmadzadeh et al. 2022). As described above, the theme which appears is that individual channels are not exclusively tied to a particular macrophage M1 or M2 phenotype but that their role reflects the totality of the macrophage's environment/signals.

Ligand-gated ion channels (LGIC) have also been the subject of a number of studies in macrophages. The P_2X_7 purinergic receptor is an ATP-gated LGIC and has been implicated in the activation of the NLRP3 inflammasome. NLRP3 activation results in the proteolytic cleavage of the pro-peptides of IL-1β and IL-18 allowing the mature cytokine to be released from the cell. The inflammasome also contributes to the activation of Gasdermin D which can initiate pyroptosis and act as conduit by which molecules can leave or enter a cell. A number of ion channels have been implicated in the activation of NLRP3 in macrophages. The P_2X_7 receptor has gained interest because it has a large K^+ ion conductance which can increase after channel activation by pore dilation and the recruitment/activation of hemichannels. It is this K^+ efflux, possibly in combination with Cl^- efflux, which is required for the canonical activation of the inflammasome (Xu et al. 2020; Xu and Núñez 2023). While a number of P_2X_7 receptor antagonists have entered clinical trials, to date none have been approved. Another LGIC implicated in macrophage function is the nicotinic acetylcholine α7 receptor (α7-nAChR). This LGIC participates in the cholinergic anti-inflammatory pathway or anti-inflammatory vagal reflex which connects the central nervous and immune system. Activation of α7-nAChR by acetylcholine (Ach) or synthetic receptor agonists results in the inhibition of NFκB translocation to the nucleus and suppression of pro-inflammatory cytokines release from activated macrophages (Fujii et al. 2017). Interestingly, this raises an important question when investigating the role of LGIC in immune cell function in physiological or disease settings; namely, does the endogenous ligand for the LGIC occur at a sufficient concentration to activate the channel? In the case of P_2X_7 and α7-nAChR, both have relatively low affinity for their natural ligands, with EC_{50} 2–4 mM for ATP and EC_{50} of 116–140 μM for Ach for their respective LGIC. It is anticipated that these concentrations of ATP and Ach cannot be found in tissues under normal physiological condition. Interestingly, the local tissue concentrations of these ligands may be sufficient to activate these LGIC in pathological setting; this may represent unique therapeutic opportunities (see below). However, little data on this is available due to spatial and temporal difficulties in the measurement of these ligands.

While ion channels have been investigated in macrophages since the 1980s, we have only scratched the surface of their role in these cells. One of the defining characteristics of macrophages is their heterogeneity and the way their local environment determines their functionality. Ion channels are well-placed as possible sensors and regulators of the macrophage-environment relationship and as such are potential therapeutic targets.

Microglia are considered resident macrophages of the brain parenchyma. Derived from erythromyeloid progenitor cells from the yolk-sac, under homeostatic conditions, they are slowly proliferating long-lived cells, but have the capability to undergo increased clonal proliferation under pathological conditions (Borst et al. 2021; Zhang et al. 2022a). In the CNS microglia play a number of important roles; they support normal neuronal function such as neurogenesis and myelination; contribute to vasculogenesis; regulate blood-brain barrier (BBB) permeability; via phagocytosis, remove cellular debris in the CNS; and participate in synapse remodelling (Borst et al. 2021). As with other macrophage types, the microglia perform their differing cellular functions by responding to extracellular cues in a cellular appropriate way, while also communicating bidirectionally with other cells of the CNS (Izquierdo et al. 2019; Prinz et al. 2019). However, while microglia have important homeostatic functions in the CNS, mounting evidence also demonstrates how these cells contribute to neuroinflammation, autoimmunity, neurodegeneration, and neoplasms (Da Fonseca et al. 2016; Colonna and Butovsky 2017; Wolf et al. 2017; Hickman et al. 2018; Borst et al. 2021). In line with other macrophage populations, ion channels have been implicated in the regulation of microglial functions and may represent possible therapeutic targets (Izquierdo et al. 2019; Luo et al. 2021; Sarkar 2022).

Intracellular Ca^{2+} plays an important role in regulating microglia function with changes in the quantitative, temporal, and spatial dynamics of the ion contributing to both physiological and pathophysiological responses of the cell (Mizoguchi and Monji 2017). Specifically, intracellular Ca^{2+} has been demonstrated to regulate migration, phagocytosis, and "pro-inflammatory" mediator release from microglial cell. As with other cells, the intracellular Ca^{2+} dynamics in microglia are regulated by Ca^{2+} transporters and channels. From the point of view of channels, SOCE; voltage-gated Ca^{2+} channels $Ca_v1.2$, $Ca_v1.3$, and $Ca_v2.2$; TRP channels; TRPV1, TRPV2, TRPV4, TRPC3, TRPM2, and TRPM7 (Pan et al. 2011); and the mechanosensitive ion channel Piezo1 (Zhang et al. 2023) have all been implicated in the regulation of microglia function. In most of the studies on Ca^{2+} channels in microglial cell, results have suggested that they contribute towards cell activation, e.g. the release of pro-inflammatory cytokines, such as TNFα and IL-1β, and the increase in generation of ROS. However, a number of studies have also suggested that Ca^{2+} channels have neuroprotective effects. For example, $Ca_v1.2$ has been reported to be involved in the development of a "neuroprotective" microglia analogous to a M2 macrophage phenotype, with microglia-specific $Ca_v1.2$ knock-down mice showing enhanced dopaminergic neuron degeneration compared to wild-type controls in the MPTP-induced model of Parkinson's disease (Wang et al. 2019). In comparison, microglia-specific $Ca_v1.2$ knock-down mice appeared to show less behavioural and motor neuron deficits in the MPTP model compared to wild-type animals (Saegusa et al. 2020), suggesting a detrimental role for $Ca_v1.2$. In a second example, it was reported that brain-derived neurotrophic factor (BDNF) reduced TNFα-induced NO production from rat microglial cells via a TRPC3-mediated increase in intracellular Ca^{2+} (Mizoguchi et al. 2014).

Considering K^+ channels, in peripheral macrophages, the resting membrane potential of the cell is primarily regulated by K^+ channels (Selezneva, Gibb, and Willis, 2022). In the case of microglial cells, the tonically active two-pore domain halothane-inhibited K^+ channel type 1 (THIK-1, $K_{2P}13.1$) appears to be involved in this important plasma membrane parameter (Luo et al. 2021). It is therefore not surprising that THIK-1 has also been reported to regulate other microglial cellular functions including cell ramification, surveillance activity, synaptic development, and the activation of the NLRP3 inflammasome (Madry et al. 2018; Izquierdo et al. 2021; Ossola et al. 2023). Microglia also express a variety of other K^+ channels including Ca^{2+}-activated ion channels, voltage-gated K^+ channels, and inward rectifier K^+ channels, and these channels have been implicated in a variety of cellular microglial functions including proliferation, migration, cytokine and ROS release, and phagocytosis (Cocozza et al. 2021).

A number of other ion channels have also been demonstrated to participate in microglia functions; these include the regulation of cell volume by volume-regulated Cl^- channels, a process important during the phagocytosis of apoptotic and necrotic cells (Furtner et al. 2007; Harl et al. 2013), the control of phagocytosis by tetrodotoxin-sensitive voltage-gated sodium channels, and ROS production by the voltage-gated proton channel. Finally, microglial cells are known to express a number of LGIC (Liu et al. 2016); these include P_2X purinergic receptors, P_2X_4 and P_2X_7 (Calovi et al. 2019); nACh receptors (Bye et al. 2023), with the microglia α7 nAChR reported to have anti-inflammatory effects (Zhou et al. 2022; Piovesana et al. 2021); the glutamate NMDA, AMPA, and kainic acid receptors (Stebbing et al. 2015; Czapski and Strosznajder 2021); and GABA-A receptors (Czapski and Strosznajder 2021; Andoh and Koyama 2023).

From a therapeutic perspective, research would suggest that microglia ion channels may play a role in the pathogenesis of a number of CNS disorders including neurodegenerative diseases, stroke, epilepsy, and brain tumours such as glioblastoma multiforme; and as a result, they may represent potential drug targets. However, as with studies on ion channels in macrophages, a number of points concerning ion channels in microglia cells need to be taken into account when considering their targeting. Firstly, a general feature of macrophages is their heterogeneity, and this is seen with microglial cells too, with different phenotypes of microglia being seen at different stages of development, in different areas of the CNS, and under different activation states (Masuda et al. 2020; Ochocka and Kaminska 2021). It should also be noted, besides microglia cells, other types of macrophages can be found in the

CNS under both physiological condition, e.g. CNS border-associated macrophages (Mundt et al. 2022; Dermitzakis et al. 2023), and pathological conditions where monocytes can be recruited from the periphery. Secondly, ion channels have been demonstrated to regulate microglial homeostatic functions as well as possible CNS pathogenic mechanisms. Therefore, it will be important to determine if the ion channel expression profiles are related to specific microglia phenotypes under physiological and pathological conditions. If so, this could allow the selective targeting of microglia populations.

8 Granulocytes and Mast Cells

Neutrophils, eosinophils, and basophils, collectively termed granulocytes owing to the presence of enzyme granules in their cytoplasm, are non-tissue-resident; such cells circulate in the bloodstream until certain signals from other innate immune cells, especially macrophages and mast cells, permit granulocyte extravasation and migration into peripheral tissues towards the source of chemotactic factors/infection. In contrast to granulocytes, mast cells are tissue-resident. Mast cells collaborate with resident tissue macrophages to promote vascular permeability and granulocyte extravasation and migration through the release of cytokines (e.g. TNFα, IL-1β, and IL-8), chemokines, and vasoactive factors like histamine. Neutrophils, constituting about 97% of granulocyte population, are highly phagocytic cells adept at intracellular killing of bacteria. Another strategy that activated neutrophils' deployment for immobilising and killing bacteria is the formation of neutrophil extracellular traps (NETs) composed of neutrophil DNA and granule proteases that act as a "spider's web" trapping targets. Eosinophils and basophils are key to fighting larger parasites, which cannot be phagocytosed by macrophages or neutrophils and instead must be attacked with a release of granule enzymatic contents directly onto parasite.

Granulocyte recruitment to site of inflammation is a cascade-like event involving the interplay of many different receptors, signalling pathways, and ion channels, such as CRAC (Schaff et al. 2010; Sogkas et al. 2015; Immler et al. 2018), TRPC1 (Lindemann et al. 2015), TRPC6 (Damann et al. 2009; Lindemann et al. 2013), TRPV4 (Yin et al. 2016), $K_{Ca}3.1$ (Henríquez et al. 2016), and $K_v1.3$ (Immler et al. 2022). TRPC6 deficiency in neutrophils was reported to affect AKT/MAPK phosphorylation downstream of CXCR2 receptor and alter remodelling of actin and neutrophil chemotaxis (Lindemann et al. 2013). A recent study identified that hydrogen peroxide guides neutrophil chemotaxis via another TRP channel, TRPM2 (Morad et al. 2021). Conflicting data about P_2X receptor regulatory functions in neutrophils exist. Lecut et al. and Maître et al. show that P_2X_1R promotes neutrophil extravasation and chemotaxis through Rho kinase activation (Lecut et al. 2009; Maître et al. 2015). Furthermore, a recent study identified a role of P_2X_1R in suicidal NETosis and its anti-cryptosporidial properties (Hasheminasab et al. 2022). Induction of NETosis depends on reactive oxygen species (ROS), the main source of which is

NADPH oxidase. However, other studies report that P_2X_1R negatively regulates neutrophil activation, chemotaxis, and ROS generation (Lecut et al. 2012; Wang et al. 2017). Another P_2X receptor, P_2X_7R, has been suggested to regulate ROS production, NLRP3 inflammasome activation, and IL-1β secretion in neutrophils (Suh et al. 2001; Karmakar et al. 2016). Granulocyte functions including FcγR-mediated phagocytosis, NADPH oxidase activity, ROS production, and degranulation were found to also depend on CRAC-mediated SOCE (Steinckwich et al. 2011), TRP channels (Bréchard et al. 2008), $K_v1.3$ (Immler et al. 2022), and H_v1 (Femling et al. 2006; Ramsey et al. 2009; el Chemaly et al. 2010; Okochi et al. 2016).

In addition to fighting infection, granulocytes and mast cells are known to play key roles in chronic inflammation and allergic/hypersensitivity reactions (Galli and Tsai 2012; Amin et al. 2016; Radermecker et al. 2018); therefore, understanding of ion channel regulation in these innate immune cells may provide new approaches to treatment. For example, inhibition of CRAC channels in mast cells was reported to attenuate mast cell degranulation and release of allergic mediators including histamine and proinflammatory cytokines, hence, reducing granulocyte recruitment and allergen-induced bronchial smooth muscle contraction, making CRAC a potential therapeutic target in asthma and other related allergic and mast cell-mediated diseases (di Capite et al. 2009; Ashmole et al. 2012; Chaki et al. 2022). Targeting P_2X_7R in mast cells has therapeutic potential to reduce intestinal inflammation in Crohn's disease (Kurashima et al. 2012; Eser et al. 2015), while $K_{Ca}3.1$ inhibition was suggested to attenuate airway remodelling and eosinophilia in models of asthma (Girodet et al. 2013; van der Velden et al. 2013). The next Sect. 9 covers therapeutic potential of targeting the various ion channels in mast and other immune cells in more detail.

9 Therapeutic Opportunities

As described above, ion channels are found in all the cells of the immune system and participate in a variety of cellular functions which are important for a normal immune response or could potentially contribute to the development of significant pathology. As would be expected, the impact of the ion channel on the cell will depend on its biochemical and electrophysiological properties, e.g. ion selectivity, gating parameters, voltage, and ligand dependency, and on its cellular location, e.g. plasma, mitochondrial, and nuclear membranes. A major determinant of ion channel properties is their subunit composition and presence of auxiliary subunits. Depending on the subunits present, ion channels from the same family can, for example, have very different channel-gating and ligand-binding properties; indeed, some ion channels have been reported to have cellular functions which are independent from their role in ion flux. Activation and gating properties of ion channels can also be influenced by the local environment, e.g. temperature, pH, mechano-stimuli, and ion concentration, and this may be an important consideration when investigating ion channels in pathological tissues, e.g. tumours and granulomas. Taking these considerations together, ion channels represent a rich biological landscape to search for new therapeutic targets.

Considering the immune system, therapeutically we would either wish to inhibit the immune system or the activation of pro-inflammatory leukocytes in the case of excessive acute inflammation, chronic inflammation, and autoinflammatory or autoimmune diseases. In comparison, in infection and some cancers, we would aim to enhance the immune response. However, a major obstacle to overcome in the design of drugs for the immune system is generalised immune suppression or activation. Too much immune suppression/anti-inflammatory activity and the individual becomes immunosuppressed and has an increased susceptibility to microbial infections and cancer; excessive immune activation can however lead to host tissue damage and the development of acute and chronic inflammatory pathologies. The goal is to develop therapies which are specific for the pathology-causing agents or mechanisms of the immune cell, while leaving their normal functionality and antimicrobial, anti-neoplastic, and homeostatic action intact. So could we exploit ion channel biology in immune cells to treat disease?

We believe there may be reasons for optimism. Firstly, ion channel function can be modulated in a number of ways; small molecular entities and biologics can be developed which block/inhibit the ion channel, increase the opening probability of the channel, or act as allosteric modulators of the channel. In addition, due to the diversity of subunit composition found in ion channels in different tissues, cells, and under different states of activation, drugs could be developed which show various levels of selectivity for ion channels. Secondly, immune cells show a high degree of heterogeneity depending on their tissue location and activation state. This heterogeneity is now being reflected in the profile of ion channels that these cells express, particularly in T lymphocytes and macrophages. Therefore, it may be possible to target immune cells with specific phenotypes, while leaving the bulk immune cell population unaffected. Finally, the environment of the pathological lesion can be significantly different to that of physiological tissue. In particular, oxygen levels, pH, temperature, extracellular matrix, and ion concentration can be significantly altered in inflammatory milieu, granuloma, and tumours. These differences can affect the properties of ion channels and their interactions with endogenous and synthetic ligands. Thus, it may be possible to develop therapies which preferentially target ion channels in a diseased/dysfunctional tissue, while having little impact on ion channels in immune cells located in non-diseased tissue.

Considering these three points, it may be feasible that therapies could be developed targeting specific ion channels in identified immune cell phenotypes and/or tissue setting. In this case it would be anticipated that therapy would specifically target pathology-associated immune cells, while leaving other immune cells to carry out their normal function. To date, only relatively few clinical trials have been conducted on drugs in inflammatory, autoimmune diseases or cancer, whose mode of action could be related to the modulation of ion channel activity in immune cells (see Table 2). Further research is required on the expression and role of ion channels in immune cells and their role in the pathogenesis of diseases if this list is to increase.

Table 2 Clinical trials in which ion channels in immune cells is predicted to be a mechanism of action of evaluated therapy

Target/drug	Type	Clinical trial phase	Notes	Reference
CRAC channels				
CM2489 (CalciMedica)	SOCE and I_{CRAC} blocker	Phase I, Ib	Oral administration to healthy cohort and moderate to severe plaque psoriasis patients. Only modest clinical effects possibly due to high protein binding. No serious adverse reactions	Stauderman (2018), Shawer et al. (2021)
Auxora (CM4620) (CalciMedica)	SOCE and I_{CRAC} blocker Inhibition ORAI1	Phase I, II	I.V. administration phase I clinical trial to healthy individuals. Phase II open label for patients with moderate to severe acute pancreatitis. Favourable safety profile and clinical efficacy	Bruen et al. (2021)
		Phase II	Phase II blind randomised placebo-controlled trials for patients with critical COVID-19 pneumonia. Favourable safety profile and clinical efficacy	Miller et al. (2020), Bruen et al. (2022)
RP3128 (Rhizen Pharmaceuticals)	SOCE and I_{CRAC} blocker	Phase I/IIb	Oral administered to healthy individuals and mild asthmatics patients. Terminated due to recruitment delays	Barde et al. (2021)
RP4010 (Rhizen Pharmaceuticals)	SOCE and I_{CRAC} blocker	Phase I/IIb	Oral administration to patients with relapsed or refractory non-Hodgkin's lymphoma. Terminated due to PK and safety issues	Stauderman (2018), Shawer et al. (2021)
Compound 36 (Lupin Ltd)	SOCE blocker	Phase I	Oral administration to healthy individuals. Expected intended patient cohort rheumatoid arthritis	Khedkar et al. (2021)
P_2X_7				
CE-224535 (Pfizer)	P_2X_7 antagonist	Phase IIb	Orally administered, randomised, double-blind, placebo-controlled clinical trial in rheumatoid arthritis patients which were refractory to methotrexate. Antagonist well-tolerated but no significant difference in clinical utility compared to control	Stock et al. (2012)

(continued)

Table 2 (continued)

Target/drug	Type	Clinical trial phase	Notes	Reference
AZD9056 (AstraZeneca)	P_2X_7 antagonist	Phase IIa	Orally administered, randomised, double-blind, placebo-controlled trial in rheumatoid arthritis patients receiving methotrexate or sulphasalazine. Significant improvement in number of swollen joints and patients achieving ACR20 level with drug compared to placebo. Drug was well-tolerated	Keystone et al. (2012)
		Phase IIb	Randomised, double-blind, placebo-controlled trial in rheumatoid arthritis patients. Trial included open-labelled group receiving etanercept. No significant difference in clinical outcomes between AZD9056 and placebo. Drug well-tolerated	Keystone et al. (2012)
		Phase IIa	Randomised, double-blind, placebo-controlled trial in active Crohn's disease patients. Drug treatment resulted in a significant decrease in Crohn's disease activity index and higher rates of remission compared to placebo group. Drug was well-tolerated	Eser et al. (2015)
K_v1.3				
Dalazatide (ShK-186) (Kv1.3 Therapeutic)	Kv1.3 inhibitor, peptide	Phase Ib	S.C. administration in patients with mild-moderate plaque psoriasis. Peptide was well-tolerated and improved global assessment of psoriatic skin lesions	Tarcha et al. (2017)

(continued)

Table 2. (continued)

Target/drug	Type	Clinical trial phase	Notes	Reference
$K_{Ca}3.1$				
Senicapoc (Icagen Inc)	$K_{Ca}3.1$ blocker	Phase II	Oral administration. Reduced late phase increase in airway resistance in response to allergen challenge in allergic asthma patients. Well-tolerated	Wulff and Castle (2010)
		Phase II	Oral administration. No improvement in lung function in patients with exercise-induced asthma	Wulff and Castle (2010)
		Phase II	Randomised open-label phase II trial. Enteral administration to COVID-19 patients with severe respiratory insufficiency. 72-h PaO_2/FiO_2 was reduced compared to control standard therapy group, but 28-day mortality rate was not significantly different between groups	Granfeldt et al. (2022)
TRPV1				
Asivatrep (PAC-14028) (Amorepacific)	TRPV1 antagonist	Phase IIb-III	Topically applied randomised, double-blind, placebo-controlled trial in atopic dermatitis patients. Significant improvement in investigators global assessment score compared to control. Well-tolerated	Lee et al. (2019), Park et al. (2022)

Clinical trials of drug in inflammatory, autoimmune disease and cancer where mechanism of action may involve ion channel blockade on the immune cells. Clinical trials targeting pain not included. Abbreviations: *CRAC* Ca^{2+} release-activated Ca^{2+} channel, K_v voltage-gated K^+ channel, K_{Ca} Ca^{2+}-dependent K^+ channel, *TRP* transient receptor potential channel

10 Conclusion

Compared to many other cell types, further research is required to investigate the role of ion channels in immune cells under physiological conditions, immune activation, and pathological conditions. The linking of cellular ion channel expression/activity profiles with disease-causing immune cell phenotypes represents a particularly interesting area from which to develop new therapies for inflammation, autoimmunity, and immunotherapy for cancer. Luckily for the immunologists, our colleagues in the fields of electrophysiology, neuroscience, and cardiovascular biology have provided us with a wealth of fundamental data on ion channel properties and robust experimental techniques and tools with which to study them. We believe it would be a missed opportunity not to use them.

References

Acharya TK, Sahu RP, Kumar S, Kumar S, Rokade TP, Chakraborty R, Dubey NK, Shikha D, Chawla S, Goswami C. Function and regulation of thermosensitive ion channel TRPV4 in the immune system. Curr Top Membr. 2022;89:155–88.

Ahmadzadeh K, Vanoppen M, Rose CD, Matthys P, Wouters CH. Multinucleated Giant cells: current insights in phenotype, biological activities, and mechanism of formation. Front Cell Dev Biol. 2022;10.

Aifantis I, Gounari F, Scorrano L, Borowski C, von Boehmer H. Constitutive pre-TCR signaling promotes differentiation through Ca2+ mobilization and activation of NF-kappaB and NFAT. Nat Immunol. 2001;2(5):403–9.

Amin K, Janson C, Bystrom J. Role of eosinophil granulocytes in allergic airway inflammation endotypes. Scand J Immunol. 2016;84(2):75–85.

Andoh M, Koyama R. Microglia and GABA: diverse functions of microglia beyond GABA-receiving cells. Neurosci Res. 2023;187:52–7.

Arya RK, Goswami R, Rahaman SO. Mechanotransduction via a TRPV4-Rac1 signaling axis plays a role in multinucleated giant cell formation. J Biol Chem. 2021;296.

Ashmole I, Duffy SM, Leyland ML, Morrison VS, Begg M, Bradding P. CRACM/Orai ion channel expression and function in human lung mast cells. J Allergy Clin Immunol. 2012;129(6):1628–35.e2.

Atcha H, Jairaman A, Holt JR, Meli VS, Nagalla RR, Veerasubramanian PK, Brumm KT, Lim HE, Othy S, Cahalan MD, Pathak MM, Liu WF. Mechanically activated ion channel Piezo1 modulates macrophage polarization and stiffness sensing. Nat Commun. 2021;12(1).

Atsumi Y, Toriyama M, Kato H, Nakamura M, Morita A, Takaishi M, Saito K, Tanaka M, Okada F, Tominaga M, Ishii KJ, Fujita F. Anti-inflammatory role of TRPV4 in human macrophages. ImmunoHorizons. 2023;7(1):81–96.

Azam P, Sankaranarayanan A, Homerick D, Griffey S, Wulff H. Targeting effector memory T cells with the small molecule Kv1.3 blocker PAP-1 suppresses allergic contact dermatitis. J Invest Dermatol. 2007;127(6):1419–29.

Barbet G, Demion M, Moura IC, Serafini N, Léger T, Vrtovsnik F, Monteiro RC, Guinamard R, Kinet JP, Launay P. The calcium-activated nonselective cation channel TRPM4 is essential for the migration but not the maturation of dendritic cells. Nat Immunol. 2008;9(10):1148–56.

Barde PJ, Viswanadha S, Veeraraghavan S, Vakkalanka SV, Nair A. A first-in-human study to evaluate the safety, tolerability and pharmacokinetics of RP3128, an oral calcium release-activated calcium (CRAC) channel modulator in healthy volunteers. J Clin Pharm Ther. 2021;46(3):677–87.

Beeton C, Wulff H, Standifer NE, Azam P, Mullen KM, Pennington MW, Kolski-Andreaco A, Wei E, Grino A, Counts DR, Wang PH, LeeHealey CJ, Andrews BS, Sankaranarayanan A, Homerick D, Roeck WW, Tehranzadeh J, Stanhope KL, Zimin P, et al. Kv1.3 channels are a therapeutic target for T cell-mediated autoimmune diseases. Proc Natl Acad Sci U S A. 2006;103(46):17414–9.

Bhakta NR, Oh DY, Lewis RS. Calcium oscillations regulate thymocyte motility during positive selection in the three-dimensional thymic environment. Nat Immunol. 2005;6(2):143–51.

Bittner S, Bobak N, Herrmann AM, Göbel K, Meuth P, Höhn KG, Stenner MP, Budde T, Wiendl H, Meuth SG. Upregulation of K2P5.1 potassium channels in multiple sclerosis. Ann Neurol. 2010;68(1):58–69.

Blériot C, Chakarov S, Ginhoux F. Determinants of resident tissue macrophage identity and function. Immunity. 2020;52(6):957–70.

Bobak N, Bittner S, Andronic J, Hartmann S, Mühlpfordt F, Schneider-Hohendorf T, Wolf K, Schmelter C, Göbel K, Meuth P, Zimmermann H, Döring F, Wischmeyer E, Budde T, Wiendl H, Meuth SG, Sukhorukov VL. Volume regulation of murine T lymphocytes relies on voltage-dependent and two-pore domain potassium channels. Biochim Biophys Acta. 2011;1808(8):2036–44.

Borst K, Dumas AA, Prinz M. Microglia: immune and non-immune functions. Immunity. 2021;54(10):2194–208.

Bréchard S, Melchior C, Plançon S, Schenten V, Tschirhart EJ. Store-operated Ca2+ channels formed by TRPC1, TRPC6 and Orai1 and non-store-operated channels formed by TRPC3 are involved in the regulation of NADPH oxidase in HL-60 granulocytes. Cell Calcium. 2008;44(5):492–506.

Bruen C, Miller J, Wilburn J, Mackey C, Bollen TL, Stauderman K, Hebbar S. Auxora for the treatment of patients with acute pancreatitis and accompanying systemic inflammatory response syndrome: clinical development of a calcium release-activated calcium channel inhibitor. Pancreas. 2021;50(4):537–43.

Bruen C, Al-Saadi M, Michelson EA, Tanios M, Mendoza-Ayala R, Miller J, Zhang J, Stauderman K, Hebbar S, Hou PC. Auxora vs. placebo for the treatment of patients with severe COVID-19 pneumonia: a randomized-controlled clinical trial. Crit Care. 2022;26(1):1–11.

Bye LJ, Finol-Urdaneta RK, Tae HS, Adams DJ. Nicotinic acetylcholine receptors: key targets for attenuating neurodegenerative diseases. Int J Biochem Cell Biol. 2023;157.

Cabral MD, Paulet PE, Robert V, Gomes B, Renoud ML, Savignac M, Leclerc C, Moreau M, Lair D, Langelot M, Magnan A, Yssel H, Mariamé B, Guéry JC, Pelletier L. Knocking down Cav1 calcium channels implicated in Th2 cell activation prevents experimental asthma. Am J Respir Crit Care Med. 2010;181(12):1310–7.

Cai G, Lu Y, Zhong W, Wang T, Li Y, Ruan X, Chen H, Sun L, Guan Z, Li G, Zhang H, Sun W, Chen M, Zhang W, Wang H. Piezo1-mediated M2 macrophage mechanotransduction enhances bone formation through secretion and activation of transforming growth factor-β1. Cell Prolif. 2023.

Calovi S, Mut-Arbona P, Sperlágh B. Microglia and the purinergic signaling system. Neuroscience. 2019;405:137–47.

Capasso M, Bhamrah MK, Henley T, Boyd RS, Langlais C, Cain K, Dinsdale D, Pulford K, Khan M, Musset B, Cherny VV, Morgan D, Gascoyne RD, Vigorito E, Decoursey TE, MacLennan ICM, Dyer MJS. HVCN1 modulates BCR signal strength via regulation of BCR-dependent generation of reactive oxygen species. Nat Immunol. 2010;11(3):265–72.

Chaki S, Alkanfari I, Roy S, Amponnawarat A, Hui Y, Oskeritzian CA, Ali H. Inhibition of orai channel function regulates mas-related G protein-coupled receptor-mediated responses in mast cells. Front Immunol. 2022;12:1.

Chauhan A, Sun Y, Sukumaran P, Quenum Zangbede FO, Jondle CN, Sharma A, Evans DL, Chauhan P, Szlabick RE, Aaland MO, Birnbaumer L, Sharma J, Singh BB, Mishra BB. M1 macrophage polarization is dependent on TRPC1-mediated calcium entry. IScience. 2018;8:85–102.

Chen K, Man Q, Miao J, Xu W, Zheng Y, Zhou X, Gao Z. Kir2.1 channel regulates macrophage polarization via the Ca2+/CaMK II/ERK/NF-κB signaling pathway. J Cell Sci. 2022a;135(13).

Chen S, Cui W, Chi Z, Xiao Q, Hu T, Ye Q, Zhu K, Yu W, Wang Z, Yu C, Pan X, Dai S, Yang Q, Jin J, Zhang J, Li M, Yang D, Yu Q, Wang Q, et al. Tumor-associated macrophages are shaped by intratumoral high potassium via Kir2.1. Cell Metab. 2022b;34(11):1843–59.e11.

Chen X, Lu W, Lu C, Zhang L, Xu F, Dong H. The CaSR/TRPV4 coupling mediates pro-inflammatory macrophage function. Acta Physiol (Oxf). 2023;237(4).

Cocozza G, Garofalo S, Capitani R, D'alessandro G, Limatola C. Microglial potassium channels: from homeostasis to neurodegeneration. Biomol Ther. 2021;11(12).

Colonna M, Butovsky O. Microglia function in the central nervous system during health and neurodegeneration. Annu Rev Immunol. 2017;35:441–68.

Concepcion AR, Wagner LE, Zhu J, Tao AY, Yang J, Khodadadi-Jamayran A, Wang YH, Liu M, Rose RE, Jones DR, Coetzee WA, Yule DI, Feske S. The volume-regulated anion channel LRRC8C suppresses T cell function by regulating cyclic dinucleotide transport and STING–p53 signaling. Nat Immunol. 2022;23(2):287–302.

Crottès D, Félix R, Meley D, Chadet S, Herr F, Audiger C, Soriani O, Vandier C, Roger S, Angoulvant D, Velge-Roussel F. Immature human dendritic cells enhance their migration through KCa3.1 channel activation. Cell Calcium. 2016;59(4):198–207.

Czapski GA, Strosznajder JB. Glutamate and GABA in microglia-neuron cross-talk in Alzheimer's disease. Int J Mol Sci. 2021;22(21).
Da Fonseca ACC, Amaral R, Garcia C, Geraldo LH, Matias D, Lima FRS. Microglia in cancer: for good or for bad? Adv Exp Med Biol. 2016;949:245–61.
Damann N, Owsianik G, Li S, Poll C, Nilius B. The calcium-conducting ion channel transient receptor potential canonical 6 is involved in macrophage inflammatory protein-2-induced migration of mouse neutrophils. Acta Physiol (Oxf). 2009;195(1):3–11.
Davies LC, Jenkins SJ, Allen JE, Taylor PR. Tissue-resident macrophages. Nat Immunol. 2013;14(10):986–95.
de Greef PC, Oakes T, Gerritsen B, Ismail M, Heather JM, Hermsen R, Chain B, de Boer RJ. The naive t-cell receptor repertoire has an extremely broad distribution of clone sizes. elife. 2020;9.
Delves PJ, Martin SJ, Burton DR, Roitt IM. Roitt's Essential Immunology. 13th ed. Chichester: John Wiley & Sons; 2017. p. 576.
Dermitzakis I, Theotokis P, Evangelidis P, Delilampou E, Evangelidis N, Chatzisavvidou A, Avramidou E, Manthou ME. CNS border-associated macrophages: ontogeny and potential implication in disease. Curr Issues Mol Biol. 2023;45(5):4285.
di Capite J, Nelson C, Bates G, Parekh AB. Targeting Ca2+ release-activated Ca2+ channel channels and leukotriene receptors provides a novel combination strategy for treating nasal polyposis. J Allergy Clin Immunol. 2009;124(5).
Di L, Srivastava S, Zhdanova O, Ding Y, Li Z, Wulff H, Lafaille M, Skolnik EY. Inhibition of the K+ channel KCa3.1 ameliorates T cell-mediated colitis. Proc Natl Acad Sci U S A. 2010;107(4):1541–6.
Di A, Xiong S, Ye Z, Malireddi RKS, Kometani S, Zhong M, Mittal M, Hong Z, Kanneganti TD, Rehman J, Malik AB. The TWIK2 potassium efflux channel in macrophages mediates NLRP3 inflammasome-induced inflammation. Immunity. 2018;49(1):56.
Du H, Bartleson JM, Butenko S, Alonso V, Liu WF, Winer DA, Butte MJ. Tuning immunity through tissue mechanotransduction. Nat Rev Immunol. 2022;23(3):174–88.
Dutta B, Goswami R, Rahaman SO. TRPV4 plays a role in matrix stiffness-induced macrophage polarization. Front Immunol. 2020;11.
el Chemaly A, Okochi Y, Sasaki M, Arnaudeau S, Okamura Y, Demaurex N. VSOP/Hv1 proton channels sustain calcium entry, neutrophil migration, and superoxide production by limiting cell depolarization and acidification. J Exp Med. 2010;207(1):129–39.
Erdogmus S, Concepcion AR, Yamashita M, Sidhu I, Tao AY, Li W, Rocha PP, Huang B, Garippa R, Lee B, Lee A, Hell JW, Lewis RS, Prakriya M, Feske S. Cavβ1 regulates T cell expansion and apoptosis independently of voltage-gated Ca2+ channel function. Nat Commun. 2022;13(1).
Eser A, Colombel JF, Rutgeerts P, Vermeire S, Vogelsang H, Braddock M, Persson T, Reinisch W. Safety and efficacy of an oral inhibitor of the purinergic receptor P2X7 in adult patients with moderately to severely active Crohn's disease A randomized placebo-controlled, double-blind, phase IIa study. Inflamm Bowel Dis. 2015;21(10):2247–53.
Félix R, Crottès D, Delalande A, Fauconnier J, Lebranchu Y, le Guennec JY, Velge-Roussel F. The Orai-1 and STIM-1 complex controls human dendritic cell maturation. PLoS One. 2013;8(5):61595.
Femling JK, Cherny VV, Morgan D, Rada B, Davis AP, Czirják G, Enyedi P, England SK, Moreland JG, Ligeti E, Nauseef WM, Decoursey TE. The antibacterial activity of human neutrophils and eosinophils requires proton channels but not BK channels. J Gen Physiol. 2006;127(6):659–72.
Feske S. CRAC channelopathies. Pflug Arch Eur J. 2010;460(2):417–35.
Feske S, Giltnane J, Dolmetsch R, Staudt LM, Rao A. Gene regulation mediated by calcium signals in T lymphocytes. Nat Immunol. 2001;2(4):316–24.
Feske S, Wulff H, Skolnik EY. Ion channels in innate and adaptive immunity. Annu Rev Immunol. 2015;33:291.
Freedman BD, Liu QH, Gaulton G, Kotlikoff MI, Hescheler J, Fleischmann BK. ATP-evoked Ca 2+ transients and currents in murine thymocytes: possible role for P2X receptors in death by neglect. Eur J Immunol. 1999;29:5.

Fujii T, Mashimo M, Moriwaki Y, Misawa H, Ono S, Horiguchi K, Kawashima K. Expression and function of the cholinergic system in immune cells. Front Immunol. 2017;8.
Furtner T, Zierler S, Kerschbaum HH. Blockade of chloride channels suppresses engulfment of microspheres in the microglial cell line, BV-2. Brain Res. 2007;1184(1):1–9.
Galli SJ, Tsai M. IgE and mast cells in allergic disease. Nat Med. 2012;18(5):693–704.
Geng J, Shi Y, Zhang J, Yang B, Wang P, Yuan W, Zhao H, Li J, Qin F, Hong L, Xie C, Deng X, Sun Y, Wu C, Chen L, Zhou D. TLR4 signalling via Piezo1 engages and enhances the macrophage mediated host response during bacterial infection. Nat Commun. 2021;12(1).
Ghanshani S, Wulff H, Miller MJ, Rohm H, Neben A, Gutman GA, Cahalan MD, Chandy KG. Up-regulation of the IKCa1 potassium channel during T-cell activation: molecular mechanism and functional consequences. J Biol Chem. 2000;275(47):37137–49.
Ghiringhelli F, Apetoh L, Tesniere A, Aymeric L, Ma Y, Ortiz C, Vermaelen K, Panaretakis T, Mignot G, Ullrich E, Perfettini JL, Schlemmer F, Tasdemir E, Uhl M, Génin P, Civas A, Ryffel B, Kanellopoulos J, Tschopp J, et al. Activation of the NLRP3 inflammasome in dendritic cells induces IL-1β–dependent adaptive immunity against tumors. Nat Med. 2009;15(10):1170–8.
Girodet PO, Ozier A, Carvalho G, Ilina O, Ousova O, Gadeau AP, Begueret H, Wulff H, Marthan R, Bradding P, Berger P. Ca(2+)-activated K(+) channel-3.1 blocker TRAM-34 attenuates airway remodeling and eosinophilia in a murine asthma model. Am J Respir Cell Mol Biol. 2013;48(2):212–9.
Goswami R, Arya RK, Biswas D, Zhu X, Rahaman SO. Transient receptor potential vanilloid 4 is required for foreign body response and Giant cell formation. Am J Pathol. 2019;189(8):1505–12.
Goswami R, Arya RK, Sharma S, Dutta B, Stamov DR, Zhu X, Rahaman SO. Mechanosensing by TRPV4 mediates stiffness-induced foreign body response and giant cell formation. Sci Signal. 2021;14(707).
Granfeldt A, Andersen LW, Vallentin MF, Hilberg O, Hasselstrøm JB, Sørensen LK, Mogensen S, Christensen S, Grejs AM, Rasmussen BS, Kristiansen KT, Strøm T, Johansen IS, Schjørring OL, Simonsen U. Senicapoc treatment in COVID-19 patients with severe respiratory insufficiency-A randomized, open-label, phase II trial. Acta Anaesthesiol Scand. 2022;66(7):838–46.
Greenberg ML, Yu Y, Leverrier S, Zhang SL, Parker I, Cahalan MD. Orai1 function is essential for T cell homing to lymph nodes. J Immunol. 2013;190(7):3197–206.
Gu Y, Gu C. Physiological and pathological functions of mechanosensitive ion channels. Mol Neurobiol. 2014;50(2):339–47.
Guilliams M, Svedberg FR. Does tissue imprinting restrict macrophage plasticity? Nat Immunol. 2021;22(2):118–27.
Harl B, Schmölzer J, Jakab M, Ritter M, Kerschbaum HH. Chloride channel blockers suppress formation of engulfment pseudopodia in microglial cells. Cell Physiol Biochem. 2013;31(2–3):319–37.
Hasheminasab SS, Conejeros I, Velásquez ZD, Borggrefe T, Gärtner U, Kamena F, Taubert A, Hermosilla C. ATP purinergic receptor P2X1-dependent suicidal NETosis induced by Cryptosporidium parvum under physioxia conditions. Biology. 2022;11(3).
He Y, Deng B, Liu S, Luo S, Ning Y, Pan X, Wan R, Chen Y, Zhang Z, Jiang J, Xu H, Xia M, Li J. Myeloid Piezo1 deletion protects renal fibrosis by restraining macrophage infiltration and activation. Hypertension. 2022;79(5):918–31.
Henríquez C, Riquelme TT, Vera D, Julio-Kalajzić F, Ehrenfeld P, Melvin JE, Figueroa CD, Sarmiento J, Flores CA. The calcium-activated potassium channel KCa3.1 plays a central role in the chemotactic response of mammalian neutrophils. Acta Physiol (Oxf). 2016;216(1):132.
Hickman S, Izzy S, Sen P, Morsett L, El Khoury J. Microglia in neurodegeneration. Nat Neurosci. 2018;21(10):1359–69.
Hu L, Gocke AR, Knapp E, Rosenzweig JM, Grishkan IV, Baxi EG, Zhang H, Margolick JB, Whartenby KA, Calabresi PA. Functional blockade of the voltage-gated potassium channel Kv1.3 mediates reversion of T effector to central memory lymphocytes through SMAD3/p21 cip1 signaling. J Biol Chem. 2012;287(2):1261–8.
Huse M. Mechanical forces in the immune system. Nat Rev Immunol. 2017;17(11):679–90.

Immler R, Simon SI, Sperandio M. Calcium signalling and related ion channels in neutrophil recruitment and function. Eur J Clin Investig. 2018;48:e12964.

Immler R, Nadolni W, Bertsch A, Morikis V, Rohwedder I, Masgrau-Alsina S, Schroll T, Yevtushenko A, Soehnlein O, Moser M, Gudermann T, Barnea ER, Rehberg M, Simon SI, Zierler S, Pruenster M, Sperandio M. The voltage-gated potassium channel KV1.3 regulates neutrophil recruitment during inflammation. Cardiovasc Res. 2022;118(5):1289–302.

Izquierdo P, Attwell D, Madry C. Ion channels and receptors as determinants of microglial function. Trends Neurosci. 2019;42(4):278–92.

Izquierdo P, Shiina H, Hirunpattarasilp C, Gillis G, Attwell D. Synapse development is regulated by microglial THIK-1 K+ channels. Proc Natl Acad Sci U S A. 2021;118(42):e2106294118.

Jairaman A, Othy S, Dynes JL, Yeromin AV, Zavala A, Greenberg ML, Nourse JL, Holt JR, Cahalan SM, Marangoni F, Parker I, Pathak MM, Cahalan MD. Piezo1 channels restrain regulatory T cells but are dispensable for effector CD4+ T cell responses. Sci Adv. 2021;7(28).

Jha MK, Badou A, Meissner M, McRory JE, Freichel M, Flockerzi V, Flavell RA. Defective survival of naive CD8+ T lymphocytes in the absence of the beta3 regulatory subunit of voltage-gated calcium channels. Nat Immunol. 2009;10(12):1275–82.

Jin J, Desai BN, Navarro B, Donovan A, Andrews NC, Clapham DE. Deletion of Trpm7 disrupts embryonic development and thymopoiesis without altering Mg2+ homeostasis. Science. 2008;322(5902):756–60.

Kan XH, Gao HQ, Ma ZY, Liu L, Ling MY, Wang YY. Kv1.3 potassium channel mediates macrophage migration in atherosclerosis by regulating ERK activity. Arch Biochem Biophys. 2016;591:150–6.

Karmakar M, Katsnelson MA, Dubyak GR, Pearlman E. Neutrophil P2X7 receptors mediate NLRP3 inflammasome-dependent IL-1β secretion in response to ATP. Nat Commun. 2016;7.

Kefauver JM, Ward AB, Patapoutian A. Discoveries in structure and physiology of mechanically activated ion channels. Nature. 2020;587(7835):567–76.

Keystone EC, Wang MM, Layton M, Hollis S, McInnes IB. Clinical evaluation of the efficacy of the P2X7 purinergic receptor antagonist AZD9056 on the signs and symptoms of rheumatoid arthritis in patients with active disease despite treatment with methotrexate or sulphasalazine. Ann Rheum Dis. 2012;71(10):1630–5.

Khedkar NR, Irlapatti NR, Dadke D, Kanoje V, Shaikh Z, Karche V, Shinde V, Deshmukh G, Patil A, Jachak S, Phukan S, Kizhakinagath PA, Gholve M, Bhankhede T, Daler J, Nemade HN, Budhe S, Pareek H, Yeshodharan R, et al. Discovery of a novel potent and selective calcium release-activated calcium channel inhibitor: 2,6-difluoro- N-(2′-methyl-3′-(4-methyl-5-oxo-4,5-dihydro-1,3,4-oxadiazol-2-yl)-[1,1′-biphenyl]-4-yl)benzamide. Structure-activity relationship and preclinical characterization. J Med Chem. 2021;64(23):17004–30.

Kis-Toth K, Hajdu P, Bacskai I, Szilagyi O, Papp F, Szanto A, Posta E, Gogolak P, Panyi G, Rajnavolgyi E. Voltage-gated sodium channel Nav1.7 maintains the membrane potential and regulates the activation and chemokine-induced migration of a monocyte-derived dendritic cell subset. J Immunol. 2011;187(3):1273–80.

Koshy S, Wu D, Hu X, Tajhya RB, Huq R, Khan FS, Pennington MW, Wulff H, Yotnda P, Beeton C. Blocking KCa3.1 channels increases tumor cell killing by a subpopulation of human natural killer lymphocytes. PLoS One. 2013;8(10).

Kulkarni OP, Lichtnekert J, Anders HJ, Mulay SR. The immune system in tissue environments regaining homeostasis after injury: is "inflammation" always inflammation? Mediat Inflamm. 2016;2016.

Kumar L, Chou J, Yee CSK, Borzutzky A, Vollmann EH, von Andrian UH, Park SY, Hollander G, Manis JP, Luigi Poliani P, Geha RS. Leucine-rich repeat containing 8A (LRRC8A) is essential for T lymphocyte development and function. J Exp Med. 2014;211(5):929–42.

Kurashima Y, Amiya T, Nochi T, Fujisawa K, Haraguchi T, Iba H, Tsutsui H, Sato S, Nakajima S, Iijima H, Kubo M, Kunisawa J, Kiyono H. Extracellular ATP mediates mast cell-dependent intestinal inflammation through P2X7 purinoceptors. Nat Commun. 2012;3.

Lecut C, Frederix K, Johnson DM, Deroanne C, Thiry M, Faccinetto C, Marée R, Evans RJ, Volders PGA, Bours V, Oury C. P2X1 ion channels promote neutrophil chemotaxis through Rho kinase activation. J Immunol. 2009;183(4):2801–9.

Lecut C, Faccinetto C, Delierneux C, van Oerle R, Spronk HMH, Evans RJ, el Benna J, Bours V, Oury C. ATP-gated P2X 1 ion channels protect against endotoxemia by dampening neutrophil activation. J Thromb Haemost. 2012;10(3):453–65.

Lee YW, Won CH, Jung K, Nam HJ, Choi G, Park YH, Park M, Kim B. Efficacy and safety of PAC-14028 cream – a novel, topical, nonsteroidal, selective TRPV1 antagonist in patients with mild-to-moderate atopic dermatitis: a phase IIb randomized trial. Br J Dermatol. 2019;180(5):1030–8.

Lee M, Du H, Winer DA, Clemente-Casares X, Tsai S. Mechanosensing in macrophages and dendritic cells in steady-state and disease. Front Cell Dev Biol. 2022;10.

Leng S, Zhang X, Wang S, Qin J, Liu Q, Liu A, Sheng Z, Feng Q, Hu X, Peng J. Ion channel Piezo1 activation promotes aerobic glycolysis in macrophages. Front Immunol. 2022;13.

Lévêque M, Penna A, Le Trionnaire S, Belleguic C, Desrues B, Brinchault G, Jouneau S, Lagadic-Gossmann D, Martin-Chouly C. Phagocytosis depends on TRPV2-mediated calcium influx and requires TRPV2 in lipids rafts: alteration in macrophages from patients with cystic fibrosis. Sci Rep. 2018;8(1):1–13.

Li C, Levin M, Kaplan DL. Bioelectric modulation of macrophage polarization. Sci Rep. 2016;6.

Lim CG, Jang J, Kim C. Cellular machinery for sensing mechanical force. BMB Rep. 2018;51(12):623–9.

Lindemann O, Umlauf D, Frank S, Schimmelpfennig S, Bertrand J, Pap T, Hanley PJ, Fabian A, Dietrich A, Schwab A. TRPC6 regulates CXCR2-mediated chemotaxis of murine neutrophils. J Immunol. 2013;190(11):5496–505.

Lindemann O, Strodthoff C, Horstmann M, Nielsen N, Jung F, Schimmelpfennig S, Heitzmann M, Schwab A. TRPC1 regulates fMLP-stimulated migration and chemotaxis of neutrophil granulocytes. Biochim Biophys Acta. 2015;1853(9):2122–30.

Link TM, Park U, Vonakis BM, Raben DM, Soloski MJ, Caterina MJ. TRPV2 plays a pivotal role in macrophage particle binding and phagocytosis. Nat Immunol. 2010;11(3):232.

Lioudyno MI, Kozak JA, Penna A, Safrina O, Zhang SL, Sen D, Roos J, Stauderman KA, Cahalan MD. Orai1 and STIM1 move to the immunological synapse and are up-regulated during T cell activation. Proc Natl Acad Sci U S A. 2008;105(6):2011–6.

Liu H, Leak RK, Hu X. Neurotransmitter receptors on microglia. Stroke Vasc. Neurol. 2016;1(2):52.

Liu CSC, Raychaudhuri D, Paul B, Chakrabarty Y, Ghosh AR, Rahaman O, Talukdar A, Ganguly D. Cutting edge: Piezo1 mechanosensors optimize human T cell activation. J Immunol. 2018;200(4):1255–60.

Liu F, Xu L, Nishi M, Ichimura A, Takeshima H. Enhanced Ca2+ handling in thioglycolate-elicited peritoneal macrophages. Cell Calcium. 2021;96.

Lo WL, Donermeyer DL, Allen PM. A voltage-gated sodium channel is essential for the positive selection of CD4+ T cells. Nat Immunol. 2012;13(9):880.

Locati M, Curtale G, Mantovani A. Diversity, mechanisms, and significance of macrophage plasticity. Annu Rev Pathol. 2020;15:123–47.

Luo L, Song S, Ezenwukwa CC, Jalali S, Sun B, Sun D. Ion channels and transporters in microglial function in physiology and brain diseases. Neurochem Int. 2021;142.

Ma J, McCarl CA, Khalil S, Lüthy K, Feske S. T-cell-specific deletion of STIM1 and STIM2 protects mice from EAE by impairing the effector functions of Th1 and Th17 cells. Eur J Immunol. 2010;40(11):3028–42.

Ma S, Dubin AE, Zhang Y, Mousavi SAR, Wang Y, Coombs AM, Loud M, Andolfo I, Patapoutian A. A role of PIEZO1 in iron metabolism in mice and humans. Cell. 2021;184(4):969.

Macian F. NFAT proteins: key regulators of T-cell development and function. Nat Rev Immunol. 2005;5(6):472–84.

Madry C, Kyrargyri V, Arancibia-Cárcamo IL, Jolivet R, Kohsaka S, Bryan RM, Attwell D. Microglial ramification, surveillance, and interleukin-1β release are regulated by the two-pore domain K+ channel THIK-1. Neuron. 2018;97(2):299–312.e6.

Maître B, Magnenat S, Heim V, Ravanat C, Evans RJ, de la Salle H, Gachet C, Hechler B. The P2X1 receptor is required for neutrophil extravasation during lipopolysaccharide-induced lethal endotoxemia in mice. J Immunol. 2015;194(2):739–49.

Masuda T, Sankowski R, Staszewski O, Prinz M. Microglia heterogeneity in the single-cell era. Cell Rep. 2020;30(5):1271–81.

Matsumoto M, Fujii Y, Baba A, Hikida M, Kurosaki T, Baba Y. The calcium sensors STIM1 and STIM2 control B cell regulatory function through interleukin-10 production. Immunity. 2011;34(5):703–14.

Maul-Pavicic A, Chiang SCC, Rensing-Ehl A, Jessen B, Fauriat C, Wood SM, Sjöqvist S, Hufnagel M, Schulze I, Bass T, Schamel WW, Fuchs S, Pircher H, McCarl CA, Mikoshiba K, Schwarz K, Feske S, Bryceson YT, Ehl S. ORAI1-mediated calcium influx is required for human cytotoxic lymphocyte degranulation and target cell lysis. Proc Natl Acad Sci U S A. 2011;108(8):3324–9.

McCarl CA, Picard C, Khalil S, Kawasaki T, Röther J, Papolos A, Kutok J, Hivroz C, LeDeist F, Plogmann K, Ehl S, Notheis G, Albert MH, Belohradsky BH, Kirschner J, Rao A, Fischer A, Feske S. ORAI1 deficiency and lack of store-operated Ca2+ entry cause immunodeficiency, myopathy, and ectodermal dysplasia. J Allergy Clin Immunol. 2009;124(6).

Melzer N, Hicking G, Göbel K, Wiendl H. TRPM2 cation channels modulate T cell effector functions and contribute to autoimmune CNS inflammation. PLoS One. 2012;7(10).

Mendu SK, Stremska ME, Schappe MS, Moser EK, Krupa JK, Rogers JS, Stipes EJ, Parker CA, Braciale TJ, Perry JSA, Desai BN. Targeting the ion channel TRPM7 promotes the thymic development of regulatory T cells by promoting IL-2 signaling. Sci Signal. 2020;13(661).

Meuth SG, Bittner S, Meuth P, Simon OJ, Budde T, Wiendl H. TWIK-related acid-sensitive K+ channel 1 (TASK1) and TASK3 critically influence T lymphocyte effector functions. J Biol Chem. 2008;283(21):14559–70.

Michalick L, Kuebler WM. TRPV4—a missing link between mechanosensation and immunity. Front Immunol. 2020;11:413.

Miller J, Bruen C, Schnaus M, Zhang J, Ali S, Lind A, Stoecker Z, Stauderman K, Hebbar S. Auxora versus standard of care for the treatment of severe or critical COVID-19 pneumonia: results from a randomized controlled trial. Crit Care. 2020;24(1).

Mizoguchi Y, Monji A. Microglial intracellular Ca2+ signaling in synaptic development and its alterations in neurodevelopmental disorders. Front Cell Neurosci. 2017;11:69.

Mizoguchi Y, Kato TA, Seki Y, Ohgidani M, Sagata N, Horikawa H, Yamauchi Y, Sato-Kasai M, Hayakawa K, Inoue R, et al. Brain-derived neurotrophic factor (BDNF) induces sustained intracellular Ca2+ elevation through the up-regulation of surface transient receptor potential 3 (TRPC3) channels in rodent microglia. J Biol Chem. 2014;289(26):18549.

Moccia F, Brunetti V, Perna A, Guerra G, Soda T, Berra-Romani R. The molecular heterogeneity of store-operated Ca2+ entry in vascular endothelial cells: the different roles of Orai1 and TRPC1/TRPC4 channels in the transition from Ca2+-selective to non-selective cation currents. Int J Mol Sci. 2023;24(4).

Morad H, Luqman S, Tan CH, Swann V, McNaughton PA. TRPM2 ion channels steer neutrophils towards a source of hydrogen peroxide. Sci Rep. 2021;11(1).

Mullen KM, Rozycka M, Rus H, Hu L, Cudrici C, Zafranskaia E, Pennington MW, Johns DC, Judge SIV, Calabresi PA. Potassium channels Kv1.3 and Kv1.5 are expressed on blood-derived dendritic cells in the central nervous system. Ann Neurol. 2006;60(1):118–27.

Mundt S, Greter M, Becher B. The CNS mononuclear phagocyte system in health and disease. Neuron. 2022;110(21):3497–512.

Nakayama T, Ueda Y, Yamada H, Shores EW, Singer A, June CH. In vivo calcium elevations in thymocytes with T cell receptors that are specific for self ligands. Science. 1992;257(5066):96–9.

Nascimento Da Conceicao V, Sun Y, Ramachandran K, Chauhan A, Raveendran A, Venkatesan M, DeKumar B, Maity S, Vishnu N, Kotsakis GA, Worley PF, Gill DL, Mishra BB, Madesh M,

Singh BB. Resolving macrophage polarization through distinct Ca2+ entry channel that maintains intracellular signaling and mitochondrial bioenergetics. IScience. 2021;24(11).

Negulescu PA, Krasieva TB, Khan A, Kerschbaum HH, Cahalan MD. Polarity of T cell shape, motility, and sensitivity to antigen. Immunity. 1996;4(5):421–30.

Ochocka N, Kaminska B. Microglia diversity in healthy and diseased brain: insights from single-cell omics. Int J Mol Sci. 2021;22(6):1–26.

Oh-hora M, Yamashita M, Hogan PG, Sharma S, Lamperti E, Chung W, Prakriya M, Feske S, Rao A. Dual functions for the endoplasmic reticulum calcium sensors STIM1 and STIM2 in T cell activation and tolerance. Nat Immunol. 2008;9(4):432.

Okochi Y, Aratani Y, Adissu HA, Miyawaki N, Sasaki M, Suzuki K, Okamura Y. The voltage-gated proton channel Hv1/VSOP inhibits neutrophil granule release. J Leukoc Biol. 2016;99(1):7–19.

Omilusik K, Priatel JJ, Chen X, Wang YT, Xu H, Choi KB, Gopaul R, McIntyre-Smith A, Teh HS, Tan R, Bech-Hansen NT, Waterfield D, Fedida D, Hunt SV, Jefferies WA. The CaV1.4 calcium channel is a critical regulator of T cell receptor signaling and naive T cell homeostasis. Immunity. 2011;35(3):349–60.

Orihara K, Odemuyiwa SO, Stefura WP, Ilarraza R, HayGlass KT, Moqbel R. Neurotransmitter signalling via NMDA receptors leads to decreased T helper type 1-like and enhanced T helper type 2-like immune balance in humans. Immunology. 2018;153(3):368–79.

Orsini EM, Perelas A, Southern BD, Grove LM, Olman MA, Scheraga RG. Stretching the function of innate immune cells. Front Immunol. 2021;12:767319–9.

Ossola B, Rifat A, Rowland A, Hunter H, Drinkall S, Bender C, Hamlischer M, Teall M, Burley R, Barker DF, et al. Characterisation of C101248: a novel selective THIK-1 channel inhibitor for the modulation of microglial NLRP3-inflammasome. Neuropharmacology. 2023;224.

Ouyang K, Leandro Gomez-Amaro R, Stachura DL, Tang H, Peng X, Fang X, Traver D, Evans SM, Chen J. Loss of IP3R-dependent Ca2+ signalling in thymocytes leads to aberrant development and acute lymphoblastic leukemia. Nat Commun. 2014;5:4814.

Pagán AJ, Ramakrishnan L. The formation and function of granulomas. Annu Rev Immunol. 2018;36:639–65.

Pan Z, Yang H, Reinach PS. Transient receptor potential (TRP) gene superfamily encoding cation channels. Hum Genomics. 2011;5(2):108.

Panyi G, Beeton C, Felipe A. Ion channels and anti-cancer immunity. Philos Trans R Soc B Biol Sci. 2014;369(1638).

Park CW, Kim BJ, Lee YW, Won C, Park CO, Chung BY, Lee DH, Jung K, Nam HJ, Choi G, Park YH, Kim KH, Park M. Asivatrep, a TRPV1 antagonist, for the topical treatment of atopic dermatitis: phase 3, randomized, vehicle-controlled study (CAPTAIN-AD). J Allergy Clin Immunol. 2022;149(4):1340–47.e4.

Piovesana R, Intriago MSS, Dini L, Tata AM. Cholinergic modulation of neuroinflammation: focus on α7 nicotinic receptor. Int J Mol Sci. 2021;22(9).

Poggi A, Zocchi MR. HIV-1 tat triggers TGF-β production and NK cell apoptosis that is prevented by pertussis toxin B. Clin Dev Immunol. 2006;13(2–4):369–72.

Prinz M, Jung S, Priller J. Microglia biology: one century of evolving concepts. Cell. 2019;179(2):292–311.

Radermecker C, Louis R, Bureau F, Marichal T. Role of neutrophils in allergic asthma. Curr Opin Immunol. 2018;54:28–34.

Rah SY, Kwak JY, Chung YJ, Kim UH. ADP-ribose/TRPM2-mediated Ca2+ signaling is essential for cytolytic degranulation and antitumor activity of natural killer cells. Sci Rep. 2015;5(1):1–10.

Ramirez GA, Coletto LA, Sciorat C, Bozzolo EP, Manunta P, Rovere-Querini P, Manfredi AA. Ion channels and transporters in inflammation: special focus on TRP channels and TRPC6. Cell. 2018;7(7).

Ramsey IS, Ruchti E, Kaczmarek JS, Clapham DE. Hv1 proton channels are required for high-level NADPH oxidase-dependent superoxide production during the phagocyte respiratory burst. Proc Natl Acad Sci U S A. 2009;106(18):7642–7.

Rangaraju S, Chi V, Pennington MW, Chandy KG. Kv1.3 potassium channels as a therapeutic target in multiple sclerosis. Expert Opin Ther Targets. 2009;13(8):909–24.

Riazanski V, Gabdoulkhakova AG, Boynton LS, Eguchi RR, Deriy LV, Hogarth DK, Loaëc N, Oumata N, Galons H, Brown ME, Shevchenko P, Gallan AJ, Yoo SG, Naren AP, Villereal ML, Beacham DW, Bindokas VP, Birnbaumer L, Meijer L, Nelson DJ. TRPC6 channel translocation into phagosomal membrane augments phagosomal function. Proc Natl Acad Sci U S A. 2015;112(47):E6486–95.

Saegusa H, Li X, Wang X, Kayakiri M, Tanabe T. Knockdown of microglial Cav2.2 N-type voltage-dependent Ca2+ channel ameliorates behavioral deficits in a mouse model of Parkinson's disease. FEBS Lett. 2020;594(17):2914–22.

Sahni J, Tamura R, Sweet IR, Scharenberg AM. TRPM7 regulates quiescent/proliferative metabolic transitions in lymphocytes. Cell Cycle. 2010;9(17):3565.

Sarkar S. Microglial ion channels: key players in non-cell autonomous neurodegeneration. Neurobiol Dis. 2022;174.

Sawada A, Takihara Y, Kim JY, Matsuda-Hashii Y, Tokimasa S, Fujisaki H, Kubota K, Endo H, Onodera T, Ohta H, Ozono K, Hara J. A congenital mutation of the novel gene LRRC8 causes agammaglobulinemia in humans. J Clin Invest. 2003;112(11):1707–13.

Schaff UY, Dixit N, Procyk E, Yamayoshi I, Tse T, Simon SI. Orai1 regulates intracellular calcium, arrest, and shape polarization during neutrophil recruitment in shear flow. Blood. 2010;115(3):657–66.

Schenk U, Frascoli M, Proietti M, Geffers R, Traggiai E, Buer J, Ricordi C, Westendorf AM, Grassi F. ATP inhibits the generation and function of regulatory T cells through the activation of purinergic P2X receptors. Sci Signal. 2011;4(162).

Scheraga RG, Abraham S, Niese KA, Southern BD, Grove LM, Hite RD, McDonald C, Hamilton TA, Olman MA. TRPV4 mechanosensitive ion channel regulates lipopolysaccharide-stimulated macrophage phagocytosis. J Immunol. 2015;196(1):428–36.

Schmidt RE, Grimbacher B, Witte T. Autoimmunity and primary immunodeficiency: two sides of the same coin? Nat Rev Rheumatol. 2017;14(1):7–18.

Schmitz C, Perraud AL, Johnson CO, Inabe K, Smith MK, Penner R, Kurosaki T, Fleig A, Scharenberg AM. Regulation of vertebrate cellular Mg2+ homeostasis by TRPM7. Cell. 2003;114(2):191–200.

Schuhmann MK, Stegner D, Berna-Erro A, Bittner S, Braun A, Kleinschnitz C, Stoll G, Wiendl H, Meuth SG, Nieswandt B. Stromal interaction molecules 1 and 2 are key regulators of autoreactive T cell activation in murine autoimmune central nervous system inflammation. J Immunol. 2010;184(3):1536–42.

Schulte-Mecklenbeck A, Bittner S, Ehling P, Döring F, Wischmeyer E, Breuer J, Herrmann AM, Wiendl H, Meuth SG, Gross CC. The two-pore domain K2P channel TASK2 drives human NK-cell proliferation and cytolytic function. Eur J Immunol. 2015;45(9):2602–14.

Selezneva A, Yoshida M, Gibb A, Willis D. Nuclear BK channels regulate CREB phosphorylation in RAW264.7 macrophages. Pharmacol Rep. 2021;73(3):881–90.

Selezneva A, Gibb AJ, Willis D. The nuclear envelope as a regulator of immune cell function. Front Immunol. 2022a;13:2765.

Selezneva A, Gibb AJ, Willis D. The contribution of ion channels to shaping macrophage behaviour. Front Pharmacol. 2022b;13.

Serafini N, Dahdah A, Barbet G, Demion M, Attout T, Gautier G, Arcos-Fajardo M, Souchet H, Jouvin M-H, Vrtovsnik F, Kinet J-P, Benhamou M, Monteiro RC, Launay P. The TRPM4 channel controls monocyte and macrophage, but not neutrophil, function for survival in sepsis. J Immunol. 2012;189(7):3689–99.

Shao Z, Makinde TO, Agrawal DK. Calcium-activated potassium channel KCa3.1 in lung dendritic cell migration. Am J Respir Cell Mol Biol. 2011;45(5):962.

Shapouri-Moghaddam A, Mohammadian S, Vazini H, Taghadosi M, Esmaeili SA, Mardani F, Seifi B, Mohammadi A, Afshari JT, Sahebkar A. Macrophage plasticity, polarization, and function in health and disease. J Cell Physiol. 2018;233(9):6425–40.

Shaw PJ, Weidinger C, Vaeth M, Luethy K, Kaech SM, Feske S. $CD4^+$ and $CD8^+$ T cell-dependent antiviral immunity requires STIM1 and STIM2. J Clin Invest. 2014;124(10):4549–63.

Shawer H, Norman K, Cheng CW, Foster R, Beech DJ, Bailey MA. ORAI1 Ca2+ channel as a therapeutic target in pathological vascular remodelling. Front Cell Dev Biol. 2021;9:653812.

Sogkas G, Vögtle T, Rau E, Gewecke B, Stegner D, Schmidt RE, Nieswandt B, Gessner JE. Orai1 controls C5a-induced neutrophil recruitment in inflammation. Eur J Immunol. 2015;45(7):2143–53.

Solis AG, Bielecki P, Steach HR, Sharma L, Harman CCD, Yun S, de Zoete MR, Warnock JN, To SDF, York AG, Mack M, Schwartz MA, Dela Cruz CS, Palm NW, Jackson R, Flavell RA. Mechanosensation of cyclical force by PIEZO1 is essential for innate immunity. Nature. 2019;573(7772):69–74.

Solovera JJ, Alvarez-Mon M, Casas J, Carballido J, Durantez A. Inhibition of human natural killer (NK) activity by calcium channel modulators and a calmodulin antagonist. J Immunol. 1987;139(3):876–80.

Sparrow EL, James S, Hussain K, Beers SA, Cragg MS, Bogdanov YD. Activation of GABA(A) receptors inhibits T cell proliferation. PLoS One. 2021;16(5):e0251632.

Spix B, Chao YK, Abrahamian C, Chen CC, Grimm C. TRPML cation channels in inflammation and immunity. Front Immunol. 2020;11.

Stauderman KA. CRAC channels as targets for drug discovery and development. Cell Calcium. 2018;74:147–59.

Stebbing MJ, Cottee JM, Rana I. The role of ion channels in microglial activation and proliferation – a complex interplay between ligand-gated ion channels, K+ channels, and intracellular Ca2+. Front Immunol. 2015;6:497.

Steinckwich N, Schenten V, Melchior C, Bréchard S, Tschirhart EJ. An essential role of STIM1, Orai1, and S100A8-A9 proteins for Ca2+ signaling and FcγR-mediated phagosomal oxidative activity. J Immunol. 2011;186(4):2182–91.

Stock TC, Bloom BJ, Wei N, Ishaq S, Park W, Wang X, Gupta P, Mebus CA. Efficacy and safety of CE-224,535, an antagonist of P2X7 receptor, in treatment of patients with rheumatoid arthritis inadequately controlled by methotrexate. J Rheumatol. 2012;39(4):720–7.

Suh B-C, Kim J-S, Namgung U, Ha H, Kim K-T. P2X7 nucleotide receptor mediation of membrane pore formation and superoxide generation in human promyelocytes and neutrophils. J Immunol. 2001;166(11):6754–63.

Sun H, Sun Z, Xu X, Lv Z, Li J, Wu R, Fei Y, Tan G, Liu Z, Liu Y, Shi D. Blocking TRPV4 ameliorates osteoarthritis by inhibiting M1 macrophage polarization via the ROS/NLRP3 signaling pathway. Antioxidants. 2022;11(12).

Szöllosi AG, Oláh A, Tóth IB, Papp F, Czifra G, Panyi G, Bíró T. Transient receptor potential vanilloid-2 mediates the effects of transient heat shock on endocytosis of human monocyte-derived dendritic cells. FEBS Lett. 2013;587(9):1440–5.

Tarcha EJ, Olsen CM, Probst P, Peckham D, Muñoz-Elías EJ, Kruger JG, Iadonato SP. Safety and pharmacodynamics of dalazatide, a Kv1.3 channel inhibitor, in the treatment of plaque psoriasis: a randomized phase 1b trial. PLoS One. 2017;12(7):e0180762–2.

Tóth BI, Benko S, Szöllosi AG, Kovács L, Rajnavölgyi É, Bíró T. Transient receptor potential vanilloid-1 signaling inhibits differentiation and activation of human dendritic cells. FEBS Lett. 2009;583(10):1619–24.

Tschumperlin DJ. Mechanotransduction. Compr Physiol. 2011;1(2):1057–73.

Vaeth M, Zee I, Concepcion AR, Maus M, Shaw P, Portal-Celhay C, Zahra A, Kozhaya L, Weidinger C, Philips J, Unutmaz D, Feske S. Ca2+ signaling but not store-operated Ca2+ entry is required for the function of macrophages and dendritic cells. J Immunol. 2015;195(3):1202–17.

van der Velden J, Sum G, Barker D, Koumoundouros E, Barcham G, Wulff H, Castle N, Bradding P, Snibson K. K(Ca)3.1 channel-blockade attenuates airway pathophysiology in a sheep model of chronic asthma. PLoS One. 2013;8(6).

Vicente R, Escalada A, Villalonga N, Texido L, Roura-Ferrer M, Martín-Satué M, López-Iglesias C, Soler C, Solsona C, Tamkun MM, Felipe A. Association of Kv1.5 and Kv1.3 contributes to the major voltage-dependent K+ channel in macrophages. J Biol Chem. 2006;281(49):37675–85.

Villalonga N, David M, Biejanska J, Vicente R, Comes N, Valenzuela C, Felipe A. Immunomodulation of voltage-dependent K+ channels in macrophages: molecular and biophysical consequences. J Gen Physiol. 2010;135(2):135–47.

Vivier E, Artis D, Colonna M, Diefenbach A, di Santo JP, Eberl G, Koyasu S, Locksley RM, McKenzie ANJ, Mebius RE, Powrie F, Spits H. Innate lymphoid cells: 10 years on. Cell. 2018;174(5):1054–66.

Vukcevic M, Spagnoli GC, Iezzi G, Zorzato F, Treves S. Ryanodine receptor activation by Cav1.2 is involved in dendritic cell major histocompatibility complex class II surface expression. J Biol Chem. 2008;283(50):34913–22.

Wang J, Lu Z-H, Gabius H-J, Rohowsky-Kochan C, Ledeen RW, Wu G. Cross-linking of GM1 ganglioside by galectin-1 mediates regulatory T cell activity involving TRPC5 channel activation: possible role in suppressing experimental autoimmune encephalomyelitis. J Immunol. 2009;182(7):4036–45.

Wang X, Qin W, Xu X, Xiong Y, Zhang Y, Zhang H, Sun B. Endotoxin-induced autocrine ATP signaling inhibits neutrophil chemotaxis through enhancing myosin light chain phosphorylation. Proc Natl Acad Sci U S A. 2017;114(17):4483–8.

Wang X, Saegusa H, Huntula S, Tanabe T. Blockade of microglial Cav1.2 Ca2+ channel exacerbates the symptoms in a Parkinson's disease model. Sci Rep. 2019;9(1).

Wang Y, Yang H, Jia A, Wang Y, Yang Q, Dong Y, Bi Y, Liu G. Dendritic cell Piezo1 integrating mechanical stiffness and inflammatory signals directs the differentiation of TH1 and Treg cells in cancer. BioRxiv. 2022;86.

Wang Y, Groeger S, Yong J, Ruf S. Orthodontic compression enhances macrophage M2 polarization via histone H3 hyperacetylation. Int J Mol Sci. 2023a;24(4).

Wang Y, Wang C, He Q, Chen G, Yu J, Cang J, Zhong M. Inhibition of sphingosine-1-phosphate receptor 3 suppresses ATP-induced NLRP3 inflammasome activation in macrophages via TWIK2-mediated potassium efflux. Front Immunol. 2023b;14.

Weber KS, Hildner K, Murphy KM, Allen PM. Trpm4 differentially regulates Th1 and Th2 function by altering calcium signaling and NFAT localization. J Immunol. 2010;185(5):2836–46.

Weidinger C, Shaw PJ, Feske S. STIM1 and STIM2-mediated Ca2+ influx regulates antitumour immunity by CD8+ T cells. EMBO Mol Med. 2013;5(9):1311–21.

Wilhelm K, Ganesan J, Müller T, Dürr C, Grimm M, Beilhack A, Krempl CD, Sorichter S, Gerlach UV, Jüttner E, Zerweck A, Gätner F, Pellegatti P, di Virgilio F, Ferrari D, Kambham N, Fisch P, Finke J, Idzko M, Zeiser R. Graft-versus-host disease is enhanced by extracellular ATP activating P2X7R. Nat Med. 2010;16(12):1434–9.

Wolf SA, Boddeke HWGM, Kettenmann H. Microglia in physiology and disease. Annu Rev Physiol. 2017;79:619–43.

Wulff H, Castle NA. Therapeutic potential of KCa3.1 blockers: recent advances and promising trends. Expert Rev Clin Pharmacol. 2010;3(3):385–96.

Wulff H, Calabresi PA, Allie R, Yun S, Pennington M, Beeton C, Chandy KG. The voltage-gated Kv1.3 K(+) channel in effector memory T cells as new target for MS. J Clin Invest. 2003;111(11):1703–13.

Wulff H, Knaus H-G, Pennington M, Chandy KG. K+ channel expression during B cell differentiation: implications for immunomodulation and autoimmunity. J Immunol. 2004;173(2):776–86.

Xu J, Núñez G. The NLRP3 inflammasome: activation and regulation. Trends Biochem Sci. 2023;48(4).

Xu R, Li C, Wu Y, Shen L, Ma J, Qian J, Ge J. Role of KCa3.1 channels in macrophage polarization and its relevance in atherosclerotic plaque instability. Arterioscler Thromb Vasc Biol. 2017;37(2):226–36.

Xu Z, Chen ZM, Wu X, Zhang L, Cao Y, Zhou P. Distinct molecular mechanisms underlying potassium efflux for NLRP3 inflammasome activation. Front Immunol. 2020;11.

Yin J, Michalick L, Tang C, Tabuchi A, Goldenberg N, Dan Q, Awwad K, Wang L, Erfinanda L, Nouailles G, Witzenrath M, Vogelzang A, Lv L, Lee WL, Zhang H, Rotstein O, Kapus A, Szaszi K, Fleming I, et al. Role of transient receptor potential vanilloid 4 in neutrophil activation and acute lung injury. Am J Respir Cell Mol Biol. 2016;54(3):370–83.

Zhang L, Cao Y, Zhang X, Gu X, Mao Y, Peng B. The origin and repopulation of microglia. Dev Neurobiol. 2022a;82(1):112–24.

Zhang X, Hou L, Li F, Zhang W, Wu C, Xiang L, Li J, Zhou L, Wang X, Xiang Y, Xiao Y, Li SC, Chen L, Ran Q, Li Z. Piezo1-mediated mechanosensation in bone marrow macrophages promotes vascular niche regeneration after irradiation injury. Theranostics. 2022b;12(4):1621.

Zhang M, Malko P, Jiang LH. Mechanosensitive Piezo1 channel regulation of microglial cell function and implications to neurodegenerative diseases and neuroinflammation. Neural Regen Res. 2023;18(11):2385.

Zhou YQ, Liu DQ, Liu C, Xu AJ, Tian YK, Mei W, Tian XB. Targeting α7 nicotinic acetylcholine receptors for chronic pain. Front Mol Neurosci. 2022;15:970040.

Zumerle S, Calì B, Munari F, Angioni R, Di Virgilio F, Molon B, Viola A. Intercellular calcium signaling induced by ATP potentiates macrophage phagocytosis. Cell Rep. 2019;27(1):1–10.e4.

Zwicker C, Bujko A, Scott CL. Hepatic macrophage responses in inflammation, a function of plasticity, heterogeneity or both? Front Immunol. 2021;12.

Ion Channel Tools and Therapeutics from Venoms and Toxins

Carol M. Trim, Abba E. Leffler, Ziyana Samanani, and Steven A. Trim

Abstract For almost as long as the ion channels in our brains have powered conscious thought, we have been fascinated by venoms and toxins. Initially thought of as magical, now we are understanding their true power and utility in drug discovery. A diverse range of venoms and toxins have been used to decipher the properties of ion channels and as a result hold a key functional place as positive controls in many assays. Such control toxins include tetrodotoxin, protoxin and charybdotoxin that are found in diverse species. The latest advances in computational chemistry and structural biology, such as cryogenic electron microscopy and free energy perturbation, are opening our drug discovery gaze to the atomic interactions of toxins and ion channels. Coupled with advances in ion channel screening and peptide biochemistry, we are on the verge of a new wave of toxin discovery and clinical application. These toxins also tackle challenging drug targets that are in desperate need of new chemical modalities. This chapter details the advances in toxin drug discovery though clinical development and new understanding of licensed drugs that feeds back to improve further therapeutic opportunities.

Keywords Venom · Toxins · Peptide therapeutics · Cryo-EM · Ziconotide · Structure-aided drug design

C. M. Trim
School of Psychology and Life Sciences, Canterbury Christ Church University, Canterbury, Kent, UK

A. E. Leffler · Z. Samanani
Schrödinger, Inc., New York, NY, USA

S. A. Trim (✉)
Venomtech Ltd., Sandwich, Kent, UK
e-mail: s.trim@venomtech.co.uk

G. Stephens, E. Stevens (eds.), *Ion Channels as Targets in Drug Discovery*,
https://doi.org/10.1007/978-3-031-52197-3_15

1 Introducing Venoms and Toxins as Ion Channel Ligands

1.1 Understanding the Evolution of Venoms as Ion Channel Tools

Venoms and toxins are intrinsically linked to ion channel drug discovery with many toxins becoming key tools for elucidating channel function, structure and pharmacology. The reason these natural compounds are so valuable to drug discovery is due to the crucial role ion channels perform in biology. Thus, when an organism evolves a toxin that disrupts ion channel homeostasis, prey becomes easier to capture, and predators learn important lessons. Microbial evolution of toxins is more complex, but similar selection pressures are likely, such as defence against competition and niche exploitation (Kumar et al. 2019). It is through the investigation of these fascinating organisms that so many standard compounds have arisen for ion channel research; plant toxins such as strychnine and curare were some of the earliest to enter scientific research. However, it is the wealth of animal toxins that has really propelled ion channel research forward; tetrodotoxin (TTX), one of the first ion channel toxins published from animal sources, is now understood to be from endosymbiotic bacteria (Lago et al. 2015). TTX-producing bacteria such as *Pseudoalteromonas tetraodonis* have been found in marine sediments such as those surrounding eel grass, thus providing an explanation as to why this potent toxin has been available for repurposing by so many aquatic species (Lee et al. 2015). With such potent microbial toxins in the marine environment, there is a strong selection pressure to evolve resistance. This is probably a precursor to weaponising saxitoxin for defence (Kvitek and Beitler 1991) and TTX for defence and predation (Hanifin 2010) as the host must not succumb to their own toxins. TTX, originally found in the liver of pufferfish (*Fugu* spp.) and batrachotoxins from the skin of the poison dart frogs (*Phyllobates terribilis*) are passive toxins for defence and thus referred to as poisons or toxins. In the study of voltage-gated sodium channels (VGSCs), TTX is well-known and has led to Na_V channels being classified as TTX-sensitive or TTX-resistant. $Na_V1.1$, $Na_V1.2$, $Na_V1.3$, $Na_V1.4$, $Na_V1.6$ and $Na_V1.7$ subtypes are considered sensitive, and $Na_V1.5$, $Na_V1.8$ and $Na_V1.9$ subtypes are considered resistant to TTX (Nieto et al. 2012). Venoms, which are actively injected, have, and continuously, produced valuable tools from α-bungarotoxin through to conotoxins and theraphotoxins. Scorpion toxins from *Leiurus quinquestriatus* were used to identify the first sodium channel alpha subunit (Beneski and Catterall 1980).

Once an animal obtains an injectable toxin, there is positive selection for gene duplication, initially increasing protein production levels, but these duplicated genes are then freed from the selection pressure. Once duplicated, these toxin genes gather random mutations, and the copies drift apart until further mutations arise that deliver new or improved activities and so become positively selected for once more. Some toxins are inevitably lost as they no longer benefit the host. Together this process of duplication, diversification and selection is the venom's own chemical lead series, and it is easy to see how we can utilise such a valuable resource in drug discovery. Convergent evolution is a common occurrence in the natural world, particularly around utility of ion channels as targets for venoms. Figure 1 displays the known toxin activities, simplified to inhibition or activation, and the corresponding ion channel target class they act upon. The simplification hides the species richness and gene duplication as there are millions of potential peptides out there to be discovered. In this simplification, the term activation includes gating modifiers that reduce thresholds or delay inactivation and, similarly, inhibitors that increase thresholds thus reducing activation. The next few decades will see many more lines drawn on this interaction web as research is currently biased for those venoms easier to collect and with greatest human envenomation effects, namely, snakes, spiders and scorpions. Neurotoxic snake venoms such as those from the Elapidae (mambas, cobras, kraits, etc.) contain a wide range of three-finger neurotoxins that evolved to block nicotinic acetylcholine receptors (nAChR) as a mechanism for prey capture. This is particularly beneficial for those hunting in complex three-dimensional habitats such as trees and underwater, where following injured prey is impractical and thus quick paralysis of prey such as with α-bungarotoxin (BTX) delivers higher

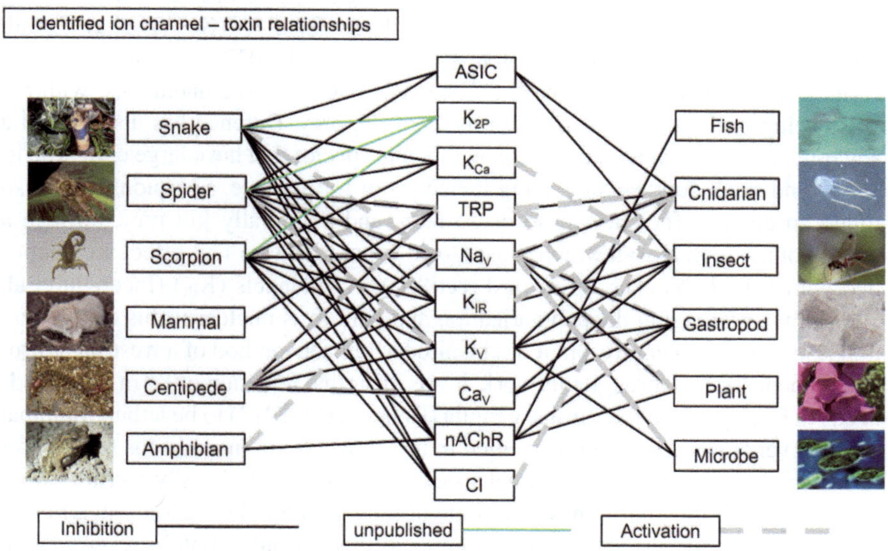

Fig. 1 Interaction web of channel families and toxin-containing species groups. Interaction web showing the known toxin interactions from the major toxin-producing organisms and major ion channel families. ASIC, acid-sensing ion channels; K_{2P}, two-pore K^+ channels; K_{Ca}, calcium-activated K^+ channels; TRP, transient receptor potential; Na_V, voltage-gated Na^+ channels; K_{IR}, inward rectifying K^+ channels; K_V, voltage-gated K^+ channels; Ca_V, voltage-gated Ca^{2+} channels; nAChR, nicotinic acetylcholine receptors; Cl, chloride channels including GABA. This interaction web demonstrates the bias of venom discovery towards snakes, spiders and scorpions and towards voltage-gated ion channels. Interactions were logged when one activity of a given animal class against a given ion channel class was found. Initial pairwise searching was performed in UniProt, and further ion channel searches in PubMed were filtered on taxonomic groups. K_{2P} data (green) is from UK customer of Venomtech Ltd. (unpublished) and represented with permission

food rewards and negates the energy expense of venom production. The need for rapid capture answers a common question of why these venoms are so potent; evolution doesn't over-engineer as this wastes energy. Another ecological niche where the need for rapid prey paralysis has led to potent ion channel blockers is with the marine gastropods, where cone snails (*Conus* spp. and *Gemmula* spp.) have evolved strategies to feed on bony fish. The snails have delicate bodies and ultra-fine envenomation apparatus, which are selected for those snails that paralysed their prey rapidly and avoided injury in capturing their meal. A very similar situation has played out with the most ancient of venomous creatures, the Cnidaria. Fish-feeding Cnidaria have the same problem as the gastropods when taking on bony or armoured prey with their delicate invertebrate arms, but they lack the defensive shells of the gastropods, with the exception of corals. Free-swimming jellyfish are at risk of damage and predation from a vast array of predators; this is the most likely selection pressure for pain-causing toxins such as the pore-forming toxins in the lethal box jellyfish *Chironex fleckeri* (Piontek et al. 2020).

Another mechanism that is apparent in sea anemones is that of analgesia which may serve to calm captured prey and thus reduce struggling which could injure the predators. Transient receptor potential (TRP) (Logashina et al. 2021), acid-sensing ion channel (ASIC) (Verkest et al. 2022), voltage-gated K^+ (K_V) (Iwakawa et al. 2019), voltage-gated Na^+ (Na_V) (Colom-Casasnovas et al. 2022) and nAChR (Kalina et al. 2022) channel blockers have all been discovered in sea anemones, with the latter being probably involved in paralysis of the prey. Defence has also played a key role in scorpion venom evolution with many species that have large claws (pedipalps) and thick cuticles only using their venom for defence. Scorpions have also found an energy-efficient way to cause pain, and potentially kill prey, through a pre-venom of high K^+ salt which causes local depolarization. This effect is exemplified with peptide blockers of inward rectifying K^+ channels (K_{IR}) (Inceoglu et al. 2003), thus prolonging the depolarization and pain. It is unclear if this depolarization is used in prey capture, but it has inspired a humane method of terrestrial arthropod euthanasia, and through this work, it was clear that injection of K^+ in arthropods needs to be targeted to the cardiac ganglia (Bennie et al. 2012) to be lethal. Mammal and fish venoms are poorly represented in the known toxin interactions (Fig. 1) for distinctly different reasons. Mammal species have only occasionally recruited venoms, and where they have, these are some of the least complex venoms so far characterised, for example, with some shrews reported to only contain seven venom components from six gene families (Hanf and Chavez 2020). Many mammalian venoms are also difficult to obtain, such as from the duck bill platypus *Ornithorhynchus anatinus* which has been shown to potently induce Ca^{2+} influx in human dorsal root ganglia (de Plater et al. 2001) and neuroblastoma cells (Kita et al. 2009). In stark contrast to the mammals, 10% of fish families contain venomous species, and these are now known to represent around 20 independent venom recruitment events (Arbuckle and Harris 2021). With fish being the most species-rich group of vertebrates and pain being a major symptom of fish envenomation, there is likely to be an explosion of new ion channel toxins discovered from fish as the techniques improve to investigate these challenging venoms. Thus, fish venoms are poorly represented due to the challenges of extracting the venoms.

1.2 Toxin Nomenclature

Whereas synthetic small molecule nomenclature was established long ago, the same is not true for most natural toxins. This is particularly important for larger molecules such as peptides because their formulae names are extremely complex due to the number of side chains. Even the so-called small molecule toxins have long rational names when using the International Union of Pure and Applied Chemistry (IUPAC) system (Table 1), but these names give rich text meaning, and that is their value. Some names for natural toxins carry a part of the taxonomy of the species they were discovered in such as TTX initially from pufferfish of the family Tetraodontidae (meaning four teeth); thus the toxin should be correctly spelt tetraodotoxin but through common use has become tetrodotoxin and reduced to TTX. The superorder Batrachia was selected for batrachotoxins; others such as vanillotoxins and wasabitoxin are derived from names of other ligands of the toxin target. A rational, systematic, nomenclature with both target pharmacology and taxonomy imparts rich meaning to the names and thus greater value. Conotoxins are an exception, with a well-designed rational nomenclature that has long been accepted and used; this consists of an initial descriptor such as a Greek letter followed by a generic name from the family of origin and lastly a four-character code for species and isoforms. Table 1 identifies the correct name for several common toxins using the preferred rational nomenclature. Shortening the family toxin name to four letters helps with ease of use, i.e. theraphotoxin as TpTx. A seminal paper in 2008 from Glenn King and colleagues set the scene for how all peptide toxins could be rationally named using the conotoxin formula (King et al. 2008), and this has been followed up in 2012 for sea anemones (Oliveira et al. 2012) and in 2014 a plea for improved snake venom toxin nomenclature (Hargreaves and Mulley 2014). Rational nomenclature will benefit all who work with these complex compounds and future-proof growth of the field as more venoms and toxins are discovered. As more toxin activities are discovered, more prefixes can be added and more families included as our understanding of venom grows. Throughout this chapter we aim to provide the rational nomenclature where possible to help with continuity and show the value of its use. However, it is still an area for improvement in toxin research, and it is often required to use the authors suggested name for clarity when referring to previous work. Therefore, we echo the plea of King, Olivera, Hargreaves and many more to challenge irrational naming when it is presented for peer review and ask editors and publishers to do the same.

Table 1 Key ion channel toxins and their origins in alphabetical order

Toxin	Rational name	Origin	Target	Key reference
Agatoxin	ω-Agatoxin-Aa1a	American funnel web spider – *Agelenopsis aperta*	$Ca_V2.1$	Adams (2004)
Batrachotoxin	[(1S)-1-[(1R,5R,6S,9R,11S,12R,14R)-9,12-dihydroxy-6,16-dimethyl-10,19-dioxa-16-azahexacyclo[12.5.3.15,9.01,14.02,11.06,11]tricosa-2,21-dien-22-yl] ethyl] 2,4-dimethyl-1H-pyrrole-3-carboxylate	Poison dart frogs – Such as *Phyllobates terribilis*	Na_V opener	Daly et al. (1978)
Bungarotoxin	α-Elapitoxintoxin-Bm1a	Banded Krait – *Bungarus multicinctus*	nAChR	Changeux et al. (1970)
Charybdotoxin	λ-Buthitoxin-Lqh1a	Deathstalker scorpion – *Leiurus quinquestriatus*	K_V's and $K_{Ca}1.1$	Galvez et al. (1990)
Curare	(1*S*,16*R*)-10,25-dimethoxy-15,15,30-trimethyl-7,23-dioxa-30-aza-15-azoniaheptacyclo[22.6.2.2^{3,6}.1^{8,12}.1^{18,22}.0^{27,31}.0^{16,34}]hexatriaconta--3(36),4,6(35),8(34),9,11,18(33),19,21,24,26,31-dodecaene-9,21-diol;chloride;pentahydrate;hydrochloride	South American vine – *Chondrodendron tomentosum*	nAChR	Pollitzer (1886)
GVIA	ω-Conotoxin-GVIA	Marine geography cone snail – *Conus geographus*	$Ca_V2.2$	Hannon and Atchison (2013)
ImI	α-Conotoxin-ImI	Marine imperial cone snail – *Conus imperialis*	α7 nAChR	Yu et al. (2011)
LvIA	α-Conotoxin-LvIA	Marine livid cone snail – *Conus lividus*	α3β2 nAChR, α3β4 nAChR	Luo et al. (2014)
ProTox II	β-Theraphotoxin-Tp2a	Green velvet spider – *Thrixopelma pruriens*	$Na_V1.7$	Xiao et al. (2010)
Psalmotoxin-1	π-Theraphotoxin-Pc1a	Trinidad chevron spider – *Psalmopoeus cambridgei*	ASIC1a	Escoubas et al. (2000)

RegIIA	α-Conotoxin-RegIIA	Marine crown cone snail – *Conus regius*	α3β2 nAChR, α3β4 nAChR	Xu et al. (2020)
Saxitoxin	[(3aS,4R,10aS)-2,6-diamino-10,10-dihydroxy-3a,4,8,9-tetrahydro-1H-pyrrolo[1,2-c]purin-4-yl]methyl carbamate	Shellfish like *Saxidomus nuttalli*, but the origin is again marine microbes, dinoflagellates and cyanobacteria	Na_V	Kao and Nishiyama (1965)
ShK	κ-Stichotoxin-She3a	Sun Sea anemone – *Stichodactyla helianthus*	$K_V1.3$ and $K_V1.1$	Pennington et al. (1995)
SNX482	ω-Theraphotoxin-Hg1a	Cameroon red baboon spider – *Hysterocrates gigas*	$Ca_V2.1$ and $Ca_V2.3$	Newcomb et al. (1998)
Strychnine	(4aR,5aS,8aR,13aS,15aS,15bR)-4a,5,5a,7,8,13a,15,15a,15b,16-decahydro-2H-4,6-methanoindolo[3,2,1-ij]oxepino[2,3,4-de]pyrrolo[2,3-h]quinolin-14-one	The strychnine tree – *Strychnos guianensis*	nAChR and glycine receptors	Woodward and Brehm (1948)
Tetrodotoxin	(1R,5R,6R,7R,9S,11S,13S,14S)-3-amino-14-(hydroxymethyl)-8,10-dioxa-2,4-diazatetracyclo[7.3.1.17,11.01,6]tetradec-3-ene-5,9,12,13,14-pentol;2-hydroxypropane-1,2,3-tricarboxylic acid	Initially pufferfish – Tetraodontidae. Correctly symbiotic bacteria – *Pseudoalteromonas tetraodonis*	Na_V	Lago et al. (2015) and Lee et al. (2015)
Vanillotoxin	τ-Theraphotoxin-Pc1a	Trinidad chevron spider – *Psalmopoeus cambridgei*	TRPV1	Siemens et al. (2006)

1.3 Ethical Sourcing of Toxin Tools

Evolution by natural selection has produced a wide variety of toxin tools for ion channels; however the act of bioprospecting is controlled by local, national and international laws, treatise and protocols. These include the Nagoya Protocol on access and benefit sharing as an addition to the Convention on Biological Diversity (Cicka and Quave 2019), commonly shortened to the Nagoya Protocol. Also the longer-established Convention on International Trade and Endangered Species of Wild Fauna and Flora (CITES) (Fukushima et al. 2019) restricts movement of organisms and their derivatives across international borders. Together with other guidance from the World Health Organisation, their aims are to protect species from over-exploitation and protect nations and cultures from biopiracy. Once a species has been discovered to contain medicinal drug-like properties, however tenuous, this can lead to high demand for collection both legitimately and in illegal trade; therefore researchers need to be aware of their moral duties and source their reagents responsibly. The western discovery of the nAChR-blocking toxins, strychnine from *Strychnos guianensis* and curare from *Chondrodendron tomentosum*, is not historically clear; however it appears different indigenous tribes have discovered similar actions from different plants to do the same task of paralysing food species. This is also confounded with some accounts of curare from *Strychnos* species (Prance 1972). Therefore, it is likely that both tribes and countries are historical victims of biopiracy in not receiving any reward for the traditional knowledge and the compounds they have discovered. The history of curare in western medicine was documented for the *Journal of the Royal College of Physicians* and is well worth a read (Lee 2005). Thankfully modern advances in peptide synthesis and refolding as well as recombinant systems with high fidelity and capacity mean that new toxins can be discovered from the natural world and then produced in the laboratory when large quantities are needed for drug development.

As with other therapeutic indications, initial uses of venom and toxin ion channel modulators were discovered through their pathological action on the human body or on their prey; classic examples include BTX and TTX. More recently, improvements in assay miniaturisation and throughput such as automated patch clamp systems have led to screening of venom libraries (bioprospecting) to uncover pharmacology not evident in envenomation pathology. Examples of such hidden ligands include ShK (κ-stichotoxin-She3a) a potent sea anemone toxin presumed to be a key defence against crustacean predators, which in humans is a picomolar potent $K_V1.3$ blocker and a powerful anti-inflammatory (Pennington et al. 2015), and m-theraphotoxin-Gs4 which is a potent inhibitor of mechanosensitive BK channels (Li et al. 2019) despite showing no cardiac symptoms from bites (Hauke and Herzig 2021).

Evolution by natural selection is the blind watchmaker of drug discovery, mutating toxins until they bring a survival benefit or redundancy to the possessor. Thus, toxins for prey capture are not necessarily those that cause a quick death; any perturbation that reduces the ability for a coordinated escape endures. Such ion channel mechanisms include openers, pore blockers, gating modifiers and even state dependence, all of which are of potential value and will be discussed within. Understanding how these toxins interact with their target channels at an atomic level allows for a deeper understanding of ion channel function and opens the door to structure-based drug design. Target pharmacology is only the first part of drug discovery; identifying and improving drug-like properties of toxins are critical to getting new therapies to patients. Small molecule drug discovery has had some ion channel successes, principally with anaesthetics, but new therapeutic modalities are needed to address the unmet needs of patients suffering from diseases driven by ion channel pathologies. Venoms and toxins have delivered some success such as the $Ca_V2.2$ blocker ziconotide, licensed as Prialt®, for severe, intractable, chronic pain which evolved in fish-hunting cone snails. This peptide is delivered intrathecally and covered in translating toxins to the clinic (Sect. 4). However, there are more venom-derived drug-like molecules making their way through in vivo development and clinical trials, so the future is bright for challenging diseases to be addressed with these new therapeutic modalities.

2 Toxin Structures and Their Targets

2.1 Venom Peptide and Ion Channel Structures in the Protein Data Bank (PDB)

How do peptides isolated from animal venoms bind to ion channels? Understanding the interactions that occur at the atomic level between venom peptides and ion channels is the key not only to deciphering the exquisite potency of these peptides for their targets but also to rationally developing them into drug leads for diseases such as pain (Wulff et al. 2019). However, obtaining structures of these peptides in complex with their ion channel targets is a challenging task (Leffler et al. 2017). The Protein Data Bank (PDB) is a valuable resource to understand the current landscape of venom peptide and ion channel structures. A search of the PDB using domain annotations was performed to identify such deposited complexes (NB, this search may not be exhaustive due to PDB structures having missing or incomplete annotations). After manual curation, a total of 40 structures of venom peptides and ion channels were identified. For each complex, the method used to solve the structure and the identity of the ion channel were extracted and summarised by deposition year to obtain a historical view of venom peptide and ion channel structural biology (Fig. 2).

From 1994 to 2006, a time period we will denote as the "first wave" of venom peptide and ion channel structural biology, solution nuclear magnetic resonance (NMR) was the primary method employed (Fig. 2a). All six structures deposited during this interval were solved using NMR. Due to the size limitations of NMR (Puthenveetil and Vinogradova 2019), these structures do not contain the full-length

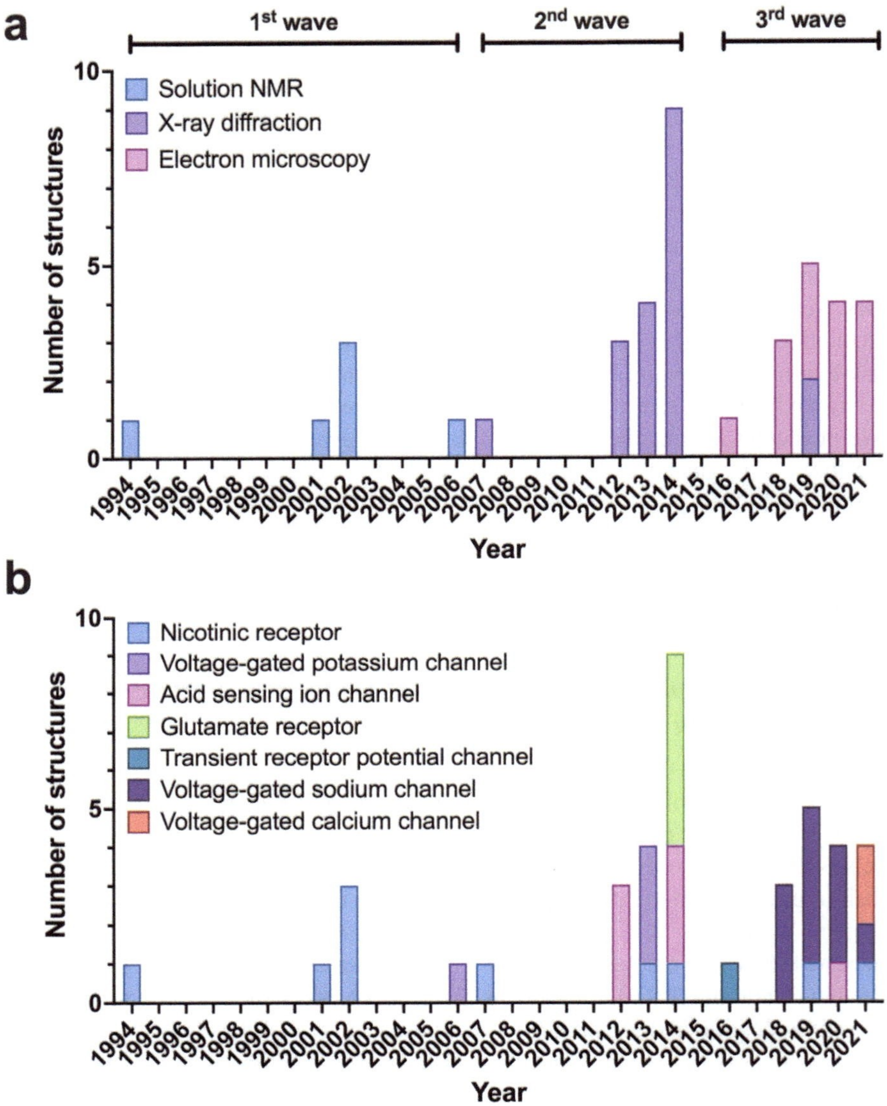

Fig. 2 Chronological deposition of ion channels complexed with venom peptides. (**a**) Venom peptide and ion channel complexes deposited in the PDB by year, coloured by the structural biology method used to obtain the complex. (**b**) Venom peptide and ion channel complexes deposited in the PDB by year, coloured by the identity of the ion channel in the complex

ion channel, but rather a short polypeptide sequence that represents the ion channel's venom peptide binding site. Examples include BTX and α-cobratoxin in complex with polypeptides representing functionally important sites on nAChRs (Fig. 2b), as well as charybdotoxin (CTX) with a subset of the pore domain of a K^+ channel (Basus et al. 1993). There are no examples of NMR being employed to solve such structures beyond 2006.

From 2007 to 2014, the "second wave" of venom peptide and ion channel structural biology, X-ray diffraction became the predominant means to solve these structures (Fig. 2a). All 17 such complexes deposited during this interval were solved using X-ray diffraction and differ in two respects from those solved earlier by NMR. First, entire subunit(s) of the channel are present in the structures. Second, a greater diversity of channel types is represented, with structures of the AMPA-type glutamate receptor and ASICs observed for the first time (Fig. 2b). The large peak of structures obtained in 2014 is not representative, as it consists primarily of five structures of the GluA2 AMPA receptor in complex with a *Conus striatus* cone snail toxin, a positive allosteric modulator, and a variety of orthosteric agonists (Chen et al. 2014). Interestingly, beyond 2014, there is a dramatic drop-off in the number of structures solved by X-ray diffraction.

Finally, the "third wave" of venom peptide and ion channel structures, from 2016 to the present, has been dominated by cryogenic electron microscopy (cryo-EM) and includes many clinically relevant targets. Of the 17 structures solved during these years, 15 utilised cryo-EM, and 2 employed X-ray diffraction (Fig. 2a). A "steady state" of approximately four structures deposited per year has taken shape. This time period has also seen the determination of the structures of two important classes of drug targets in complex with venom peptides for the first time: voltage-gated Na^+ channels and voltage-gated Ca^{2+} channels (Fig. 2b). Of the 11 sodium channel structures, 10 were solved with cryo-EM, and 1 was solved using X-ray diffraction. Amongst these structures were several of $Na_V1.7$, a widely pursued pain target (Dib-Hajj et al. 2013), with gating-modifier toxins (GMTs) (Xu et al. 2019). Finally, in 2021, the structure of the non-opioid analgesic drug, ziconotide, was solved in complex with the calcium channel $Ca_V2.2$, providing the first example of how a venom peptide that is an FDA-approved drug interacts with its target (Gao et al. 2021).

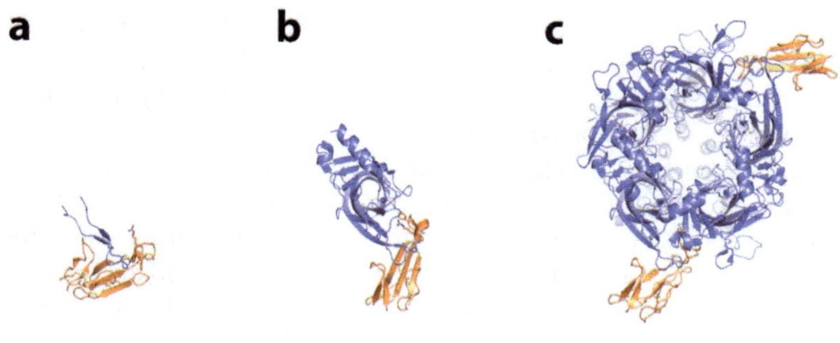

Fig. 3 Increasing resolution of BTX binding to the nAChR. (**a**) NMR structure of BTX in complex with the *Tetronarce californica* α1 residues 182 to 202 (PDB code: 1LJZ), (**b**) X-ray diffraction structure of BTX in complex with the *Mus musculus* α1 subunit (PDB code: 2QC1), (**c**) BTX in complex with *Tetronarce californica* muscle subtype pentamer (PDB code: 6UWZ). Cryogenic electron microscopy structure of BTX is depicted as an orange cartoon, while the nAChR is shown as the blue cartoon. For clarity the PDB entry 1LJZ uses the basionym *Torpedo californica for Tetronarce californica*

The "three waves" of venom peptide and ion channel structural biology – NMR, X-ray diffraction and cryo-EM – can be best summarised with a concrete example, BTX and the muscle-subtype nAChR. Although the fact that BTX interacts with this nAChR was known since the 1960s and 1970s (Changeux et al. 1970), a structure showing how the two proteins bind remained elusive for decades. In 2003, during the "first wave", NMR was used to solve a structure of BTX in complex with a 20-residue polypeptide (residues 182 to 202 of the *Torpedo californica* α1 subunit) corresponding to a functionally relevant binding site on the α subunit of the nAChR (Samson et al. 2002) (Fig. 3a). This structure clearly shows that BTX binds to the nAChR by placing a small helical segment below the vicinal disulphide bond on loop C of the nAChR. Interestingly, however, the cation-π sandwich formed between residues R36 and F32 of BTX (Infield et al. 2021) is not present in this structure (Fig. 3a). In 2007, the X-ray diffraction structure of BTX with the extracellular domain (ECD) of the mouse nAChR α1 subunit was solved (Fig. 3b). In addition to revealing additional details about the interaction between BTX and the α1 nAChR ECD, it included ordered water molecules (Fig. 3b) (Dellisanti et al. 2007). Finally, in 2020 the closed state structure of the muscle-type nAChR was determined with cryo-EM, revealing the binding of BTX at two subunit interfaces (Fig. 3c). Notably, in the pentameric assembly, the orientation of BTX relative to the channel clearly differs from the NMR and X-ray diffraction structures, highlighting the contribution of the complementary subunit to the binding mode of BTX (Fig. 3c). This series of three structures demonstrates how each successive technology expanded and refined the knowledge of the BTX and nAChR interaction.

2.2 *Venom Peptide and Ion Channel Structures Explain Mutagenesis Data*

Structures of venom peptides in complex with their ion channel targets are highly useful for visualising the overall peptide binding mode, rationalising structure-activity relationships (SAR) from mutagenesis experiments and guiding the development of new mutants. For example, CTX is a potent blocker of voltage-gated K^+ channels, and it has been shown that mutation of lysine (K27) to methionine (K27M) leads to a marked reduction in the affinity of CTX (Banerjee et al. 2013). Visual inspection of the CTX and K^+ channel structure clearly explains this SAR. The amino group of K27 forms a network of four hydrogen bonds to the backbone carbonyls of the residues that line the mouth of the channel's selectivity filter; these interactions are eliminated with the mutation to methionine, which has no hydrogen bond donors (Fig. 4a).

Similarly, while the key SAR of ziconotide for the N-type voltage-gated Ca^{2+} channel $Ca_V2.2$ had been deduced via mutagenesis studies (Nadasdi et al. 1995), the structural features underlying these interactions were unclear. For example, it was reported that mutation of tyrosine (Y13) on ziconotide to phenylalanine (Y13F)

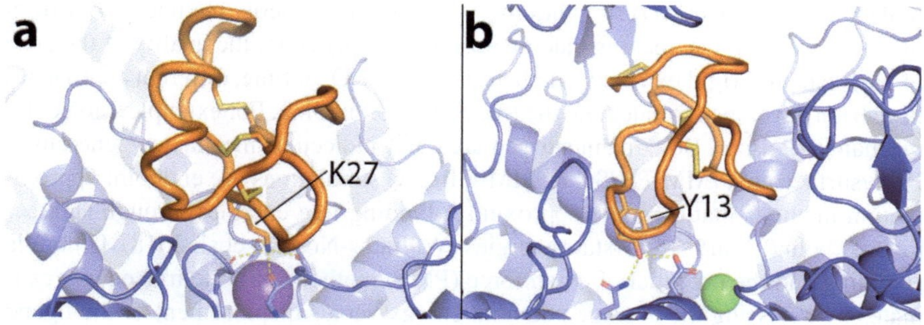

Fig. 4 High-resolution structures play a key role in understanding atomic level interactions. (**a**) Structure of CTX in complex with a eukaryotic voltage-gated K^+ channel (PDB code: 4JTA). (**b**) Structure of ziconotide in complex with human $Ca_V2.2$ (PDB code: 7MIX). In both panels, the venom peptide is shown as an orange cartoon and the ion channel as a blue cartoon. K^+ is shown as a purple sphere and Ca^{2+} is shown as a green sphere. Hydrogen bonds are shown as yellow, dashed lines. Disulphide bonds are depicted as yellow sticks

resulted in >99% loss in potency. Interestingly, mutation of Y13 to monoiodotyrosine only had a 40% loss in potency, highlighting the importance of the phenolic group of Y13 (Nadasdi et al. 1995). Why is this specific hydroxyl group so essential for ziconotide's block of $Ca_V2.2$? Examination of the structure reveals that the hydroxyl group is perfectly positioned by ziconotide to form strong hydrogen bonds with the side chains of a pair of glutamine and aspartate residues on $Ca_V2.2$ (Fig. 4b). Overall, these two examples highlight how venom peptide and ion channel structures can play a significant role in understanding the nuanced interactions occurring at these complexes.

2.3 *Free Energy Methods for Venom Peptide and Ion Channel Structures*

The structures of venom peptides in complex with ion channels have limitations for structure-based design. Although they are highly useful for understanding the enthalpic interactions that occur between the two molecules in the bound state, other important contributions to the binding free energy, such as entropy, cannot be deduced from them by observation. As a result, there is growing interest in combining these structures with computational approaches to maximise their utility (Shah et al. 2020).

Free energy methods are a category of calculations that work by considering the energies of the "free" (ligand in solution interacting with solvent) and "bound" (ligand bound to target through non-covalent interactions) states (Williams-Noonan et al. 2018). They require accurate modelling of protein (i.e. ion channel), ligand (i.e. venom peptide) and solvent structures and energies (Clark et al. 2017). Multiple states of the ligand and receptor are generally sampled as well, typically with molecular dynamics (MD) simulations. As a result, free energy methods differ from the more well-known docking calculations that simply score the quality of the pose of the ligand in a rigid receptor (Kitchen et al. 2004). Over time, different approaches have been developed to calculate the free energy of binding. For example, molecular mechanics-Poisson Boltzmann/surface area and molecular mechanics-generalised Born/surface area (MM-PB/SA or MM-GB/SA) simulations are endpoint approximation methods that compute approximate binding free energies through subtraction of "bound"- and "free"-state energies (Williams-Noonan et al. 2018). Umbrella sampling and the potential of mean force (PMF) methods provide free energy estimates by assessing the average potential energy versus displacement of the ligand over a series of restrained positions (Williams-Noonan et al. 2018). Finally, free energy perturbation (FEP) calculates the relative free energy of binding by summing the free energy differences between sequential transformations of the ligand as it is mutated from an initial to a final state Katz et al. (2021a).

These free energy methods have been applied with success to predicting the potency of venom peptides for their ion channel targets (Table 2). In 2002, Eriksson and Roux established that free energy methods could be readily applied to study venom peptide and ion channel complexes in their work on deriving a model of the AgTx2-K^+ channel complex (Eriksson and Roux 2002). Yu et al. used MM-GB/SA to validate their models of the α7 nAChR and α-conotoxin-ImI complex, finding an R^2 of 0.74 between the experimental and calculated relative free energies of 30 ImI point mutants (Yu et al. 2011). Most recently, Katz et al. used the FEP implementation FEP+ to retrospectively predict the relative potencies of 47 point mutants of the GMTs ProTx-II and huwentoxin (HwTx) for $Na_V1.7$ with 64% accuracy (Katz et al. 2021a). Future studies that use these methods to prospectively identify mutants that enhance potency, as opposed to only validating models with existing mutational data, will be particularly valuable for enhancing their impact.

Another key topic, the selectivity of venom peptides, has been studied in the context of the HsTX1-$K_V1.3$/$K_V1.1$ (Rashid et al. 2014) and RegIIA α3β2/α3β4 nAChR (Xu et al. 2020) systems. In the former case, umbrella sampling and PMF calculations were employed to identify the mutant HsTX1[R14A] that is picomolar potent and highly selective for $K_V1.3$ (Rashid et al. 2014). In the latter case, MM-GB/SA calculations using an ensemble of MD simulation frames led to the identification of three RegIIA analogues with a 10- to 37-fold increase in selectivity for the α3β2 nAChR (Xu et al. 2020). More recently, Katz et al. used FEP+ to retrospectively predict the selectivity's of 11 LvIA conotoxin point mutants for the α3β2 nAChR over the α3β4 nAChR with 91% accuracy (Katz et al. 2021b).

3 Drug-Like Properties of Ion Channel Toxins

Evolution by natural selection is not anthropocentric, and thus it is not obvious that venom toxins can be therapeutically useful. But yet, they can be, as evidenced by captopril, exenatide and ziconotide. So, what is driving the selection of

Table 2 Studies of free energy methods applied to venom peptides and ion channels

Venom peptide(s)	Ion channel(s)	Summary	Reference
AgTx2	Shaker K^+	A model for the AgTx2-Shaker complex was derived using high-temperature MD simulations to identify candidate complexes, followed by pose stability and relative binding free energy calculations of mutants to identify the highest-quality complexes	Eriksson and Roux (2002)
ImI	α7 nAChR	More than 30 single-point mutations to either the α7 nAChR or ImI were studied using comparative modelling and MD simulations. Methods combining different conformational refinement procedures and binding energy calculations were explored, with the best results obtained from explicit water energy minimisation and MM-GB/SA	Yu et al. (2011)
HsTX1	$K_V1.3$, $K_V1.1$	Docking and MD simulations were used to create complexes of HsTX1 with $K_V1.3$/$K_V1.1$, followed by umbrella sampling simulations and PMF calculations. PMF results predicted that HsTX1[R14A] would have a greater than 2 kcal/mol gain in selectivity for $K_V1.3$ relative to the wild type	Rashid et al. (2014)
RegIIA	α3β2 nAChR, α3β4 nAChR	MM-GB/SA was applied for the rational design of RegIIA analogues selectively targeting the α3β2 versus the α3β4 nAChR. Through in silico mutagenesis applied to an ensemble of frames from MD simulations, three RegIIA analogues with 10- to 37-fold selectivity improvement for the α3β2 subtype were found	Xu et al. (2020)
LvIA	α3β2 nAChR, α3β4 nAChR	FEP+ was used to retrospectively predict the selectivity's of 11 LvIA point mutants for the α3β2 over the α3β4 nAChR with 91% accuracy	Katz et al. (2021b)
ProTx-II, HwTx	$Na_V1.7$	FEP+ was employed to retrospectively predict the relative potencies of 47 point mutants of the $Na_V1.7$ GMTs ProTx-II and HwTx with 64% accuracy. Additionally, FEP+ and MM-GB/SA identified 15 non-standard ProTx-II[W24] mutants predicted to have a gain in potency	Katz et al. (2021a)

therapeutically useful peptides in venoms? Part of the answer will be serendipity, as the targets in the prey have also evolved and thus human therapeutic targets are distantly related to prey species but closer to that of some predators. The evolution of pain-causing venom in spitting cobras and the act of spraying venom are compellingly likely to be due to the cobras defence against upright hominins (Kazandjian et al. 2021). In this instance the pain-causing three-finger toxins are dependant on cytotoxic phospholipase A2 enzymes, which are not very drug-like, but does exemplify the potential. In the same way that antimicrobial peptides were discovered long before the venom microbiome explanation for their evolution (Esmaeilishirazifard et al. 2022), the same is likely to be possible with ion channel peptides. There are over 10,000 published papers linking venoms to ion channel target pharmacology, and the list is growing; although the rate appeared to peak in the year 2000, there are still over 200 papers per year being published. What needs to happen to turn this stream of new ligands into new therapeutics is the development of a greater understanding of the other drug-like properties of these peptides. Characterisation of properties like bioavailability and stability, combined with the growing interest in new therapeutic modalities, will drive this new era. New technology is proving opportunities to improve on nature where there has been no natural selection, i.e. that of desired pharmacokinetics (PK) and pharmacodynamics (PD) in patients. There is already a good precedent for peptide therapeutics with over 80 on the market already and a further 150 in clinical development (Muttenthaler et al. 2021). Those venom peptide drugs that have made it to the clinic and beyond are just a small subsection of the chemical space available, the same as with small molecules.

3.1 Modern Bioprospecting

The purpose of bioprospecting for new ion channel ligands is changing. The vast majority of the work on discovering new ion channel tools from venoms and toxins was exactly that, discovery, the hope that a drug-ready peptide could be found and fast-tracked to the clinic. Many papers end with statements like "this peptide could be a new therapeutic for disease X that is poorly controlled by current therapies" with only target engagement data, thus lacking bioavailability, PK and PD. However, this discovery approach has been fruitful, and the discovery of ω-conotoxins inhibiting $Ca_V2.2$ though pore blocking (Fig. 4a) led to ziconotide. Evolution has provided far more toxins than we so far screened in the last 100 years of systematic drug discovery. It is worth bearing in mind that 96-well and 384-well plates were only invented in the 1990s and high-throughput 384-well automated patch clamp technology has only been practical since 2013 (Obergrussberger et al. 2021). With these new technologies, screening thousands of new venom peptides against ion channel targets is quick and practical; serendipity may deliver other ready-made drugs, but there is another way. The modern method of bioprospecting that is gaining traction in the literature and within several pharmaceutical companies represents screening for a venom peptide ligand as a suitable starting point for drug discovery and development. In the same way, the practice of finding a small molecule hit then proceeds to building a lead series around it, followed by SAR and rational drug design to improve the physical properties. These starting points need to be potent, such as the picomolar active $K_V1.3$ toxins HsTx1 (κ-scorpiotoxin-Hs1a) (Huang et al. 2012) and ShK (κ-stichotoxin-Sh1a) (Iwakawa et al. 2019) and the low nanomolar active $Na_V1.7$-blocking peptides ProTox II (β/ω-theraphotoxin-Tp2a) and GpTx1 (μ-theraphotoxin-Gp1a) (Diochot 2021). As $Na_V1.7$ is such a popular target for analgesic drug discovery, many theraphosid toxins have been found to target this channel including those from *Ceratogyrus cornuatus* (Bosmans et al. 2006) and more famously ProTx-II from *Thrixopelma pruriens* (Flinspach et al. 2016). CcoTx toxins are from the theraphosid spider *Ceratogyrus cornuatus*, which has been reclassified as *Ceratogyrus marshalli*, and therefore the toxins should be correctly referred to as β-theraphotoxins-Cm1a, Cm1b and Cm1c. These are particularly interesting as there are three closely related peptides identified, and naturally occurring amino acid substitutions modify their respective selectivity to $Na_V1.7$ (Bosmans et al. 2006). Another peptide pair was identified for blocking homomeric and heteromeric ASIC1a channels from the black mamba snakes *Dendroaspis polylepis*, mambalgin-1 and mambalgin-2 (Diochot et al. 2012). It would be logical to use a rational name of π-elapitoxin-Dp1a and Dp1b for these toxins, but 10 years of mambalgin use will be hard to change. In this case, the two peptides only differ at position four, either having a phenylalanine (F) or tyrosine (Y) at this location; thus it is only the presence of the oxygen on the aromatic ring that differs, giving atomic-level power for SAR. Such is the convergent nature of venom evolution; another potent ASIC1a blocker was discovered from the Trinidad chevron spider, *Psalmopoeus cambridgei*, PcTx1 (π-theraphotoxin-Pc1a), about 3 years earlier (Qadri et al. 2009). Both mambalgins and PcTx-1 are common structures for the species they were discovered in; however there is no commonality between the inhibitor cystine knot structure of PcTx-1 and the three-finger structure of mambalgins, thus giving multiple options for structural starting points. Since then, several other toxins have been found to modulate ASIC channels, proving the deeper we look the more toxins we find. Selectivity is a benefit but not a necessity in these starting points, as many groups have shown this to be tuneable. The $K_V1.3$-blocking peptide called HsTx1 by the authors (but κ-scorpiotoxin-Hs1a is preferred) was discovered in the venom of the Asian forest scorpion *Heterometrus spinnifer*. This is a typical inhibitor cystine knot motif of 34 amino acids with 4 stabilising disulphide bonds and a C-terminal amide (Lebrun et al. 1997). Although picomolar potent at $K_V1.3$, it still had nanomolar potency at the neuronal $K_V1.1$ channel. Through in silico potential mean force modelling, arginine 14 (R14) was identified as binding to $K_V1.1$ but having no effect in the binding pocket of $K_V1.3$; thus mutating R14 to alanine [R14A] produced a 2000-fold selectivity for $K_V1.3$ (Rashid et al. 2014).

3.2 *Understanding Mechanisms of Action*

When investigating the mechanism of action of toxins on ion channels, the ArachnoServer database was unique in that the toxins could be browsed by taxonomic relationships or target pharmacology. This gave the drug discovery community an incredibly useful resource; although focused on spider toxins, viewing all the ligands for your target of interest in one place was a great resource. ArachnoServer was also a champion platform for rational nomenclature which displays mechanism of action through its rich text format (see Sec. 1.2). Unfortunately, at the time of print, the ArachnoServer database (Pineda et al. 2018) is undergoing significant code revision and requiring a major input of funding and thus is unavailable (2022 conversation with G.F. King, V. Herzig and lead author, unreferenced). However, there are other ways of getting this target information, such as through using the rational nomenclature prefix which is particularly powerful for Na_V channel tools and searching the UniProt database (Bateman et al. 2021). Beta (β) toxins shift voltage dependence of Na_V channels, delta (δ) toxins delay inactivation, iota (ι) are agonist, and mu (μ) toxins directly inhibit these ion channels. In October 2022 UniProt database (www.uniprot.org) held 446 Na_V β-toxins, 49 δ-toxins, 0 ι-toxins and 94 μ-toxins, although there are ι-toxins such as the $Na_V1.6$-activating conotoxin ι-RXIA published, which shifts the voltage dependence of activation similar to scorpion β-toxins. Several of these β-toxins reduce the voltage threshold of activation to a point where the channels are open at the resting potential of the cell and therefore producing physiological agonist activity. This supports the dogma of toxin research that binding to parts of the voltage sensor domain is a very common mode of action for venom toxins.

State- and use-dependant modulation of ion channels is becoming a key area of focus for drug discovery as it is hoped that this may alleviate some of the unwanted effects of ion channel drugs. TTX and saxitoxin have been found to have some state and use dependency, with the latter having increased affinity immediately after repolarization (Huang et al. 2012; Makielski et al. 1993).

Figure 5 demonstrates the slow on-rate of ProTx-II, compared with TTX, at $Na_V1.7$. Whereas TTX has engaged and blocked the pore by the time of the first voltage pulse, ProTx-II has had minimal effect at the first pulse, and with each voltage train, there is a decrease in current flow. The progression of inhibition with repeated voltage cycling does not appear to be affected by the second compound addition, implying that the rate is not concentration dependant. As ProTx-II binds to the membrane-embedded voltage sensor (Henriques et al. 2016), it is likely that channel opening is required for toxin engagement to the voltage sensor, thus explaining the use dependence observed. The use dependence of TTX published elsewhere was with submaximal concentrations (Huang et al. 2012).

Data from an automated patch clamp screen of 960 venom fractions against $Na_V1.4$ and $Na_V1.7$ identified novel venom peptides from theraphosid spiders which also demonstrated a slow on-rate (Fig. 5). On further investigation, this was seen to demonstrate the same use-dependant blockade of the channels as seen in ProTx-II (Fig. 5). This effect was seen in both selective and non-selective ligands (Fig. 6).

Toxin such as GaTx1 from the deathstalker scorpion *Leiurus quinquestriatus* (which should have the rational name of Σ-buthotoxin-Lq1a) is a state-dependant blocker of the cystic fibrosis transmembrane conductance regulator (CFTR) chloride channel (Fuller et al. 2005, 2007); this peptide increases the number of channels in the closed state after repeated cycling with ATP.

3.3 Improving on Nature

Peptides, including toxins, are great starting points for building a drug discovery lead series as there are a number of relatively easy ways of making structural changes. One way is to use the natural mutations acquired during speciation;

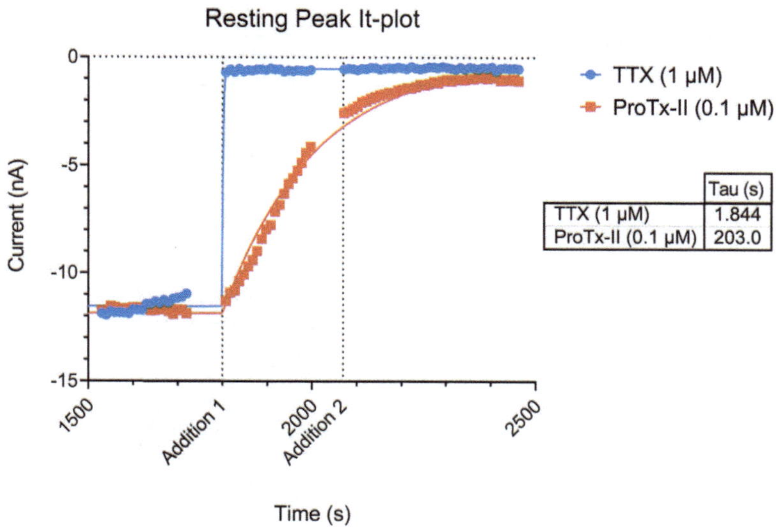

Fig. 5 Evidence of potential use dependence in ProTx-II inhibition of $Na_V1.7$. Originally presented at the European Laboratory Research Interest Group (ELRIG) Drug Discovery 2021 conference by Duncan, E., Williams, S. and Kammonen, J. This current vs time (It) plot of TTX (blue) and ProTx-II (orange) was generated with CHO T-REX $Na_V1.7$ expressing cells on the Qube automated patch clamp system (Sophion) with each dot a separate voltage pulse. At the maximal concentrations tested, TTX has blocked the $Na_V1.7$ pore fully by the first voltage train; however ProTx-II inhibition builds with subsequent voltage pulses and is not affected by a second compound addition

although the mutations were originally random, those that were not effective have been lost through natural selection.

Many toxin classes are present in a wide variety of related species, such as inhibitor cystine knot (ICK) peptides in theraphosid spiders. This presents the drug discovery scientist with a pre-made lead series with structural and potency differences ripe for SAR (Fig. 7). There are many other $Na_V1.7$-blocking toxins not included here as they are less similar to HxTx-4 but with equally interesting structure and pharmacology, such as β/μ-theraphotoxin-Pe1b (Pe1b) from the Malaysian purple earth spider *Phormingochilus everetti* (Rupasinghe et al. 2020).

Another method is to build the lead series with synthetic peptide chemistry where systematic mutations can be added along with side group modifications and non-natural amino acids chosen. This method has worked well for Janssen Pharmaceuticals, where screening of systematic mutations in HxTx1 revealed that isoleucine (I5), lysine (K27) and threonine (T28) were critical to blocking $Na_V1.7$ (Neff et al. 2020). These mutations were then combined with non-natural amino acids to improve the metabolic stability. Natural mutations at these positions have also seen improvements in $Na_V1.7$ potency across particularly those associated with the *Grammostola* genus from the Americas (Fig. 7).

Lastly there is an option combining human ingenuity with the power of natural selection, a process called directed evolution. Many labs are working with these

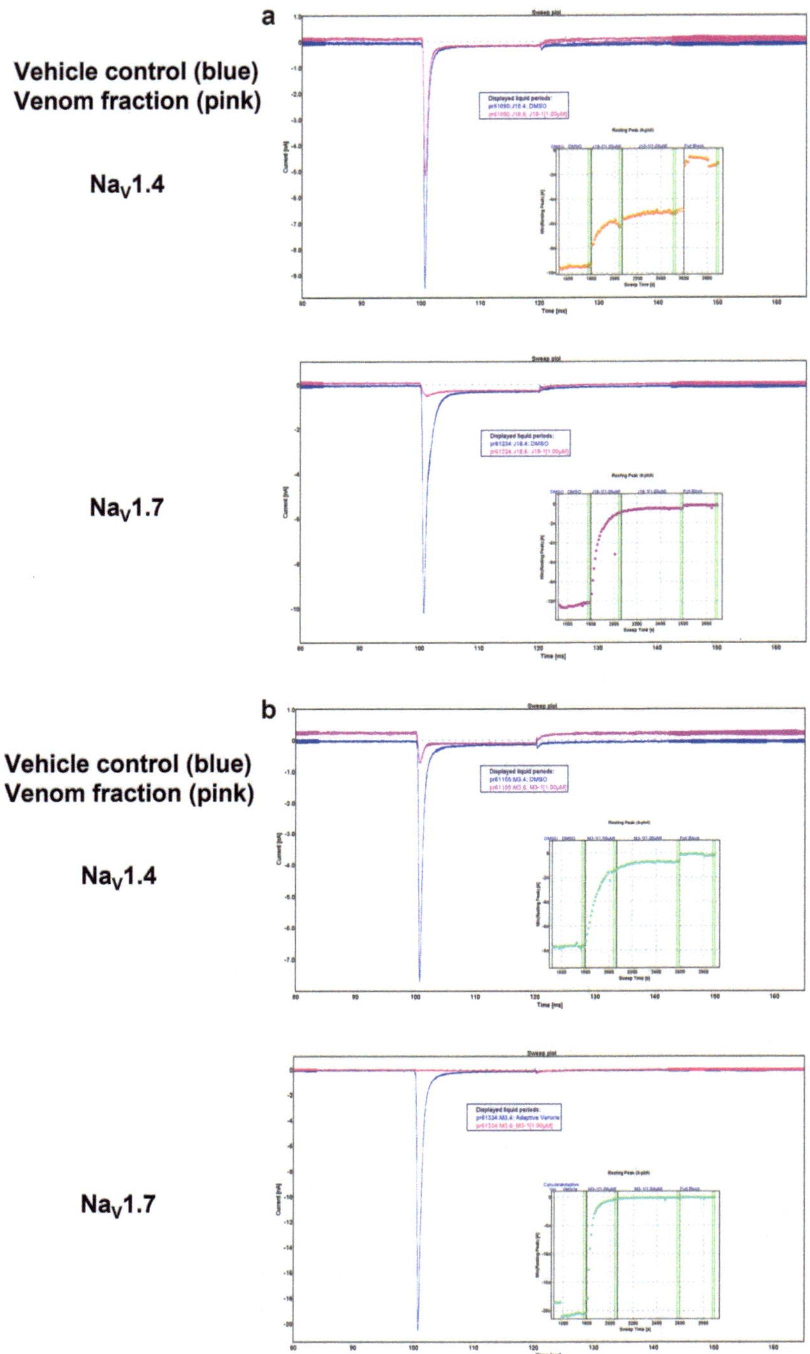

Fig. 6 Slow use-dependant blockage of Na_V channels with novel toxins. A collaboration between Venomtech Ltd. (UK) and Charles River Laboratories (UK) presented as poster number 1031 at SfN Neuroscience 2021 (virtually) and orally at ELRIG Discovery Technologies 2021. Screening a Targeted Venom Discovery Array® (Venomtech) containing 960 fractions generated a hit rate of 3.6% many of which are new toxins. The examples shown here of a partial selective 3.1a and a non-selective 3.1b peptide both exhibit the slow, pulse-dependant onset of current inhibition

						peptide alignment																																					
UniProt ID	short name	species	origin	identity	NaV1.7 potency	0	1	2	3	4	5	6	7	8	9	10	11	12	13	14	15	16	17	18	19	20	21	22	23	24	25	26	27	28	29	30	31	32	33	34	35	36	37
P08426	U6-TRTX-Hs1a	*Cyriopagopus schmidti*	Asia	63	untested	R	E	C	L	G	I	F	K	A	C	T	P	G	K	N	E	C	C	S	-	-	N	R	V	C	S	G	K	T	R	W	S	K	W	K	L	-	-
P0DP95	u-TRTX-Osp1a	*Orphnaecus sp.*	Asia	59	none published	-	G	C	K	G	F	G	K	A	C	K	Y	G	A	D	E	C	C	K	-	-	N	L	V	C	S	K	K	H	K	W	C	K	Y	T	L	-	-
B3EWN2	u-TRTX-Hlv1a	*Cyriopagopus lividus*	Asia	80	none published	-	A	C	L	G	F	G	E	K	C	N	P	S	N	D	K	C	C	K	S	S	S	L	V	C	S	Q	K	H	K	W	C	K	Y	G	W	-	-
A0A0J9XIW9	u-TRTX-Hd1a	*Cyriopagopus doriae*	Asia	68	111nM	-	A	C	L	G	F	G	K	S	C	N	P	S	N	D	Q	C	C	K	S	S	S	L	A	C	S	T	K	H	K	W	C	K	Y	E	L	-	-
P83303	u-TRTX-HxTx4	*Cyriopagopus schmidti*	Asia	100	9.6nM	R	E	C	L	E	I	F	K	A	C	N	P	S	N	D	Q	C	C	K	S	S	K	L	V	C	S	R	K	T	R	W	C	K	Y	Q	I	G	K
P60975	u-TRTX-Hhn1a	*Cyriopagopus hainanus*	Asia	77	42.3nM	-	E	C	L	G	F	G	K	G	C	N	P	S	N	D	Q	C	C	K	S	A	N	L	V	C	S	R	K	H	R	W	C	K	Y	E	I	-	-
D2Y234	u-TRTX-Hhn1c	*Cyriopagopus hainanus*	Asia	77	not published	R	E	C	L	G	F	G	K	G	C	N	P	S	N	D	Q	C	C	K	S	S	N	L	V	C	S	R	K	H	R	R	C	K	Y	E	I	G	K
D2Y207	u-TRTX-Hhn1b	*Cyriopagopus hainanus*	Asia	80	not published	R	E	C	L	G	F	G	K	G	C	N	P	S	N	D	Q	C	C	K	S	S	N	L	V	C	S	R	K	H	R	W	C	K	Y	E	I	G	K
D2Y235	u-TRTX-Hhn1d	*Cyriopagopus hainanus*	Asia	77	not published	R	E	C	L	G	F	G	K	G	C	N	P	S	S	D	Q	C	C	K	S	S	N	L	V	C	S	R	K	H	R	W	C	K	Y	E	I	G	K
P84510	u-TRTX-Ps1a	*Paraphysa scrofa*	Americas	63	not published	-	D	C	L	G	F	L	W	K	C	N	P	S	N	D	K	C	C	R	P	-	N	L	V	C	S	R	K	D	K	W	C	K	Y	Q	I	-	-
P0DL73	identical to othe	*Paraphysa scrofa*	Americas	52	0.58nM	-	D	C	L	G	F	M	R	K	C	I	P	D	N	D	K	C	C	R	P	-	N	L	V	C	S	R	T	H	K	W	C	K	Y	V	F	-	-
P0DL72	GpTx1?	*Grammostola porteri*	Americas	52	0.58nM	-	D	C	L	G	F	M	R	K	C	I	P	D	N	D	K	C	C	R	P	-	N	L	V	C	S	R	T	H	K	W	C	K	Y	V	F	-	-
P0DJA9	GTx1-15	*Grammostola rosea*	Americas	52	0.58nM	-	D	C	L	G	F	M	R	K	C	I	P	D	N	D	K	C	C	R	P	-	N	L	V	C	S	R	T	H	K	W	C	K	Y	V	F	G	K
P0DL64	u-TRTX-Tp1a	*Thrixopelma pruriens*	Americas	52	2.1nM	-	D	C	L	K	F	G	W	K	C	N	P	R	N	D	K	C	C	S	-	-	G	L	K	C	G	S	N	H	N	W	C	K	L	H	I	-	-
A0A1S4NYE8	b-TRTX-Pre1a	*Psalmopoeus reduncus*	Americas	51	not published	E	D	C	L	G	W	F	S	R	C	S	P	K	N	D	K	C	C	P	-	-	N	Y	K	C	S	S	K	D	L	W	C	K	Y	K	I	W	-
P0DL84	b-TRTX-Cd1a	*Ceratogyrus darlingi*	Africa	54	16nM	-	D	C	L	G	W	F	K	S	C	D	P	K	N	D	K	C	C	K	-	-	N	Y	S	C	S	R	R	D	R	W	C	K	Y	D	L	-	-
P84507	b-TRTX-Cm1a	*Ceratogyrus marshalli*	Africa	54	129.1nM	-	D	C	L	G	W	F	K	S	C	D	P	K	N	D	K	C	C	K	-	-	N	Y	T	C	S	R	R	D	R	W	C	K	Y	D	L	-	-
P84508	b-TRTX-Cm1b	*Ceratogyrus marshalli*	Africa	58	95.5nM	-	D	C	L	G	W	F	K	S	C	D	P	K	N	D	K	C	C	K	-	-	N	Y	T	C	S	R	R	D	R	W	C	K	Y	Y	L	-	-
P0DL74	VSTX3	*Paraphysa scrofa*	Americas	52	770nM	-	D	C	L	G	W	F	K	G	C	D	P	D	N	D	K	C	C	E	-	-	G	Y	K	C	N	R	R	D	K	W	C	K	Y	K	L	W	-
P0C2P5	VSTX3	*Grammostola rosea*	Americas	52	430nM	-	D	C	L	G	W	F	K	G	C	D	P	D	N	D	K	C	C	E	-	-	G	Y	K	C	N	R	R	D	K	W	C	K	Y	K	L	W	-
P0DP93	Ccy1a	*Chromatopelma cyaneopubescens*	Americas	64	not published	D	D	C	L	G	I	F	K	S	C	N	P	D	N	D	K	C	C	E	-	-	S	Y	K	C	S	R	R	D	K	W	C	K	Y	V	L	-	-
P0DP94	Ccy1b	*Chromatopelma cyaneopubescens*	Americas	55	not published	D	D	C	L	G	F	F	K	S	C	N	P	D	N	D	K	C	C	E	-	-	N	Y	K	C	N	R	R	D	K	W	C	K	Y	V	L	-	-
P83745	K-TRTX-Tb1a	*Theraphosa blondi*	Americas	50	not published	A	A	C	L	G	M	F	E	S	C	D	P	N	N	D	K	C	C	P	-	-	N	R	E	C	N	R	K	H	K	W	C	K	Y	K	L	W	-
P0DQO3	u-TRTX-Tap1a	*Theraphosa apophysis*	Americas	54	not published	D	D	C	L	G	M	F	S	S	C	D	P	N	N	D	K	C	C	P	-	-	N	R	K	C	S	R	K	D	Q	W	C	K	Y	Q	L	W	-
B1P1F1	U16-TRTX-Cg1a	*Chilobrachys guangxiensis*	Asia	50	not published	D	D	C	L	G	M	F	S	S	C	N	P	D	N	D	K	C	C	E	-	-	G	R	K	C	D	R	R	D	Q	W	C	K	W	N	P	W	-
B3EWN0	b-TRTX-Pmr1a	*Pterinochilus murinus*	Africa	52	not published	D	D	C	L	G	M	F	S	S	C	D	P	D	N	D	K	C	C	E	-	-	G	R	K	C	N	R	K	D	K	W	C	K	Y	V	L	-	-
B1P1F2	U16-TRTX-Cg1a	*Chilobrachys guangxiensis*	Asia	50	not published	D	D	C	L	G	L	F	S	S	C	N	P	D	N	D	K	C	C	E	-	-	G	R	K	C	N	R	R	D	K	W	C	K	L	K	L	W	-

Fig. 7 Peptide BLAST limited to Theraphosidae showing species and potency of related sequences. Sequence presented from NCBI BLASTP (*National Centre for Biotechnology Information. Basic Local Alignment Search Tool – Protein (BLASTP)*, 2022) demonstrating the conserved cystine knot motif of six cysteine from voltage-gated Na^+ channel-inhibiting peptides. Only sequences with = > 50% identity to huwentoxin HxTx4 were included, potency reported as the lowest published for the UniProt entry. The diversity of biological starting points here demonstrates over 1000-fold difference in recorded potency and originates from three major continents. Several β-toxins have a shorter loop 3 and truncated C-termini. Strongly conserved residues are in blue and those for $Na_V1.7$ binding in green

techniques to improve on the great starting points of venom peptides and shape them into drug-like molecules. One such example published by Pfizer and Amunix utilises gene shuffling directed evolution from a starting point of CcoTx-1 (correctly named β-theraphotoxin-Cm1a) to selectively and potently block $Na_V1.7$ as a pain therapeutic (Shcherbatko et al. 2016). The resultant selection and follow-up saturation mutagenesis increased the potency at $Na_V1.7$ to 2.6 nM and 1000-fold selectivity over muscular $Na_V1.4$ and $Na_V1.5$, as well as eliminating K^+ and Ca^+ activity that had been seen in the native protein (Shcherbatko et al. 2016).

Although oral bioavailability is still challenging with peptides, it is not impossible. Of the possible opportunities, full cyclisation has been shown to produce oral bioavailability in conotoxin Vc1.1 from *Conus victoriae* (Clark et al. 2010). In this study, adding a 6-amino acid linker to the C-terminus and then cyclising by native chemical ligation, the authors demonstrated increased stability in synthetic gastric and intestinal fluids (>24 h) before oral dosing in a pain model (Clark et al. 2010). Although target analgesia through $Ca_V2.2$ was achieved in the chronic constriction injury model, no plasma samples were recorded for evidencing bioavailability and total exposure. Even some whole venoms, such as those from shrews, with long plasma half-lives of 16 h (Liao et al. 2022), and HsTx-1(R19A), maintained 90% bioavailability at 5 h (Jin et al. 2016a, b).

Techniques such as amino acid modifications are useful to avoid protease degradation, so too are the use of D-amino acids and C-terminal amides all of which are expertly reviewed by Muttenthaler et al. 2021. However, such modifications also naturally occur in venom peptides, probably due to the powerful selection pressure for stability. HsTx-1 (Jin et al. 2016a, b) has a C-terminal amide, and so far, platypus, spider and cone snail venoms have been discovered with D-amino acids (Mast et al. 2021). As our technology to decipher these modifications in nature improves, so will the knowledge about how nature has provided drug-like stability in venom peptides. Snake venoms, particularly those from rattlesnakes, have been known to contain L-amino acid oxidase (LAAO) for decades, and it is thus highly likely that many new D-amino acid-containing peptides will be discovered in their venom as a natural response to this enzyme.

In summary, the selection pressures for venoms, to be ready for use at a moment's notice, have led to useful modifications of weaponised peptides that further improves their utility in human therapeutics.

4 Translating Toxins to the Clinic

4.1 *Autoimmune Disease*

The autoimmune diseases multiple sclerosis and rheumatoid arthritis are challenging to treat. Current therapies are non-specific acting broadly, so producing a number of off-target effects (Fugger et al. 2020). More specific therapies are needed and targeting ion channels could be the answer. $K_v1.3$ channels overexpressed in activated effector memory (T_{EM}) lymphocytes are implicated in the development of these types of autoimmune diseases (Tanner et al. 2017). $K_v1.3$ channels are needed for chronically activated memory T cells to function. In multiple sclerosis and rheumatoid arthritis, the $K_v1.3$ potassium channel is a key regulator of $CCR7^-$ effector memory T lymphocyte (T_{EM}) cell activation, and in these conditions, T_{EM} cells that lack CCR7 expression are key drivers of inflammation (Tanner et al. 2017).

The peptide (ShK) from the sea anemone *Stichodactyla helianthus* blocks Kv1.3 channels with high affinity but poor selectivity. From this was developed ShK-186 (called dalazatide) with superior selectivity, which has been through human phase I clinical trials utilising subcutaneous injection for the treatment of plaque psoriasis (Tajti et al. 2020). The early clinical trial in humans has shown that dalazatide is generally well-tolerated and can improve psoriatic skin lesions, but larger studies are needed (Tarcha et al. 2017).

Following on from ShK and its analogues, new, selective $K_V1.3$ inhibitors have been discovered from other species. One of these improved peptides, HsTx1[R14A] from the Asian forest scorpion *Heterometrus spinnifer* discussed earlier, is a promising therapeutic candidate (Rashid et al. 2014).

Venom peptide delivery mechanisms are also being investigated, as although oral bioavailability of peptides is a long-standing challenge, progress is being made. Pulmonary administration of HsTx1[R14A] could be a good alternative for peptide delivery instead of oral delivery, with 44% bioavailability in the rat plasma peaking at 2 h post dose (Jin et al. 2016a, b). Another method of delivery was investigated utilising treatment via the buccal mucosa which, in mice, delivered the peak peptide concentrations to the plasma at 4 h (Jin et al. 2016a, b). The non-invasive systemic delivery of the $K_V1.3$-blocking peptide HsTx1[R14A] showed that this could be a promising delivery mechanism via this route. Invasive intrathecal delivery such as that used with ziconotide is not thought to be well-tolerated by this patient group.

Rat intravenous and subcutaneous injections of labelled HsTx1[R14A] were used to understand the distribution and pharmacokinetics. HsTx1[R14A] was conjugated with NOTA (1,4,7-triazacyclononane-1,4,7-triacetic acid), and the peptide was also radiolabelled with copper-64 to undertake biodistribution and positron emission tomography (PET) studies (Bergmann et al. 2017).

A study by Jin et al. 2021, investigated another method of HsTX1[R14A] peptide delivery by utilising a controlled release microsphere formulation. The peptide was encapsulated in microspheres made using different poly(lactic-co-glycolic acid) (PLGA) materials. PLGA breaks down to lactic and glycolic acid which are excreted as carbon dioxide and water, and its use has also been approved by the FDA. The peptide/PLGA encapsulations were subcutaneously delivered to Sprague-Dawley rats, and the RG503H PLGA microsphere formulation was found to be the most promising (Jin et al. 2021).

4.2 Pain Therapeutics

Prialt® (ziconotide) (as discussed previously) was developed from *Conus magus*, a fish-hunting cone snail venom peptide called ω-conotoxin-MVIIA, and was approved by the FDA in 2004 to treat severe chronic pain related to AIDS, cancer and neuropathies (Safavi-Hemami et al. 2019). This drug blocks voltage-gated calcium channel $Ca_V2.2$. Ziconotide is administered intrathecally, has good efficacy and doesn't have the opioid-related side effects. For long-term chronic pain patients that require intrathecal drugs to control their pain, ziconotide is recommended as a first-line therapy (Chalil et al. 2021).

From the venom of the sea anemone *Anthopleura elegantissima*, the peptide APETx2 was discovered (Diochot et al. 2004) which inhibits ASIC3 and is selective over ASIC1a, ASIC1b and ASIC2a. APETx2 was successful in inhibiting inflammatory pain in rats (Karczewski et al. 2010). APETx2 has quite a complex pharmacology; Lee et al. showed that it potentiated rat ASIC1b at concentrations 30–100-fold higher than inhibition at ASIC3, and this needs to be considered if APETx2 is to be used in the clinic as an ASIC inhibitor (Lee et al. 2018). Another sea anemone peptide from *Urticina grebelnyi* called π-AnmTX Ugr 9a-1 (Ugr 9–1) which inhibits ASIC3 also has analgesic activity, as shown using in vivo tests in mice (Osmakov et al. 2013). The mambalgins that target ASICs also have analgesic activity and knock-down, and knockout mice were used to investigate the effects on ASIC1a and ASIC1b channels (Diochot et al. 2016).

Huwentoxin-IV (HwTx-IV) a $Na_v1.7$ inhibitor (above) has been shown to have analgesic effects on mice and rats in an inflammatory pain model (Liu et al. 2014a, b). Another theraphosid peptide that acts on $Na_v1.7$ is from *Pamphobeteus nigricolor*. The peptide, called μ-theraphotoxin-Pn3a peptide, blocks $Na_v1.7$ and shows analgesic activity in rodents when co-administered with a sub-therapeutic dose of opioids but not on its own (Deuis et al. 2017).

Despite its lack of lack of selectivity, the potent voltage-gated Na^+ channel blocker TTX has been published as producing analgesia in Swiss Webster mice (Marcil et al. 2006). Studies have also been undertaken in rats which also show TTX to have an analgesic effect (Zhang et al. 2006). It is surprising that these studies have been performed with ethical approval, as the involvement of $Na_V1.7$ in pain has been confirmed since 2004 (Nassar et al. 2004) followed by the human data on congenital insensitivity to pain in 2006 (Cox et al. 2006); also TTX is not therapeutic material with its well-known human toxicity.

A peptide from the venom of the armed spider *Phoneutria nigriventer* called Phα1β blocks the TRPA1 channel selectively, which causes an analgesic effect, and a recombinant form of the peptide called CTK was shown to work in vitro and in vivo (Tonello et al. 2017).

Centipede venoms are a relatively untapped source of ion channel tools that interact with Na_v, K_v, Ca_v and TRPV1 channels (Chu et al. 2020). Two peptides from centipedes have been characterised as specific Na^+ channel blockers, namely, μ-$SLPTX_3$-Ssm2a and μ-$SLPTX_3$-Ssm3a. In a rodent pain model, μ-$SLPTX_3$-Ssm2a, via targeting $Na_v1.7$, has shown to be more effective to block pain than morphine and is highly stable in human plasma (Dib-Hajj et al. 2013).

Another *Phoneutria nigriventer* peptide Tx3–3 (ω-ctenitoxin-Pn3), which blocks the activity of voltage-gated Ca^{2+} channels, has been shown to reduce pain symptoms in reserpine-induced fibromyalgia (Pedron et al. 2021). Given intrathecally at 30 picomoles/injection, Tx3–3 was slightly more effective than pregabalin and presumably could act on $Ca_V2.2$ (like ziconotide). However, as fibromyalgia is a chronic and diffuse pain condition, it was unclear from the study if the analgesic effect was restricted to the L4-L5 spinal nerves or could give systemic analgesia.

Pain-causing toxins are also useful tools in drug discovery; the plant compound capsaicin is probably the most common. An excellent review on a range of algesic tools has been produced by a collaboration of several Australian universities (Jami et al. 2018). Several scorpion venoms appear to cause pain through activating $Na_V1.7$ channels (Rowe et al. 2011). OD1 is a toxin from the Iranian scorpion *Odonthobuthus doriae* which increases $Na_V1.7$ current through delayed inactivation (Maertens et al. 2006). This has led to the publication of a novel in vivo mouse model of intra-plantar pain mediated by $Na_V1.7$ following injection of OD1 (Deuis et al. 2016). Several research groups tested their new venom peptides and compounds in this model by co-injecting the inhibitor with the OD1 (Rahnama et al. 2017; Sousa et al. 2017). Follow-up studies were performed with intra-peritoneal injection 10 min before injection of OD1 (Deuis et al. 2017); thus, it appears to be a suitable target-specific model of pain, like capsaicin is for TRPV1.

4.3 Central Nervous System (CNS) Disorders

Another armed spider peptide PhKV (previously Tx3-1, which is a good example of why rational nomenclature is needed) is demonstrating interesting results in the fight against Alzheimer's disease. PhKV has been shown to block A-type K^+ currents (I_A). Mice were given a novel object recognition task in wooden chambers after being administered PhKV; the task showed that the mice had enhanced both short- and long-term memory consolidation after dosing (Gomes et al. 2013).

Ion channels are dysregulated in epilepsy, and so potential targets for treatment include voltage-gated Na^+, K^+, Ca^{2+} channels, Cys-loop receptors, ionotropic glutamate receptors and P2X receptors and can be targeted with venoms in the search for antiepileptic drugs (Chow et al. 2020). SCN1A encodes the voltage-gated sodium channel $Na_V1.1$; however, in most people with Dravet syndrome, in one allele of SCN1A, there is a dominant loss of function mutation. This causes a reduced activity of the channel. Hm1a venom peptide isolated from the African arboreal theraphosid *Heteroscodra maculata* targets $Na_V1.1$ through inhibition of inactivation and rescues mice from seizures and premature death (Richards et al. 2018).

Micrurotoxins 1 and 2 (MmTX1 and MmTX2) from *Micrurus mipartitus* coral snake venom enhance ionotropic γ-aminobutyric acid type A ($GABA_A$) agonist sensitivity, although the in vivo effect is a reduction in basal activity followed by busts of intense seizures (Rosso et al. 2015). Other labs have found both agonist and antagonist activity of *Micrurus* spp. venoms at GABA (Kleiz-Ferreira et al. 2022).

4.4 Cancer

Ion channel expression alteration and/or activity can cause oncochannels, ion channels with oncogenic potential (Huber 2013); as a result of these discoveries, ion channels are now considered novel therapeutic targets in cancer research. Overexpression of $K_V1.3$ channels is observed in breast, colon, smooth muscle, skeletal muscle and lymph node cancers (Teisseyre et al. 2019). Scorpion venom peptides have been shown to block voltage-gated K^+ and Na^+ channels in cancer (Díaz-García and Varela 2020).

A human glioma U251-MG rodent xenograft model has been used to test the effect of the scorpion venom BmK from *Buthus martensii* (W.-X. Wang and Ji 2005). The venom induced glioma cell apoptosis, and the authors hypothesise that the apoptosis and inhibition of growth of the tumour are due to effect on ion channels. Anti-tumour analgesic peptide (AGAP) from the venom of the Chinese scorpion *Buthus martensii* inhibits the voltage-gated proton channel H_v1 (Tang et al. 2020) via the Zn^{2+} binding pocket and voltage-gated Ca^{2+} channels in DRGs (Liu et al. 2014a, b).

This peptide has also been shown to inhibit Na^+ channels. When AGAP and $AGAP^{W38G}$ (glycine replacing tryptophan at position 38) were investigated (which both inhibit $Na_v1.7$ and $Na_v1.8$), $AGAP^{W38G}$ had greater therapeutic utility with the ability to block pain with less toxicity to skeletal and cardiac muscles when investigated in vivo. This is because it had greater selectivity over $Na_v1.4$ and $Na_v1.5$ (Xu et al. 2017).

Another scorpion peptide has been shown to have therapeutic and diagnostic value for cancer patients. The 36-amino acid peptide chlorotoxin (CTX or CLTX) from the scorpion *Leiurus quinquestriatus* interacts with voltage-dependant Cl^- channels and was first used as a tool to characterise Cl^- channels but was found to target the most common primary brain malignancy, gliomas (Dardevet et al. 2015). CLTX binds these tumours but not to healthy brain cells and can penetrate the blood-brain tumour barrier (BBTB). CLTX has been shown to target tumours in mice using a transgenic glioma model (Akcan et al. 2011). CLTX-directed chimeric antigen receptor (CAR) T cells have been developed to target glioblastoma (Wang et al. 2020). The combination of CLTX and CAR T (CLTX-CAR T cells) is superior to CAR T alone. Tumour regression was shown in orthotopic xenograft glioblastoma tumour models and was without off-target effects on healthy brain cells. CLTX has been successfully evaluated in pre-clinical and clinical trials for cancer (Cohen et al. 2018). Safety and feasibility studies in humans look promising (Brown et al. 2021).

4.5 Infectious Disease

The incredible scientific efforts undertaken during the worse of the coronavirus pandemic saw many scientists investigating their field of expertise in relation to COVID-19, and this was no different for the ion channel experts.

Calcium signalling has long been known to play a vital role in virus infection (Berlansky et al. 2022). Ca^{2+} channel blockers have been used in epithelial lung cells to inhibit SARS-CoV-2 infectivity. TRP channels could be potential targets to prevent or treat COVID-19 (Jaffal and Abbas 2021). A number of animal venoms have been used to block TRP channels, and this could be investigated to see if it would be useful in preventing or treating SARS-CoV-2. α7-AChR was thought to be part of SARS-CoV2 pathology through the nicotinic hypothesis; however this is now thought unlikely (Godellas et al. 2022).

The field of bacterial ion channels as therapeutic targets of infection has not really taken off yet, but it is possible that venoms will yield toxins to bacterial ion channels as a result of venom microbiome and the large number of venom toxins targeting ion channels.

5 Techniques for Identifying Novel Venom Peptide Ligands

Before the invention of high-performance liquid chromatography (HPLC), venoms were often studied as whole venom, sometimes referred to as crude venom in reference to its complexity, like crude oil. In this state, the observed pharmacology is the sum of all active compounds or just those of the most potent or most abundant. With suitable assays and with target venoms that lack cytotoxicity, crude venoms can be successfully screened to find new drug-like activities. A fluorescent screen with SHS-Y5Y cells endogenously expressing $Na_V1.7$ was used to screen 205 crude theraphosid venoms, as they are a rich source of Na_V ligands (see previously); 40% demonstrated some inhibition of veratridine-induced Ca^{2+} flux, with a further 14.6% activators identified (Klint et al. 2015). From the 82 inhibitor hits, Klint and colleagues separated the venoms by HPLC and rescreened to uncover 7 new selective $Na_V1.7$ inhibitors, the best being μ-theraphotoxin-Hd1a, which has 68% homology to HxTx4 (Fig. 7). Screens like these demonstrate the potential for new molecules and new starting material for many challenging targets in drug discovery. Finding that ~40% of Theraphosidae species contain some $Na_V1.7$ inhibitors, and that these were unrelated at the sub-family level (Klint et al. 2015), means this could extrapolate to other untested species of Theraphosidae. With over 1000 species of Theraphosidae currently described (Sherwood et al. 2021), there are potentially twice as many new venoms inhibiting $Na_V1.7$ than were screened by Klint et al. (and this was probably one of the largest screens of theraphosid venoms against $Na_V1.7$ published). These complex mixtures are now understood to contain more than just what is produced by the venom gland cells. The origin of TTX from pufferfish has turned out to be symbiotic bacteria (Lago et al. 2015), probably similar to those in the blue-ringed octopus (Whitelaw et al. 2019); this hints at a wider co-evolution of endosymbiotic bacteria influencing venom composition and evolution of venomous animals. Endosymbiotic microbes and parasites are being discovered in a wide range of species, and now we have the tools to look for them. The Initiative for Venom Associated Microbes and Parasites (iVAMP) is a community of venom researchers that are finding this symbiosis wherever they look (Ul-Hasan et al. 2019), and the latest evidence is showing co-evolution, as expected, between venomous host and colonising bacteria explaining the selection pressure for antimicrobial peptides (Esmaeilishirazifard et al. 2022). Selection for peptides that modulate microbial and parasitic ion channels may be an underlying mechanism of such diversity yet to be discovered. In order to access this complexity and identify the drug-like peptides within these complex mixtures, they need to be simplified. Advances in HPLC including ultra- and nano-fractionation (Neumann et al. 2020), when coupled with high-throughput screening (HTS) systems such as automated patch clamp, are paving the way for rapid growth in the discovery of venom compounds as ligands for ion channels. A process for building targeted venom fraction libraries using two-dimensional HPLC has been recently described (Trim et al. 2021), and the use of these libraries has demonstrated new ligands for well-characterised targets such as $Na_V1.7$ and $Na_V1.4$ (Poster 1031, Neuroscience 2021). Many other platforms for high-throughput peptide drug discovery and development are being published, such as those reviewed in 2021 (Muttenthaler et al. 2021).

In addition to proteomics, genomics and transcriptomics are frequently used to elucidate the sequences of active peptides from venoms. Although transcriptomics delivers powerful sequence data, this often involves euthanasia of the venomous species in order to dissect out the venom gland for RNA extraction. Occasionally research is even published where the authors were not able to dissect out the venom glands so whole body segments were processed leading to contaminating transcripts from cuticle, muscle and other possible tissues; this problem has been reviewed recently (Vonk et al. 2021). As technology improves for isolating nucleic acids from smaller samples, we are discovering that RNA is not just present in snake venom (Currier et al. 2012) but also scorpion venom (Vonk et al. 2021) where there is sufficient quantity and quality to be used for transcriptomics. There is no common venomous ancestor between snakes and scorpions, and thus the presence of useable RNA in both groups' venom implies it may be ubiquitous and therefore could lead to a new generation of euthanasia-free venom transcriptomics. One plausible mechanism for persistence of these transcripts could be explained by specialised extracellular vesicles called exosomes (Wilson et al. 2014), which are commonly released from secretary cells. Exosomes have been detected in venoms (Ogawa et al. 2008; Souza-Imberg et al. 2017) and been shown to transport nucleic acids (Ohshima et al. 2010); further work is on-going by several laboratories in this exciting field. Generation of cDNA libraries from venomous animals produces great opportunities to expand and improve on nature by producing recombinant peptide display libraries; in these large libraries, $>10^{10}$ peptides can then be screened in assays and enriched to discover the pharmacology of interest. So far, we are still only scratching the surface of what nature has to offer, and combining millions of years of evolution with the accelerating pace of human ingenuity is a very exciting time to be looking at venom peptide drug discovery.

6 Conclusion

"Endless forms most beautiful and most wonderful have been and are being evolved", said Darwin and many others since. This applies to venom peptides as much as it does to the whole organism scale that Charles Darwin was referring to. Initial interest and drug discovery with venom peptides and toxins focused on the effects of envenomation, but with high-throughput screening techniques, drug discovery is starting to inform what we know about the evolution of venoms. The large interaction surface of venom peptides allows for greater selectivity between closely related ion channels that have very different functions. Understanding these interactions will allow us to develop new therapeutics from venom peptide starting points. Venom peptide and ion channel structural biology efforts over the last 25 years can be divided into three waves, each driven by a specific technology that progressively yielded more insight into how venom peptides bind to ion channels (Fig. 2). In the first wave (1994–2006), NMR provided a glimpse into these interactions via solution structures of venom peptides with polypeptides representing short sequences of ion channels (Fig. 3). In the second wave (2007–2014), X-ray diffraction yielded structures of entire ion channels in complex with venom peptides for the first time. Finally, in the third and current wave (2016–present), cryo-EM has played a dominant role, bringing into focus clinically relevant classes of ion channels such as voltage-gated Na^+ and Ca^{2+} channels (Fig. 2b and Fig. 4). The technologies that will build on this remarkable progress to catalyse the "fourth wave" of venom peptide and ion channel structural biology remain to be seen.

Computational methods are likely to play a major role in this next wave (Table 2). One possibility is that deep learning-based structure prediction methods such as AlphaFold (Jumper et al. 2021) will vastly expand the repertoire of venom peptide and ion channel complexes. While this approach has achieved great success at predicting the structures of single-protein domains to high resolution, significant progress at modelling the multiple conformational states of proteins (Del Alamo et al. 2022), the disulphide frameworks of venom peptides (Gulsevin and Meiler 2022) and the geometries of heteromultimers (Evans et al. 2022) will facilitate similar progress at predicting the structures of venom peptide and ion channel complexes. Another possibility is that free energy methods will be more broadly adopted to dramatically increase the identification of potency- and selectivity-enhancing mutations of venom peptides, a central step in the drug discovery process. One of these free energy methods, FEP+, appears to accurately predict the free energy of mutation even when using comparatively low-resolution electron microscopy structures, suggesting it may have broad utility (Katz et al. 2021a, b). Ultimately, the combination of computational methods with the considerable and growing body of structures will accelerate the development of venom peptides into drugs that target ion channels (Holford et al. 2018).

Venoms are such powerful organ systems that bring significant survival benefits to their possessors that we should not be surprised they have evolved so many times. This overarching view fails to demonstrate the scale of the opportunity. There are potentially 16 million unique venom peptides to investigate in spiders alone, and another 100,000 in the known scorpions (King et al. 2008) without counting thousands of other venomous species. This is why rational nomenclature is crucial for continuity and will also provide a structure for naming modified peptides from SAR and structure-guided drug design. There is a world of opportunity to explore, and it needs to be done ethically and responsibly for the benefit of the patients. Venom peptide drugs have already improved patients' lives, with several more in clinical and pre-clinical research. Peptide chemistry is starting to open up the possibilities of oral delivery, but the boom in biologicals such as antibodies has also increased injectable delivery as acceptable ways to take medication. Finding novel therapeutics in venoms adds previously unknown value to these species which are often persecuted and thus could encourage people to value all species as we need them to advance human, and animal, disease treatment.

References

Adams ME. Agatoxins: ion channel specific toxins from the american funnel web spider, Agelenopsis aperta. Toxicon. 2004;43(5):509–25.

Akcan M, Stroud MR, Hansen SJ, Clark RJ, Daly NL, Craik DJ, Olson JM. Chemical re-engineering of chlorotoxin improves bioconjugation properties for tumor imaging and targeted therapy. J Med Chem. 2011;54(3):782–7.

Arbuckle K, Harris RJ. Radiating pain: venom has contributed to the diversification of the largest radiations of vertebrate and invertebrate animals. BMC Ecol Evol. 2021;21(1):1–12.
Banerjee A, Lee A, Campbell E, Mackinnon R. Structure of a pore-blocking toxin in complex with a eukaryotic voltage-dependent K(+) channel. Elife. 2013;2:e00594.
Basus VJ, Song G, Hawrot E. NMR solution structure of an alpha-bungarotoxin/nicotinic receptor peptide complex. Biochemistry. 1993;32(46):12290–8.
Bateman A, Martin MJ, Orchard S, Magrane M, Agivetova R, Ahmad S, Alpi E, Bowler-Barnett EH, Britto R, Bursteinas B, Bye-A-Jee H, Coetzee R, Cukura A, da Silva A, Denny P, Dogan T, Ebenezer TG, Fan J, Castro LG, et al. UniProt: the universal protein knowledgebase in 2021. Nucleic Acids Res. 2021;49(D1):D480–9.
Beneski DA, Catterall WA. Covalent labeling of protein components of the sodium channel with a photoactivable derivative of scorpion toxin. Proc Natl Acad Sci U S A. 1980;77(1):639–43.
Bennie NAC, Loaring CD, Bennie MMG, Trim SA. Methods & techniques: an effective method for terrestrial arthropod euthanasia. J Exp Biol. 2012;215(24):4237–41.
Bergmann R, Kubeil M, Zarschler K, Chhabra S, Tajhya RB, Beeton C, Pennington MW, Bachmann M, Norton RS, Stephan H. Distribution and kinetics of the Kv1.3-blocking peptide HsTX1[R14A] in experimental rats. Sci Rep. 2017;7(1):1–11.
Berlansky S, Sallinger M, Grabmayr H, Humer C, Bernhard A, Fahrner M, Frischauf I. Calcium signals during SARS-CoV-2 infection: assessing the potential of emerging therapies. Cell. 2022;11(2)
Bosmans F, Rash L, Zhu S, Diochot S, Lazdunski M, Escoubas P, Tytgat J. Four novel tarantula toxins as selective modulators of voltage-gated sodium channel subtypes. Mol Pharmacol. 2006;69(2):419–29.
Brown C, Blanchard MS, Aftabizadeh M, Hibbard J, Dodia R, Lingaraju CR, Starr R, D'Apuzzo M, Forman SJ, Portnow J, Ressler JA, Stiller T, Barva B, Kilpatrick J, McNamara P, Wagner JR, Wang D, Barish M, Badie B. CTIM-29. Clinical evaluation of chlorotoxin-directed car t cells for patients with recurrent glioblastoma. Neuro-Oncology. 2021;23(Supplement_6):vi57–vi57.
Chalil A, Staudt MD, Harland TA, Leimer EM, Bhullar R, Argoff CE. A safety review of approved intrathecal analgesics for chronic pain management. Expert Opin Drug Saf. 2021;20(4):439–51.
Changeux J-P, Kasai M, Lee C-Y. Use of a snake venom toxin to characterize the cholinergic receptor protein. Proc Natl Acad Sci. 1970;67(3):1241–7.
Chen L, Durr KL, Gouaux E. X-ray structures of AMPA receptor-cone snail toxin complexes illuminate activation mechanism. Science. 2014;345(6200):1021–6.
Chow CY, Absalom N, Biggs K, King GF, Ma L. Venom-derived modulators of epilepsy-related ion channels. Biochem Pharmacol. 2020;181:114043.
Chu Y, Qiu P, Yu R. Centipede venom peptides acting on ion channels. Toxins. 2020;12(4)
Cicka D, Quave C. Bioprospecting for pharmaceuticals: an overview and vision for future access and benefit sharing. Med Plants From Farm Pharm. 2019;17–34
Clark RJ, Jensen J, Nevin ST, Callaghan BP, Adams DJ, Craik DJ. The engineering of an orally active conotoxin for the treatment of neuropathic pain. Angewandte Chemie Int Ed. 2010;49(37):6545–8.
Clark AJ, Gindin T, Zhang B, Wang L, Abel R, Murret CS, Xu F, Bao A, Lu NJ, Zhou T, Kwong PD, Shapiro L, Honig B, Friesner RA. Free energy perturbation calculation of relative binding free energy between broadly neutralizing antibodies and the gp120 glycoprotein of HIV-1. J Mol Biol. 2017;429(7):930–47.
Cohen G, Burks SR, Frank JA. Chlorotoxin – a multimodal imaging platform for targeting glioma tumors. Toxins. 2018;10(12)
Colom-Casasnovas A, Garay E, Cisneros-Mejorado A, Aguilar MB, Lazcano-Pérez F, Arellano RO, Sánchez-Rodríguez J. Sea anemone Bartholomea annulata venom inhibits voltage-gated Na+ channels and activates GABAA receptors from mammals. Sci Rep. 2022;12(1):5352.
Cox JJ, Reimann F, Nicholas AK, Thornton G, Roberts E, Springell K, Karbani G, Jafri H, Mannan J, Raashid Y, Al-Gazali L, Hamamy H, Valente EM, Gorman S, Williams R, McHale DP, Wood

JN, Gribble FM, Woods CG. An SCN9A channelopathy causes congenital inability to experience pain. Nature. 2006;444(7121):894–8.

Currier RB, Calvete JJ, Sanz L, Harrison RA, Rowley PD, Wagstaff SC. Unusual stability of messenger RNA in snake venom reveals gene expression dynamics of venom replenishment. PLoS One. 2012;7(8):1–10.

Daly JW, Brown GB, Mensah-Dwumah M, Myers CW. Classification of skin alkaloids from neotropical poison-dart frogs (Dendrobatidae). Toxicon. 1978;16(2):163–88.

Dardevet L, Rani D, El Aziz TA, Bazin I, Sabatier J-M, Fadl M, Brambilla E, De Waard M. Chlorotoxin: a helpful natural scorpion peptide to diagnose glioma and fight tumor invasion. Toxins. 2015;7(4):1079–101.

de Plater GM, Milburn PJ, Martin RL. Venom from the platypus, *Ornithorhynchus anatinus,* induces a calcium-dependent current in cultured dorsal root ganglion cells. J Neurophysiol. 2001;85(3):1340–5.

Del Alamo D, Sala D, McHaourab HS, Meiler J. Sampling alternative conformational states of transporters and receptors with AlphaFold2. Elife. 2022;11

Dellisanti CD, Yao Y, Stroud JC, Wang ZZ, Chen L. Crystal structure of the extracellular domain of nAChR alpha1 bound to alpha-bungarotoxin at 1.94 A resolution. Nat Neurosci. 2007;10(8):953–62.

Deuis JR, Wingerd JS, Winter Z, Durek T, Dekan Z, Sousa SR, Zimmermann K, Hoffmann T, Weidner C, Nassar MA, Alewood PF, Lewis RJ, Vetter I. Analgesic effects of GpTx-1, PF-04856264 and CNV1014802 in a mouse model of NaV1.7-mediated pain. Toxins. 2016;8(3)

Deuis JR, Dekan Z, Wingerd JS, Smith JJ, Munasinghe NR, Bhola RF, Imlach WL, Herzig V, Armstrong DA, Rosengren KJ, Bosmans F, Waxman SG, Dib-Hajj SD, Escoubas P, Minett MS, Christie MJ, King GF, Alewood PF, Lewis RJ, et al. Pharmacological characterisation of the highly Na v 1.7 selective spider venom peptide Pn3a. Sci Rep. 2017;7(January):1–18.

Díaz-García A, Varela D. Voltage-gated K+/Na+ channels and scorpion venom toxins in cancer. Front Pharmacol. 2020;11:913.

Dib-Hajj SD, Yang Y, Black JA, Waxman SG. The Na v 1.7 sodium channel: from molecule to man. Nat Rev Neurosci. 2013;14(1):49–62.

Diochot S. Pain-related toxins in scorpion and spider venoms: a face to face with ion channels. J Venom Anim Toxins Incl Trop Dis. 2021;27(March 2021):1–33.

Diochot S, Baron A, Rash LD, Deval E, Escoubas P, Scarzello S, Salinas M, Lazdunski M. A new sea anemone peptide, APETx2, inhibits ASIC3, a major acid-sensitive channel in sensory neurons. EMBO J. 2004;23(7):1516–25.

Diochot S, Baron A, Salinas M, Douguet D, Scarzello S, Dabert-Gay AS, Debayle D, Friend V, Alloui A, Lazdunski M, Lingueglia E. Black mamba venom peptides target acid-sensing ion channels to abolish pain. Nature. 2012;490(7421):552–5.

Diochot S, Alloui A, Rodrigues P, Dauvois M, Friend V, Aissouni Y, Eschalier A, Lingueglia E, Baron A. Analgesic effects of mambalgin peptide inhibitors of acid-sensing ion channels in inflammatory and neuropathic pain. Pain. 2016;157(3):552–9.

Eriksson MA, Roux B. Modeling the structure of agitoxin in complex with the Shaker K+ channel: a computational approach based on experimental distance restraints extracted from thermodynamic mutant cycles. Biophys J. 2002;83(5):2595–609.

Escoubas P, De Weille JR, Lecoq A, Diochot S, Waldmann R, Champigny G, Moinier D, Menez A, Lazdunski M. Isolation of a tarantula toxin specific for a class of proton-gated Na+ channels. J Biol Chem. 2000;275(33):25116–21.

Esmaeilishirazifard E, Usher L, Trim C, Denise H, Sangal V, Tyson GH, Barlow A, Redway KF, Taylor JD, Kremyda-Vlachou M, Davies S, Loftus TD, Lock MMG, Wright K, Dalby A, Snyder LAS, Wuster W, Trim S, Moschos SA. Bacterial adaptation to venom in snakes and Arachnida. Microbiol Spectr. 2022;10(3):e0240821.

Evans R, O'Neill M, Pritzel A, Antropova N, Senior A, Green T, Žídek A, Bates R, Blackwell S, Yim J, Ronneberger O, Bodenstein S, Zielinski M, Bridgland A, Potapenko A, Cowie A,

Tunyasuvunakool K, Jain R, Clancy E, et al. Protein complex prediction with AlphaFold-Multimer. BioRxiv; 2022.
Flinspach M, Xu Q, Piekarz AD, Fellows R, Hagan R, Gibbs A, Liu Y, Neff RA, Freedman J, Eckert WA, Zhou M, Bonesteel R, Pennington MW, Eddinger KA, Yaksh TL, Hunter M, Swanson RV, Wickenden AD. Insensitivity to pain induced by a potent selective closed-state Nav1.7 inhibitor. 2016;
Fugger L, Jensen LT, Rossjohn J. Challenges, progress, and prospects of developing therapies to treat autoimmune diseases. Cell. 2020;181(1):63–80.
Fukushima C, Mendoza JI, West RC, Longhorn SJ, Rivera E, Cooper EWT, Hénaut Y, Henriques S, Cardoso P. Species conservation profiles of tarantula spiders (araneae, theraphosidae) listed on CITES. Biodivers Data J. 2019;7:e39342.
Fuller MD, Zhang ZR, Cui G, McCarty NA. The block of CFTR by scorpion venom is state-dependent. Biophys J. 2005;89(6):3960–75.
Fuller MD, Thompson CH, Zhang Z-R, Freeman CS, Schay E, Szakács G, Bakos E, Sarkadi B, McMaster D, French RJ, Pohl J, Kubanek J, McCarty NA. State-dependent inhibition of cystic fibrosis transmembrane conductance regulator chloride channels by a novel peptide toxins. J Biol Chem. 2007;
Galvez A, Gimenez-Gallego G, Reuben JP, Roy-Contancin L, Feigenbaum P, Kaczorowski GJ, Garcia ML. Purification and characterization of a unique, potent, peptidyl probe for the high conductance calcium-activated potassium channel from venom of the scorpion Buthus tamulus. J Biol Chem. 1990;265(19):11083–90.
Gao S, Yao X, Yan N. Structure of human Cav2.2 channel blocked by the painkiller ziconotide. Nature. 2021;596(7870):143–7.
Godellas NE, Cymes GD, Grosman C. An experimental test of the nicotinic hypothesis of COVID-19. Proc Natl Acad Sci. 2022;119(44)
Gomes GM, Dalmolin GD, Cordeiro M, Gomez MV, Ferreira J, Rubin MA. The selective A-type K+ current blocker Tx3-1 isolated from the Phoneutria nigriventer venom enhances memory of naïve and Aβ(25-35)-treated mice. Toxicon. 2013;76:23–7.
Gulsevin A, Meiler J. Benchmarking peptide structure prediction with AlphaFold2. BioRxiv; 2022.
Hanf ZR, Chavez AS. A comprehensive multi-omic approach reveals a relatively simple venom in a diet generalist, the northern short-tailed shrew, Blarina brevicauda. Genome Biol Evol. 2020;12(7):1148–66.
Hanifin CT. The chemical and evolutionary ecology of tetrodotoxin (TTX) toxicity in terrestrial vertebrates. Mar Drugs. 2010;8(3):577–93.
Hannon HE, Atchison WD. Omega-conotoxins as experimental tools and therapeutics in pain management. Mar Drugs. 2013;11(3):680–99.
Hargreaves AD, Mulley JF. A plea for standardized nomenclature of snake venom toxins. Toxicon. 2014;90(August):351–3.
Hauke TJ, Herzig V. Muscle spasms – a common symptom following theraphosid spider bites? Toxicon. 2021;192(December 2020):74–7.
Henriques ST, Deplazes E, Lawrence N, Cheneval O, Chaousis S, Inserra M, Thongyoo P, King GF, Mark AE, Vetter I, Craik DJ, Schroeder CI. Interaction of tarantula venom peptide ProTx-II with lipid membranes is a prerequisite for its inhibition of human voltage-gated sodium channel NaV1.7. J Biol Chem. 2016;291(33):17049–65.
Holford M, Daly M, King GF, Norton RS. Venoms to the rescue. Science. 2018;361(6405):842–4.
Huang CJ, Schild L, Moczydlowski EG. Use-dependent block of the voltage-gated Na+ channel by tetrodotoxin and saxitoxin: effect of pore mutations that change ionic selectivity. J Gen Physiol. 2012;140(4):435–54.
Huber SM. Oncochannels. Cell Calcium. 2013;53(4):241–55.
Inceoglu B, Lango J, Jing J, Chen L, Doymaz F, Pessah IN, Hammock BD. One scorpion, two venoms: prevenom of Parabuthus transvaalicus acts as an alternative type of venom with distinct mechanism of action. Proc Natl Acad Sci U S A. 2003;100(3):922–7.
Infield DT, Rasouli A, Galles GD, Chipot C, Tajkhorshid E, Ahern CA. Cation-pi interactions and their functional roles in membrane proteins. J Mol Biol. 2021;433(17):167035.

Iwakawa N, Baxter NJ, Wai DCC, Fowler NJ, Morales RAV, Sugase K, Norton RS, Williamson MP. Conformational exchange in the potassium channel blocker ShK. Sci Rep. 2019;9(1):19307.

Jaffal SM, Abbas MA. TRP channels in COVID-19 disease: potential targets for prevention and treatment. Chem Biol Interact. 2021;345:109567.

Jami S, Erickson A, Brierley SM, Vetter I. Pain-causing venom peptides: insights into sensory neuron pharmacology. Toxins. 2018;10(1) MDPI AG

Jin L, Boyd BJ, Larson IC, Pennington MW, Norton RS, Nicolazzo JA. Enabling noninvasive systemic delivery of the Kv1.3-blocking peptide HsTX1[R14A] via the buccal mucosa. J Pharm Sci. 2016a;105(7):2173–9.

Jin L, Zhou Q, Chan HK, Larson IC, Pennington MW, Morales RAV, Boyd BJ, Norton RS, Nicolazzo JA. Pulmonary delivery of the Kv1.3-blocking peptide HsTX1[R14A] for the treatment of autoimmune diseases. J Pharm Sci. 2016b;105(2):650–6.

Jin L, Pan Y, Pham AC, Boyd BJ, Norton RS, Nicolazzo JA. Prolonged plasma exposure of the Kv1.3-inhibitory peptide HsTX1[R14A] by subcutaneous Administration of a poly(lactic-co-glycolic acid) (PLGA) microsphere formulation. J Pharm Sci. 2021;110(3):1182–8.

Jumper J, Evans R, Pritzel A, Green T, Figurnov M, Ronneberger O, Tunyasuvunakool K, Bates R, Zidek A, Potapenko A, Bridgland A, Meyer C, Kohl SAA, Ballard AJ, Cowie A, Romera-Paredes B, Nikolov S, Jain R, Adler J, et al. Highly accurate protein structure prediction with AlphaFold. Nature. 2021;596(7873):583–9.

Kalina RS, Kasheverov IE, Koshelev SG, Sintsova OV, Peigneur S, Pinheiro-Junior EL, Popov RS, Chausova VE, Monastyrnaya MM, Dmitrenok PS, Isaeva MP, Tytgat J, Kozlov SA, Kozlovskaya EP, Leychenko EV, Gladkikh IN. Nicotinic acetylcholine receptors are novel targets of APETx-like toxins from the sea anemone Heteractis magnifica. Toxins. 2022;14(10)

Kao CY, Nishiyama A. Actions of saxitoxin on peripheral neuromuscular systems. J Physiol. 1965;180(1):50.

Karczewski J, Spencer RH, Garsky VM, Liang A, Leitl MD, Cato MJ, Cook SP, Kane S, Urban MO. Reversal of acid-induced and inflammatory pain by the selective ASIC3 inhibitor, APETx2. Br J Pharmacol. 2010;161(4):950–60.

Katz D, Sindhikara D, DiMattia M, Leffler AE. Potency-enhancing mutations of gating modifier toxins for the voltage-gated sodium channel NaV1.7 can be predicted using accurate free-energy calculations. Toxins (Basel). 2021a;13(3)

Katz D, DiMattia MA, Sindhikara D, Li H, Abraham N, Leffler AE. Potency- and selectivity-enhancing mutations of conotoxins for nicotinic acetylcholine receptors can be predicted using accurate free-energy calculations. Mar Drugs. 2021b;19(7)

Kazandjian TD, Petras D, Robinson SD, van Thiel J, Greene HW, Arbuckle K, Barlow A, Carter DA, Wouters RM, Whiteley G, Wagstaff SC, Arias AS, Albulescu L-O, Plettenberg Laing A, Hall C, Heap A, Penrhyn-Lowe S, McCabe CV, Ainsworth S, et al. Convergent evolution of pain-inducing defensive venom components in spitting cobras. Science. 2021;371(6527):386–90.

King GF, Gentz MC, Escoubas P, Nicholson GM. A rational nomenclature for naming peptide toxins from spiders and other venomous animals. Toxicon. 2008;52(2):264–76.

Kita M, Black DS, Ohno O, Yamada K, Kigoshi H, Uemura D. Duck-billed platypus venom peptides induce Ca2+ influx in neuroblastoma cells. J Am Chem Soc. 2009;131(50):18038–9.

Kitchen DB, Decornez H, Furr JR, Bajorath J. Docking and scoring in virtual screening for drug discovery: methods and applications. Nat Rev Drug Discov. 2004;3(11):935–49.

Kleiz-Ferreira JM, Bernaerts H, Pinheiro-Junior EL, Peigneur S, Zingali RB, Tytgat J. Pharmacological screening of venoms from five Brazilian Micrurus species on different ion channels. Int J Mol Sci. 2022;23(14)

Klint JK, Smith JJ, Vetter I, Rupasinghe DB, Er SY, Senff S, Herzig V, Mobli M, Lewis RJ, Bosmans F, King GF. Seven novel modulators of the analgesic target NaV1.7 uncovered using a high-throughput venom-based discovery approach. Br J Pharmacol. 2015;172(10):2445–58.

Kumar R, Feltrup TM, Kukreja RV, Patel KB, Cai S, Singh BR. Evolutionary features in the structure and function of bacterial toxins. Toxins. 2019;11(1)

Kvitek R, Beitler M. Relative insensitivity of butter clam neurons to saxitoxin; a pre-adaptation for sequestering paralytic shellfish poisoning toxins as a chemical defense. Mar Ecol Prog Ser. 1991;69:47–54.
Lago J, Rodriguez LP, Blanco L, Vieites JM, Cabado AG. Tetrodotoxin, an extremely potent marine neurotoxin: distribution, toxicity, origin and therapeutical uses. Marine Drugs. 2015;13(10):6384.
Lebrun B, Romi-Lebrun R, Martin-Eauclaire MF, Yasuda A, Ishiguro M, Oyama Y, Pongs O, Nakajima T. A four-disulphide-bridged toxin, with high affinity towards voltage-gated K+ channels, isolated from Heterometrus spinnifer (Scorpionidae) venom. Biochem J. 1997;328(1):321–7.
Lee MR. Curare: the South American arrow poison. J R Coll Physicians Edinb. 2005;35(1):83–92.
Lee RD, Jospin G, Lang JM, Eisen JA, Coil DA. Draft genome sequence of Pseudoalteromonas tetraodonis strain UCD-SED8 (phylum Gammaproteobacteria). Genome Announc. 2015;3(6)
Lee JYP, Saez NJ, Cristofori-Armstrong B, Anangi R, King GF, Smith MT, Rash LD. Inhibition of acid-sensing ion channels by diminazene and APETx2 evoke partial and highly variable antihyperalgesia in a rat model of inflammatory pain. Br J Pharmacol. 2018;175(12):2204–18.
Leffler AE, Kuryatov A, Zebroski HA, Powell SR, Filipenko P, Hussein AK, Gorson J, Heizmann A, Lyskov S, Tsien RW, Poget SF, Nicke A, Lindstrom J, Rudy B, Bonneau R, Holford M. Discovery of peptide ligands through docking and virtual screening at nicotinic acetylcholine receptor homology models. Proc Natl Acad Sci U S A. 2017;114(38):E8100–9.
Li H, Xu J, Shen ZS, Wang GM, Tang M, Du XR, Lv YT, Wang JJ, Zhang FF, Qi Z, Zhang Z, Sokabe M, Tang QY. The neuropeptide GsMTx4 inhibits a mechanosensitive BK channel through the voltage-dependent modification specific to mechano-gating. J Biol Chem. 2019;294(31):11892–909.
Liao Z, Tang X, Chen W, Jiang X, Chen Z, He K, Li Q, Duan Z, He X, Kamau PM, Lv L, Zhang Z, Rong M, Lv Q, Lai R. Shrew's venom quickly causes circulation disorder, analgesia and hypokinesia. Cell Mol Life Sci CMLS. 2022;79(1):35.
Liu X, Li C, Chen J, Du J, Zhang J, Li G, Jin X, Wu C. AGAP, a new recombinant neurotoxic polypeptide, targets the voltage-gated calcium channels in rat small diameter DRG neurons. Biochem Biophys Res Commun. 2014a;452(1):60–5.
Liu Y, Wu Z, Tang D, Xun X, Liu L, Li X, Nie D, Xiang Y, Yi J, Yi J. Analgesic effects of Huwentoxin-IV on animal models of inflammatory and neuropathic pain. Protein Pept Lett. 2014b;21(2):153–8.
Logashina YA, Palikova YA, Palikov VA, Kazakov VA, Smolskaya SV, Dyachenko IA, Tarasova NV, Andreev YA. Anti-inflammatory and analgesic effects of TRPV1 polypeptide modulator APHC3 in models of osteo- and rheumatoid arthritis. Mar Drugs. 2021;19(1)
Luo S, Zhangsun D, Schroeder CI, Zhu X, Hu Y, Wu Y, Weltzin MM, Eberhard S, Kaas Q, Craik DJ, McIntosh JM, Whiteaker P. A novel α4/7-conotoxin LvIA from Conus lividus that selectively blocks α3β2 vs. α6/α3β2β3 nicotinic acetylcholine receptors. FASEB J. 2014;28(4):1842–53.
Maertens C, Cuypers E, Amininasab M, Jalali A, Vatanpour H, Tytgat J. Potent modulation of the voltage-gated sodium channel Nav1.7 by OD1, a toxin from the scorpion Odonthobuthus doriae. Mol Pharmacol. 2006;70(1):405–14.
Makielski JC, Satin J, Fan Z. Post-repolarization block of cardiac sodium channels by saxitoxin. Biophys J. 1993;65(2):790–8.
Marcil J, Walczak J-S, Guindon J, Ngoc AH, Lu S, Beaulieu P. Antinociceptive effects of tetrodotoxin (TTX) in rodents. Br J Anaesth. 2006;96(6):761–8.
Mast DH, Checco JW, Sweedler JV. Advancing d-amino acid-containing peptide discovery in the metazoan. Biochim Biophys Acta Proteins Proteomics. 2021;1869(1):140553.
Muttenthaler M, King GF, Adams DJ, Alewood PF. Trends in peptide drug discovery. Nat Rev Drug Discov. 2021;20(4):309–25.
Nadasdi L, Yamashiro D, Chung D, Tarczy-Hornoch K, Adriaenssens P, Ramachandran J. Structure-activity analysis of a Conus peptide blocker of N-type neuronal calcium channels. Biochemistry. 1995;34(25):8076–81.

Nassar MA, Stirling LC, Forlani G, Baker MD, Matthews EA, Dickenson AH, Wood JN. Nociceptor-specific gene deletion reveals a major role for Nav1.7 (PN1) in acute and inflammatory pain. Proc Natl Acad Sci U S A. 2004;101(34):12706–11.

National Centre for Biotechnology Information. Basic Local Alignment Search Tool – Protein (BLASTP). (2022). Retrieved 15 September 2022, from

Neff RA, Flinspach M, Gibbs A, Shih AY, Minassian NA, Liu Y, Fellows R, Libiger O, Young S, Pennington MW, Hunter MJ, Wickenden AD. Comprehensive engineering of the tarantula venom peptide huwentoxin-IV to inhibit the human voltage-gated sodium channel hNav1.7. J Biol Chem. 2020;295(5):1315–27.

Neumann C, Slagboom J, Somsen GW, Vonk F, Casewell NR, Cardoso CL, Kool J. Development of a generic high-throughput screening assay for profiling snake venom protease activity after high-resolution chromatographic fractionation. Toxicon. 2020;178:61–8.

Newcomb R, Szoke B, Palma A, Wang G, Chen X, Hopkins W, Cong R, Miller J, Urge L, Tarczy-Hornoch K, Loo JA, Dooley DJ, Nadasdi L, Tsien RW, Lemos J, Miljanich G. Selective peptide antagonist of the class E calcium channel from the venom of the tarantula Hysterocrates gigas. Biochemistry. 1998;37(44):15353–62.

Nieto FR, Cobos EJ, Tejada MÁ, Sánchez-Fernández C, González-Cano R, Cendán CM. Tetrodotoxin (TTX) as a Therapeutic Agent for Pain. Marine Drugs. 2012;10(2):281–305

Obergrussberger A, Friis S, Brüggemann A, Fertig N. Automated patch clamp in drug discovery: major breakthroughs and innovation in the last decade. Expert Opin Drug Discovery. 2021;16(1):1–5.

Ogawa Y, Kanai-Azuma M, Akimoto Y, Kawakami H, Yanoshita R. Exosome-like vesicles in Gloydius blomhoffii blomhoffii venom. Toxicon. 2008;51(6):984–93.

Ohshima K, Inoue K, Fujiwara A, Hatakeyama K, Kanto K, Watanabe Y, Muramatsu K, Fukuda Y, Ogura SI, Yamaguchi K, Mochizuki T. Let-7 microRNA family is selectively secreted into the extracellular environment via exosomes in a metastatic gastric cancer cell line. PLoS One. 2010;5(10):1–10.

Oliveira JS, Fuentes-Silva D, King GF. Development of a rational nomenclature for naming peptide and protein toxins from sea anemones. Toxicon. 2012;60(4):539–50.

Osmakov DI, Kozlov SA, Andreev YA, Koshelev SG, Sanamyan NP, Sanamyan KE, Dyachenko IA, Bondarenko DA, Murashev AN, Mineev KS, Arseniev AS, Grishin EV. Sea anemone peptide with uncommon β-hairpin structure inhibits acid-sensing ion channel 3 (ASIC3) and reveals analgesic activity. J Biol Chem. 2013;288(32):23116–27.

Pedron C, Antunes FTT, Rebelo IN, Campos MM, Correa ÁP, Klein CP, de Oliveira IB, do Nascimento Cordeiro M, Gomez MV, de Souza AH. Phoneutria nigriventer Tx3-3 peptide toxin reduces fibromyalgia symptoms in mice. Neuropeptides. 2021;85:102094.

Pennington MW, Byrnes ME, Zaydenberg I, Khaytin I, de Chastonay J, Krafte DS, Hill R, Mahnir VM, Volberg WA, Gorczyca W. Chemical synthesis and characterization of ShK toxin: a potent potassium channel inhibitor from a sea anemone. Int J Pept Protein Res. 1995;46(5):354–8.

Pennington MW, Chang SC, Chauhan S, Huq R, Tajhya RB, Chhabra S, Norton RS, Beeton C. Development of highly selective Kv1.3-blocking peptides based on the sea anemone peptide ShK. Mar Drugs. 2015;13(1):529–42.

Pineda SS, Chaumeil PA, Kunert A, Kaas Q, Thang MWC, Le L, Nuhn M, Herzig V, Saez NJ, Cristofori-Armstrong B, Anangi R, Senff S, Gorse D, King GF. ArachnoServer 3.0: an online resource for automated discovery, analysis and annotation of spider toxins. Bioinformatics. 2018;34(6):1074–6.

Piontek M, Seymour JE, Wong Y, Gilstrom T, Potriquet J, Jennings E, Nimmo A, Miles JJ. The pathology of Chironex fleckeri venom and known biological mechanisms. Toxicon. 2020;X(6):100026.

Pollitzer S. On Curare. J Physiol. 1886;7(3):274.

Prance GT. An ethnobotanical comparison of four tribes of Amazonian Indians. Acta Amazon. 1972;2(2):7–27.

Puthenveetil R, Vinogradova O. Solution NMR: a powerful tool for structural and functional studies of membrane proteins in reconstituted environments. J Biol Chem. 2019;294(44):15914–31.

Qadri YJ, Berdiev BK, Song Y, Lippton HL, Fuller CM, Benos DJ. Psalmotoxin-1 docking to human acid-sensing ion channel-1. J Biol Chem. 2009;284(26):17625–33.

Rahnama S, Deuis JR, Cardoso FC, Ramanujam V, Lewis RJ, Rash LD, King GF, Vetter I, Mobli M. The structure, dynamics and selectivity profile of a NaV 1.7 potency-optimised huwentoxin-IV variant. PLoS One. 2017;12(3)

Rashid MH, Huq R, Tanner MR, Chhabra S, Khoo KK, Estrada R, Dhawan V, Chauhan S, Pennington MW, Beeton C, Kuyucak S, Norton RS. A potent and Kv1.3-selective analogue of the scorpion toxin HsTX1 as a potential therapeutic for autoimmune diseases. Sci Rep. 2014;4:4509.

Richards KL, Milligan CJ, Richardson RJ, Jancovski N, Grunnet M, Jacobson LH, Undheim EAB, Mobli M, Chow CY, Herzig V, Csoti A, Panyi G, Reid CA, King GF, Petrou S. Selective NaV1.1 activation rescues Dravet syndrome mice from seizures and premature death. Proc Natl Acad Sci U S A. 2018;115(34):E8077–85.

Rosso JP, Schwarz JR, Diaz-Bustamante M, Céard B, Gutiérrez JM, Kneussel M, Pongs O, Bosmans F, Bougis PE. MmTX1 and MmTX2 from coral snake venom potently modulate receptor activity modulate GABAa receptor activity. Proc Natl Acad Sci U S A. 2015;112(8):E891–900.

Rowe AH, Xiao Y, Scales J, Linse KD, Rowe MP, Cummins TR, Zakon HH. Isolation and characterization of CvIV4: a pain inducing α- scorpion toxin. PLoS One. 2011;6(8)

Rupasinghe DB, Herzig V, Vetter I, Dekan Z, Gilchrist J, Bosmans F, Alewood PF, Lewis RJ, King GF. Mutational analysis of ProTx-I and the novel venom peptide Pe1b provide insight into residues responsible for selective inhibition of the analgesic drug target NaV1.7. Biochem Pharmacol. 2020;181:114080.

Safavi-Hemami H, Brogan SE, Olivera BM. Pain therapeutics from cone snail venoms: from Ziconotide to novel non-opioid pathways. J Proteome. 2019;190:12–20.

Samson A, Scherf T, Eisenstein M, Chill J, Anglister J. The mechanism for acetylcholine receptor inhibition by alpha-neurotoxins and species-specific resistance to alpha-bungarotoxin revealed by NMR. Neuron. 2002;35(2):319–32.

Shah B, Sindhikara D, Borrelli K, Leffler AE. Water thermodynamics of peptide toxin binding sites on ion channels. Toxins (Basel). 2020;12(10)

Shcherbatko A, Rossi A, Foletti D, Zhu G, Bogin O, Galindo Casas M, Rickert M, Hasa-Moreno A, Bartsevich V, Crameri A, Steiner AR, Henningsen R, Gill A, Pons J, Shelton DL, Rajpal A, Strop P. Engineering highly potent and selective microproteins against Nav1.7 sodium channel for treatment of pain. J Biol Chem. 2016;291(27):13974–86.

Sherwood D, Blick T, Gabriel R, Lucas SM. On the authorship and publication dates of the nomina Theraphosa and Theraphosidae (Araneae: Mygalomorphae). Zootaxa. 2021;4950(2):395–7.

Siemens J, Zhou S, Piskorowski R, Nikai T, Lumpkin EA, Basbaum AI, King D, Julius D. Spider toxins activate the capsaicin receptor to produce inflammatory pain. Nature. 2006;444(7116):208–12.

Sousa SR, Wingerd JS, Brust A, Bladen C, Ragnarsson L, Herzig V, Deuis JR, Dutertre S, Vetter I, Zamponi GW, King GF, Alewood PF, Lewis RJ. Discovery and mode of action of a novel analgesic β-toxin from the African spider Ceratogyrus darlingi. PLoS One. 2017;12(9)

Souza-Imberg A, Carneiro SM, Giannotti KC, Sant'Anna SS, Yamanouye N. Origin and characterization of small membranous vesicles present in the venom of Crotalus durissus terrificus. Toxicon. 2017;136:27–33.

Tajti G, Wai DCC, Panyi G, Norton RS. The voltage-gated potassium channel KV1.3 as a therapeutic target for venom-derived peptides. Biochem Pharmacol. 2020;181:114146.

Tang D, Yang Y, Xiao Z, Xu J, Yang Q, Dai H, Liang S, Tang C, Dong H, Liu Z. Scorpion toxin inhibits the voltage-gated proton channel using a Zn2+ –like long-range conformational coupling mechanism. Br J Pharmacol. 2020;177(10):2351–64.

Tanner MR, Tajhya RB, Huq R, Gehrmann EJ, Rodarte KE, Atik MA, Norton RS, Pennington MW, Beeton C. Prolonged immunomodulation in inflammatory arthritis using the selective Kv1.3 channel blocker HsTX1[R14A] and its PEGylated analog. Clin Immunol. 2017;180:45–57.

Tarcha EJ, Olsen CM, Probst P, Peckham D, Muñoz-Elías EJ, Kruger JG, Iadonato SP. Safety and pharmacodynamics of dalazatide, a Kv1.3 channel inhibitor, in the treatment of plaque psoriasis: a randomized phase 1b trial. PLoS One. 2017;12(7):e0180762.

Teisseyre A, Palko-Labuz A, Sroda-Pomianek K, Michalak K. Voltage-gated potassium channel Kv1.3 as a target in therapy of cancer. Front Oncol. 2019;9:933.

Tonello R, Fusi C, Materazzi S, Marone IM, De Logu F, Benemei S, Gonçalves MC, Coppi E, Castro-Junior CJ, Gomez MV, Geppetti P, Ferreira J, Nassini R. The peptide Phα1β, from spider venom, acts as a TRPA1 channel antagonist with antinociceptive effects in mice. Br J Pharmacol. 2017;174(1):57–69.

Trim CM, Byrne LJ, Trim SA. Utilisation of compounds from venoms in drug discovery. Prog Med Chem. 2021;60:1–66.

Ul-Hasan S, Rodríguez-Román E, Reitzel AM, Adams RMM, Herzig V, Nobile CJ, Saviola AJ, Trim SA, Stiers EE, Moschos SA, Keiser CN, Petras D, Moran Y, Colston TJ. The emerging field of venom-microbiomics for exploring venom as a microenvironment, and the corresponding initiative for venom associated microbes and parasites (iVAMP). Toxicon. 2019;X:4.

Verkest C, Salinas M, Diochot S, Deval E, Lingueglia E, Baron A. Mechanisms of action of the peptide toxins targeting human and rodent acid-sensing ion channels and relevance to their in vivo analgesic effects. Toxins. 2022;14(10)

Vonk FJ, Bittenbinder MA, Kerkkamp HMI, Grashof DGB, Archer JP, Afonso S, Richardson MK, Kool J, Van der Meijden A. A non-lethal method for studying scorpion venom gland transcriptomes, with a review of potentially suitable taxa to which it can be applied. PLoS One. 2021;16(11 November):1–16.

Wang W-X, Ji Y-H. Scorpion venom induces glioma cell apoptosis in vivo and inhibits glioma tumor growth in vitro. J Neuro-Oncol. 2005;73(1):1–7.

Wang D, Starr R, Chang W-C, Aguilar B, Alizadeh D, Wright SL, Yang X, Brito A, Sarkissian A, Ostberg JR, Li L, Shi Y, Gutova M, Aboody K, Badie B, Forman SJ, Barish ME, Brown CE. Chlorotoxin-directed CAR T cells for specific and effective targeting of glioblastoma. Sci Transl Med. 2020;12(533)

Whitelaw BL, Cooke IR, Finn J, Zenger K, Strugnell JM. The evolution and origin of tetrodotoxin acquisition in the blue-ringed octopus (genus Hapalochlaena). Aquat Toxicol (Amsterdam, Netherlands). 2019;206:114–22.

Williams-Noonan BJ, Yuriev E, Chalmers DK. Free energy methods in drug design: prospects of 'alchemical perturbation' in medicinal chemistry. J Med Chem. 2018;61(3):638–49.

Wilson CM, Naves T, Vincent F, Melloni B, Bonnaud F, Lalloué F, Jauberteau MO. Sortilin mediates the release and transfer of exosomes in concert with two tyrosine kinase receptors. J Cell Sci. 2014;127(18):3983–97.

Woodward RB, Brehm WJ. The structure of strychnine. Formulation of the neo bases. J Am Chem Soc. 1948;70(6):2107–15.

Wulff H, Christophersen P, Colussi P, Chandy KG, Yarov-Yarovoy V. Antibodies and venom peptides: new modalities for ion channels. Nat Rev Drug Discov. 2019;18(5):339–57.

Xiao Y, Blumenthal K, Jackson JO, Liang S, Cummins TR. The tarantula toxins ProTx-II and Huwentoxin-IV differentially interact with human Nav1.7 voltage sensors to inhibit channel activation and inactivation. Mol Pharmacol. 2010;78(6):1124–34.

Xu Y, Meng X, Hou X, Sun J, Kong X, Sun Y, Liu Z, Ma Y, Niu Y, Song Y, Cui Y, Zhao M, Zhang J. A mutant of the Buthus martensii Karsch antitumor-analgesic peptide exhibits reduced inhibition to hNav1.4 and hNav1.5 channels while retaining analgesic activity. J Biol Chem. 2017;292(44):18270–80.

Xu H, Li T, Rohou A, Arthur CP, Tzakoniati F, Wong E, Estevez A, Kugel C, Franke Y, Chen J, Ciferri C, Hackos DH, Koth CM, Payandeh J. Structural basis of Nav1.7 inhibition by a gating-modifier spider toxin. Cell. 2019;176(4):702–715 e14.

Xu Q, Tae HS, Wang Z, Jiang T, Adams DJ, Yu R. Rational design of alpha-conotoxin RegIIA analogues selectively inhibiting the human alpha3beta2 nicotinic acetylcholine receptor through computational scanning. ACS Chem Neurosci. 2020;11(18):2804–11.

Yu R, Craik DJ, Kaas Q. Blockade of neuronal alpha7-nAChR by alpha-conotoxin ImI explained by computational scanning and energy calculations. PLoS Comput Biol. 2011;7(3):e1002011.

Zhang HL, Han R, Chen ZX, Chen BW, Gu ZL, Reid PF, Raymond LN, Qin ZH. Opiate and acetylcholine-independent analgesic actions of crotoxin isolated from crotalus durissus terrificus venom. Toxicon. 2006;48(2):175–82.

Index

G. Stephens, E. Stevens (eds.), *Ion Channels as Targets in Drug Discovery*,
https://doi.org/10.1007/978-3-031-52197-3

G

H

I

K

L

M

N

P

S

MIX
Papier aus verantwortungsvollen Quellen
Paper from responsible sources
FSC® C105338

If you have any concerns about our products,
you can contact us on
ProductSafety@springernature.com

In case Publisher is established outside the EU,
the EU authorized representative is:
Springer Nature Customer Service Center GmbH
Europaplatz 3, 69115 Heidelberg, Germany

Printed by Libri Plureos GmbH
in Hamburg, Germany